S0-EJQ-667

Differential Diagnosis of Soft Tissue and Bone Tumors

"Classifications are ideal and mutable. Observation and experience are solid and unchangeable"

MALPIGHI (1628–1694)

Differential Diagnosis of Soft Tissue and Bone Tumors

STEVEN I. HAJDU, M.D.
Attending Pathologist,
Surgical Pathology
and Autopsy Services,
Department of Pathology
Chief, Cytology Service
Memorial Sloan-Kettering Cancer Center
Professor of Pathology
Cornell University Medical College
New York, New York

with the assistance of
EVA M. HAJDU, B.S.

LEA & FEBIGER · 1986 · PHILADELPHIA

LEA & FEBIGER
600 Washington Square
Philadelphia, Pa. 19106-4198
U.S.A.
(215) 922-1330

Library of Congress Cataloging in Publication Data

Hajdu, Steven I.
Differential diagnosis of soft tissue and bone tumors.

Bibliography: p.
Includes index.
1. Sarcoma—Diagnosis. 2. Diagnosis, Cytologic.
3. Diagnosis, Differential. I. Hajdu, Eva M. II. Title.
[DNLM: 1. Sarcoma—Diagnosis. 2. Diagnosis, Differential.
QZ 345 H154d]
RC270.H333 1984 616.99′4075 83-19990
ISBN 0-8121-0895-7

Printed in the United States of America

Print Number: 5 4 3 2 1

Dedicated to the Pathologist

Preface

THIS BOOK is the outgrowth of more than twenty-five years of my diagnostic experience with the histology and cytology of soft tissue and bone tumors. In my presentation of tumors I have followed a sequence consistent with that followed by pathologists in their examinations. All soft tissue and bone neoplasms, as well as non-neoplastic lesions, are defined and classified, regardless of their histogenesis or organ of origin, according to *histologic pattern* (arranged, spreading, lacy, epithelioid, alveolar, and disarranged) and *cell morphology* (slender spindle cells, plump spindle cells, granular epithelioid cells, clear epithelioid cells, isomorphic giant cells, and pleomorphic giant cells), with consideration of the *appearance of stroma* and the *products of cells*. Some tumors may fit into more than one of these histologic categories and may be listed more than once.

After a brief consideration of the histogenesis of soft tissue and bone tumors (Table 5), in the introductory chapter readers will find the definitions and the synonyms of the most common non-neoplastic lesions (Table 10), benign neoplasms (Table 11), and malignant neoplasms (Table 12) listed along with the most pertinent references.

The clinical presentation, size, site, and gross and radiologic appearance of reactive and neoplastic lesions are presented with reference to age and sex of the patient to augment diagnostic accuracy. To obtain optimal results, the dissection and handling of soft tissue and bone specimens are described.

The tumors presented in depth are those that are most common and those that present the greatest diagnostic difficulties to pathologists. The main portion of the book is devoted to illustrations of the growth pattern, cell morphology, appearance of stroma, and products of cells. All photomicrographs have been taken from hematoxylin and eosin stained sections unless specified. A limited number of photomicrographs illustrate diagnostic features of tissue sections stained with special, histochemical, or immunochemical stains; in addition, ultrastructural features are included to call attention to instances when such studies may contribute to the diagnosis. All photographs have been cropped to show only pertinent features. The pages were laid out with the resemblances of various entities in mind, and I have made a point of selecting illustrations that show features important to the differential diagnosis.

The thousands of cross-references, tables, and illustrations with special markings are designed to save time and facilitate an accurate differential diagnosis. In the last chapters the grading and staging procedure of sarcomas is outlined with consideration of prognosis and differential diagnosis. At the end of the book, in the form of an appendix, the cytologic appearance of soft tissue and bone tumors is illustrated as seen in smears prepared from exfoliative and aspiration specimens and stained by using Papanicolaou's method or hematoxylin and eosin.

In publishing this book my main purpose was to integrate the histologic appearance of soft tissue and bone tumors, and to improve the skills of pathologists in diagnosing intra- and extraskeletal connective tissue tumors by pattern recognition. To render the various terms and classifications suitable for inclusion I had to exercise considerable freedom in selecting the most appropriate ones. I realize that some of the concepts presented in this book may not be accepted by everyone, but I feel as William Cooke did in 1822, when he wrote the following: "The daily observation of disease which baffles the utmost skill, will present to the feeling mind an adequate inducement to cultivate a spirit of diligent research; and where this benevolent principle is the actuating motive, the individual need not be anxious about an apology for the publication of his ideas."

1275 York Avenue
New York, New York 10021

STEVEN I. HAJDU

Acknowledgments

I OWE MY most appreciative thanks to Myron R. Melamed, M.D., Chairman, Department of Pathology of Memorial Sloan-Kettering Cancer Center, and Eva O. Hajdu, M.D., my wife, for their encouragement and support. My thanks are also due to my many colleagues and associates, particularly those in the Department of Pathology of the Memorial Sloan-Kettering Cancer Center, all of whom have contributed in some way to the development and realization of the concepts presented in this book.

Many of the thoughts presented in this book are deeply rooted in the works of our predecessors in pathology and took shape after innumerable formal and informal discussions in study groups such as the Soft Tissue Task Force of the American Joint Committee and the Eastern Cooperative Oncology Group. I wish to express my warmest appreciation to the thousands of pathologists who attended my various workshops, seminars, and symposia and sought my opinion as a consultant through the years.

The skillful printing of photographs from my often less-than-well-exposed negatives was carried out by Ms. M. Ryon and Mr. K. Kong. Some of the ultrastructural illustrations were selected and annotated by R. Erlandson, Ph.D., and printed by Mr. Roy Keppie. The technical skills of Ms. Trudy Bodak, Mrs. Maryann Gangi, Mrs. Daisy Jimenez-Joseph, Ms. Lucille Mercer, and Mr. Antonio Scorza are acknowledged with appreciation.

My special thanks to my secretary and laboratory coordinator, Ms. R. Nager, who typed the whole manuscript, including nearly 100 tables, 1,500 legends, and an endless number of cross-references, with professional dedication and cheerful attentiveness.

And finally I wish to thank members of the editorial and production staff of Lea & Febiger, particularly Mr. R. Kenneth Bussy, Executive Editor, and Mr. Thomas J. Colaiezzi, Production Manager, for their many suggestions and artful production of this book.

Contents

1. *Histogenesis and Classification* 3
   - Embryology
   - Differentiation
   - Cell of Origin
   - Books and Monographs
   - Terminology
   - Incidence
   - Role of the Pathologist
   - Histogenetic Classification
   - Growth Phases of Sarcomas
   - Non-neoplastic Lesions
     - Definitions, Synonyms, and References
   - Benign Neoplasms
     - Definitions, Synonyms, and References
   - Malignant Neoplasms
     - Definitions, Synonyms, and References
   - Diagnosis by Growth Pattern, Cell Morphology, Appearance of Stroma, and Products of Cells

2. *Clinical Presentation* 35
   - Non-neoplastic Lesions
   - Benign Neoplasms
   - Malignant Neoplasms
   - Superficial Soft Tissue Lesions
   - Multifocal Bone Lesions
   - Solitary Bone Lesions

3. *Age Distribution* 50
   - Non-neoplastic Lesions
   - Benign Neoplasms
   - Malignant Neoplasms

4. *Sex Distribution* 54
   - Non-neoplastic Lesions
   - Benign Neoplasms
   - Malignant Neoplasms

5. *Anatomic Site* 56
   - Non-neoplastic Soft Tissue Lesions
   - Benign Soft Tissue Neoplasms
   - Malignant Soft Tissue Neoplasms
   - Non-neoplastic Bone Lesions
   - Benign Bone Neoplasms
   - Malignant Bone Neoplasms

6. *Size* 79
   - Non-neoplastic Lesions
   - Benign Neoplasms
   - Malignant Neoplasms

7. *Radiologic Appearance* 82
   - Non-neoplastic Lesions
   - Benign Neoplasms
   - Malignant Neoplasms

8. *Dissection of Specimen* 113

9. *Gross Appearance* 114

Non-neoplastic Lesions
Benign Neoplasms
Malignant Neoplasms

10. *Growth Pattern* 136

Arranged Pattern
Spreading Pattern
Lacy Pattern
Epithelioid Pattern
Alveolar Pattern
Disarranged Pattern

11. *Cell Morphology* 214

Slender Spindle Cells
Plump Spindle Cells
Granular Epithelioid Cells
Clear Epithelioid Cells
Isomorphic Giant Cells
Pleomorphic Giant Cells

12. *Appearance of Stroma* 265

Fibrillar Stroma
Sclerosed Stroma
Myxoid Stroma
Vascular Stroma
Inflamed Stroma
Necrotic Stroma
Chondrified Stroma
Ossified Stroma
Calcified Stroma

13. *Products of Cells* 340

Collagen
Glycogen
Polysaccharides
Fat
Melanin
Secretory Granules
Crystals
Fine Structure
Tissue Antigens

14. *Histologic Grade of Sarcomas* 402

Low Grade Sarcomas
High Grade Sarcomas
Either Low Grade or High Grade Sarcomas

15. *Stage of Sarcomas* 405

Stage 0
Stage I
Stage II
Stage III
Stage IV

16. *Prognosis* 408

Non-neoplastic Lesions
Benign Neoplasms
Malignant Neoplasms

17. *Differential Diagnosis* 411

Non-neoplastic Lesions
Benign Neoplasms
Malignant Neoplasms
Tumors with Arranged Pattern
Tumors with Spreading Pattern
Tumors with Lacy Pattern
Tumors with Epithelioid Pattern
Tumors with Alveolar Pattern
Tumors with Disarranged Pattern

Appendix 446

Exfoliative and Aspiration Cytology

Index 459

Differential Diagnosis of Soft Tissue and Bone Tumors

"Sarcomatous tumors are of very various kinds, and consequently every attempt to devise appropriate names to distinguish them is at least laudable"

SAMUEL COOPER (1780–1848)

1. Histogenesis and Classification

SOFT tissue and bone tumors arise from derivatives of the embryonic mesoderm. Cytologically, the cells of the embryonic disc have no distinctive features, but once they are organized into specific tissues and organs the microscopic similarity disappears. Most, if not all, organs are derived from at least two of the three germ layers, ectoderm, mesoderm, and endoderm, one of which is very commonly the mesoderm (Fig. 1). The embryonic mesoderm is segmented into a series of somites and split into somatic and splanchnic layers that are the source of the mesenchymal or connective tissues (Fig. 2). Both skeletal and soft tissues are composed of living, constantly changing cells fulfilling specific structural and functional roles. The difference between skeletal and soft tissues depends not so much on the dissimilarity of cellular elements, but on their degree of differentiation, maturation, genetically determined assembly, products, and preordered functional role.

In the course of differentiation, mesenchymal cells become specialized and may assume the cytologic characteristics of fibroblasts, myoblasts, lipoblasts, chondroblasts, osteoblasts, or an endless number of other primitive forms (Fig. 3). The fact that mesenchymal cells are ubiquitous and highly versatile is further complicated by the observation that any well-defined mesenchymal cell may either undergo maturation arrest, so-called "dedifferentiation," or at the conclusion of the mitotic cycle reach a higher level of differentiation by acquiring complex cytoplasmic organelles, depositing biochemically and immunologically active products, and assuming a phenotype different from that of the parent cell. For example, a noncollagenous fibrous histiocytic cell may emerge as a collagenous fibroblast, a fibroblast as a bone-producing osteoblast, or an undifferentiated pericyte as a leiomyoblast.

The recent reappearance of the term "dedifferentiation" is an unfortunate one, and its use should be discouraged. Cells, mesenchymal cells in particular, do not dedifferentiate, but may undergo maturation arrest or fail to differentiate. Often, it seems that there was a barrier to differentiation. There is ample clinical and experimental evidence that maturation arrest, for example, in embryonal rhabdomyo-

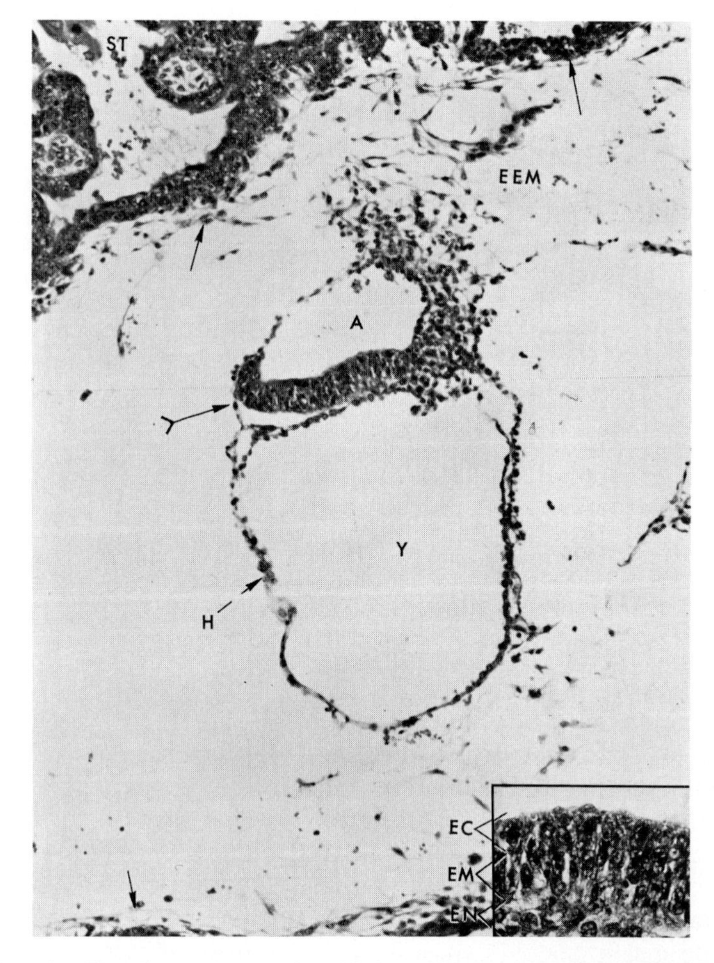

FIG. 1. Histologic section of an 18-day-old human embryo embedded in the endometrium. The endometrium almost completely surrounds the embryo. The cytotrophoblasts *(arrows)* form an irregular solid inner lining. Beyond this are irregular lacunae containing syncytiotrophoblasts *(ST)*. The embryonic side of the cytotrophoblastic lining is covered by extraembryonic mesoderm *(EEM)* that is condensed on its inner aspect to form Heuser's membrane *(H)* around the yolk sac *(Y)*. The amniotic cavity *(A)* is separated from the yolk sac by the embryonic disc *(arrow with tails)*. The embryonic disc is composed of three distinct linings (see insert): ectoderm *(EC)*, embryonic mesoderm, *(EM)*, and endoderm *(EN)*. (From Hajdu, S.I., and Hajdu, E.O.: *Cytopathology of Sarcomas and Other Nonepithelial Malignant Tumors.* Philadelphia, W.B. Saunders, 1976.)

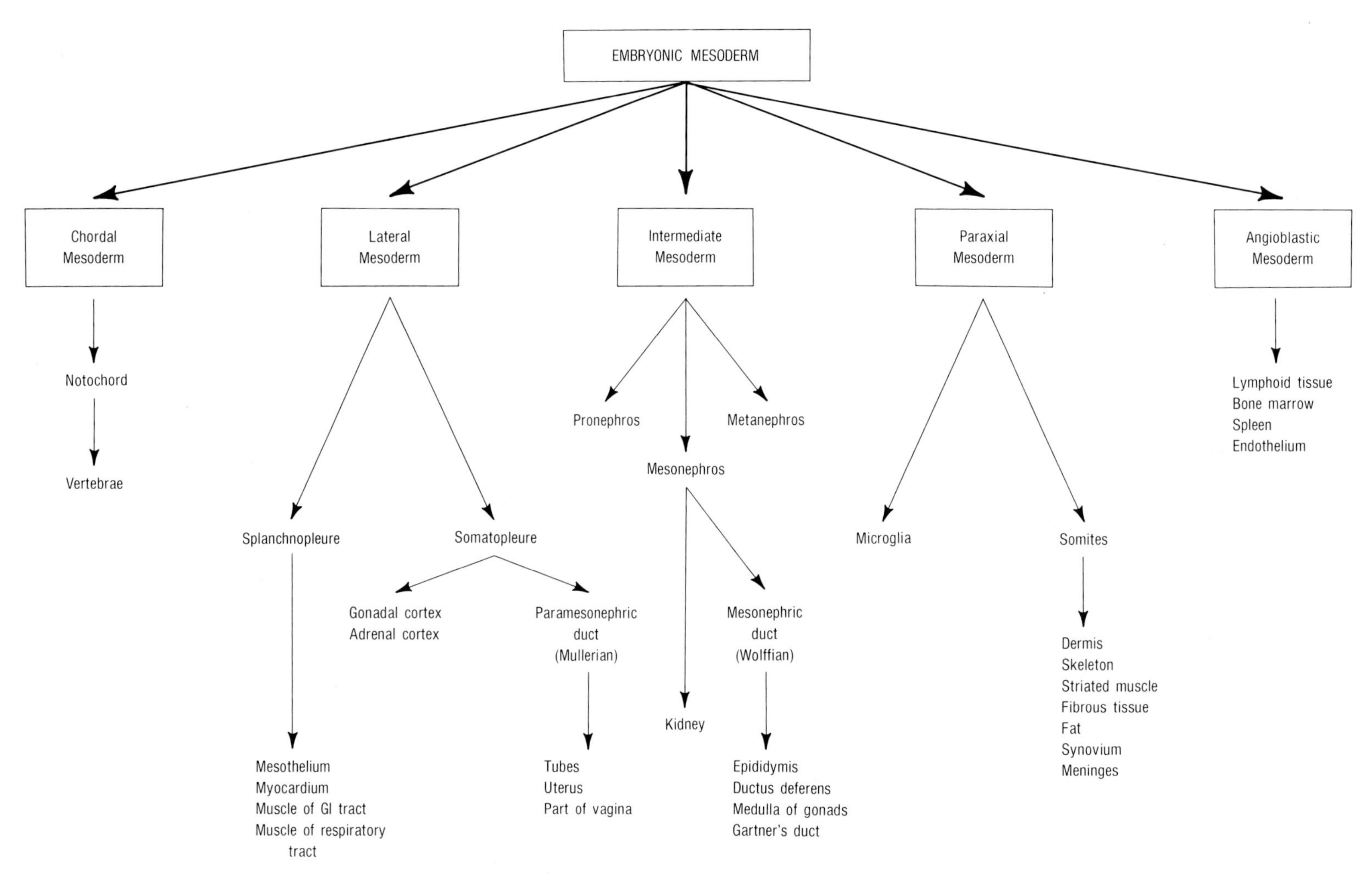

Fig. 2. Schematic illustration of the origin of various organs and organ systems from the embryonic mesoderm through a series of intermediate structures. Note the role of the *lateral mesoderm* in the development of the mesothelial lining, the smooth muscle, and female genitalia. Practically all soft tissues and the skeleton develop from the *paraxial mesoderm;* the entire lymphoreticular system originates from the *angioblastic mesoderm.* (From Hajdu, S.I., and Hajdu, E.O.: *Cytopathology of Sarcomas and Other Nonepithelial Malignant Tumors.* Philadelphia, W.B. Saunders, 1976.)

sarcoma, primitive neuroectodermal tumor, and Ewing's sarcoma, is analogous to that in acute leukemia and some lymphocytic neoplasms (Fig. 4). Similarly, there is no such thing as "malignant degeneration."

Soft tissues and bones are composed of clones of mesenchymal cells such as adipocytes, fibroblasts, osteoblasts, chondroblasts, and hematopoietic elements. The diversity and dissimilarity of these cells are expressed in the tissue pattern of the organs they build. Both differentiated and undifferentiated tumors may be composed of a combination of cellular elements at various stages of differentiation and may grow in a variety of tissue patterns. Once it is recognized that cell morphology and tissue patterns are subject to modulation and changes, and are influenced by local tissue conditions and a host of other factors, it is not difficult to understand that mesenchymal tissues and tumors (primary, recurrent, and metastatic) may assume, permanently or temporarily, dangerously misleading, overlapping, tissue patterns and cell morphology (Fig. 5). Therefore, different pathologists may label primary, recurrent, and metastatic soft tissue and bone neoplasms differently, and occasionally the same pathologist may call the same tumor by different names.

In many tumors, for example, in epithelial or glial tumors, the differences in growth patterns exhibited by the tumor cells are a reliable guide to the tissue of origin. In soft tissue and bone tumors, however, the relationship between differentiation and cell of origin is often blurred by the nonspecific assembly of undifferentiated cellular elements that may show no apparent structural differences between reactive and neoplastic growths. In 1919, James Ewing wrote that "the capacity of connective tissue to indulge in exuberant reactive or reparative growth is remarkable." No one would deny that actively growing reactive lesions, for example, granuloma, fasciitis, myositis ossificans, or callus, may contain all, or nearly all, the cellular elements that embryonic mesenchyme can produce and may mimic neoplastic growth.

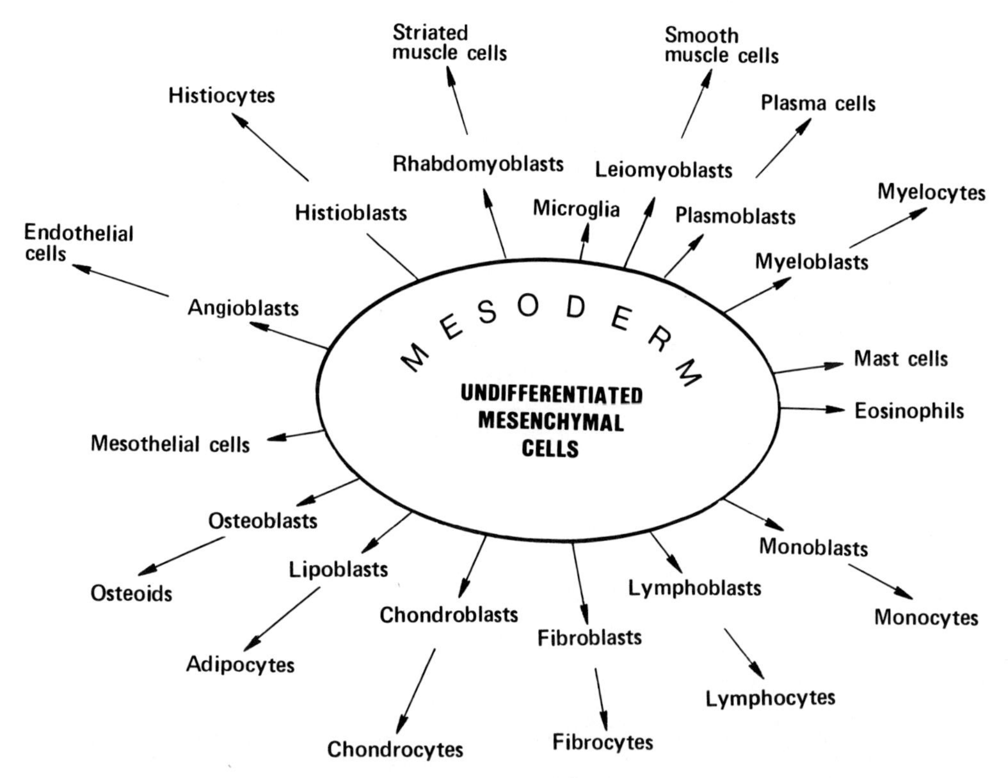

Fig. 3. Primitive mesenchymal cells may produce any of these forms. Undifferentiated, blastic or poorly differentiated, cells may mature and become well differentiated forms (Modified from Hajdu, S.I.: *Pathology of Soft Tissue Tumors.* Philadelphia, Lea & Febiger, 1979.)

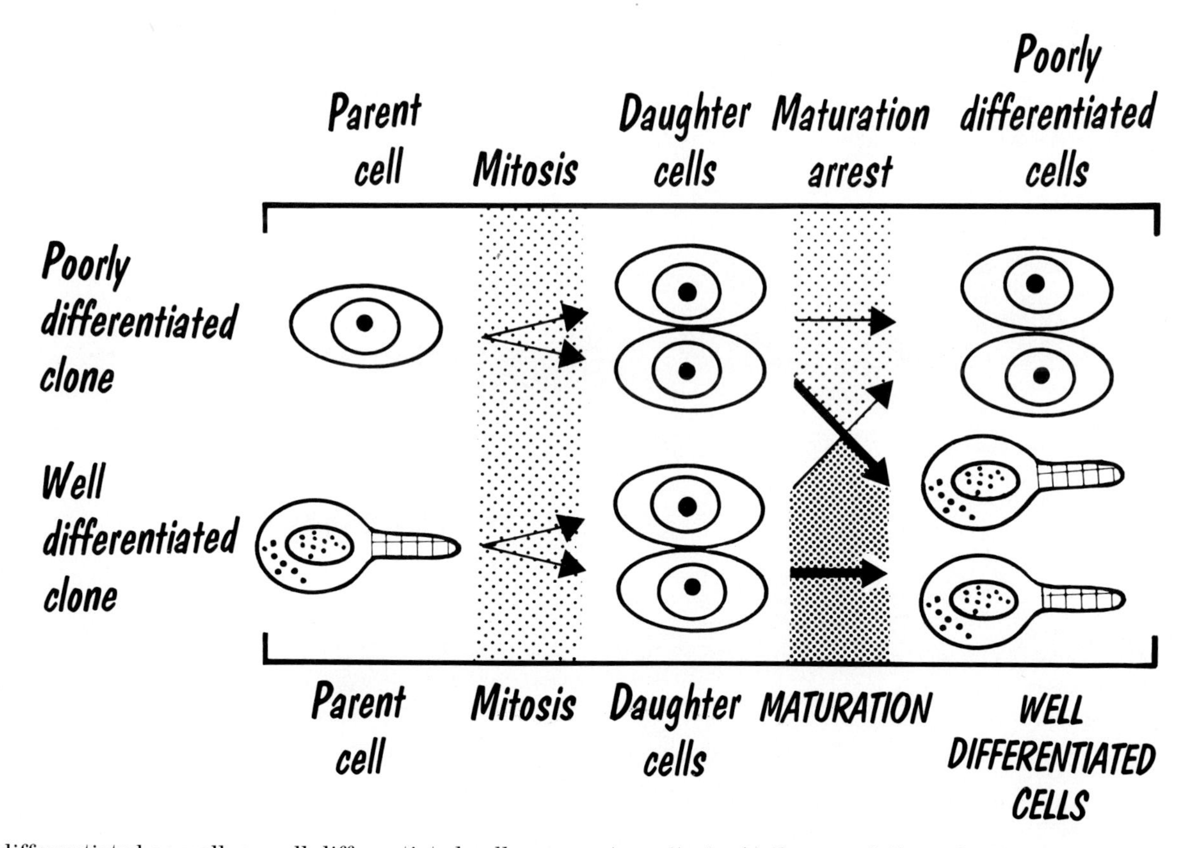

Fig. 4. Poorly differentiated as well as well-differentiated cells may enter mitosis. At the completion of mitosis, the daughter cells may remain undifferentiated, arrested (see right upper half), or may mature and become well-differentiated forms (see right lower half).

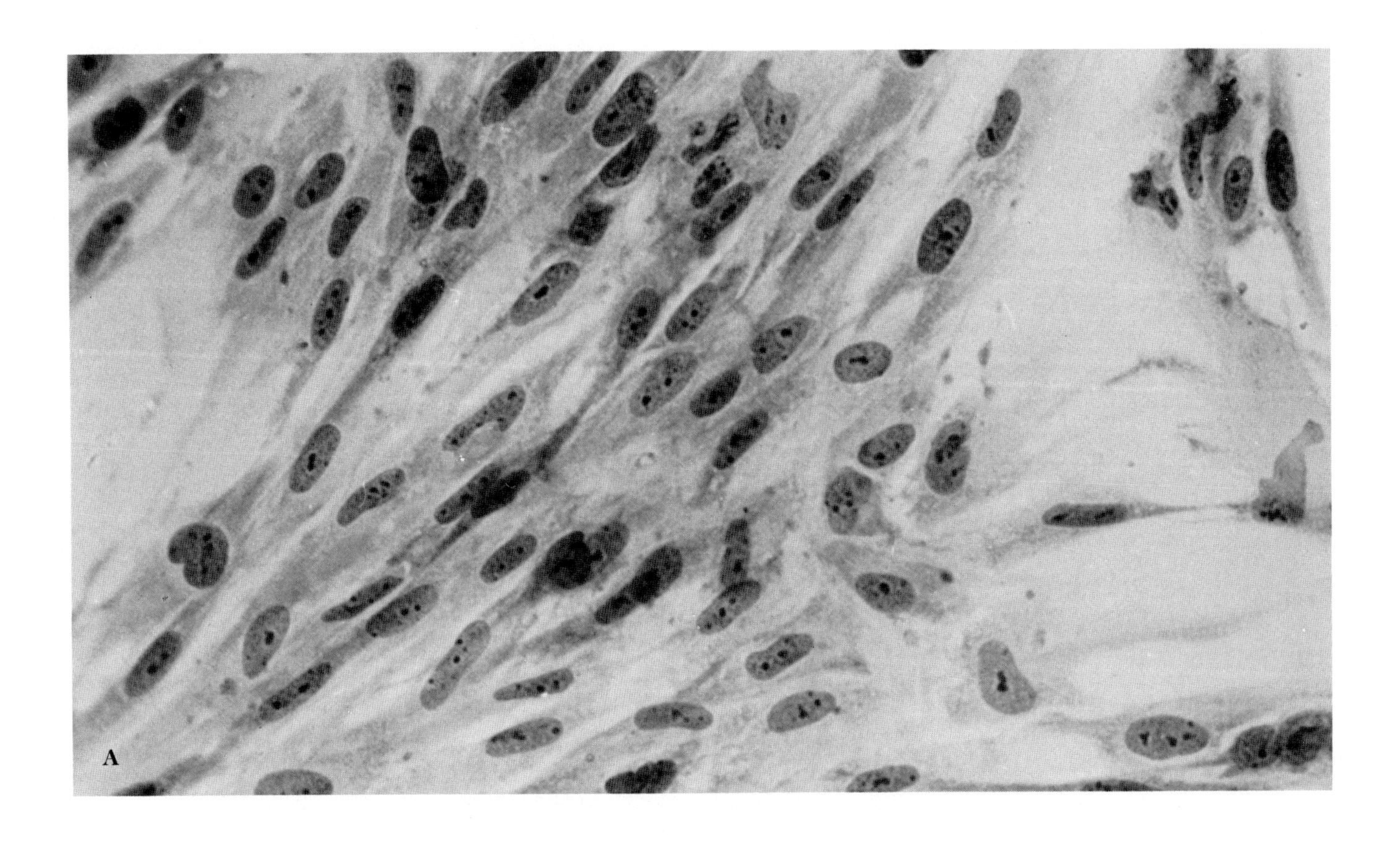

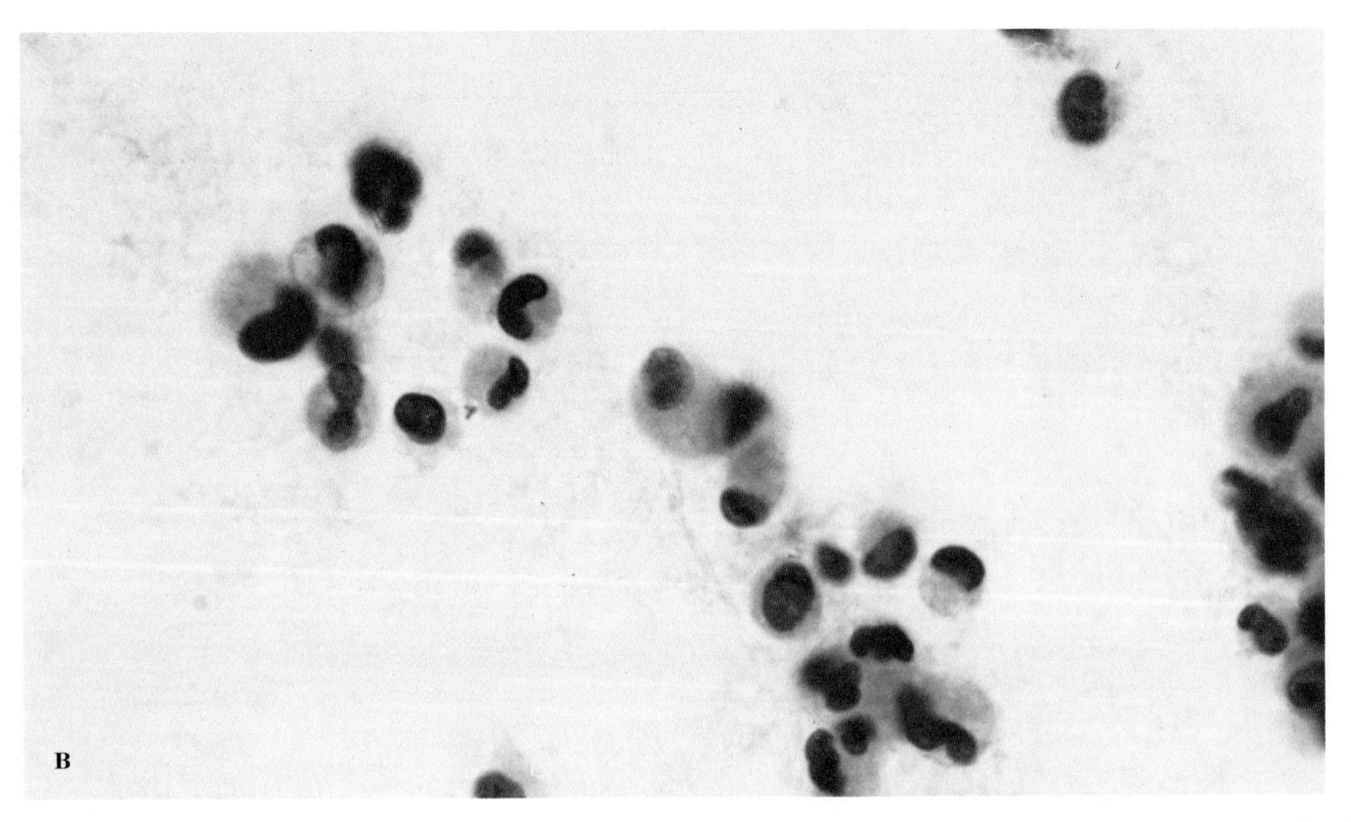

Fig. 5A. Tissue culture of human fibroblasts growing as a monolayer after alcohol fixation and staining. The uniform spindly fibroblasts with regular oval nuclei show the characteristic appearance of benign fibroblasts in tissue culture (Papanicolaou stain, ×470). B. Human fibroblasts identical in origin with that in Figure 5A in a smear prepared from suspension of a trypsinized monolayer tissue culture. Note the striking difference in microscopic appearance of fibroblasts from that of Figure 5A as the result of trypsinization.

Ascertaining the cell of origin for many soft tissue and bone tumors is still problematic, but it seems that there is a general agreement among investigators that soft tissue and bone tumors originate from primitive pluripotential mesenchymal cells. Traditionally, soft tissue and bone tumors have been discussed in different texts, and bone pathology, especially bone tumor pathology, has enjoyed an exclusive status for more than a century and has been promoted in more than three dozen books and monographs. Not until 1979 was the first comprehensive text on the pathology of soft tissue tumors published (Tables 1 and 2). The artificial separation of mesenchymal tumors according to anatomic boundaries (that is, intraskeletal, bone, and extraskeletal, soft tissue) served its purpose, but also produced setbacks and disappointment. Who has not felt frustrated from time to time with the ambiguous, constantly changing, and confusing terminology? And how many pathologists have misused or misunderstood the various names and definitions of soft tissue and bone tumors while evaluating their pathologic and clinical behavior?

Due to a sectarian or exclusive club approach, the present generation of pathologists inherited a dozen or more different names of lesions that look microscopically similar or identical in soft tissues and bone (Table 3). Virchow recognized in the mid-1800s that "osteosarcoma is an ossified fibrosarcoma," and Stout stated in 1953 that "extraskeletal

Table 1. *Books and Monographs on Soft Tissue Tumors*

Pack, G.T., and Ariel, I.M.: Tumors of Soft Somatic Tissues. New York, Paul Hoebner Inc., 1958.

Saavedra, J.A.: Sarcomas y Lesiones Seudosarcomatosas de Partes Blandas. Mexico, La Prenna Medica Mexicana, 1967.

Stout, A.P., and Lattes, R.: Tumors of the Soft Tissues. Atlas of Tumor Pathology. 2nd Ed. Washington, D.C., Armed Forces Institute of Pathology, 1967.

Mackenzie, D.M.: The Differential Diagnosis of Fibroblastic Disorders. Oxford, Blackwell, 1970.

Berlin, S.J.: Soft Somatic Tumors of the Foot. Mount Kisco, New York, Futura Publishing Co., 1976.

Ashbury, A.K., and Johnson, P.C.: Pathology of Peripheral Nerve. Philadelphia, W.B. Saunders, 1978.

Hajdu, S.I.: Pathology of Soft Tissue Tumors. Philadelphia, Lea & Febiger, 1979.

Allen, P.W.: Tumors and Proliferations of Adipose Tissue. A Clinicopathologic Approach. New York, Masson Publishing, 1981.

Enzinger, F.M., and Weiss, W.S.: Soft Tissue Tumors. St. Louis, C.V. Mosby, 1983.

Table 2. *Books and Monographs on Bone Tumors*

Coley, B.L.: Neoplasms of Bone. Hagerstown, Hoebner Medical Division, 1948.

Geschickter, C.F., and Copeland, M.M.: Tumors of Bone. 3rd Ed. Philadelphia, J.B. Lippincott, 1949.

Jaffee, H.L.: Tumors and Tumorous Conditions of the Bone and Joints. Philadelphia, Lea & Febiger, 1958.

Barry, H.C.: Paget's Disease of Bone. Baltimore, Williams and Wilkins, 1969.

Dahlin, D.C.: Bone Tumors. Springfield, Charles C Thomas, 1970.

Jelliffe, A.M., and Strickland, B. (eds.): Symposium Ossium. Edinburgh, E & S Livingstone, 1970.

Lichtenstein, L.: Disorders of Bone and Joints. St. Louis, C.V. Mosby, 1970.

Waldenstrom, J.: Diagnosis and Treatment of Multiple Myeloma. New York, Grune & Stratton, 1971.

Boseman, E.F. (ed.): The Human Spine in Health and Disease. New York, Grune & Stratton, 1971.

Hirohata, K. and Morimoto, K.: Ultrastructure of Bone and Joint Diseases. New York, Grune & Stratton, 1971.

Lodwick, G.S.: The Bone and Joints: An Atlas of Tumor Radiology. Chicago, Year Book Medical Publishers, 1971.

Spjut, H.J., Dorfman, H.D., Fechner, R.E., and Ackerman, L.V.: Tumors of Bone and Cartilage. Atlas of Tumor Pathology. 2nd Series. Washington, D.C., Armed Forces Institute of Pathology, 1971.

Jaffe, H.L.: Metabolic, Degenerative and Inflammatory Diseases of Bone and Joints. Philadelphia, Lea & Febiger, 1972.

Lichtenstein, L.: Bone Tumors, 4th Ed. St. Louis, C.V. Mosby, 1972.

Price, C.H.G., and Ross, F.G.M.: Bone—Certain Aspects of Neoplasia. London, Butterworth, 1972.

Netherlands Committee on Bone Tumors. Radiological Atlas of Bone Tumors. Vol. 2. Baltimore, Williams & Wilkins, 1973.

Vinogradova, T.: Bone Neoplasms. Moscow, Izdatelstvo Meditsina, 1973.

Rasmussen, H., and Bordier, P.: The Physiological and Cellular Basis of Metabolic Bone Disease. Baltimore, Williams & Wilkins, 1974.

Aegerter, E., and Kirkpatric, J.A.: Orthopedic Diseases. Physiology, Pathology, Radiology. 4th Ed. Philadelphia, W.B. Saunders, 1975.

Ackerman, L.V., Spjut, H.J., and Abell, M.R.: Bones and Joints. International Academy of Pathology Monograph. Baltimore, Williams & Wilkins, 1976.

Grundman, E. (ed.): Malignant Bone Tumors. New York, Springer-Verlag, 1976.

Hajdu, S.I., and Hajdu, E.O.: Cytopathology of Sarcomas and Other Nonepithelial Malignant Tumors. Philadelphia, W.B. Saunders, 1976.

Steinbock, R.T.: Paleopathological Diagnosis and Interpretation. Bone Diseases in Ancient Human Populations. Springfield, Charles C Thomas, 1976.

Murray, R.O., and Jacobson, E.H.: The Radiology of Skeletal Disorders. 2nd Ed. New York, Churchill Livingstone, 1977.

Huvos, A.G.: Bone Tumors. Diagnosis, Treatment and Prognosis. Philadelphia, W.B. Saunders, 1979.

Mirra, J.M.: Bone Tumors. Diagnosis and Treatment. Philadelphia, J.B. Lippincott, 1980.

Schajowicz, F.: Tumors and Tumor-Like Lesions of Bones. New York, Springer-Verlag, 1982.

Wilner, D. (ed.): Radiology of Bone Tumors and Allied Disorders. Philadelphia, W.B. Saunders, 1982.

Bogumill, G.P., and Schwamm, H.A.: Orthopaedic Pathology. Philadelphia, W.B. Saunders, 1984.

Bullough, P.G., and Vigorita, V.J.: Atlas of Orthopaedic Pathology with Clinical and Radiologic Correlations. Baltimore, University Park Press, 1984.

TABLE 3. *Microscopically Identical Lesions Known by Different Names in Soft Tissue and Bone Pathology*

SOFT TISSUES		BONE
Tendosynovitis	=	Aneurysmal bone cyst
Benign fibrous histiocytoma	=	Nonossifying fibroma
Xanthogranuloma	=	Nonossifying fibroma
Fibromatosis	=	Fibrous dysplasia
Fibroma	=	Desmoplastic fibroma
Desmoid tumor	=	Desmoplastic fibroma
Histiocytic fibrous histiocytoma	=	Giant cell tumor
Osseous metaplasia	=	Desmoplastic bone formation
Hemangiosarcoma	=	Hemangioendothelioma
Scar	=	Callus
Fat necrosis	=	Bone infarct
Abscess	=	Osteomyelitis
Synovial chondromatosis	=	Chondroblastoma
Granulocytic sarcoma	=	Granulocytic leukemia

chondrosarcomas show the histologic characteristics that are generally accepted as characterizing chondrosarcoma of bone." The concept that giant cell tumor of bone is microscopically different from giant cell tumor of soft tissues has no foundation.

A peculiar difference between soft tissue and bone tumors is that a host of reactive lesions and benign tumors are not known to occur in bone, possibly because they are asymptomatic and remain undetected. Consequently, the introduction of new names and terms that do not indicate the tumor's pathophysiologic or histogenetic roots is useless and confusing. Not to realize that osteoid is a specialized form of collagen, or that there is no such thing as synovioma, or that the so-called "osteoclasts" in bone are multinucleated histiocytic forms in soft tissues will lead to the continuation of misuse of terms, propagation of misunderstanding, and mismanagement of patients.

There is no statistically reliable, all-inclusive figure for soft tissue and bone tumors. However, it is estimated that over 6,000 new cases of soft tissue sarcomas and fewer than 2,000 bone sarcomas are diagnosed annually in the United States. While soft tissue sarcomas represent about 1% of all malignant neoplasms in adults and 7% in children, skeletal sarcomas are responsible for less than 0.2% of malignant tumors in adults and 5% in children. Because of their rarity, soft tissue and bone sarcomas represent a rather minor part of the diagnostic experience of pathologists; about a dozen centers are large enough to see a sufficient number of soft tissue and bone tumors to be familiar with the microscopic appearance of all variants. In addition, soft tissue and bone tumors are heterologous lesions, and their wide morphologic range reflects the complexity of mesenchymal tissues from which they stem; there are over 200 more or less well-defined microscopic forms of soft tissue and bone tumors, of which 72 are malignant neoplasms, 69 benign neoplasms, and 85 reactive, non-neoplastic lesions that may resemble neoplasms (Fig. 6).

We must also be cognizant that benign non-neoplastic and benign neoplastic lesions outnumber malignant neoplasms by a margin of about 100 to 1. Due to overlapping morphology of many benign and malignant lesions and the inability of pathologists to recognize "borderline connective tissue lesions" and "sarcoma in situ," connective tissue neoplasms traditionally are called either benign or malignant. This type of classification is fueled by the view held by some workers that most soft tissue and bone sarcomas are malignant de novo, and very few have benign precursors. While it is tempting to take issue with such thinking (Table 4), it is perhaps sufficient to point out that there was a time when lesions, for example, carcinoma in situ of the uterine cervix, the urinary bladder, and the stomach, were not recognized. Some physicians have held on to their views for decades, believing that adenocarcinoma of the colon never develops in a polyp, or that a mammary lesion such as lobular carcinoma in situ does not exist.

Without doubt, a number of problems in regard to the histogenesis and pathology of sarcomas remain to be solved. Whatever the histogenesis, pathologists may enhance the diagnosis of soft tissue and bone neoplasms by knowing the clinical presentation, size, site, radiologic appearance, and age of the patient. It is needless to say that in most cases close cooperation between the surgeon, radiologist, and pathologist is essential in order to arrive at an accurate microscopic diagnosis. It must also be recognized that the role of the pathologist is not ancillary but crucial in the diagnosis of soft tissue and bone tumors, a difficult and complex task with serious therapeutic ramifications. Those who doubt the complexity of diagnosis should remember Osler's advice to his clinical colleagues, "You are as good as your pathologist."

No pathologist should render final diagnosis without having access to clinical history and radiologic findings, for as Lauren Ackerman said, "The pathologist can make enough errors with all available information." The more information the pathologist has the more information the pathologist can give. The proper time to evaluate the clinical information and radiologic findings is after, and not before, the histology has been assessed, but prior to issuing a definitive diagnosis. If a pathologist has been forced to make a diagnostic decision without accurate clinical information, or has been given misinformation and has committed an error in diagnosis, the blame should be placed on those who misled the pathologist. Also, the practice of seeking an unbiased second opinion, that is, asking the pathologist to render a diagnosis without clinical and other information pertinent to the case, should be discouraged. One does well to remember the view held by Boyd: "It is the high function of the pathologist not merely to attach correct labels to lesions, but to reconstruct the course of events from the earliest inception of disease to the final moment of life."

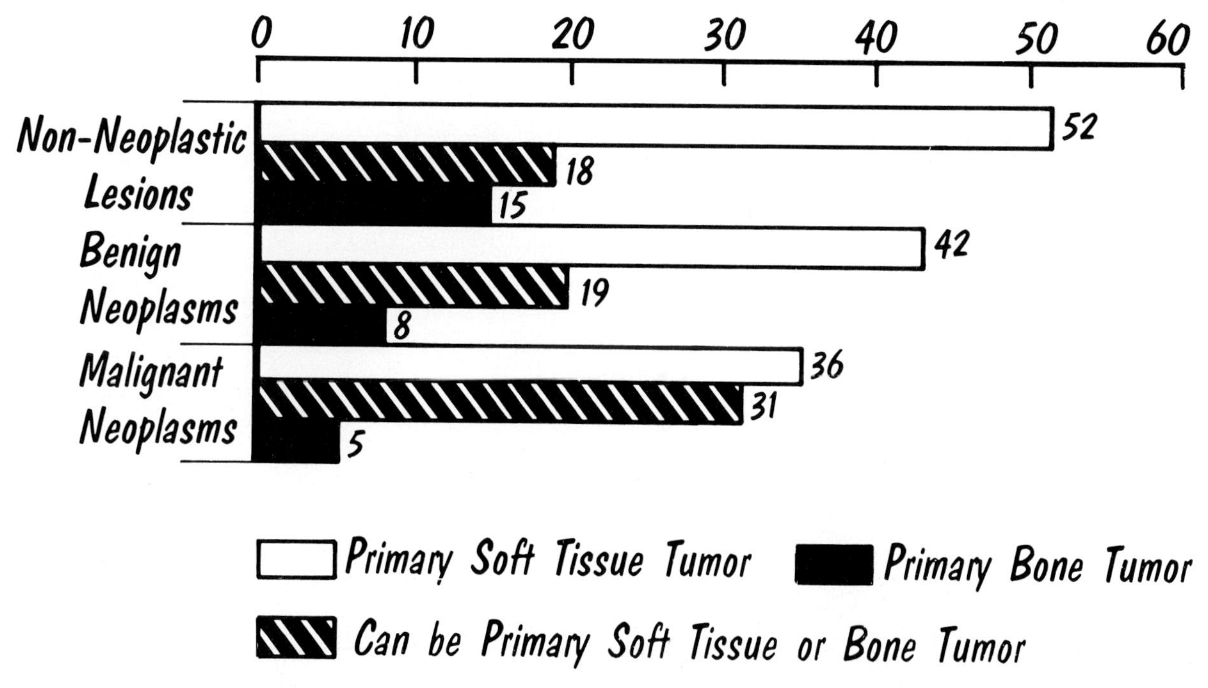

Fig. 6. Comparison of approximate number of soft tissue and bone tumors.

The better understanding of sarcomas was held back for centuries by, among other things, misuse and misunderstanding of various names and definitions that were often due to the unavailability of accurate information or to plain ignorance. Galen defined sarcomas in such a way that many forms of inflammation or infectious swelling, as well as all sorts of benign neoplasms, were called "sarcomas." As late as the mid-1800s, Rokitansky stated that "Sarcomata represent benign new-growths, they are always purely local affections, they are curable by complete extirpation: that is, they do not recur at the same spot, and still less do they multiply in other localities."

Current classification schemes of soft tissue and bone tumors separate tumors whose cell of origin is thought to be known and those whose histogenesis is not known. These schemes are the result of the work of many pathologists, to mention just a few, Johannes Muller, Virchow, James Ewing, Henry Jaffe, and Stout. As new diagnostic techniques develop, it will be possible to establish the histogenesis of soft tissue and bone tumors more and more accurately; and in the great majority of cases, the present classification schemes will need revision. It is hoped though that frequent revision will not take place.

Histogenetic classification, based on characteristic microscopic features, permits subdivision of tumors regardless of the anatomic site of the tumor or age of the patient. A classification that departs from such a scheme would be equivalent to comparing apples with oranges or to classifying dogs as the gypsies did in former times—according to

Table 4. *Some Benign Lesions Known to Undergo Malignant Transformation*

Benign fibrous histiocytoma
Scar
Fibromatosis
Fibrous dysplasia
Tendosynovitis
Lipoma
Angiomyolipoma
Angiomyoma
Myositis ossificans
Leiomyoma
Benign glomus tumor
Infarct
Lymphedema
Neurofibroma
Benign schwannoma
Benign paraganglioma
Benign mesothelioma
Chondroblastoma
Enchondroma
Osteochondroma
Osteoblastoma
Paget's disease
Osteomyelitis
Benign thymoma
Benign granular cell tumor
Meningioma
Benign cystosarcoma phyllodes
Benign mesenchymoma

TABLE 5. *Histogenetic Classification of Soft Tissue and Bone Tumors**

	NON-NEOPLASTIC LESIONS	BENIGN NEOPLASMS	MALIGNANT NEOPLASMS
Undifferentiated Connective Tissue Tumors	● Aneurysmal bone cyst ● Giant cell granuloma ● Hyperparathyroidism ○ Xanthogranuloma ◑ Foreign body granuloma	◑ Benign fibroblastic fibrous histiocytoma ◑ Benign histiocytic fibrous histiocytoma ◑ Benign pleomorphic fibrous histiocytoma ● Nonossifying fibroma	◑ Malignant fibroblastic fibrous histiocytoma ◑ Malignant histiocytic fibrous histiocytoma ◑ Malignant pleomorphic fibrous histiocytoma
Fibrous Tissue Tumors	○ Fasciitis ○ Fasciitis ossificans ○ Scar ○ Keloid ○ Fibromatosis ○ Fibromatosis ossificans ● Fibrous dysplasia ○ Elastofibroma ○ Collagenoma	○ Fibroma ● Desmoplastic fibroma ● Ossifying fibroma	◑ Desmoid tumor ◑ Fibroblastic fibrosarcoma ◑ Pleomorphic fibrosarcoma
Tendosynovial Tumors	○ Tendosynovitis ◑ Tendosynovial cyst ○ Tendosynovial chondromatosis ○ Tendosynovitis ossificans ○ Granulomatous tendosynovitis ○ Rheumatoid arthritis ◑ Osteoarthritis ○ Gouty arthritis ○ Pseudogout ○ Carpal tunnel syndrome	○ Fibroma ○ Lipoma ○ Benign histiocytic fibrous histiocytoma	○ Biphasic tendosynovial sarcoma ○ Monophasic tendosynovial sarcoma, spindle cell type ○ Monophasic tendosynovial sarcoma, pseudoglandular type ○ Epithelioid sarcoma ○ Clear cell sarcoma ○ Chordoid sarcoma ○ Malignant histiocytic fibrous histiocytoma
Adipose Tissue Tumors	◑ Fat necrosis ◑ Lipogranuloma ○ Proliferative panniculitis ○ Panniculitis ossificans ○ Lipodystrophia ○ Adiposis dolorosa ○ Steatopygia ○ Piezogenic papule	◑ Well-differentiated lipoma ◑ Myxoid lipoma ◑ Fibroblastic lipoma ○ Lipoblastoma ◑ Pleomorphic lipoma ◑ Angiolipoma ○ Angiomyolipoma ○ Myelolipoma ○ Hibernoma	◑ Well-differentiated liposarcoma ◑ Myxoid liposarcoma ◑ Lipoblastic liposarcoma ◑ Fibroblastic liposarcoma ◑ Pleomorphic liposarcoma
Muscle Tumors	○ Proliferative myositis ○ Myositis ossificans ○ Fibromatosis ○ Atrophy ○ Dystrophy ○ Polymyositis ○ Rhabdomyolysis	◑ Leiomyoma ○ Leiomyomatosis ○ Rhabdomyoma ○ Lipoma ◑ Angiomyoma	◑ Leiomyoblastoma ◑ Leiomyosarcoma ○ Embryonal rhabdomyosarcoma ○ Alveolar rhabdomyosarcoma ○ Myxoid rhabdomyosarcoma ○ Rhabdomyoblastoma ○ Pleomorphic rhabdomyosarcoma
Tumors of Vessels	○ Pyogenic granuloma ○ Angifollicular lymphoid hyperplasia ◑ Arteriovenous malformation ◑ Hereditary hemorrhagic telangiectasia ◑ Vasculitis ◑ Infarct ○ Lymphedema ○ Cystic hygroma	◑ Capillary hemangioma ◑ Cavernous hemangioma ◑ Arteriovenous hemangioma ○ Venous hemangioma ○ Hypertrophic hemangioma ◑ Hemangiomatosis ◑ Papillary endothelial hyperplasia ○ Hemangioblastoma ○ Angiofibroma ○ Angiomyoma ◑ Angiolipoma ○ Angiomyolipoma ◑ Benign glomus tumor ○ Lymphangioma ○ Lymphangiomatosis ○ Lymphangiomyoma ○ Lymphangiomyomatosis	◑ Hemangiopericytoma ◑ Hemangiosarcoma ○ Lymphangiosarcoma ◑ Leiomyosarcoma ◑ Malignant glomus tumor ○ Malignant angiomyolipoma

	NON-NEOPLASTIC LESIONS	BENIGN NEOPLASMS	MALIGNANT NEOPLASMS
Tumors of Peripheral Nerve	○ Traumatic neuroma ○ Hypertrophy ○ Degeneration	◑ Neurofibroma Pacinian neurofibroma Myxoid neurofibroma Plexiform neurofibroma Benign Triton tumor Neurofibromatosis ◑ Benign schwannoma Benign glandular schwannoma Benign nevoid schwannoma Benign pigmented schwannoma ◑ Benign undifferentiated peripheral nerve tumor Benign neuroepithelioma	◑ Neurofibrosarcoma Malignant Triton tumor ◑ Malignant schwannoma Malignant glandular schwannoma Malignant nevoid schwannoma Malignant pigmented schwannoma ◑ Malignant undifferentiated peripheral nerve tumor Primitive neuroectodermal tumor Malignant neuroepithelioma
Tumors of Autonomic Nerve		◑ Neuroma ◑ Ganglioneuroma ○ Benign paraganglioma	○ Neuroblastoma ○ Ganglioneuroblastoma ○ Malignant paraganglioma
Mesothelial Tumors	○ Mesothelial hyperplasia ○ Hydrocele ○ Mesothelial cyst	○ Benign epithelioid mesothelioma ○ Benign fibrous mesothelioma ○ Adenomatoid tumor	○ Malignant epithelioid mesothelioma ○ Malignant fibrous mesothelioma
Cartilage-producing Tumors	◑ Chondroid metaplasia ○ Tendosynovial chondromatosis ● Callus ● Prolapsed intervertebral disc	● Chondroblastoma ● Chondromyxoid fibroma ◑ Chondroma ● Osteochondroma	◑ Well-differentiated chondrosarcoma ◑ Poorly differentiated chondrosarcoma ◑ Myxoid chondrosarcoma ◑ Mesenchymal chondrosarcoma
Osteoid-producing Tumors	● Paget's disease ● Enostosis ◑ Osteoid metaplasia ● Callus ◑ Periostitis ○ Fibromatosis ossificans ○ Fasciitis ossificans ○ Panniculitis ossificans ○ Myositis ossificans ○ Fibrodysplasia ossificans progressiva	● Osteoblastoma ● Osteoid osteoma ○ Osteoma ● Ossifying fibroma	◑ Osteosarcoma Fibrous histiocytic osteosarcoma Fibrosarcomatous osteosarcoma Chondrosarcomatous osteosarcoma Osteoblastic osteosarcoma Telangiectatic osteosarcoma Sclerosing osteosarcoma ◑ Parosteal osteosarcoma ● Paget's sarcoma
Lymphoreticular Tumors	◑ Eosinophilic granuloma ◑ Histiocytosis ◑ Lymphoid hyperplasia ○ Lymphocytoma cutis ● Leukocytosis ○ Mastocytosis ● Osteomyelitis ◑ Plasma cell granuloma ○ Thymic cyst ○ Angiofollicular lymphoid hyperplasia ○ Angiomatous lymphoid hamartoma	○ Benign thymoma	◑ Granulocytic sarcoma ● Leukemia ○ Mycosis fungoides ◑ Malignant lymphoma ○ Hodgkin's disease ◑ Plasmacytoma ● Plasma cell myeloma ○ Malignant thymoma
Miscellaneous Tumors	◑ Amyloidoma ◑ Gaucher's disease ◑ Osteogenesis imperfecta ◑ Mucopolysaccharidoses ● Hypophosphatasemia ● Hyperphosphatasemia ● Osteoporosis ● Osteomalacia	○ Benign granular cell tumor ○ Meningioma ○ Benign cystosarcoma phyllodes ◑ Benign mesenchymoma ● Ecchordosis physaliphora	○ Malignant granular cell tumor ○ Alveolar soft part sarcoma ○ Kaposi's sarcoma ○ Meningeal sarcoma ○ Malignant cystosarcoma phyllodes ◑ Ewing's sarcoma ● Chordoma ● Adamantinoma ◑ Angioendotheliomatosis ◑ Radiation induced sarcoma ◑ Chemotherapy induced sarcoma ◑ Malignant mesenchymoma ◑ Undifferentiated sarcoma

*○ Primary soft tissue tumor; ● primary bone tumor; ◑ can be primary soft tissue or bone tumor.

the length of the tail, the size of the body, the shape of the ears, and whether the dog belonged to the king. No one would deny that soft tissue and bone tumors are often difficult to classify, especially in limited material, according to their histogenesis. The histogenesis of a tumor is, after all, a conclusion arrived at by inference and deduction based on experience, not by seeing the cell of origin. Although determining the histogenesis of a tumor may remain problematic and controversial in many instances, it should not be ignored and must be included in therapeutic planning.

Contrary to the wishes of some writers to create more and more new labels, it is perhaps time to concentrate on the comparison, organization, and redefinition of existing entities. Lumping histogenetically related but microscopically and prognostically different lesions is not done for the sake of "convenience," but because the lesions are common in origin. Furthermore, it is logical, scientific, and practical: who would deny that uterine stromal sarcoma, leiomyosarcoma, and mesodermal mixed tumor, despite different histologic appearance and behavior, are histogenetically related uterine neoplasms? Consequently, it should not take too much effort to recognize that histogenetically such diverse neoplasms may originate in a common organ or an organ system, although they look unrelated in microscopic appearance: who would deny the microscopic and behavioral dissimilarity of various gliomas, pulmonary neoplasms, bone neoplasms, skin tumors, or peripheral nerve tumors?

Some writers do not attempt to distinguish between benign neoplasms and non-neoplastic, reactive lesions because they believe "it has no practical value." So simplistic an approach cannot be anything but the result of detachment from clinical reality. Is it practical to distinguish inflammatory pseudopolyps of the colon from villous adenoma, parathyroid hyperplasia from parathyroid adenoma, giant cell granuloma from giant cell tumor, or mastitis from duct hyperplasia? If the answer is affirmative, I propose that the subdivision of soft tissue and bone tumors for non-neoplastic lesions, benign neoplasms, and malignant neoplasms is to be retained until there is satisfactory proof that such a classification is obsolete (Table 5).

Soft tissue and bone tumors often pose a diagnostic challenge and, no matter how well experienced the pathologist is, there will be a certain number of tumors that cannot be placed in existing categories because a number of tumors are characterized by microscopic similarities. There is no perfect classification and none should be written in stone. In 1939, Bucy and Gustafson said that "Classification must be regarded as providing merely arbitrary pockets into which we can place tumors in order that they may be more easily considered." It would be wrong to call every pleomorphic neoplasm that contains a fibrous histiocytic, undifferentiated mesenchymal component a fibrous histiocytoma, just as it would be a mistake to label all neoplasms that have pericytic areas hemangiopericytomas.

Like other neoplasms, soft tissue and bone neoplasms must have a beginning, developing from the preneoplastic and incipient (borderline or equivocal) phases to the established (or in-situ) and invasive phases (Fig. 7). It must be remembered that a biopsy from a tumor, at a given phase, is like a single frame from a movie film and the time it takes to progress from one phase to another may vary from tumor to tumor, for as James Ewing said, "Beyond the autonomy of growth, it is difficult to add any element that will apply to all tumors."

Our inability to detect the submicroscopic changes that take place in the DNA during the preneoplastic or induction phase hinder early detection. It may require years before the neoplasm passes through the preneoplastic phase and enters the incipient (proliferative cellular, atypical or borderline) phase. A neoplasm may remain for months or years in the incipient phase and may be characterized by an unclearly defined cytologic and histologic mutation; as a rule, it usually remains small and asymptomatic, and it is seldom diagnostic. Most sarcomas in the established (in-situ) phase are symptomatic, reach variable size, and can be diagnosed, though not without difficulty. Sarcomas may remain in the invasive phase for weeks or months prior to entering the metastatic phase.

Most malignant soft tissue and bone neoplasms are diagnosed in the invasive phase because they are symptomatic. In general, they are about 5 cm in size, exhibit characteristic growth patterns, and show identifiable cytologic abnormalities (Table 6). The earlier the lesion, the greater the margin of uncertainty in histologic diagnosis. The fact that sarcomas can be mistaken for benign lesions is proof of their deceptively harmless microscopic appearance during the initial phases of growth. This complicates matters further, for as Mackenzie said, "Innocent morphology is not always accompanied by innocent behavior." Experience may reduce the margin of uncertainty but does not abolish it. If a clear distinction between a benign and a malignant neoplasm cannot be made, the use of the term "borderline" is advisable to express our uncertainty as to its potential behavior (Table 7). The interpretation of the borderline or histologically equivocal neoplasm is a difficult one, and may vary from pathologist to pathologist because borderline neoplasms show the earliest structural changes. Once the neoplasm has entered the metastatic phase, the local tissue conditions, resistance of the host, aggressiveness of the neoplastic clone, and effectiveness of the therapy are what decide the duration and the outcome of the disease.

The cause of soft tissue and bone tumors is, with a few exceptions, unknown. The list of forms and types non-neoplastic lesions may assume is endless. Most known benign neoplasms have known malignant counterparts. Those few benign neoplasms that do not are listed in Table 8. Likewise, there are only a dozen malignant neoplasms without corresponding benign forms (Table 9).

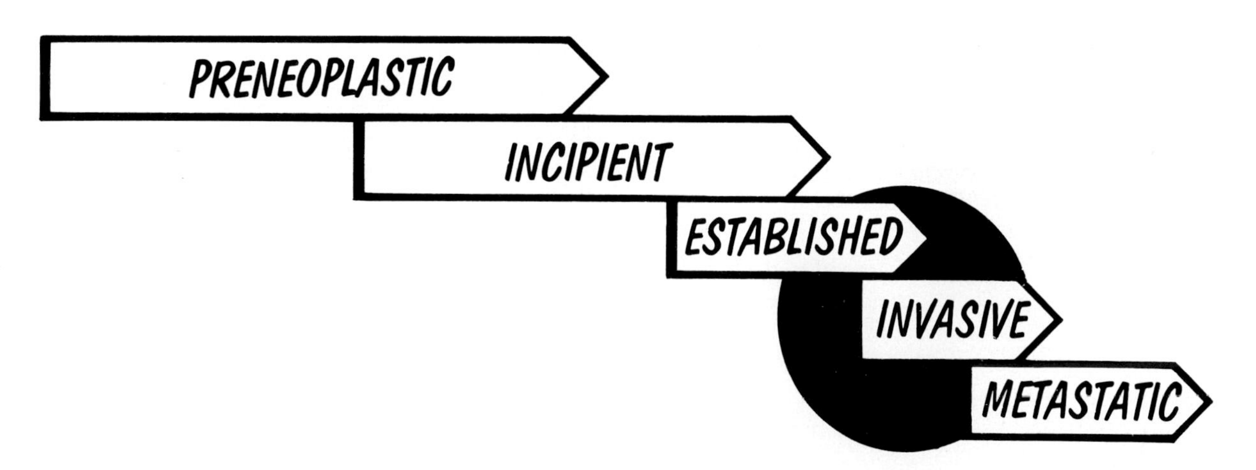

Fig. 7. Theoretical growth phases of sarcomas. The incipient or borderline phase is the link between the preneoplastic (nondiagnostic) phase and the established (earliest diagnostic or sarcoma in-situ) phase. The shadowed circle indicates when most sarcomas are diagnosed (see Table 6).

Ideally, a system of classification lists tumors according to histogenesis, site, cell of origin, dominant pattern, and the probable clinical outcome. The diverse cell morphology and overlapping histologic patterns of soft tissue and bone tumors make exact classification hazardous at times and require thorough microscopic evaluation often coupled with both special studies and consideration of pertinent clinicopathologic findings. Not realizing that neoplasms as well as reactive lesions occasionally show flagrant departures from the usual patterns may lead to misclassification, a danger warranting more reliance on cell morphology, for example, in certain muscle and nerve tumors. Reliable assessment and identification of intraskeletal and extraskeletal connective tissue tumors depend on the pathologist's ability to distinguish reactive, non-neoplastic, lesions from benign and malignant neoplasms. Such an evaluation requires, among other things, knowledge of the names, definitions, and synonyms of various entities, as well as up to date familiarity with the most significant publications. Tables 10, 11, and 12 are designed to give such information with easy to follow cross-references to pertinent figures and tables.

Despite John Hunter's statement that "Of all things on the face of the earth definitions are the most cursed," just like classifications definitions are needed to put what we know in order and to place our knowledge in perspective. Definitions for bone tumors have been in existence for a long time, but for soft tissue tumors definitions do not exist. The attempt presented here should be viewed with the understanding that definitions of tumors can remain consistent only by changing to reflect new developments and discoveries.

The various malignant neoplasms, benign neoplasms, and especially the reactive, non-neoplastic, lesions are known to many pathologists by different synonyms. The use of synonyms and nebulous terms should be discouraged, but to prevent misunderstanding a list of the most commonly mentioned synonyms is presented in Tables 10, 11, 12, and 13.

No text can be all inclusive, especially if it is designed as a practical diagnostic guide to the surgical pathologist, much like a cookbook to the cook. In Table 14, readers will find a list of entities that are not considered in detail in this text, and are urged to consult some of the standard texts on these topics listed in Tables 1 and 2.

A diagnosis that is made on the basis of microscopic appearance largely depends on the pathologist's visual memory. The tissue pattern, cell morphology, appearance of the stroma, and products of the cells should be studied as objectively as possible, with consideration of the patient's age, and the site and size of the tumor (Tables 15–18). Admittedly, the classification of soft tissue and bone tumors is histogenetic; pathologists who do not see a substantial number of connective tissue tumors may find the histogenetic classification difficult, but they are able to recognize similar or identical growth patterns and cells by microscopic comparison that will lead to the identification and proper classification of a given tumor. Roger Williams said that "Comparison is the greatest organon of biological and consequently also of pathological research." Undoubtedly, most histologic examinations are a matter of pattern and cell recognition by comparison.

Unfortunately there are very few criteria, other than histologic pattern and cell morphology, by which malignant soft tissue and bone neoplasms can be diagnosed. Special stains, histochemical stains, electron microscopy, immunofluorescence microscopy, and immunohistochemistry have not fulfilled every expectation. All of these techniques are undoubtedly helpful in establishing the histogenesis and the cell of origin of a given neoplasm, but none is helpful in distinguishing a malignant neoplasm from a benign one.

The diagnostic problems soft tissue and bone tumors present to the pathologist are many. Often the tissue ob-

TABLE 6. *Estimated Time Sarcomas Remain in Various Growth Phases**

	PRENEOPLASTIC PHASE	INCIPIENT PHASE	ESTABLISHED PHASE	INVASIVE PHASE	METASTATIC PHASE
Malignant fibroblastic fibrous histiocytoma	yrs	yrs	yrs	mos	mos–yrs
Malignant histiocytic fibrous histiocytoma	yrs	yrs	mos	mos	mos–yrs
Malignant pleomorphic fibrous histiocytoma	yrs	yrs–mos	mos–wks	mos–wks	mos
Desmoid tumor	yrs	yrs	yrs	yrs–mos	yrs
Fibroblastic fibrosarcoma	yrs	yrs	yrs–mos	mos	mos–yrs
Pleomorphic fibrosarcoma	yrs	yrs–mos	yrs–mos	mos	mos
Biphasic tendosynovial sarcoma	yrs	yrs	yrs–mos	mos	mos–yrs
Monophasic tendosynovial sarcoma, spindle cell type	yrs–mos	yrs–mos	mos–wks	wks	mos–wks
Monophasic tendosynovial sarcoma, pseudoglandular type	yrs	yrs	yrs–mos	mos	mos–yrs
Epithelioid sarcoma	yrs	yrs–mos	mos	wks	mos
Clear cell sarcoma	yrs–mos	yrs–mos	mos	wks	mos
Chordoid sarcoma	yrs	yrs	yrs	yrs–mos	yrs
Well-differentiated liposarcoma	yrs	yrs	yrs	yrs	yrs
Myxoid liposarcoma	yrs	yrs	yrs	yrs	yrs
Lipoblastic liposarcoma	yrs–mos	yrs–mos	mos	wks	mos
Fibroblastic liposarcoma	yrs	yrs–mos	mos	mos	yrs
Pleomorphic liposarcoma	yrs–mos	yrs–mos	mos–wks	wks	mos
Leiomyoblastoma	yrs	yrs–mos	mos	mos–wks	mos–yrs
Leiomyosarcoma	yrs	yrs–mos	mos	mos–wks	mos–yrs
Embryonal rhabdomyosarcoma	yrs	yrs–mos	mos	mos–wks	mos
Alveolar rhabdomyosarcoma	yrs–mos	yrs–mos	mos–wks	mos–wks	mos
Myxoid rhabdomyosarcoma	yrs	yrs–mos	mos	mos	mos
Rhabdomyoblastoma	yrs	yrs–mos	mos	mos	mos
Pleomorphic rhabdomyosarcoma	yrs–mos	yrs–mos	mos–wks	wks	mos
Hemangiopericytoma	yrs	yrs	yrs–mos	mos	mos–yrs
Hemangiosarcoma	yrs–mos	yrs–mos	mos–wks	wks	mos
Lymphangiosarcoma	yrs	yrs–mos	mos–wks	wks	mos
Malignant glomus tumor	yrs	yrs	yrs	mos	yrs
Neurofibrosarcoma	yrs	yrs	yrs–mos	mos	mos–yrs
Malignant schwannoma	yrs	yrs	yrs	mos	mos–yrs
Malignant undifferentiated peripheral nerve tumor	yrs	yrs–mos	mos	mos–wks	mos
Neuroblastoma	yrs–mos	yrs–mos	mos–wks	wks	mos
Malignant paraganglioma	yrs	yrs	yrs	mos–yrs	yrs
Malignant epithelioid mesothelioma	yrs	yrs–mos	mos	wks	mos
Malignant fibrous mesothelioma	yrs	yrs	yrs	mos	mos–yrs
Well-differentiated chondrosarcoma	yrs	yrs	yrs–mos	mos	yrs
Poorly differentiated chondrosarcoma	yrs	yrs–mos	mos–wks	mos–wks	mos
Mesenchymal chondrosarcoma	yrs	yrs–mos	mos	mos–wks	mos
Osteosarcoma	yrs–mos	yrs–mos	mos–wks	wks	mos
Parosteal osteosarcoma	yrs	yrs	yrs	yrs	yrs
Paget's sarcoma	yrs	yrs	yrs–mos	mos	mos
Granulocytic sarcoma	yrs–mos	mos	mos–wks	wks	mos–wks
Malignant lymphoma	yrs–mos	mos	mos–wks	wks	mos–yrs
Plasma cell myeloma	yrs–mos	mos	mos–wks	wks	mos
Malignant granular cell tumor	yrs–mos	yrs–mos	mos	mos–wks	mos
Alveolar soft part sarcoma	yrs	yrs	mos	mos	yrs
Kaposi's sarcoma	yrs	yrs	yrs	yrs	yrs
Malignant cystosarcoma phyllodes	yrs–mos	mos	mos	mos	yrs
Ewing's sarcoma	yrs–mos	mos	mos–wks	wks	mos
Chordoma	yrs	yrs–mos	mos	wks	mos–yrs
Chemotherapy induced sarcoma	yrs	yrs	mos–yrs	mos	mos
Adamantinoma	yrs	yrs	yrs	mos	yrs
Radiation induced sarcoma	yrs	yrs	yrs–mos	mos–wks	mos

*yrs = years; mos = months; wks = weeks.

TABLE 7. *Borderline or Atypical Forms That Can Be Difficult to Differentiate from Cellular Benign and Well-differentiated Malignant Neoplasms**

Borderline fibroblastic fibrous histiocytoma
Borderline histiocytic fibrous histiocytoma
Borderline pleomorphic fibrous histiocytoma
Borderline fibrous tissue neoplasm
Borderline adipose tissue neoplasm
Borderline smooth muscle neoplasm
Borderline vascular neoplasm
Borderline glomus tumor
Borderline peripheral nerve tumor
Borderline paraganglioma
Borderline mesothelioma
Borderline cartilaginous tumor
Borderline meningioma
Borderline cystosarcoma phyllodes
Borderline mesenchymoma

*See Fig. 7.

TABLE 8. *Benign Neoplasms That Have No Known Malignant Counterpart*

Angiolipoma
Hibernoma
Papillary endothelial hyperplasia
Hemangioblastoma
Lymphangiomyoma

TABLE 9. *Malignant Neoplasms That Have No Known Benign Counterpart**

Monophasic tendosynovial sarcoma
Epithelioid sarcoma
Clear cell sarcoma
Chordoid sarcoma
Hemangiopericytoma
Primitive neuroectodermal tumor
Neuroblastoma
Hodgkin's disease
Alveolar soft part sarcoma
Kaposi's sarcoma
Ewing's sarcoma
Chordoma
Myxopapillary ependymoma
Adamantinoma
Angioendotheliomatosis

*See Table 4.

tained is not sufficient in amount and not representative of the lesion. Aspiration biopsy techniques, cytologic or tissue, are less than optimal diagnostic techniques for the definitive assessment of primary soft tissue and bone tumors.

The claim of the novice, unexperienced, and unscrupulous, that aspiration biopsy is safe, accurate, and a substitute for surgical biopsy of primary soft tissue and bone lesions is untenable. Needle biopsy and aspirate can be useful in evaluating recurrent and metastatic neoplasms (see Appendix), but the value of these techniques is very limited in the primary diagnosis of soft tissue and bone tumors because the material obtained is (1) very often not representative of the lesion and (2) insufficient for the unequivocal and precise histologic typing and grading of neoplasms on which therapeutic decisions depend. Similarly, frozen section diagnosis of soft tissue and bone tumors should be practiced with full knowledge of the limitation of the technique, and with the awareness that it is prudent to defer a diagnostic decision until paraffin sections are available. The pathologist who relies on suboptimal biopsy material is like the fellow who tries to guess the identity of the bird in a covered cage from a plucked feather.

It is hoped that the pathologist who has a full knowledge of age (see Chapter 3) and sex (see Chapter 4) of the patient, site (see Chapter 5), size (see Chapter 6), and radiologic (see Chapter 7) and gross (see Chapters 8 and 9) appearance of the lesion will be able to reach a diagnosis with the microscope by examining generous and technically adequate tissue samples. In the case of malignant neoplasms the microscopic examination must end with assessment of the histologic grade (see Chapter 14). It is most important that the pathologist, surgeon, and oncologist agree on the best course of action and advise the patient accordingly, but it must also be recognized that the opinion of all involved parties is based on the pathologist's examination. If the pathologist is unable to render a definitive diagnosis, the case should be referred to a consultant with special expertise. Similarly, if the clinician, or the institution, is not prepared to carry out definitive treatment, patients with suspected soft tissue or bone tumors should be referred to another institution prior to biopsy attempts.

The role of the pathologist in diagnosing soft tissue and bone tumors is to determine whether the lesion is reactive, non-neoplastic, or neoplastic, and if neoplastic, whether the lesion is benign or malignant. Such a diagnosis cannot be done without familiarity with the histogenetic classification, yet knowledge of histogenetic classification does not necessarily lead to accurate diagnosis. An accurate diagnosis can be made only by accurate recognition of the lesion's growth pattern, familiarity with the cell morphology, the appearance of the stroma, and ability to analyze the product of the cells by using special techniques (see Chapters 10–13).

The chapters that follow are designed to ensure accurate and dependable diagnostic assessment of the most common reactive and neoplastic soft tissue and bone tumors, and to minimize the inconsistency as well as the danger of misdiagnosis.

Table 10. *Non-neoplastic Lesions*

DEFINITIONS	SYNONYMS AND RELATED TERMS
Aneurysmal bone cyst *is an osteolytic and expansile intramedullary non-neoplastic lesion. It is composed of blood-filled cystic spaces, partly lined by granular epithelioid cells, and solid areas containing epulis type giant cells and slender spindle cells.*	Hemorrhagic bone cyst, atypical giant cell tumor, intraskeletal villonodular synovitis, angiomatoid fibrous histiocytoma
Giant cell granuloma *is a non-neoplastic lesion of maxillofacial and cranial bones. It is composed of granular epithelioid and small isomorphic epulis type giant cells in hemorrhagic nonfibrillar stroma.*	Giant cell reparative granuloma
Hyperparathyroidism *presents in bone as a tender swelling of slow growth and is associated with elevated serum calcium. Microscopically, it resembles benign histiocytic fibrous histiocytoma (benign giant cell tumor).*	Brown tumor
Xanthogranuloma *is a reactive cutaneous lesion of infants and young children. It is formed by short and slender spindle cells in arranged pattern and granular epithelioid and Touton giant cells in finely reticular and myxoid stroma that may contain eosinophiles.*	Atypical juvenile xanthoma, nevoxanthoendothelioma, congenital xanthoma multiplex
Fasciitis *is a self-limited, infiltrative or nodular, non-neoplastic lesion of superficial soft tissues. It is composed of slender spindle cells in a spreading or arranged pattern and an inflammatory, vascular, or myxoid matrix; it may contain numerous mitotic figures, extravasated red blood cells, and metaplastic cartilage and bone.*	Nodular fasciitis, infiltrative fasciitis, subcutaneous pseudosarcomatous fasciitis, paraosteal fasciitis, proliferative fasciitis, fasciitis with metaplasia, metaplastic fasciitis, fasciitis ossificans
Keloid *is the product of non-neoplastic and disarranged proliferation of dermal and subcutaneous fibroblasts and fibrocytes, coupled with formation of scar tissue, and heavy deposits of collagen Type I.*	Cheloid, keloid fibroma, folliculitis keloidalis
Fibromatosis *is an ubiquitous, non-neoplastic, proliferative, and infiltrative soft tissue lesion. It is predominantly composed of spreading and disarranged slender fibroblasts in fibrillar or sclerosing stroma and may contain metaplastic cartilage and bone.*	See Table 13
Fibrous dysplasia *is a non-neoplastic, monostotic or polyostotic, sharply delineated lesion. Histologically, it is characterized by plump, spreading, and disarranged spindle cells in collagenous and woven bony stroma.*	Ossifying fibroma, focal osteitis fibroma, fibrosis of bone
Elastofibroma *is a non-neoplastic, self-limited lesion composed of tangled collagenous and elastic fibers in a disarranged myxomatous and necrotic matrix. It occurs almost exclusively in the soft tissues of the back.*	Elastofibroma dorsi, elastodysplasia, pre-elastofibroma
Tendosynovitis *is a non-neoplastic lesion involving tendosynovial tissues and composed of granular and clear epithelioid cells and epulis type giant cells in sheets and in papillary or villous arrangement. The matrix is often pigmented and rich in fine capillary vessels.*	Villonodular synovitis, pigmented villonodular synovitis, tendosynovitis, tendonitis, bursitis, synovial hyperplasia, benign synovioma, proliferative synovitis, angiomatoid fibrous histiocytoma
Tendosynovial cyst *is a non-neoplastic, tumor-like lesion of tendosynovial tissues. It contains colorless, viscous fluid and is found most commonly in soft tissues of hands and feet, however, it may present intraskeletally.*	Ganglion, synovial cyst, cystic myxoma, Baker's cyst, bursal cyst

SELECTED REFERENCES*	ILLUSTRATIONS†
Tillman, B.P., Dahlin, D.C., Lipscomb P.R., et al.: Aneurysmal bone cyst: An analysis of ninety-five cases. Mayo Clin. Proc. *43*:478, 1968. Biesecker, J.L., Marcove, R.C., Huvos, A.G., et al.: Aneurysmal bone cysts: A clinicopathological study of 66 cases. Cancer *26*:615, 1970.	**Figs. 96, 125, 150, 163, 509, 785, 1015, 1058**
Anderson, L., Fejerkow, O., Philipsen, H.P.: Giant cell granuloma. A clinical and histological study of 129 new cases. Acta Pathol. Microbiol. Scand. (A) *81*:600, 1973. Hirschl, S., Katz, A.: Giant cell reparative granuloma outside the jaw bone. Hum. Pathol. *5*:171, 1974.	**Figs. 119, 632, 787, 1000, 1069**
Woolner, L.B., Keating, F.R., Black, B.M.: Tumors and hyperplasia of the parathyroid glands. Review of pathological findings in 140 cases of primary hyperparathyroidism. Cancer *5*:1069, 1952.	
Helwig, E.B., Hackney, V.C.: Juvenile xanthogranuloma (nevoxanthoendothelioma). Am. J. Pathol. *30*:625, 1954.	**Figs. 607, 734, 779, 1031, 1127, 1201**
Allen, P.W.: Nodular fasciitis. Pathology *4*:9, 1972. Bernstein, K., Lattes, R.: Nodular (pseudosarcomatous) fasciitis, a non-recurrent lesion. Clinicopathologic study of 134 cases. Cancer *49*:1668, 1982.	**Figs. 82, 345, 368, 389, 403, 408, 578, 605, 621, 834, 851, 855, 858, 947, 956, 971, 991, 1020, 1030, 1045, 1080**
King, G.D., Salzman, F.A.: Keloid scars. Analysis of 89 patients. Surg. Clin. North Am. *50*:595, 1970.	**Figs. 877, 1153**
Allen, P.W.: The fibromatoses: A clinicopathologic classification based on 140 cases. Am. J. Surg. Pathol. *1*:255, 1977. Richards, R.C., Rogers, S.W., Gardner, E.J.: Spontaneous mesenteric fibromatosis in Gardner's syndrome. Cancer *47*:597, 1981.	**Figs. 17, 104, 280, 367, 385, 387, 419, 581, 582, 616, 860, 861, 873, 874, 878, 882, 1022, 1025, 1035, 1036, 1050, 1081, 1095, 1121, 1122**
Harris, W.H., Dudley, H.R., Barry, R.J.: The natural history of fibrous dysplasia. An orthopaedic, pathological and roentgenographic study. J. Bone Joint Surg. *49A*:207, 1962. Bejui-Thivolet, J., Patricot, L.M., and Vauzelle, J.L.: Transformation sarcomateuse sur dysplasie fibreuse. A propos d'un cas, revue de la littérature. Sem. Hop. Paris *58*:1329, 1982.	**Fig. 104**
Rasmussen, J., Jensen, H., Henschel, A.: Elastofibroma dorsi: two cases and a review of literature. Pathologe *3*:104, 1982. Nagamine, N., Nohara, Y., Ito, E.: Elastofibroma in Okinawa. A clinicopathologic study of 170 cases. Cancer *50*:1794, 1982.	**Figs. 914, 1123, 1140, 1206**
Myers, B.W., Masi, A.T., Feigenbaum, S.L.: Pigmented villonodular synovitis and tenosynovitis. A clinical epidemiologic study of 166 cases and literature review. Medicine *59*:223, 1980.	**Figs. 70, 786, 867, 1023, 1029, 1136**
Soren, A.: Pathogenesis and treatment of ganglion. Clin. Orthop. *48*:173, 1966. Sim, F.H., Dahlin, D.C.: Ganglion cyst of bone. Mayo Clin. Proc. *46*:484, 1971.	**Figs. 144, 186, 508, 1014**

*For additional references see texts listed in Tables 1 and 2.

†See Tables 3–5, 13, 19, 22, 23b, 24, 27, 30, 33, 36, 39, 42, 45, 48, 51, 54, 57, 58, 66, 72, 81, 84, 87–92.

Table 10. *Non-neoplastic Lesions (continued)*

DEFINITIONS	SYNONYMS AND RELATED TERMS
Tendosynovial chondromatosis *is a non-neoplastic, monoarticular arthropathy. It is characterized by multiple fibrocartilagenous nodules in inflamed and villous synovial tissues.*	Synovial chondrometaplasia, loose bodies, articular chondrosis, synovial chondromatosis
Lipogranuloma *is a self-limited lesion of adipose tissue containing granular and clear epithelioid cells, as well as isomorphic giant cells in an inflammatory, vascular, and necrotic background.*	Organized fat necrosis, nodular panniculitis, reticulohistiocytic granuloma, lipoid dermatoarthritis
Proliferative panniculitis *is a fasciitis-like, non-neoplastic infiltrative and nodular lesion of the interlobular and perivascular connective tissues of mature fat. It is composed of spindle cells and epithelioid cells in a vascular and inflammatory stroma, and may contain calcified nests and metaplastic bone.*	Proliferative fasciitis, organized fat necrosis, fasciitis involving fat, nodular panniculitis, Weber-Christian disease, inflammation of adipose tissue
Piezogenic papule *is a cystic and nodular herniation of subcutaneous fat into the dermal collagen. It is a disease of obese patients and is seen almost exclusively on the medial and lateral aspects of the heels on weight-bearing.*	Dermatocele
Proliferative myositis *is a non-neoplastic, ill-defined, infiltrative lesion of skeletal muscle. It is composed of granular epithelioid cells and pleomorphic giant histiocytic forms with prominent nuclei and nucleoli resembling ganglion cells.*	Myositis proliferans
Myositis ossificans *is a self-limited, non-neoplastic lesion of skeletal muscle. Histologically, it is composed of partly calcified, orderly, bony trabeculae and diverse fibrohistiocytic and inflammatory elements in disarranged pattern.*	Myositis ossificans progressiva
Pyogenic granuloma *is a localized, non-neoplastic cutaneous or submucosal lesion. It is composed of nests of minute capillary vessels and inflammatory cells in fibrillar stroma.*	Granuloma pyogenicum, granuloma gravidarum, granuloma teleangiecticum, lobular capillary hemangioma, granulation-tissue-type hemangioma
Angiofollicular lymphoid hyperplasia *is a non-neoplastic cutaneous lesion, predominantly of the head and neck areas, mostly affecting middle-aged women. It is composed of proliferating capillary vessels, aggregates of lymphocytes, and eosinophils in solitary or multiple nodules.*	Kimura's disease, angiolymphoid hyperplasia with eosinophila, atypical pyogenic granuloma, angioblastic lymphoid hyperplasia, epithelioid hemangioma
Arteriovenous malformation *is a non-neoplastic, self-limited vascular abnormality. It is found most commonly in muscles of the extremities and is composed of tortuous, thick-walled, anastomosing vessels and sparsely cellular intervascular stroma.*	Vascular hamartoma, arteriovenous shunt, arteriovenous hemangioma
Traumatic neuroma *is a non-neoplastic, disorderly proliferation of axon-nerve sheath complex and often presents in multinodular fashion.*	Amputation neuroma, Morton's neuroma
Paget's disease *is a non-neoplastic, polyostotic proliferative disease of the elderly. It is characterized by intraskeletal deposits of woven bone, increase of osteoid, and elevation of serum alkaline phosphatase.*	Osteitis deformans
Amyloidoma *is a tumor-like localized deposit of amyloid. It can be found in soft tissues, parenchymal organs, as well as in bone, frequently without any associated disease.*	Amyloid tumor, localized amyloid formation, localized amyloidosis

SELECTED REFERENCES*	ILLUSTRATIONS†
Villacin, A.B., Brigham, L.N., Bullough, P.G.: Primary and secondary chondrometaplasia. Human Pathol. *10*:439, 1979.	**Fig. 83**
Tredwell, S.J., Morton, K.S., Vassar, P.S.: Lipogranuloma of bone. Can. J. Surg. *12*:420, 1969. Castillo-Oertel, Y., Johnson, F.B.: Sclerosing lipogranuloma of male genitalia. Arch. Pathol. *101*:321, 1977.	**Figs. 820, 907, 1043**
Steinberg, B.: Systemic nodular panniculitis. Am. J. Pathol. *29*:1059, 1953. Churg, E.B., Enzinger, F.M.: Proliferative fasciitis. Cancer *36*:1450, 1975.	**Figs. 430, 552, 721, 910**
Grant, J.: Piezogenic pedal papules. Arch. Dermatol, *101*:619, 1970. Hajdu, S.I.: Pathology of Soft Tissue Tumors. Philadelphia, Lea & Febiger, 1979, pp. 234–235.	
Kern, W.H.: Proliferative myositis. A pseudosarcomatous reaction to injury. Arch. Pathol. *69*:209, 1960.	**Figs. 597, 738, 863, 951**
Angervall, L., Stener, B., Stener, I., et al.: Pseudomalignant osseous tumor of soft tissue. J. Bone Joint Surg. *51B*:654, 1969. Mitra, M., Sen, A.K., Deb, H.K.: Myositis ossificans traumatica. A complication of tetanus. Report of a case and review of the literature. J. Bone Joint Surg. *58A*:885, 1976.	**Figs. 145, 384, 620, 913, 1083–1085**
Bhaskar, S.N., Jacoway, J.R.: Pyogenic granuloma—clinical features, incidence, histology and result of treatment. Report of 242 cases. J. Oral Surg. *24*:391, 1966. Mills, S.E., Cooper, P.H., Fechner, R.E.: Lobular capillary hemangioma. The underlying lesion of pyogenic granuloma. Am. J. Surg. Pathol. *4*:471, 1980.	**Figs. 331, 332, 981**
Eady, R.A.J., Jones, E.W.: Pseudopyogenic granuloma. Enzyme histochemical and ultrastructural study. Human Pathol. *8*:653, 1977.	**Figs. 12, 13, 979, 1040**
Girard, E., Graham, J.H., Johnson, W.C.: Arteriovenous hemangioma (arteriovenous shunt). A clinicopathological and histochemical study. J. Clin. Pathol. *1*:73, 1974.	**Figs. 139, 197, 911, 1006, 1009**
Das Gupta, T.W., Brasfield, R.D.: Amputation neuroma in cancer patients. N.Y. J. Med. *69*:2129, 1969.	**Fig. 1264**
Miller, A.S., Cuttino, C.L., Elzay, R.P., et al.: Giant cell tumor of the jaws associated with Paget's disease of bone: report of two cases and review of the literature. Arch. Otolaryngol. *100*:223, 1974.	**Fig. 105**
Brownstein, M.H., Helwig, E.B.: The cutaneous amyloidoses. Arch. Dermatol. *102*:8, 1978. Lipper, S., Kahn, L.B.: Amyloid tumor. Am. J. Surg. Pathol. *2*:141, 1978.	

*For additional references see texts listed in Tables 1 and 2.

†See Tables 3–5, 13, 19, 22, 23b, 24, 27, 30, 33, 36, 39, 42, 45, 48, 51, 54, 57, 58, 66, 72, 81, 84, 87–92.

TABLE 11. *Benign Neoplasms*

DEFINITIONS	SYNONYMS AND RELATED TERMS
Benign fibroblastic fibrous histiocytoma *is a small, benign, and ill-defined neoplasm most commonly found in the dermis of adults. It is composed of thin-walled capillary vessels surrounded by slender fibroblasts forming a storiform pattern.*	Sclerosing hemangioma, benign fibrous histiocytoma, myxofibroma, dermatofibroma
Benign histiocytic fibrous histiocytoma *is a poorly demarcated benign neoplasm of soft tissues and bone. It contains a large number of isomorphic, epulis-type giant cells in a matrix composed of monomorphic granular epithelioid cells and fine capillary vessels.*	Benign giant cell tumor, fibrous histiocytoma, villonodular synovitis, synovioma, giant cell tumor of tendon sheath, myeloplexoma, myeloxanthoma, osteoclastoma, giant cell tumor of soft tissues, nodular synovitis, pigmented villonodular synovitis
Benign pleomorphic fibrous histiocytoma *is a poorly outlined benign neoplasm, usually less than 5 cm in size, of dermal, subcutaneous or deep tissues. Histologically, it contains an admixture of foamy histocytes and isomorphic Touton giant cells, as well as occasional epulis-type giant cells. The stroma is vascular, finely reticular, and loosely cellular.*	Xanthoma, juvenile xanthoma, atypical fibroxanthoma, xanthofibroma, fibroxanthoma, reticulohistiocytoma, tendinous xanthoma, xanthoma of bone, xanthelasma, tuberous xanthoma, angiomatoid fibrous histiocytoma
Nonossifying fibroma *is a well-delineated, multiloculated, and lytic metaphyseal benign fibrous histiocytoma. Histologically, it is composed of slender spindle cells in arranged pattern and epulis-type giant cells and foam cells in finely reticular stroma.*	Metaphyseal fibrous defect, subcortical nonossifying fibroma, fibrous cortical defect, xanthic giant cell tumor, xanthogranuloma of bone, benign fibrous histiocytoma
Fibroma *is an ubiquitous benign neoplasm, often ill-defined with infiltrative edges. It is composed of fibroblastic elements at various stages of differentiation, in spreading pattern, and the stroma may appear myxoid or densely sclerotic.*	Dermatofibroma, sclerosing fibroma, xanthofibroma, synovial fibroma, myxofibroma, juvenile aponeurotic fibroma, fascial fibroma, fibroblastoma, adenofibroma, fibroma durum
Desmoplastic fibroma *is a rare and locally destructive benign neoplasm of bone. It is composed of fibroblastic elements in spreading pattern and in richly collagenous and sparsely cellular matrix.*	Desmoid tumor of bone
Lipoma *is a benign neoplasm of adipose tissue and may occur in any organ, including bone. It is usually an encapsulated globoid lesion. The consistency and cellularity vary according to histologic type.*	Well-differentiated lipoma, myxoid lipoma, fibroblastic lipoma, myxolipoma, myxofibrolipoma, myelolipoma, lipoma arborescens, angiolipoma, lipoma dolorosa, infiltrating lipoma, myxoma, spindle cell lipoma, intraneural lipoma, lipomatosis, angiomyxoma
Lipoblastoma *is a benign neoplasm of lipoblasts and occurs almost exclusively in the subcutis of the extremities of young children.*	Embryonal lipoma, lipoma foetocellular, lipoblastomatosis
Pleomorphic lipoma *is a benign neoplasm of adipose tissue found most commonly in the posterior neck and in the retroperitoneum. It is characterized by pleomorphic, multinucleated histiocytic giant cells dispersed in sparsely cellular myxoid matrix.*	Myxoma with giant cells, pleomorphic myxoma
Angiomyolipoma *is a benign neoplasm, seen predominantly in adult women and found most commonly in the kidney. It is composed of thick-walled, hamartomatous capillary vessels, plump smooth muscle cells, and adipocytes.*	Renal hamartoma
Hibernoma *is a benign neoplasm of adults found most commonly in the subcutaneous tissues of the back and shoulder. It is composed of granular, lipid-rich adipocytes with round or polyhedral contour, is arranged in lobules, and contains centrally positioned dark nuclei.*	Lipoma of brown fat, atypical lipoma, lipoma granulare

SELECTED REFERENCES*	ILLUSTRATIONS†
Gross, R.W., Wolback, S.B.: Sclerosing hemangiomas, their relationship to dermatofibroma, histiocytoma, xanthoma and to certain pigmented lesions of the skin. Am. J. Pathol. *19*:533, 1943. Font, R.L., Hidayat, A.A.: Fibrous histiocytoma of the orbit. A clinicopathologic study of 150 cases. Human Pathol. *13*:199, 1982.	**Figs. 333, 334, 352, 960, 1021, 1044**
Hutter, R.V.P., Worcester, J.N., Frances, K.C., et al.: Benign and malignant giant cell tumors of bone. A clinicopathologic analysis of the natural history of the disease. Cancer *15*:653, 1962. Jones, F.E., Soule, E.H., Coventry, M.B.: Fibrous xanthoma of synovium (giant cell tumor of tendon sheath, pigmented nodular synovitis). J. Bone Joint Surg. *51A*:76, 1969.	**Figs. 71, 115, 174, 178, 183, 222, 325, 490, 501, 774, 778**
Fretzin, D.F., Helwig, E.B.: Atypical fibroxanthoma of the skin—a clinicopathologic study of 144 cases. Cancer *31*:1541, 1973.	**Figs. 135, 346, 417, 422, 459, 722, 782**
Steiner, G.C.: Fibrous cortical defect and nonossifying fibroma of bone. Arch. Pathol. *97*:206, 1974. Campanacci, M., Laus, M., Boriani, S.: Multiple non-ossifying fibromata with extraskeletal anomalies: A new syndrome? J. Bone Joint Surg. *65*(*B*): 627, 1983.	**Figs. 112, 117, 631, 852**
Churg, E.B., Enzinger, F.M.: Fibroma of tendon sheath. Cancer *44*:1945, 1979.	**Figs. 336, 859, 884, 886, 904, 915, 917, 1027, 1109, 1204**
Sugiura, I.: Desmoplastic fibroma. Case report and review of the literature. J. Bone Joint Surg. *58A*:126, 1976.	**Figs. 106, 633**
Angervall, L., Dahl, I., Kindbloom, L.G.: Spindle cell lipoma. Acta Pathol. Microbiol. Scand. *84*:477, 1976. Milgram, J.W.: Intraosseous lipomas with reactive ossification in the proximal femur. Skeletal Radiol. *7*:1, 1981.	**Figs. 24, 73, 198, 243, 455, 635, 847, 908, 939–950, 1041, 1168, 1181, 1187**
Stringel, G., Shandling, B., Mancer, K., et al.: Lipoblastoma in infants and children. J. Pediatr. Surg. *17*:277, 1982. Churg, E.B., Enzinger, F.M.: Benign lipoblastomatosis: an analysis of 35 cases. Cancer *32*:482, 1983.	**Figs. 412, 413, 437, 996**
Hajdu, S.I.: Pathology of Soft Tissue Tumors. Philadelphia, Lea & Febiger, 1979, pp. 240–243. Shmookler, B.M., Enzinger, F.M.: Pleomorphic lipoma. A benign tumor simulating liposarcoma. Cancer *47*:126, 1981.	**Figs. 73, 613, 781, 821, 909, 957, 1182, 1183**
Hajdu, S.I., Foote, F.W.: Angiomyolipoma of the kidney. Report of 27 cases and review of the literature. J. Urol. *102*:396, 1969.	**Fig. 959, 1124**
Mesara, B.W., Batsakis, J.G.: Hibernoma of the neck. Arch. Otolaryngol. *85*:95, 1967.	**Figs. 432, 717, 1162**

*For additional references see texts listed in Tables 1 and 2.

†See also Tables 3–5, 7, 8, 20, 22, 23a, 23b, 25, 28, 31, 34, 37, 40, 43, 46, 49, 52, 57, 59, 62, 64, 65, 67, 70, 72, 82, 85, 87–92.

Table 11. *Benign Neoplasms (continued)*

DEFINITIONS	SYNONYMS AND RELATED TERMS
Leiomyoma *is an ubiquitous neoplasm of adults and may be multifocal. It is composed of interlacing bundles of well-differentiated, nonmitotic, muscle cells in fibrillar, sclerosed, or myxoid stroma.*	Vascular leiomyoma, leiomyomatosis, angiomyoma, leiomyoma cutis, peritoneal leiomyomatosis, benign leiomyoblastoma, intravenous leiomyomatosis
Rhabdomyoma *is a benign neoplasm of striated muscle. It is seen most commonly in the oral cavity and larynx in adults, and in the vagina and heart in infants, usually as a glistening well-outlined nodule. It is composed of closely packed, well-defined, polygonal granular cytoplasmic rhabdomyocytes in disarranged pattern. In occasional cells crystals can be seen.*	Myoblastoma, adult rhabdomyoma, fetal rhabdomyoma, rhabdomyoma of the heart, cardiac rhabdomyoma
Hemangioma *is a benign neoplasm of blood vessels. It is an ubiquitous lesion, but it is found most commonly in the subcutis and in the dermis. Microscopically, it may appear in several forms.*	Capillary hemangioma, cavernous hemangioma, venous hemangioma, hypertrophic hemangioma, juvenile hemangioma, arteriovenous hemangioma, intramuscular hemangioma, hemangiomatosis, synovial hemangioma, perineural hemangioma, histiocytic hemangioma
Angiofibroma *is a benign neoplasm composed of a network of angiomatous vessels in fibromyxoid stroma. It occurs almost exclusively in the nasopharynx of boys.*	Juvenile nasopharyngeal angiofibroma, juvenile angiofibroma, hemangiofibroma
Benign glomus tumor *is an ubiquitous vascular neoplasm most commonly found in the fingers and forearm. Histologically it is composed of thin and thick-walled cavernous capillary vessels surrounded by ribbons and sheets of uniform, granular epithelioid cells.*	Glomangioma, benign hemangiopericytoma, angioneuromyoma
Lymphangiomyoma *is an uncommon benign neoplasm manifested by proliferation of endothelial and smooth muscle cells of tortuous lymphatic vessels. It most commonly affects the abdominal and thoracic lymphatics of adult women.*	Lymphangiopericytoma, angiomyoma, angiomatous hyperplasia, lymphangiomatosis
Neurofibroma *is a benign fibroblastic neoplasm of peripheral nerve and can be solitary or multiple. Its consistency and histologic appearance vary according to differentiation of neoplastic elements from myxoid to fibrous. The slender spindle cells are often arranged in parallel bundles and the nuclei may form palisades. Virtually no mitotic figures can be seen.*	Benign fibroblastic peripheral nerve tumor, benign collagenous peripheral nerve tumor, plexiform neuroma, angioneuroma, neurofibromatosis, myxoid neurofibroma, pleomorphic neurofibroma, benign collagenous schwannoma, neurothekoma, pigmented neurofibroma, granular cell neurofibroma, pacinian neurofibroma, glandular neurofibroma, von Recklinghausen's disease, pigmented neurofibromatosis, plexiform neurofibromatosis, diffuse neurofibromatosis, localized neurofibromatosis, ancient neurofibroma, angioneurofibroma
Benign schwannoma *is a nonfibroblastic, self-limited, and solitary neoplasm of peripheral nerve. Its consistency and histologic appearance vary according to the differentiation and composition of its neoplastic elements, from myxoid to fibrous histiocytic or nevoid. The neoplastic cells are often arranged in palisades and storiform pattern. Mitotic figures are seldom seen.*	Benign nonfibroblastic peripheral nerve tumor, myxoid schwannoma, cystic schwannoma, molluscum fibrosum, benign noncollagenous peripheral nerve tumor, myxoid neurilemma, benign peripheral nerve sheath tumor, glandular schwannoma, nevoid schwannoma, pigmented schwannoma, epithelioid benign peripheral nerve tumor, ancient neurilemoma, ancient schwannoma
Benign paraganglioma *is an encapsulated neoplasm of the autonomic nervous system most commonly found in the neck, retroperitoneum, and mediastinum. The histologic appearance is influenced by site of origin, and the types of cells that may range from indistinct, clear or granular epithelioid cells to pleomorphic polygonal cells in organoid nests.*	Benign carotid body tumor, chemodectoma, benign nonchromaffin paraganglioma, benign glomus jugulare tumor, benign vagal body paraganglioma, benign aortic body tumor, benign mediastinal paraganglioma, benign retroperitoneal paraganglioma

SELECTED REFERENCES*	ILLUSTRATIONS†
Williams, L.J., Pavlick, F.T.: Leiomyomatosis peritonealis disseminata. Cancer *45*:1726, 1980. Hachisuga, T., Hashimoto, H., Enjoji, M.: Angioleiomyoma. A clinicopathologic study of 562 cases. Cancer *54*:126, 1984	**Figs. 380, 1010**
Fenoglio, J.J., McAllister, H.A., Ferrans, V.J.: Cardiac rhabdomyoma: A clinicopathologic and electron microscopic study. Cancer *37*:2283, 1976. Walter, P., Guerbaou, M.: Rhabdomyome foetal. Virchows Arch (Pathol. Anat.) *371*:59, 1976.	**Figs. 497, 737, 793**
Conners, J.F., Khan, G.: Hemangioma of the striated muscle. South. Med. J. *70*:1423, 1977. Pandey, S., Pandey, A.K.: Osseous hemangiomas. Arch. Orthop. Traumat. Surg. *99*:23, 1981.	**Figs. 18, 83, 98, 549, 553, 554, 559, 560, 974, 987, 988, 1004, 1221**
English, G.M., Hemenway, W.G., Cundy, R.L.: Surgical treatment of invasive angiofibroma. Arch. Otolaryngol. *96*:312, 1972.	**Fig. 862, 912, 1254**
Carroll, R.E., Berman, A.T.: Glomus tumors of the hand. Review of the literature and report of twenty-eight cases. J. Bone Joint Surg. *54A*:691, 1972.	**Figs. 475, 481, 961**
Kitzsteiner, K.A., Mallen, R.G.: Pulmonary lymphangiomyomatosis. Cancer *46*:2248, 1980.	**Figs. 562, 978**
Bird, C.C., Willis, R.A.: The histogenesis of pigmented neurofibroma. J. Pathol. *97*:631, 1969. Brasfield, R.D., Das Gupta, T.K.: Von Recklinghausen's disease. A clinicopathologic study. Ann. Surg. *175*:86, 1972.	**Figs. 11, 44, 74, 204, 245, 253, 363, 365, 487, 527, 627, 652, 840, 842, 871, 942, 955, 993, 1013, 1114, 1142, 1143**
Das Gupta, T.K., Brasfield, K.D., Strong, E.W., Hajdu, S.I.: Benign solitary schwannoma (neurilemomas). Cancer *24*:355, 1969. Erlandson, R.A., Woodruff, J.M.: Peripheral nerve sheath tumors: An electron microscopic study of 43 cases. Cancer *49*:273, 1982.	**Figs. 257, 356, 421, 471, 486, 516, 517, 846, 850, 1228, 1268**
Lack, E.E., Cubilla, A.L., Woodruff, J.M.: Paragangliomas of the head and neck region. A pathologic study of tumors from 70 patients. Human Pathol. *10*:199, 1979. Lack, E.E., Cubilla, A.L., Woodruff, J.M., et al.: Extra-adrenal paragangliomas of the retroperitoneum. A clinicopathologic study of 12 tumors. Am. J. Surg. Pathol. *4*:109, 1980.	**Figs. 276, 483, 544, 714, 790, 794, 853, 1196, 1220**

*For additional references see texts listed in Tables 1 and 2.

†See also Tables 3–5, 7, 8, 20, 22, 23a, 23b, 25, 28, 31, 34, 37, 40, 43, 46, 49, 52, 57, 59, 62, 64, 65, 67, 70, 72, 82, 85, 87–92.

TABLE 11. ***Benign Neoplasms (continued)***

DEFINITIONS	SYNONYMS AND RELATED TERMS
Chondroblastoma *is a rare and highly cellular neoplasm of bone. It is composed of immature and mature chondroid elements with distinct outlines in a vascular, isomorphic epulis-type giant cell containing focally calcified matrix.*	Codman's tumor, epiphyseal chondroblastoma, calcified giant cell tumor, chondromatous osteoclastoma, epiphyseal giant cell tumor.
Chondromyxoid fibroma *is a benign bone tumor of long bones composed of chondroid and fibromyxoid tissues, in variable proportion, in a lacy pattern.*	Myxoid chondroma, fibromyxoid chondroma
Chondroma *is a well-demarcated, benign neoplasm that may be primary in bone or in soft tissues. It is composed of well-differentiated and distended mononucleated chondrocytes in a noncrowded and orderly arrangement.*	Enchondroma, synovial chondroma, Ollier's disease
Osteochondroma *is a bony trabeculae containing mature, cartilage-capped projection protruding into soft tissues from cortical bone. It is a benign neoplasm most often located at the metaphysis of long bone.*	Exostosis, osteocartilagenous exostosis
Osteoblastoma *is a benign, osteoid-producing neoplasm. Histologically it is composed of osteoblasts surrounding trabeculae of woven bone. The matrix is myxofibrillar and contains epulis type giant cells, slender spindle cells, and distended capillaries and devoid of chondrocytes.*	Giant osteoid osteoma, cementoblastoma
Osteoid osteoma *is a solitary, benign neoplasm of bone usually less than 1 cm. Histologically, it is characterized by a central nidus and peripheral orderly reactive bone formation.*	Osteoma, periosteal osteoma
Benign granular cell tumor *is an ubiquitous, usually solitary, benign neoplasm of unknown histogenesis. It is composed of fairly uniform ill-defined granular epithelioid cells containing round nuclei and arranged in epithelioid patterns.*	Granular cell myoblastoma, granular cell neurofibroma, granular cell schwannoma, granular neuroma
Benign cystosarcoma phyllodes *is a circumscribed, firm, and lobular mammary neoplasm of middle-aged women. It is composed of sparsely cellular, papillary intracanalicular and intracystic projections of myxofibrillar connective tissue tongues lined with benign epithelial cells.*	Giant fibroadenoma, cystosarcoma, intracanaliculare fibroadenoma

SELECTED REFERENCES*	ILLUSTRATIONS†
Dahlin, D.C., Ivins, J.C.: Benign chondroblastoma. A study of 125 cases. Cancer *30*:401, 1972. Huvos, A.G., Marcove, R.C.: Chondroblastoma of bone. A critical review. Clin. Orthop. *95*:300, 1973.	**Figs. 116, 128, 506, 1103, 1104, 1111**
Rahimi, A., Beabout, J.W., Ivins, J.C.: Chondromyxoid fibroma. A clinicopathologic study of 76 cases. Cancer *27*:726, 1972. Standefer, M., Hardy, R.W., Marks, K., et al.: Chondromyxoid fibroma of the cervical spine. Neurosurgery *11*:288, 1982.	**Fig. 114**
Takigawa, K.M.D.: Chondroma of the bones of the hand. A review of 110 cases. J Bone Joint Surg. *53(A)*:591, 1971. Dahlin, D.C., Salvador, H.H.: Cartilagenous tumor of the soft tissues of the hands and feet. Mayo Clin. Proc. *49*:725, 1974.	**Figs. 94, 191, 772, 1066, 1152**
Nosanchuk, J.K., Kaufer, H.: Recurrent periosteal chondroma. Report of two cases and review of the literature. J. Bone Joint Surg. *51(A)*:375, 1969. Hershey, S.L., Landen, F.T.: Osteochondromas as a cause of false popliteal aneurysms. Review of the literature and report of two cases. J. Bone Joint Surg *54(A)*:1765, 1977.	**Figs. 113, 155, 167, 184, 302, 303, 772, 1063, 1066, 1090, 1152**
Marsh, B.N., Bonfiglio, M., Brady, L.P., et al.: Benign osteoblastoma. Range of manifestations. J. Bone Joint Surg. *57(A)*:1, 1975.	**Figs. 97, 211, 234, 1092, 1155**
Dunlap, J.A.Y., Morton, K.S., Elliott, G.B.: Recurrent osteoid osteoma. Report of a case with review of the literature. J. Bone Joint Surg. *52(B)*:128, 1980. Cohen, M.D., Harrington, T.M., Ginsburg, W.W.: Osteoid osteoma: 95 cases and a review of the literature. Seminars in Arthritis and Rheumatism *12*:265, 1983.	**Figs. 101, 1089**
Strong, E.W., McDivitt, R.W., Brasfield, R.: Granular cell myoblastoma. Cancer *25*:415, 1970.	**Figs. 75, 261, 498, 539, 736, 1160, 1235**
Hajdu, S.I., Espinosa, M.H., Robbins, G.F.: Recurrent cystosarcoma phyllodes. A clinicopathologic study of 32 cases. Cancer *38*:1402, 1976.	**Figs. 43, 48, 247, 414, 921, 938**

*For additional references see texts listed in Tables 1 and 2.
†See also Tables 3–5, 7, 8, 20, 22, 23a, 23b, 25, 28, 31, 34, 37, 40, 43, 46, 49, 52, 57, 59, 62, 64, 65, 67, 70, 72, 82, 85, 87–92.

Table 12. *Malignant Neoplasms*

DEFINITIONS	SYNONYMS AND RELATED TERMS
Malignant fibroblastic fibrous histiocytoma *is a neoplasm that is usually considerable in size and is most often found in the subcutis of the shoulder area. It is composed of slender fibroblasts in storiform pattern.*	Dermatofibrosarcoma protuberans, malignant fibrous histiocytoma, Darier's tumor, myxoid fibrous histiocytoma, intermediate fibrohistiocytic tumor, myxofibrosarcoma
Malignant histiocytic fibrous histiocytoma *is an uncommon neoplasm. It can be primary in soft tissues, as well as in bone. Histologically, it consists of short spindle cells, granular epithelioid cells, and sparse, isomorphic, epulis-type giant cells.*	Malignant giant cell tumor of bone, malignant giant cell tumor of tendon sheath, malignant giant cell tumor of soft tissues, malignant nevoxanthoendothelioma, malignant osteoclastoma
Malignant pleomorphic fibrous histiocytoma *is a neoplasm of soft tissues and bone most commonly found in the extremities. It is composed of pleomorphic and slender fibroblastic cells, as well as mononucleated and multinucleated histiocytic forms in disarranged pattern and in a richly vascular matrix.*	Xanthosarcoma, spindle and giant cell sarcoma, malignant xanthoma, fibroxanthosarcoma, angiomatoid fibrous histiocytoma, inflammatory fibrous histiocytoma, fibrous histiocytoma of bone in association with bone infarct, malignant nevoxanthoendothelioma, malignant xanthogranuloma
Desmoid tumor *is a low grade malignant neoplasm with scar-like consistency found most often in deep soft tissues. It is composed of uniform collagenous fibrocytes in spreading pattern and abundant intercellular collagenous matrix.*	Aggressive fibromatosis, desmoid fibromatosis, desmoid, desmoplastic fibroma, deep fibromatosis, abdominal desmoid, extra-abdominal desmoid
Fibrosarcoma *is a malignant tumor of soft tissues and bone. It is composed of collagenous fibroblasts in broad bands and herringbone pattern but the microscopic appearance may vary according to histologic grade.*	Fibroblastic fibrosarcoma, pleomorphic fibrosarcoma, myxofibrosarcoma, spindle and giant cell sarcoma, spindle cell sarcoma, cicatricial fibrosarcoma, congenital fibrosarcoma, infantile fibrosarcoma, adult fibrosarcoma
Biphasic tendosynovial sarcoma *is usually of considerable size and is predominantly found in the proximity of joints. It is composed of granular epithelioid cells in pseudoglandular arrangement and uniform slender spindle cells in spreading fascicles.*	Synovial sarcoma, malignant synovioma, biphasic synovial sarcoma, synovioma
Monophasic tendosynovial sarcoma, spindle cell type, *is one of the most aggressive and fastest growing soft tissue sarcomas. It is a disease of young adults and is found predominantly in the extremities. Microscopically, it is characterized by uniform, non-giant-cell-forming, slender spindle cells, with short cytoplasmic processes in fascicles and about tissue slits.*	Monophasic synovial sarcoma, synovial sarcoma, synovial fibrosarcoma, synovial spindle cell sarcoma
Epithelioid sarcoma *is usually present in the form of multinodular deposits along tendons of distal parts of the extremities, and has a tendency to metastasize to regional lymph nodes. Histologically, it is comprised of short oval and round epithelioid cells with finely granular cytoplasm in cords and clusters.*	Epitheliosarcoma, aponeurotic sarcoma
Clear cell sarcoma *is a high grade malignant neoplasm and is found predominantly in the distal part of the extremities along tendon-aponeurotic tissues. Microscopically, it is composed of clusters and fascicles of well-defined and uniform clear epithelioid cells containing clear nuclei, prominent nucleoli and possess nonspecific intracytoplasmic organelles and glycogen.*	Clear cell sarcoma of tendons, aponeurotic clear cell sarcoma
Well differentiated liposarcoma *usually presents as a bulky, pseudoencapsulated, low grade malignant neoplasm of adults. The predominant microscopic elements are mono- and multivacuolated adipocytes and preadipocytes, some of them widely distended, in the background of a fine capillary network and sparsely sclerosed stroma.*	Lipoma-like liposarcoma
Myxoid liposarcoma *is a bulky, pseudoencapsulated, low grade malignant neoplasm of adults. It is comprised of multivacuolated preadipocytes and occasional lipoblasts held together by branching capillary vessels in a lacy pattern.*	Myxosarcoma, malignant myxoma, lipomyxosarcoma, myxoliposarcoma

SELECTED REFERENCES*	ILLUSTRATIONS†
Taylor, H.B., Helwig, E.B.: Dermatofibrosarcoma protuberans. A study of 115 cases. Cancer *15*:717, 1962. Zaatari, G., Hajdu, S.I., Shiu, M.H.: Soft tissue sarcoma of the trunk. In Press.	**Figs. 37, 40, 54, 76, 111, 143, 241, 301, 339, 340, 350, 373, 645, 672, 710, 760, 833, 1133–1135, 1147, 1178, 1200**
Kahn, L.B.: Malignant giant cell tumor of tendon sheath. Ultrastructural study and review of the literature. Arch. Pathol. *95*:203, 1973. Angervall, L., Hagmar, B., Kindbloom, L.G., et al.: Malignant giant cell tumor of soft tissues. Cancer *47*:736, 1981.	**Figs. 57, 156, 161, 300, 452, 534, 592, 600, 711, 756, 775, 776, 784, 791, 816, 872, 892, 893, 930, 1018B, 1048, 1149, 1172, 1246, 1247**
Hajdu, S.I.: Pathology of Soft Tissue Tumors. Philadelphia, Lea & Febiger, 1979, pp. 94–121. Capanna, R., Bertoni, F., Bacchini, P., et al.: Malignant fibrous histiocytoma of bone. Cancer *54*:177, 1984.	**Figs. 25, 88, 111, 130, 138, 140, 148, 201, 203, 232, 237, 262, 288, 293, 344, 577, 598, 679, 780, 788, 789, 792, 798, 805, 937, 962, 1003, 1148, 1179, 1180, 1253**
Das Gupta, T.K., Brasfield, R.D., O'Hara, J.: Extra-abdominal desmoids. Ann. Surg. *170*:109, 1969. Shiu, M.H., Flanebaum, L., Hajdu, S.I., et al.: Malignant soft tissue tumors of the anterior abdominal wall. Arch. Surg. *115*:152, 1980.	**Figs. 9, 36, 52, 77, 185, 202, 209, 244, 282, 376, 386, 397, 662, 686, 865, 899, 1118**
Castro, E.B., Hajdu, S.I., Fortner, J.G.: Surgical therapy of fibrosarcoma of extremities. Arch. Surg. *107*:784, 1973. Huvos, A.G., Higinbotham, N.L.: Primary fibrosarcoma of bone. A clinicopathologic study of 130 patients. Cancer *35*:837, 1975.	**Figs. 33, 60, 91, 117, 131, 165, 170, 240, 260, 297, 371, 393, 410, 601, 608, 625, 647, 660, 671, 809, 825, 832, 900, 901, 1117, 1205**
Lee, S.M., Hajdu, S.I., Exelby, P.R.: Synovial sarcoma in children. Surg. Gynecol. Obstet. *138*:701, 1974. Potter, G.K., Walkes, M.H., Penny, T.R.: Tendosynovial sarcoma. A clinicopathologic review of foot cases with a case report. J. Am. Podiatry Assoc. *74*:312, 1984.	**Figs. 85, 122, 152, 153, 320, 323, 514, 520, 522, 535, 536, 547, 574, 586, 727, 765, 906, 1099, 1100, 1173, 1202**
Hajdu, S.I., Shiu, M.H., Fortner, J.G.: Tendosynovial sarcoma—a clinicopathologic study of 136 cases. Cancer *39*:120, 1977. Krall, R.A., Kostianovsky, M., Patchefsky, A.S.: Synovial sarcoma. A clinical, pathological and ultrastructural study of 26 cases supporting the recognition of monophasic variant. Am. J. Surg. Pathol. *5*:135, 1981.	**Figs. 39, 64, 85, 179, 190, 315, 326, 391, 392, 396, 398, 399, 467, 491, 492, 542, 622, 642, 646, 648, 650, 675, 683, 684, 854, 864, 870, 894, 902, 903, 1144, 1145**
Miettinen, M., Lehto, V.P., Vartio, T., Virtanen, I.: Epithelioid sarcoma. Ultrastructural and immunohistologic features suggesting a synovial origin. Arch. Pathol. Lab. Med. *106*:620, 1982. Owens, J.C., Shiu, M.H., Smith, R., Hajdu, S.I.: Soft Tissue Sarcomas of the Hand and Foot. Cancer *55*:2010, 1985.	**Figs. 62, 85, 181, 187, 258, 259, 322, 326, 462, 464, 468, 491, 676, 830, 866, 895, 1052, 1059, 1146, 1170, 1244, 1245**
Hajdu, S.I.: Pathology of Soft Tissue Tumors. Philadelphia, Lea & Febiger, 1979, pp. 204–210. Eckardt, J.J., Pritchard, D.J., Soule, E.H.: Clear cell sarcoma. A clinicopathologic study of 27 cases. Cancer *52*:1482, 1983.	**Figs. 62, 85, 322, 447, 492, 526, 759, 761, 896, 927, 1145, 1223, 1244, 1245**
Hajdu, S.I.: Pathology of Soft Tissue Tumors. Philadelphia, Lea & Febiger, 1979, pp. 261–267.	**Figs. 58, 86, 219, 431, 435, 822, 999, 1184, 1233**
Cody, S.I., Turnbull, A.D., Fortner, J.G., Hajdu, S.I.: The continuing challenge of retroperitoneal soft tissue sarcoma. Cancer *47*:2147, 1981.	**Figs. 49, 56, 59, 86, 188, 208, 269, 290, 438, 439, 636, 783, 933, 948, 997, 1046, 1113, 1169, 1232**

*For additional references see texts listed in Tables 1 and 2.

†See also Tables 5, 7, 9, 21–23B, 26, 29, 32, 35, 38, 41, 44, 47, 50, 53, 56, 57, 60, 62, 64, 65, 68, 70, 75, 76, 79, 80, 83, 86, 87–92.

TABLE 12. *Malignant Neoplasms (continued)*

DEFINITIONS	SYNONYMS AND RELATED TERMS
Lipoblastic liposarcoma *is a high grade malignant neoplasm that can vary in size. Histologically, it is comprised of uniform granular epithelioid lipoblasts in a background of a network of thin-walled capillary vessels.*	Round cell liposarcoma, epithelioid liposarcoma
Fibroblastic liposarcoma *is predominantly composed of slender spindle cells, prolipoblasts, and rare lipoblasts held together, often in storiform pattern, by branching fine capillary vessels.*	Spindle cell liposarcoma, sclerosing liposarcoma, fibroblastic liposarcoma
Pleomorphic liposarcoma *is a high grade, commonly bulky and necrotic, malignant neoplasm of deep soft tissues. It is characterized by a mixture of microscopic elements, ranging from slender and plump spindle cells to vacuolated adipocytes and multinucleated giant forms.*	Giant cell liposarcoma, spindle and giant cell liposarcoma
Leiomyoblastoma *is a malignant smooth muscle neoplasm most commonly occurring in the wall of the stomach and intestines in the form of a bulging, cystic, and exophytic mass. Microscopically, it is characterized by vacuolated epithelioid myoblasts in various stages of differentiation, from mononucleated round forms to multinucleated pleomorphic elements.*	Epithelioid leiomyosarcoma, round cell leiomyosarcoma, malignant leiomyoblastoma, clear cell leiomyosarcoma, bizarre leiomyosarcoma
Leiomyosarcoma *is an ubiquitous pseudoencapsulated malignant neoplasm of smooth muscle. It is composed of bundles of spindly leiomyocytes and fibrovascular stroma. The grade of malignancy depends on mitotic count, cellularity, differentiation, vascularity, and amount of necrosis.*	Malignant smooth muscle tumor, peripheral leiomyosarcoma, vascular leiomyosarcoma, bizarre leiomyosarcoma
Embryonal rhabdomyosarcoma *is a small cell, ill-defined malignant neoplasm of striated muscle. It is found most often in children; microscopically the growth pattern may vary from spreading to alveolar and may appear epithelioid or myxoid.*	Alveolar rhabdomyosarcoma, botryoid sarcoma, tubular rhabdomyosarcoma, solid rhabdomyosarcoma, epithelioid rhabdomyosarcoma
Rhabdomyoblastoma *is a rare and primitive high grade malignant neoplasm. It is composed of sharply outlined globoid rhabdomyoblasts resembling ganglion cells in minimal stroma. The neoplastic cells stain pink with hematoxylin and eosin and exhibit a positive reaction for myoglobin.*	Rhabdoid tumor, sarcoma with intracytoplasmic filamentous inclusions
Pleomorphic rhabdomyosarcoma *is a highly malignant neoplasm of striated muscle that is composed of spindle, round, and bizarre pleomorphic cells in a disarranged pattern. Cross-striation can be detected by a careful search, and isolated cells may stain positive for myoglobin.*	Spindle and giant cell sarcoma
Hemangiopericytoma *is an ill-defined rare malignant neoplasm of pericytes. Histologically, it is composed of thin-walled and branching vessels surrounded by ribbons of mostly round granular cells or sheets of short and uniform slender spindle cells.*	Perithelioma, juvenile hemangiopericytoma, malignant hemangiopericytoma
Hemangiosarcoma *is a malignant neoplasm of blood vessels. It is composed of tightly packed nests of epithelioid cells and an intricate network of anastomosing small- and medium-size blood vessels containing polygonal endothelial cells in single or multiple lining.*	Malignant hemangioendoethelioma, hemangioendothelioma, angiosarcoma, angioendothelioma, epithelioid hemangioendothelioma, hemangioendotheliomatosis
Neurofibrosarcoma *is a malignant fibroblastic neoplasm of peripheral nerve. It is composed of slender collagenous fibroblasts at various stages of maturation. The mitotic count is usually more than 5 per 10 high power fields.*	Malignant collagenous peripheral nerve tumor, malignant neurofibromatosis, malignant pigmented neurofibromatosis, malignant cutaneous neurofibroma, malignant schwannoma, malignant fibroblastic peripheral nerve tumor, neurofibroma associated with Recklinghausen's disease, Triton tumor

SELECTED REFERENCES*	ILLUSTRATIONS†
Hajdu, S.I.: Pathology of Soft Tissue Tumors. Philadelphia, Lea & Febiger, Philadelphia, 1979, pp. 274–280.	**Figs. 55, 86, 199, 415, 448, 450, 543, 687, 753, 934, 973, 1231**
Hajdu, S.I.: Pathology of Soft Tissue Tumors. Philadelphia, Lea & Febiger, 1979, pp. 271–274.	**Figs. 30, 86, 343, 440, 641, 668, 837, 848, 926, 998**
Spittle, M.F., Newton, K.A., Mackenzie, D.H.: Liposarcoma: a review of 60 cases. Br. J. Cancer *24*:696, 1970. Reitan, J.B., Kaalhus, O., Brennhovd, I.O., et al.: Prognostic factors in liposarcoma. Cancer *55*:2482, 1985.	**Figs. 86, 141, 284, 294, 458, 590, 799, 823, 928, 1019A, 1049**
Lavin, P., Hajdu, S.I., Foote, F.W.: Gastric and extragastric leiomyoblastoma. Cancer *29*:305, 1972. Shiu, M.H., Farr, G.H., DiMetrios, N., Hajdu, S.I.: Myosarcomas of the stomach. Cancer *49*:177, 1982.	**Figs. 196, 270, 274, 277, 278, 420, 429, 503, 513, 691, 713, 733, 744, 745, 747, 750, 795, 814, 828, 868, 964, 1011, 1053, 1139, 1158, 1213, 1214, 1248**
Fields, J.P., Helwig, E.B.: Leiomyosarcoma of the skin and subcutaneous tissue. Cancer *47*:156, 1981. Hochstetter, A.R., Eberle, H., Ruttner, J.R.: Primary leiomyosarcoma of extragnathic bones. Case report and review of literature. Cancer *53*:2194, 1984.	**Figs. 72, 110, 193, 220, 231, 273, 341, 347, 360, 362, 366, 381, 404, 588, 618, 659, 664, 666, 667, 670, 680, 681, 685, 746, 808, 841, 844, 849, 856, 875, 890, 994, 1132, 1163, 1211, 1212, 1249, 1274**
Maurer, H.M., Moon, T., Donaldson, M., et al.: The intergroup rhabdomyosarcoma study. Cancer *40*:2015, 1977. Lloyd, R.V., Hajdu, S.I., Knapper, W.H.: Embryonal rhabdomyosarcoma in adults. Cancer. *51*:557, 1983.	**Figs. 10, 20, 28, 66–68, 78, 120, 207, 263, 281, 291, 364, 372, 378, 416, 427, 443, 446, 472, 515, 529, 530, 567, 570, 571, 637, 644, 706, 708, 751, 804, 889, 923, 935, 944, 946, 949, 952, 1033, 1129, 1164, 1251**
Hajdu, S.I.: Pathology of Soft Tissue Tumors. Philadelphia, Lea & Febiger, 1979, pp. 352–356. Haas, J.E., Palmer, N.F., Weinborg, A.G., et al.: Ultrastructure of malignant rhabdoid tumor of the kidney. Hum. Pathol. *12*:646, 1981.	**Figs. 473, 477, 500, 528, 569, 594, 695, 709, 715, 739, 796, 936, 1034, 1159, 1260**
Hajdu, S.I.: Pathology of Soft Tissue Tumors. Philadelphia, Lea & Febiger, 1979, pp. 336–352. Hajdu, S.I., Bell, D., Botet, J., et al.: Intraskeletal presentation of pleomorphic rhabdomyosarcoma of soft tissues. In Press.	**Figs. 29, 87, 121, 292, 295, 296, 299, 313, 599, 604, 609, 612, 800, 804, 810, 813, 1130, 1167, 1216, 1250, 1255, 1273**
McMaster, M.J., Soule, E.H., Ivins, J.C.: Hemangiopericytoma: A clinicopathologic study and long-term follow-up of 60 patients. Cancer *36*:2232, 1975. Wold, L.E., Unni, K.K., Cooper, K.L., et al.: Hemangiopericytoma of bone. Am. J. Surg. Pathol. *6*:53, 1982.	**Figs. 93, 470, 479, 484, 525, 533, 693, 696, 972, 984, 995, 1203**
Sordillo, E.M., Sordillo, P.O., Hajdu, S.I.: Primary hemangiosarcoma of spleen. Med. Pediatr. Oncol. *9*:319, 1981. Wold, L.E., Unni, D.K., Beabout, J.W., et al.: Hemangioendothelial sarcoma of bone. Am. J. Surg. Pathol. *6*:59, 1982.	**Figs. 16, 32, 42, 46, 92, 102, 316, 523, 555, 557, 563, 564, 576, 720, 731, 905, 975, 976, 989, 990, 1005, 1209, 1210**
Guccion, J.G., Enzinger, F.M.: Malignant schwannoma associated with von Recklinghausen's neurofibromatosis. Virchow's Arch (Pathol. Anat.) *383*:43, 1979. Storm, F.K., Eilber, F.R., Mirra, J., et al.: Neurofibrosarcoma. Cancer *45*:126, 1980.	**Figs. 22, 23, 26, 34, 38, 53, 79, 236, 256, 264, 271, 275, 289, 369, 370, 401, 584, 593, 606, 624, 626, 732, 812, 813, 838, 1116, 1229, 1263, 1259, 1266**

*For additional references see texts listed in Tables 1 and 2.

†See also Tables 5, 7, 9, 21–23B, 26, 29, 32, 35, 38, 41, 44, 47, 50, 53, 56, 57, 60, 62, 64, 65, 68, 70, 75, 76, 79, 80, 83, 86, 87–92.

Table 12. *Malignant Neoplasms (continued)*

DEFINITIONS	SYNONYMS AND RELATED TERMS
Malignant schwannoma *is a nonfibroblastic peripheral nerve tumor. It is usually solitary and histologically is composed of noncollagenous slender spindle cells, often arranged in storiform pattern, and pleomorphic forms in fibrillar stroma. The mitotic count is usually more than 5 per 10 high power fields.*	Malignant nonfibroblastic peripheral nerve tumor, neurogenic sarcoma, noncollagenous malignant peripheral nerve tumor, malignant neurilemoma, ectomesenchymoma, Triton tumor, malignant nevoid schwannoma, malignant melanoma of soft tissues, clear cell sarcoma containing melanin, malignant peripheral nerve tumor with glandular inclusion
Undifferentiated malignant peripheral nerve tumor *is a rare form of peripheral nerve tumor. It is ubiquitous and may present in a number of poorly differentiated microscopic forms ranging from predominantly cluster-forming neuroblastic to tubule-forming neuroepithelial neoplasms.*	Primitive neuroectodermal tumor, peripheral neuroblastoma, malignant neuroepithelioma, peripheral medulloepithelioma, olfactory neuroepithelioma, malignant pigmented neuroectodermal tumor of infancy, malignant retinal anlage tumor
Chondrosarcoma *is a usually bulky, malignant, cartilagenous neoplasm. It is composed of neoplastic chondrocytes at various stages of maturation, in myxoid stroma, and can be primary in bone as well as soft tissues.*	Chondroblastic sarcoma, malignant chondroblastoma, myxoid chondrosarcoma, dedifferentiated chondrosarcoma, clear cell chondrosarcoma, well-differentiated chondrosarcoma, poorly differentiated chondrosarcoma, extraskeletal chondrosarcoma
Mesenchymal chondrosarcoma *is a malignant cartilagenous neoplasm of bone and soft tissues. It is composed of small, poorly differentiated cartilagenous nests embedded in sarcomatous tissue with sparsely cellular matrix that is commonly undifferentiated.*	Chondrosarcoma, dedifferentiated chondrosarcoma, poorly differentiated chondrosarcoma, primitive multipotential primary sarcoma of bone, polyhistioma
Osteosarcoma *is a highly malignant neoplasm of bone and may also occur in soft tissues. It is composed of malignant mesenchymal cells at various stages of maturation in disarranged pattern and is characterized by the deposit of malignant osteoid.*	Osteogenic sarcoma, sclerosing osteosarcoma, telangiectatic osteosarcoma, small cell osteosarcoma, chondroblastic osteosarcoma, large cell predominant osteosarcoma, giant cell predominant osteosarcoma, fibrous histiocytic osteosarcoma, fibrosarcomatous osteosarcoma, chondrosarcomatous osteosarcoma, osteoblastic osteosarcoma
Parosteal osteosarcoma *is a slowly growing, usually bulky, malignant bone tumor arising in the juxtacortical portion of long bones. It is composed of well-differentiated fibroblastic elements commonly in arranged pattern and sparsely cellular malignant osteoid.*	Juxtacortical osteosarcoma, juxtacortical osteogenic sarcoma, ossifying parosteal sarcoma, parosteal osteogenic sarcoma, periosteal osteosarcoma, low grade osteosarcoma
Paget's sarcoma *is a highly malignant polyostotic neoplasm. It develops in less than 1% of the patients with Paget's disease and is predominantly composed of osteosarcoma and fibrosarcoma components in a disarranged pattern.*	Bone sarcoma in Paget's disease.
Alveolar soft part sarcoma *is a slowly progressive malignant neoplasm of unknown histogenesis. It is characterized by distinct polygonal large cells that are arranged in alveolar pattern and have granular cytoplasm and prominent nuclei. In occasional cells protein-carbohydrate crystals are detectable with PAS stain and electron microscope.*	Malignant granular cell myoblastoma with organoid structure, malignant angioreninoma
Kaposi's sarcoma *is an idiopathic hemorrhagic sarcoma. It is composed of slender and plump spindle cells in a richly vascular background. It is a multifocal disease of the elderly but it can be found at any age in association with acquired immunodeficiency.*	Kaposi's disease, idiopathic hemorrhagic sarcoma, angiosarcoma multiplex, acquired immunodeficiency disease, K-sarcoma
Ewing's sarcoma *is an undifferentiated malignant neoplasm of bone and may present in soft tissues. It is characterized by densely packed, uniform, and small epithelioid cells with round nuclei and indistinct, glycogen containing, granular cytoplasm. The stroma is minimal and finely vascular.*	Ewing's tumor, extraskeletal Ewing's sarcoma
Chordoma *is an uncommon neoplasm arising from the primitive notochord at the two ends of the axial skeleton. It is composed of vacuolated physaliferous cells in a lobular lacy arrangement and mucoid intercellular matrix.*	Sacrococcygeal chordoma, cervical chordoma

SELECTED REFERENCES*	ILLUSTRATIONS†
Sordillo, P.P., Helson, L., Hajdu, S.I., et al.: Malignant schwannoma—clinical characteristics, survival and response to therapy. Cancer *47*:2503, 1981.	**Figs. 21, 38, 51, 80, 200, 238, 242, 252, 355, 357, 382, 521, 580, 583, 591, 603, 619, 628, 661, 701, 735, 815, 836, 839, 845, 967, 1192, 1207, 1208**
Tang, C.K., Hajdu, S.I.: Neuroblastoma in adolescence and adulthood. N.Y. State J. Med. *75*:1434, 1975. Harper, P.G., Pringle, J., Souhami, R.L.: Neuroepithelioma. A rare malignant peripheral nerve tumor of primitive origin. Report of two cases and a review of the literature. Cancer *48*:2282, 1981.	**Figs. 309, 342, 478, 489, 519, 561, 565, 568, 572, 585, 596, 690, 724, 827, 982, 1115, 1238, 1239**
Dahlin, D.C., Henderson, E.D.: Chondrosarcoma, a surgical and pathologic problem. Review of 212 cases. J. Bone Joint Surg. *38(A)*:1025, 1956. Hajdu, S.I., Shiu, M., Sordillo, P.: Extra-osseous chondrosarcoma. In Press.	**Figs. 81, 95, 123, 124, 126, 136, 137, 147, 182, 213, 214, 216, 223, 285, 308, 407, 434, 441, 445, 638, 766, 773, 932, 1061, 1064, 1065, 1097, 1101, 1102, 1110, 1222, 1275**
Salvador, A.H., Beabout, J.W., Dahlin, D.C.: Mesenchymal chondrosarcoma. Observations on 30 new cases. Cancer *38*:605, 1971.	**Figs. 103, 151, 175, 180, 318, 390, 444, 507, 595, 674, 707, 920, 1062, 1265**
Dahlin, D.C., Conventry, M.B.: Osteogenic sarcoma. A study of six hundred cases. J. Bone Joint Surg. *49(A)*:101, 1967. Caron, A.S., Hajdu, S.I., Strong, E.W.: Osteogenic sarcoma of the facial and cranial bones. A review of 43 cases. Am. J. Surg. *122*:719, 1973. Sordillo, P., Hajdu, S.I., Magill, G.B., et al.: Extraosseous osteogenic sarcoma. Review of 48 patients. Cancer *51*:727, 1983.	**Figs. 90, 108, 109, 127, 146, 149, 157, 158, 162, 164, 168, 176, 230, 248, 251, 272, 304, 305, 306, 312, 319, 324, 395, 405, 411, 461, 602, 610, 611, 643, 730, 754, 769, 777, 801, 807, 817, 819, 831, 876, 897, 898, 958, 968, 1001, 1070–1079, 1086–1088, 1091, 1105–1108, 1150, 1156**
Ahuja, S.C., Villacin, A.B., Smith, J., et al.: Juxtacortical (parosteal) osteogenic sarcoma. Histological grading and prognosis. J. Bone Joint Surg. *59(A)*:632, 1977.	**Figs. 129, 217, 224, 249, 287, 353, 354, 656, 1082, 1154**
Greditzer, H.G., McLeod, R.A., Unni, K.K., et al.: Bone sarcomas in Paget's disease. Radiology *146*:327, 1983. Huvos, A.G., Butler, A., Bretsky, S.S.: Osteogenic sarcoma associated with Paget's disease of bone. Cancer *52*:1489, 1983.	**Figs. 107, 134, 234B and C, 918, 1103, 1128**
Lieberman, P.H., Foote, F.W., Stewart, F.W., et al.: Alveolar soft part sarcoma. JAMA *198*:1047, 1966. Deschryver-Kecskemeti, K., Kraus, F.T., Engleman, W., et al: Alveolar soft part sarcoma—a malignant angioreninoma. Am. J. Surg. Pathol. *6*:5, 1982.	**Figs. 89, 538, 697, 718, 1161, 1199, 1227**
Urmacher, C., Myskowsi, P., Ochoa, M., et al.: Outbreak of Kaposi's sarcoma with cytomegalovirus in young homosexual males. Am. J. Med. *72*:569, 1982. Lobenthal, S.W., Hajdu, S.I., Urmacher, C.: Cytologic findings in homosexual males with acquired immunodeficiency. Acta Cytol. *27*:597, 1983.	**Figs. 63, 388, 629, 655, 658, 977, 987, 1126**
Dahlin, D.C., Conventry, M.B., Scanlon, P.W.: Ewing's sarcoma. A critical analysis of 165 cases. J. Bone Joint Surg. *48(A)*:185, 1961. Angervall, L., Enzinger, F.M.: Extraskeletal neoplasm resembling Ewing's sarcoma. Cancer *36*:240, 1975.	**Figs. 100, 172, 173, 210, 214, 227, 229, 250, 286, 307, 314, 330, 449, 453, 476, 573, 698, 1157, 1230**
Higinbotham, H.L., Phillips, R.F., Farr, H.W., et al.: Chordoma. Cancer *20*:1841, 1967. Wang, C.C., James, A.E.: Chordoma with a brief review of the literature and report of a case and widespread metastases. Cancer *22*:162, 1968.	**Figs. 69, 215, 310, 505, 729, 767, 771, 1174, 1226**

*For additional references see texts listed in Tables 1 and 2.

†See also Tables 5, 7, 9, 21–23B, 26, 29, 32, 35, 38, 41, 44, 47, 50, 53, 56, 57, 60, 62, 64, 65, 68, 70, 75, 76, 79, 80, 83, 86, 87–92.

Table 13. *The Major Types of Fibromatoses and Their Synonyms**

	SYNONYMS AND RELATED TERMS
Subdermal fibromatosis	Fibrous hamartoma of infancy; subdermal fibromatous tumor of infancy; Reye's disease
Fibromatosis colli	Torticollis, wry neck, fibrous sternomastoid tumor of infancy; congenital muscular torticollis; congenital sternocleidomastoid tumor
Digital fibromatosis	Recurring digital fibrous tumor of infancy; infantile dermal fibromatosis
Fascial fibromatosis	Congenital localized fibromatosis; myositis fibrosa; diffuse infantile fibromatosis
Multifocal fibromatosis	Congenital generalized fibromatosis; aggressive infantile fibromatosis
Gingival fibromatosis	Hereditary gingival fibromatosis
Plantar fibromatosis	Ledderhose's disease
Aponeurotic fibromatosis	Juvenile aponeurotic fibroma, fibroblastoma; calcifying fibroma of Keasbey
Mesenteric fibromatosis	Sclerosing mesenteritis; retractile mesenteritis; sclerosing lipogranulomatosis; subperitoneal sclerosis; Gardner's syndrome
Palmar fibromatosis	Dupuytren's contracture; fibroplasia of palmar fascia; knuckle pads
Penile fibromatosis	Peyronie's disease
Mediastinal fibromatosis	Chronic mediastinitis; idiopathic mediastinitis; fibrous mediastinitis
Retroperitoneal fibromatosis	Ormond's disease; periureteral fibromatosis; periureteritis; periureteritis plastica; pelvic fibrosis; idiopathic retroperitoneal fibrosis
Miscellaneous fibromatosis	Myofibromatosis; cicatricial fibromatosis; hyalin fibromatosis; post-radiation fibromatosis

*Modified from Hajdu, S.I.: Pathology of Soft Tissue Tumors. Philadelphia, Lea & Febiger, 1979.

Table 14A. *Non-neoplastic Soft Tissue and Bone Lesions Not Considered in Detail**

SOFT TISSUE LESIONS	BONE LESIONS
Hyperparathyroidism	Infarct
Scar	Chondroid metaplasia
Fibromatosis ossificans	Enostosis
Fasciitis ossificans	Osteoid metaplasia
Tendosynovitis ossificans	Periostitis
Granulomatous tendosynovitis	Lymphoid hyperplasia
Rheumatoid arthritis	Leukocytosis
Osteoarthritis	Osteomyelitis
Gout	Plasma cell granuloma
Carpal tunnel syndrome	Prolapsed intervertebral disc
de Querrain's disease	Osteogenesis imperfecta
Lipodystrophia	Hypophosphatasemia
Panniculitis ossificans	Hyperphosphatasemia
Adiposis dolorosa	Osteoporosis
Steatopygia	Osteomalacia
Foreign body granuloma	
Atrophy of muscle	
Dystrophy of muscle	
Polymyositis	
Rhabdomyolysis	
Hereditary hemorrhagic telangiectasia	
Vasculitis	
Infarct	
Lymphedema	
Cystic hygroma	
Traumatic neuroma	
Hypertrophy of nerve	
Degeneration of nerve	
Mesothelial hyperplasia	
Hydrocele	
Mesothelial cyst	
Chondroid metaplasia	
Osteoid metaplasia	
Periostitis	
Fibrodysplasia ossificans progressiva	
Lymphoid hyperplasia	
Lymphocytoma cutis	
Histiocytosis	
Mastocytosis	
Plasma cell granuloma	
Thymic cyst	
Angiomatous lymphoid hamartoma	
Amyloidoma	
Gaucher's disease	
Mucopolysaccharidoses	

*For detail see texts listed in Tables 1 and 2.

Table 14B. *Benign Soft Tissue and Bone Neoplasms Not Considered in Detail**

SOFT TISSUE LESIONS	BONE LESIONS
Angiolipoma	Benign mesenchymoma
Myelolipoma	Ecchordosis physaliphora
Myxoid lipoma	
Fibroblastic lipoma	
Leiomyomatosis	
Papillary endothelial hyperplasia	
Venous hemangioma	
Hemangiomatosis	
Lymphangiomyomatosis	
Hemangioblastoma	
Lymphangioma	
Lymphangiomatosis	
Pacinian neurofibroma	
Myxoid neurofibroma	
Benign nevoid schwannoma	
Benign undifferentiated peripheral nerve tumor	
Neuroma	
Benign glandular schwannoma	
Benign epithelioid mesothelioma	
Adenomatoid tumor	
Osteoma	
Benign thymoma	
Benign mesenchymoma	

*For detail see texts listed in Tables 1 and 2.

Table 14C. *Malignant Soft Tissue and Bone Neoplasms Not Considered in Detail**

SOFT TISSUE LESIONS	BONE LESIONS
Malignant glomus tumor	Leukemia
Malignant angiomyolipoma	Malignant lymphoma
Angioendotheliomatosis	Plasmacytoma
Malignant pigmented schwannoma	Adamantinoma
Neuroblastoma	Chemotherapy induced sarcoma
Ganglioneuroblastoma	Radiation induced sarcoma
Mycosis fungoides	
Malignant lymphoma	
Hodgkin's disease	
Plasmacytoma	
Meningeal sarcoma	
Malignant thymoma	
Chemotherapy induced sarcoma	
Radiation induced sarcoma	
Malignant mesenchymoma	
Undifferentiated sarcoma	

*For detail see texts listed in Tables 1 and 2.

Table 15. *Growth Pattern*

1. Arranged
2. Spreading
3. Lacy
4. Epithelioid
5. Alveolar
6. Disarranged

Table 16. *Cell Morphology*

1. Slender spindle
2. Plump spindle
3. Granular epithelioid
4. Clear epithelioid
5. Isomorphic giant
6. Pleomorphic giant

Table 17. *Appearance of Stroma*

1. Fibrillar
2. Sclerosed
3. Myxoid
4. Vascular
5. Inflamed
6. Necrotic
7. Chondrified
8. Ossified
9. Calcified

Table 18. *Product of Cells*

1. Collagen
2. Glycogen
3. Mucopolysaccharides
4. Fat
5. Melanin
6. Secretory granules
7. Crystals
8. Fine structure
9. Tissue antigens

"Sarcoma is a soft swelling without pain"

WILLIAM CULLEN (1710–1790)

2. Clinical Presentation

Non-neoplastic Lesions (Tables 19, 22, and 23; Figures 12, 13, 17, 19)

Benign Neoplasms (Tables 20, 22, and 23; Figures 8, 11, 18, 24, 43, 44, 48, 50)

Malignant Neoplasms (Tables 21, 22, and 23; Figures 9, 10, 14–16, 20–23, 25–42, 45–47, 49, 51–69)

IN general, the presence of a mass, tenderness, pain in resting position, and pressure on adjacent structures constitute the most common presenting complaints of patients with soft tissue and bone tumors. Limitation of motion usually does not occur until the tumor has reached considerable size. The cause of the overwhelming majority of soft tissue and bone tumors is unknown. However, it is known that long-standing lymphedema, congenital or acquired, may lead to development of lymphangiosarcoma, and exposure to incidental or therapeutic irradiation may produce various neoplasms, including fibrous histiocytoma, peripheral nerve tumor, and osteosarcoma. Chronic exposure to vinylchloride may cause hepatic angiosarcoma, and exposure to asbestos may give rise to epithelioid or fibrous mesothelioma, as well as pulmonary carcinoma. Recently, alleged increased incidence of soft tissue sarcomas in men exposed to herbicides containing phenoxy acids and chlororphenols or so-called "Agent-Orange" was publicized in a handful of poorly documented case reports that could not be substantiated on review. It is known that about 50% of patients with acquired immunodeficiency syndrome develop Kaposi's sarcoma. With the exception of Ewing's sarcoma and fibromatosis, which are extremely rare in blacks, and keloid, which is far less common in whites than in blacks, soft tissue and bone tumors do not have any apparent racial propensity.

The clinical presentation of the most common non-neoplastic lesions is summarized in Table 19. In Table 20, the clinical presentation of benign neoplasms is listed, and in Table 21 the clinical presentation of the most common malignant neoplasms is presented. Superficial, multifocal, and solitary lesions are listed in Tables 22 and 23, respectively.

Familial linkage has been established in neurofibromatosis, in palmar and plantar fibromatosis, and in some other forms of fibromatosis as part of Gardner's syndrome. Rare familial tendency is suspected in elastofibromas, lipomas, cutaneous leiomyomas, and paragangliomas.

Spontaneous transformation of up to one-third of multiple neurofibromas to neurofibrosarcoma, 1% of Paget's disease to Paget's sarcoma, and 5% of benign giant cell tumors of bone to malignant giant cell tumor is recognized. The role of trauma in inducing sarcoma, for example, the occasional transformation of myositis ossifications to sarcoma, is a subject of continuous debate but undoubtedly trauma can occasionally induce tissue changes that may lead to appearance of sarcoma.

There are only a limited number of clinical and laboratory "markers" that can be used to facilitate diagnosis of mesenchymal tumors. As a rule, serum alkaline phosphatase is elevated in Paget's disease, Paget's sarcoma, osteosarcoma, and plasma cell myeloma. Elevated serum calcium can be found in hyperparathyroidism and plasma cell myeloma. Urine levels of hydroxyproline are elevated in Paget's disease, and urine and serum levels of epinephrine and norepinephrine, as well as their catabolic products (metanephrines and VMA), are elevated in patients with functionally active paraganglioma and neuroblastoma.

Patients with plasma cell myeloma may have elevated levels of IgG, IgA, IgD, and IgE coupled with Bence-Jones proteinuria. IgM can be occasionally detected in increased quantities in serum of malignant lymphoma patients.

The cryptic and often insidious nature of soft tissue and bone tumors calls for alertness and close cooperation between patients and physicians to assure early detection.

TABLE 19. ***Clinical Presentation of Non-neoplastic Lesions***

	SWELLING	PAIN	TENDERNESS	LIMITATION OF MOTION	SUPERFICIAL	DEEP	MULTIFOCAL	RATE OF GROWTH	RADIOLOGIC FINDING
Aneurysmal bone cyst	X	X		X				Mos, yrs	X
Giant cell granuloma	X	X						Wks	X
Hyperparathyroidism (1)	X		X		X			Mos	X
Xanthogranuloma (2)								Wks	
Fasciitis (3)	X		X		X		X[*]	Wks	
Keloid (4)								Mos, yrs	
Fibromatosis (5)					X	X	X	Mos	
Fibrous dysplasia (6)	X	X	X					Mos	X
Elastofibroma (7)	X		X		X[†]	X		Mos	
Collagenoma (8)	X		X		X		X	Mos, yrs	
Tendosynovitis	X		X	X	X[†]	X		Mos	
Tendosynovial cyst			X	X				Mos	X[‡]
Tendosynovial chondromatosis	X	X	X	X		X	X	Mos, yrs	X
Gouty arthritis (9)	X	X	X	X	X	X	X	Mos, yrs	X
Proliferative panniculitis			X		X			Wks	
Fat necrosis	X		X		X	X[†]		Wks	
Lipogranuloma	X		X		X	X		Mos	X
Proliferative myositis	X		X	X		X		Wks	X
Myositis ossificans (10)	X			X		X		Mos	X
Pyogenic granuloma (11)			X		X			Wks, mos	
Angiofollicular lymphoid hyperplasia (12)			X		X		X	Mos	
Arteriovenous malformation	X		X			X	X[*]	Mos, yrs	X[‡]
Traumatic neuroma	X	X	X	X	X[†]	X	X[*]	Mos	
Paget's disease (13)	X	X	X	X			X	Mos, yrs	

[*]Can be multifocal. [†]Can be deep. [‡]If it is in bone.

1. Elevated serum calcium
2. Cutaneous nodule, asymptomatic
3. Arthralgia
4. Cutaneous nodule, familial, asymptomatic
5. Familial, can be associated with Gardner's syndrome
6. Cutaneous pigmentation, asymmetria of skull
7. Familial tendency
8. Multiple, familial
9. Elevated uric acid
10. Microdactylia in the multifocal progressive form
11. Papular
12. Eosinophilia
13. Fractures, bone deformity, deafness, elevated serum alkaline phosphatase and urinary hydroxyproline, familial tendency

Table 20. *Clinical Presentation of Benign Neoplasms*

	SWELLING	PAIN	TENDERNESS	LIMITATION OF MOTION	SUPERFICIAL	DEEP	MULTIFOCAL	RATE OF GROWTH	RADIOLOGIC FINDING
Benign fibroblastic fibrous histiocytoma	X		X		X	X*		Mos, yrs	X†
Benign histiocytic fibrous histiocytoma	X	X	X		X	X*		Mos	X†
Benign pleomorphic fibrous histiocytoma (1)	X		X		X	X*		Mos, yrs	X†
Nonossifying fibroma (2)						X		Mos, yrs	X
Fibroma (3)	X		X	X	X	X*	X‡	Mos, yrs	X†
Desmoplastic fibroma		X				X		Mos, yrs	X
Lipoma (4)	X				X	X	X	Yrs	X
Lipoblastoma	X				X		X‡	Mos	
Angiomyolipoma (5)								Yrs	X
Hibernoma	X		X		X	X*		Yrs	
Leiomyoma	X		X		X§	X	X	Mos, yrs	X
Rhabdomyoma	X				X	X*		Mos, yrs	
Glomus tumor	X	X	X		X	X*	X‡	Mos	
Lymphangiomyoma (6)						X	X‡	Yrs	X
Neurofibroma (7)	X	X	X	X	X	X	X	Mos, yrs	
Benign schwannoma	X	X	X	X	X§	X	X‡	Mos, yrs	
Benign paraganglioma	X		X			X		Mos, yrs	
Chondroblastoma (8)		X		X		X		Mos, yrs	X
Chondromyxoid fibroma	X	X				X		Mos	X
Chondroma (9)	X	X	X	X	X§	X	X‡	Mos, yrs	X
Osteochondroma (10)	X	X	X		X§	X	X‡	Mos, yrs	X
Osteoblastoma	X	X				X		Mos	X
Osteoid osteoma		X	X			X		Mos	X
Benign granular cell tumor	X				X		X‡	Mos, yrs	
Benign cystosarcoma phyllodes	X		X			X	X‡	Yrs	X

*Can be deep. †If it is in bone. ‡Can be multifocal. §Can be superficial.

1. Familial tendency, cutaneous pigmentation
2. Asymptomatic
3. Location influences symptoms
4. Familial tendency
5. Location and size influence symptoms
6. Disease of women
7. Familial tendency, café-au-lait spots
8. Joint effusion in 30% of cases
9. Location influences symptoms
10. Location influences symptoms, familial tendency

Table 21. *Clinical Presentation of Malignant Neoplasms*

	SWELLING	PAIN	TENDERNESS	LIMITATION OF MOTION	SUPERFICIAL	DEEP	MULTIFOCAL	RATE OF GROWTH	RADIOLOGIC FINDING
Malignant fibroblastic fibrous histiocytoma	X				X	X*		Mos, yrs	X
Malignant histiocytic fibrous histiocytoma	X	X	X		X	X		Mos	X
Malignant pleomorphic fibrous histiocytoma	X	X†	X		X	X*		Mos	X
Desmoid tumor (1)			X			X		Yrs, mos	X
Fibrosarcoma (2)	X	X	X	X	X	X		Mos, wks	X
Tendosynovial sarcoma	X	X		X	X	X		Mos, wks	X
Liposarcoma	X		X			X	X‡	Mos, yrs	X
Leiomyosarcoma	X	X†	X		X	X		Mos	X
Embryonal rhabdomyosarcoma (3)	X		X	X		X		Mos, wks	X
Myxoid rhabdomyosarcoma (4)	X				X	X		Mos, wks	X
Rhabdomyoblastoma	X		X			X		Mos, wks	X
Pleomorphic rhabdomyosarcoma	X		X	X		X		Mos, wks	X
Hemangiopericytoma	X	X	X			X		Mos	X
Hemangiosarcoma	X	X	X		X	X*	X‡	Mos	X
Lymphangiosarcoma (5)	X	X			X	X	X	Yrs	
Neurofibrosarcoma (6)	X	X			X	X	X‡	Mos, yrs	X
Malignant schwannoma	X	X	X	X		X	X‡	Mos, yrs	X
Malignant mesothelioma (7)			X			X	X‡	Mos, yrs	X
Chondrosarcoma of bone (8)	X	X	X			X		Mos, yrs	X
Extraskeletal chondrosarcoma	X		X		X	X		Mos, yrs	X
Mesenchymal chondrosarcoma (9)	X	X	X		X	X		Mos	X
Osteosarcoma (10)	X	X	X			X	X†	Mos	X
Parosteal osteosarcoma	X	X	X	X	X	X		Yrs	X
Paget's sarcoma (11)		X	X			X		Mos	X
Granulocytic sarcoma	X		X		X	X	X‡	Wks	X
Plasma cell myeloma (12)		X	X		X	X	X	Mos	X
Malignant granular cell tumor	X				X		X‡	Mos	
Alveolar soft part sarcoma	X				X	X		Mos	X
Kaposi's sarcoma (13)	X				X	X*	X	Yrs, mos	
Ewing's sarcoma (14)		X	X			X		Mos	X
Chordoma		X	X			X		Mos	X

*Can be deep. †If it is in bone. ‡Can be multifocal.

1. 1% associated with Gardner's syndrome
2. History of pre-existing bone lesion, if it is in bone. Can be congenital
3. Can be congenital
4. Can be congenital
5. Lymphedema
6. Café-au-lait spots and neurofibromatosis
7. Hemorrhagic effusion
8. One-third associated with benign intraskeletal lesion
9. One-third extraskeletal
10. 0.3% multifocal. Elevated serum alkaline phosphatase
11. Elevated serum alkaline phosphatase
12. Hypercalcemia, immunoglobulins. Bence-Jones protein, amyloidosis, elevated serum alkaline phosphatase and serum calcium
13. Can be associated with acquired immune deficiency
14. Fever, seldom seen in blacks

Table 22. *The Most Common Superficial Soft Tissue Lesions*

Non-neoplastic Lesions	
	Xanthogranuloma
	Fasciitis
	Fasciitis ossificans
	Scar
	Keloid
	Fibromatosis
	Fibromatosis ossificans
	Elastofibroma
	Collagenoma
	Tendosynovitis
	Tendosynovial cyst
	Fat necrosis
	Lipogranuloma
	Panniculitis
	Adiposis dolorosa
	Steatopygia
	Piezogenic papule
	Pyogenic granuloma
	Angioblastic lymphoid hyperplasia
	Hereditary hemorrhagic telangiectasia
	Cystic hygroma
	Lymphocytoma cutis
	Mastocytosis
Benign Neoplasms	
	Benign fibroblastic fibrous histiocytoma
	Benign pleomorphic fibrous histiocytoma
	Fibroma
	Lipoma
	Lipoblastoma
	Angiolipoma
	Hibernoma
	Leiomyoma
	Rhabdomyoma
	Angiomyoma
	Hemangioma
	Hypertrophic hemangioma
	Hemangiomatosis
	Glomus tumor
	Neurofibroma
	Benign schwannoma
	Osteoma
	Benign granular cell tumor
Malignant Neoplasms	
	Malignant fibroblastic fibrous histiocytoma
	Epithelioid sarcoma
	Leiomyosarcoma
	Hemangiosarcoma
	Lymphangiosarcoma
	Malignant glomus tumor
	Neurofibrosarcoma
	Mycosis fungoides
	Malignant granular cell tumor
	Kaposi's sarcoma

TABLE 23A. *The Most Common Multifocal Bone Lesions*

NON-NEOPLASTIC LESIONS	BENIGN NEOPLASMS	MALIGNANT NEOPLASMS
Aneurysmal bone cyst	Nonossifying fibroma	Hemangiosarcoma
Hyperparathyroidism	Hemangioma	Paget's sarcoma
Fibrous dysplasia	Chondroblastoma	Granulocytic sarcoma
Arteriovenous malformation	Chondroma	Plasma cell myeloma
Paget's disease	Osteochondroma	
Eosinophilic granuloma		
Osteomyelitis		

TABLE 23B. *Lesions Usually Present as Solitary Bone Lesions*

NON-NEOPLASTIC LESIONS	BENIGN NEOPLASMS	MALIGNANT NEOPLASMS
Giant cell granuloma	Benign fibroblastic fibrous histiocytoma	Malignant fibroblastic fibrous histiocytoma
Tendosynovial cyst	Benign histiocytic fibrous histiocytoma	Malignant histiocytic fibrous histiocytoma
Infarct	Benign pleomorphic fibrous histiocytoma	Malignant pleomorphic fibrous histiocytoma
Callus	Desmoplastic fibroma	Fibrosarcoma
Enostosis	Chondromyxoid fibroma	Chondrosarcoma
	Osteoblastoma	Osteosarcoma
	Osteoid osteoma	Parosteal osteosarcoma
		Ewing's sarcoma
		Chordoma
		Adamantinoma

Fig. 8. Benign teratoma of the neck of an infant (see Fig. 235).

Fig. 9. (Top center) Desmoid tumor of the neck of an adult woman (see Figs. 36, 52, 77, 185, 202, 244, 376, 662).

Fig. 10. Embryonal rhabdomyosarcoma (see Figs. 20, 28, 66, 67, 68, 78, 226, 263, 364).

Fig. 11. Neurofibroma of the vagus nerve (see Figs. 74, 245, 253, 363, 487, 652, 840, 1142).

Fig. 12. (Bottom center) Multiple papulomacular lesions of angiofollicular lymphoid hyperplasia (see Figs. 13, 979, 1040).

Fig. 13. Solitary nodule of angiofollicular lymphoid hyperplasia (see Figs. 12, 979, 1040).

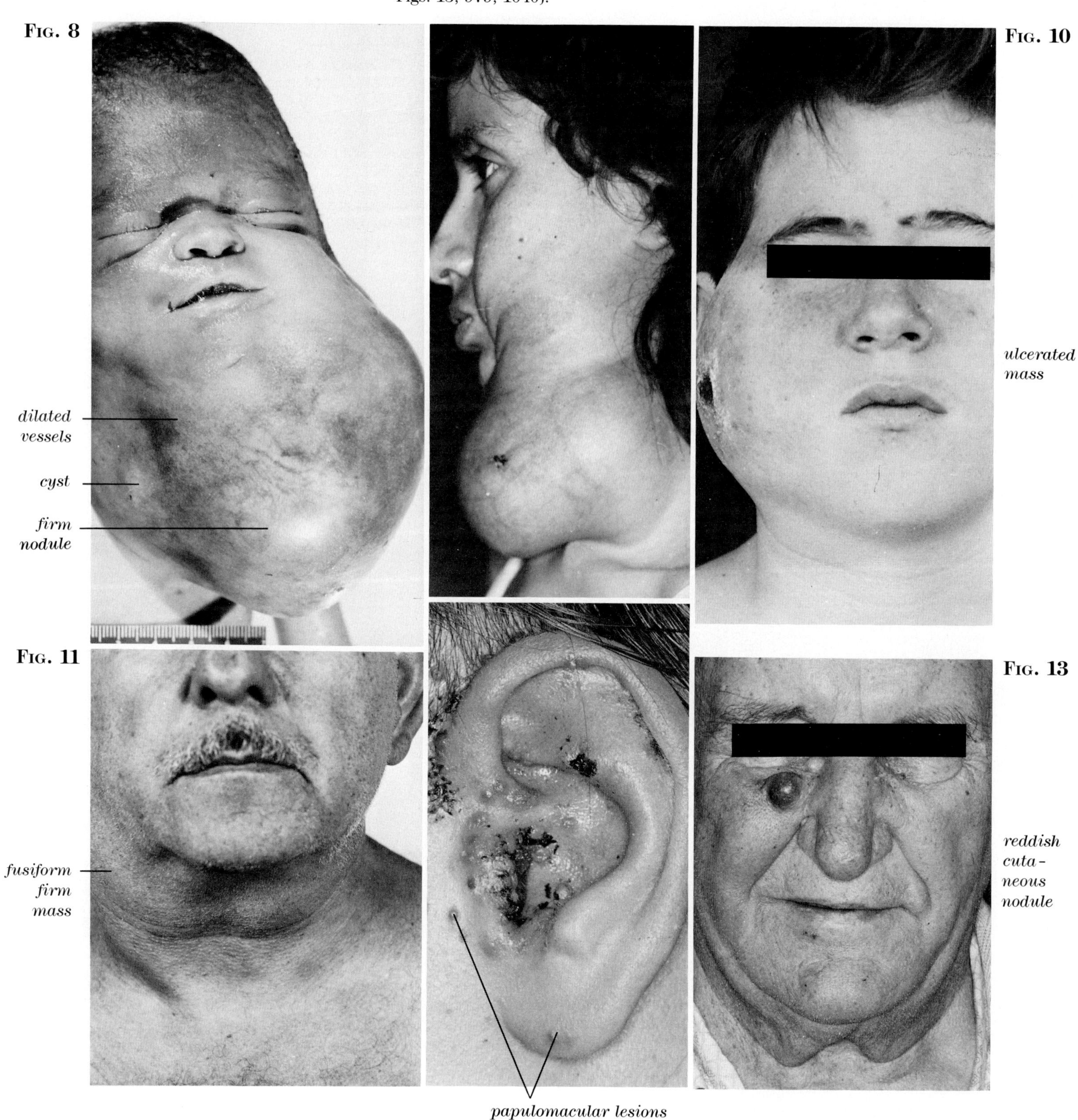

Fig. 14. Seminoma of testis metastatic to neck (see Figs. 566, 757, 1224).

Fig. 15. (Top center) Burkitt's lymphoma (see Figs. 132, 169, 423, 463, 494).

Fig. 16. Primary cutaneous hemangiosarcoma (see Figs 32, 42, 46, 92, 102).

Fig. 17. Gingival fibromatosis (see Figs. 280, 367, 385, 387, 581, 616).

Fig. 18. (Center) Congenital capillary hemangioma of the tip of the tongue (see Figs. 549, 559, 988).

Fig. 19. Cystic hygroma (see Fig. 1016).

Fig. 20. Embryonal rhabdomyosarcoma of an adult (see Figs. 10, 28, 66–68, 78, 226, 263, 364, 637).

Fig. 21. (Bottom center) Malignant schwannoma of the face (see Figs. 38, 51, 80, 200, 238).

Fig. 22. Neurofibrosarcoma with stigmata of von Recklinghausen's disease (see Fig. 23, 26, 34, 53, 79).

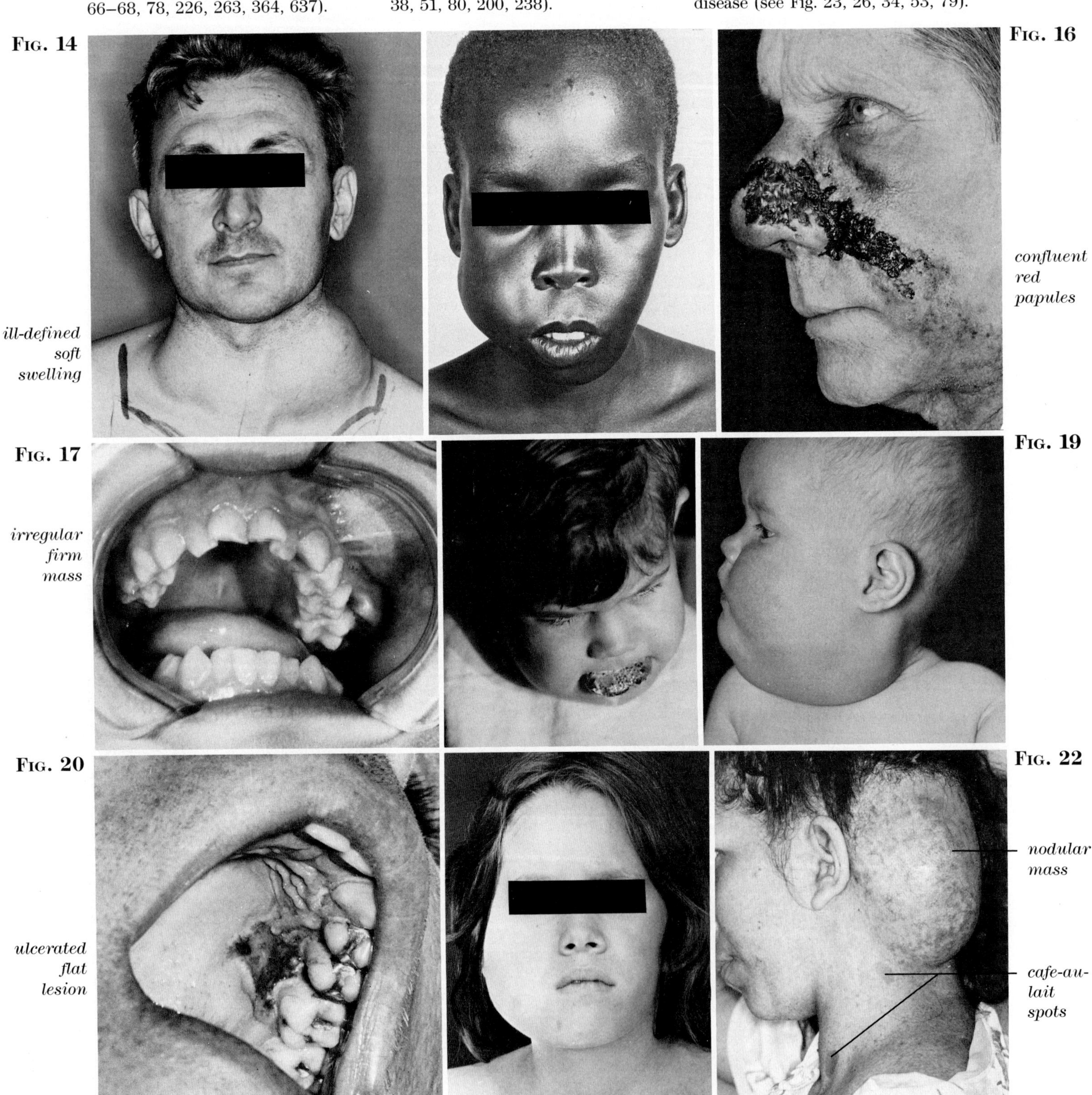

Fig. 23. Neurofibrosarcoma with cafe-au-lait spots (see Figs. 22, 26, 34, 53, 79, 236, 254, 256, 369).

Fig. 24. (Top center) Pedunculated lipoma (Courtesy of Dr. W. Knapper, Memorial Sloan-Kettering Cancer Center) (see Figs. 24, 73, 198, 243).

Fig. 25. Recurrent malignant pleomorphic fibrous histiocytoma (see Figs. 31, 88, 111, 130, 138, 140, 148, 237, 255, 262, 344).

Fig. 26. Recurrent neurofibrosarcoma of the neck (see Figs. 22, 23, 34, 53, 256, 264, 271, 275, 370).

Fig. 27. (Center) Hodgkin's disease with axillary presentation (see Figs. 742, 885, 1024).

Fig. 28. Embryonal rhabdomyosarcoma of the neck of a youngster (Courtesy of Dr. M. Shiu, Memorial Sloan-Kettering Cancer Center) (see Figs. 10, 20, 66–68).

Fig. 29. Pleomorphic rhabdomyosarcoma of the shoulder with erosion of the scapula (see Figs. 87, 121, 292, 295, 296).

Fig. 30. (Bottom center) Fibroblastic liposarcoma of an elderly man (Courtesy of Dr. W. Knapper, Memorial Sloan-Kettering Cancer Center) (see Figs. 86, 343, 440, 641).

Fig. 31. Recurrent malignant pleomorphic fibrous histiocytoma (see Figs. 25, 88, 111, 130, 138, 201, 203, 232, 288, 293, 460).

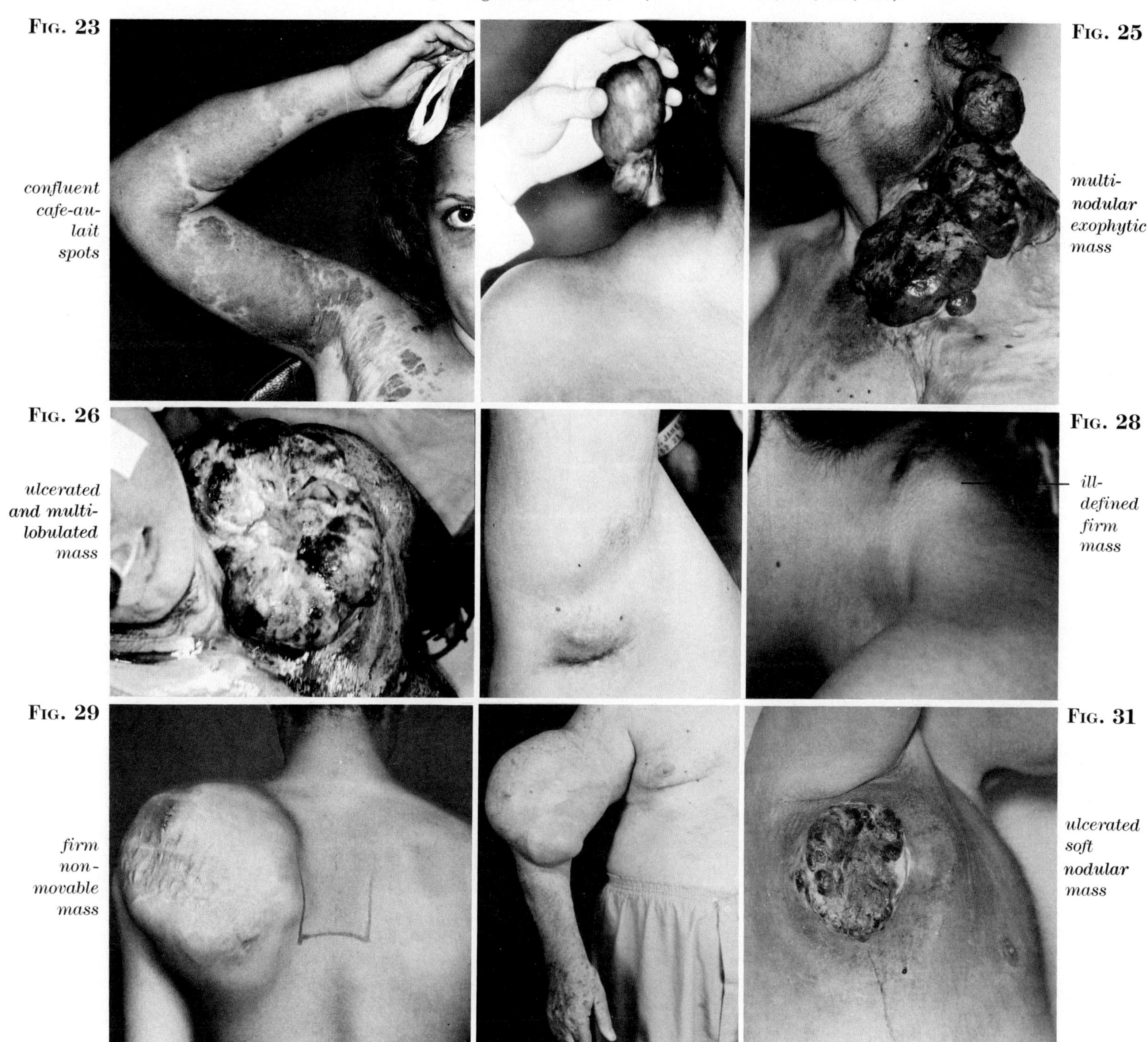

Fig. 32. Recurrent hemangiosarcoma after disarticulation (see Figs. 16, 42, 46, 92, 102, 316, 523, 541).

Fig. 35. Post-mastectomy lymphangiosarcoma (see Figs. 41, 61, 84, 154, 159, 321, 424, 551, 649).

Fig. 38. Recurrent malignant schwannoma. (Courtesy of Dr. W. Knapper, Memorial Sloan-Kettering Cancer Center) (see Figs. 21, 51, 80, 200, 238, 242, 246).

Fig. 33. (Top center) A neglected case of fibrosarcoma (see Figs. 60, 91, 117, 131, 165, 170, 260, 410).

Fig. 36. (Center) Desmoid tumor that required disarticulation (see Figs. 9, 52, 77, 185, 202, 209, 244).

Fig. 39. (Bottom center) Subcutaneous nodules of metastatic tendosynovial sarcoma (see Figs. 64, 85, 179, 190, 315, 326, 391, 392, 396).

Fig. 34. Multifocal neurofibrosarcoma (see Figs. 23, 26, 53, 79, 236).

Fig. 37. Malignant fibroblastic fibrous histiocytoma, so-called "dermatofibrosarcoma protuberans" (see Figs. 37, 40, 54, 76, 111, 143, 241).

Fig. 40. Recurrent malignant fibroblastic fibrous histiocytoma (Courtesy of Dr. M. Shiu, Memorial Sloan-Kettering Cancer Center) (see Figs. 37, 54, 76, 111, 143, 241).

Fig. 32

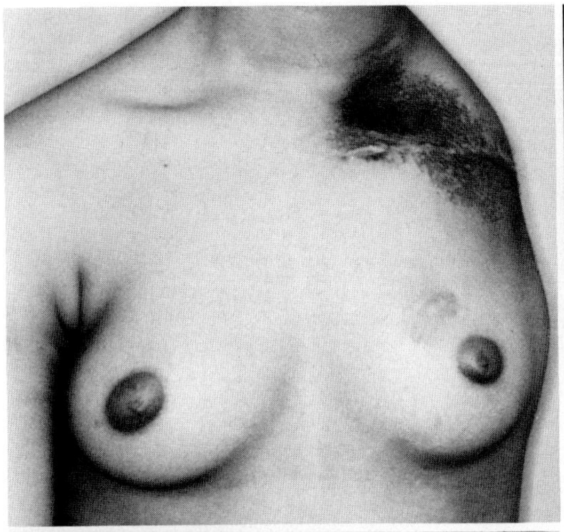

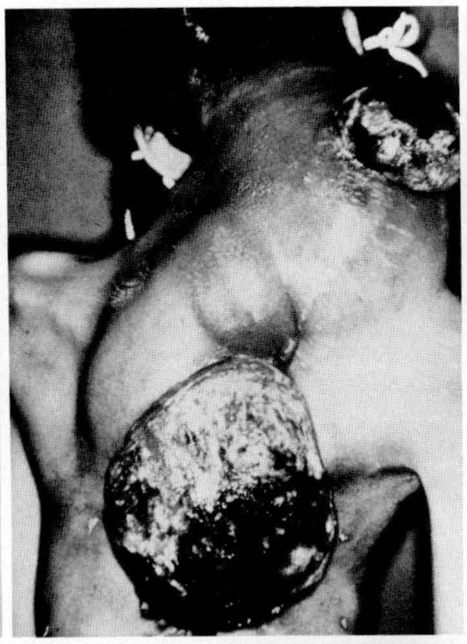

Fig. 34

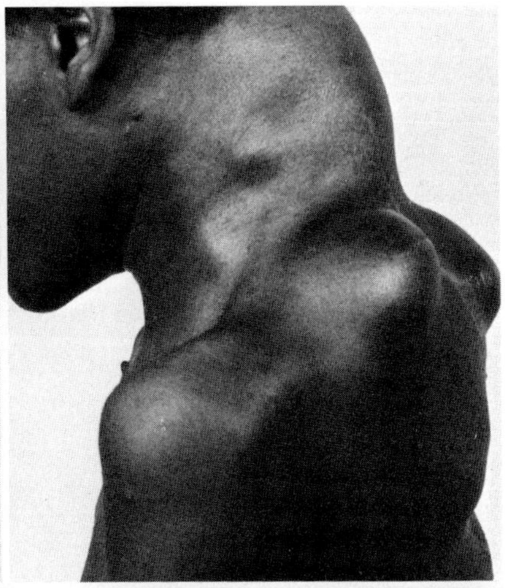

Fig. 35

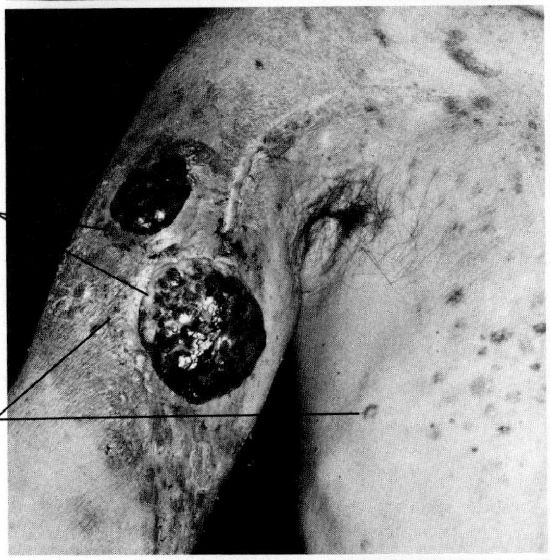

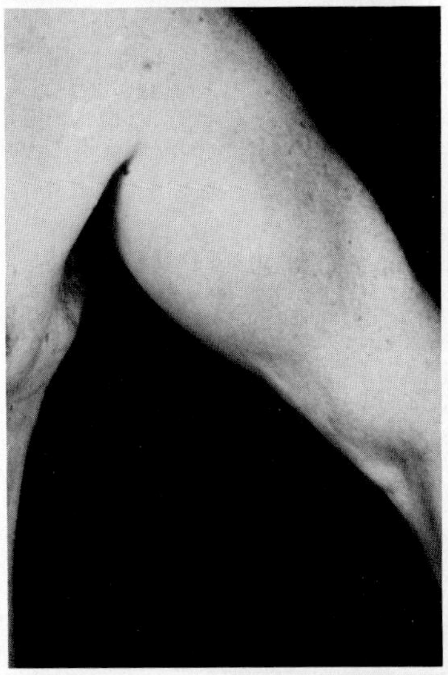

Fig. 37

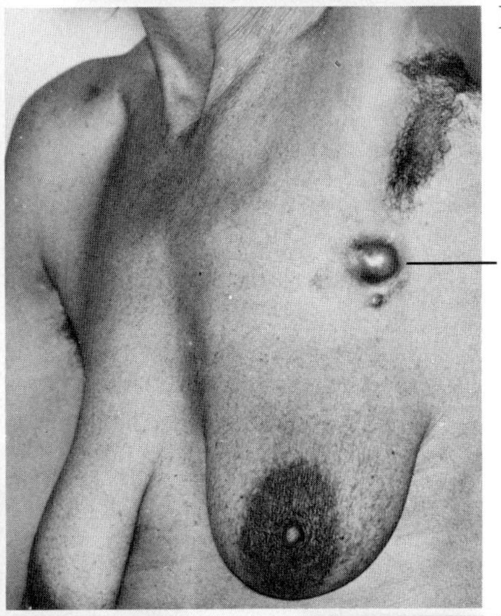

Fig. 38

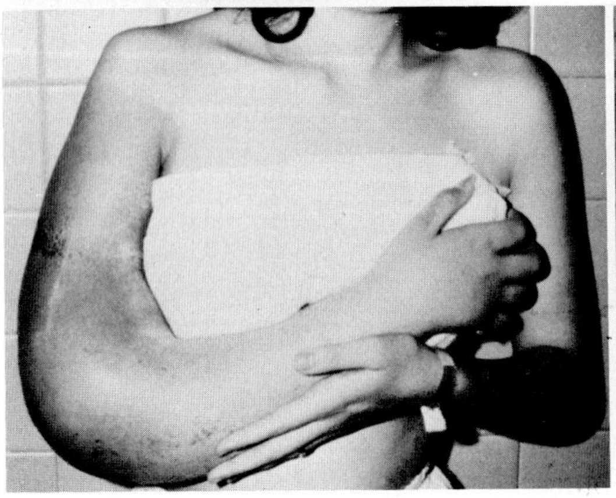

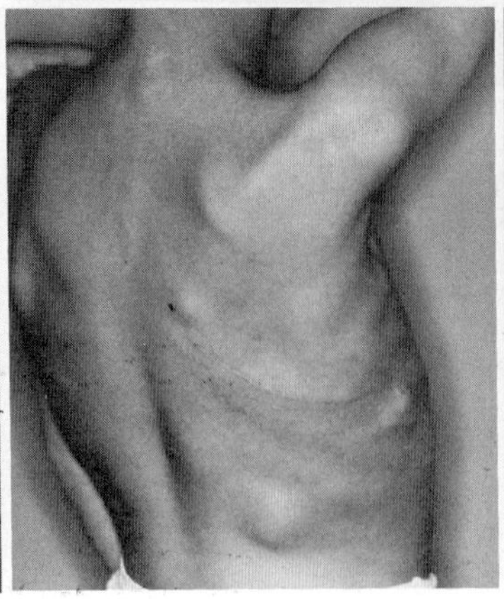

Fig. 40

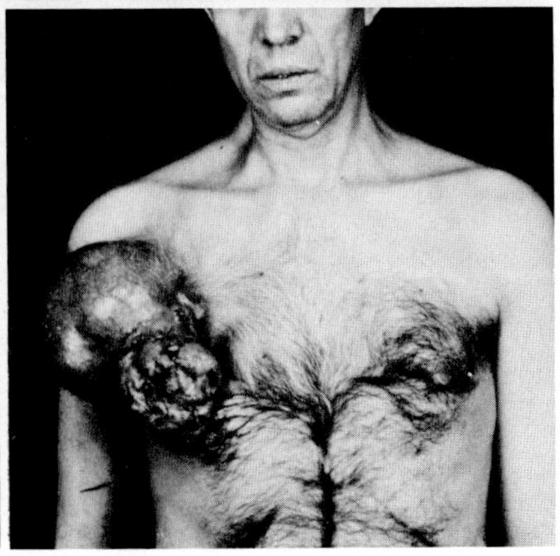

Fig. 41. Lymphangiosarcoma associated with post-mastectomy lymphedema (see Figs. 35, 61, 84, 154, 159, 321, 424, 551, 558, 649).

Fig. 44. Neurofibromatosis (see Fig. 204, 245, 253, 363, 487)

Fig. 47. Malignant cystosarcoma phyllodes (see Figs. 351, 377, 657, 663).

Fig. 42. (Top center) Hemangiosarcoma of breast (see Figs. 16, 32, 46, 92, 102, 316, 523, 541, 555, 557).

Fig. 45. (Center) Wilms' tumor (Courtesy of Dr. P. Exelby, Memorial Sloan-Kettering Cancer Center) (see Figs. 383, 835).

Fig. 43. Benign cystosarcoma phyllodes (see Figs. 48, 247, 414, 921, 938).

Fig. 46. Recurrent hemangiosarcoma of breast (see Figs. 32, 42, 92, 102, 316, 523, 555, 563, 564).

Fig. 48. Recurrent benign cystosarcoma phyllodes (see Figs. 43, 247, 414, 921, 938).

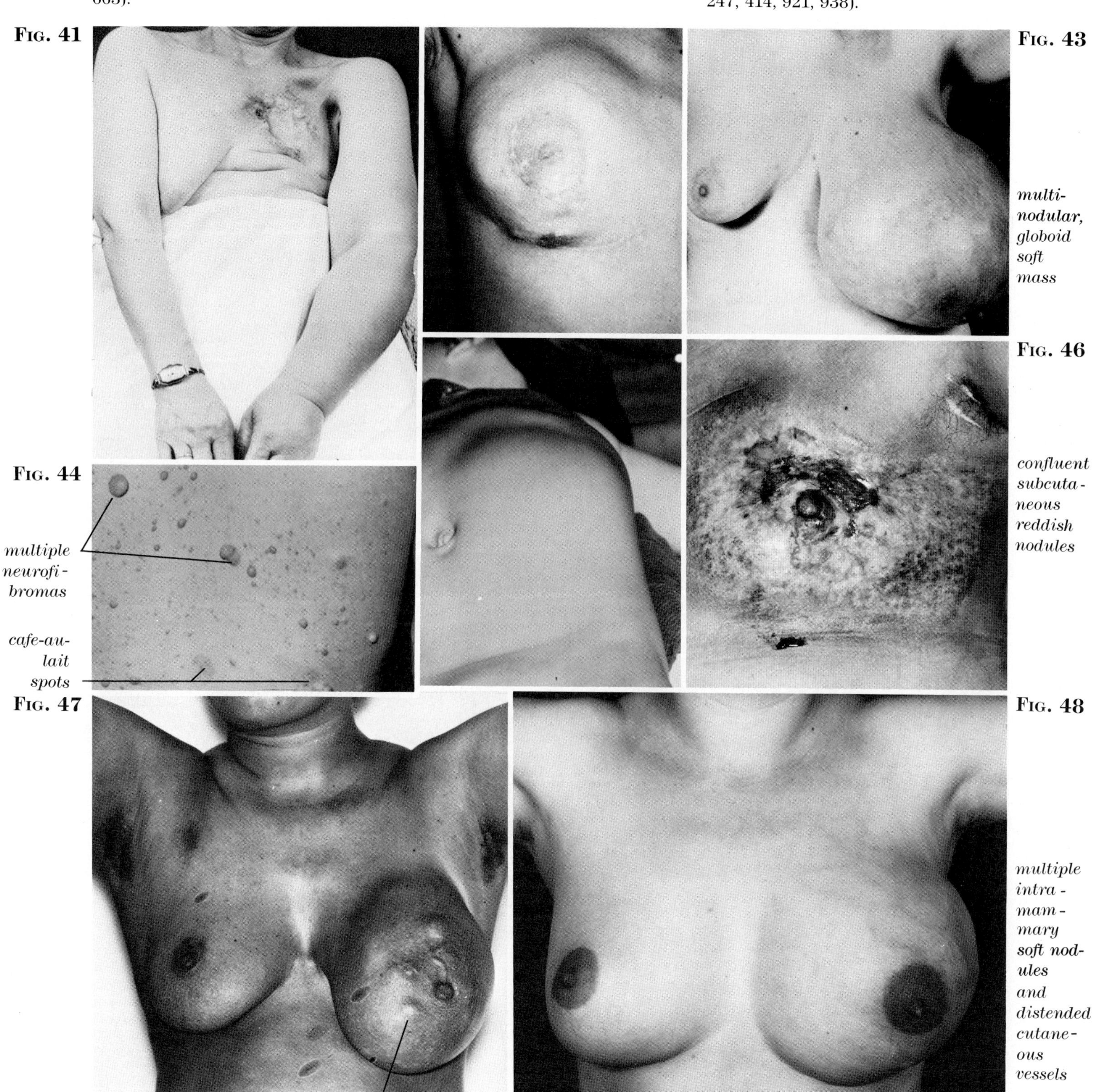

Fig. 49. Myxoid liposarcoma (Courtesy of Dr. W. Knapper, Memorial Sloan-Kettering Cancer Center) (see Figs. 56, 59, 86, 188, 208, 269, 290, 438, 439).

Fig. 51. Retroperitoneal malignant peripheral nerve tumor (see Figs. 21, 38, 80, 200, 238, 242, 246, 252, 298, 355, 357).

Fig. 50. Peritoneal leiomyomatosis. Multiple tumor nodules are visible through an abdominal incision (see Figs. 358, 359, 361, 402, 654).

Fig. 52. Recurrent desmoid tumor of the abdominal wall (see Figs. 9, 36, 77, 185, 202, 209, 244, 282, 376, 386

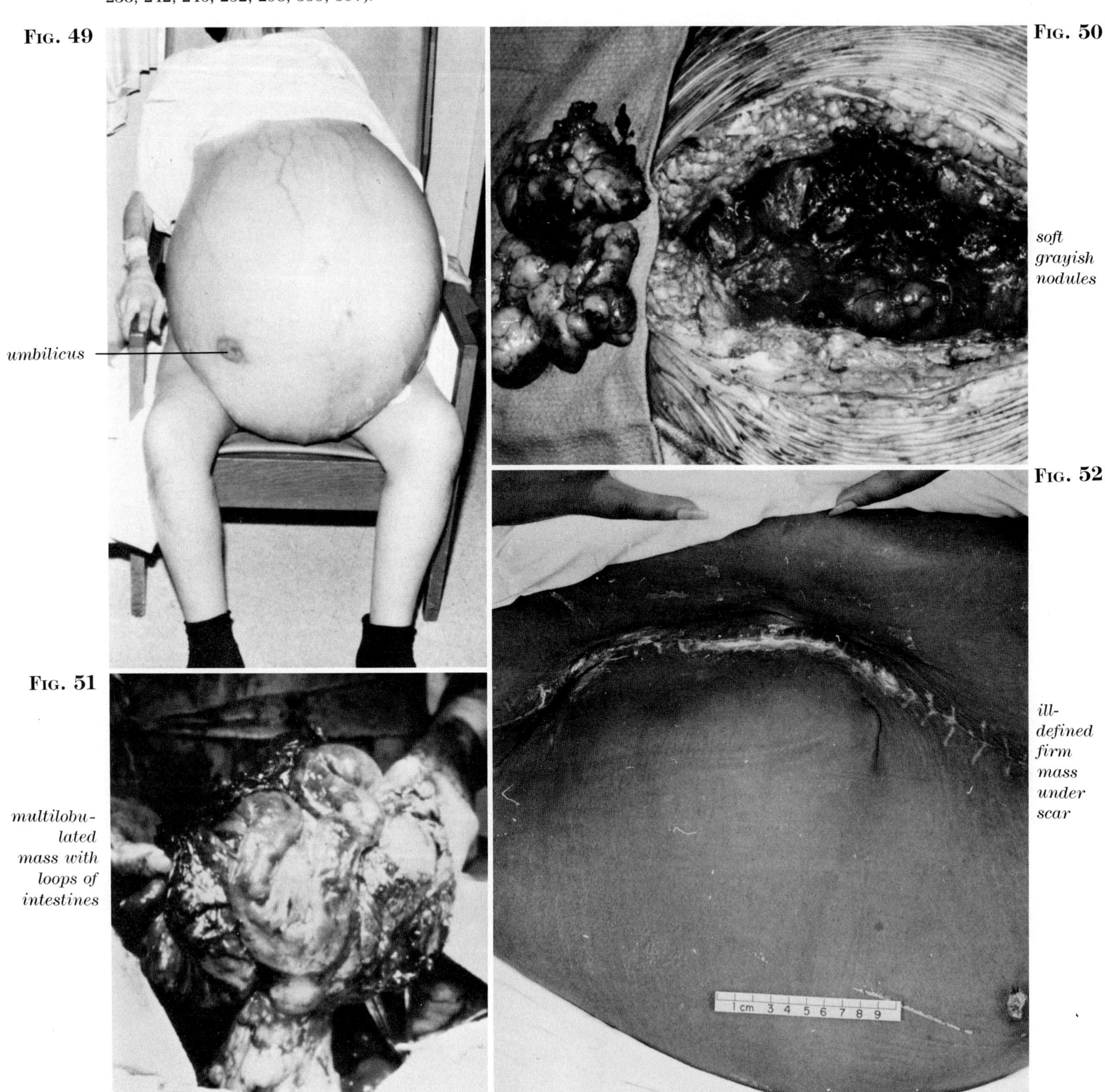

Fig. 53. Neurofibrosarcoma of the right thigh and neurofibromatosis with cafe-au-lait spots (see Figs. 26, 34, 79, 236, 254, 256, 264, 271).

Fig. 54. (Top center) Recurrent malignant fibroblastic fibrous histiocytoma (see Figs. 37, 40, 76, 111, 143, 241, 301, 339, 340, 350).

Fig. 55. Lipoblastic liposarcoma (see Figs. 86, 199, 415, 448, 450).

Fig. 56. Recurrent myxoid liposarcoma (see Figs. 49, 59, 86, 188, 208, 269, 290, 438, 439, 636).

Fig. 57. (Bottom center) Recurrent malignant histiocytic fibrous histiocytoma (see Figs. 156, 161, 300, 452, 534, 592, 600, 711).

Fig. 58. Well-differentiated liposarcoma of the right thigh (see Figs. 86, 219, 431, 435, 822, 999, 1184).

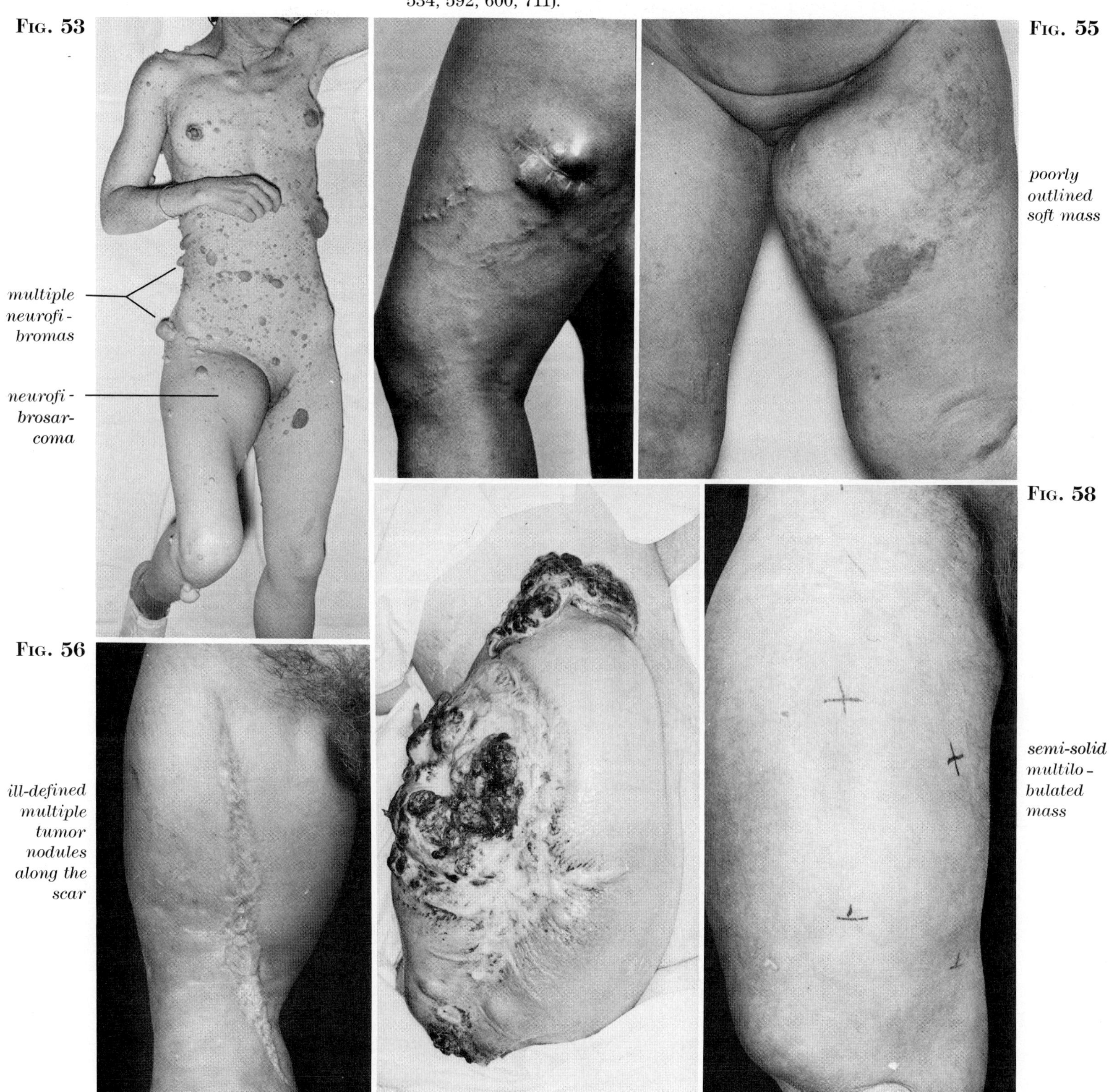

Fig. 59. Myxoid liposarcoma (see Figs. 49, 56, 86, 188, 208, 269, 290).

Fig. 60. (Top center) Fibrosarcoma (see Figs. 33, 91, 117, 131, 165).

Fig. 61. Lymphangiosarcoma 20 years after filarial infection induced lymphedema (see Figs. 35, 41, 84, 154, 159, 321, 424, 551, 558).

Fig. 62. Ill-defined subcutaneous nodules of tendosynovial sarcoma, epithelioid and clear cell type (see Figs. 85, 181, 187, 258, 259, 322).

Fig. 63. (Center) Kaposi's sarcoma (see Figs. 388, 629, 655, 658).

Fig. 64. Large tendosynovial sarcoma, monophasic spindle cell type (see Figs. 39, 85, 179, 190, 315, 326).

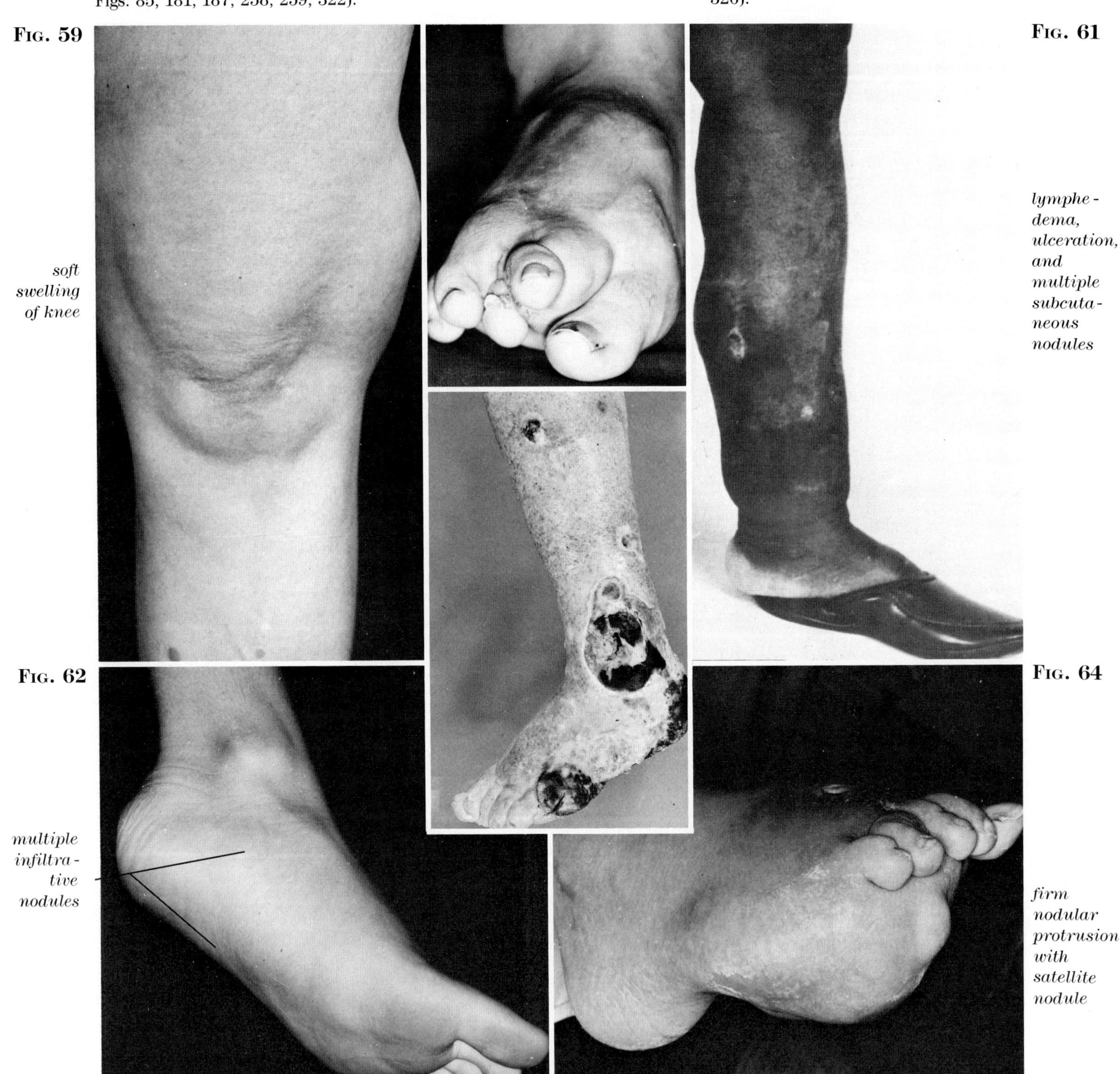

Fig. 65. Mesodermal mixed tumor of the uterus protruding into the vagina (see Figs. 502, 589, 763, 969, 1165).

Fig. 67. Embryonal rhabdomyosarcoma of the urinary bladder (Courtesy of Dr. P. Exelby, Memorial Sloan-Kettering Cancer Center) (see Figs. 28, 66, 68, 78, 226, 207, 263, 281, 291, 364).

Fig. 66. Embryonal rhabdomyosarcoma of the perineum and buttock of an infant (see Figs. 20, 28, 67, 68, 78, 207, 226, 263, 281, 291, 372).

Fig. 68. Paratesticular embryonal rhabdomyosarcoma with bilateral groin and pelvic metastases (see Figs. 66, 67, 78, 226, 263, 281, 291, 364, 372, 378).

Fig. 69. Sacrococcygeal chordoma (see Figs. 215, 310, 505, 729, 767, 771, 1174, 1226).

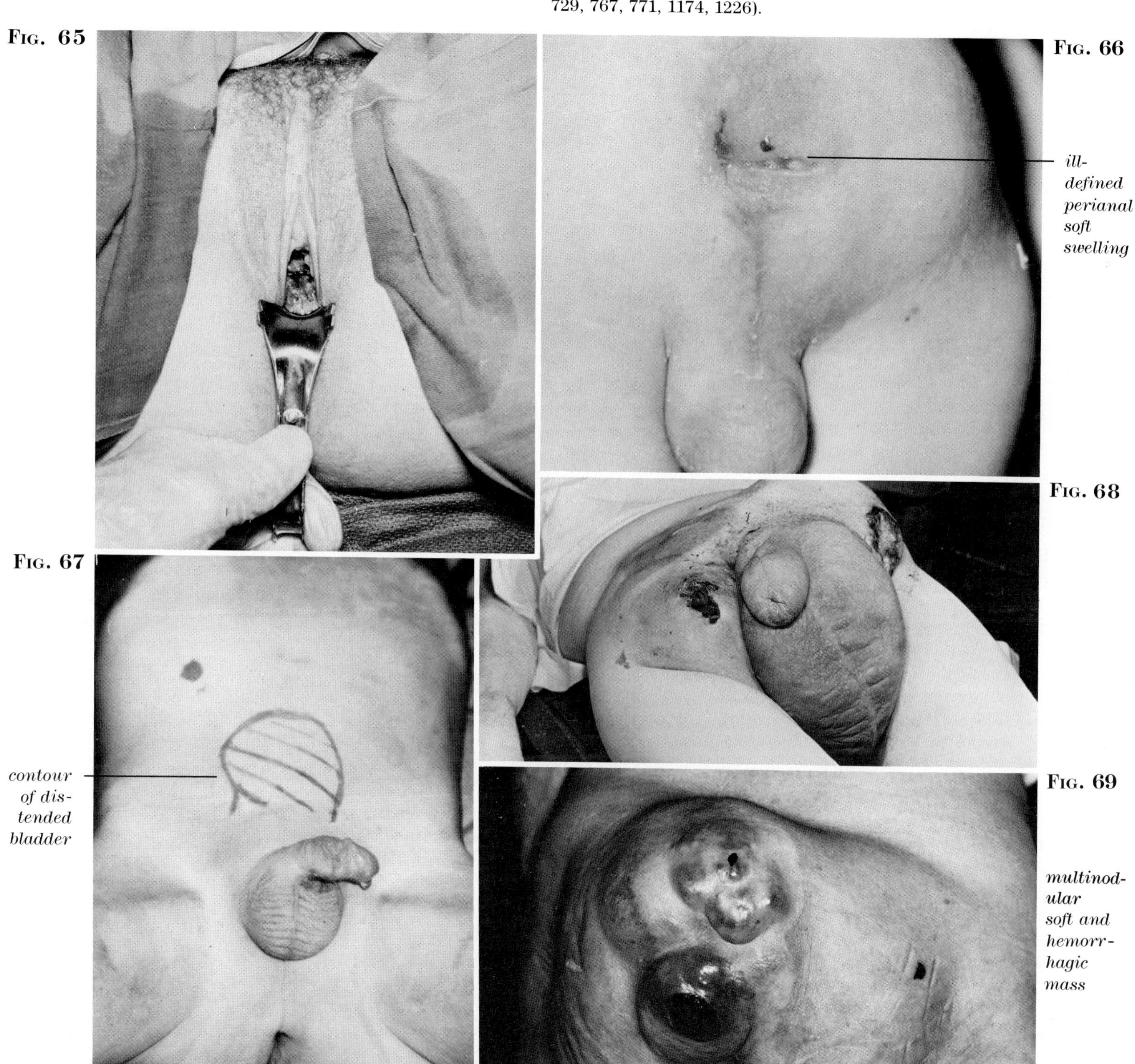

"Sarcomas vary in prognosis and natural history depending on cell type, anatomic site and the patient's age"
ARTHUR PURDY STOUT (1885–1967)

3. Age Distribution

Non-neoplastic Lesions (Table 24)

Benign Neoplasms (Table 25)

Malignant Neoplasms (Table 26)

PRACTICALLY all malignant neoplasms occurring in patients under 15 years of age are nonepithelial in origin. Rokitansky, more than 150 years ago, pointed out that "Sarcomata are the most common neoplasms of children." Soft tissue sarcomas comprise 7% and bone sarcomas 5% of childhood malignant neoplasms. The three most common malignant soft tissue and bone neoplasms that account for 90% of the sarcomas in patients under 15 years of age are embryonal rhabdomyosarcoma, Ewing's sarcoma, and intraskeletal osteosarcoma. The three most common non-neoplastic lesions are fibromatosis, aneurysmal bone cyst, and fibrous dysplasia; the three most common benign neoplasms are lipoblastoma, nonossifying fibroma, and osteoid osteoma. It is estimated that one-half of all so-called "solid tumors" in children occur in the soft tissues and bones.

On the other end of the age scale, among patients 60 years old and older, the most common non-neoplastic lesions include elastofibroma, gout, lipogranuloma, and Paget's disease. Primary intraskeletal benign neoplasms in this age group are extremely rare, but benign soft tissue neoplasms such as benign fibrous histiocytomas, various lipomas, and leiomyoma are not uncommon. In elderly patients, the most commonly seen malignant neoplasms are liposarcoma, leiomyosarcoma, plasma cell myeloma, Paget's sarcoma, extraskeletal chondrosarcoma, and extraskeletal osteosarcoma.

The influence of age on the microscopic appearance and prognosis of soft tissue and bone tumors is well documented. In 1959, Stout wrote that "One can feel no assurance that tumors with a comparable histological structure will behave in the same way in children as they do in adults." One needs only to consider the histologically excellent prognosis of childhood fibrosarcoma, tendosynovial sarcoma, and such epithelial neoplasms as thyroid and salivary gland carcinomas and germ cell tumors. Due to recent progress in multimodal therapy, embryonal rhabdomyosarcoma, Ewing's sarcoma, and osteosarcoma should be added to the list of sarcomas that have a far better prognosis in children and adolescents than in adults. As a fact, so called childhood neoplasms, for example, embryonal rhabdomyosarcoma, Ewing's sarcoma, osteosarcoma and neuroblastoma, when they occur in adults, carry extremely grave prognosis.

Admittedly, knowledge of the age of the patient often influences the microscopic diagnosis in equivocal cases. However, knowing that malignant neoplasms such as liposarcoma, pleomorphic fibrous histiocytoma, pleomorphic rhabdomyosarcoma, mesothelioma, plasma cell myeloma, extraskeletal osteo- and chondrosarcoma practically never occur in children should not influence the pathologist to not consider the microscopic appearance of the neoplasm above all other aspects.

Table 24. *Age Distribution of Non-neoplastic Lesions*

Soft Tissue Lesions — Average Age, Years (65+ 60 55 50 45 40 35 30 25 20 15 10 5 0)		Bone Lesions — Average Age, Years (0 5 10 15 20 25 30 35 40 45 50 55 60 65+)
	Aneurysmal bone cyst	———
	Giant cell granuloma	———
———	Xanthogranuloma	
———	Fasciitis	
———	Keloid	
———	Fibromatosis	
	Fibrous dysplasia	———
———	Elastofibroma	
———	Tendosynovitis	
———	Tendosynovial cyst	———
———	Tendosynovial chondromatosis	
———	Gouty arthritis	
———	Lipogranuloma	———
———	Proliferative panniculitis	
———	Proliferative myositis	
———	Myositis ossificans	
———	Pyogenic granuloma	
———	Angiofollicular lymphoid hyperplasia	
———	Arteriovenous malformation	———
	Paget's disease	———

Table 25. *Age Distribution of Benign Neoplasms*

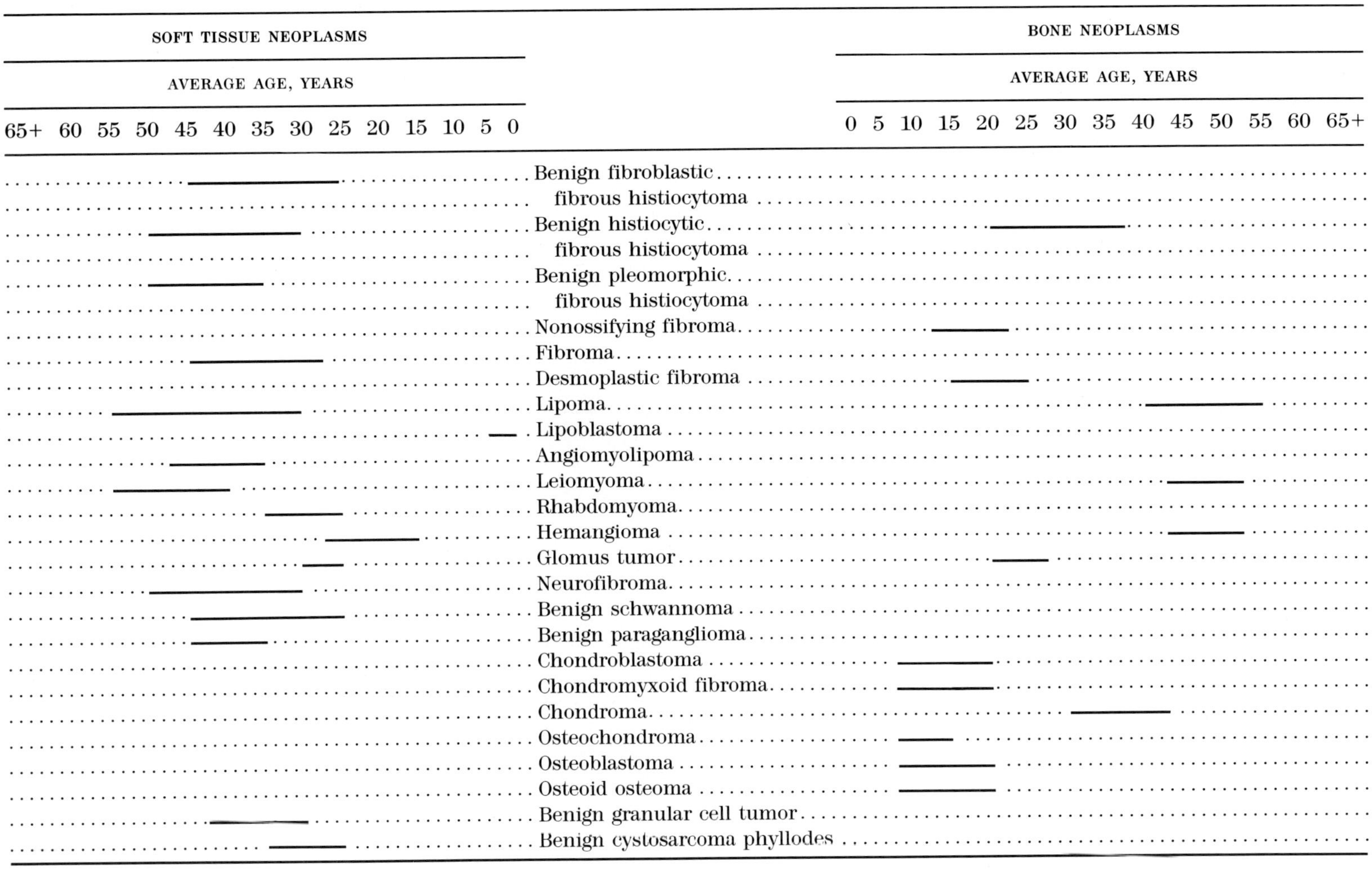

Soft tissue neoplasms: average age, years (65+ 60 55 50 45 40 35 30 25 20 15 10 5 0)		Bone neoplasms: average age, years (0 5 10 15 20 25 30 35 40 45 50 55 60 65+)
~45–25	Benign fibroblastic fibrous histiocytoma	
~50–30	Benign histiocytic fibrous histiocytoma	~20–38
~50–35	Benign pleomorphic fibrous histiocytoma	
	Nonossifying fibroma	~15–22
~45–28	Fibroma	
	Desmoplastic fibroma	~15–25
~55–30	Lipoma	~40–55
~3–0	Lipoblastoma	
~47–35	Angiomyolipoma	
~55–40	Leiomyoma	~43–53
~35–25	Rhabdomyoma	
~27–15	Hemangioma	~43–53
~30–25	Glomus tumor	~20–27
~50–30	Neurofibroma	
~45–25	Benign schwannoma	
~45–35	Benign paraganglioma	
	Chondroblastoma	~10–20
	Chondromyxoid fibroma	~10–20
	Chondroma	~30–43
	Osteochondroma	~10–15
	Osteoblastoma	~10–20
	Osteoid osteoma	~10–20
~42–29	Benign granular cell tumor	
~35–25	Benign cystosarcoma phyllodes	

Table 26. *Age Distribution of Malignant Neoplasms*

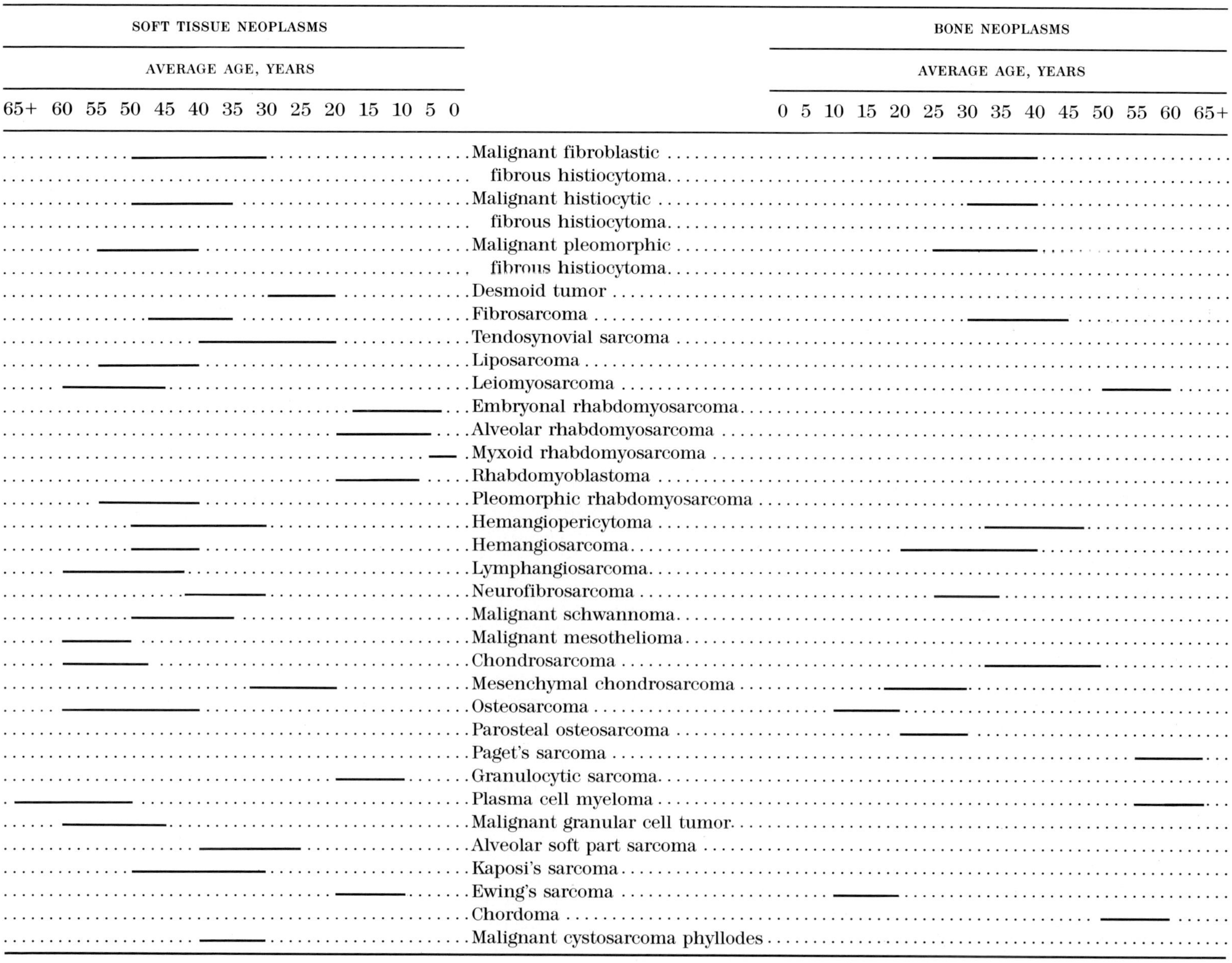

Soft tissue neoplasms: average age, years	Neoplasm	Bone neoplasms: average age, years
50–30	Malignant fibroblastic fibrous histiocytoma	25–40
50–35	Malignant histiocytic fibrous histiocytoma	30–40
55–40	Malignant pleomorphic fibrous histiocytoma	25–40
30–20	Desmoid tumor	
47–35	Fibrosarcoma	30–45
40–20	Tendosynovial sarcoma	
55–40	Liposarcoma	
60–45	Leiomyosarcoma	50–60
17–3	Embryonal rhabdomyosarcoma	
20–5	Alveolar rhabdomyosarcoma	
5–0	Myxoid rhabdomyosarcoma	
20–7	Rhabdomyoblastoma	
55–40	Pleomorphic rhabdomyosarcoma	
50–30	Hemangiopericytoma	32–47
50–40	Hemangiosarcoma	20–40
60–42	Lymphangiosarcoma	
42–30	Neurofibrosarcoma	25–35
50–35	Malignant schwannoma	
60–50	Malignant mesothelioma	
60–47	Chondrosarcoma	32–50
32–20	Mesenchymal chondrosarcoma	17–30
60–40	Osteosarcoma	10–20
	Parosteal osteosarcoma	20–30
	Paget's sarcoma	55–65+
20–10	Granulocytic sarcoma	
65+–50	Plasma cell myeloma	55–65+
60–45	Malignant granular cell tumor	
40–25	Alveolar soft part sarcoma	
50–30	Kaposi's sarcoma	
20–10	Ewing's sarcoma	10–20
	Chordoma	50–60
40–30	Malignant cystosarcoma phyllodes	

"Cancer is the most grave, and most incurable disease and one of the diseases which most frequently affect human beings"

JEAN CRUVEILHIER (1791–1874)

4. Sex Distribution

Non-neoplastic Lesions (Table 27)

Benign Neoplasms (Table 28)

Malignant Neoplasms (Table 29)

UNTIL recently, no particular attention was paid to sex distribution of soft tissue and bone tumors. Consideration of the overall sex predominance occasionally blurs age-related differences. For example, intraskeletal histiocytic fibrous histiocytoma, or so-called "giant cell tumor of bone," in patients younger than 18 years is seen predominantly in women; however, due to its predominance in men in their third and fourth decades there is no apparent overall sex predominance.

It is generally accepted that soft tissue and bone tumors are found most commonly in men. Tables 27 to 29 list the most common soft tissue and bone tumors with apparent sex predominance. It is shown that there is a 2:1 overall male predominance that increases to 3:1 if sarcomas are considered only (see Table 29). However, it should be pointed out that benign and malignant peripheral nerve tumors, neuroendocrine tumors, and vascular and smooth muscle neoplasms are found more commonly in women than in men. On the other hand, tendosynovial tumors, striated muscle tumors, malignant adipose tissue neoplasms, intra- and extraskeletal fibrous and fibrohistiocytic tumors, as well as most osteoid and cartilage-producing tumors, are seen most commonly in men.

TABLE 27. ***Sex Distribution of Non-neoplastic Lesions***

SOFT TISSUE LESIONS	BONE LESIONS
Male Predominance	
Fasciitis	Arteriovenous malformation
Fibromatosis	Paget's disease
Elastofibroma	
Tendosynovial chondromatosis	
Gouty arthritis	
Proliferative panniculitis	
Proliferative myositis	
Myositis ossificans	
Arteriovenous malformation	
Female Predominance	
Tendosynovial cyst	Aneurysomal bone cyst
Angiofollicular lymphoid hyperplasia	Giant cell granuloma
	Fibrous dysplasia
	Tendosynovial cyst

Table 28. *Sex Distribution of Benign Neoplasms*

SOFT TISSUE NEOPLASM	BONE NEOPLASMS
Male Predominance	
Benign histiocytic fibrous histiocytoma	Nonossifying fibroma
Benign pleomorphic fibrous histiocytoma	Chondroblastoma
Lipoblastoma	Chondromyxoid fibroma
Rhabdomyoma	Osteochondroma
	Osteoblastoma
	Osteoid osteoma
Female Predominance	
Lipoma	Hemangioma
Angimyolipoma	
Leiomyoma	
Hemangioma	
Lymphangiomyoma	
Neurofibroma	
Benign paraganglioma	
Benign granular cell tumor	
Benign schwannoma	

Table 29. *Sex Distribution of Malignant Neoplasms*

SOFT TISSUE NEOPLASMS	BONE NEOPLASMS
Male Predominance	
Malignant fibroblastic fibrous histiocytoma	Fibrosarcoma
Malignant histiocytic fibrous histiocytoma	Leiomyosarcoma
Malignant pleomorphic fibrous histiocytoma	Chondrosarcoma
Desmoid tumor, extra-abdominal	Mesenchymal chondrosarcoma
Fibrosarcoma	Osteosarcoma
Tendosynovial sarcoma	Paget's sarcoma
Liposarcoma	Granulocytic sarcoma
Embryonal rhabdomyosarcoma	Ewing's sarcoma
Alveolar rhabdomyosarcoma	
Rhabdomyoblastoma	
Pleomorphic rhabdomyosarcoma	
Malignant mesothelioma	
Chondrosarcoma	
Mesenchymal chondrosarcoma	
Granulocytic sarcoma	
Plasma cell myeloma	
Kaposi's sarcoma	
Ewing's sarcoma	
Female Predominance	
Desmoid tumor, abdominal	Hemangiosarcoma
Leiomyosarcoma	Lymphangiosarcoma
Neurofibrosarcoma	Parosteal osteosarcoma
Malignant granular cell tumor	Chordoma
Alveolar soft part sarcoma	

"There is such a great difference in the behavior of tumors bearing the same name but developing in different parts of the body"

ARTHUR PURDY STOUT (1885–1967)

5. Anatomic Site

Non-neoplastic Soft Tissue Lesions (Table 30; Figures 70, 82, 83)

Benign Soft Tissue Neoplasms (Table 31; Figures 71, 73–75)

Malignant Soft Tissue Neoplasms (Table 32; Figures 72, 76–81, 84–93)

Non-neoplastic Bone Lesions (Tables 33, 36; Figures 96, 104, 105)

Benign Bone Neoplasms (Tables 34, 37; Figures 94, 97, 98, 101, 106, 112–116)

Malignant Bone Neoplasms (Tables 35, 38; Figures 95, 99, 100, 102, 103, 107–111, 117)

SOFT tissue and bone tumors vary significantly in site of origin. Soft tissue tumors can arise in a number of locations from the head to the toes, but most of them are found in the thigh, the arm, the trunk, and the head and neck areas (Tables 30–32 and Figs. 70–93). Primary bone tumors are almost as ubiquitous, but most occur in the long bones of the extremities (Tables 33–38 and Figs. 94–117).

The site of origin of soft tissue and bone lesions must be considered along with the age and sex of the patient but cannot serve as a major component for classification or as a sure lead to the diagnosis. Although many soft tissue and bone sarcomas are ubiquitous in their anatomic distribution, most show a predilection for certain sites. Approximately one-half of the soft tissue and bone sarcomas have an unique site preference for the extremities. A definite affinity to certain sites or preferential sites of origin within specific soft tissue compartments or part of bones is often difficult to establish due to the bulky nature of many soft tissue and bone sarcomas, but, in general, tumors of a particular cell type occur in the area where the corresponding primitive mesenchymal cells are most prevalent. For example, tendosynovial sarcomas are commonly found in the vicinity of tendons, bursae, and synovial membranes; and histiocytic fibrous histiocytomas (giant cell tumors) of bone are almost always located in the epulis-type, giant-cell-rich fields of the metaphysis.

One can only speculate about the factors that control and determine the site preference of many tumors. Without doubt, regional variations in the blood supply, composition, and mitotic activity of connective tissue elements, and metabolic and structural requirements, as well as extrinsic humoral and physical stimuli, play a role in the tumor's preference for specific sites.

TABLE 30. *Sites of Non-neoplastic Soft Tissue Lesions**

	HEAD	NECK	ARM AND SHOULDER	ELBOW AND FOREARM	THIGH AND BUTTOCK	KNEE AND LEG	HAND AND FOOT	CHEST WALL	ABDOMINAL WALL	INTERNAL THORAX	RETROPERITONEUM	GASTRO-INTESTINAL TRACT	INTERNAL CRANIUM
Xanthogranuloma	++	+		–		–		+			+		
Fasciitis[†]			+	++	–		–	+					
Keloid	–	++						+	–				
Fibromatosis	–	+		+			++				+		
Elastofibroma		+			–			++					
Collagenoma								++					
Tendosynovitis	–		–	+		+	++						
Tendosynovial chondromatosis[‡]			–	+	+	++	+						
Gouty arthritis			–	–		+	++						
Fat necrosis			–		+			++[§]	–		+		
Lipogranuloma[‡]				+	++	+		+			–	–	
Panniculitis[†]			+	++	+	+		+					
Piezogenic papule[†]						+	++"						
Proliferative myositis		–	+	–	++	–		+					
Myositis ossificans			+		++	–			–				
Pyogenic granuloma[†]	++	+					–	–				–	
Angiofollicular lymphoid hyperplasia[†‡]	++	+	–					–					
Arteriovenous malformation		+	++		+						–	–	–
Hereditary hemorrhagic telangiectasia[†‡]	+	+	–	–	–	–		++	+			+	–
Lymphedema		–	–	+		++	+				+		

*++ = most common; + = common; – = rare. [†]Commonly superficial. [‡]Commonly multifocal. [§]Breast. "Foot.

TABLE 31. *Sites of Benign Soft Tissue Neoplasms**

	HEAD	NECK	ARM AND SHOULDER	ELBOW AND FOREARM	THIGH AND BUTTOCK	KNEE AND LEG	HAND AND FOOT	CHEST WALL	ABDOMINAL WALL	INTERNAL THORAX	RETROPERITONEUM	GASTRO-INTESTINAL TRACT	INTERNAL CRANIUM
Benign fibroblastic fibrous histiocytoma[†]	+	−	++	+		+		−					
Benign histiocytic fibrous histiocytoma	−		+		+	+	++						
Benign pleomorphic fibrous histiocytoma[†,‡]	+	+	++	+		+	−	+			−		−
Lipoma[†,‡]		−		−	+			++			+	−	
Lipoblastoma[†]		−	+		++	+		−					
Angiomyolipoma					−						++	+	
Hibernoma		++	+		−			+					
Leiomyoma[†]					−	+					−	++	
Rhabdomyoma[†]	++									+[§]	−[″]		
Hemangioma[†,‡]	+	+			+	+		++			−	−	−
Hemangioblastoma										−	−		++
Glomus tumor[†,‡]			+		+		++	+		−	−	+	
Lymphangioma[‡]		+	−							+	++	+	
Lymphangiomyoma		−								++	+	−	
Neurofibroma[†,‡]		+	−		+			++		−	+		
Benign schwannoma		−	+		+	−		++			−		−
Ganglioneuroma		−								++	+	+	−
Benign paraganglioma		+								+	++	−	
Chondroma[‡]	−			−		+	++	−					
Benign granular cell tumor[†]	−	−		+	+			++					
Meningioma	+									−	+		++

*++ = most common; + = common; − = rare. [†]Commonly superficial. [‡]Commonly multifocal. [§]Heart. [″]Vagina.

Table 32. *Sites of Malignant Soft Tissue Neoplasms**

	HEAD	NECK	ARM AND SHOULDER	ELBOW AND FOREARM	THIGH AND BUTTOCK	KNEE AND LEG	HAND AND FOOT	CHEST WALL	ABDOMINAL WALL	INTERNAL THORAX	RETROPERITONEUM	GASTRO-INTESTINAL TRACT	INTERNAL CRANIUM
Malignant fibroblastic fibrous histiocytoma†	−		++		+			+					
Malignant histiocytic fibrous histiocytoma		−	+		++	+							
Malignant pleomorphic fibrous histiocytoma		+	+	+	++	−		−			−		
Desmoid tumor		+	+		+	+		+	++		−		
Fibrosarcoma			+	+	++						−		
Biphasic tendosynovial sarcoma		−	+	+		++	−						
Monophasic tendosynovial sarcoma, spindle cell type		−	+	−	++	+	+	−					
Epithelioid sarcoma†,‡			+	++		+	+	−					
Clear cell sarcoma		−	−	+		++	+						
Chordoid sarcoma				+		−	++						
Liposarcoma			+		++	+		−			+		
Leiomyoblastoma			−		−						+	++	
Leiomyosarcoma			−		+	+					+	++	
Embryonal rhabdomyosarcoma	+	+	+		++		−	+					
Rhabdomyoblastoma		−	+		++			−			−		
Pleomorphic rhabdomyosarcoma		−	+		++	−		−					
Hemangiopericytoma		+	−	−	++			+			−		
Hemangiosarcoma	+		−		++			+			−		
Lymphangiosarcoma†,‡	−		++	−	+	+		−					
Neurofibrosarcoma†,‡		+	−		+			++		−	+		
Malignant schwannoma		+	+	−	++	−		−			−		−
Malignant paraganglioma		+								+	++		
Chondrosarcoma		+	+		++	+		−					
Osteosarcoma			+		++	−		+	−		−		
Malignant granular cell tumor†,‡	−	−	+		++			+					
Alveolar soft part sarcoma			+		++	+					−		
Kaposi's sarcoma†,‡	+			+	−	++	+	+	−				

*++ = most common; + = common; − = rare. †Commonly superficial. ‡Commonly multifocal.

Table 33. ***Sites of Non-neoplastic Bone Lesions***

	Cranial Bones	Facial Bones	Clavicle	Humerus	Ulna and Radius	Femur	Tibia and Fibula	Bones of Hand and Foot	Bones of Fingers and Toes	Ribs	Sternum	Vertebrae	Pelvic Bones	Sacrum	Scapula
Aneurysmal bone cyst				+	−	−	+	+‡				++	−		
Giant cell granuloma	−	++						−	+	−					
Fibrous dysplasia	+	+		+		+	++(tibia)		−	−			+		
Paget's disease†	+		−	+		+	+			−			++		
Eosinophilic granuloma†	++		+		+			−		+			−		−

*++ = most common; + = common; − = rare. †Commonly multifocal. ‡Metatarsal.

Table 34. ***Sites of Benign Bone Neoplasms****

	Cranial Bones	Facial Bones	Clavicle	Humerus	Ulna and Radius	Femur	Tibia and Fibula	Bones of Hand and Foot	Bones of Fingers and Toes	Ribs	Sternum	Vertebrae	Pelvic Bones	Sacrum	Scapula
Benign histiocytic fibrous histiocytoma				−	+	+	++			−		−		+	
Nonossifying fibroma			−	+	+	+	++			−					
Desmoplastic fibroma		+		++	−	+	−						+		+
Hemangioma†	+	+	−	−		+	+				−	++			
Chondroblastoma			−	+		++	+(tibia)	+(foot)		−			+		
Chondromyxoid fibroma			−			++	+	+(foot)		−			−		
Chondroma†			−	+		+		+	++	−					
Osteochondroma				+	−	++	+(tibia)						−		
Osteoblastoma			−	−		+	+	+	+			++			
Osteoid osteoma				+	−	+	++					+			

*++ = most common; + = common; − = rare. †Commonly multifocal.

TABLE 35. ***Sites of Malignant Bone Neoplasms****

	CRANIAL BONES	FACIAL BONES	CLAVICLE	HUMERUS	ULNA AND RADIUS	FEMUR	TIBIA AND FIBULA	BONES OF HAND AND FOOT	BONES OF FINGERS AND TOES	RIBS	STERNUM	VERTEBRAE	PELVIC BONES	SACRUM	SCAPULA
Malignant fibroblastic fibrous histiocytoma	–			+		++	+			+			–		–
Malignant histiocytic fibrous histiocytoma				–	–	+	++(tibia)			–					
Malignant pleomorphic fibrous histiocytoma	–			+		++	+	–		+			–		–
Fibrosarcoma		+		+		++	+(tibia)			–			–		+
Leiomyosarcoma				+		++	+								
Hemangiosarcoma			–	+	–	++	+			–		+	+		
Chondrosarcoma		–		+		+	+			+	–		++		
Mesenchymal chondrosarcoma	–	+				++		–		+		–	+		–
Osteosarcoma		–		+		+	++(tibia)			–			+		
Parosteal osteosarcoma				+		++	+(tibia)			–					
Paget's sarcoma	+	–		+		++	–						–		
Plasma cell myeloma[†]	++		–	–		+						+	+		
Ewing's sarcoma			–	+	–	+	++(tibia)			–			+		
Chordoma	–											+		++	

*++ = most common; + = common; – = rare. [†]Commonly multifocal.

Table 36. *Most Common Location of Non-neoplastic Lesions in Long Bones**

	EPIPHYSIS	EPIPHYSIS-METAPHYSIS	METAPHYSIS	METAPHYSIS-DIAPHYSIS	DIAPHYSIS	ANYWHERE	JUXTACORTICAL COMPONENT
Aneurysmal bone cyst	–	+	++	+		–	–
Fibrous dysplasia			–	+	++	–	–
Paget's disease			–	+	+	++	
Callus			–	+	++	–	+
Eosinophilic granuloma			–	+	++	+	–
Osteomyelitis		–	++	+	–		–
Synovial cyst	++	+	–				–
Simple cyst		–	++	+	+		

*++ = most common; + = common; – = rare.

Table 37. *Most Common Location of Benign Neoplasms in Long Bones**

	EPIPHYSIS	EPIPHYSIS-METAPHYSIS	METAPHYSIS	METAPHYSIS-DIAPHYSIS	DIAPHYSIS	ANYWHERE	JUXTACORTICAL COMPONENT
Benign histiocytic fibrous histiocytoma	++	+	+	–			
Nonossifying fibroma	–	–	++	+	+		
Desmoplastic fibroma	–	–	++	+	+		–
Chondroblastoma	++	+	–				
Chondromyxoid fibroma	–	–	++	+	+	+	
Osteochondroma		–	++	+		–	++
Osteoblastoma	++	+	+	–	–		
Osteoid osteoma		–	++	+	+		–

*++ = most common; + = common; – = rare.

Table 38. *Most Common Location of Malignant Neoplasms in Long Bones**

	EPIPHYSIS	EPIPHYSIS-METAPHYSIS	METAPHYSIS	METAPHYSIS-DIAPHYSIS	DIAPHYSIS	ANYWHERE	JUXTACORTICAL COMPONENT
Malignant fibroblastic fibrous histiocytoma		–	++	+	+		
Malignant histiocytic fibrous histiocytoma	++	+	–				
Malignant pleomorphic fibrous histiocytoma		–	++	+	+	–	
Fibrosarcoma		+	++	+	+		–
Leiomyosarcoma		–	++	+			
Hemangiosarcoma	–	–	+	+	++		–
Chondrosarcoma		–	++	+			+
Mesenchymal chondrosarcoma			++	+	+		–
Osteosarcoma	–	–	++	+	+		
Parosteal osteosarcoma			+				++
Paget's sarcoma						++	
Plasma cell myeloma		–	+	+	++	+	
Ewing's sarcoma			–	+	++		
Adamantinoma			–	+	++		

*++ = most common; + = common; – = rare.

Fig. 70. Tendosynovitis (see Figs. 786, 867, 1023, 1029, 1136).

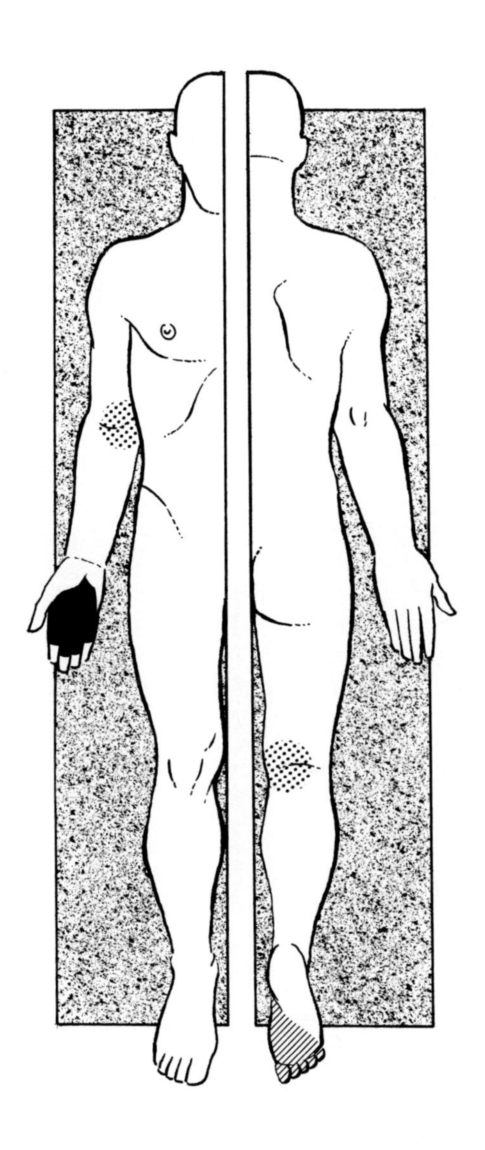

Fig. 71. Benign histiocytic fibrous histiocytoma (Benign giant cell tumor) (see Figs. 115, 174, 178, 183, 325, 490, 501, 774, 778).

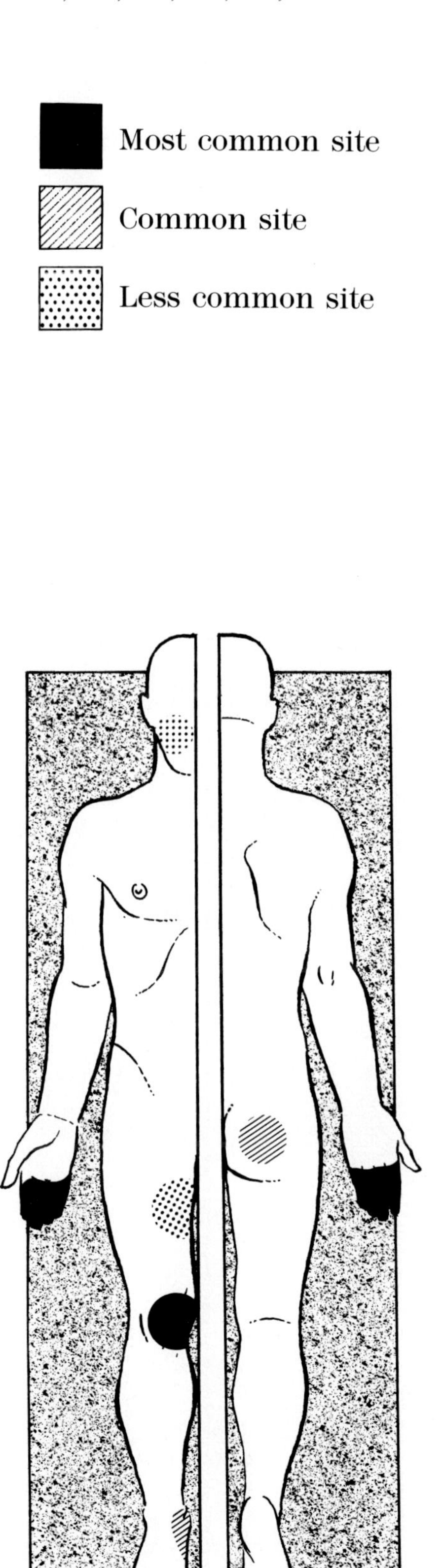

Fig. 72. Leiomyosarcoma/also retroperitoneum ■ major vessels ▦ (see Figs. 110, 193, 218, 220, 231).

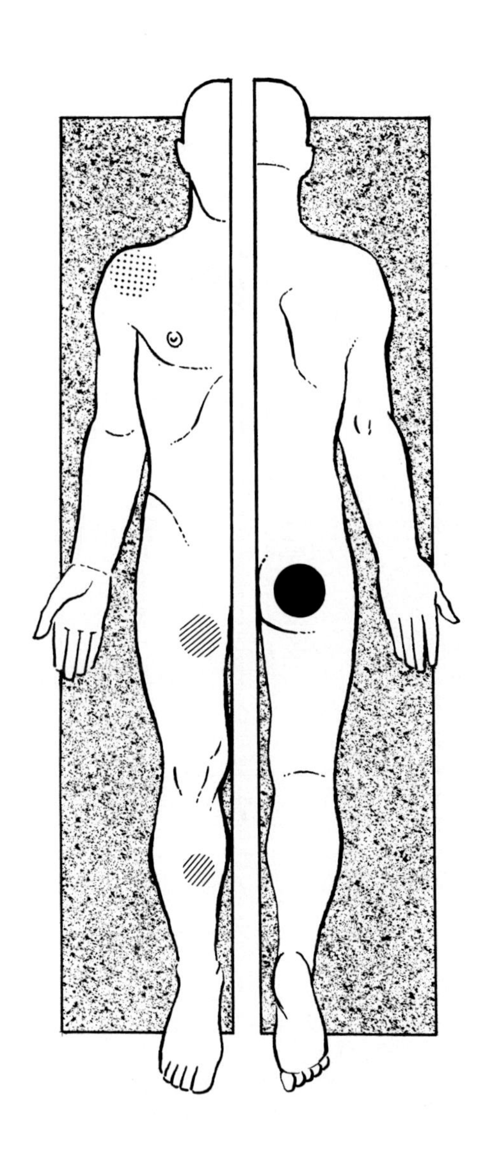

Fig. 73. Lipoma/also retroperitoneum ■ (see Figs. 24, 198, 243, 455).

Fig. 74. Neurofibroma/also retroperitoneum ▨ pelvis ▦ (see Figs. 11, 245, 253, 363, 487).

Fig. 75. Benign granular cell tumor/also tongue ■ (see Figs. 75, 261, 498, 539, 736, 1160, 1235).

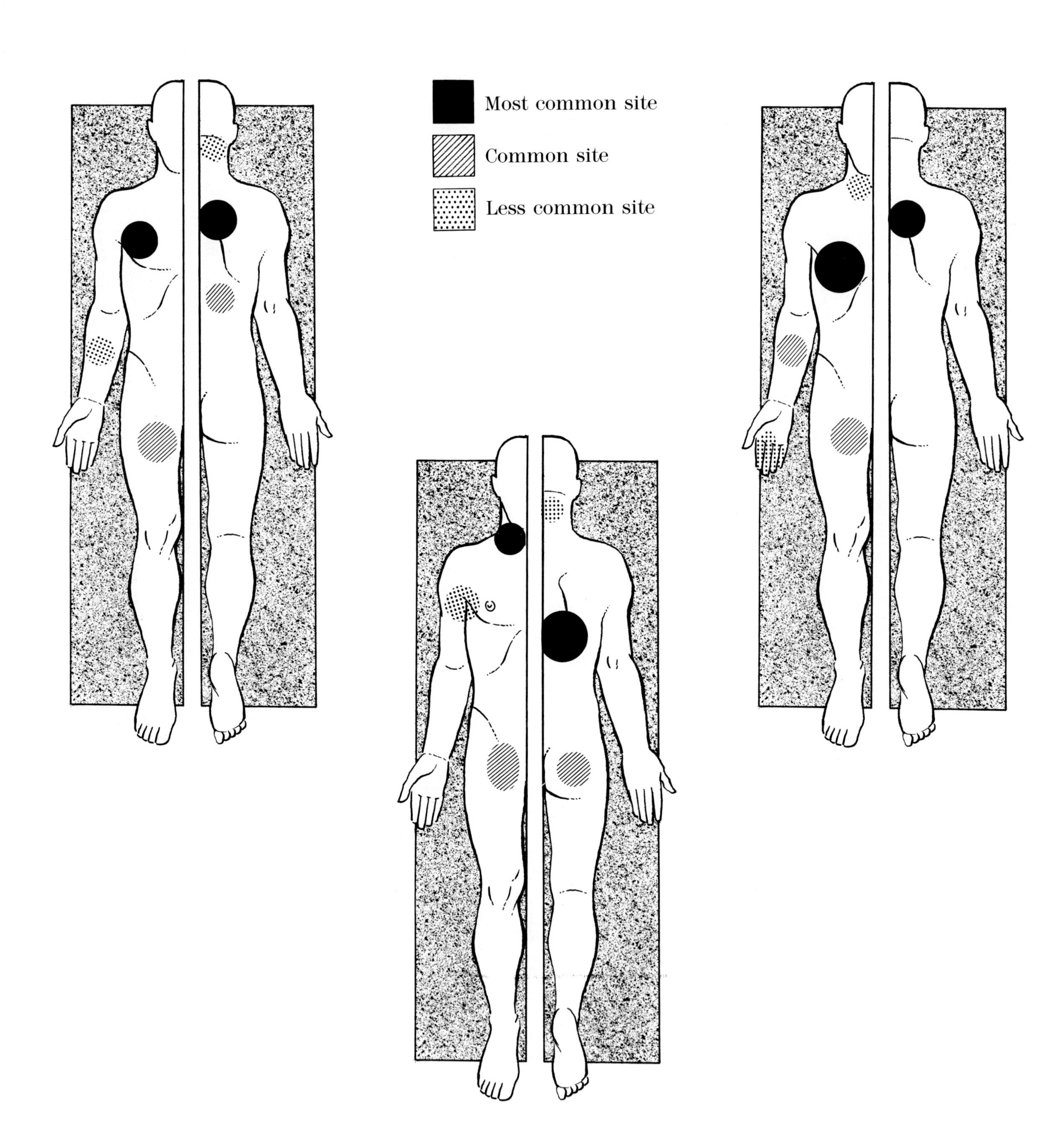

Fig. 76. Malignant fibroblastic fibrous histiocytoma (Dermatofibrosarcoma protuberans) (see Figs. 111, 143, 241, 301, 339, 340, 350).

Fig. 77. Desmoid tumor/also retroperitoneum and pelvis ▨ (see Figs. 52, 185, 202, 209, 244, 282, 376, 386).

Fig. 78. Embryonal rhabdomyosarcoma/also bladder and vagina ▨ paratestis ⊞ (see Figs. 67, 68, 120, 206, 207, 226, 263, 281, 291, 364).

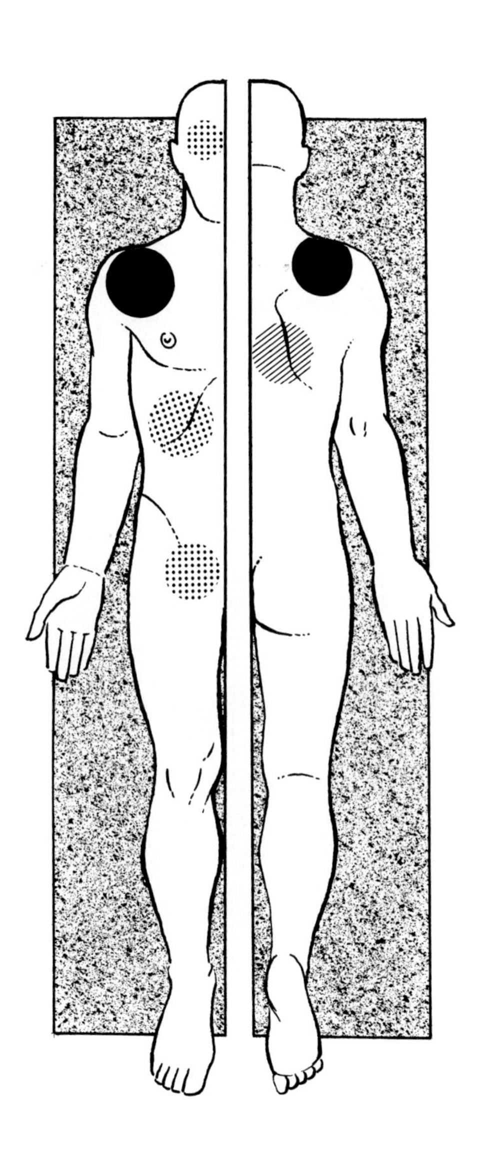

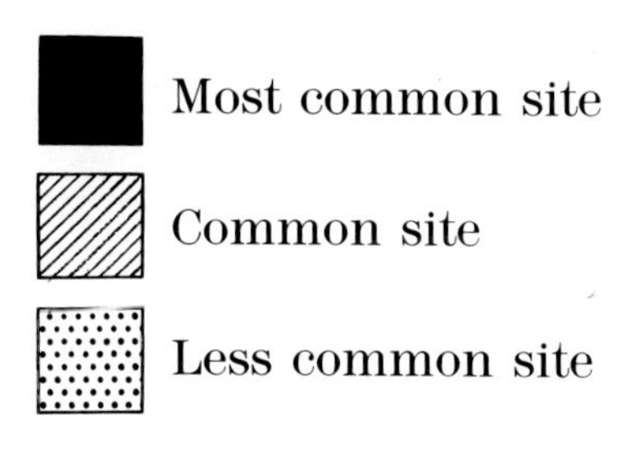

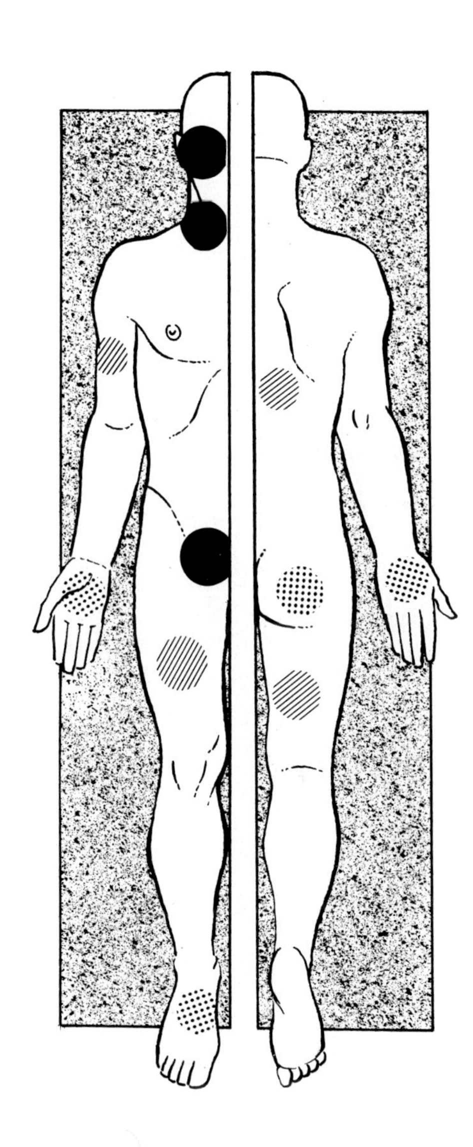

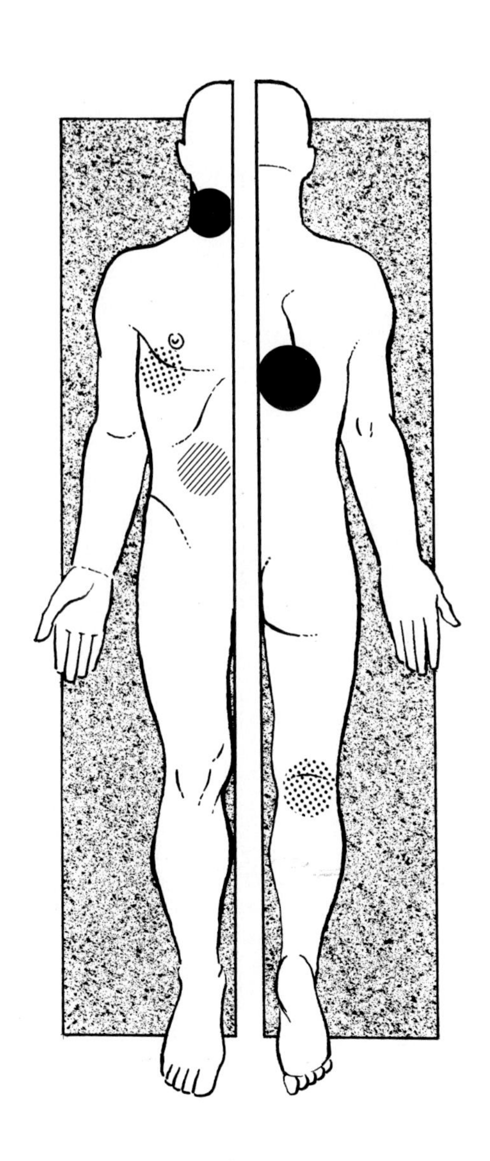

Fig. 79. Neurofibrosarcoma/also retroperitoneum ▨ (see Figs. 34, 53, 236, 254, 256, 264, 271, 275).

Fig. 80. Malignant schwannoma (see Figs. 51, 200, 238, 242, 246).

Fig. 81. Chondrosarcoma (see Figs. 95, 123, 124, 126, 136, 137, 147, 213).

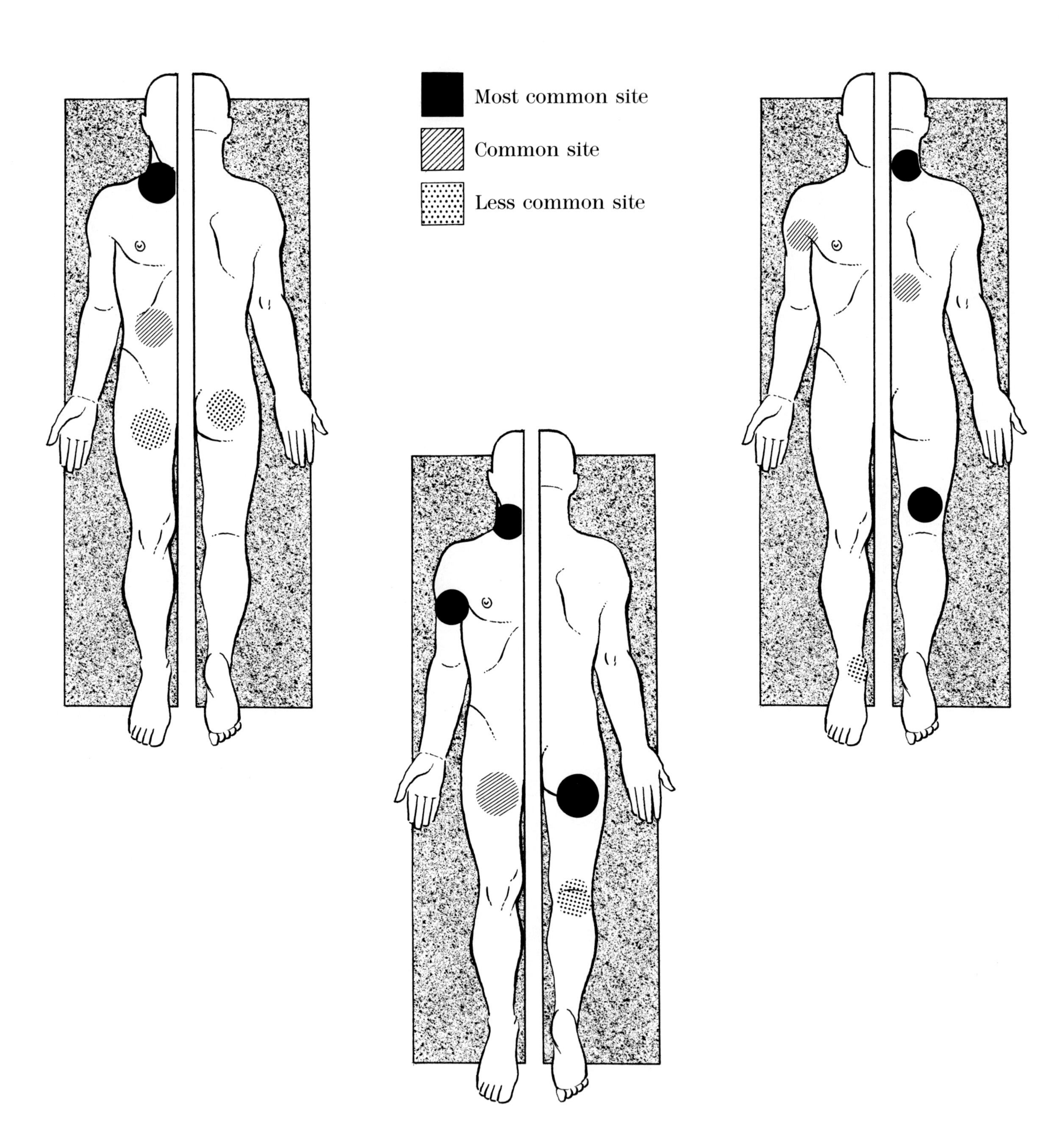

Fig. 82. Fasciitis (see Figs. 345, 368, 389, 403, 408, 578, 605).

Fig. 83. Tendosynovial chondromatosis (see Figs. 786, 1060).

Fig. 84. Lymphangiosarcoma (see Figs. 41, 61, 154, 159, 321).

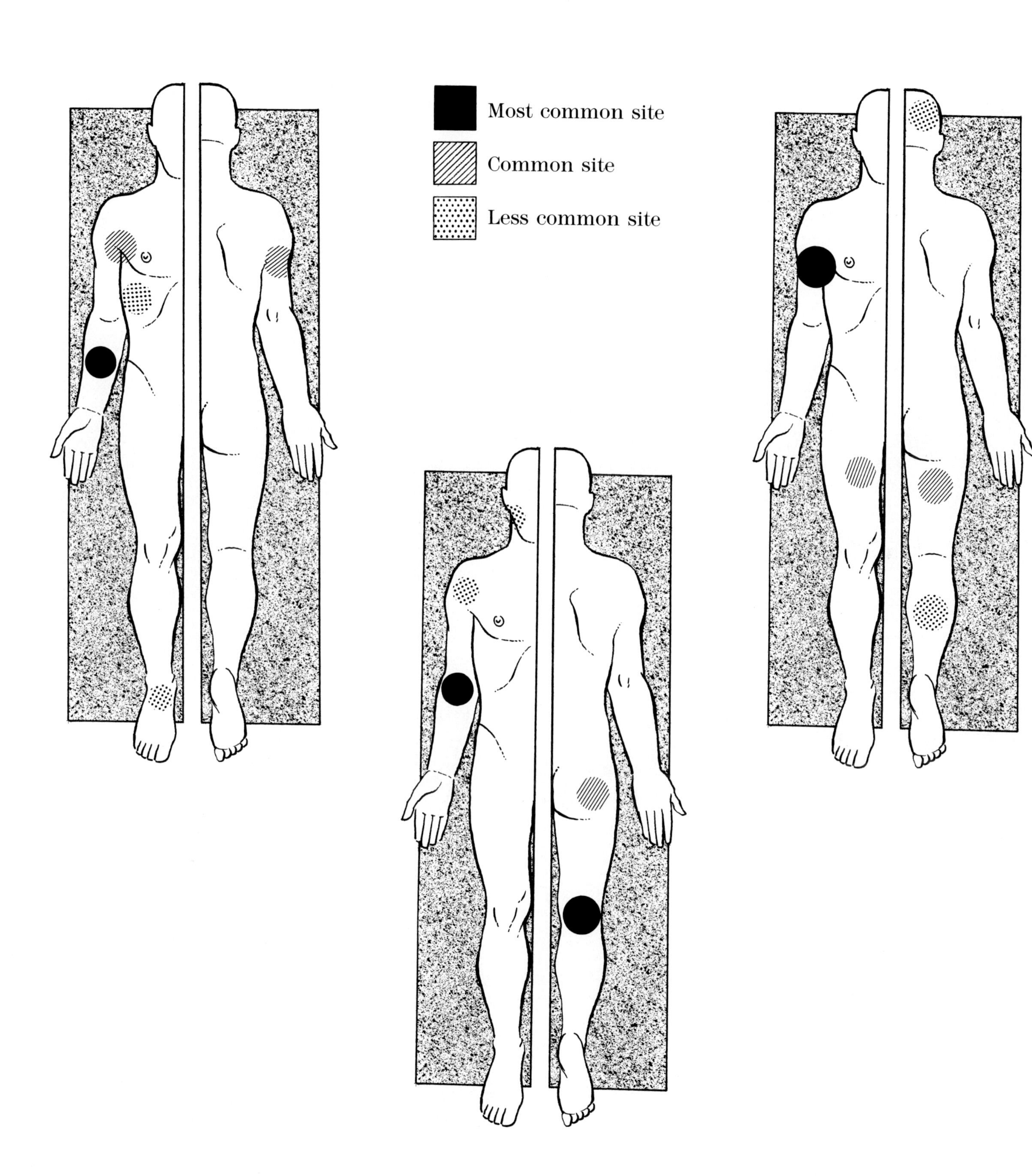

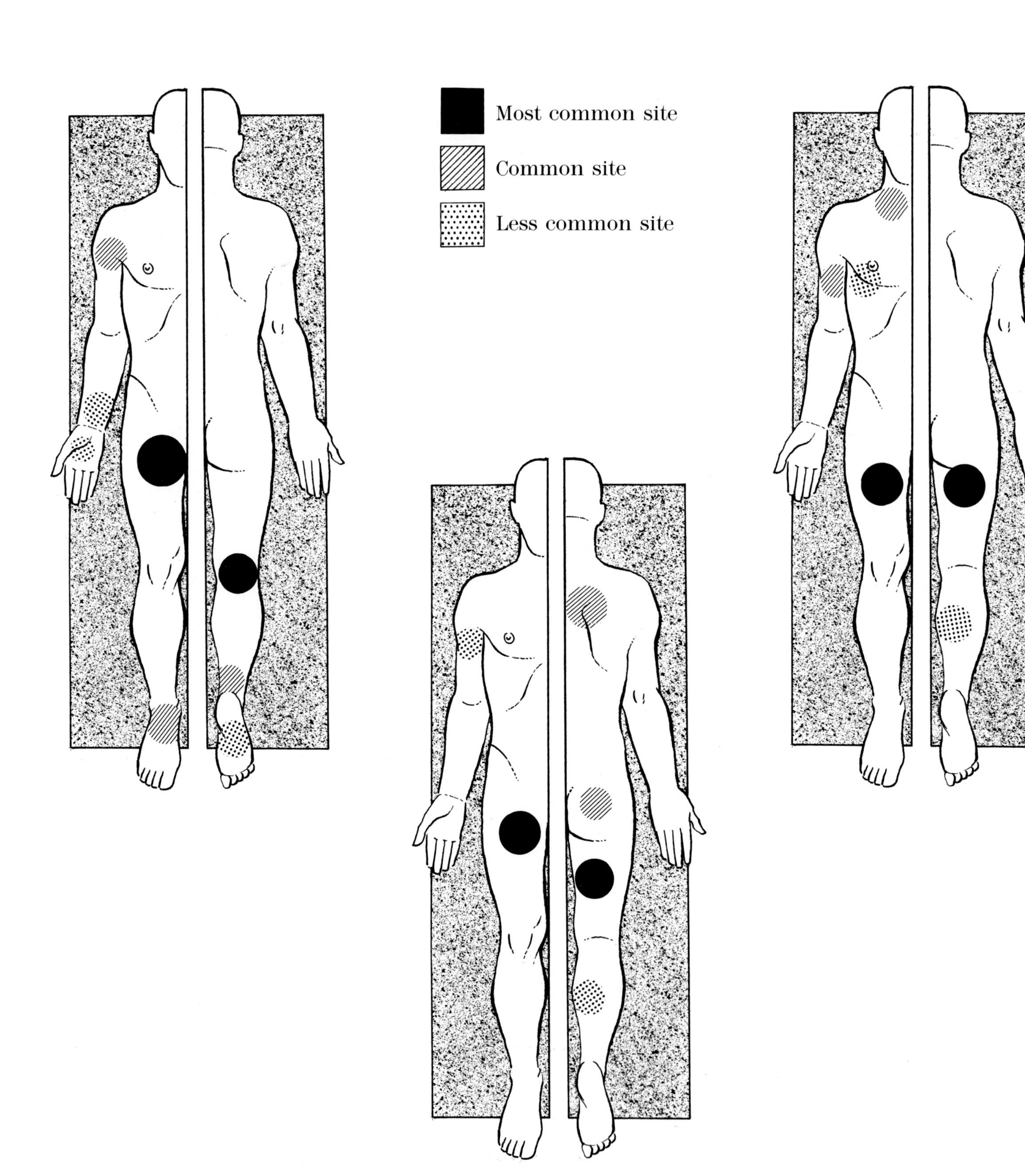

Fig. 85. Tendosynovial sarcoma (see Figs. 64, 179, 190, 315, 320, 323, 326, 391, 392, 396, 398, 514).

Fig. 86. Liposarcoma/also retroperitoneum ▨ (see Figs. 58, 208, 219, 269, 290, 294, 343, 415, 431, 438).

Fig. 87. Pleomorphic rhabdomyosarcoma (see Figs. 29, 121, 292, 295, 296, 299, 313, 599, 604, 609).

Fig. 88. Malignant pleomorphic fibrous histiocytoma/also pelvis and retroperitoneum ⚃ (see Figs. 111, 130, 138, 140, 148, 237, 255, 262, 288).

Fig. 89. Alveolar soft part sarcoma (see Figs. 538, 697, 718, 1161, 1199).

Fig. 90. Osteosarcoma (see Figs. 108, 127, 142, 146, 149, 157, 158, 162, 164, 168, 230, 233, 248).

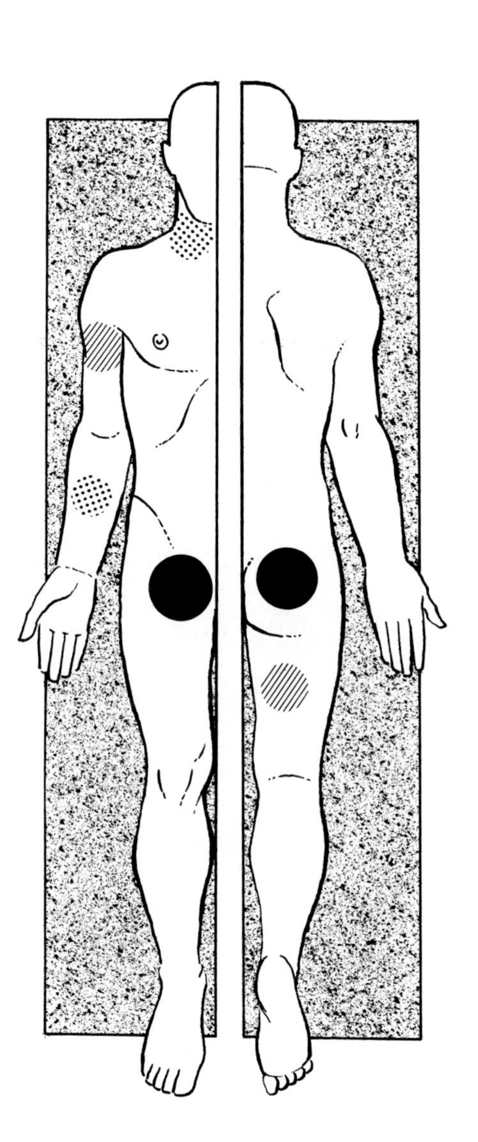

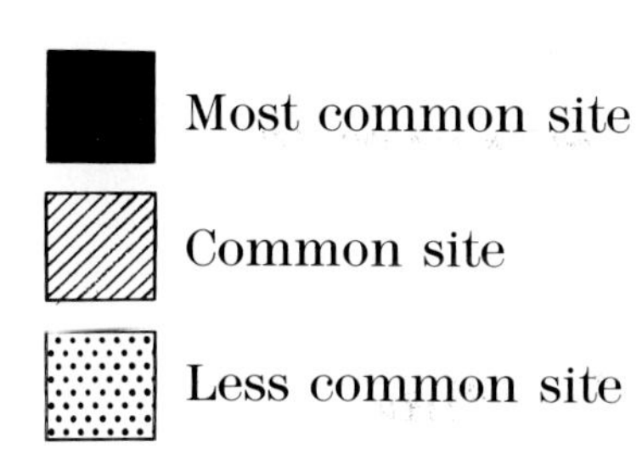

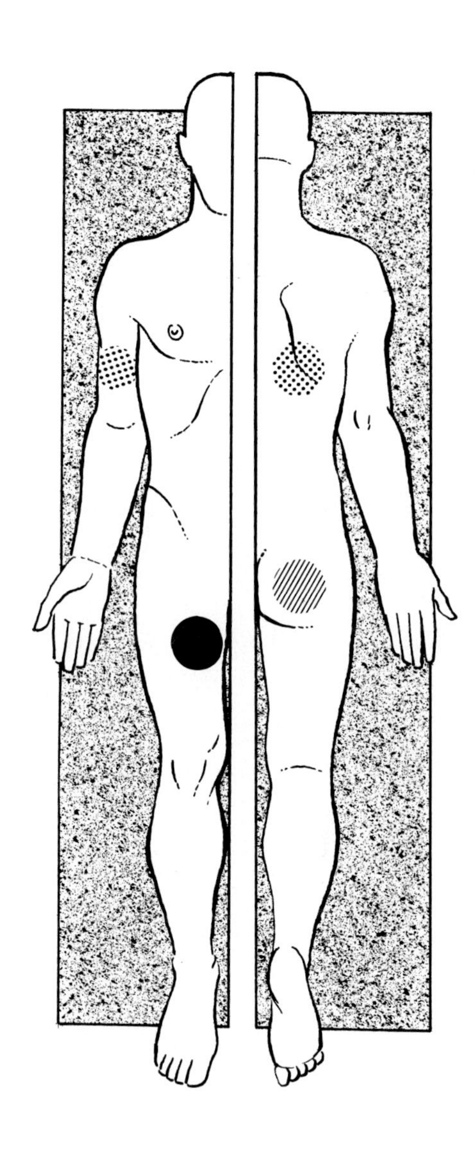

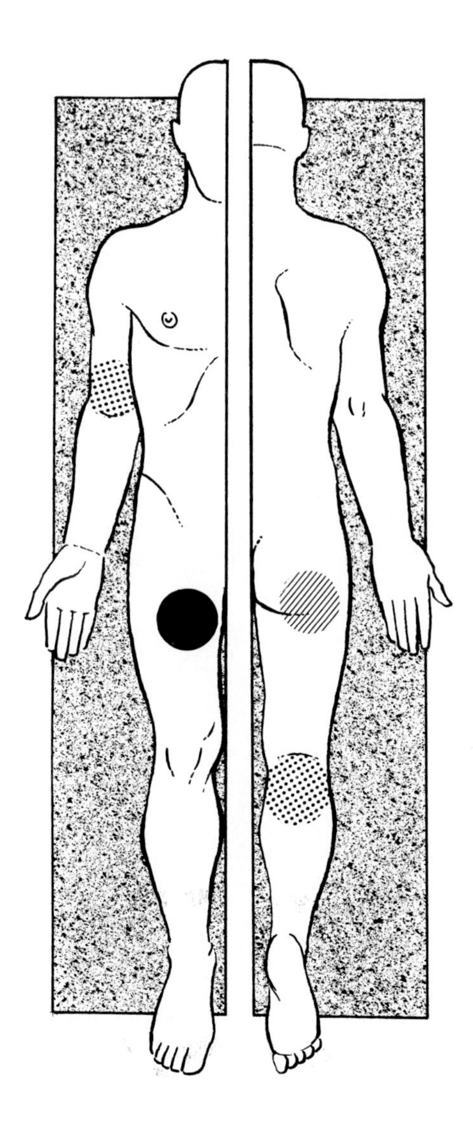

Fig. 91. Fibrosarcoma (see Figs. 60, 117, 131, 165, 240, 260, 297).

Fig. 92. Hemangiosarcoma/also liver ▦ (see Figs. 46, 102, 316, 523, 541, 555, 557, 563, 564, 576).

Fig. 93. Hemangiopericytoma/also retroperitoneum and pelvis ▨ (see Figs. 470, 479, 484, 525, 533).

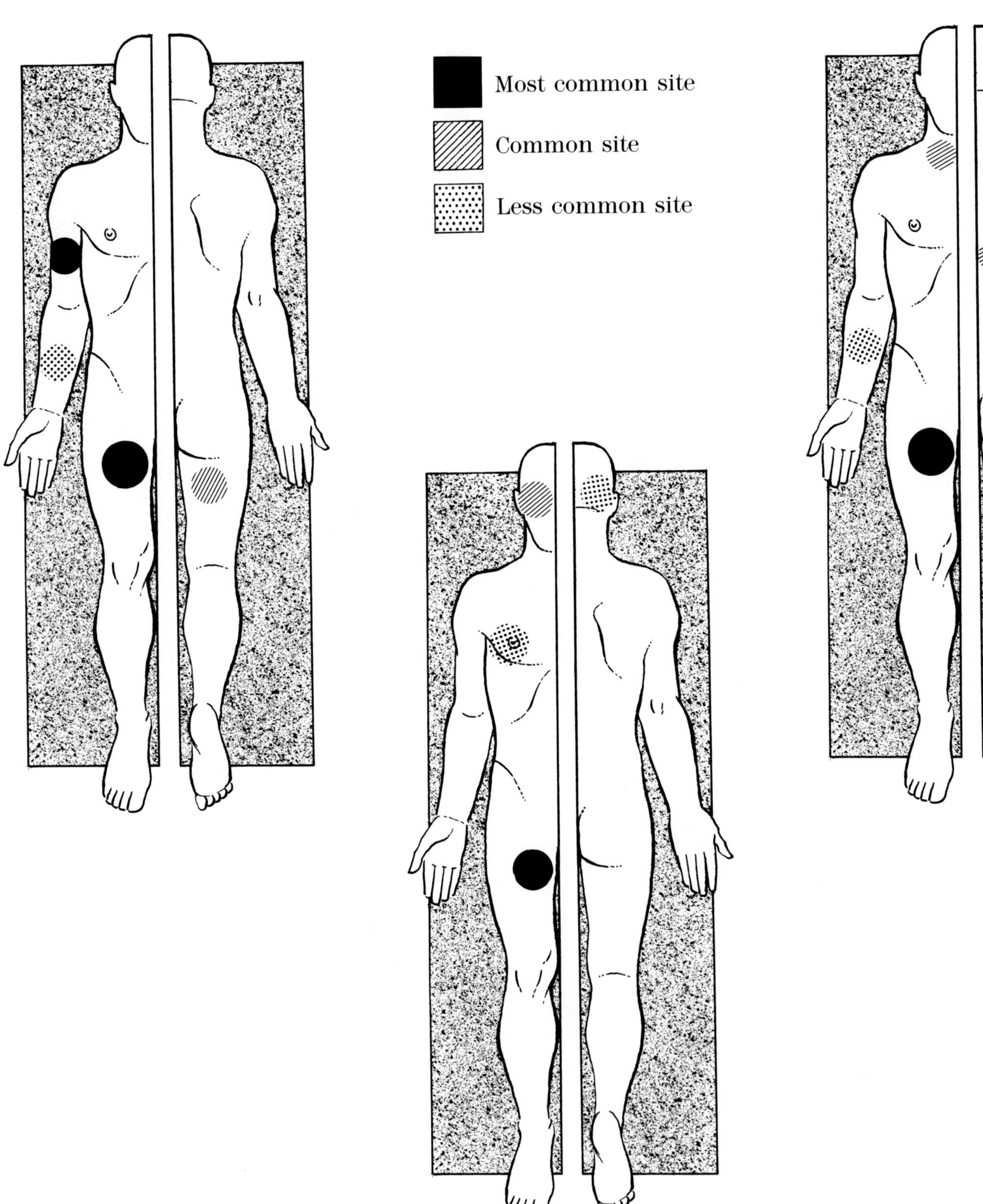

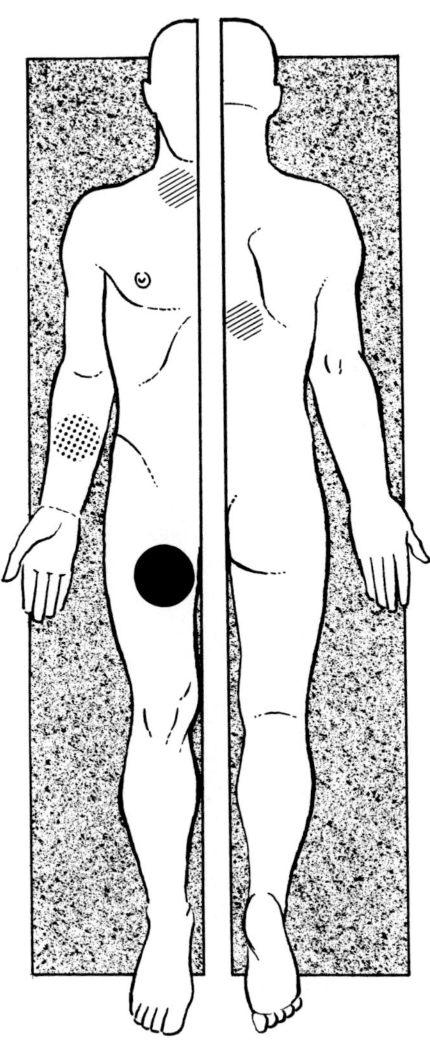

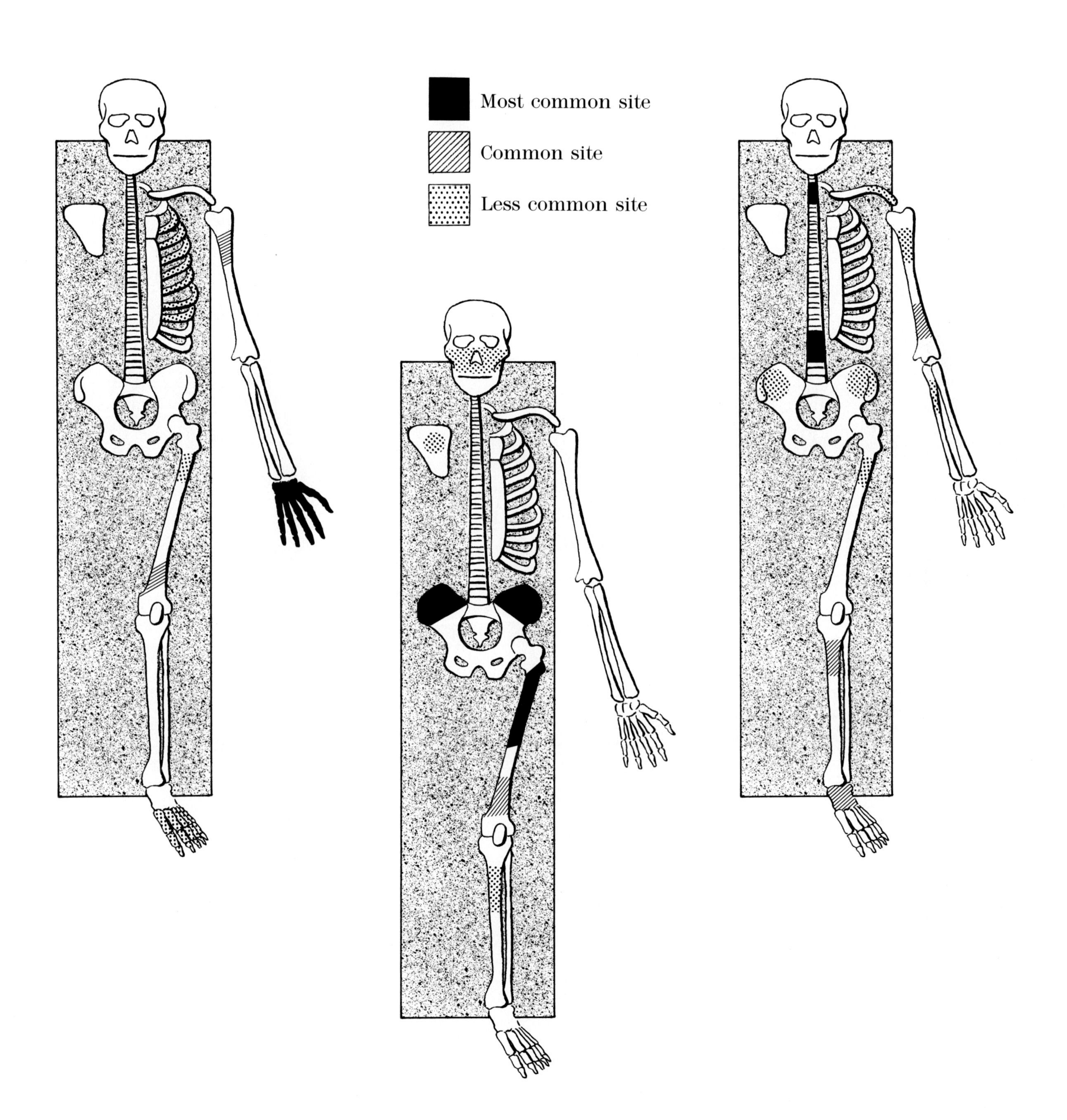

Fig. 94. Chondroma (see Fig. 191).

Fig. 95. Chondrosarcoma (see Figs. 81, 123, 124, 136, 216, 223, 285).

Fig. 96. Aneurysmal bone cyst (see Figs. 125, 150, 163, 509, 785, 1015).

Fig. 97. Osteoblastoma (see Figs. 211, 234, 1092, 1155).

Fig. 98. Hemangioma (see Figs. 18, 98, 549, 553, 559, 988).

Fig. 99. Plasma cell myeloma (see Figs. 454, 692, 741).

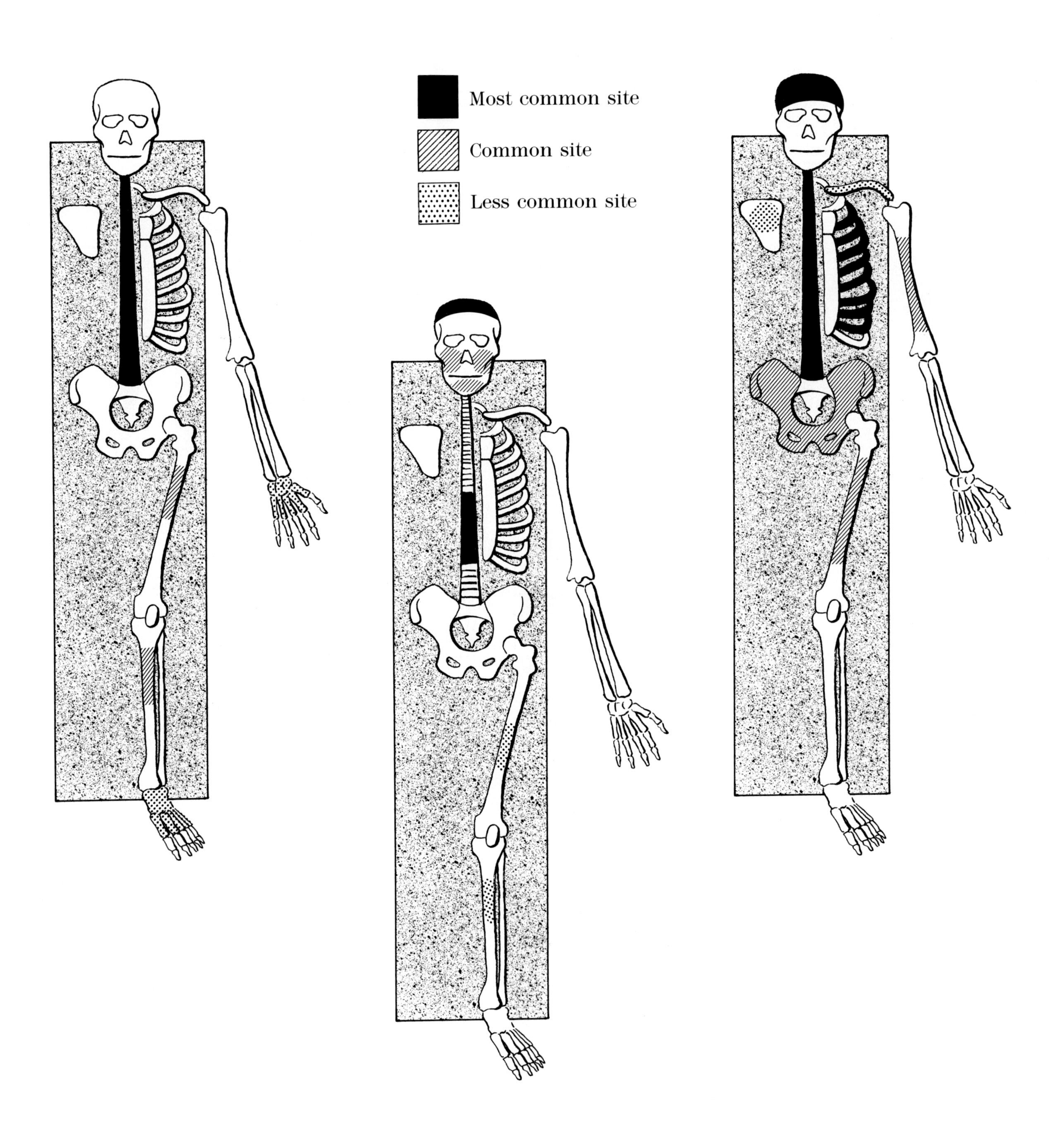

Fig. 100. Ewing's sarcoma (see Figs. 172, 173, 210, 214, 227).

Fig. 101. Osteoid osteoma (see Fig. 1089).

Fig. 102. Hemangiosarcoma (see Figs. 42, 92, 316, 523, 541).

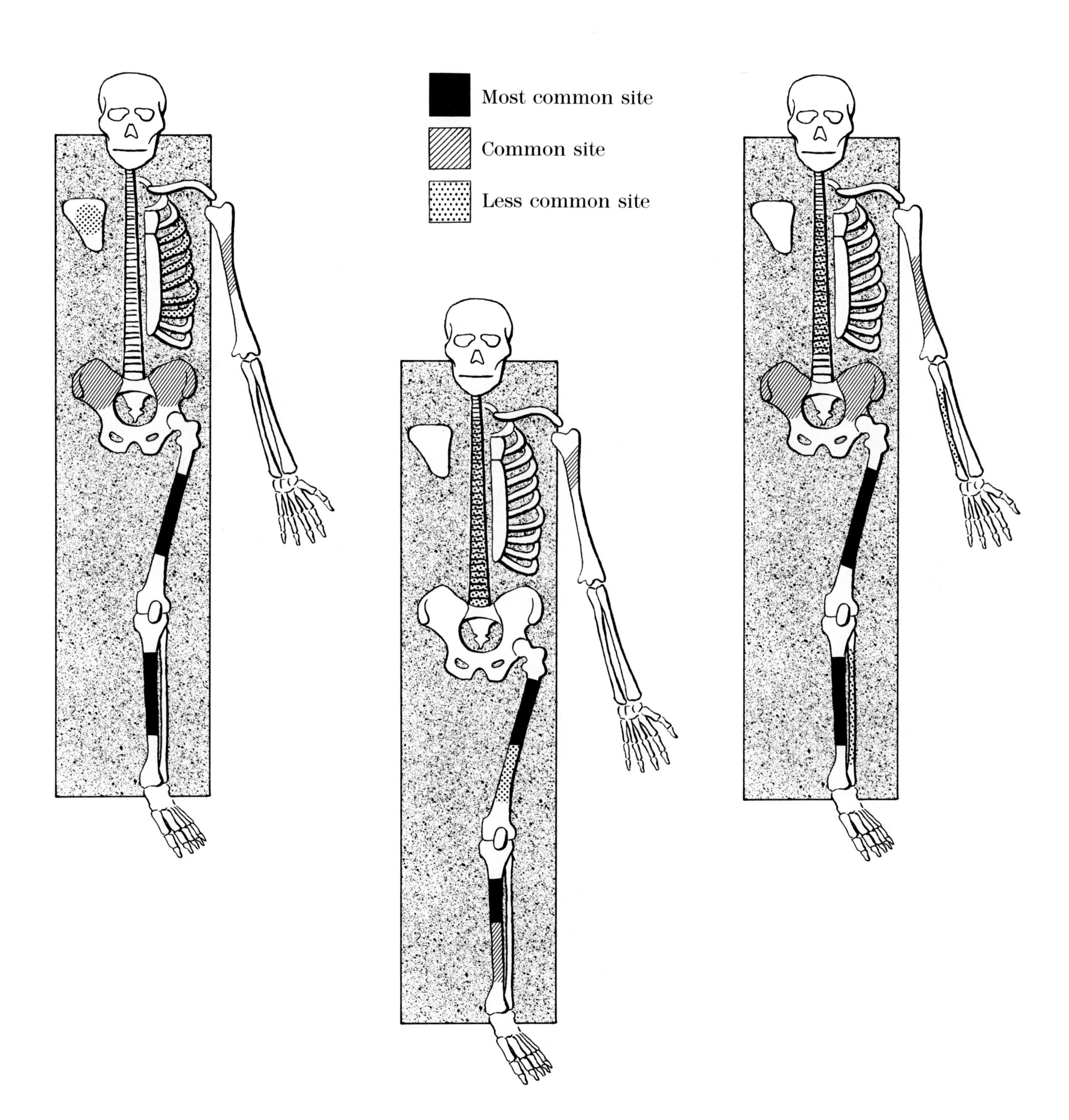

Fig. 103. Mesenchymal chondrosarcoma (see Figs. 151, 175, 180, 318, 390, 444, 507, 595, 674).

Fig. 104. Fibrous dysplasia (see Figs. 1050, 1081, 1095).

Fig. 105. Paget's disease (see Fig. 107).

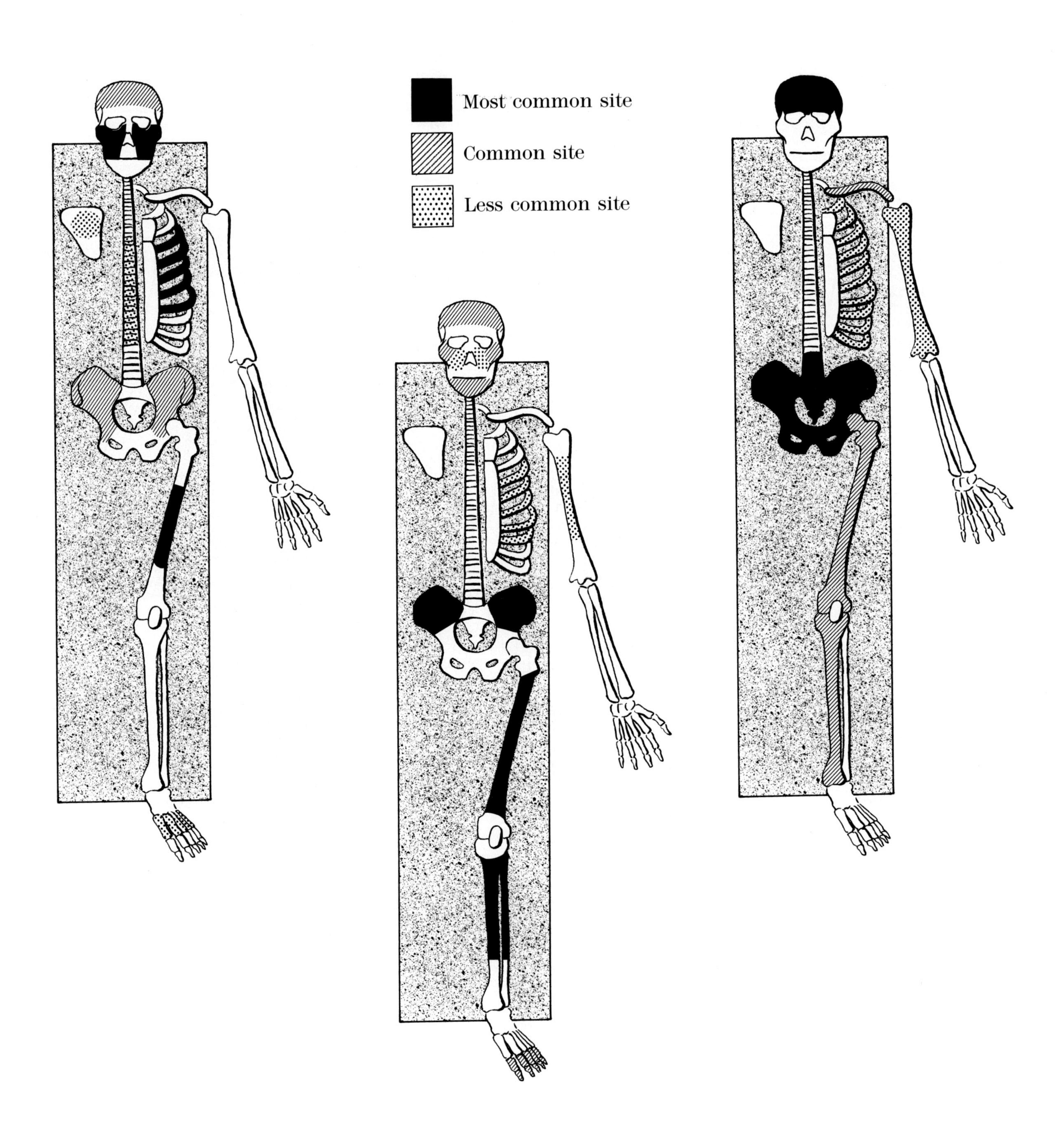

Fig. 106. Desmoplastic fibroma (see Fig. 633).

Fig. 107. Paget's sarcoma (see Figs. 105, 134, 918, 1103, 1128).

Fig. 108. Osteosarcoma (see Figs. 90, 109, 127, 142, 146, 149, 157, 158, 162, 164, 168, 176, 221, 225, 228).

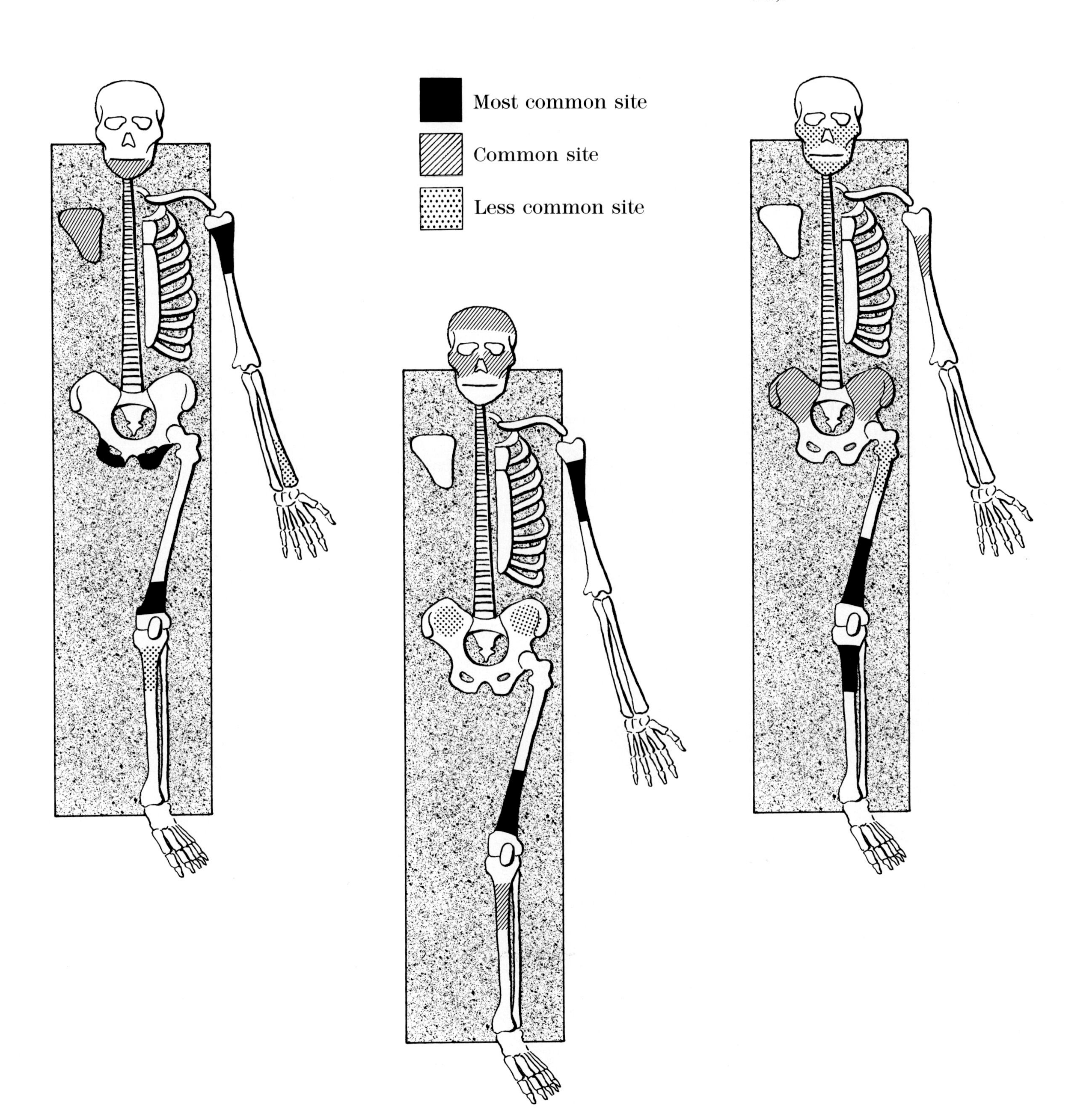

Fig. 109. Parosteal osteosarcoma (see Figs. 108, 129, 217, 224, 249).

Fig. 110. Leiomyosarcoma (see Figs. 72, 193, 218, 220, 231).

Fig. 111. Malignant fibroblastic and pleomorphic fibrous histiocytoma (see Figs. 76, 143, 241, 301, 339).

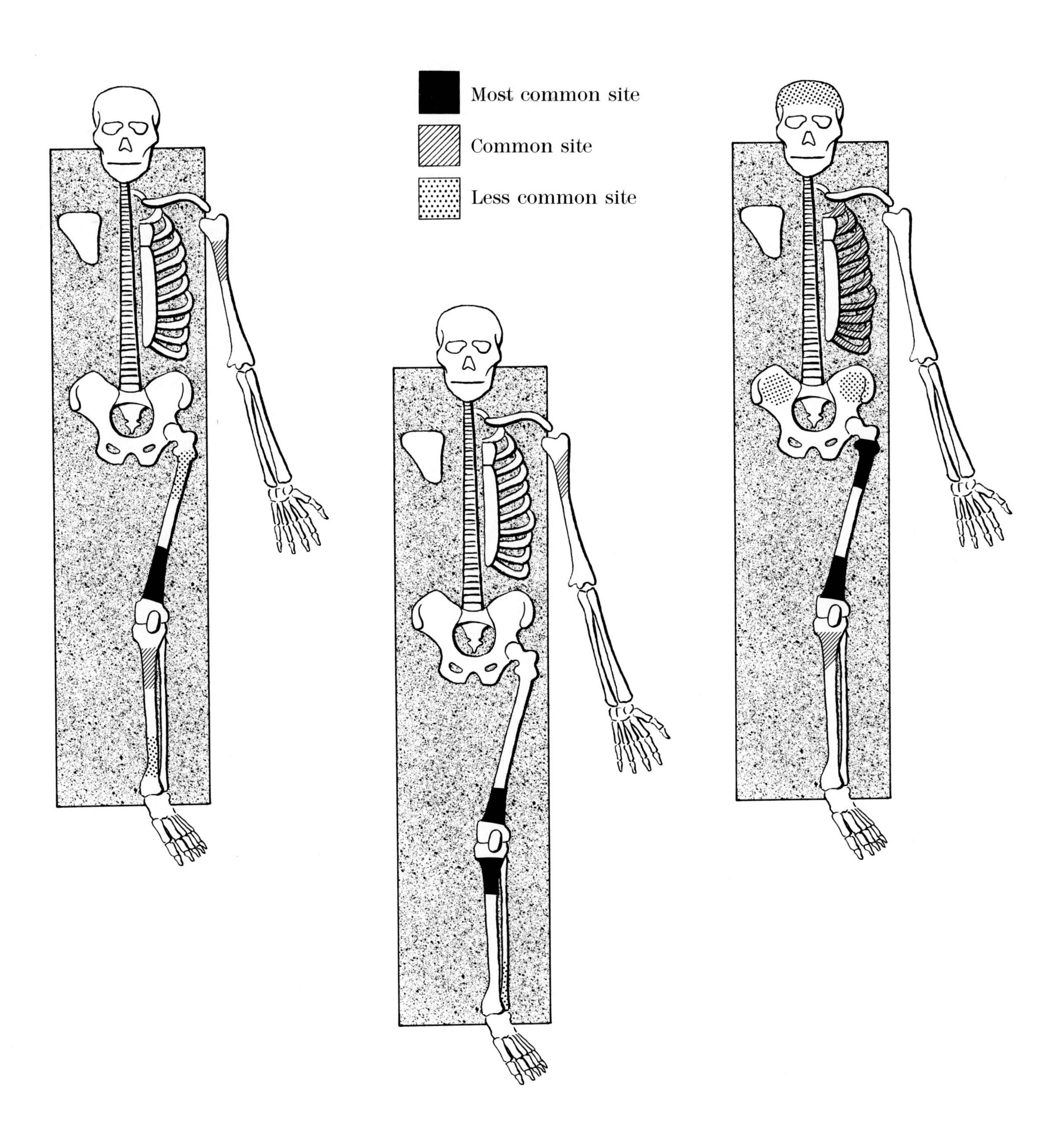

Fig. 112. Nonossifying fibroma (see Figs. 177, 631, 852).

Fig. 113. Osteochondroma (see Figs. 167, 302, 303, 772).

Fig. 114. Chondromyxoid fibroma (see Fig. 106).

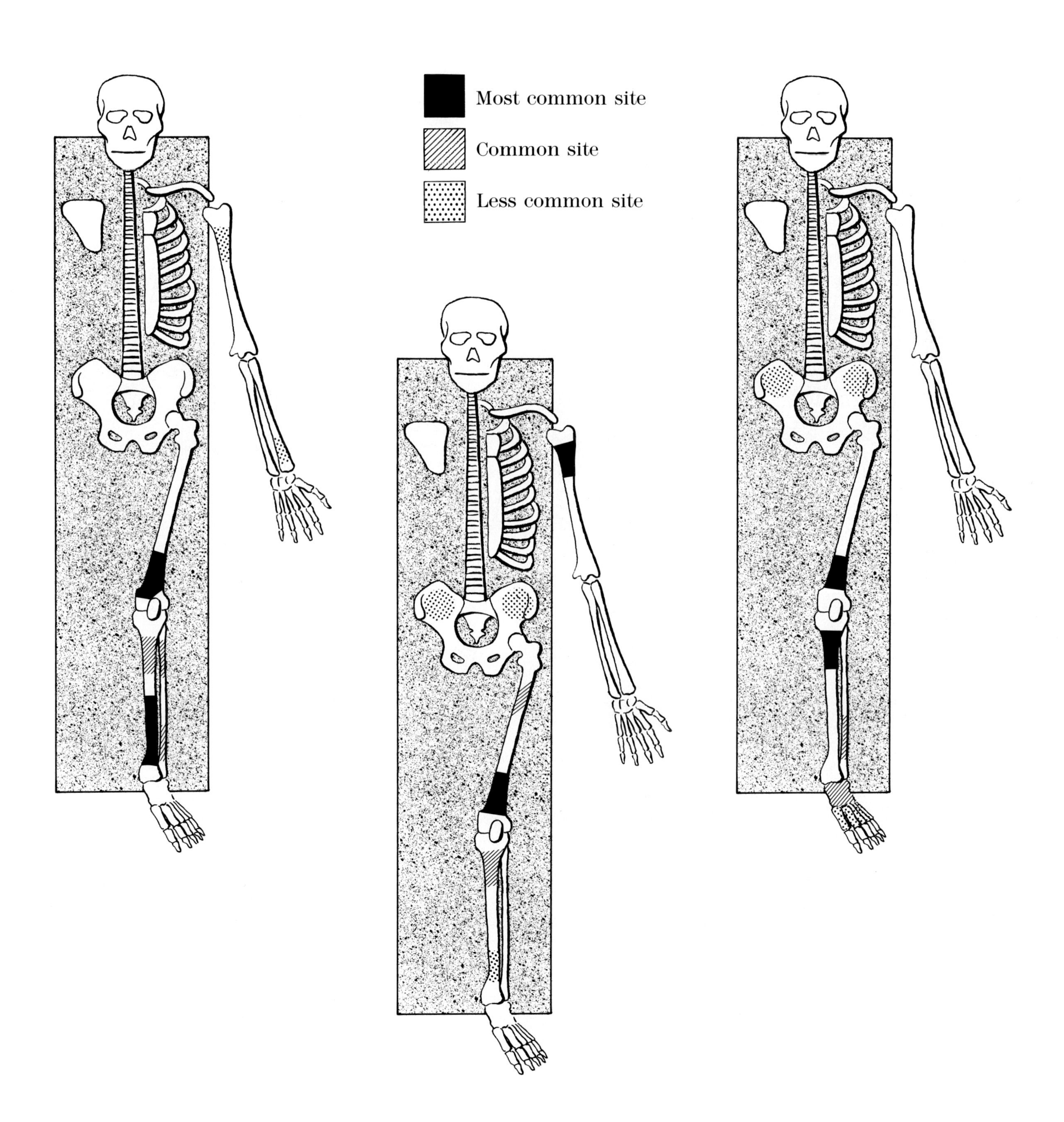

Fig. 115. Benign histiocytic fibrous histiocytoma (Benign giant cell tumor) (see Figs. 71, 174, 178, 183, 325, 490, 501, 774, 778).

Fig. 116. Chondroblastoma (see Figs. 128, 506, 1103, 1104, 1111).

Fig. 117. Fibrosarcoma (see Figs. 91, 131, 165, 170, 240, 260, 297).

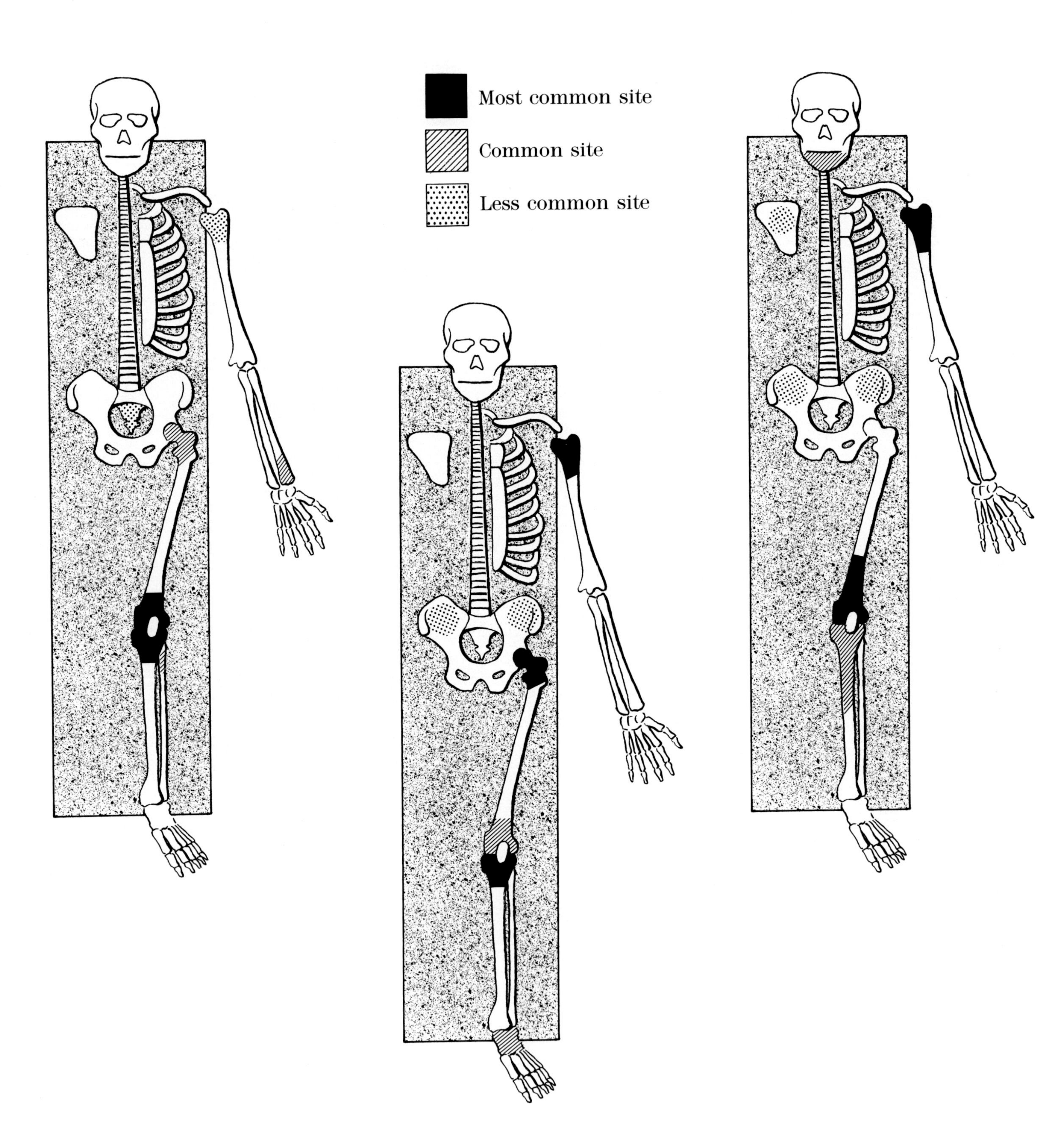

"Large size of a tumor is just as ominous as high cellularity"
HENRY L. JAFFE (1896–1979)

6. Size

Non-neoplastic Lesions (Table 39)

Benign Neoplasms (Table 40)

Malignant Neoplasms (Table 41)

IN general, the larger the soft tissue or a bone tumor is, the greater the likelihood that it is malignant. Conversely, as a rule the smaller the tumor, the bigger the chance that it is benign. Less than a handful of non-neoplastic lesions are on the average bigger than 5 cm: elastofibroma, lipogranuloma, proliferative myositis, myositis ossificans, and arteriovenous malformation (Table 39). About a dozen benign soft tissue and bone neoplasms are on the average 5 cm or bigger: nonossifying fibroma, desmoplastic fibroma, lipoma, angiomyolipoma, leiomyoma, neurofibroma, benign schwannoma, benign paraganglioma, chondroma, osteochondroma, and benign cystosarcoma phyllodes (Table 40); however, there are at least three dozen sarcomas that are bigger than 5 cm. In fact, of the sarcomas listed in Table 41, only three sarcomas are usually less than 5 cm in size.

Size can indicate the nature and potential behavior of a neoplasm, and it can also challenge the surgeon, who must resect the neoplasm without violating it, and with clear margins. Sarcomas larger than 5 cm may necessitate amputation or disarticulation because a clear margin cannot be obtained without mutilating surgery. The statement that "It is better to be alive with one leg off than to die with two legs on" perhaps is not a source of solace to the patient, but in the case of bulky, high grade, and unresectable neoplasms it is nonetheless as true today as it was in the past.

Size is an important consideration in various staging schemes. Admittedly, the classification of sarcomas according to whether they are greater or less than 5 cm in size is an arbitrary and crude practice of oncology, but in the absence of a better criterion we must adhere to it and continue to use it. It seems that most soft tissue sarcomas and bone sarcomas are diagnosed when they have reached, more or less, the size of 5 cm, and just about the time when they are passing from the established phase to the invasive phase and are becoming symptomatic (see Figure 7).

Table 39. *Average Size of Non-neoplastic Lesions*

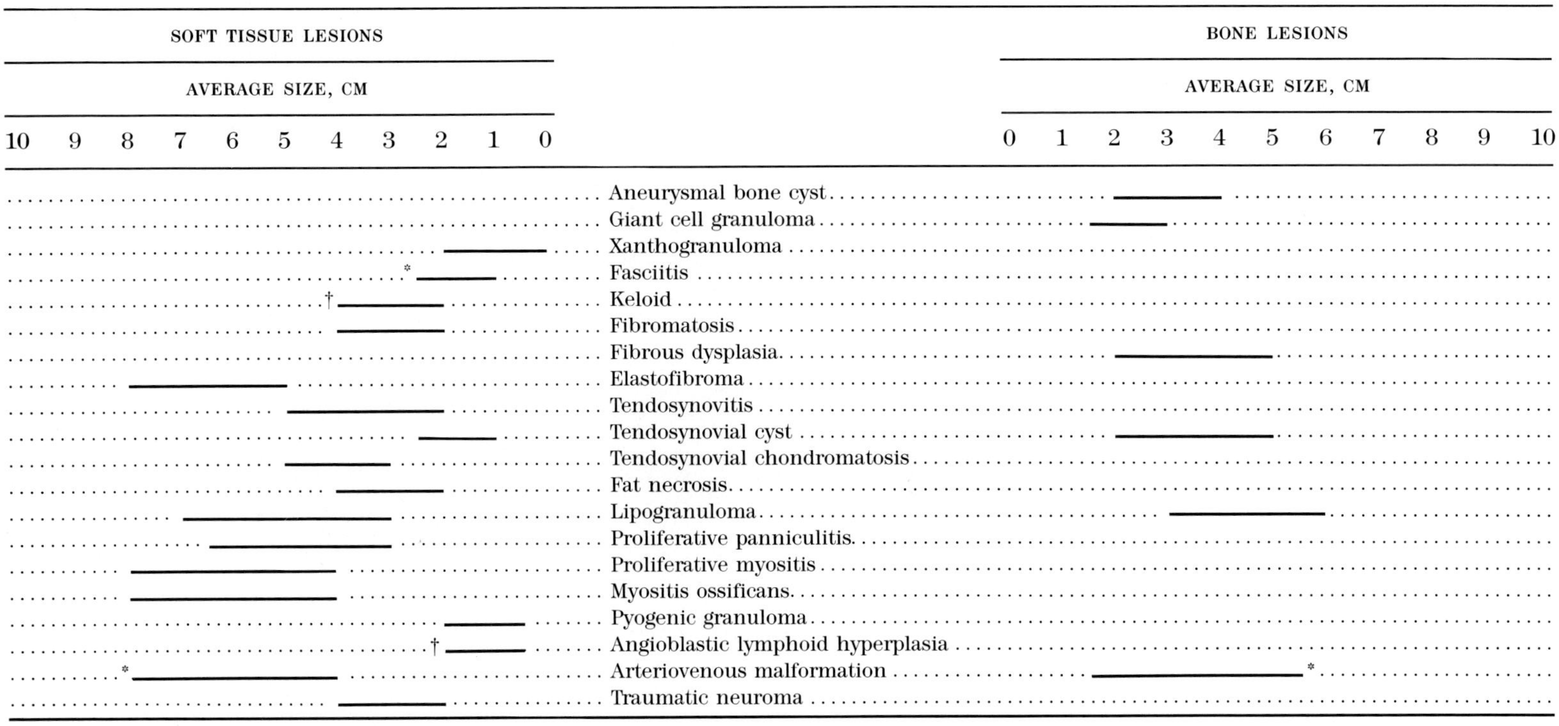

Soft Tissue Lesions: Average Size, cm (10 9 8 7 6 5 4 3 2 1 0)		Bone Lesions: Average Size, cm (0 1 2 3 4 5 6 7 8 9 10)
	Aneurysmal bone cyst	—
	Giant cell granuloma	—
—	Xanthogranuloma	
*—	Fasciitis	
†—	Keloid	
—	Fibromatosis	
	Fibrous dysplasia	—
—	Elastofibroma	
—	Tendosynovitis	
—	Tendosynovial cyst	—
—	Tendosynovial chondromatosis	
—	Fat necrosis	
—	Lipogranuloma	—
—	Proliferative panniculitis	
—	Proliferative myositis	
—	Myositis ossificans	
—	Pyogenic granuloma	
†—	Angioblastic lymphoid hyperplasia	
*—	Arteriovenous malformation	—*
—	Traumatic neuroma	

*Can be multifocal. †Often multifocal.

Table 40. *Average Size of Benign Neoplasms*

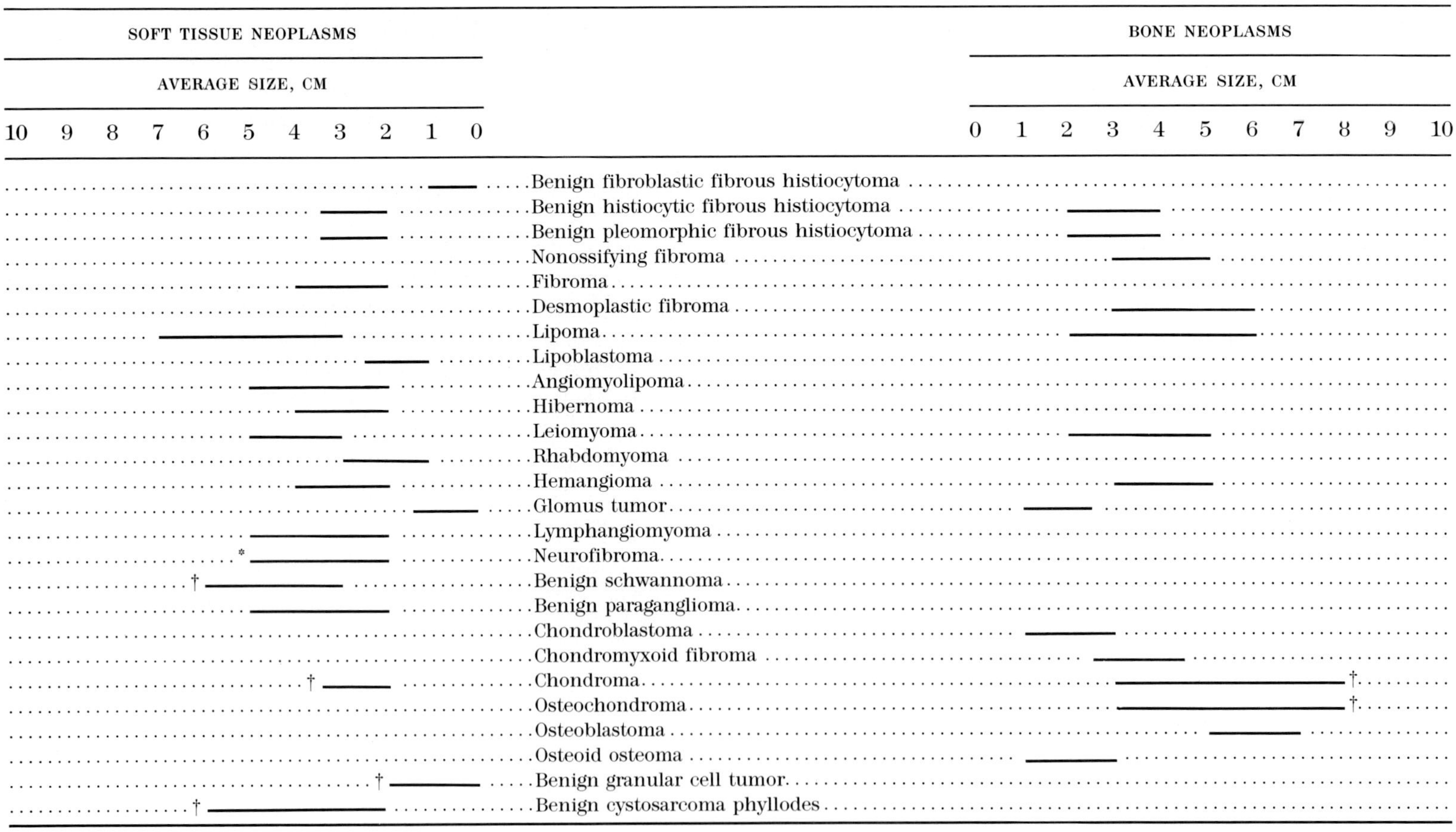

Soft Tissue Neoplasms: Average Size, cm (10 9 8 7 6 5 4 3 2 1 0)		Bone Neoplasms: Average Size, cm (0 1 2 3 4 5 6 7 8 9 10)
—	Benign fibroblastic fibrous histiocytoma	
—	Benign histiocytic fibrous histiocytoma	—
—	Benign pleomorphic fibrous histiocytoma	—
	Nonossifying fibroma	—
—	Fibroma	
	Desmoplastic fibroma	—
—	Lipoma	—
—	Lipoblastoma	
—	Angiomyolipoma	
—	Hibernoma	
—	Leiomyoma	—
—	Rhabdomyoma	
—	Hemangioma	—
—	Glomus tumor	—
—	Lymphangiomyoma	
*—	Neurofibroma	
†—	Benign schwannoma	
—	Benign paraganglioma	
	Chondroblastoma	—
	Chondromyxoid fibroma	—
†—	Chondroma	—†
	Osteochondroma	—†
	Osteoblastoma	—
	Osteoid osteoma	—
†—	Benign granular cell tumor	
†—	Benign cystosarcoma phyllodes	

*Often multifocal. †Can be multifocal.

Table 41. *Average Size of Malignant Neoplasms*

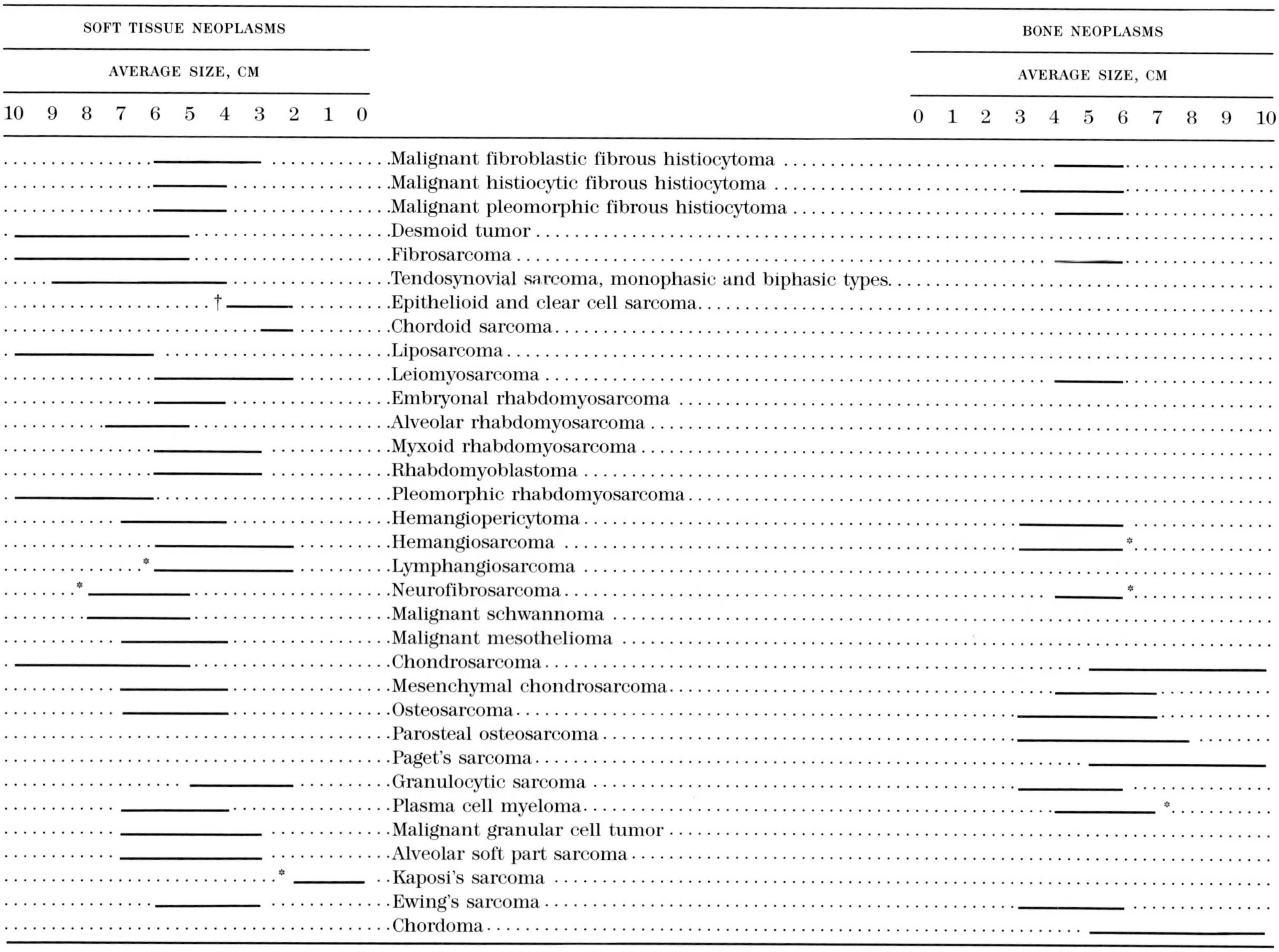

Soft tissue neoplasms: average size, cm (10 9 8 7 6 5 4 3 2 1 0)		Bone neoplasms: average size, cm (0 1 2 3 4 5 6 7 8 9 10)
6–3	Malignant fibroblastic fibrous histiocytoma	4–6
6–4	Malignant histiocytic fibrous histiocytoma	3–6
6–4	Malignant pleomorphic fibrous histiocytoma	4–6
10–5	Desmoid tumor	
10–5	Fibrosarcoma	4–6
9–4	Tendosynovial sarcoma, monophasic and biphasic types	
†4–2	Epithelioid and clear cell sarcoma	
3–2	Chordoid sarcoma	
10–6	Liposarcoma	
6–2	Leiomyosarcoma	4–6
6–4	Embryonal rhabdomyosarcoma	
7.5–5	Alveolar rhabdomyosarcoma	
6–3	Myxoid rhabdomyosarcoma	
6–3	Rhabdomyoblastoma	
10–6	Pleomorphic rhabdomyosarcoma	
7–4	Hemangiopericytoma	3–6
6–2	Hemangiosarcoma	3–6*
*6–2	Lymphangiosarcoma	
*8–5	Neurofibrosarcoma	4–6*
8–5	Malignant schwannoma	
7–4	Malignant mesothelioma	
10–5	Chondrosarcoma	5–10
7–4	Mesenchymal chondrosarcoma	4–7
7–4	Osteosarcoma	3–7
	Parosteal osteosarcoma	3–8
	Paget's sarcoma	5–10
5–2	Granulocytic sarcoma	3–6
7–4	Plasma cell myeloma	4–7*
7–3	Malignant granular cell tumor	
7–3	Alveolar soft part sarcoma	
*2–0	Kaposi's sarcoma	
6–3	Ewing's sarcoma	3–6
	Chordoma	5–10

*Often multifocal. †Can be multifocal.

"The pathologist can make enough errors with all available information"
LAUREN V. ACKERMAN (1905–)

7. Radiologic Appearance

Non-neoplastic Lesions (Table 42; Figures 118, 119, 125, 139, 144, 145, 150, 163, 186, 194, 197)

Benign Neoplasms (Table 43; Figures 128, 135, 155, 166, 167, 174, 177, 178, 183, 184, 191, 198, 204, 211, 222, 234A)

Malignant Neoplasms (Table 44; Figures 120–124, 126, 127, 129–134, 136–138, 140–143, 146–149, 151–154, 156–162, 164, 165, 168–173, 175, 176, 179–182, 185, 187–190, 192, 193, 195, 196, 199–203, 205–210, 212–221, 223–233, 234B and C)

INITIAL evaluation of soft tissue and bone tumors usually requires evaluation of the precise location, size, density, and extent of the tumor. Radiologic examination in most cases is followed by incisional biopsy, or by excisional biopsy if the tumor is a small soft tissue lesion less than 5 cm, and by biopsy or curettage if it is a bone tumor.

The radiologic finding and impression should be fully disclosed to the pathologist before a definitive diagnosis on the basis of microscopic examination is reached. While pathologists are not trained in diagnostic radiology and should not assume the role of diagnostic radiologists, they must know whether there is a specific radiologic finding. If the pathologist's microscopic finding is supported by radiologic and clinical findings, a definitive diagnosis can be made. On the other hand, if there is a lack of concordance between the pathologist's finding and the radiologist's impression, the biopsy material may not be representative of the lesion, or the radiologist's assessment of the lesion may be inaccurate. In this situation the surgeon and the radiologist must review their findings with the pathologist before reaching histologic conclusion. They may reassess their view and accept the pathologist's diagnosis or perform new studies, including another biopsy of the lesion.

Surgeons, radiologists, and pathologists must be jointly aware that some of the most obvious soft tissue and bone lesions may possess a dangerously misleading appearance. One needs only to consider the ominous clinical appearance of proliferative myositis or myositis ossificans, the occasionally benign appearance of the radiologic features of osteosarcoma, the unimpressive histology of parosteal osteosarcoma, the frisky and unpredictable microscopy of fasciitis, or the dangerously malignant appearance, under the microscope, of osteoblastoma or ordinary callus. The pathologist must take into account clinical and radiologic findings, but it would be a mistake to tend to a particular diagnosis because of knowledge of the lesion's clinical or radiologic appearance; foreknowledge of clinical circumstances and radiologic findings, like complete ignorance of them, can occasionally lead to errors in pathologic diagnosis, and the best hope lies with the well-informed pathologist who assesses the histology objectively.

Clinically and on conventional radiographic film, soft tissue and bone tumors may present without specific features. Prior to present day radiography, the role of radiology in the evaluation of intra- and extraskeletal tumors was limited. Tomography, computed tomography, xeroradiography, and ultrasound techniques may detect density differences in the tumor and disclose the relationship of the tumor to the surrounding tissues, information hardly possible to get with conventional radiographic techniques. Arteriography and, in certain cases, lymphangiography, radionuclide scans, and scintigrams are usually helpful in determining the precise location of the tumor and choosing the best biopsy site. The use of these techniques in evaluation of soft tissue and bone tumors permits the following in addition to diagnostic assessment: (1) evaluation of tumor size and extent; (2) determination of whether a tumor is well defined or poorly defined; (3) determination of which bones, compartments, or organs are involved either by direct extension or extrinsic pressure; and (4) evaluation of the relation of the tumor to major vessels, nerves, and other vital structures.

Many radiologists feel that magnetic resonance imaging complements computed tomography in the evaluation of many tumors by not only producing an image of density, but also by indicating various chemical and biochemical parameters. Despite promising early results in detecting brain tumors, magnetic resonance imaging is still in the developmental stage, and its full potential and role in detection of soft tissue and bone tumors has only begun to be understood.

Tables 42, 43, and 44 were designed and Figures 118 to 234 were selected to give an overview of certain specific and suggestive radiologic features of some of the most common soft tissue and bone tumors. Readers are urged to consult some of the specific texts for more detail (see Tables 1 and 2).

TABLE 42. *Radiologic Appearance of Non-neoplastic Lesions*

	LYTIC	BLASTIC OR SCLEROTIC	EXPANSILE	DESTRUCTIVE	THINNED CORTEX	MULTI-LOCULATED	VASCULAR	CALCIFIED	WELL DEMARCATED	INFILTRATIVE	SCLEROSING EDGE	MULTI-FOCAL	SITES IN LONG BONES
Aneurysmal bone cyst (1)	X		X	X	X	X	X		X		X	X†	
Giant cell granuloma* (2)	X		X	X	X			X	X		X		
Hyperparathyroidism (3)	X	X	X	X			X			X		X	Diaphysis
Fasciitis ossificans		X					X			X		X†	
Fibrous dysplasia* (4)	X	X	X	X		X					X		
Tendosynovial cyst*	X			X	X	X		X	X		X		Epiphysis
Tendosynovial chondromatosis		X						X	X		X	X†	
Panniculitis ossificans		X						X		X		X†	
Myositis ossificans* (5)		X						X	X		X		
Chondroid metaplasia		X						X	X				
Paget's disease of bone (6)	X	X								X		X	
Arteriovenous malformation	X					X	X	X	X			X†	
Osteoid metaplasia		X						X	X				
Callus		X	X	X			X			X			
Eosinophilic granuloma (7)	X		X	X	X	X	X		X			X	Diaphysis
Osteomyelitis (8)	X	X										X†	Epiphysis

*Solitary. †Can be multifocal.

1. Codman's triangle
2. Commonly in mandible
3. Sclerotic bands away from the lesion
4. Ground glass appearance
5. Bony trabeculae
6. Fracture, thickened trabeculae
7. Onion skin appearance
8. Codman's triangle, onion skin appearance

Table 43. *Radiologic Appearance of Benign Neoplasms*

	Lytic	Blastic or Sclerotic	Expansile	Destructive	Thinned Cortex	Multi-loculated	Vascular	Calcified	Well Demarcated	Infiltrative	Sclerosing Edge	Multi-focal	Sites in Long Bones
Benign histiocytic fibrous histiocytoma of bone*	X		X	X		X		X		X			Epiphysis
Nonossifying fibroma* (1)	X		X	X		X				X	X		Metaphysis
Desmoplastic fibroma* (2)	X	X		X	X				X		X		Metaphysis
Intraskeletal lipoma*	X								X				
Angiolipoma*							X		X				
Angiomyolipoma*	X					X	X	X	X				
Hemangioma of bone (3)	X	X		X		X	X					X†	
Lymphangioma	X		X			X	X			X			
Neurofibroma of bone (4)	X			X						X		X	
Benign paraganglioma*			X				X	X	X				
Chondroblastoma	X	X		X				X‡			X	X‡	Epiphysis
Chondromyxoid fibroma*	X		X		X				X		X		Metaphysis
Chondroma	X		X	X				X	X			X†	
Osteochondroma		X						X	X			X†	Metaphysis
Osteoblastoma* (5)	X	X	X	X				X	X				Epiphysis
Osteoid osteoma* (6)		X						X	X				Metaphysis

*Solitary †Can be multifocal ‡Scattered in "chicken-wire" type

1. Lobulated edge
2. Irregular edge
3. Irregular edge, thickened trabeculae
4. Pseudoarthrosis
5. Periosteal reaction and scattered calcification
6. Central nidus and central calcification

TABLE 44. *Radiologic Appearance of Malignant Neoplasms*

	LYTIC	BLASTIC OR SCLEROTIC	EXPANSILE	DESTRUCTIVE	THINNED CORTEX	MULTI-LOCULATED	VASCULAR	CALCIFIED	WELL DEMARCATED	INFILTRATIVE	SCLEROSING EDGE	MULTI-FOCAL	SITES IN LONG BONES
Malignant fibroblastic fibrous histiocytoma of bone*	X		X	X	X		X			X			Metaphysis
Malignant histiocytic fibrous histiocytoma of bone* (1)	X		X	X		X	X			X			Epiphysis
Malignant pleomorphic fibrous histiocytoma of bone* (2)	X			X			X			X			Metaphysis
Desmoid tumor*		X		X				X‡		X			
Fibrosarcoma of bone	X	X		X	X					X		X‡	Metaphysis
Liposarcoma	X		X			X	X		X				
Leiomyosarcoma of bone*	X			X			X	X†		X			Metaphysis
Pleomorphic rhabdomyosarcoma* (3)	X			X						X			
Hemangiopericytoma of bone	X		X	X		X	X			X		X‡	
Hemangiosarcoma of bone (4)	X		X	X		X	X			X			Diaphysis
Malignant mesothelioma				X		X		X†		X			
Chondrosarcoma	X		X	X		X				X		X‡	Epiphysis
Mesenchymal chondrosarcoma			X	X		X				X			
Osteosarcoma* (5)	X	X		X						X			Metaphysis
Parosteal osteosarcoma* (6)		X		X	X					X			
Paget's sarcoma of bone		X	X	X			X			X		X‡	
Granulocytic sarcoma	X		X	X	X					X		X	
Malignant lymphoma of bone*	X		X	X						X		X‡	Diaphysis
Plasma cell myeloma	X		X	X	X	X			X	X		X	
Ewing's sarcoma (7)	X			X			X			X			Diaphysis
Chordoma*	X		X	X					X				Diaphysis
Myxopapillary ependymoma*			X			X	X			X			
Alveolar soft part sarcoma*			X	X		X	X	X		X			
Adamantinoma*	X			X			X			X			Diaphysis

*Solitary. †Can be calcified. ‡Can be multifocal.

1. Eccentric
2. Eccentric
3. Infiltrates bone
4. One-third multifocal
5. 0.3% multifocal
6. Juxtacortical
7. Periosteal reaction

Fig. 118A, B. Untreated eosinophilic granuloma of the skull of a child showing progression in a period of 13 months (see Fig. 1028).

Fig. 119. Giant cell granuloma of the mandible showing lytic destruction of the mandible (see Figs. 632, 787, 1000, 1069).

Fig. 120. Myxoid embryonal rhabdomyosarcoma of the nasopharynx with extension into the left maxillary sinus, orbit, and ethmoid (see Figs. 78, 206, 226, 263, 281, 416, 418, 587).

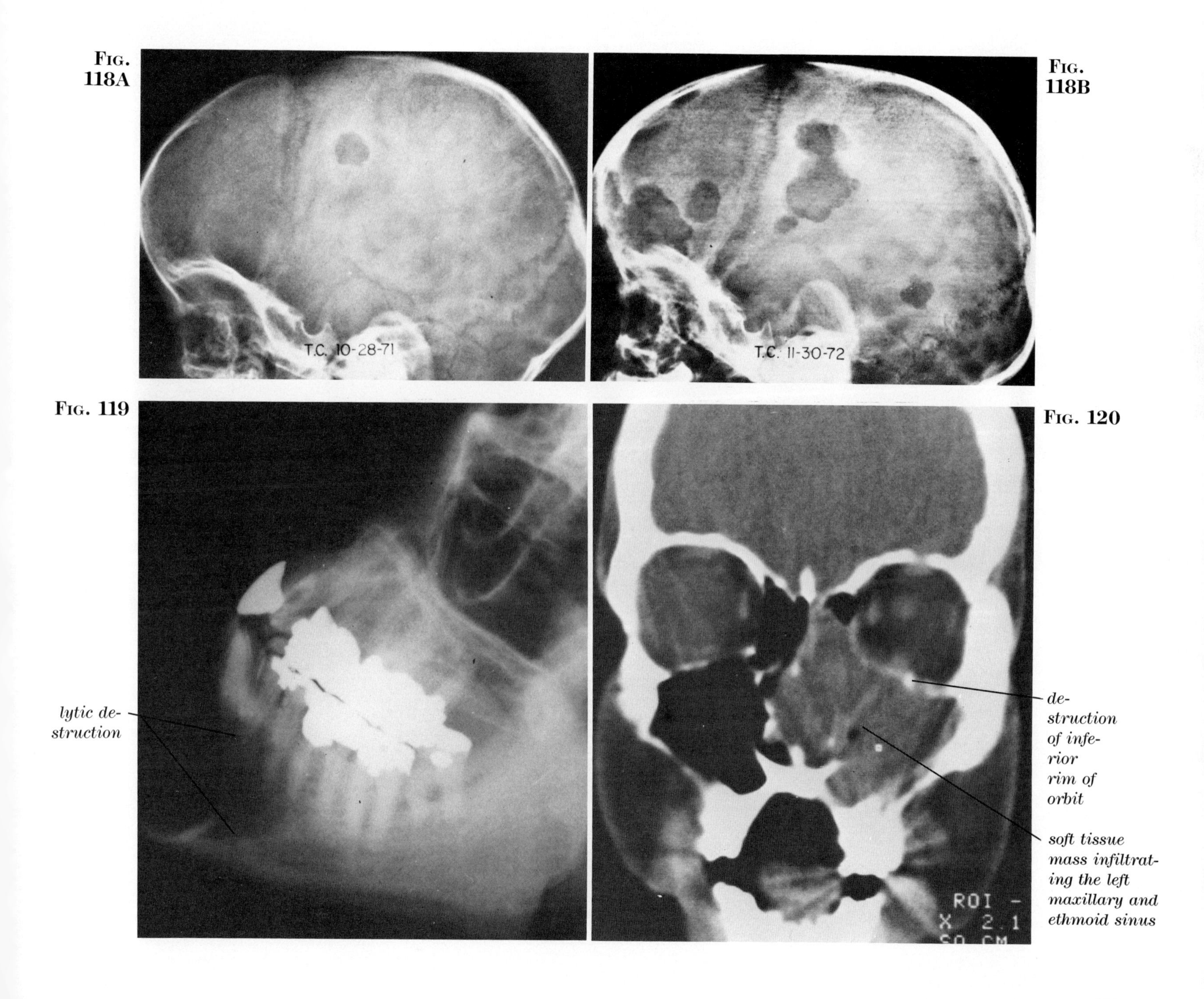

Fig. 121. Pleomorphic rhabdomyosarcoma of the left shoulder with invasion of scapula. A. An irregular lytic area in the glenoid fossa of the scapula. B. Hypervascularity of the sarcoma is demonstrated by arteriography (see Figs. 87, 292, 295, 299, 313, 599, 604, 609, 612).

Fig. 122. Hypervascularity of tendosynovial sarcoma of the right shoulder is seen by arteriography (see Figs. 152, 153, 179, 181, 187, 190, 320, 323).

Fig. 123. Chondrosarcoma of the right scapula shows hypovascularity by arteriography (see Figs. 95, 124, 126, 136, 216, 223).

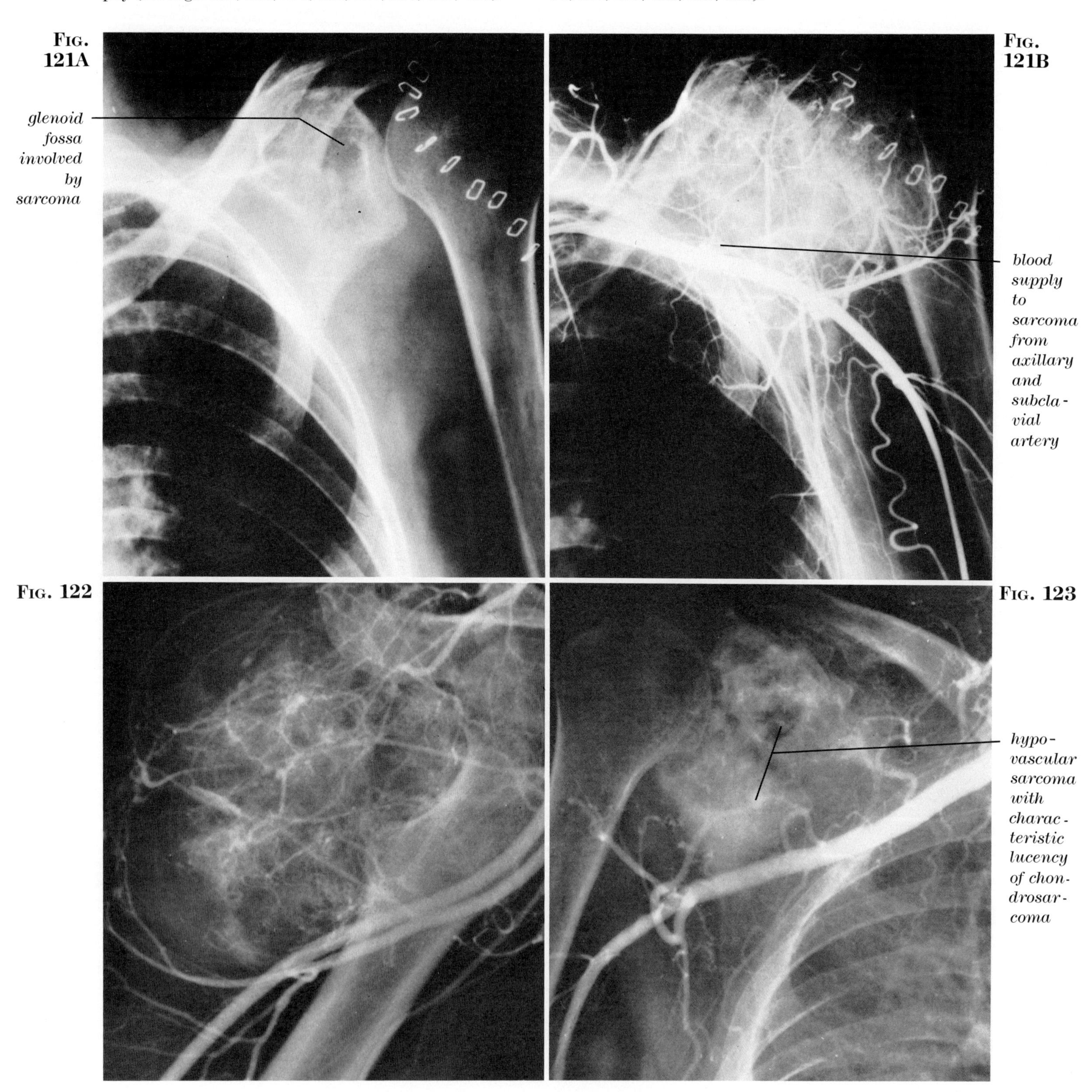

Fig. 124. Chondrosarcoma of the right humerus, showing pathologic fracture (see Figs. 123, 136, 137, 213, 216, 223, 285, 317, 327, 329).

Fig. 125. (Top center) Aneurysmal bone cyst of the right humerus (see Figs. 96, 150, 163, 509, 785, 1015).

Fig. 126. Chondrosarcoma of the left humerus (see Figs. 81, 95, 123, 124, 136, 147, 182, 213, 216).

Fig. 127. Osteosarcoma of the left humerus (see Figs. 108, 109, 142, 146, 149, 157, 158, 162, 164, 168).

Fig. 128. (Bottom center) Chondroblastoma of the left humerus showing stipled calcification (see Figs. 116, 506, 1103, 1104, 1111).

Fig. 129. Parosteal osteosarcoma of the humerus (Case of Dr. T. Simon, Syracuse) (see Figs. 217, 224, 249, 287, 353, 354, 656, 1082).

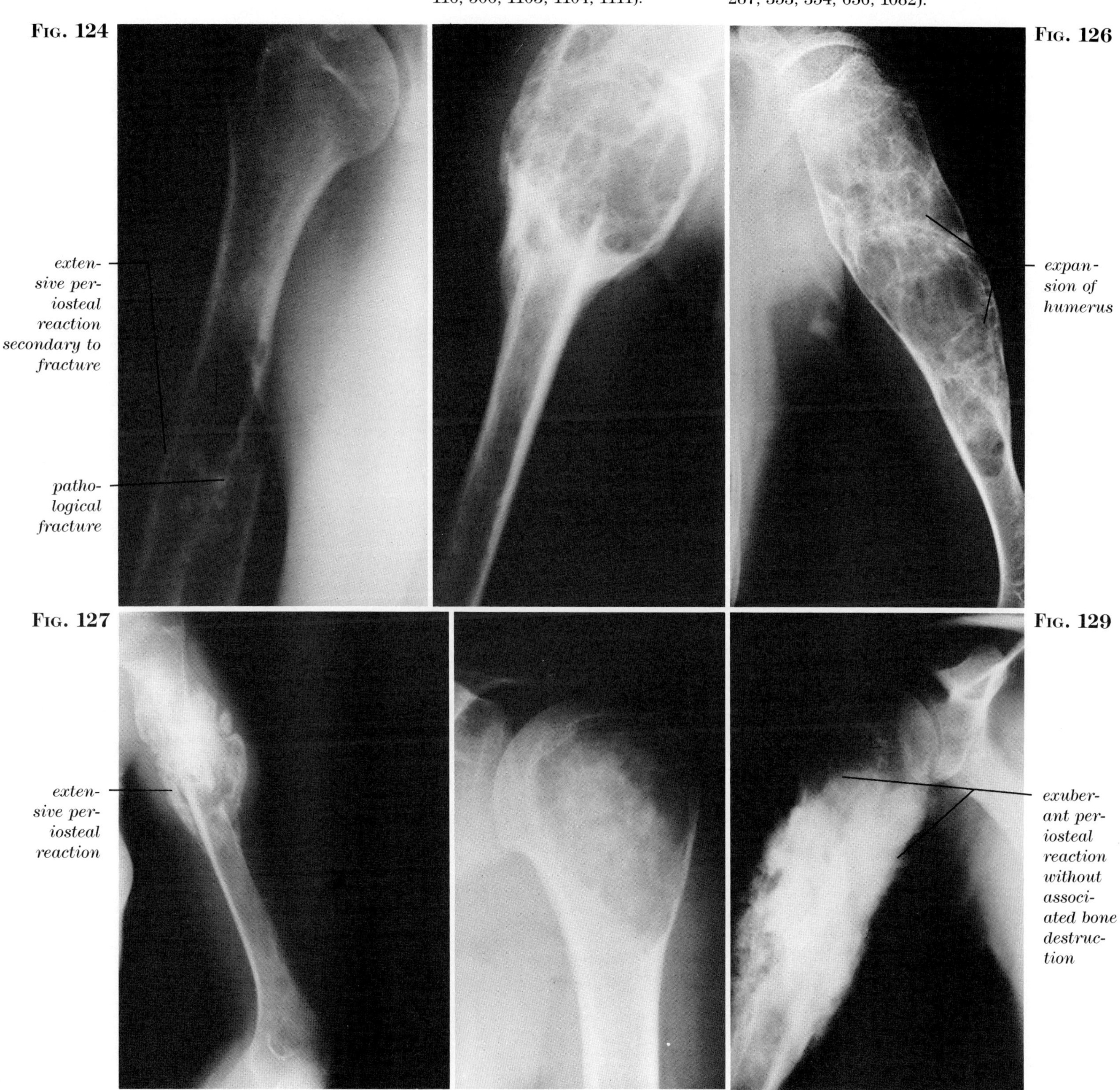

Fig. 130. Malignant pleomorphic fibrous histiocytoma, of the elbow showing invasion of the ulna (see Figs. 111, 138, 140, 148, 201, 203, 232, 237, 255).

Fig. 131. Intramedullary fibrosarcoma of the humerus (see Figs. 91, 117, 165, 170, 240, 260, 297, 410).

Fig. 132. Malignant lymphoma presents as dense, hardly visible nodule in the soft tissues of the cubital fossa (see Figs. 15, 169, 423, 463, 494, 702, 704, 705, 712, 743).

Fig. 133. Plasmacytoma of the right humerus (see Figs. 192, 726).

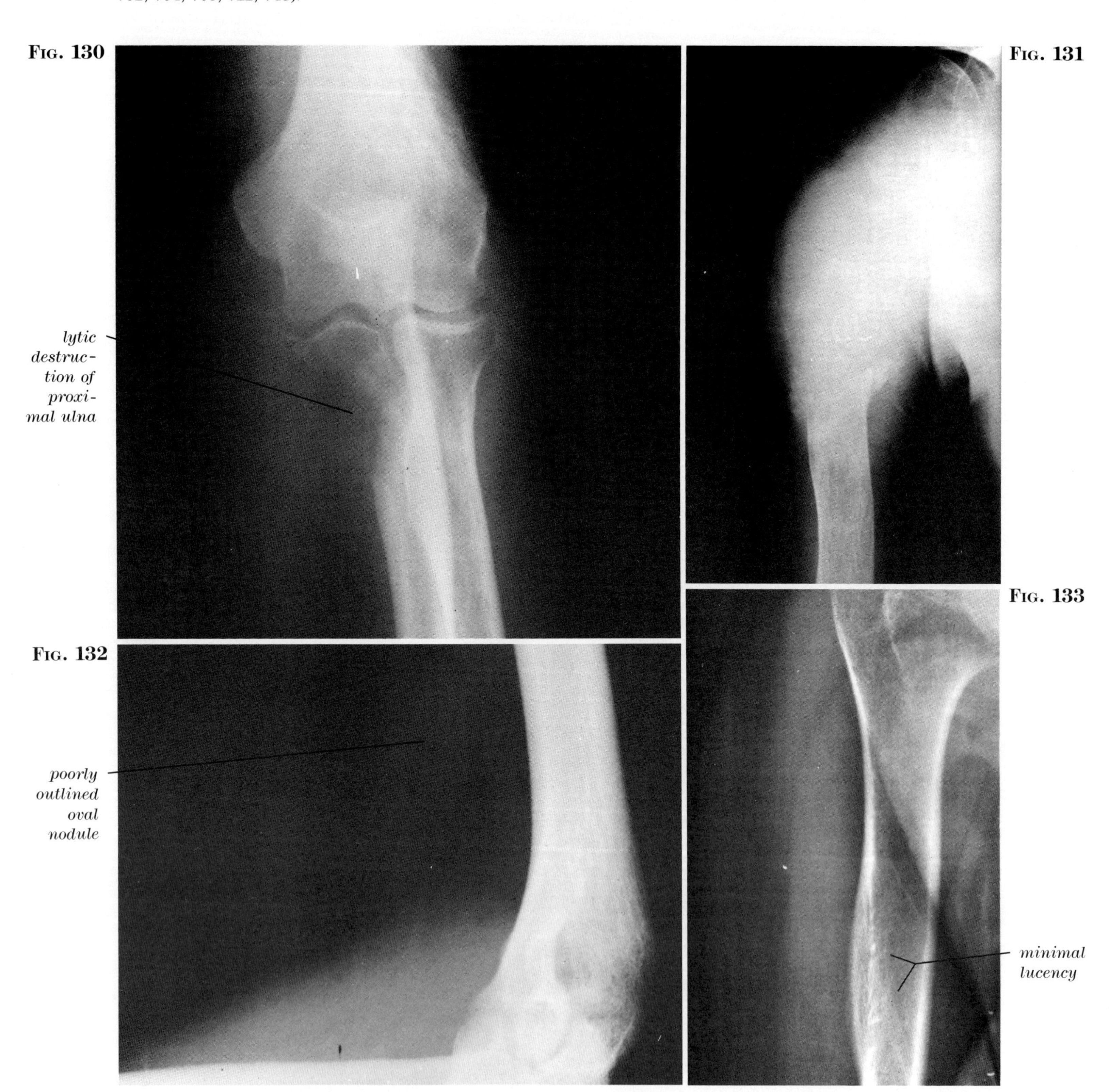

Fig. 134. Paget's sarcoma of the right ilium (Courtesy of Dr. R. Heelan, Memorial Sloan-Kettering Cancer Center, New York) (see Figs. 107, 234A and B, 918, 1103, 1128).

Fig. 136. Chondrosarcoma of the left iliac bone (see Figs. 95, 124, 126, 137, 147, 182, 213, 223, 285, 308, 317, 327, 329).

Fig. 135. Benign pleomorphic fibrous histiocytoma of proximal femur (see Figs. 346, 417, 422, 459, 722).

Fig. 137. Chondrosarcoma of proximal femur (see Figs. 124, 136, 182, 213, 223, 285, 308, 317, 327).

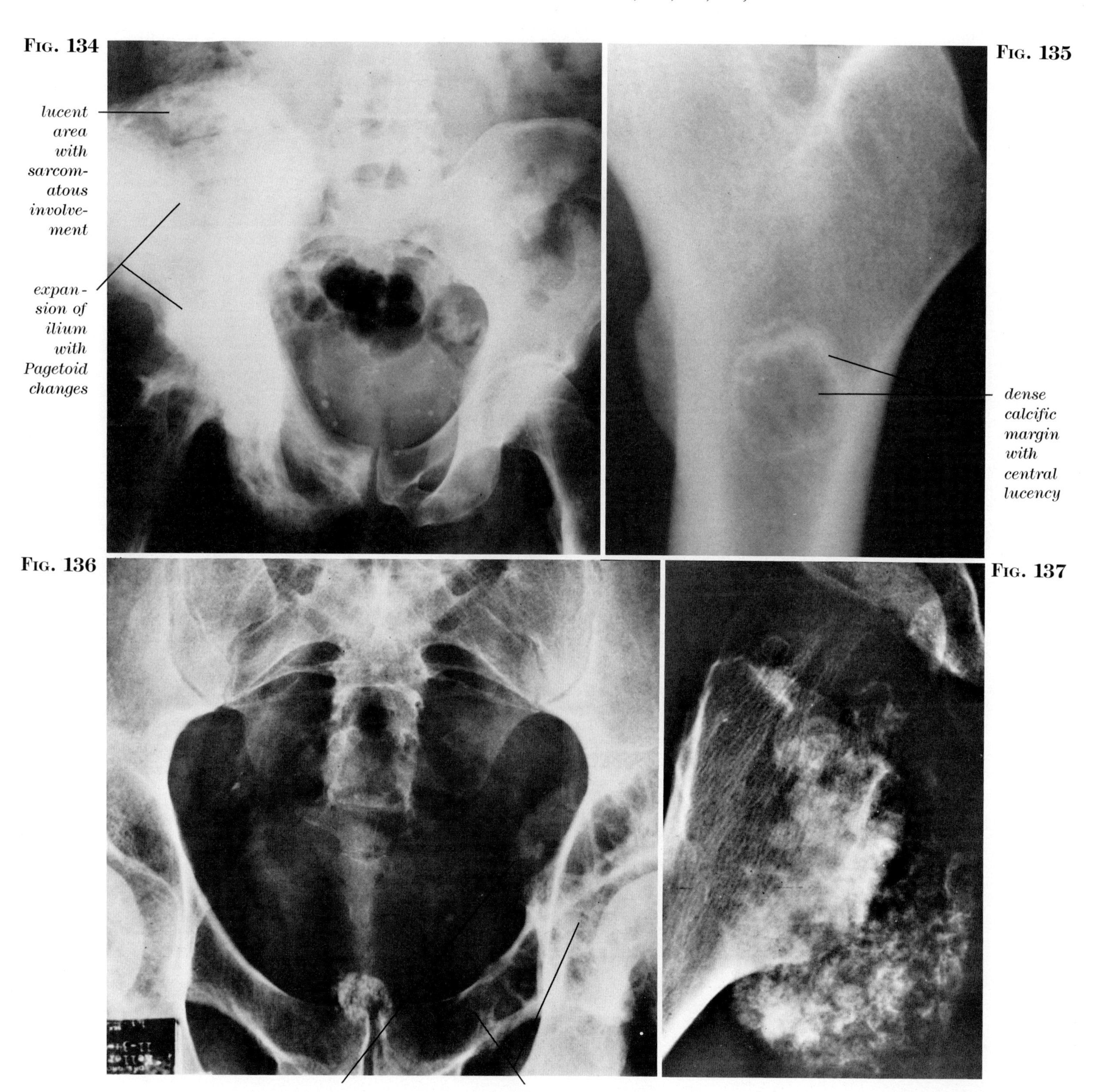

Fig. 138. Malignant pleomorphic fibrous histiocytoma of the inner thigh showing erosion of the ischial tuberosity (see Figs. 130, 140, 148, 201, 203, 232, 237, 255, 262, 288).

Fig. 140. Malignant pleomorphic fibrous histiocytoma of proximal thigh (see Figs. 31, 88, 138, 148, 201, 203, 232, 237, 255, 262, 288, 293).

Fig. 139. Arteriovenous malformation of the proximal thigh seen after angiography (Case of Dr. J. Mongeau, Canada) (see Figs. 197, 911, 1006, 1009).

Fig. 141. Liposarcoma of mid-thigh (see Figs. 86, 188, 199, 208, 219, 284, 294, 458, 590).

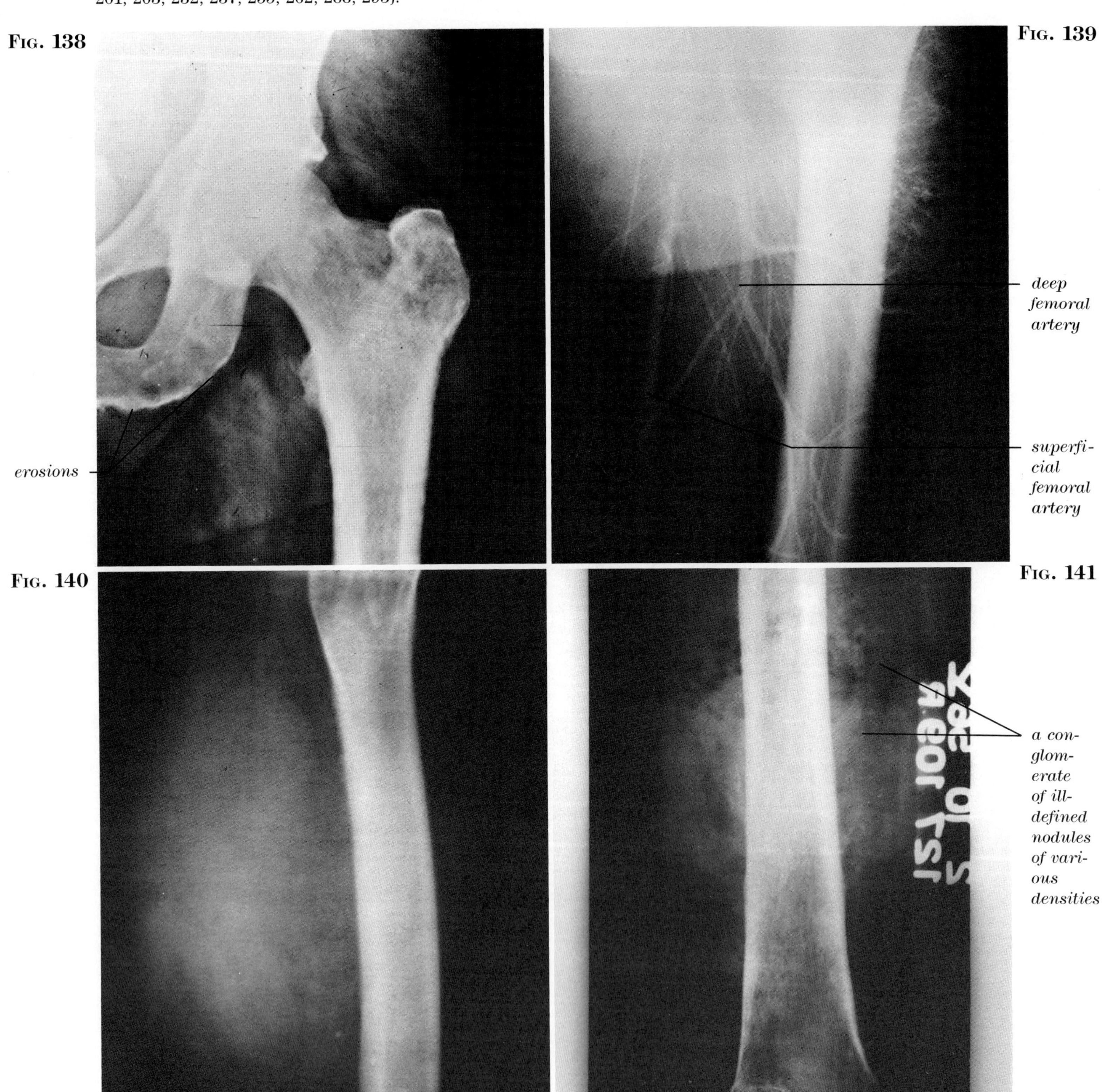

Fig. 142. Osteosarcoma of the left distal femur with parosteal extension (see Figs. 127, 146, 149, 157, 158, 162, 164, 168, 176, 221, 230, 304).

Fig. 143. (Top center) Malignant fibrous histiocytoma of distal femur. A. Frontal view. B. (center) Lateral view (see Figs. 111, 203, 232, 241, 301, 339, 340, 350, 373, 460).

Fig. 144. Intraosseous synovial cyst of the right distal femur of a 14-year-old (see Figs. 186, 508, 1014).

Fig. 145. Myositis ossificans of lateral mid-thigh (see Figs. 384, 620, 913, 1083).

Fig. 146. (Bottom center) Extraskeletal osteosarcoma of the left buttock (see Figs. 90, 142, 149, 157, 158, 162, 164, 168, 176, 221, 225, 228).

Fig. 147. Extraskeletal chondrosarcoma of the popliteal space (see Figs. 81, 95, 123, 124, 126, 137, 182, 213, 216, 223, 285, 308).

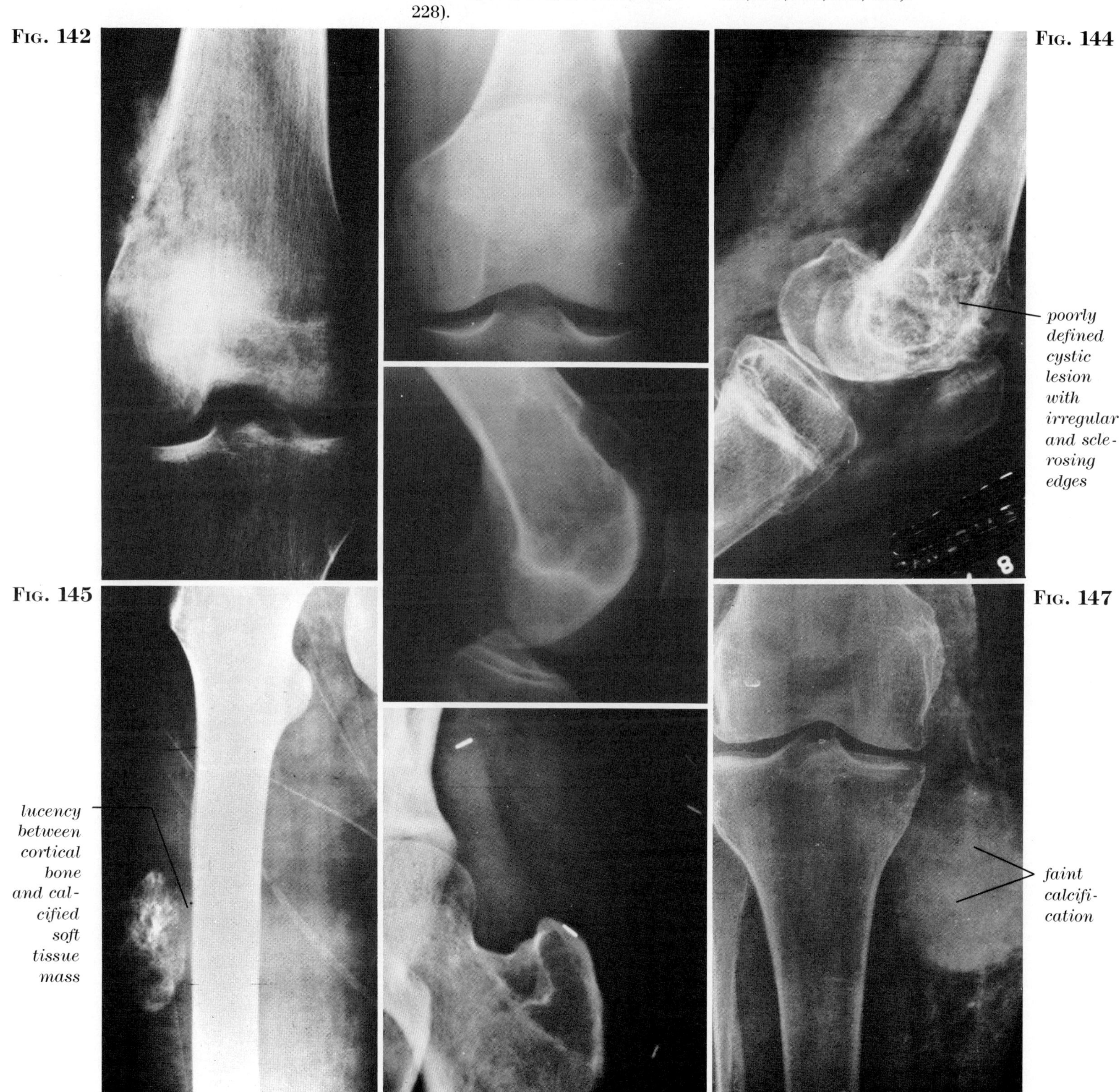

Fig. 148. Malignant fibrous histiocytoma of distal femur with extension into soft tissues (see Figs. 111, 140, 148, 201, 203, 232, 237, 255, 293, 301).

Fig. 149. Osteosarcoma of distal femur with extension into the popliteal space (see Figs. 108, 109, 146, 157, 158, 162, 164, 168, 176, 221, 225, 228, 305, 306).

Fig. 150. Aneurysmal bone cyst of distal femur (see Figs. 96, 125, 163, 509, 785, 1015, 1058).

Fig. 151. Mesenchymal chondrosarcoma of distal femur and soft tissues of thigh (see Figs. 103, 175, 180, 318, 390, 444, 507, 595, 674, 707).

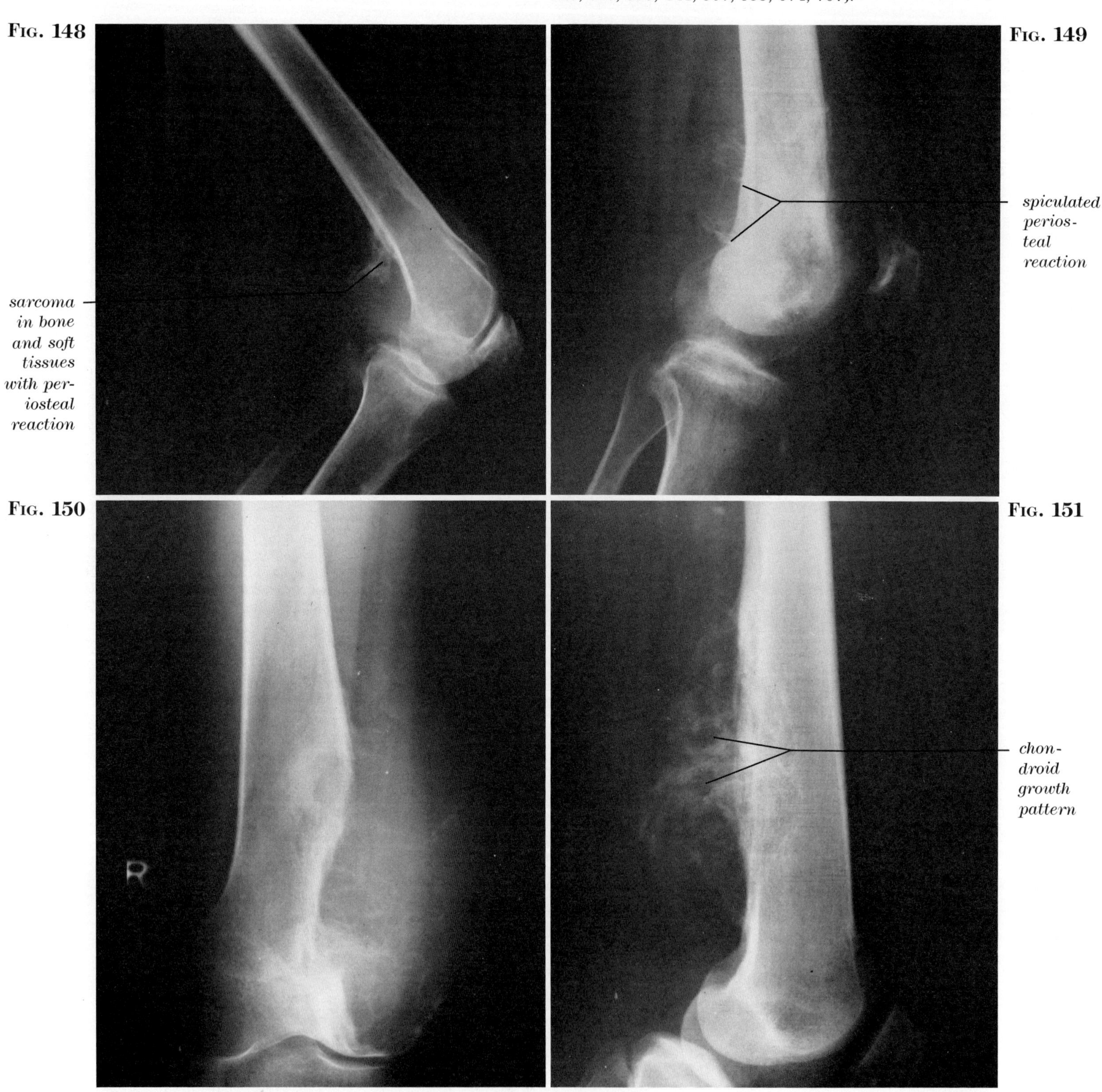

Fig. 152. Biphasic tendosynovial sarcoma of knee of a 2-year-old boy (Case of Dr. G. Sanchez, Panama) (see Figs. 85, 122, 153, 179, 320, 323, 514, 520, 522, 535, 536).

Fig. 153. Biphasic tendosynovial sarcoma of popliteal fossa of an adult (see Figs. 85, 152, 320, 323, 514, 520, 522, 535, 536, 547, 574).

Fig. 154. Multifocal lymphangiosarcoma of the right leg following two decades of filariasis-induced lymphedema (see Figs. 61, 84, 159, 321, 424, 551, 558, 649, 954).

Fig. 155. Osteochondroma of proximal fibula of a teenager (see Figs. 113, 167, 184, 302, 303, 772, 1063, 1066).

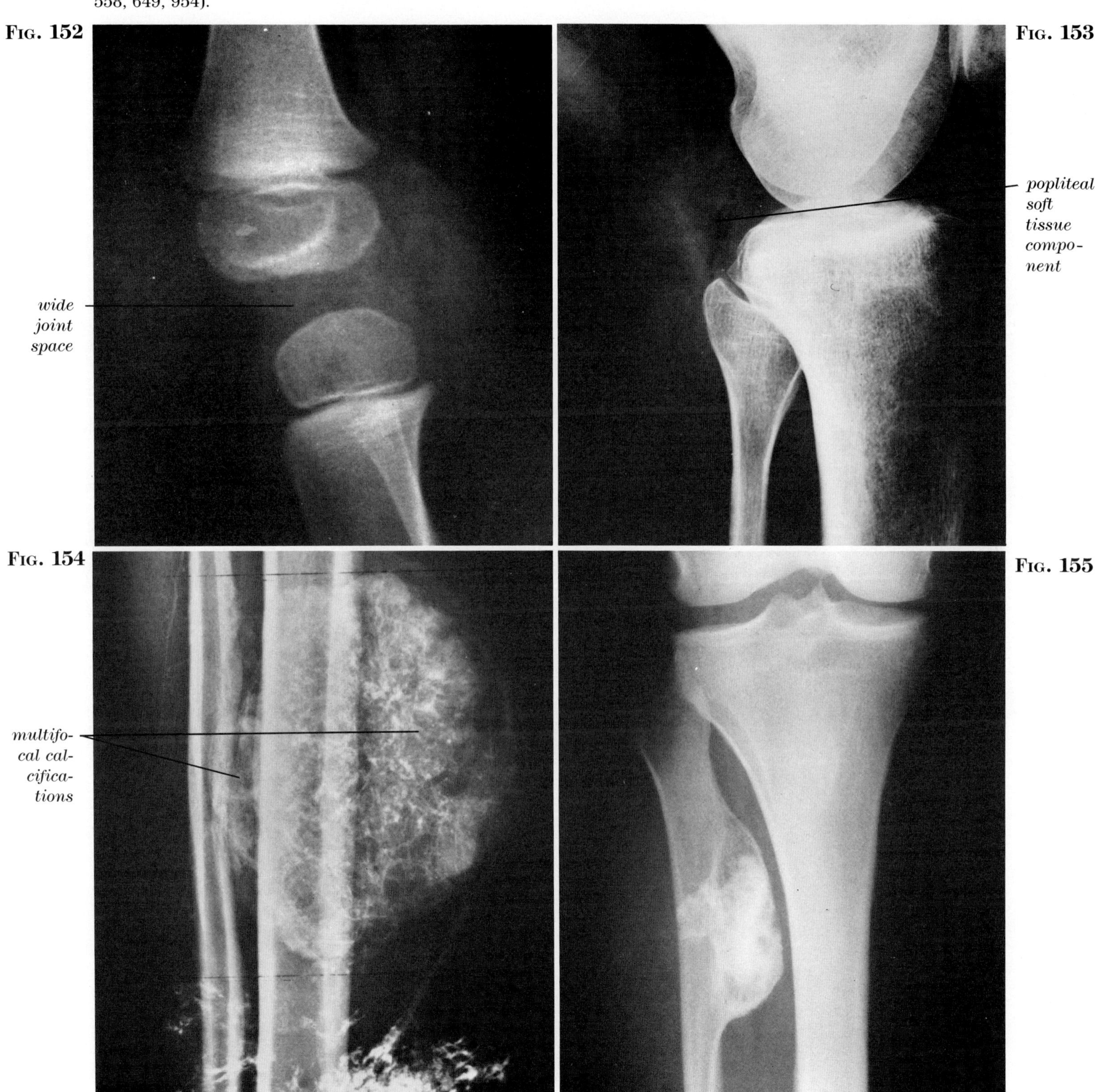

Fig. 156. Popliteal view of malignant histiocytic fibrous histiocytoma of the distal right femur (see Figs. 57, 111, 140, 143, 148, 161, 201, 203, 288, 300, 301, 452, 534).

Fig. 157. Osteosarcoma of the distal femur demonstrated by angiography (see Figs. 108, 149, 158, 162, 164, 168, 176, 221, 225, 304).

Fig. 158. Osteosarcoma of distal femur after arteriography (see Figs. 108, 157, 162, 164, 168, 176, 221, 225, 228, 304–306, 312).

Fig. 159. Hypervascularity of filariasis-induced lymphangiosarcoma of the leg demonstrated by arteriography (see Figs. 61, 84, 154, 321, 424, 551, 558).

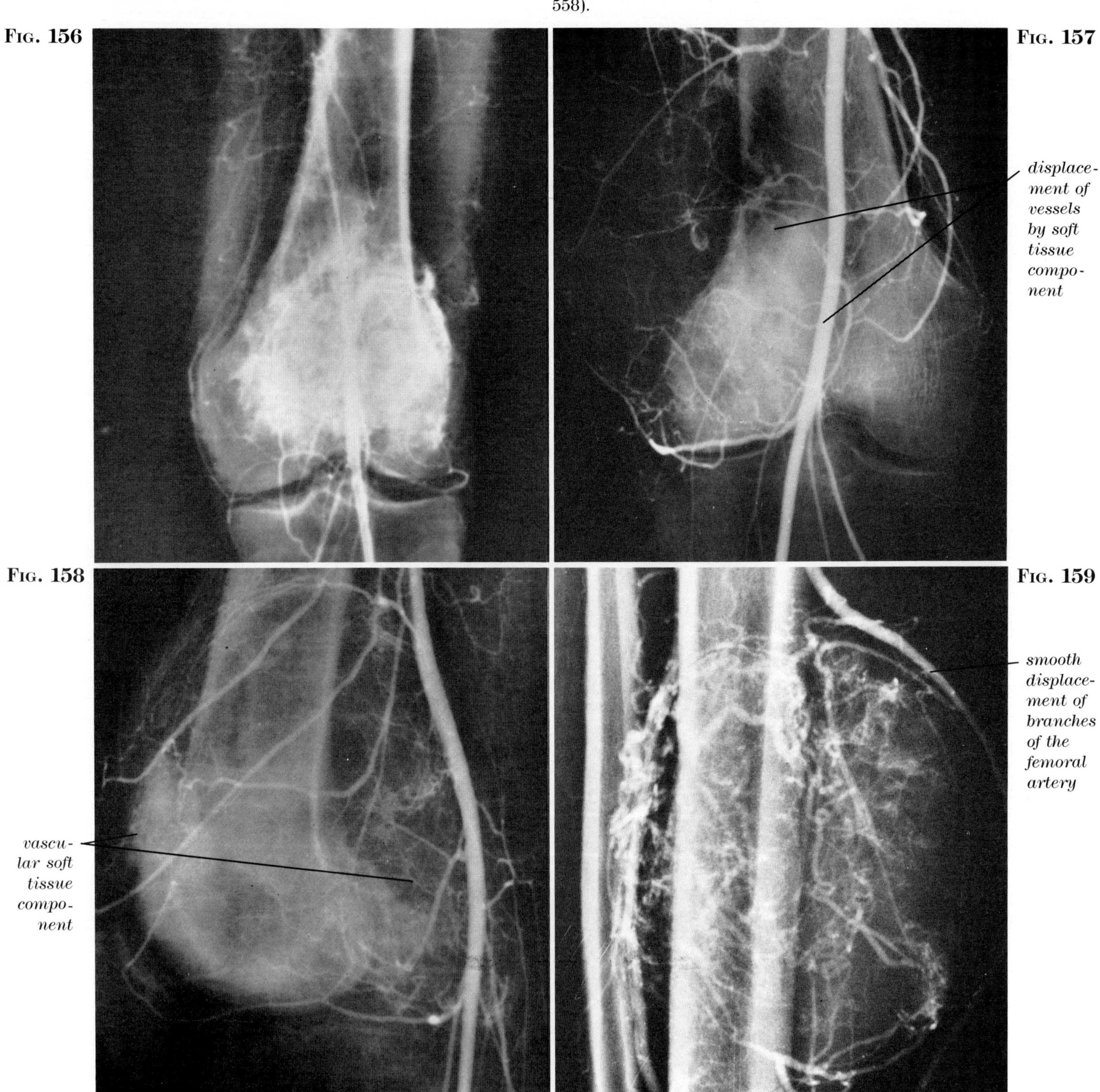

Fig. 160. Mammary carcinoma metastatic to the metaphysis of the tibia (see Figs. 682, 1259).

Fig. 161. Malignant histiocytic fibrous histiocytoma of proximal tibia (see Figs. 57, 156, 300, 452, 534).

Fig. 162. Sclerosing osteosarcoma of proximal tibia of a child (see Figs. 108, 158, 164, 168, 176, 221, 225, 233, 305, 306, 312, 319).

Fig. 163. Aneurysmal bone cyst of proximal tibia (see Figs. 96, 125, 150, 509, 785, 1015, 1058).

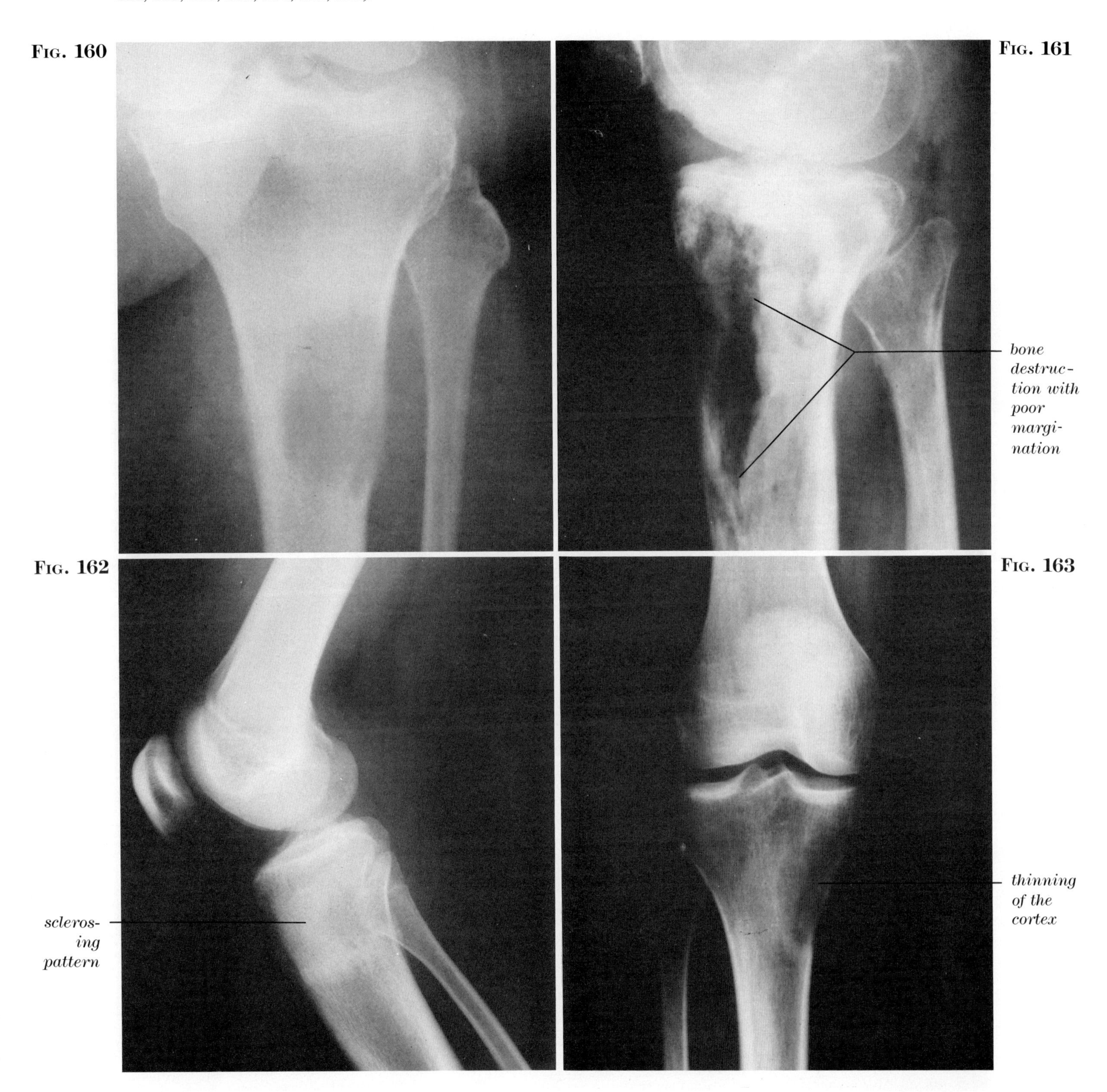

Fig. 164. Telangiectatic osteosarcoma of proximal tibia of a 13-year-old girl (see Figs. 108, 109, 127, 142, 146, 157, 168, 176, 221, 233, 328).

Fig. 165. (Top center) Low grade intramedullary fibrosarcoma of proximal tibia of a 16-year-old girl (see Figs. 121, 170, 260, 410, 601).

Fig. 166. Ganglioneuroma of proximal tibia of a 7-month-old baby (see Figs. 486, 630, 811, 824, 857).

Fig. 167. Osteochondroma of tibia and fibula (see Figs. 113, 302, 303, 772, 1063, 1066, 1090, 1152).

Fig. 168. Osteosarcoma of fibula of a 60-year-old man, predominantly growing as chondrosarcoma (see Figs. 108, 164, 176, 221, 225, 228, 233, 305).

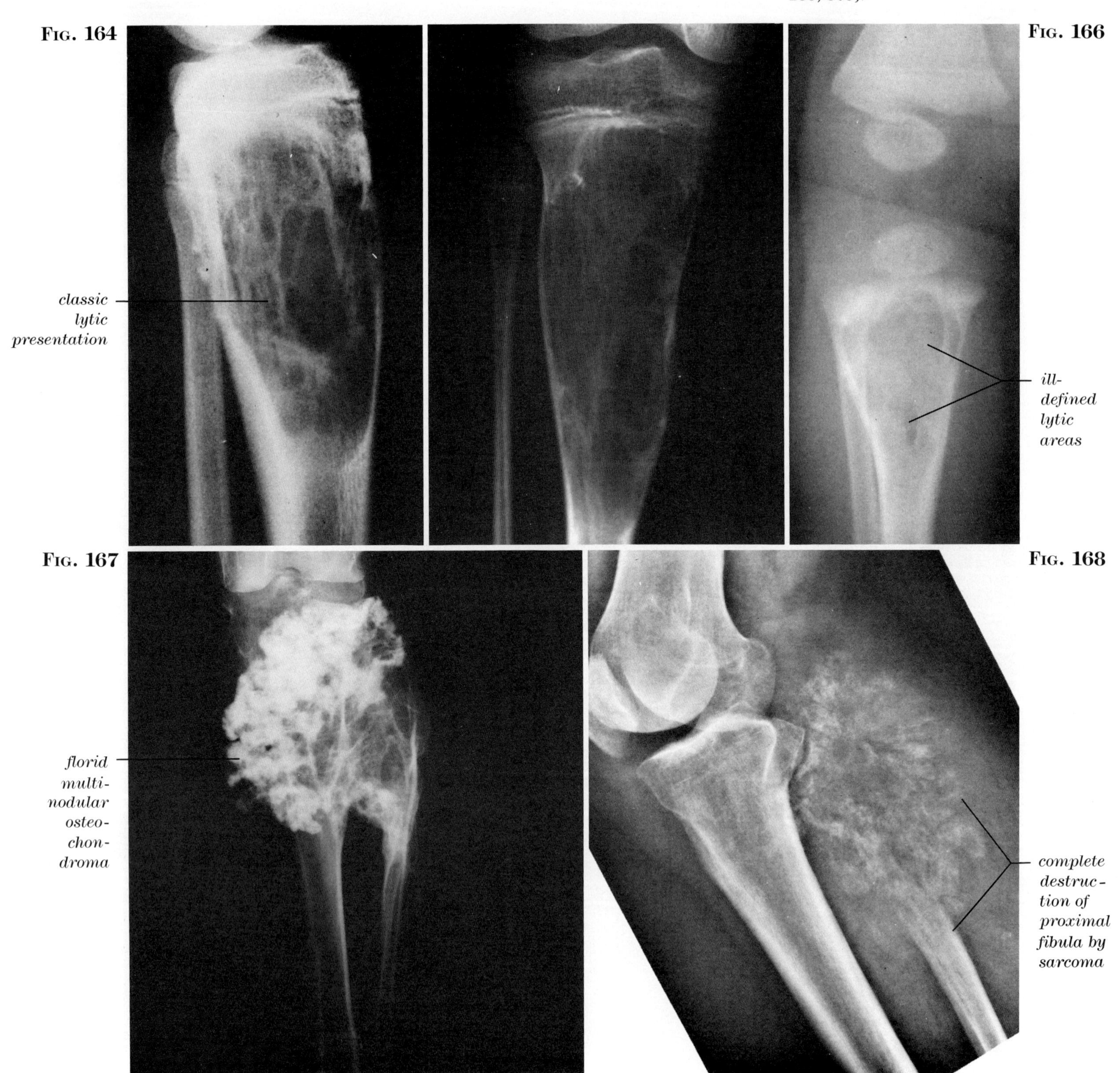

Fig. 169. Malignant lymphoma of tibia (see Figs. 15, 132, 423, 463, 494, 702, 704, 705, 712, 743).

Fig. 170. Fibrosarcoma of the leg of a 16-year-old girl with invasion of the fibula (Case of Dr. J. Ryan, Detroit) (see Figs. 60, 117, 131, 165, 240, 260, 297, 371).

Fig. 171. Adamantinoma of tibia of an adult woman (see Figs. 493, 634).

Fig. 172. Ewing's sarcoma of the right tibia of a child (see Figs. 100, 173, 210, 214, 227, 229, 250, 286, 307).

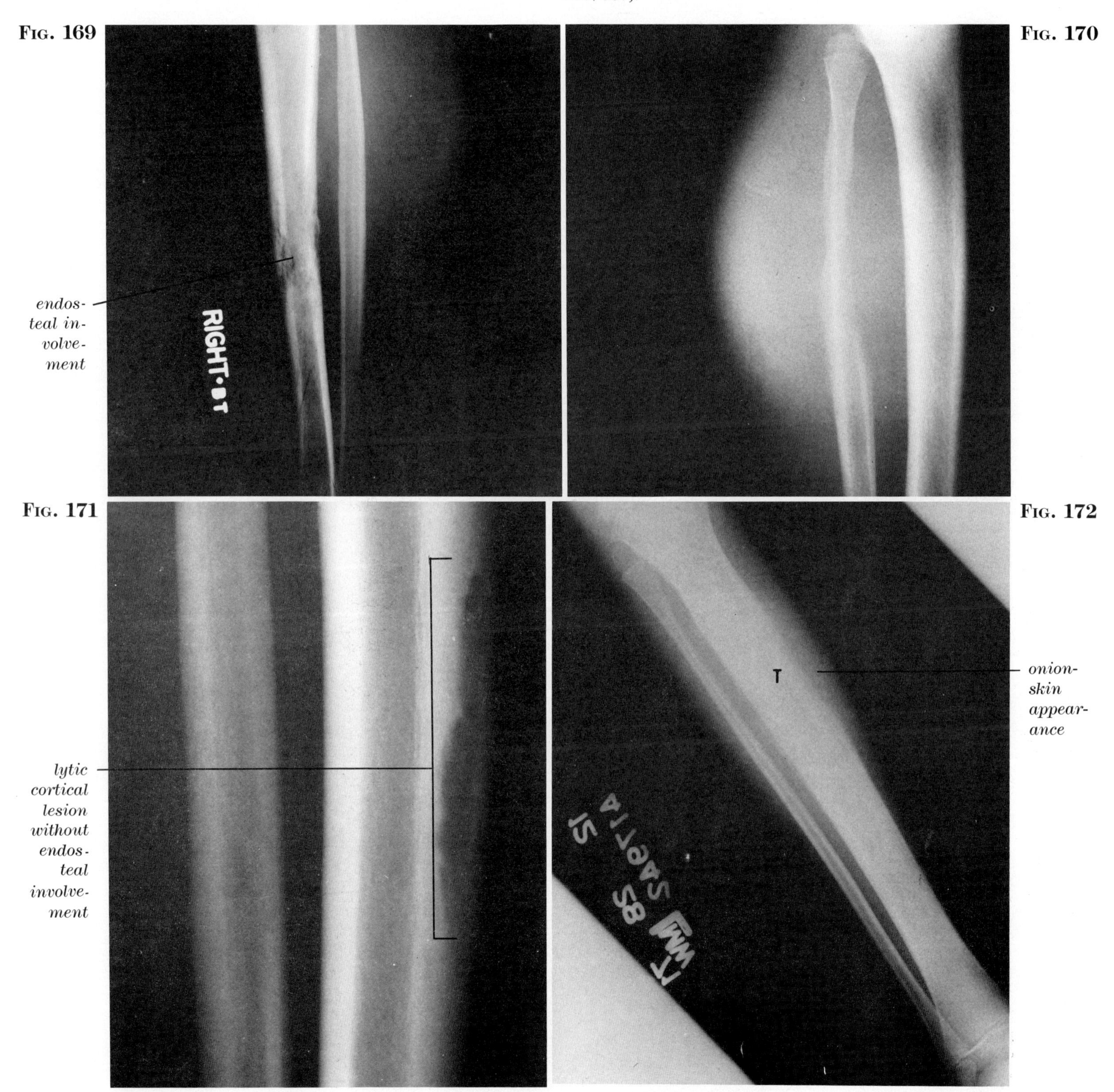

Fig. 173. Ewing's sarcoma of the right distal fibula (see Figs. 100, 172, 210, 214, 227, 229, 250, 286, 307, 314, 330).

Fig. 174. Benign histiocytic fibrous histiocytoma of distal fibula of a 13-year-old, partly growing as nonossifying fibroma (see Figs. 115, 178, 183, 222, 325, 490).

Fig. 175. Recurrent mesenchymal chondrosarcoma of distal tibia (see Figs. 103, 151, 180, 318, 390).

Fig. 176. (Bottom center) Telangiectatic osteosarcoma of distal tibia (see Figs. 108, 162, 168, 221, 225, 228, 233, 328).

Fig. 177. Nonossifying fibroma of proximal fibula of an 11-year-old (see Figs. 112, 631, 852).

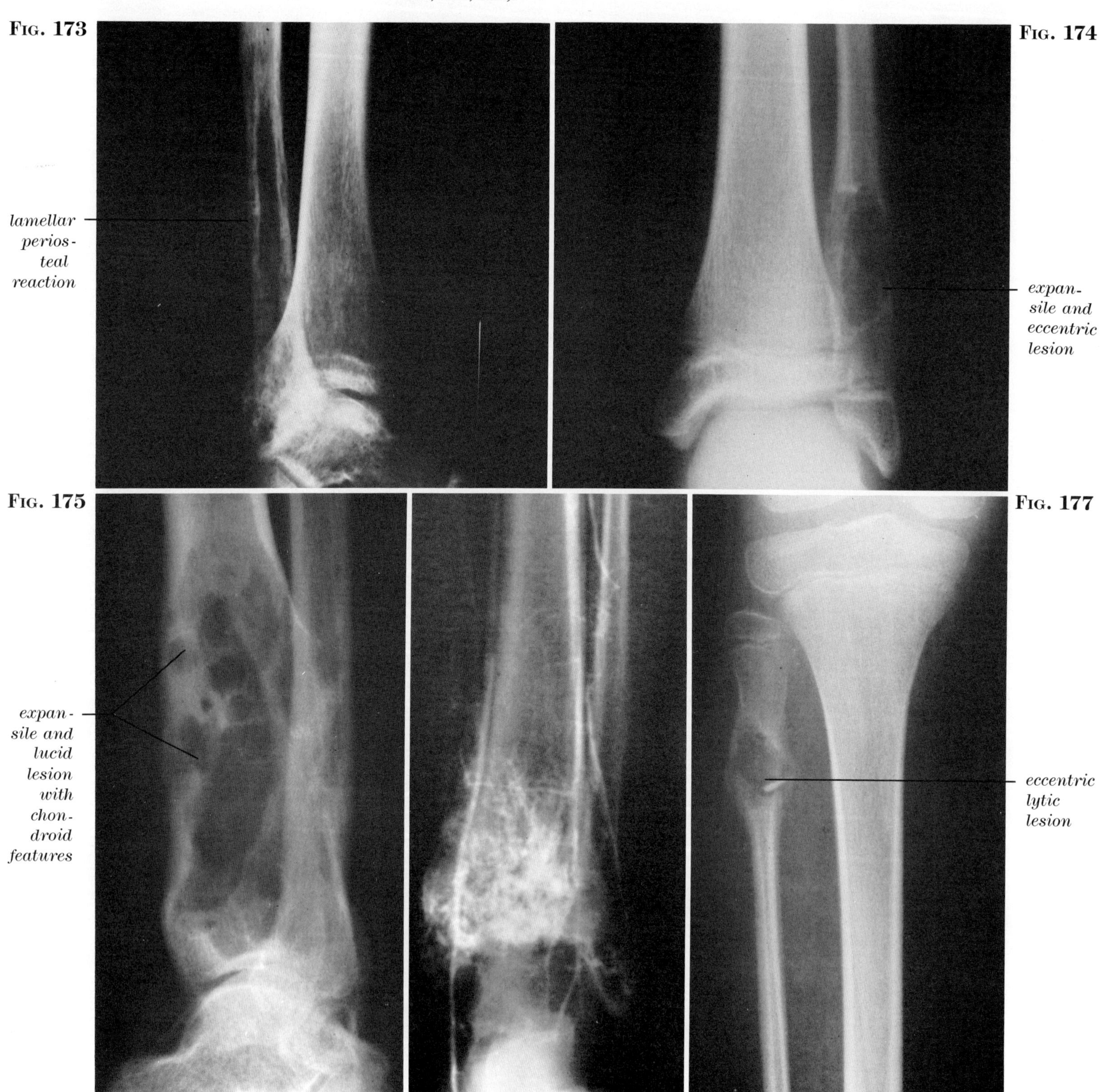

Fig. 178. Benign histiocytic fibrous histiocytoma of tendon sheath with erosion of (A) lateral side of cuboid and (B) calcaneus (see Figs. 71, 174, 183, 222, 325, 490, 501, 774, 778).

Fig. 179. Monophasic tendosynovial sarcoma of foot (see Figs. 85, 190, 315, 320, 326, 391).

Fig. 181. Tendosynovial sarcoma of ankle partly growing as epithelioid sarcoma with erosion of medial aspect of the talus (see Figs. 85, 187, 258, 259, 320, 322).

Fig. 180. Mesenchymal chondrosarcoma of os calcis (see Figs. 103, 151, 175, 318, 390, 444, 507, 595).

Fig. 182. Chondrosarcoma of multiple bones of foot (see Figs. 95, 124, 136, 137, 213, 216, 223).

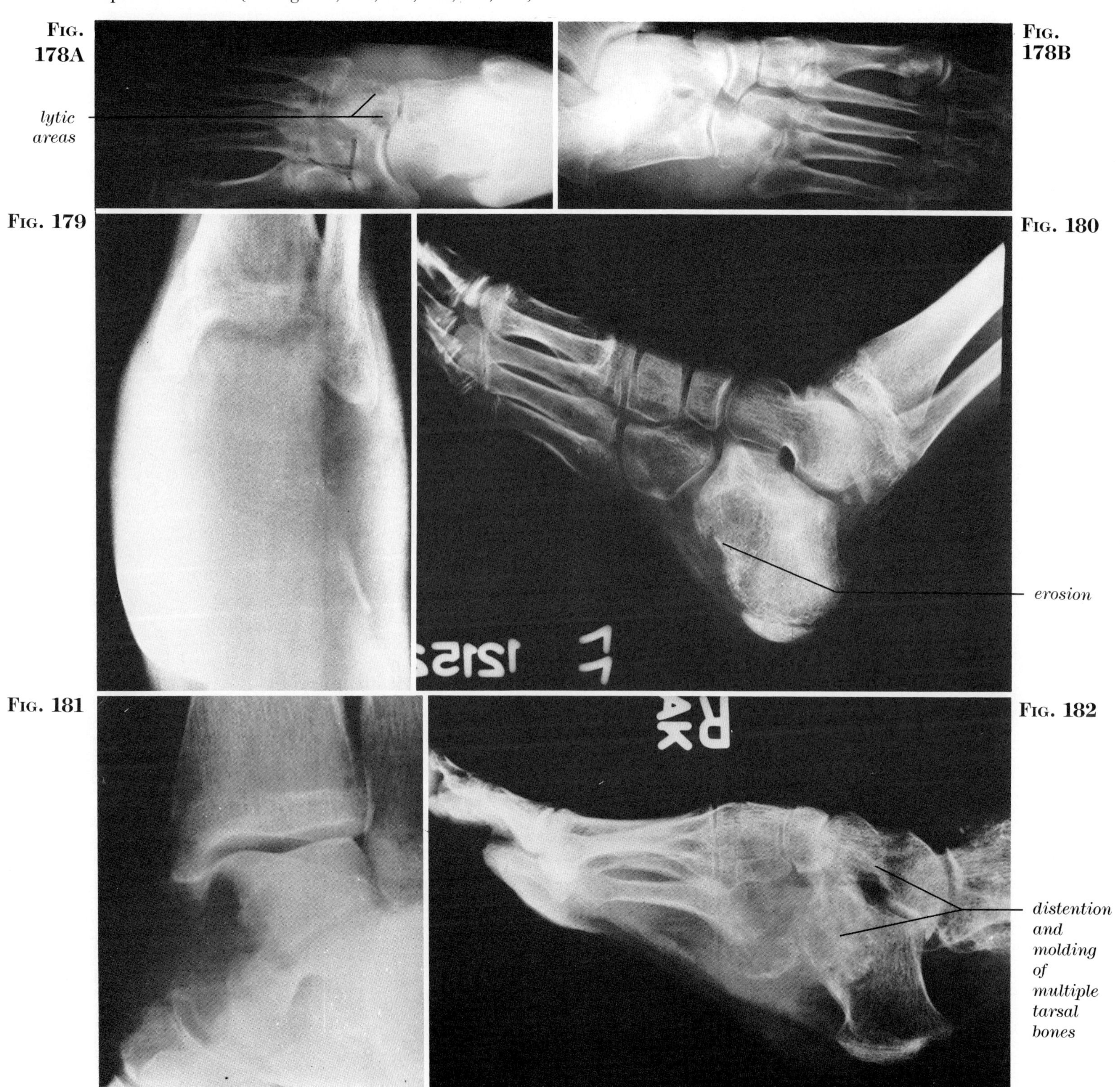

Fig. 183. Benign histiocytic fibrous histiocytoma with erosion of the proximal phalange of the index finger (see Figs. 71, 178, 222, 325, 490, 501, 778).

Fig. 184. (Top center) Osteochondroma of ulna (see Figs. 113, 167, 302, 303).

Fig. 185. Desmoid tumor with invasion of ulna (see Figs. 9, 36, 52, 77, 202, 209, 244, 282).

Fig. 186. Intraosseous synovial cyst of proximal aspect of the metacarpal of the thumb (see Figs. 144, 508, 1014).

Fig. 187. Epithelioid sarcoma invading 4th metacarpal bone (see Figs. 85, 181, 258, 259, 322, 326).

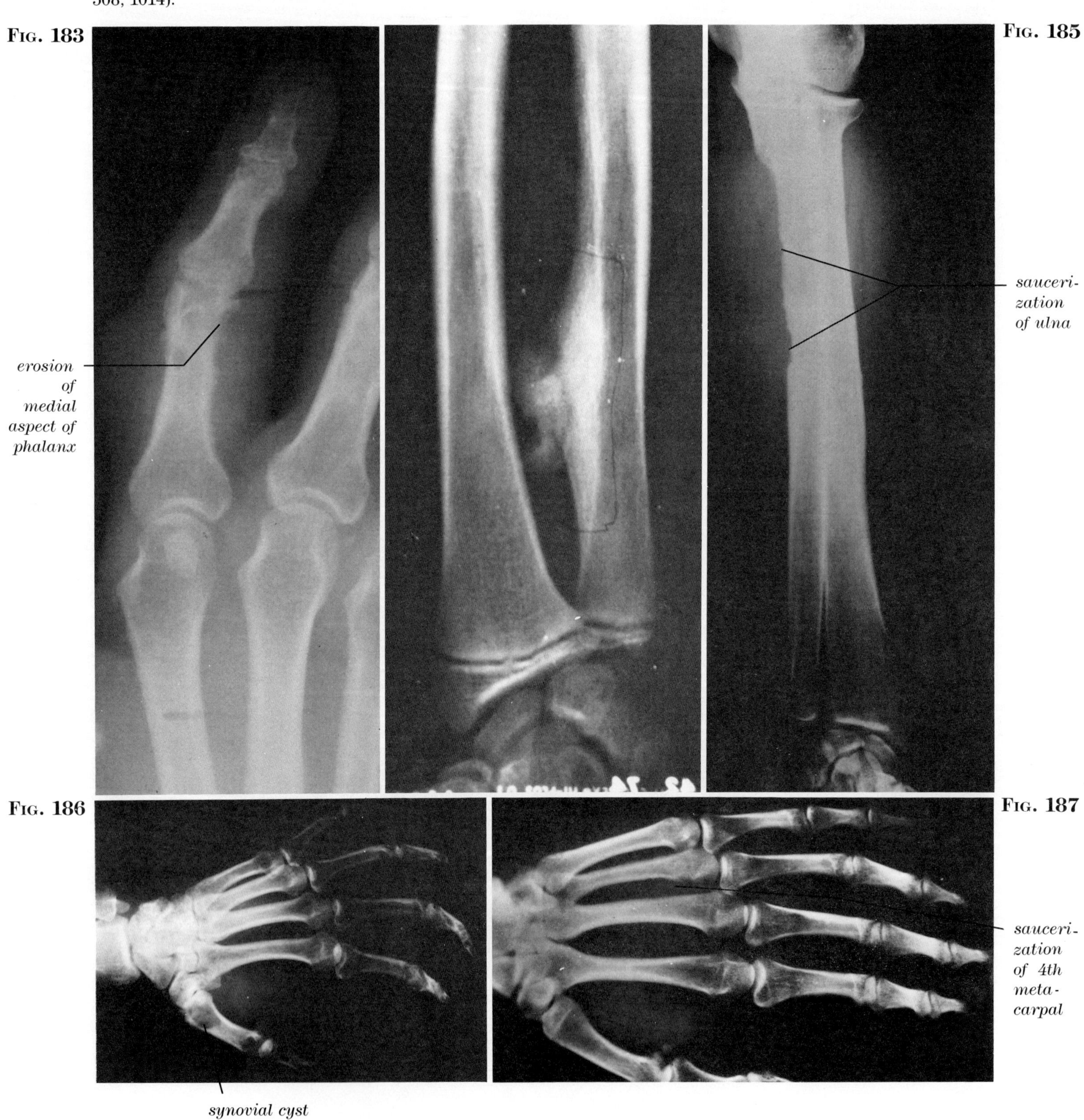

Fig. 188. Myxoid liposarcoma metastatic to the right lung (see Figs. 86, 208, 269, 290, 438).

Fig. 189. Malignant fibrous mesothelioma of the right lung (see Figs. 265, 266, 268, 406, 614).

Fig. 190. Monophasic tendosynovial sarcoma, spindle cell type, metastatic to the lung (see Figs. 179, 315, 326, 391, 392).

Fig. 191. (Bottom center) Multiple enchondromas of ribs (see Fig. 94).

Fig. 192. Plasmacytoma of the sternum (see Figs. 133, 192, 208A, 726).

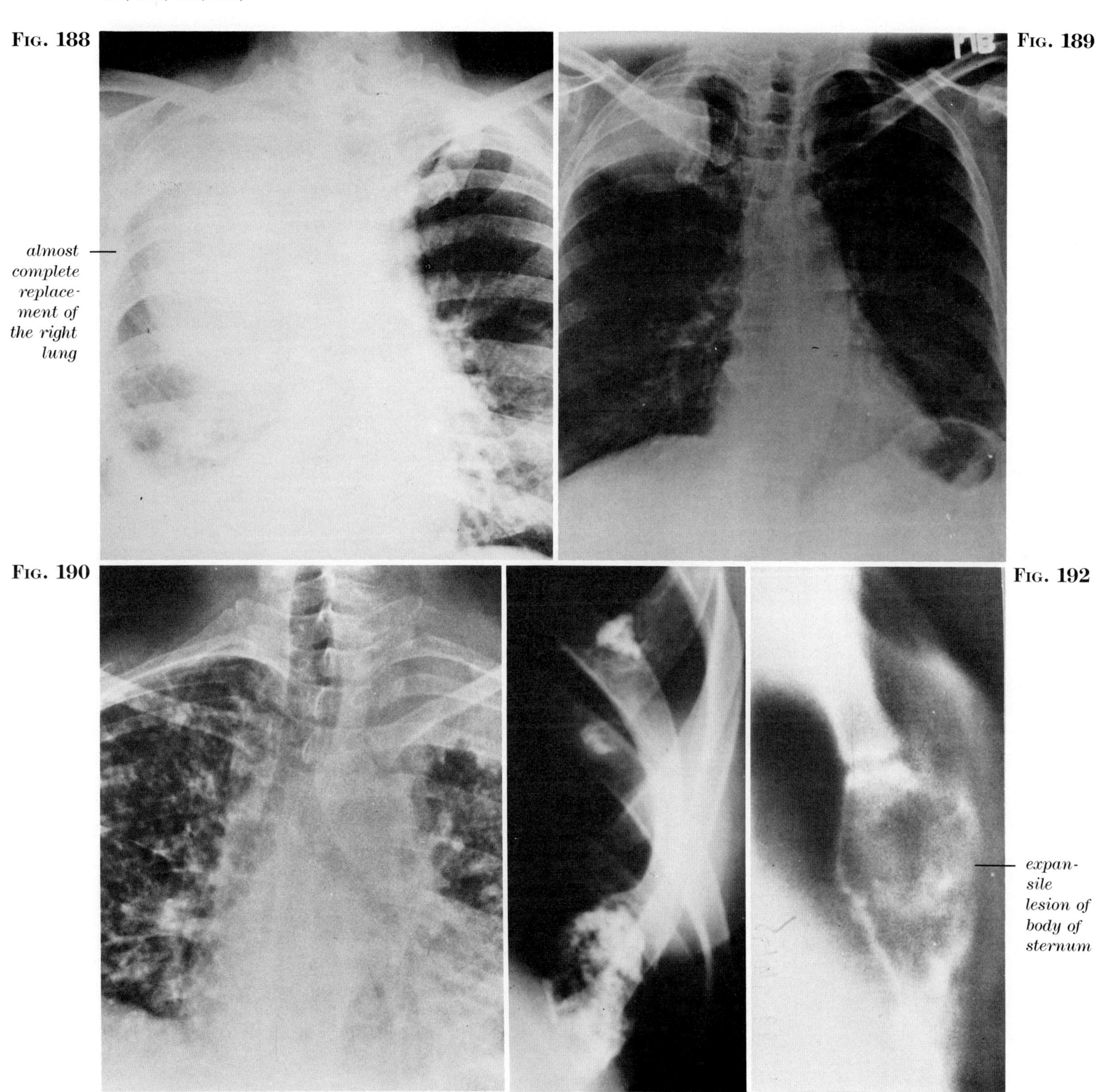

Fig. 193. Leiomyosarcoma involving loops of intestines, mesentery, and retroperitoneum (see Figs. 72, 110, 218, 220, 231, 273).

Fig. 194. Collagenoma of the right retroperitoneum shows almost complete obliteration of blood vessels.

Fig. 195. Extragonadal choriocarcinoma of the left pelvis and retroperitoneum with displacement of the left ureter to the right (see Figs. 829, 1055).

Fig. 196. Leiomyoblastoma of the stomach (case of Dr. E. Hajdu, Long Island Jewish Hospital, New Hyde Park) (see Figs. 270, 274, 277, 278, 420).

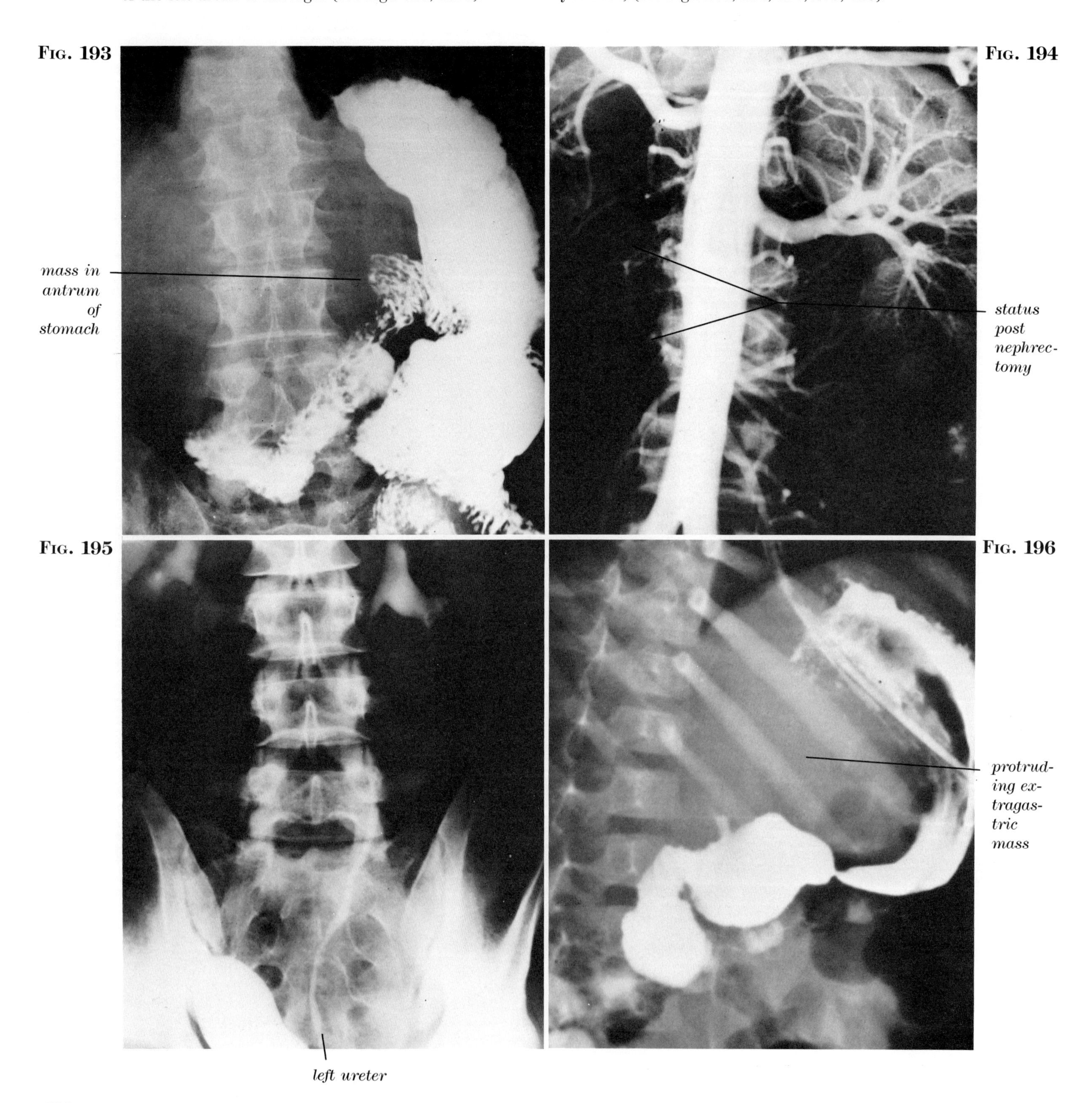

Fig. 197. Arteriovenous malformation of anterior aspect of upper right thigh (Case of Dr. J. Mongeau, Canada) (see Figs. 139, 911, 1006, 1009).

Fig. 199. Liposarcoma infiltrating the adductor muscles of the left thigh (see Figs. 86, 141, 188, 208, 294, 415, 448, 450, 543).

Fig. 201. Malignant fibrous histiocytoma of the left anterior thigh (Case of Dr. S. Kumari, Long Island Jewish Hospital, New Hyde Park, New York) (see Figs. 88, 148, 203, 232, 237, 255, 262, 288, 293).

Fig. 198. Lipoma of the left posterior thigh (Courtesy of Dr. S. Kumari, Long Island Jewish Hospital, New Hyde Park, New York) (see Figs. 73, 243, 455).

Fig. 200. Malignant schwannoma of the posterior aspect of the right thigh (Courtesy of Dr. R. Heelan, Memorial Sloan-Kettering Cancer Center, New York) (see Figs. 80, 238, 242, 246, 252, 298, 355).

Fig. 202. Poorly defined desmoid tumor of the left calf. (see Figs. 77, 185, 209, 244, 282, 376).

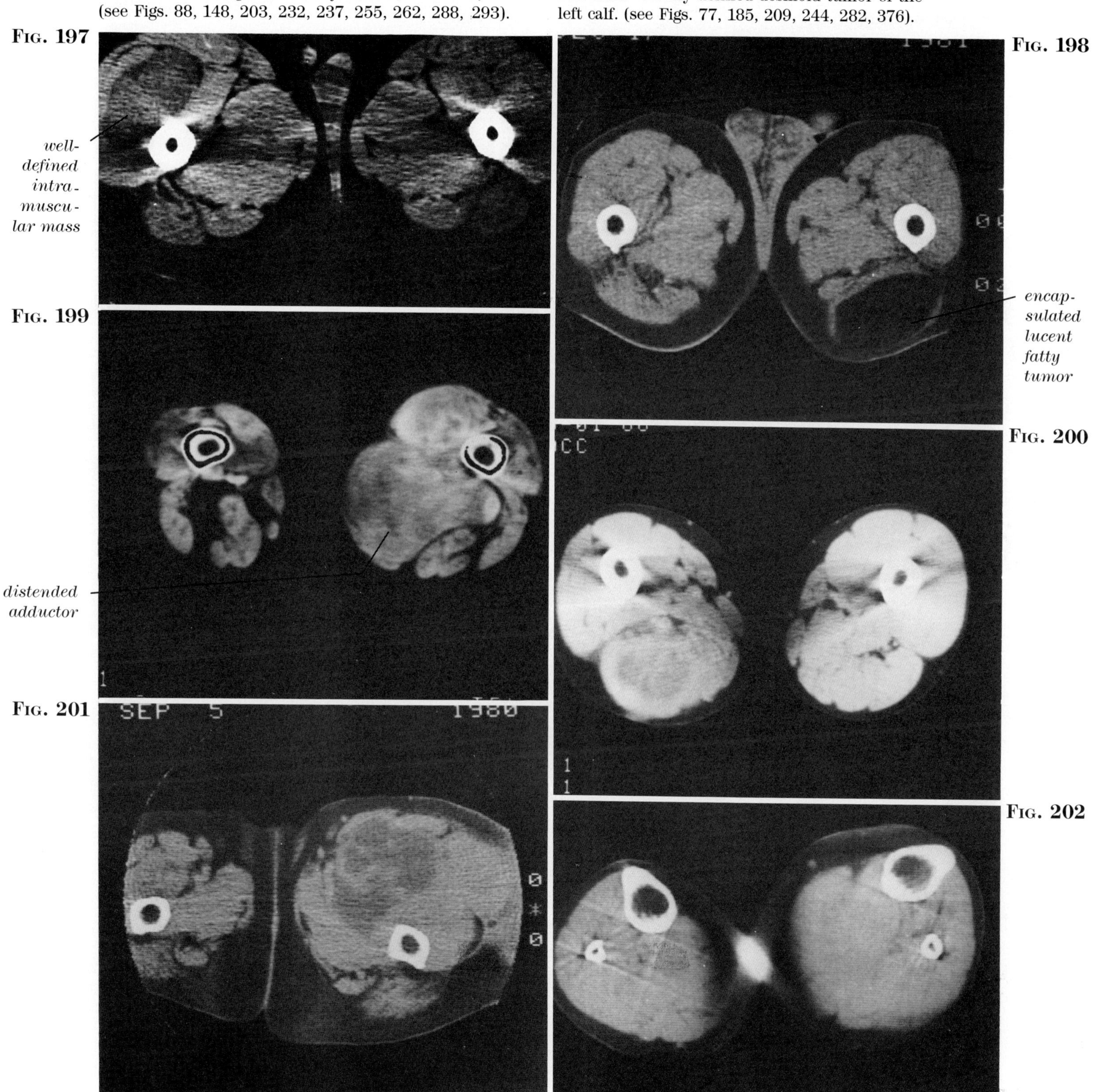

Fig. 203. Post-radiation malignant fibrous histiocytoma of the anterior right chest wall extending into the right axilla and mediastinum (see Figs. 76, 88, 130, 138, 140, 148, 201, 232, 237, 255, 262).

Fig. 204. Neurofibromatosis of the left axilla obliterating the normal neurovascular structures (Courtesy of Dr. R. Heelan, Memorial Sloan-Kettering Cancer Center, New York) (see Figs. 44, 74, 245, 253).

Fig. 206. Embryonal rhabdomyosarcoma metastatic to the mediastinum (see Figs. 78, 120, 207, 226, 263, 281, 291, 364, 372, 378, 379).

Fig. 205A, B. Post-radiation fibrosarcoma of left chest wall with intrathoracic and axillary extension (see Figs. 91, 374, 484).

Fig. 207. Embryonal rhabdomyosarcoma metastatic to the right breast (Courtesy of Dr. R. Heelan, Memorial Sloan-Kettering Cancer Center, New York) (see Figs. 78, 206, 226, 263, 281, 291, 364).

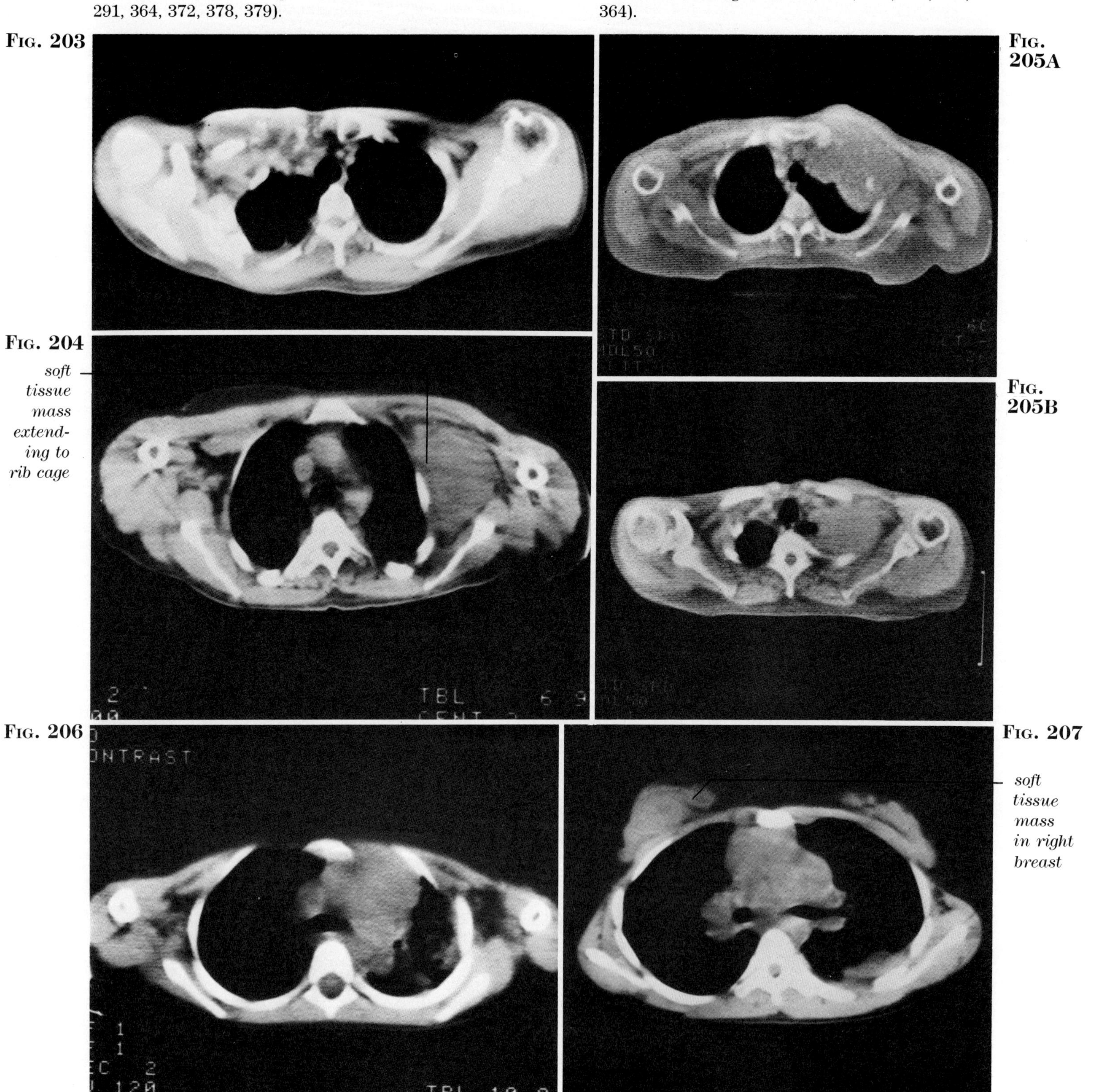

Fig. 208. Myxoid liposarcoma of the anterior abdominal wall (see Figs. 86, 188, 269, 290, 438, 439).

Fig. 208A. Plasma cell myeloma of the left scapula (see Figs. 99, 133, 192, 454).

Fig. 210. Paget's sarcoma of the left ilium (Courtesy of Dr. R. Heelan, Memorial Sloan-Kettering Cancer Center, New York) (see Figs. 107, 134, 918, 1103, 1128).

Fig. 209. Recurrent pelvic desmoid tumor with invasion of the right iliac bone (see Figs. 77, 185, 202, 244, 282, 376, 386).

Fig. 211. Atypical osteoblastoma of the sacrum (see Figs. 97, 234, 1092, 1155).

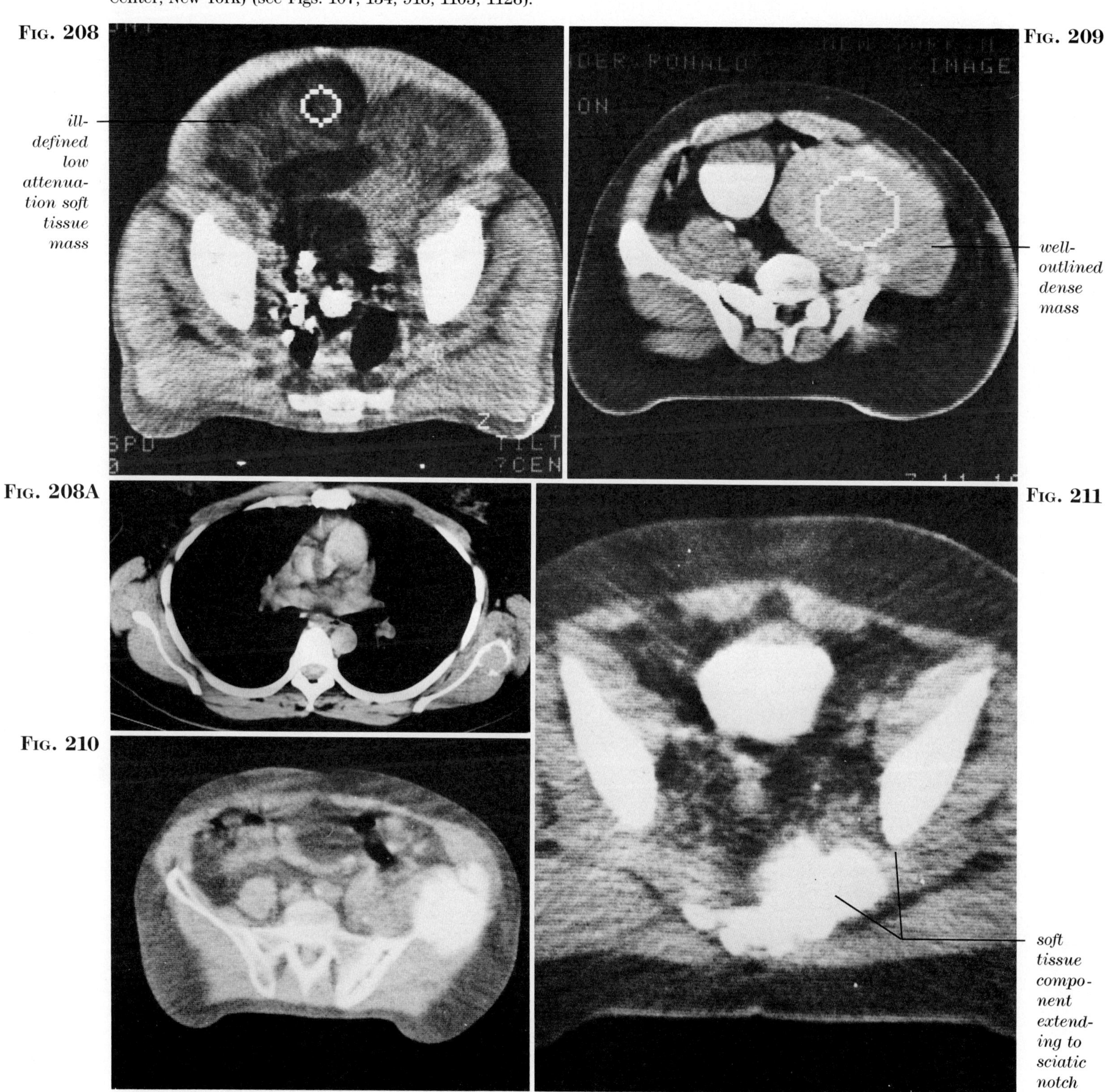

Fig. 212. Osteoblastic metastasis of prostatic carcinoma to right ilium (Courtesy of Dr. J. Botet, Memorial Sloan-Kettering Cancer Center, New York) (see Fig. 1272).

Fig. 214. Ewing's sarcoma of the right ilium and sacrum (see Figs. 100, 172, 173, 210, 227, 229, 250, 286, 307, 314, 330).

Fig. 216. Low grade chondrosarcoma arising in osteochondroma of the right ilium (Courtesy of Dr. J. Botet, Memorial Sloan-Kettering Cancer Center, New York) (see Figs. 95, 136, 137, 182, 213, 223, 285, 308, 327, 329).

Fig. 213. Chondrosarcoma of rib (see Figs. 95, 137, 182, 216, 223, 285, 308, 327, 329).

Fig. 215. Chordoma of the sacrum (Courtesy of Dr. J. Botet, Memorial Sloan-Kettering Cancer Center, New York) (see Figs. 69, 310, 505, 729).

Fig. 217. Parosteal osteosarcoma of the left humerus (see Figs. 109, 129, 224, 249, 287, 353, 354, 656).

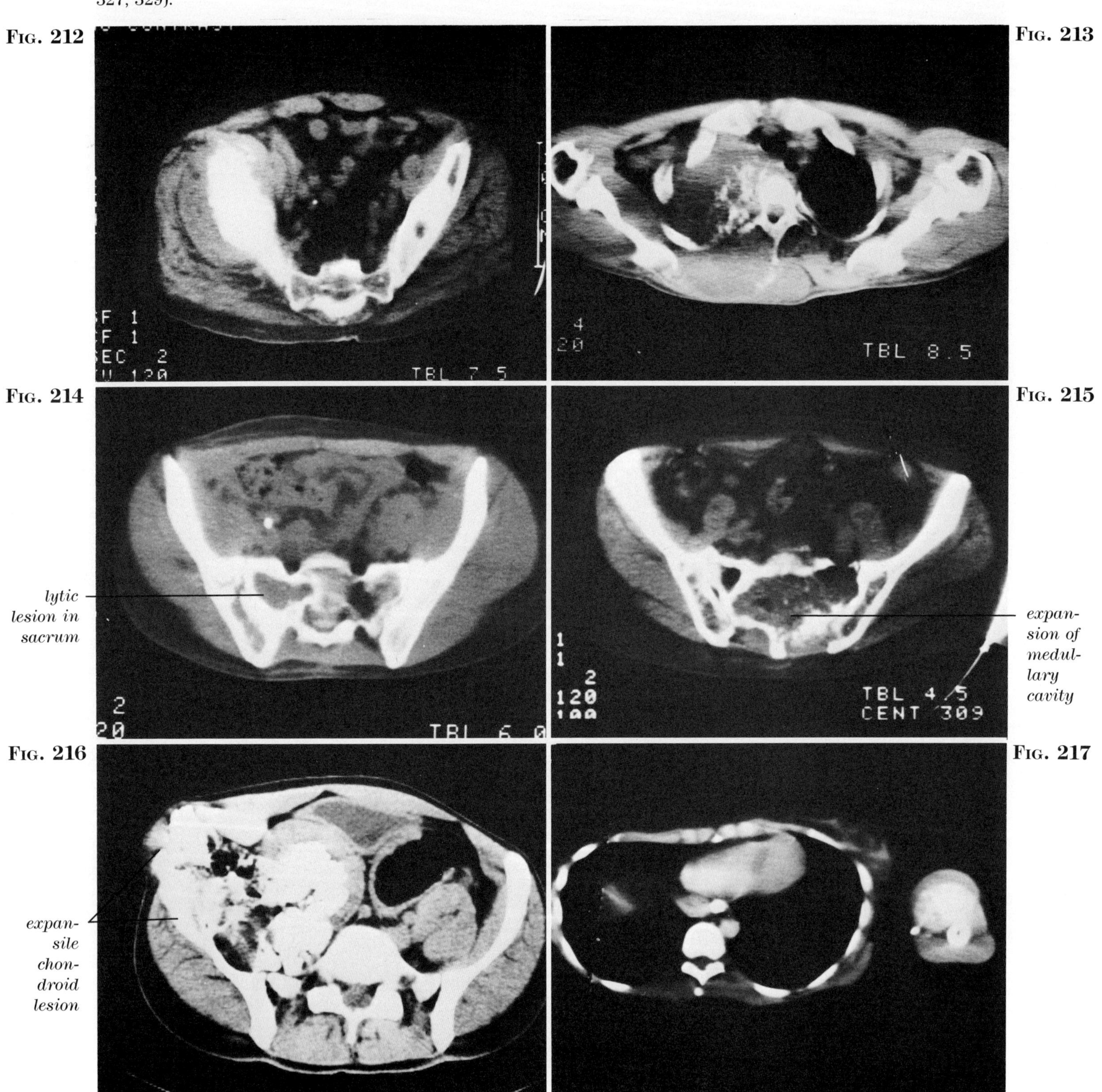

Fig. 218. Low grade uterine leiomyosarcoma metastatic to the sternum 2 years status post hysterectomy (see Figs. 72, 193, 196, 220, 231, 273).

Fig. 219. Well-differentiated liposarcoma of right distal thigh (Case of Dr. A. Jacobson, San Antonio, Texas) (see Figs. 58, 86, 431, 435, 822, 999).

Fig. 220. Leiomyosarcoma of the right diaphragm (see Figs. 72, 218, 231, 273, 341, 347).

Fig. 221. Osteosarcoma of the right ilium (Courtesy of Dr. J. Lane, Memorial Sloan-Kettering Cancer Center, New York) (see Figs. 108, 142, 149, 158, 164, 176, 225, 233).

Fig. 222. Benign histiocytic fibrous histiocytoma (benign giant cell tumor) of the right proximal humerus (see Figs. 71, 115, 174, 178, 183, 222, 325).

Fig. 223. Low grade chondrosarcoma of the left ilium (see Figs. 95, 124, 136, 213, 216, 285, 308, 327, 329).

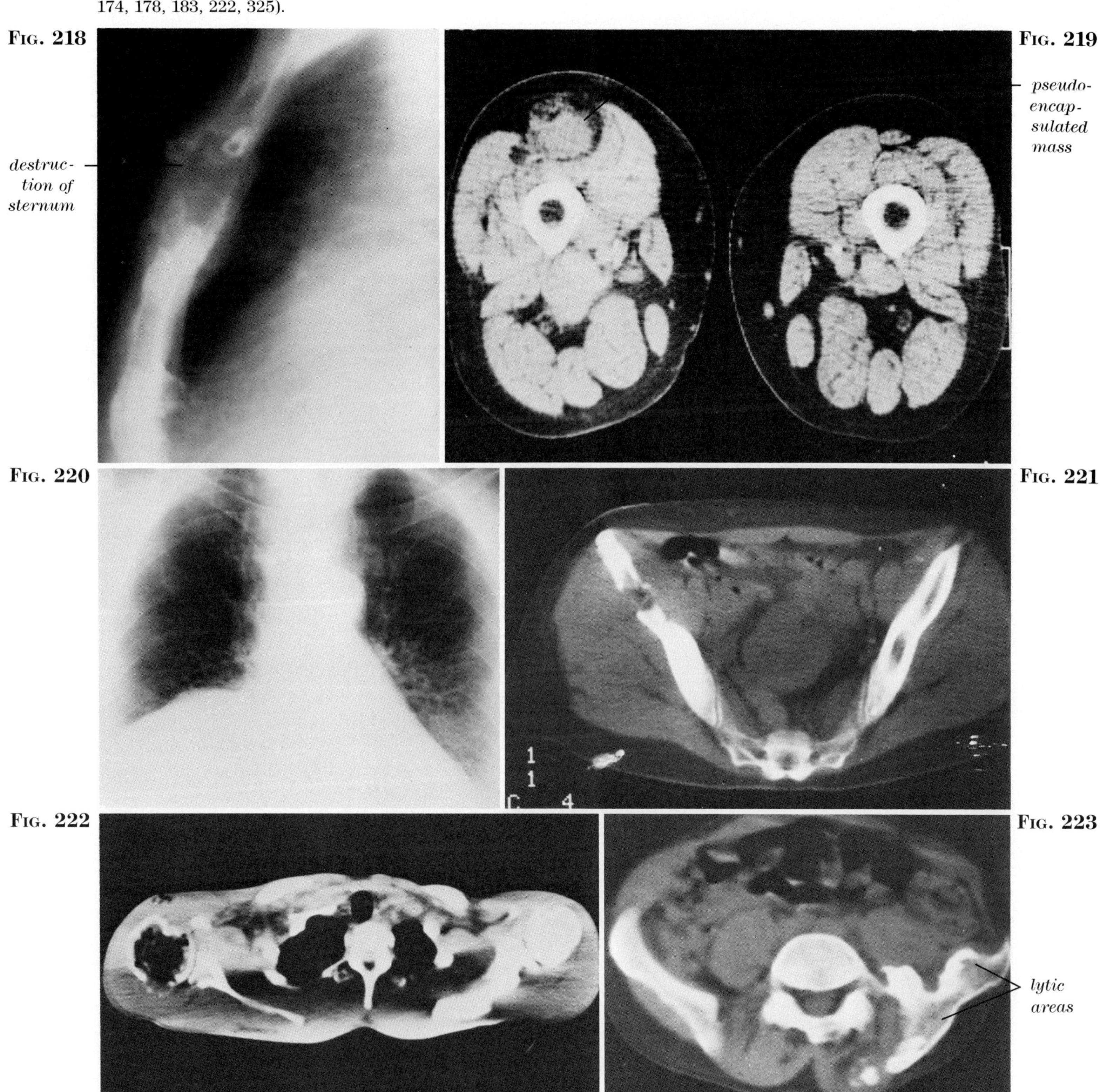

Fig. 224A–C. Parosteal osteosarcoma of distal femur of 27-year-old woman (Case of Dr. Davies, Johannesburg, South Africa) (see Figs. 109, 129, 217, 224, 249, 287, 353, 354).

Fig. 225. Osteosarcoma of distal metaphysis of the left femur (see Figs. 108, 127, 146, 157, 162, 176, 221, 228, 304, 312).

Fig. 226. Embryonal rhabdomyosarcoma of the left leg of a 2-year-old after multiple biopsy attempts (see Figs. 78, 120, 206, 263, 281, 291).

Fig. 227. Ewing's sarcoma of the right midfibula (see Figs. 100, 172, 173, 210, 214, 229, 307, 314).

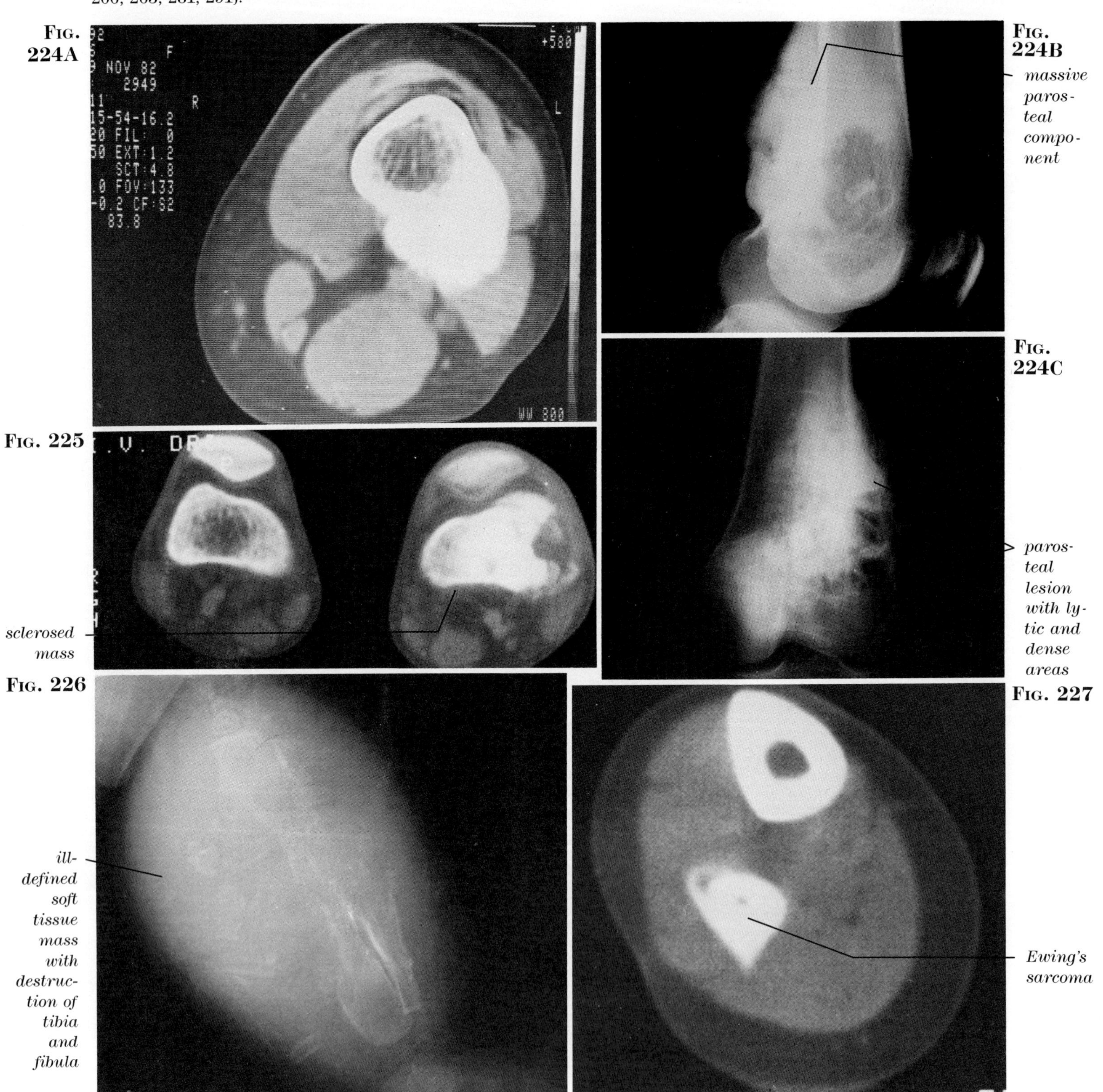

Fig. 228. Bone scan with ^{18}F shows abnormal uptake in osteosarcoma of the right distal femur (Courtesy of Dr. S. Yeh, Memorial Sloan-Kettering Cancer Center, New York) (see Figs. 127, 142, 149, 158, 162, 176, 221, 225, 230, 304).

Fig. 229. (Top center) Bone scan with Technetium shows abnormal uptake in Ewing's sarcoma of the left distal fibula (see Figs. 173, 210, 214, 227, 250, 314).

Fig. 230. Bone scan with Technetium shows abnormal uptake in osteosarcoma of the left distal femur (see Figs. 149, 158, 176, 221, 225, 228, 233, 304).

Fig. 231. (Center) Leiomyosarcoma metastatic to the liver is demonstrated with Technetium (Courtesy of Dr. S. Yeh, Memorial Sloan-Kettering Cancer Center, New York) (see Figs. 220, 273, 341, 347, 360).

Fig. 232. Bone scan of malignant fibrous histiocytoma of the left distal femur (see Figs. 203, 237, 255, 262, 288, 293, 301).

Fig. 233. (Bottom center) Bone scan of telangiectatic osteosarcoma of left proximal tibia (see Figs. 149, 158, 164, 176, 228, 230, 248, 328).

Fig. 234A. Bone scan shows abnormal uptake in atypical osteoblastoma of the sacrum (see Figs. 97, 211, 1092, 1155).

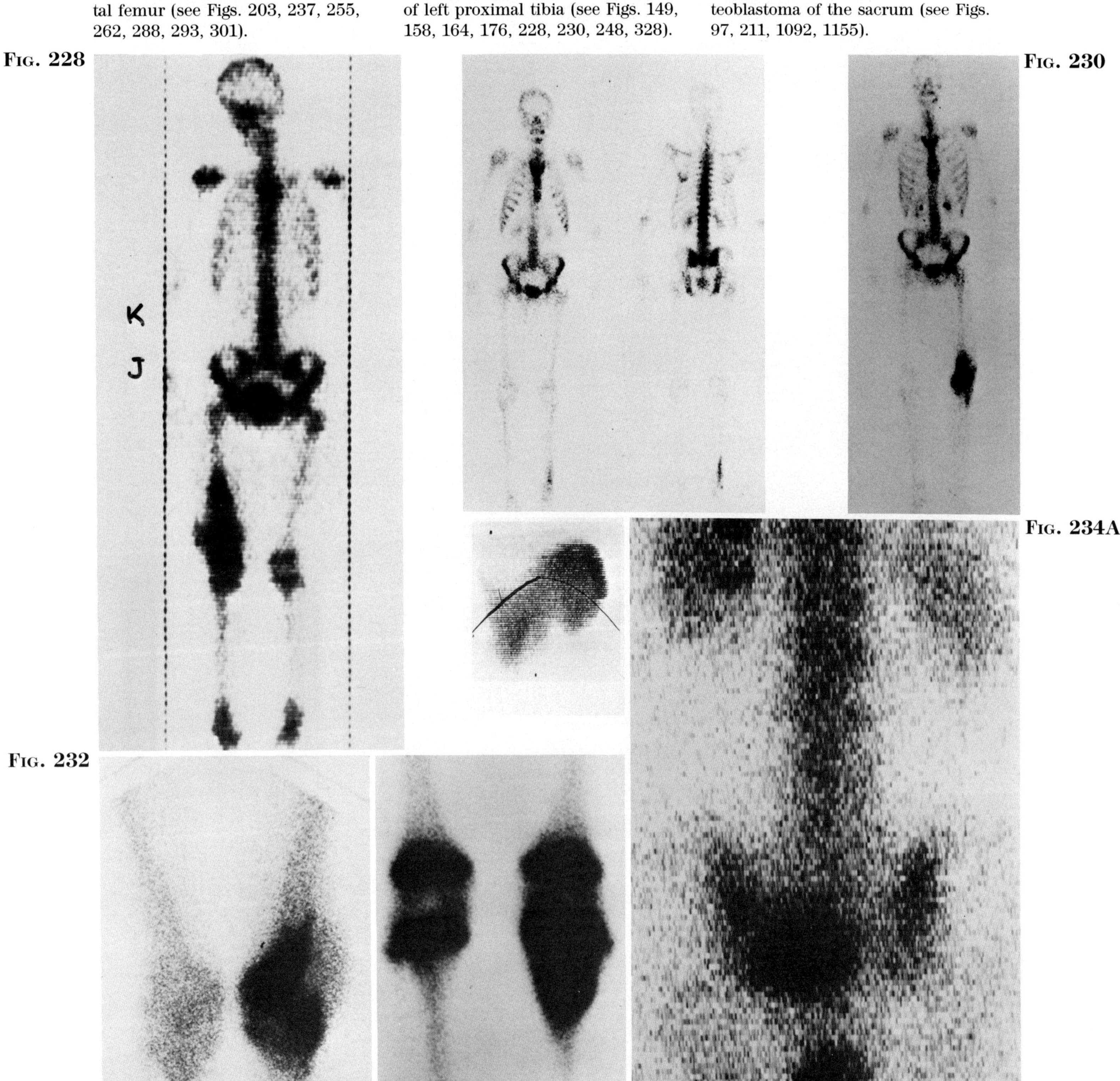

Fig. 234B and C. Magnetic resonance image of lumbo-sacral Paget's sarcoma predominantly growing as malignant histiocytic fibrous histiocytoma (Courtesy of J. Botet, M.D., Memorial Sloan-Kettering Cancer Center) (see Figs. 107, 134, 918, 1103, 1128).

Fig. 234B

Fig. 234C

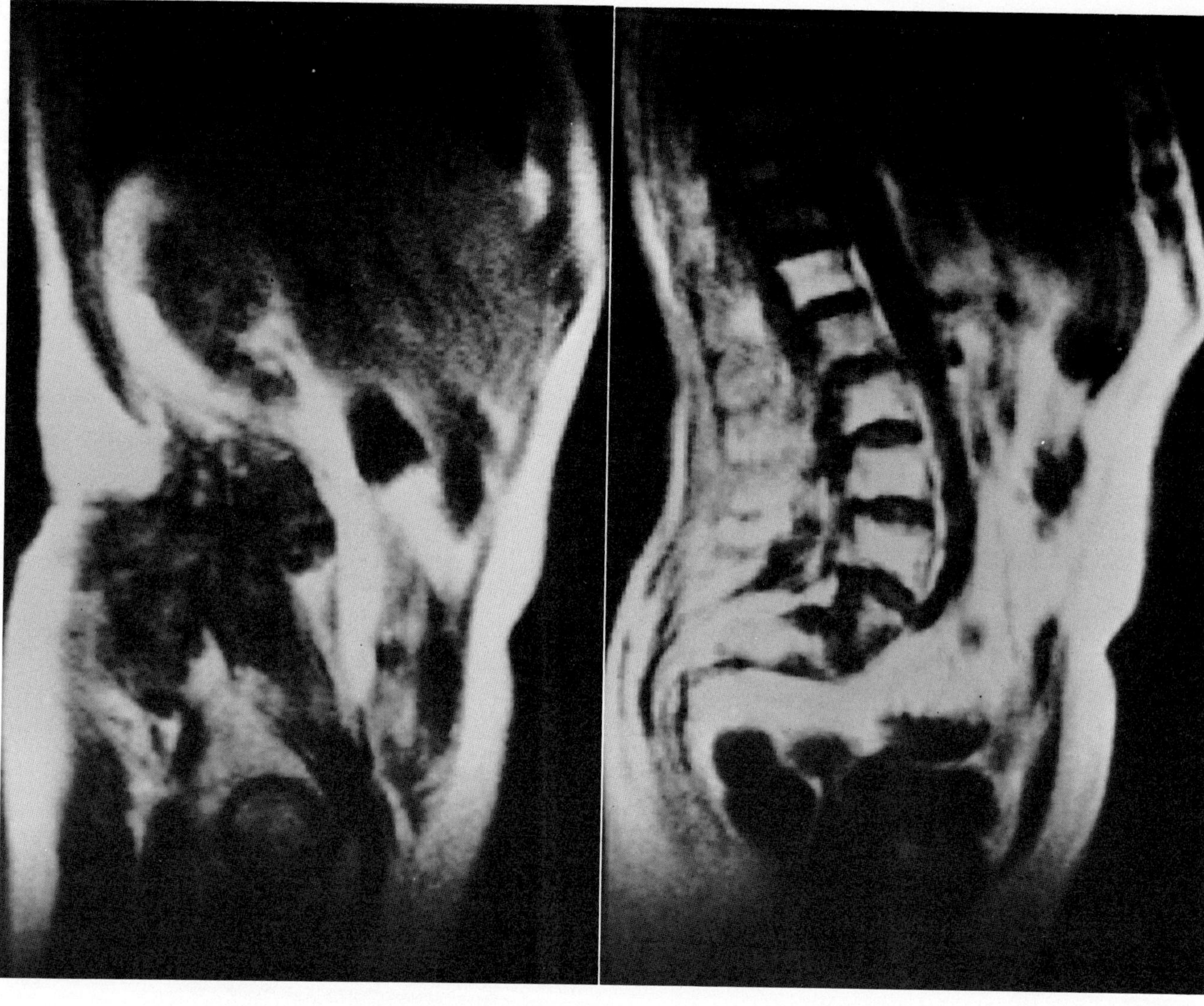

"It is the high function of the pathologist not merely to attach correct labels to lesions, but to reconstruct the course of events from the earliest inception of disease to the final moment of life"
WILLIAM BOYD (1885–1979)

8. Dissection of Specimen

THE bone marrow is the only organ that has been historically studied more extensively in smears than in tissue sections. Although smears prepared from aspirates can occasionally be substituted under specific conditions for tissue biopsy (see Appendix), in most instances if soft tissue or bone tumors are suspected formal incisional or excisional tissue biopsies are warranted.

Incisional biopsies or excisional biopsies less than 5 cm should be sectioned and blocked in toto for microscopic examination. It is advisable to place small and representative sections in special fixatives for cell marker and other studies (for example, electron microscopy). Excisional biopsy specimens should be dissected with consideration of anatomic structures (for example, nerves, vessels, tendons, skin, bone, etc.), and the size, color, and consistency of the tumor and whether the excisional margins are free of tumor should be recorded. As much soft tissue as possible must be removed from incisional bone biopsies and bone curettings for microscopic examination. The bone should be decalcified and processed separately.

Major surgical specimens resulting from procedures such as en bloc resection or amputation should be inspected for the presence of scars, swellings, ulcers, fractures, and for suture markers left in by the surgeon. If possible, the specimens should then be photographed. Each specimen should be dissected and described according to anatomic landmarks, and the relation of the tumor to skin, major vessels, nerves, soft tissues, periosteum, or bone should be recorded. Whether the tumor has involved or infiltrated any of these structures, and whether the tumor is involving one or multiple muscle groups should be recorded. The three dimensions of the tumor must be recorded, and the cut surface of the tumor should be described, including contour, consistency, and color and extent of infiltration of surrounding structures, with special attention given to perivascular and perineural soft parts. Major blood vessels should be opened, and fragments of blood clots removed from veins should be submitted for microscopic examination. In case of bone tumors, the specimen should be dissected according to anatomic landmarks, and the bone should be cut with a saw. The description must state the size, color, consistency, and location of the tumor in the bone, and whether it extends along the periosteum, medullary cavity, and joint space. In both soft tissue and bone tumors, areas proximal and distal to the tumor should be dissected in order to search for satellite or metastatic tumor nodules or skip areas. The distance of the edges of the tumor from the resection margins must be measured.

There is no rule as to the number of sections the pathologist must submit for microscopic examination, but it is important that the tumor is generously sampled and the sections taken from representative areas. Sections of the tumor, pertinent resection margins (including bone margins), and all lymph nodes proximal to the tumor must be studied by microscopic examination. It is advisable to place a representative sample of the specimen in special fixative for ultrastructural examination, and it is perhaps wise to keep another sample fresh for special histochemical and immunochemical studies.

The quality of the result of microscopic examination is directly proportional to the quality of the material that is submitted for the study. Poorly preserved and inappropriately processed tissue is of no value. At the conclusion of the studies, the pathologist must be prepared to answer questions that go beyond diagnostic labelling, including the grade, stage, and overall potential behavior of the tumor, and whether the surgical margins were free of tumor. The pathologist's answers to these questions will influence or decide the clinicians' therapeutic strategy. Drawings or photographs taken during dissection of the specimen are an absolute necessity for permanent recording and for unequivocal answers to these and other questions that may arise at a later time.

"We call sarcoma fleshy excrescence"
GALEN (A.D. 130–200)

9. Gross Appearance

Non-neoplastic Lesions (Table 45; Figures 235, 280)

Benign Neoplasms (Table 46; Figures 243, 245, 247, 253, 257, 261, 302, 303, 325)

Malignant Neoplasms (Table 47; Figures 236–242, 244, 246, 248–252, 254–256, 258–260, 262–279, 281–301, 304–324, 326–330)

THE gross appearance of soft tissue and bone tumors, in the majority of cases, is unreliable and not helpful in diagnosis. The difficulty in distinguishing tumors that are benign from those that are malignant is partly due to their similarity in composition of cellular elements and deposit of intercellular matrix. One needs only to consider the gross appearance of fibromatoses, desmoid tumor, and fibrosarcoma, or the resemblance of leiomyoma to leiomyosarcoma. The gross appearance of callus and osteosarcoma or giant cell tumor and telangiectatic osteosarcoma may look alike. Hemorrhage and cystic and myxomatous changes can be seen in large benign tumors, as well as malignant neoplasms. The gross similarity of lipomas to liposarcomas or schwannomas to tendosynovial sarcomas is so striking that even the most experienced pathologist can be fooled—every pathologist is familiar with the dangerously misleading gross appearance of benign granular cell tumor of the breast.

Intraskeletal and extraskeletal benign and malignant neoplasms grow by expansion and infiltration of surrounding tissues. The expansile growth leads to compression of adjacent tissues and organs, resulting in pseudoencapsulation, which often gives a harmless, self-limited gross appearance to some of the high grade malignant neoplasms such as pleomorphic fibrous histiocytoma, monophasic tendosynovial sarcoma, pleomorphic rhabdomyosarcoma, and poorly differentiated chondrosarcoma.

Infiltrative edge or poor demarcation, on the other hand, cannot be taken at gross cutting as a sure sign of malignancy. Fasciitis, panniculitis, fibromatosis, desmoplastic fibroma, and nonossifying fibroma may appear to be more infiltrative than most sarcomas. Although the gross appearance of connective tissue tumors cannot be taken for granted, an experienced pathologist may be able to discern among a spectrum of nonspecific gross features a few that can be taken as suggestive features. A list of such suggestive features is given in Tables 45, 46, and 47, and illustrated in Figures 235 to 330.

Table 45. *Gross Appearance of Non-neoplastic Lesions*

Aneurysmal bone cyst: Sponge-like fragments, dark reddish with cysts containing clear fluid
Giant cell granuloma: Sponge-like, brownish-gray
Hyperparathyroidism: Sponge-like, reddish with sclerotic bands
Xanthogranuloma: Yellow, well-outlined, cutaneous nodule that may protrude into subcutaneous tissue
Fasciitis: Gray-white, semi-firm with infiltrative edges
Keloid: Ill-defined, firm, and scar-like
Fibromatosis: Poorly outlined, firm, nodular, and grayish
Fibrous dysplasia: Gritty with cystic areas
Elastofibroma: Nonencapsulated and lobulated semisolid mass
Tendosynovitis: Brownish gray, granular, and cystic with villous projections
Tendosynovial cyst: Cystic with sclerosing wall and viscous fluid in lumen
Synovial chondromatosis: Villous and cystic with dense chondroid, "loose" bodies
Gouty arthritis: Gritty, grayish-yellow, and necrotic
Fat necrosis: Pseudoencapsulated, semisolid, grayish-yellow, and lobular
Lipogranuloma: Well-outlined, yellowish, and cheesy
Proliferative panniculitis: Yellowish semisolid, ill-defined with grayish fibrous trabeculae
Proliferative myositis: Yellowish-gray and ill-defined mass in skeletal muscle
Myositis ossificans: Well-delineated, semi-firm, reddish gray mass with bony trabeculae
Pyogenic granuloma: Reddish-brown, mucosal or cutaneous nodule
Angioblastic lymphoid hyperplasia: Reddish-brown, single or multiple cutaneous nodules
Arteriovenous malformation: Hemorrhagic and solid with Swiss cheese appearance
Traumatic neuroma: Semi-firm micronodules in soft tissues
Eosinophilic granuloma: Gritty, reddish-gray, and amorphous
Amyloidoma: Semisolid, grayish, and cheesy

Table 46. *Gross Appearance of Benign Neoplasms*

Benign fibroblastic fibrous histiocytoma: Semi-firm and whitish yellow cutaneous nodule or deeper mass
Benign histiocytic fibrous histiocytoma (Benign giant cell tumor): Well-demarcated cystic or solid trabecular and friable mass of yellowish-gray or reddish-gray color
Benign pleomorphic fibrous histiocytoma: Yellowish-tan, ill-defined, soft cutaneous nodule of soft tissues or bone
Nonossifying fibroma: Soft and yellow with grayish solid areas
Desmoplastic fibroma: Well-outlined, globoid, dense and gray tumor with irregular edges
Lipoma: Encapsulated lobular yellow mass from soft to fibrous consistency
Angiomyolipoma: Pseudoencapsulated, semisolid, yellowish-gray mass with soft and cystic areas
Leiomyoma: Globoid and encapsulated yellowish-gray firm mass with fasciculated cut surface
Hemangioma: Ill-defined hemorrhagic, cystic, and trabecular nodule, often subcutaneous
Benign glomus tumor: Ill-defined grayish-red nodule with minute cystic areas
Neurofibroma: Encapsulated, oval, and semisolid or firm mass with fascicular cut surface
Benign schwannoma: Encapsulated, oval mass with cystic areas, arising from nerve
Benign undifferentiated peripheral nerve tumor: Encapsulated and fusiform, from soft to semisolid consistency, often arising from a major nerve
Benign mesothelioma: Solid, grayish-yellow, solitary mesothelial nodule
Chondroblastoma: Ill-defined, grayish-pink, with hemorrhagic areas from mucoid to firm consistency
Chondroma: Well-outlined, whitish glistening nodule or mass with gritty consistency
Osteochondroma: Cartilage-capped, well-outlined bony protuberance
Osteoblastoma: Well-demarcated hemorrhagic and gritty nodule, 2 cm or larger
Osteoid osteoma: Firm bony nodule, less than 2 cm, with reddish-white central nidus
Benign thymoma: Encapsulated, reddish-brown semisolid mass
Benign granular cell tumor: Semi-firm, subcutaneous or submucosal, ill-defined yellowish-gray nodule
Benign cystosarcoma phyllodes: Multilobular, yellowish-gray, well-defined intramammary mass

Table 47. *Gross Appearance of Malignant Neoplasms*

Malignant fibroblastic fibrous histiocytoma: Semi-firm, homogenous grayish tumor with stellate fascicles and irregular edges
Malignant histiocytic fibrous histiocytoma (Malignant giant cell tumor): Semi-firm, grayish tumor with irregular or infiltrative edges
Malignant pleomorphic fibrous histiocytoma: Solid, yellowish-gray, ill-defined tumor with hemorrhagic or cystic areas
Desmoid tumor: Yellowish-white tumor with scar-like consistency and infiltrative edges
Fibrosarcoma: Ill-defined, firm, and homogenous tumor
Tendosynovial sarcoma: Pseudoencapsulated, glistening yellowish-gray, semisolid mass, solitary or multiple, with cystic and hemorrhagic areas
Liposarcoma: Pseudoencapsulated, glistening yellow, and nodular-containing cysts with consistency ranging from soft to firm
Leiomyosarcoma: Pseudoencapsulated, solid or firm yellowish-gray tumor with trabecular cut-surface
Leiomyoblastoma: Pseudoencapsulated, semisolid tumor with cystic and grayish solid areas
Embryonal rhabdomyosarcoma: Ill-defined, yellowish-gray, and solid tumor, usually in muscle
Myxoid rhabdomyosarcoma: Multinodular, commonly submucosal, soft, and grape-like
Pleomorphic rhabdomyosarcoma: Bulky, infiltrative, yellowish-gray mass, usually in muscle
Hemangiopericytoma: Semi-firm, nodular, yellowish-gray, and well-demarcated
Hemangiosarcoma: Dark red, friable and rubbery, well-outlined tumor with blood clots in cystic spaces
Lymphangiosarcoma: Gray and semi-firm rubbery nodules with irregular edges, often subcutaneous
Neurofibrosarcoma: Well-outlined fusiform and firm mass, often multifocal
Malignant schwannoma: Encapsulated, semi-firm, and fusiform mass, usually adjacent to or in a peripheral nerve
Malignant paraganglioma: Pseudoencapsulated, grayish-red, semisolid lobular mass
Malignant mesothelioma: Multinodular, semisolid, yellowish-gray mesothelial mass
Chondrosarcoma: Shiny, bluish, well-defined tumor with gray-white nodules and empty cysts
Osteosarcoma: Solid or firm ill-defined tumor with bony or cartilagenous consistency
Parosteal osteosarcoma: Firm, lobulated tumor attached to cortical bone with consistency ranging from semisolid to firm
Granulocytic sarcoma: Semisolid, grayish tumor that may appear green in white light and often fluoresces red with ultraviolet light
Plasma cell myeloma: Reddish gray, soft friable, and amorphous tissue
Kaposi's sarcoma: Semisolid gray, hemorrhagic nodule or mass
Ewing's sarcoma: Friable, reddish brown tumor
Chordoma: Lobular and cystic ill-defined tumor with solid areas

Fig. 235. Congenital multinodular and cystic teratoma of the neck (see Fig. 8).

Fig. 236. Neurofibrosarcoma of the vagus nerve (see Figs. 22, 23, 34, 53, 79, 254, 256, 264, 271).

Fig. 237. Malignant pleomorphic fibrous histiocytoma of the face infiltrating the parotid (see Figs. 25, 31, 88, 138, 255, 262, 288, 293).

Fig. 238. Hemimandibulectomy specimen of a 10-year-old with an invasive malignant schwannoma (see Figs. 21, 38, 51, 80, 242, 246, 252, 298).

Fig. 239. Ependymoma of the brain metastatic to the axilla (see Figs. 466, 518).

Fig. 240. Fibrosarcoma invading scapula (see Figs. 33, 60, 91, 297, 371).

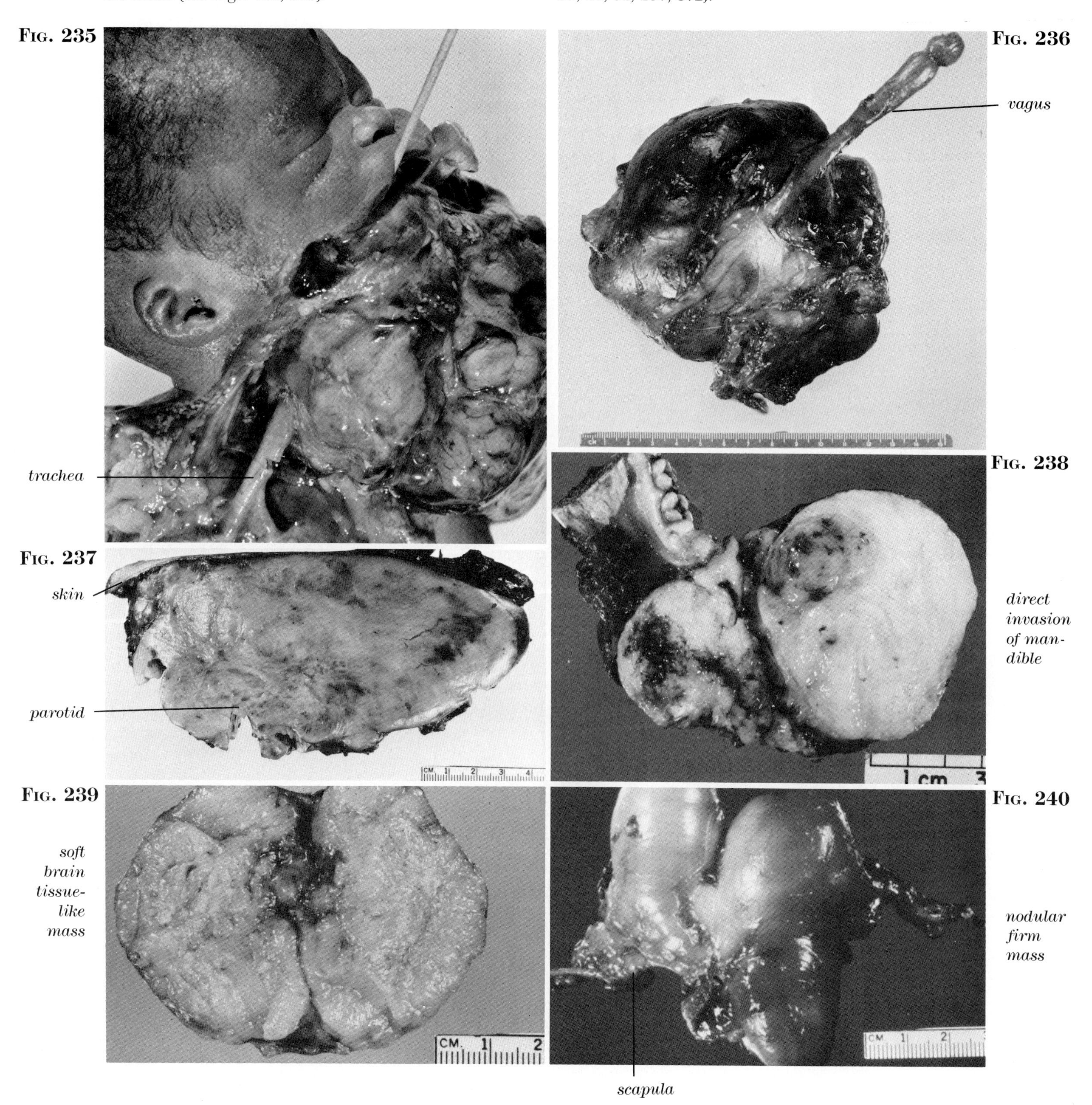

Fig. 241. Malignant fibroblastic fibrous histiocytoma (see Figs. 37, 40, 54, 76, 301, 339, 340).

Fig. 242. (Top center) Malignant schwannoma of the back. Note the multinodular portion of the tumor (see Figs. 38, 51, 80, 238, 246, 252, 298).

Fig. 243. Pedunculated lipoma of the shoulder (see Figs. 24, 73, 198, 455).

Fig. 244. Desmoid tumor of the abdominal wall developed after appendectomy (see Figs. 9, 36, 52, 77, 209, 282, 376).

Fig. 245. Solitary neurofibroma of the chest wall (see Figs. 11, 74, 253, 257, 363).

Fig. 246. Conglomerate of multiple nodules of malignant schwannoma of the shoulder (see Figs. 21, 38, 80, 242, 252, 298, 355, 357).

Fig. 247. Crossected benign cystosarcoma phyllodes of the breast (see Figs. 43, 48, 414).

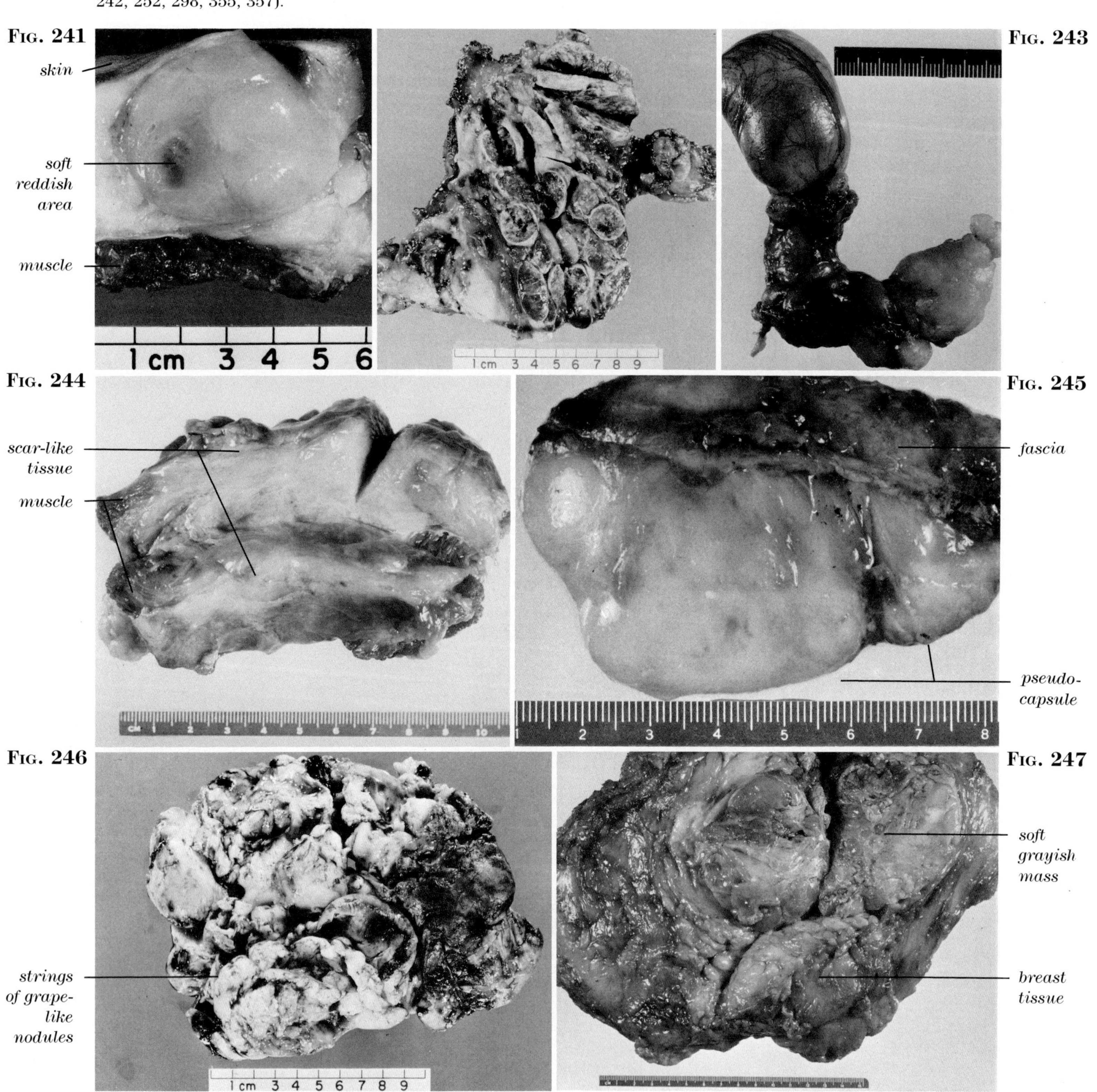

Fig. 248. Osteosarcoma of proximal humerus (see Figs. 90, 127, 142, 251, 272, 306, 312, 319, 324).

Fig. 249. Parosteal osteosarcoma of proximal humerus (Case of Dr. T. Simon, Syracuse, New York) (see Figs. 109, 129, 217, 224, 287, 353).

Fig. 250. Ewing's sarcoma of ulna with extension into soft tissues (see Figs. 100, 172, 173, 210, 214, 286, 307).

Fig. 251. Osteosarcoma of distal humerus (see Figs. 90, 108, 142, 146, 248, 272, 304, 305, 306, 312).

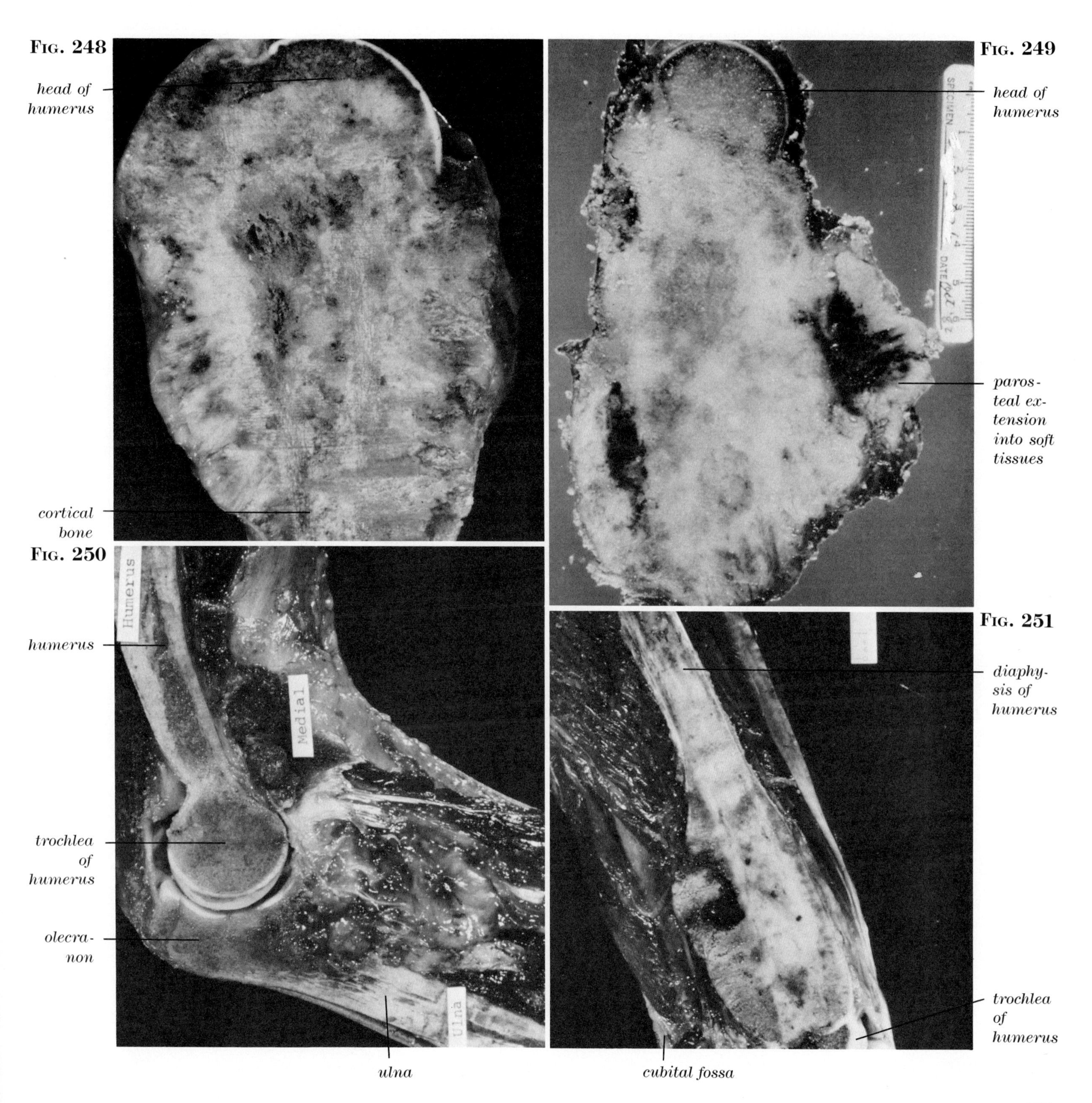

Fig. 252. Malignant schwannoma. Note the multinodular thickening of branches of the brachial plexus (see Figs. 80, 200, 238, 242, 246, 298).

Fig. 253. (Top center) Neurofibroma of the hand (see Figs. 11, 74, 245, 363, 365, 487, 527).

Fig. 254. Cross-sectioned solitary neurofibrosarcoma of the median nerve (see Figs. 23, 26, 79, 236, 256, 264, 271, 275, 289).

Fig. 256. Neurofibrosarcoma of the axilla and arm found in association with congenital birth marks (see Figs. 26, 34, 236, 254, 264, 271).

Fig. 255. (Top center) Malignant pleomorphic fibrous histiocytoma metastatic to the axilla (see Figs. 25, 31, 88, 237, 262, 288, 293).

Fig. 257. Solitary benign schwannoma of the sciatic nerve (see Figs. 356, 556, 846, 850, 1228).

Fig. 252

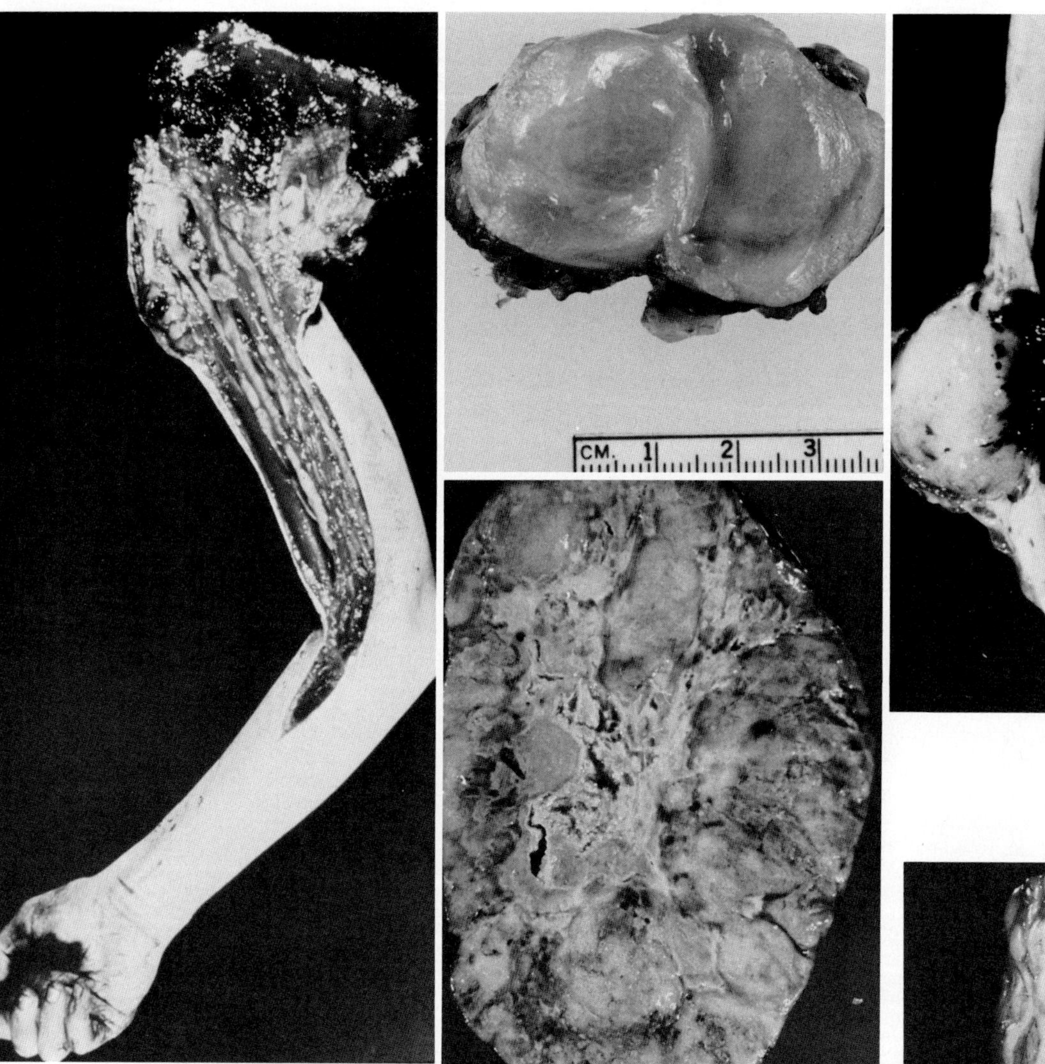

Fig. 254

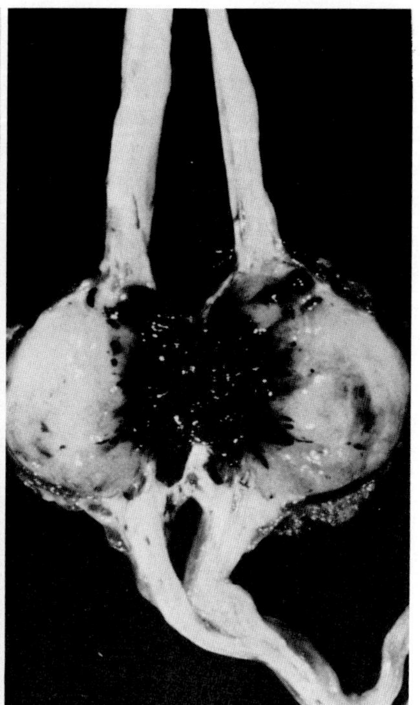

Fig. 257

encapsulated soft nodule

Fig. 256

ulnar nerve

fusiform thickening of nerve

firm mass

Fig. 258. Tendosynovial epithelioid sarcoma (see Figs. 62, 85, 181, 187, 259, 322, 326, 462, 464).

Fig. 259. Epithelioid sarcoma of the hand (see Figs. 62, 181, 187, 258, 322).

Fig. 260. An ill-defined fibrosarcoma of the forearm (see Figs. 33, 91, 131, 165, 170, 240, 297, 371, 410).

Fig. 261. Subcutaneous granular cell tumor of the arm (see Figs. 75, 498, 539, 736, 1160, 1235).

Fig. 262. Superficial malignant pleomorphic fibrous histiocytoma of the forearm (see Figs. 25, 31, 88, 148, 237, 255, 288, 293).

Fig. 263. Embryonal rhabdomyosarcoma of the hand of a child (see Figs. 20, 28, 66, 67, 68, 78, 281, 291).

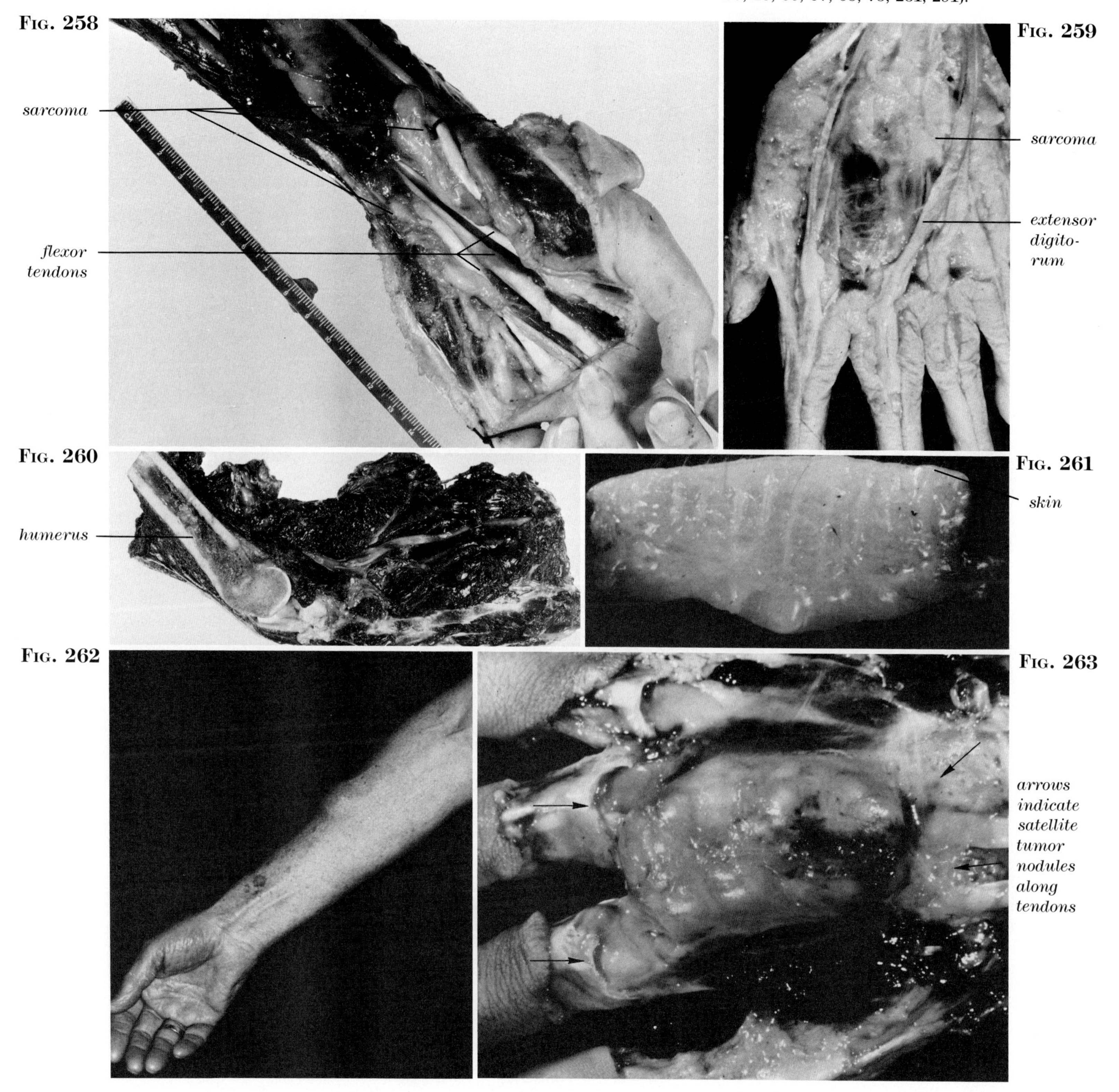

Fig. 264. Multiple nodules of neurofibrosarcoma along the intercostal nerves (see Figs. 26, 34, 79, 236, 254, 256, 271, 275, 289).

Fig. 265. Malignant fibroepithelial mesothelioma of the chest wall (see Figs. 267, 511, 531, 869).

Fig. 266. Malignant fibrous mesothelioma of the abdominal wall (see Figs. 189, 268, 406, 614).

Fig. 267. Multinodular epithelioid mesothelioma after pleural stripping (see Figs. 265, 511, 531, 869, 1242, 1243).

Fig. 268. Malignant fibrous mesothelioma of the pericardium (see Figs. 189, 406, 614, 678).

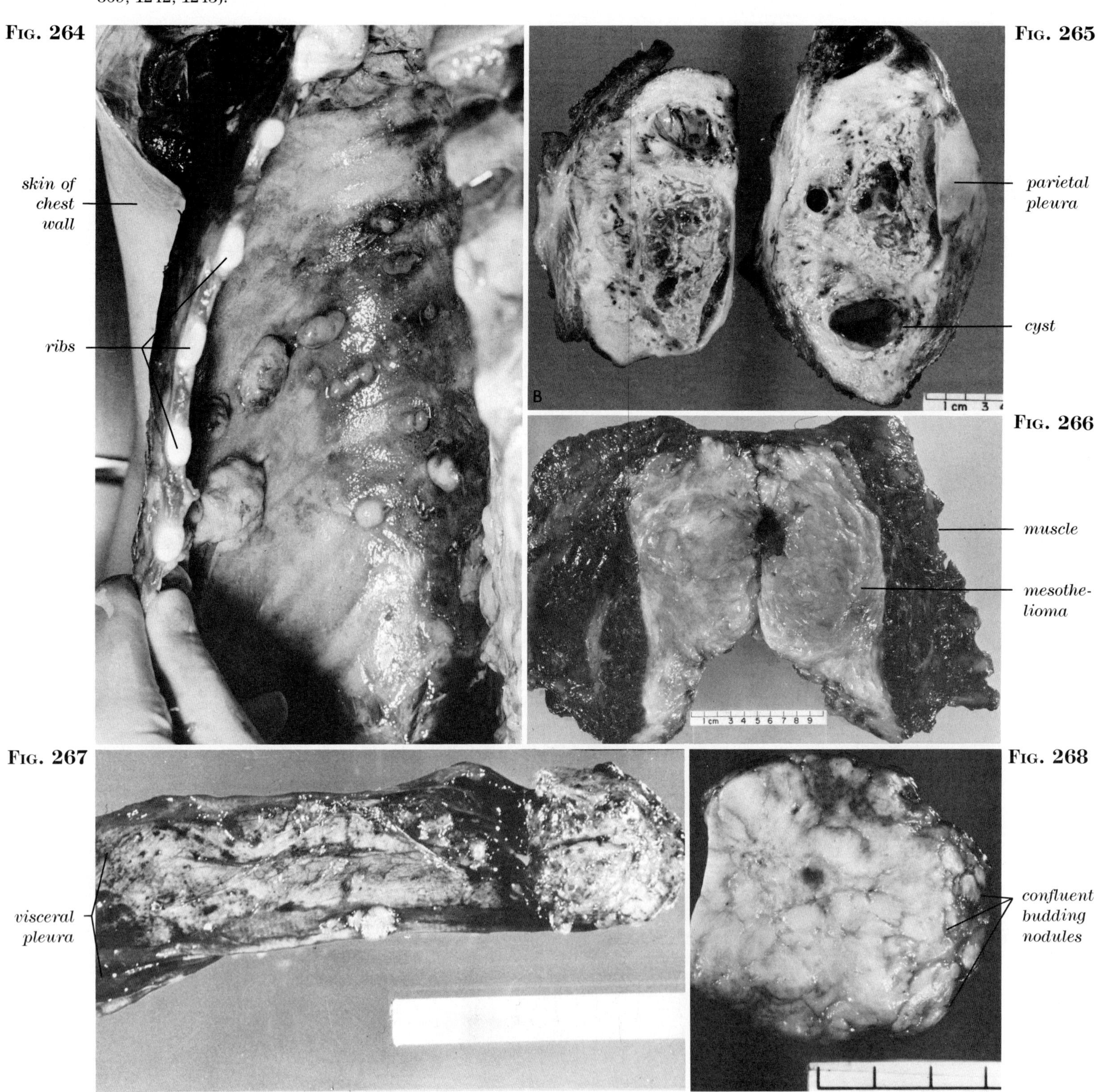

Fig. 269. Gigantic retroperitoneal liposarcoma (Courtesy of Dr. W. Knapper, Memorial Sloan-Kettering Cancer Center, New York) (see Figs. 56, 58, 86, 188, 208, 219, 290).

Fig. 271. Neurofibrosarcoma of the retroperitoneum (see Figs. 34, 53, 79, 254, 264, 271, 275, 289).

Fig. 273. Intramural and polypoid leiomyosarcoma of the colon (see Figs. 72, 110, 218, 220, 231, 341).

Fig. 270. Cross-section of leiomyoblastoma of the stomach (see Figs. 196, 274, 277, 278, 420, 429, 503).

Fig. 272. Cross-section of extraosseous osteosarcoma of the retroperitoneum with invasion of the colon (see Figs. 90, 108, 164, 248, 251, 272, 304, 305, 306, 312).

Fig. 274. Exophytic leiomyoblastoma of the small intestine (see Figs. 196, 270, 277, 278, 420, 429, 503).

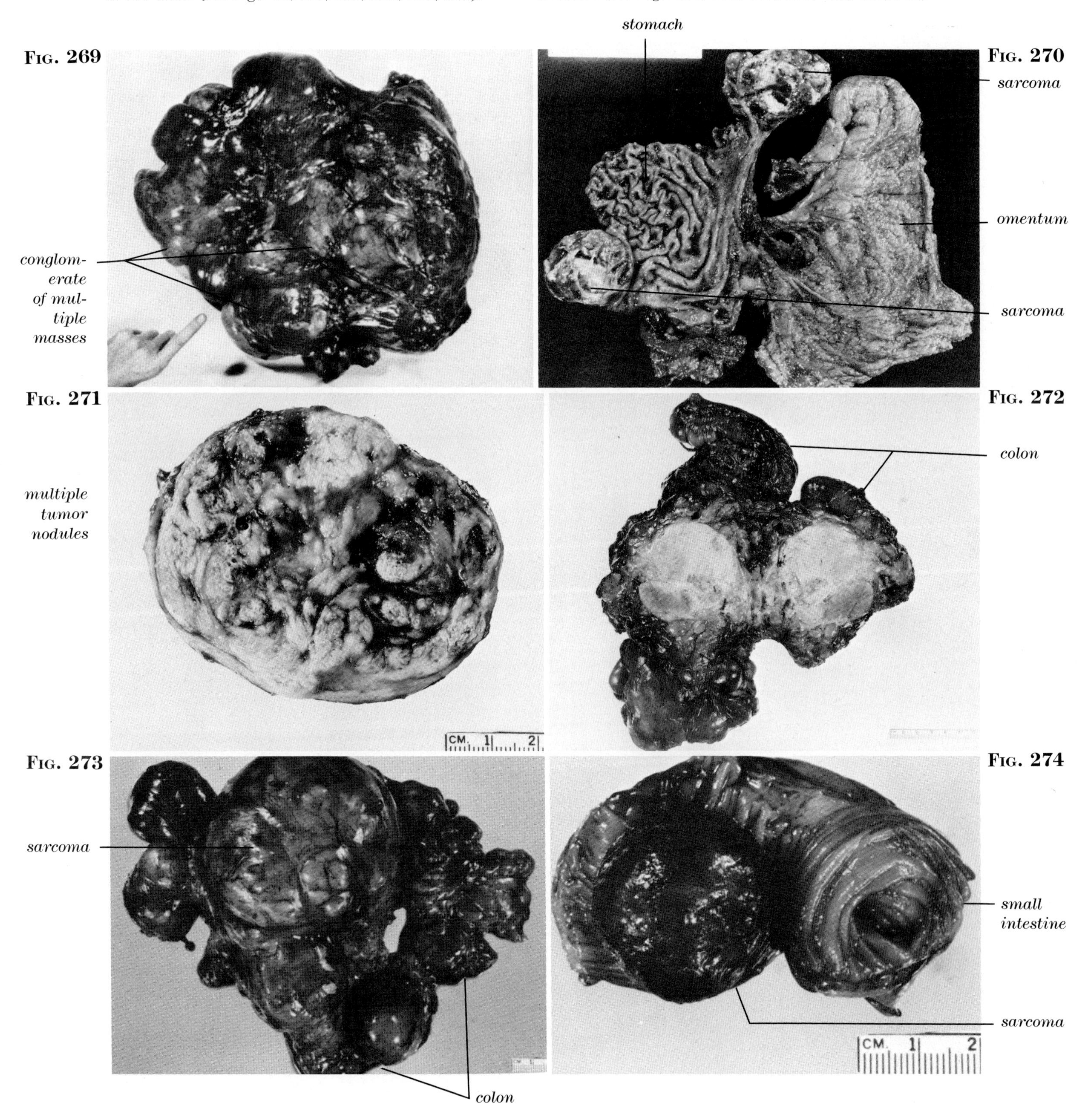

Fig. 275. Neurofibrosarcoma of the retroperitoneum in association with neurofibromatosis (see Figs. 79, 236, 254, 256, 264, 271, 289, 369).

Fig. 277. Extramural and cystic leiomyoblastoma of the stomach (see Figs. 196, 270, 274, 278, 420, 429, 503, 513).

Fig. 279. Multiple nodules of malignant teratoma of testis metastatic to retroperitoneum (see Figs. 349, 688, 922).

Fig. 276. Multinodular malignant paraganglioma of the pelvis (see Figs. 483, 544, 714, 790, 794).

Fig. 278. Gastric leiomyoblastoma with nodular extension into mesentery and retroperitoneum (see Figs. 196, 270, 274, 277, 420, 429, 503).

Fig. 280. Fused nodules of mesenteric fibromatosis (see Figs. 17, 367, 385, 387, 419, 581, 582).

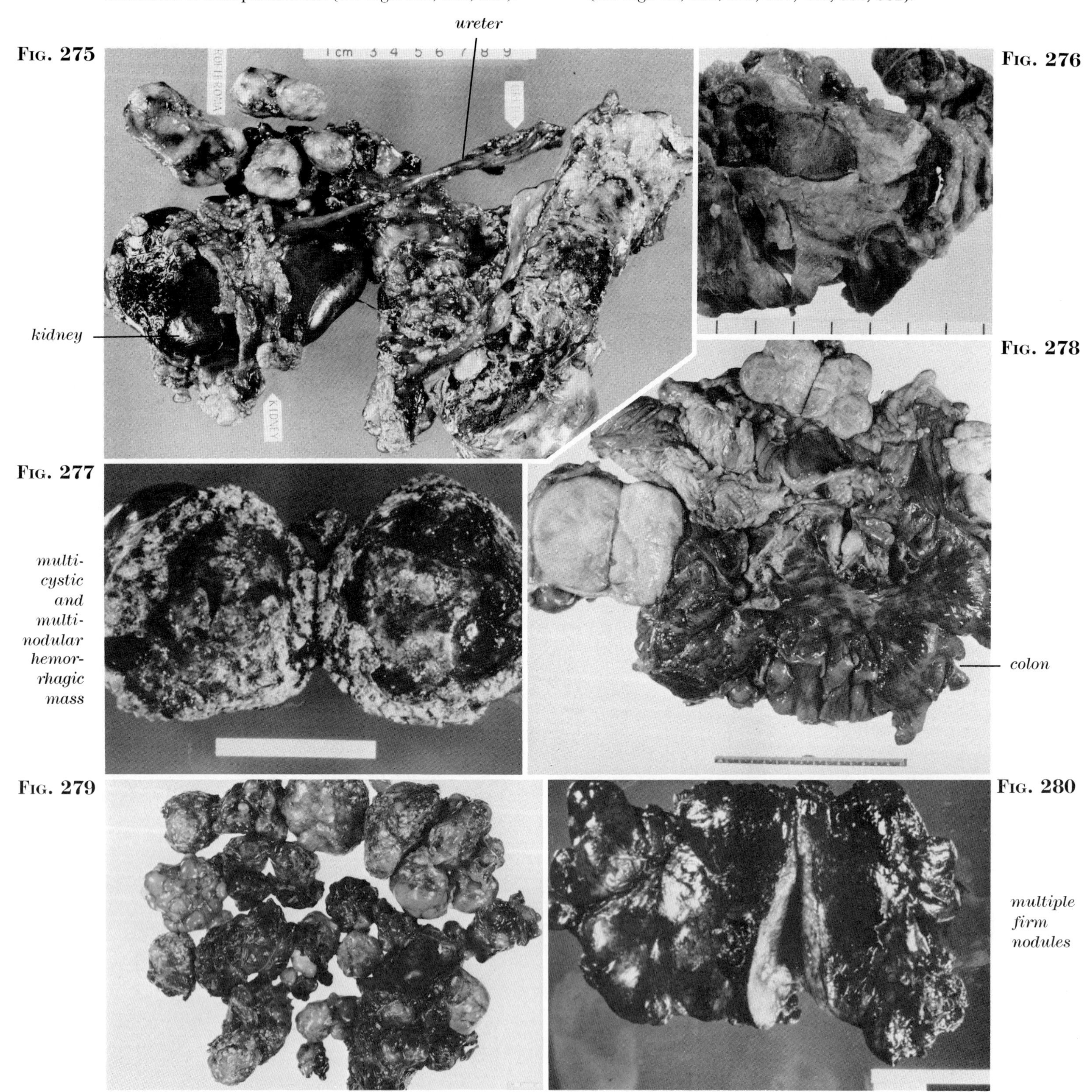

Fig. 281. Cross-section of paratesticular embryonal rhabdomyosarcoma (see Figs. 28, 66, 67, 68, 78, 263, 291).

Fig. 282. Desmoid tumor of the buttock. Note the ill-defined edges. (see Figs. 52, 77, 244, 376).

Fig. 283. Retroperitoneal meningeal sarcoma metastatic to the liver 10 years after excision of the primary (see Figs. 495, 806, 887).

Fig. 284. Liposarcoma of the flank. Note the cross-sected kidney (see Figs. 30, 56, 59, 86, 141, 269, 290, 294).

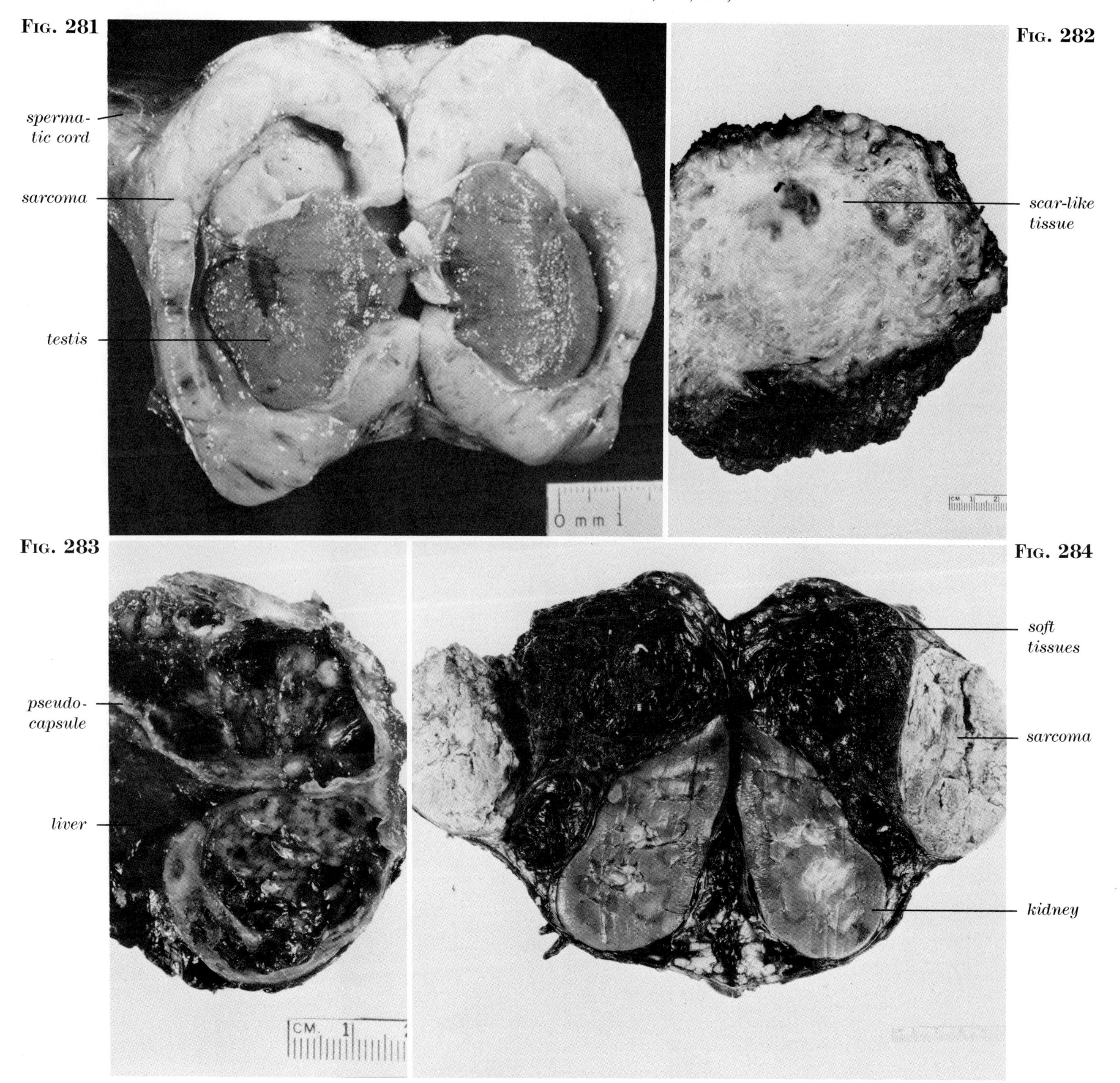

Fig. 285. Chondrosarcoma of the right pelvis and thigh (see Figs. 95, 182, 213, 216, 223, 308, 317, 327, 442).

Fig. 286. Ewing's sarcoma of the right femur with a prominent soft tissue component (see Figs. 100, 214, 227, 229, 250, 307, 314, 330).

Fig. 287. Parosteal osteosarcoma of the left distal femur (see Figs. 109, 129, 217, 224, 249, 353, 354, 656).

Fig. 288. Malignant pleomorphic fibrous histiocytoma of distal femur with intramedullary and parosteal components (see Figs. 31, 111, 140, 148, 201, 203, 232, 237, 255, 262, 293).

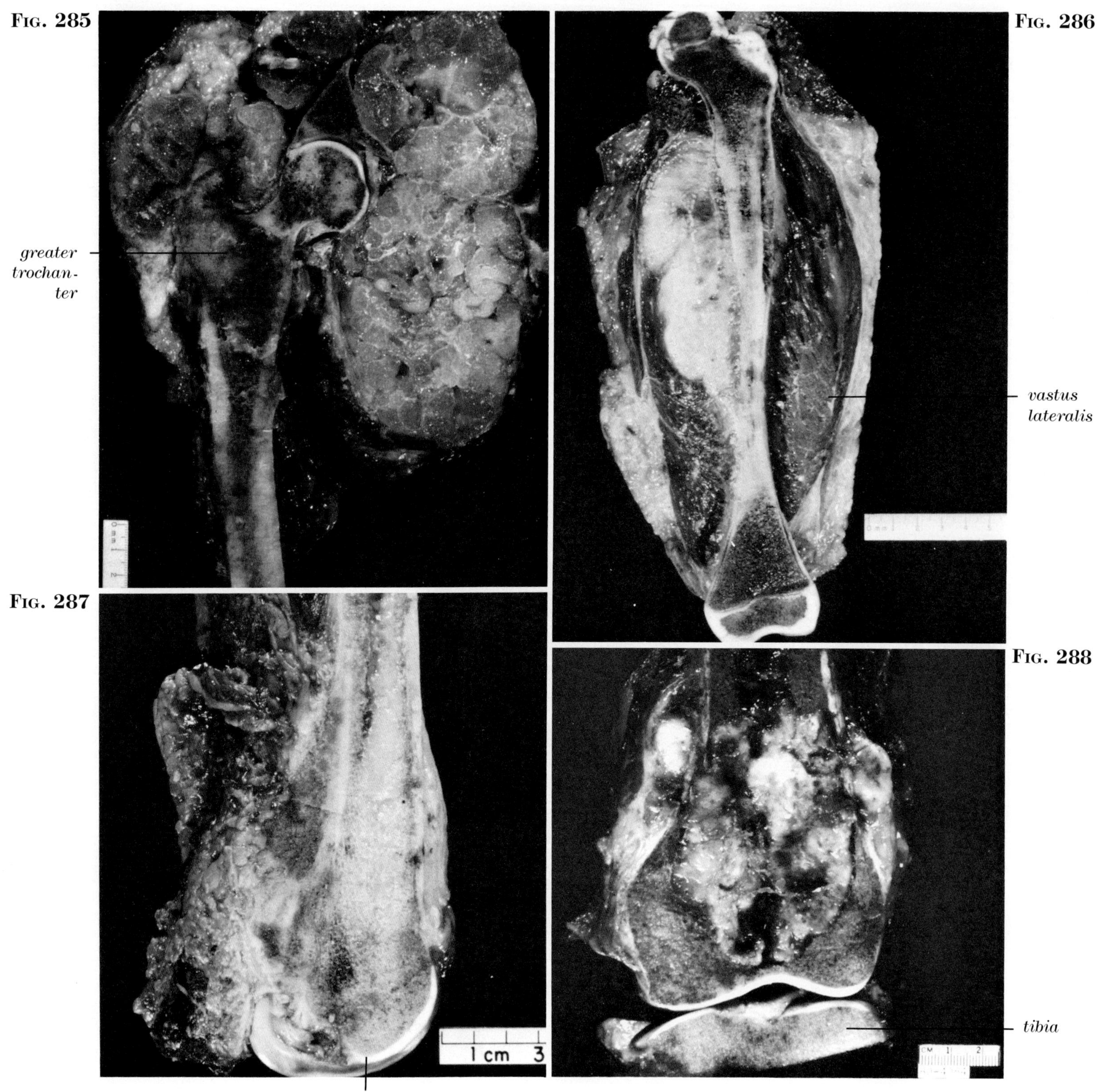

Fig. 289. Neurofibrosarcoma of the right thigh (see Figs. 34, 53, 79, 271).

Fig. 290. (Top center) Myxoid liposarcoma of thigh encircling the femur (see Figs. 59, 208, 269).

Fig. 291. Embryonal rhabdomyosarcoma of thigh in an adult (see Figs. 67, 68, 263, 281).

Fig. 292. En bloc resection of pleomorphic rhabdomyosarcoma of thigh (see Figs. 29, 87, 295, 296).

Fig. 293. (Bottom center) Malignant pleomorphic fibrous histiocytoma of thigh (see Figs. 25, 88, 237, 255, 262).

Fig. 294. Pleomorphic liposarcoma of thigh, en bloc resection (see Figs. 30, 55, 141, 284).

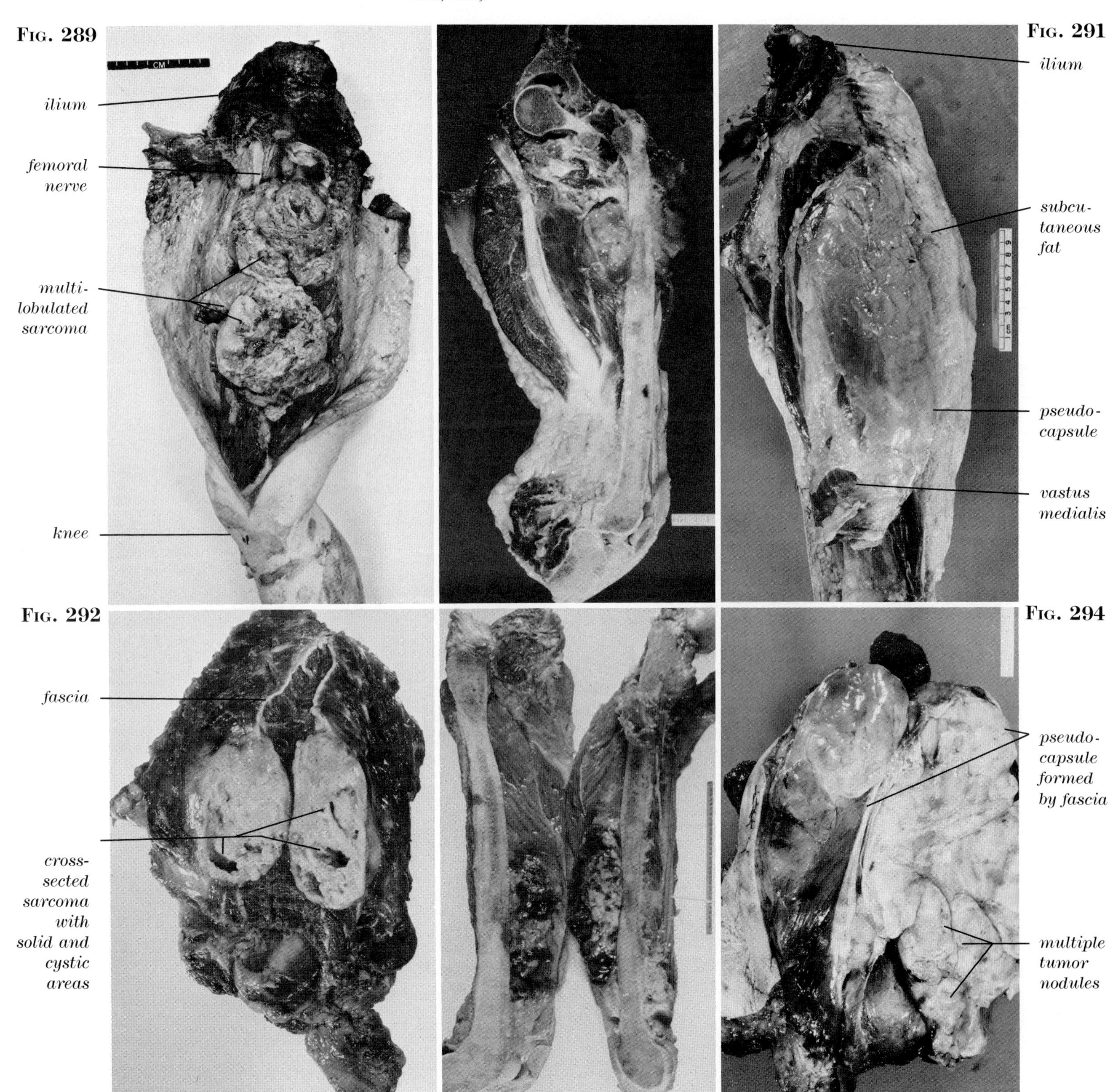

Fig. 295. Pleomorphic rhabdomyosarcoma of knee with ulceration (see Figs. 87, 121, 292, 296, 299, 313).

Fig. 296. (Top center) Bulky pleomorphic rhabdomyosarcoma of distal posterior thigh (see Figs. 87, 292, 295, 599).

Fig. 297. Popliteal fibrosarcoma (see Figs. 91, 240, 371, 374, 393, 410, 601, 608, 625).

Fig. 298. Malignant schwannoma of sciatic nerve (see Figs. 51, 80, 238, 246, 252, 355, 357, 382, 619).

Fig. 299. (Bottom center) Pleomorphic rhabdomyosarcoma of the left posterior thigh (see Figs. 87, 292, 295, 296).

Fig. 300. Malignant histiocytic fibrous histiocytoma of anterior thigh (see Figs. 57, 156, 452, 534).

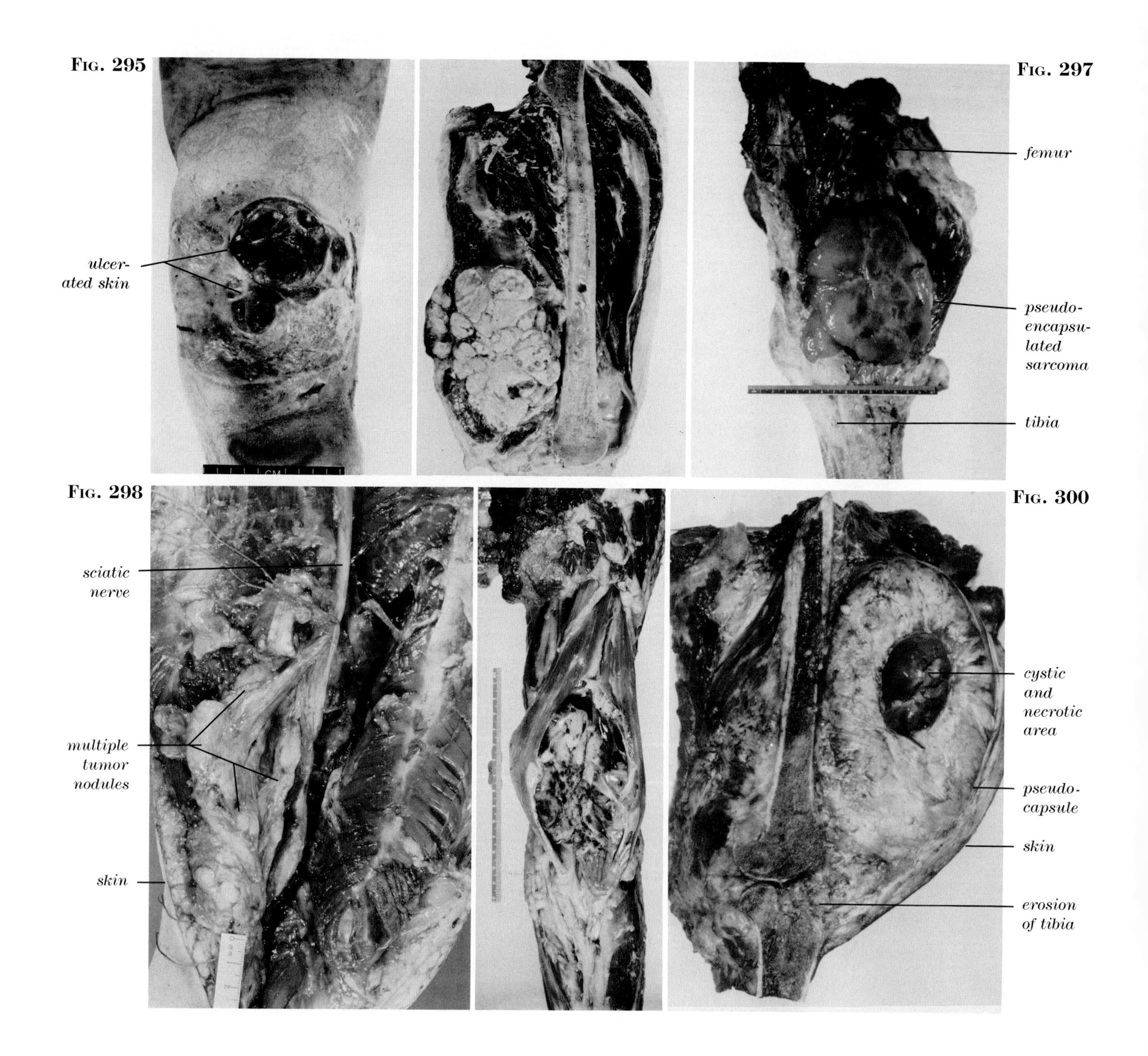

Fig. 301. Malignant fibroblastic fibrous histiocytoma of the right thigh (see Figs. 37, 40, 54, 76, 143, 241, 339, 340).

Fig. 302. Osteochondroma of the left proximal femur (see Figs. 113, 155, 167, 184, 303, 772, 1063, 1066, 1090).

Fig. 303. Cross-section of an osteochondroma of distal femur (see Figs. 113, 155, 167, 302, 772, 1063, 1066, 1090, 1152).

Fig. 304. Osteosarcoma of distal femur with prominent extramedullary component (see Figs. 108, 109, 142, 149, 176, 225, 272, 305, 306, 312, 319).

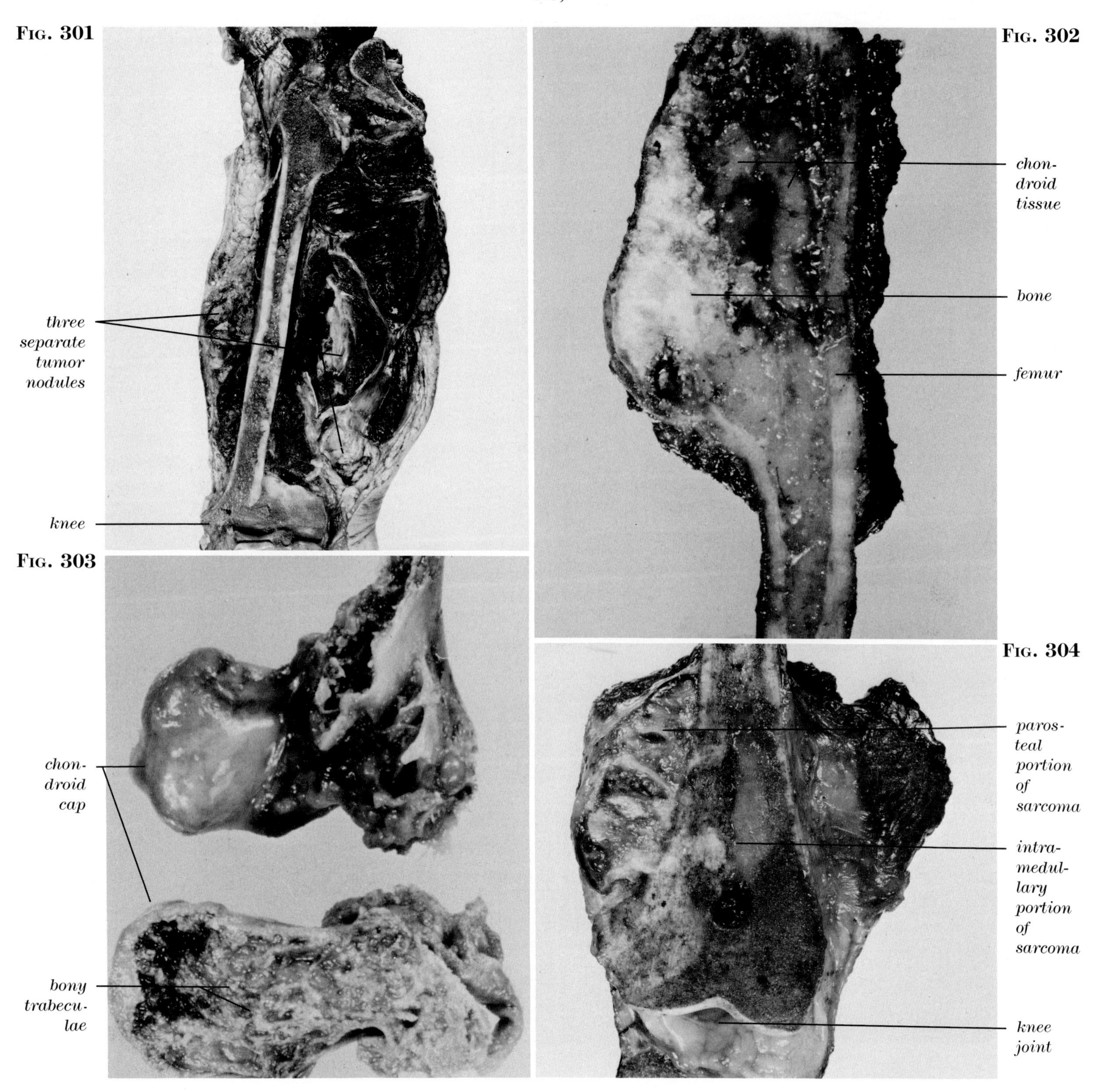

Fig. 305. Osteosarcoma of proximal tibia (see Figs. 108, 109, 221, 233, 262, 304, 306).

Fig. 306. Sclerosing osteosarcoma of the proximal metaphysis of tibia (see Figs. 108, 109, 162, 164, 228, 248, 251, 305, 312).

Fig. 307. Ewing's sarcoma of tibia (see Figs. 100, 172, 214, 227, 229, 250, 286, 314, 330).

Fig. 308. Chondrosarcoma of proximal fibula (see Figs. 95, 213, 216, 223, 285, 308, 317, 327, 329).

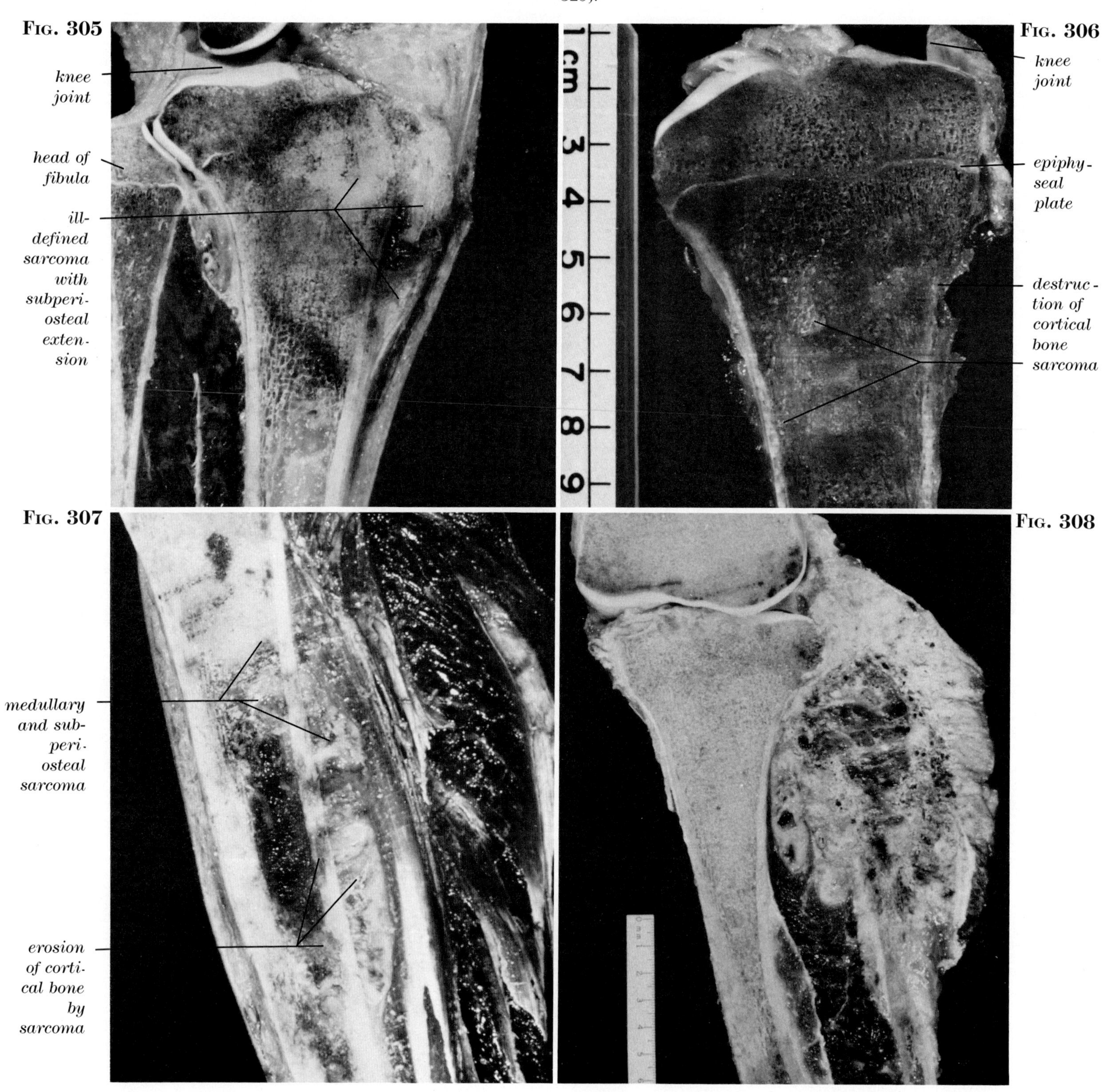

Fig. 309. Malignant neuroepithelioma of sciatic nerve above the popliteal bifurcation. A. Nodular surface. B. (Top center) Longitudinal section. C. Close-up of cut surface (see Figs. 478, 489, 568, 585, 690, 1115, 1238, 1239).

Fig. 310. Sacral chordoma with lumbar vertebrae at autopsy (see Figs. 69, 215, 505, 729, 767, 771, 1174, 1226).

Fig. 311. Medulloblastoma spreading subdurally around spinal cord (see Figs. 469, 723).

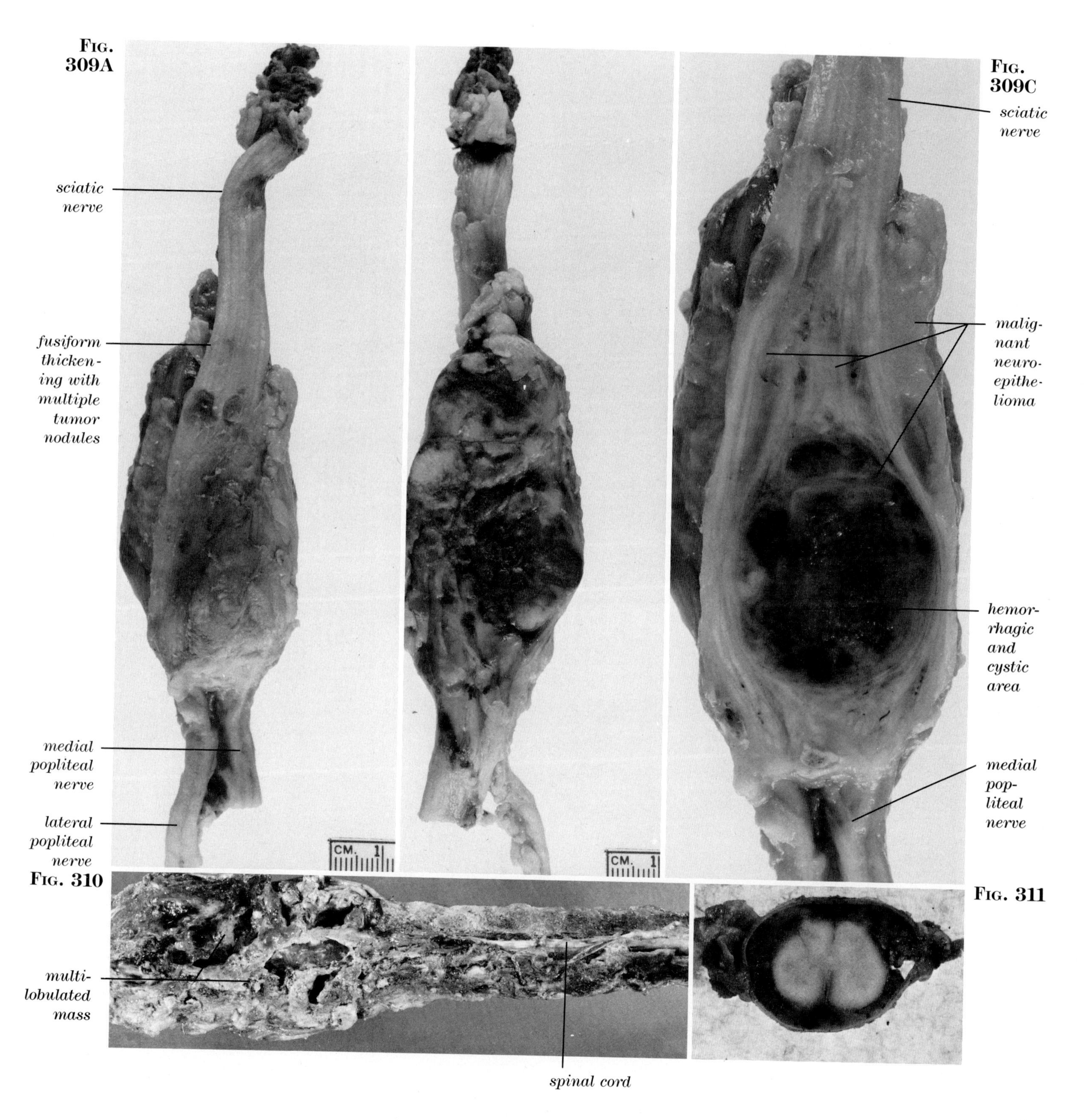

Fig. 312. Osteosarcoma of distal femur (see Figs. 108, 109, 157, 158, 217, 225, 228, 230, 306, 319, 324).

Fig. 313. Pleomorphic rhabdomyosarcoma of thigh invading distal femur (see Figs. 29, 87, 121, 292, 295, 296, 299, 599).

Fig. 314. Ewing's sarcoma of distal fibula (see Figs. 100, 173, 227, 229, 250, 286, 307, 330, 449).

Fig. 315. Monophasic tendosynovial sarcoma of the foot with invasion of bones (see Figs. 39, 64, 85, 179, 190, 320, 326, 391, 392).

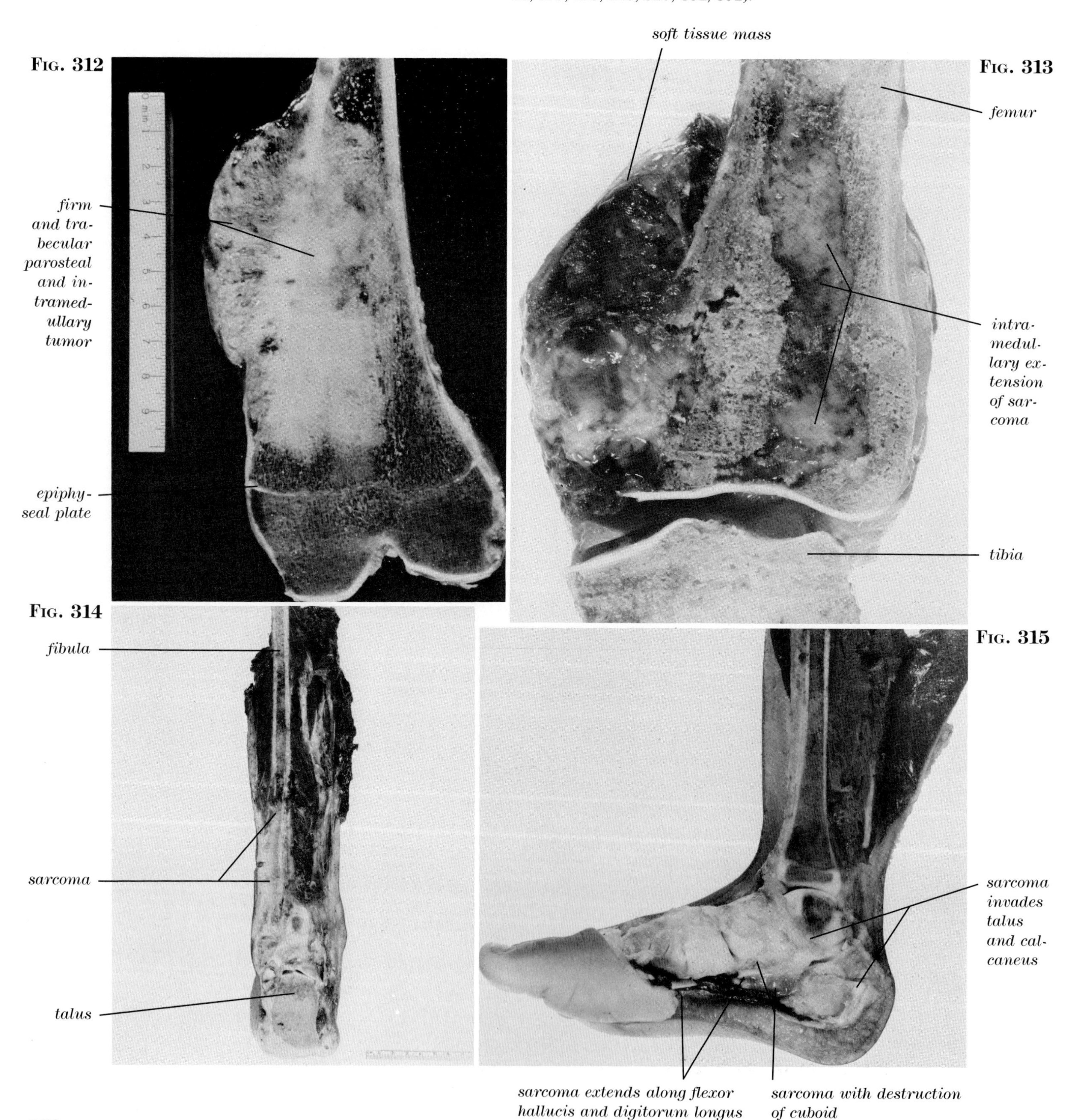

FIG. 316. Scapulohumeral amputation specimen showing multifocal hemangiosarcoma (see Figs. 32, 42, 92, 102, 523, 541, 555, 557).

FIG. 318. Mesenchymal chondrosarcoma of foot (see Figs. 103, 151, 175, 180, 390, 444, 507, 595).

FIG. 317. Chondrosarcoma of proximal tibia (see Figs. 95, 137, 147, 182, 213, 216, 223, 285, 308, 327, 329, 407).

FIG. 319. Osteosarcoma of tibia (see Figs. 108, 109, 162, 164, 225, 233, 248, 251, 272, 304–306, 312, 324, 328).

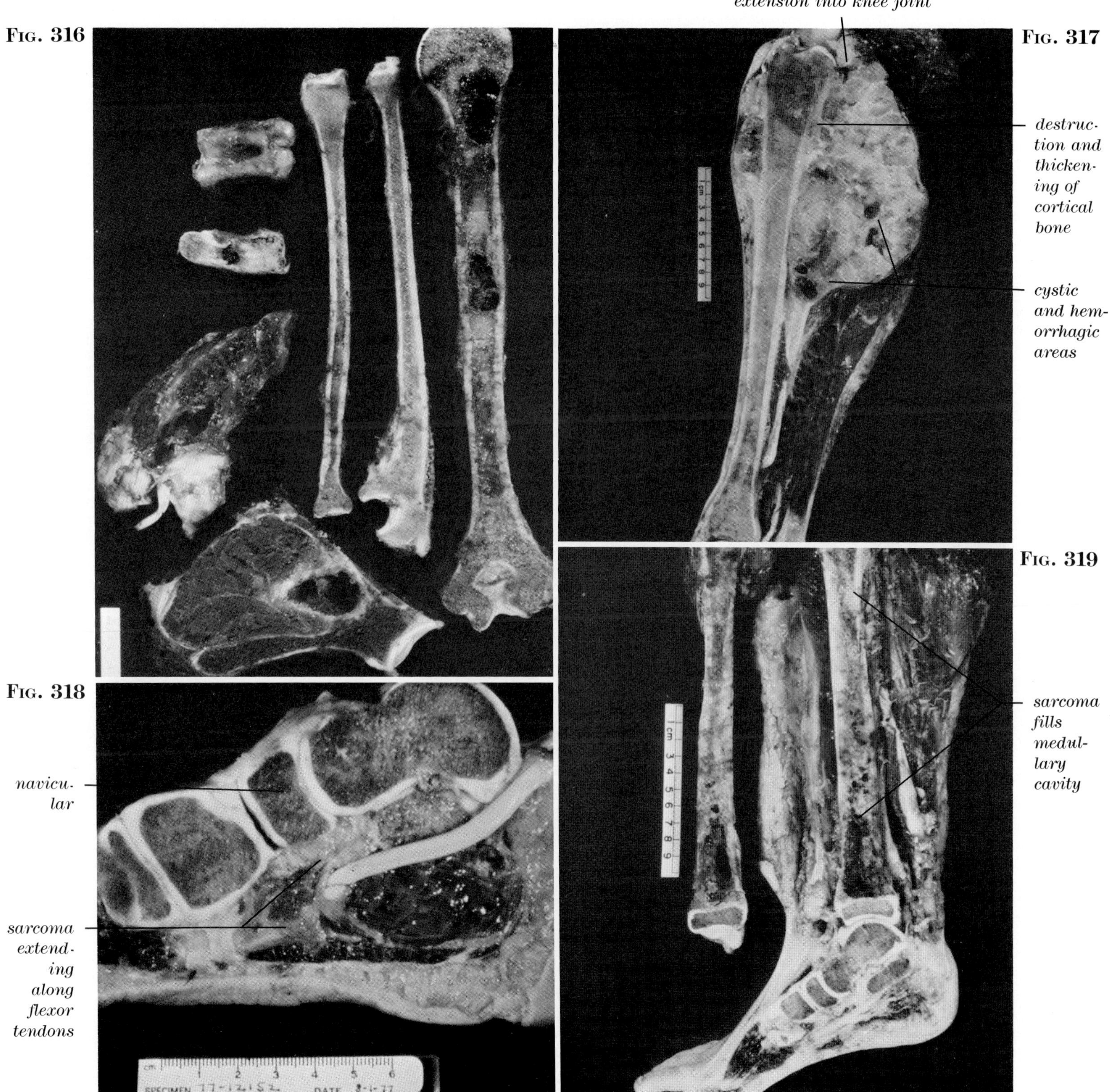

Fig. 320. Tendosynovial sarcoma of foot (see Figs. 39, 64, 85, 122, 152, 153, 179, 181, 323).

Fig. 321. (Top center) Filarial-infection-induced, lymphedema-associated lymphangiosarcoma of leg (see Figs. 35, 41, 61, 84, 154, 159, 424).

Fig. 322. Tendosynovial sarcoma of ankle, predominantly growing as epithelioid and clear cell sarcoma (see Figs. 62, 179, 181, 187, 259).

Fig. 323. Popliteal tendosynovial sarcoma encircling tendons and nerves (see Figs. 152, 153, 320, 514).

Fig. 324. Pseudoencapsulated extraosseous osteosarcoma involving gastrocnemius (see Figs. 90, 228, 248, 251, 305, 319, 328, 395).

Fig. 325. Benign histiocytic fibrous histiocytoma of popliteal space (see Figs. 71, 222, 490, 501, 778).

Fig. 326. Multinodular tendosynovial sarcoma of foot (see Figs. 39, 62, 64, 179, 190, 315, 322, 326).

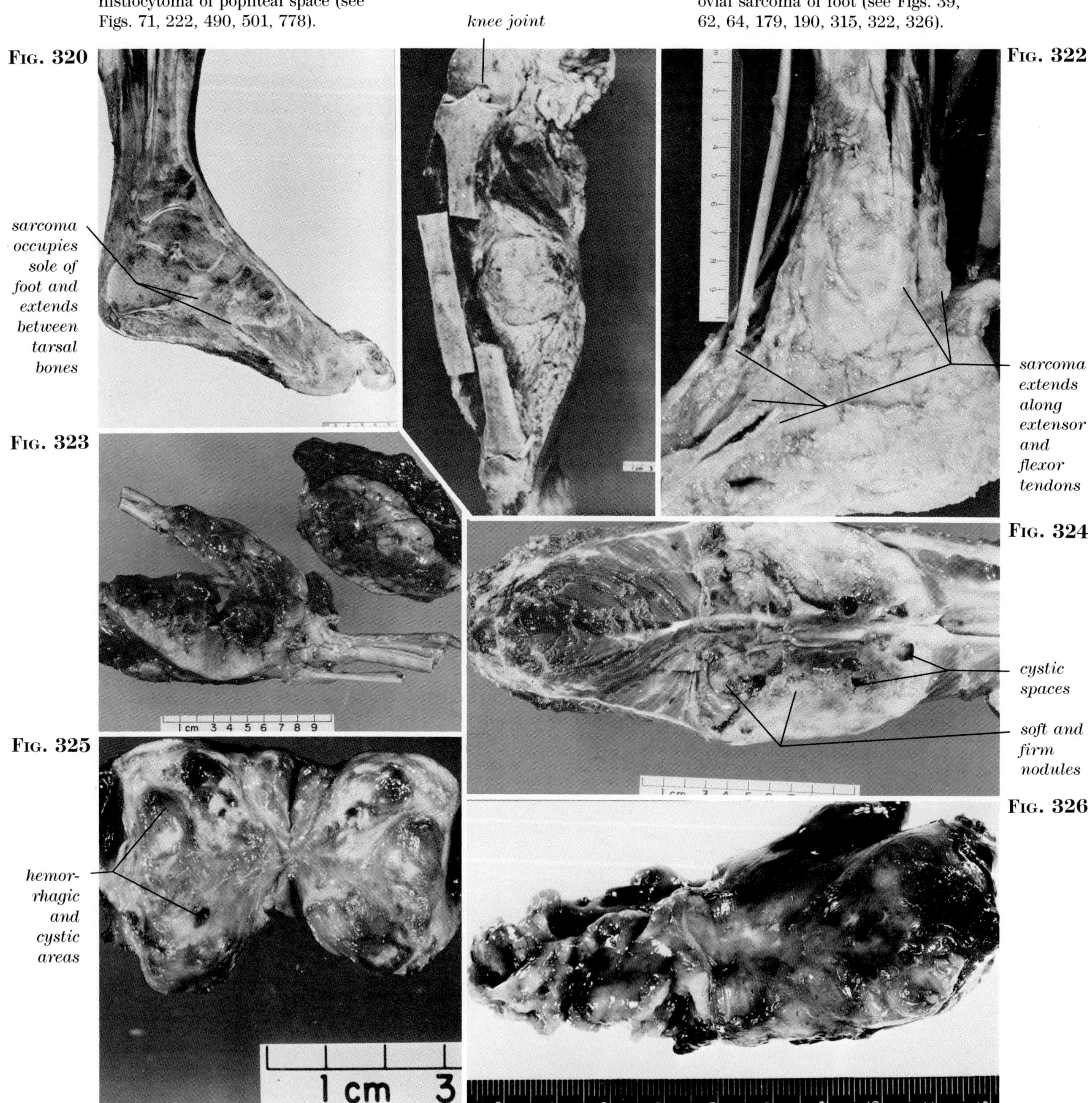

Fig. 327. Cross-section of chondrosarcoma of pelvis (see Figs. 81, 95, 136, 147, 216, 223, 285, 317, 329, 442).

Fig. 329. Chondrosarcoma of ilium (see Figs. 136, 213, 216, 223, 308, 318, 327, 390, 407, 434, 441, 442, 638).

Fig. 328. Telangiectatic osteosarcoma of proximal tibia (see Figs. 108, 109, 164, 217, 228, 324, 395).

Fig. 330. Ewing's sarcoma metastatic to lung (see Figs. 100, 172, 173, 229, 250, 286, 314, 449).

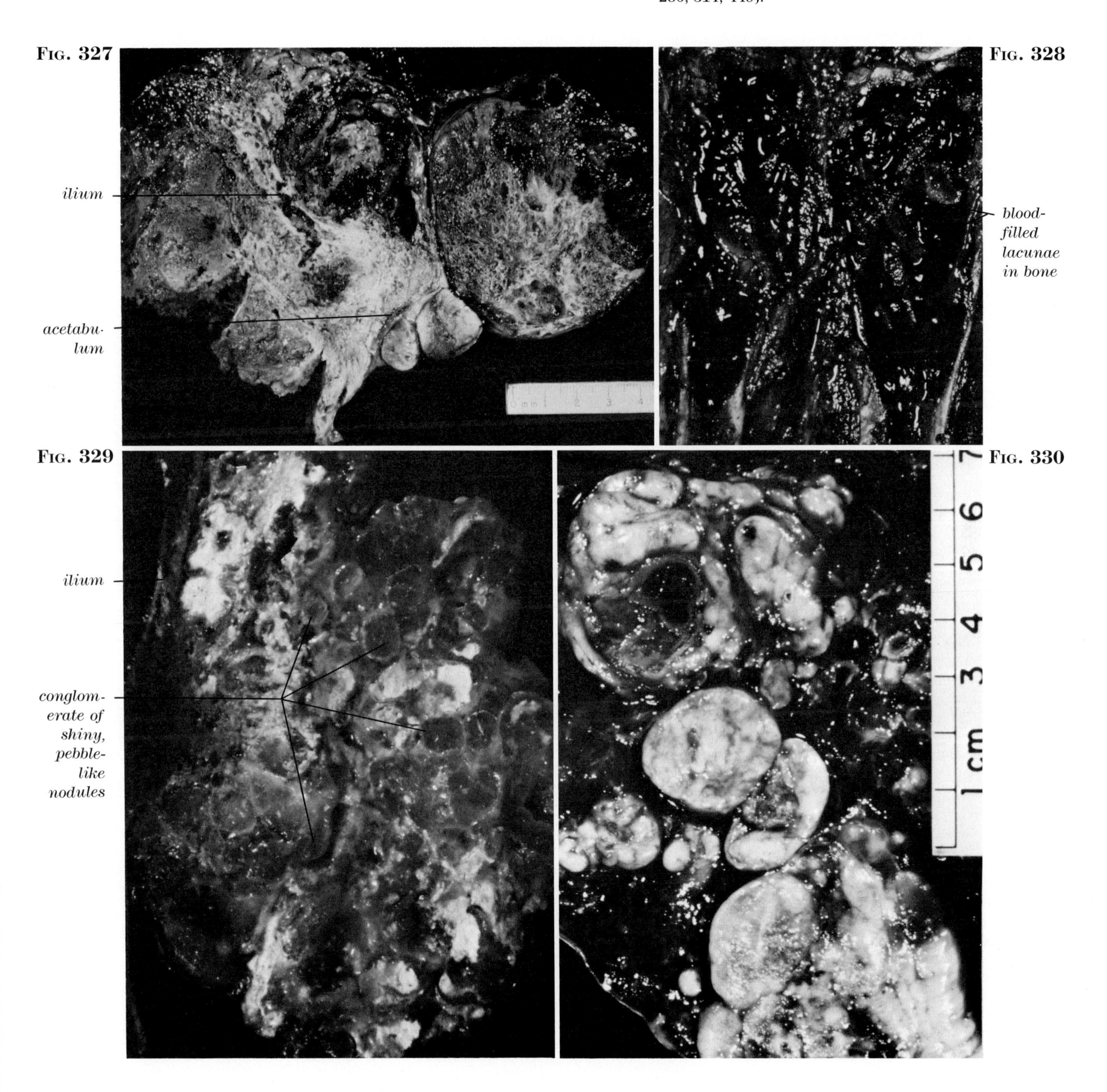

"In applying a diagnosis to a particular tumor, one is guided by the dominant histologic tissue pattern"

HENRY L. JAFFE (1896–1979)

10. Growth Pattern

	NON-NEOPLASTIC LESIONS	BENIGN NEOPLASMS	MALIGNANT NEOPLASMS
1. Arranged Pattern. See Tables 48 to 50.			
a. Storiform pattern. See Figures:	331, 332, 345, 348	333, 334, 336, 346, 352	335, 337–344, 347, 349–351, 353, 354
b. Palisading pattern. See Figures:	367, 368	356, 358, 359, 361, 363, 365	355, 357, 360, 362, 364, 366
c. Herring-bone pattern. See Figures:			369–374
2. Spreading Pattern. See Tables 48 to 50.			
See Figures:	384, 385, 387, 389, 403, 408	380, 400, 402	375–379, 381–383, 386, 388, 390–399, 401, 404–407, 409–411
3. Lacy Pattern. See Tables 48 to 50.			
See Figures:	419, 425, 430, 433	412–414, 417, 421, 422, 432, 437, 455, 459	415, 416, 418, 420, 423, 424, 426–429, 431, 434–436, 438–454, 456–458, 460, 461
4. Epithelioid Pattern. See Tables 48 to 50.			
See Figures:	508, 509	471, 474, 475, 481, 483, 485–487, 490, 496–498, 501, 506, 516, 517	462–470, 472, 473, 476–480, 482, 484, 488, 489, 491–495, 499, 500, 502–505, 507, 510–515, 518–521

	NON-NEOPLASTIC LESIONS	BENIGN NEOPLASMS	MALIGNANT NEOPLASMS
5. Aveolar Pattern. See Tables 48 to 50.			
See Figures:	539, 552	527, 537, 549, 553, 554, 556, 559, 560, 562	522–526, 528–536, 538, 540–548, 550, 551, 555, 557, 558, 561, 563–576
6. Disarranged Pattern. See Tables 48 to 50.			
See Figures:	578, 581, 582, 597, 605, 607	613	577, 579, 580, 583–596, 598–604, 606, 608–612, 614, 615

The pathologist may rely more heavily on the pattern of growth than on any other criteria in the microscopic diagnosis of soft tissue and bone tumors. Most soft tissue and bone tumors exhibit considerable variation in cell morphology and differentiation, but many of them present with a characteristic growth pattern. In the diagnosis of tumors any growth pattern that is significantly different from the usual should be taken seriously as a signal that the presumptive diagnosis may not be a correct one.

Each soft tissue and bone tumor may exhibit one or more of the six growth patterns listed in Tables 48 to 50, but in most tumors there is only one pattern that is predominant. However, due to growth patterns of benign and malignant lesions that overlap in similarity, the growth pattern alone, without consideration of the cell morphology, the appearance of stroma, and the products of cells, does not necessarily indicate whether a neoplasm is benign or malignant.

The correct identification of growth pattern is the cornerstone in tumor diagnosis. The predominant pattern of growth is easily recognizable with a low power microscopic lens in hematoxylin- and eosin-stained sections of high technical quality. With consideration of cellular elements and knowledge of the patient's age and the site and size of the tumor, the recognition of the pattern should lead to the correct diagnosis in the majority of the cases.

In the *arranged pattern* the cells exhibit a preordered and repetitive, easily recognizable growth tendency. In the storiform pattern the cells form avascular twisted nebulae. Nuclear palisading is the result of periodical and parallel arrangement of a symmetrically aligned stack of cells producing a wavy effect. The herring-bone pattern shows cells wedged against each other in sharp angles in interweaving bundles.

The *spreading pattern* shows cells in uninterrupted, monotonous, and one-directional alignment.

The interconnected cells in the *lacy pattern* form a filigree or spiderweb-like network.

The *epithelioid pattern* is the product of densely packed and adherent cells in clusters and sheets.

Tightly linked cells forming solid alveoli, rings, or tubules are responsible for the *alveolar pattern.*

Tumors that grow without appreciable pattern and whose cellular elements show complete disarray belong to the *disarranged pattern.*

TABLE 48. *Growth Pattern of Non-neoplastic Lesions**

	ARRANGED PATTERN	SPREADING PATTERN	LACY PATTERN	EPITHELIOID PATTERN	ALVEOLAR PATTERN	DISARRANGED PATTERN
● Aneurysmal bone cyst				X		X
● Giant cell granuloma				X		X
● Hyperparathyroidism				X		
○ Xanthogranuloma	X		X			X
○ Fasciitis	X	X				X
○ Keloid		X				X
○ Fibromatosis		X				X
● Fibrous dysplasia		X				
○ Elastofibroma						X
○ Collagenoma		X				X
○ Tendosynovitis			X	X		
◑ Tendosynovial cyst			X	X		
○ Tendosynovial chondromatosis			X			
○ Gouty arthritis			X			X
◑ Fat necrosis			X			X
◑ Lipogranuloma			X			X
○ Proliferative panniculitis		X		X		X
○ Proliferative myositis		X				X
○ Myositis ossificans						X
○ Pyogenic granuloma					X	
○ Angifollicular lymphoid hyperplasia					X	
◑ Arteriovenous malformation			X		X	X
○ Traumatic neuroma	X					
● Callus						X
◑ Eosinophilic granuloma			X	X		

*○ Primary soft tissue tumor. ● Primary bone tumor. ◑ Can be primary soft tissue or bone tumor.

Table 49. *Growth Pattern of Benign Neoplasms**

	Arranged Pattern	Spreading Pattern	Lacy Pattern	Epithelioid Pattern	Alveolar Pattern	Disarranged Pattern
◑ Benign fibroblastic fibrous histiocytoma	X					
◑ Benign histiocytic fibrous histiocytoma			X	X		
◑ Benign pleomorphic fibrous histiocytoma			X			X
● Nonossifying fibroma	X					
○ Fibroma		X				
● Desmoplastic fibroma		X				
◑ Well differentiated lipoma			X			
◑ Myxoid lipoma			X			
○ Fibroblastic lipoma	X		X			
○ Lipoblastoma			X	X		
○ Pleomorphic lipoma	X		X			X
○ Angiomyolipoma	X		X			X
○ Hibernoma			X	X		
◑ Leiomyoma	X	X				
○ Rhabdomyoma				X		X
◑ Hemangioma			X		X	
◑ Benign glomus tumor			X	X	X	
◑ Neurofibroma	X	X				
○ Plexiform neurofibroma	X					
◑ Benign schwannoma	X	X				
○ Benign nevoid schwannoma			X	X		
◑ Benign undifferentiated peripheral nerve tumor	X	X		X	X	
◑ Ganglioneuroma		X				X
○ Benign paraganglioma	X		X	X		
○ Benign fibrous mesothelioma		X				
● Chondroblastoma			X	X		
● Chondromyxoid fibroma		X	X			
◑ Chondroma			X			
● Osteochondroma			X			
● Osteoblastoma			X			X
● Osteoid osteoma						X
○ Benign granular cell tumor				X		
○ Meningioma	X			X		
○ Benign cystosarcoma phyllodes	X	X				

*○ Primary soft tissue tumor. ● Primary bone tumor. ◑ Can be primary soft tissue or bone tumor.

TABLE 50. *Growth Pattern of Malignant Neoplasms**

	ARRANGED PATTERN	SPREADING PATTERN	LACY PATTERN	EPITHELIOID PATTERN	ALVEOLAR PATTERN	DISARRANGED PATTERN
◑ Malignant fibroblastic fibrous histiocytoma	X					
◑ Malignant histiocytic fibrous histiocytoma	X	X		X		
◑ Malignant pleomorphic fibrous histiocytoma	X					X
◑ Desmoid tumor		X				
◑ Fibroblastic fibrosarcoma	X	X				
◑ Pleomorphic fibrosarcoma		X				X
○ Biphasic tendosynovial sarcoma		X			X	
○ Monophasic tendosynovial sarcoma, spindle cell type		X				
○ Monophasic tendosynovial sarcoma, pseudoglandular type					X	
○ Epithelioid sarcoma				X		
○ Clear cell sarcoma			X	X		
○ Chordoid sarcoma			X			
◑ Well differentiated liposarcoma			X		X	
◑ Myxoid liposarcoma	X		X			
○ Lipoblastic liposarcoma			X	X		
○ Fibroblastic liposarcoma	X		X			
○ Pleomorphic liposarcoma						X
◑ Leiomyoblastoma			X	X		
◑ Leiomyosarcoma	X	X				X
○ Embryonal rhabdomyosarcoma		X		X		X
○ Alveolar rhabdomyosarcoma				X	X	
○ Myxoid rhabdomyosarcoma			X			
○ Rhabdomyoblastoma				X		X
○ Pleomorphic rhabdomyosarcoma						X
◑ Hemangiopericytoma		X		X	X	
◑ Hemangiosarcoma				X	X	
○ Lymphangiosarcoma				X	X	
◑ Neurofibrosarcoma	X	X				
◑ Malignant schwannoma	X	X		X		
◑ Malignant undifferentiated peripheral nerve tumor			X	X	X	
○ Malignant epithelioid mesothelioma				X		
○ Malignant fibrous mesothelioma		X				
◑ Chondrosarcoma			X	X		
◑ Mesenchymal chondrosarcoma	X	X	X			
◑ Osteosarcoma		X		X		X
◑ Parosteal osteosarcoma	X	X				X
● Paget's sarcoma		X		X		X
◑ Granulocytic sarcoma			X	X		
● Plasma cell myeloma			X	X		
○ Malignant granular cell tumor						X
○ Alveolar soft part sarcoma					X	
○ Kaposi's sarcoma		X				
◑ Ewing's sarcoma				X		
● Chordoma			X			
○ Myxopapillary ependymoma			X	X	X	

*○ Primary soft tissue tumor. ● Primary bone tumor. ◑ Can be primary soft tissue or bone tumor.

Fig. 331. Pyogenic granuloma (×140) (see Figs. 332, 981).

Fig. 332. Pyogenic granuloma (×85) (see Figs. 331, 981).

Fig. 333. Benign fibroblastic fibrous histiocytoma (×140) (see Figs. 334, 352, 960, 1021, 1044).

Fig. 334. Benign fibroblastic fibrous histiocytoma, deeply infiltrating (×140) (see Figs. 333, 352, 960, 1021, 1044).

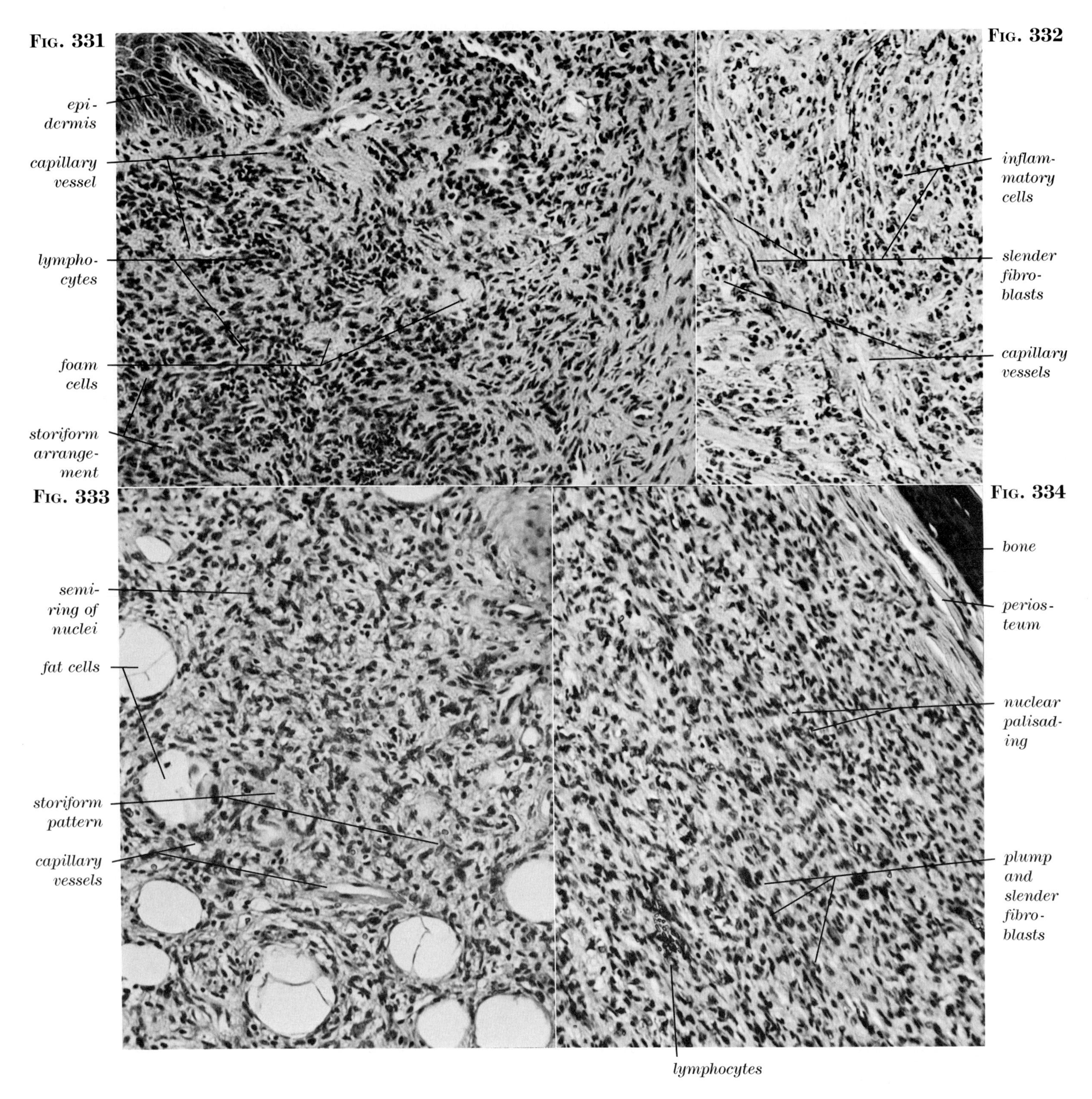

Fig. 335. Stromal sarcoma of uterus (×140) (see Figs. 465, 969).

Fig. 336. Fibroma, so-called "dermatofibroma" (×140) (see Figs. 859, 884, 886, 904, 915, 917, 1027, 1109).

Fig. 337. Malignant thymoma (×85) (see Figs. 375, 546, 617, 966).

Fig. 338. Desmoplastic epidermoid carcinoma (×350) (see Fig. 673).

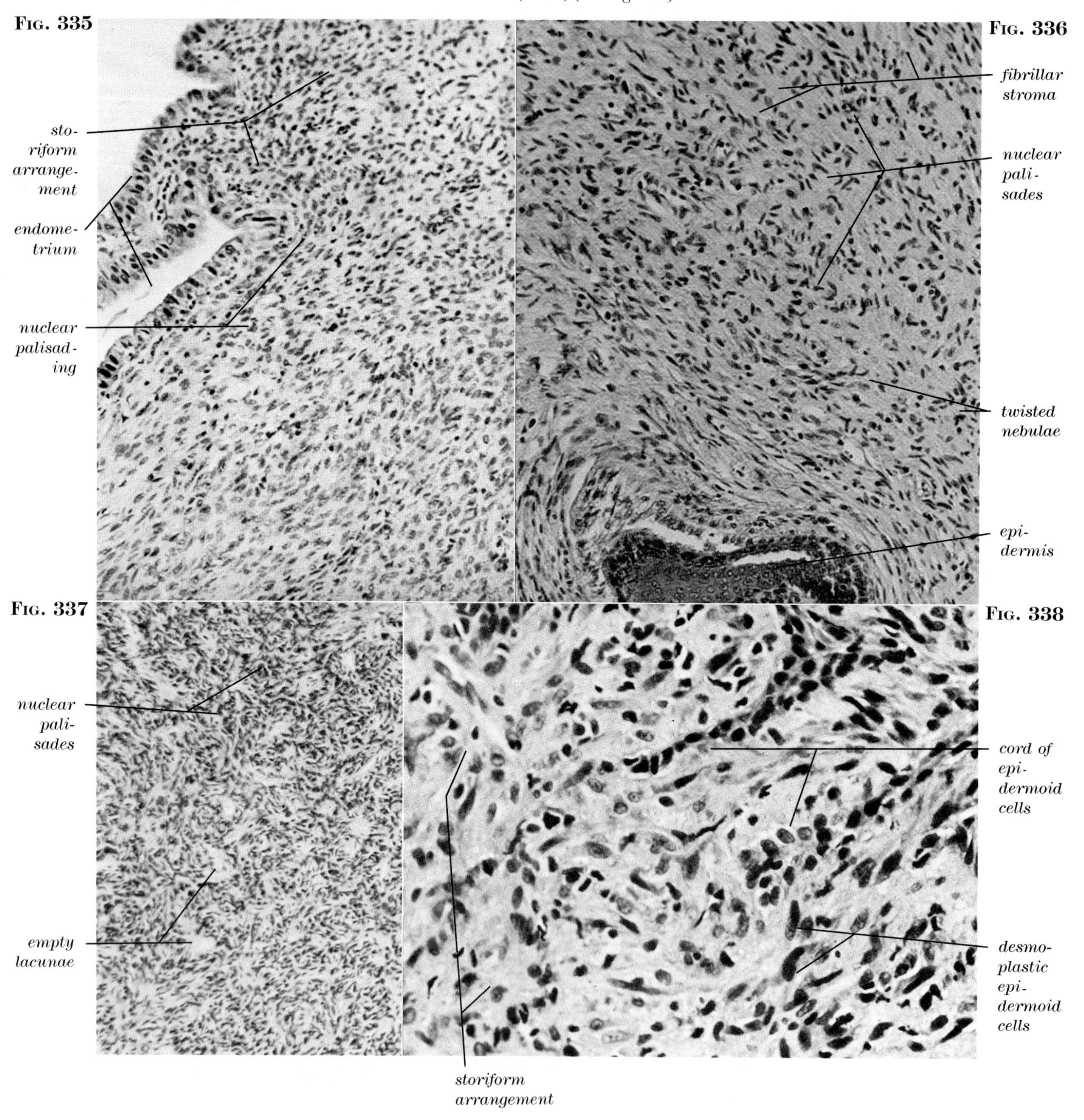

Fig. 339. Malignant fibroblastic fibrous histiocytoma, so-called "dermatofibrosarcoma protuberans" (×350) (see Figs. 241, 301, 340, 350, 373, 645, 672, 710, 760, 833).

Fig. 340. Malignant fibroblastic fibrous histiocytoma (×350) (see Figs. 241, 339, 350, 373).

Fig. 341. Low grade leiomyosarcoma (×350) (see Figs. 72, 110, 193, 218, 273, 347, 360, 362, 366, 381, 404, 588, 618, 659, 664, 666, 667).

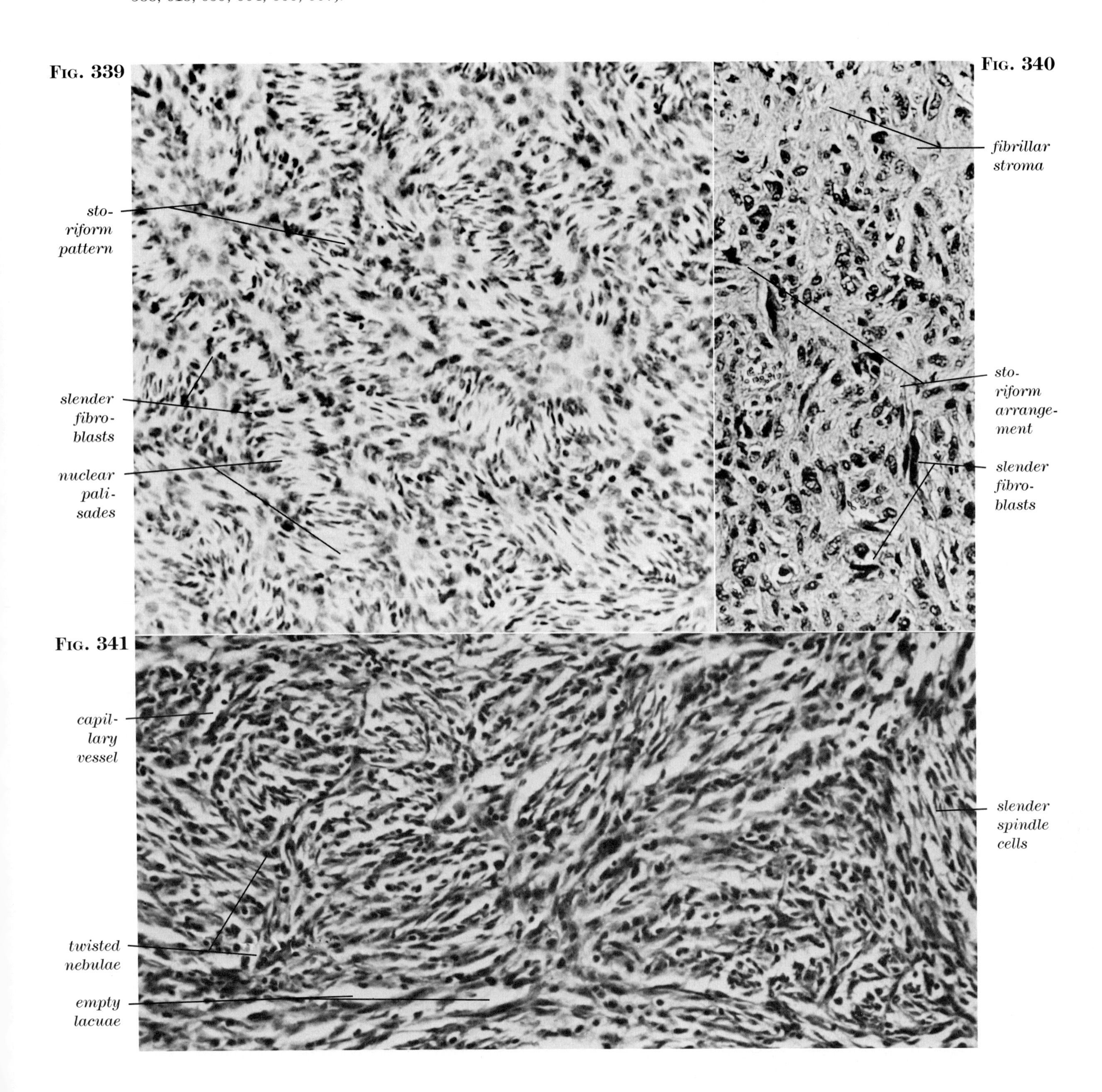

Fig. 342. Malignant undifferentiated peripheral nerve tumor growing as malignant fibroblastic fibrous histiocytoma (×350) (see Figs. 519, 596, 827, 982).

Fig. 343. Fibroblastic liposarcoma (×350) (see Figs. 30, 86, 440, 641, 668, 837, 848, 926, 998).

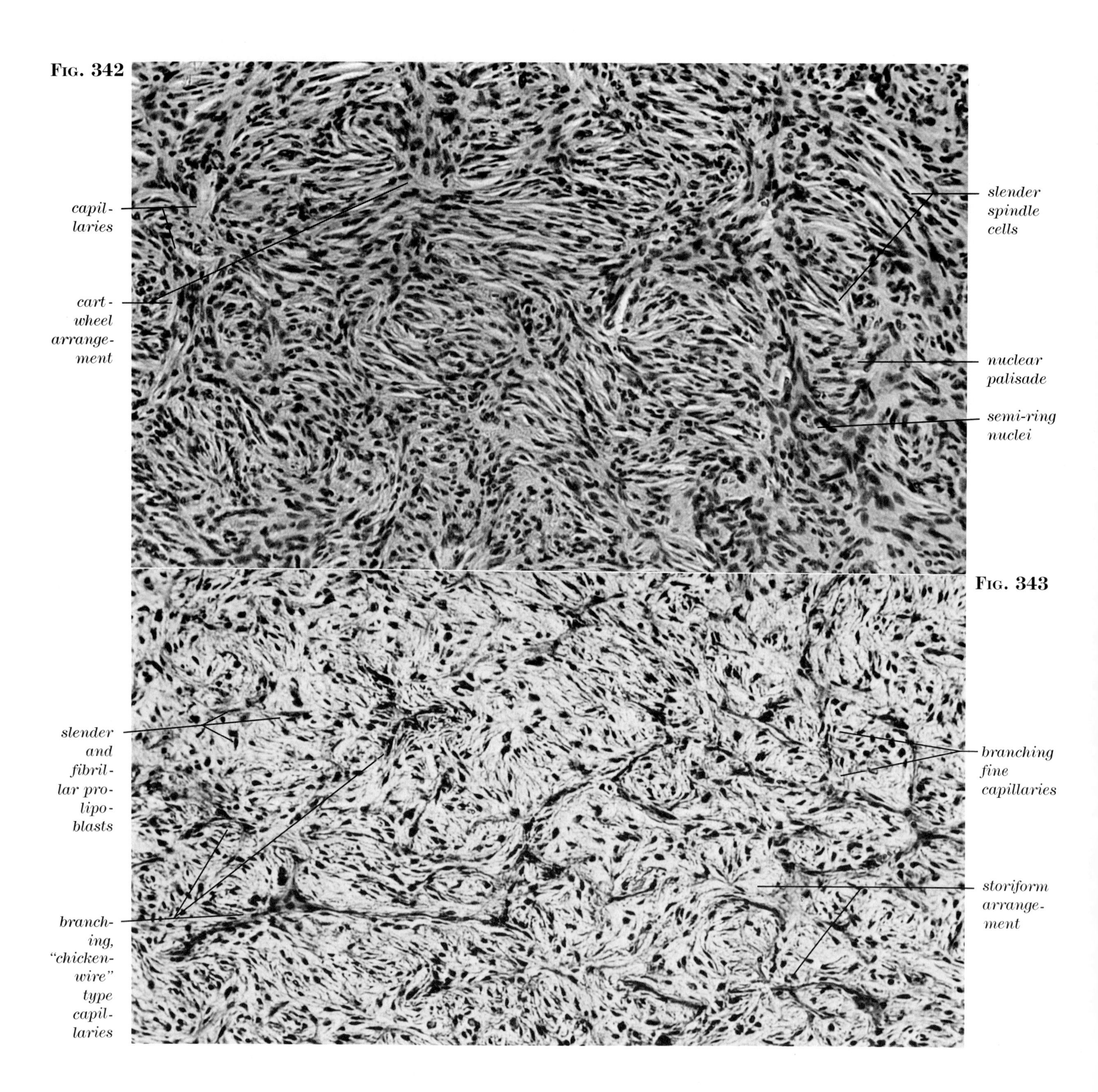

Fig. 344. Low grade malignant pleomorphic fibrous histiocytoma (×350) (see Figs. 88, 237, 460, 577, 579, 598, 679, 780, 788, 789).

Fig. 345. Fasciitis (×350) (see Figs. 82, 368, 389, 403).

Fig. 346. Borderline or atypical pleomorphic fibrous histiocytoma (×350) (see Figs. 135, 417, 422, 459, 722, 782).

Fig. 347. Low grade leiomyosarcoma of stomach (×350) (see Figs. 72, 110, 193, 220, 273, 341, 360, 362, 366, 381, 404, 588).

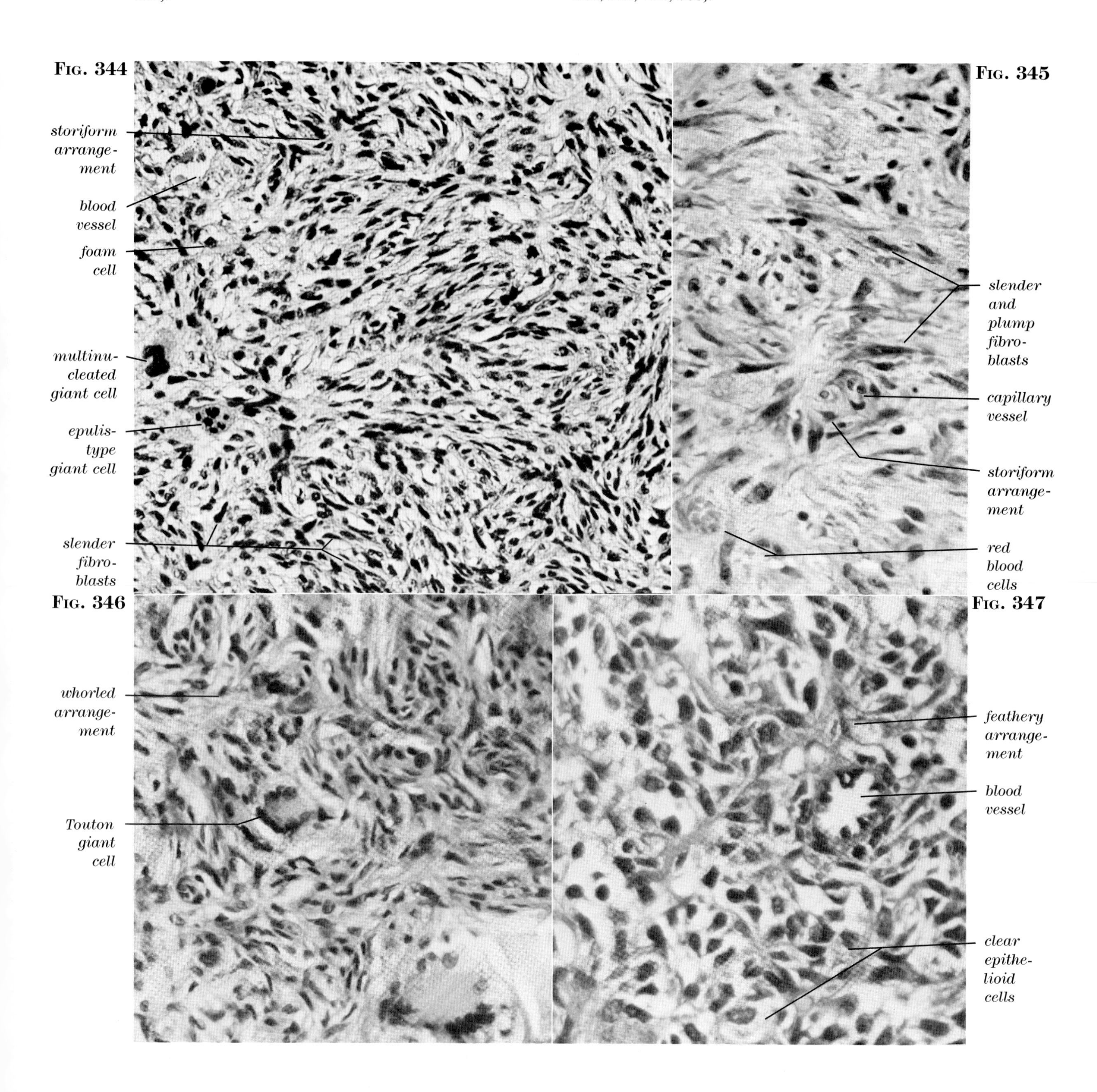

Fig. 348. Benign ovarian stroma (×350) (see Fig. 623).

Fig. 349. Immature, malignant teratoma of testis (×350) (see Figs. 279, 688, 922).

Fig. 350. Malignant fibroblastic fibrous histiocytoma of tibia (×350) (see Figs. 111, 143, 339, 340, 373, 645, 672, 710, 760).

Fig. 351. Malignant cystosarcoma of breast metastatic to bone (×350) (see Figs. 47, 377, 657, 663).

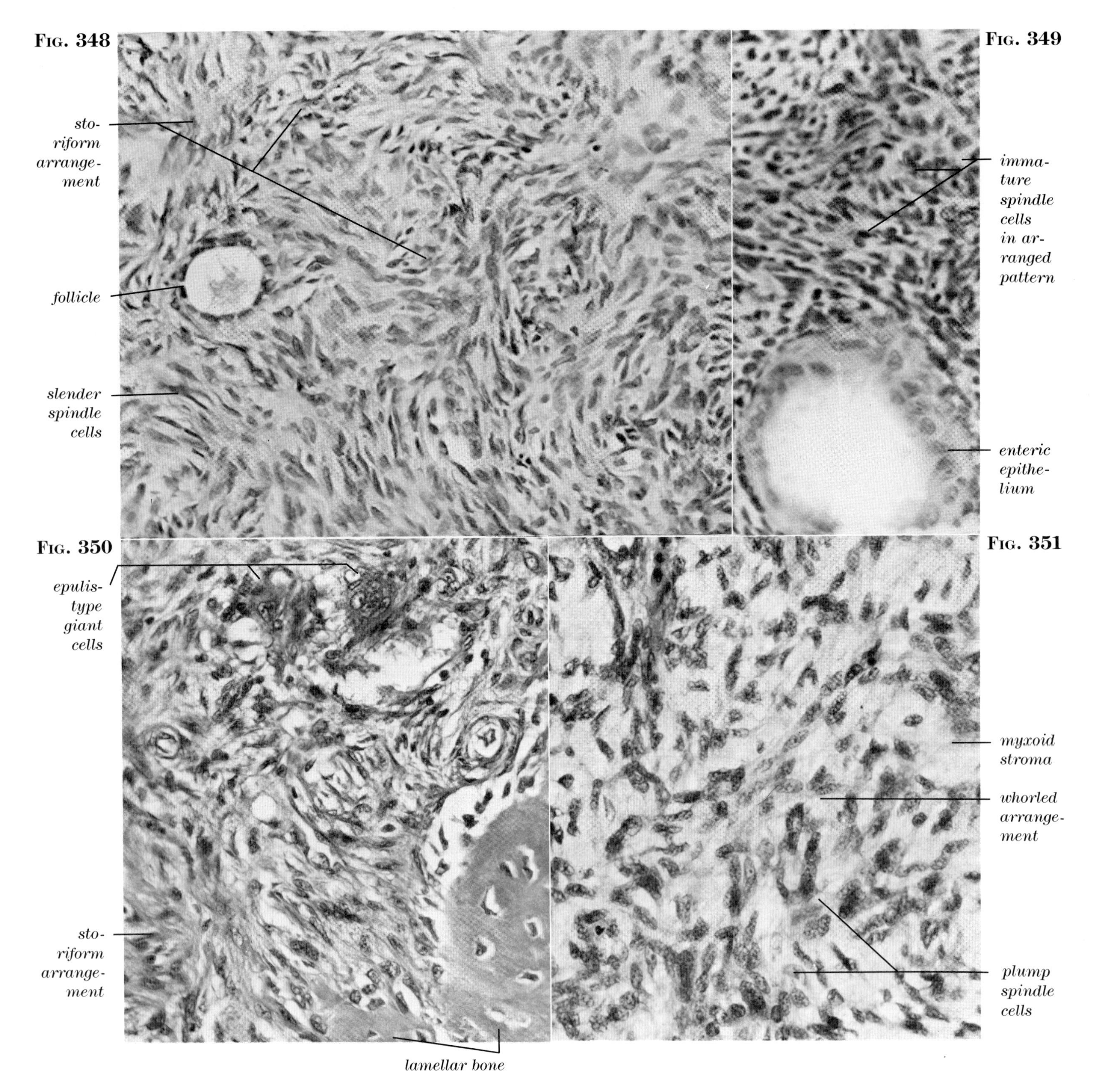

Fig. 352. Benign fibrous histiocytoma, so-called "sclerosing hemangioma," of lung (×35) (see Figs. 333, 334, 960, 1021).

Fig. 353. Parosteal, low grade osteosarcoma of femur (×35) (see Figs. 129, 217, 224, 249, 287, 354).

Fig. 354. Parosteal, low grade osteosarcoma (×350) (see Figs. 109, 129, 217, 224, 287, 353, 656, 1082, 1154).

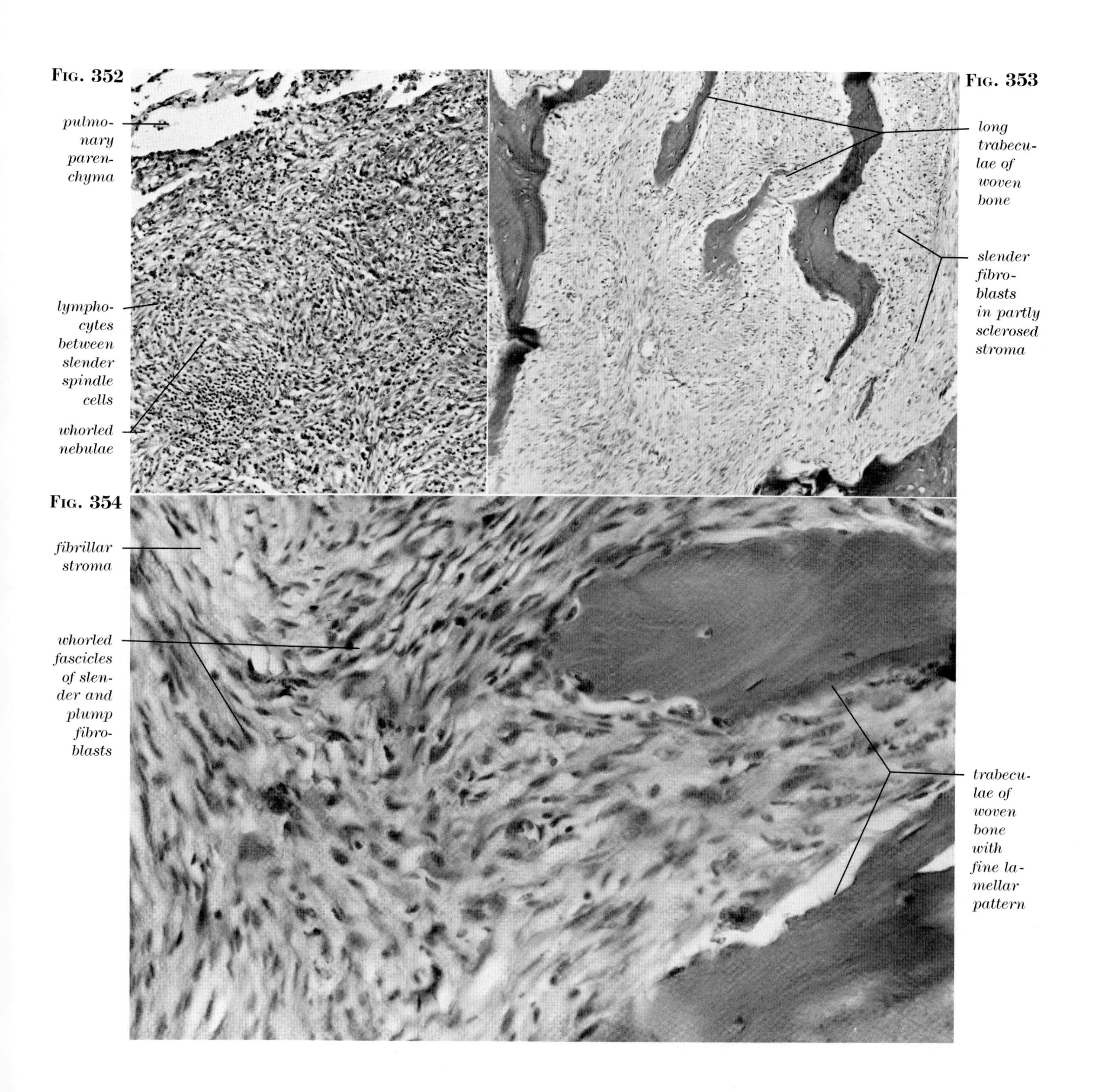

Fig. 355. Low grade malignant schwannoma (×140) (see Figs. 21, 38, 80, 200, 238, 242, 246, 252, 357, 382, 580, 583).

Fig. 357. High grade, poorly differentiated, malignant schwannoma (×140) (see Figs. 51, 238, 355, 382, 580, 583, 591, 603, 619).

Fig. 356. Borderline or atypical schwannoma (×140) (see Figs. 257, 556, 846, 850, 1228).

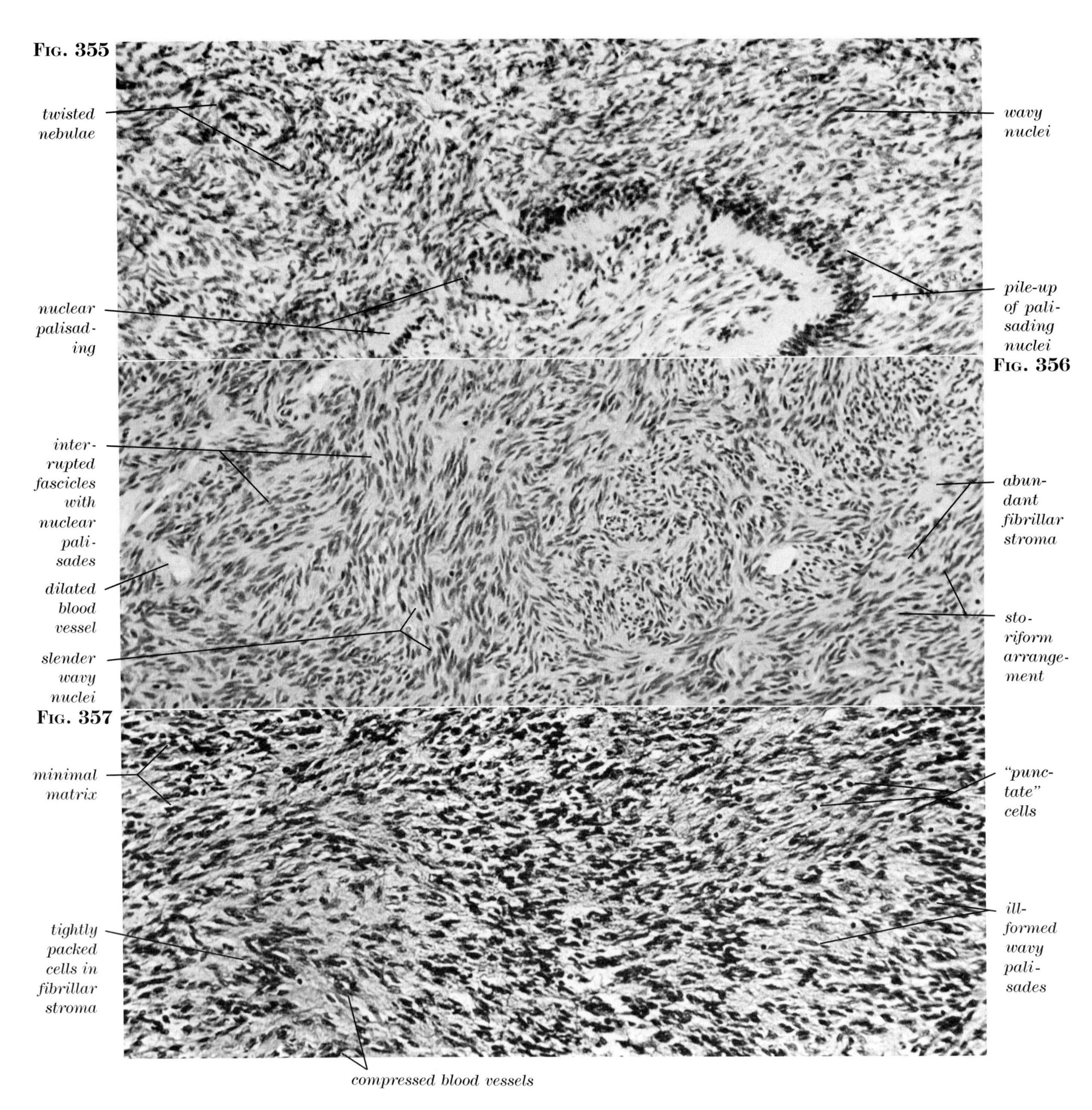

Fig. 358. A detached pseudoencapsulated nodule of peritoneal leiomyomatosis (×10) (see Figs. 50, 359, 361, 402, 654).

Fig. 359. Peritoneal leiomyomatosis (×140) (see Figs. 50, 358, 361, 380, 402, 654, 1010).

Fig. 360. Low grade peripheral leiomyosarcoma (×140) (see Figs. 72, 193, 218, 220, 231, 273, 341, 347, 360, 362, 366, 381, 404, 588).

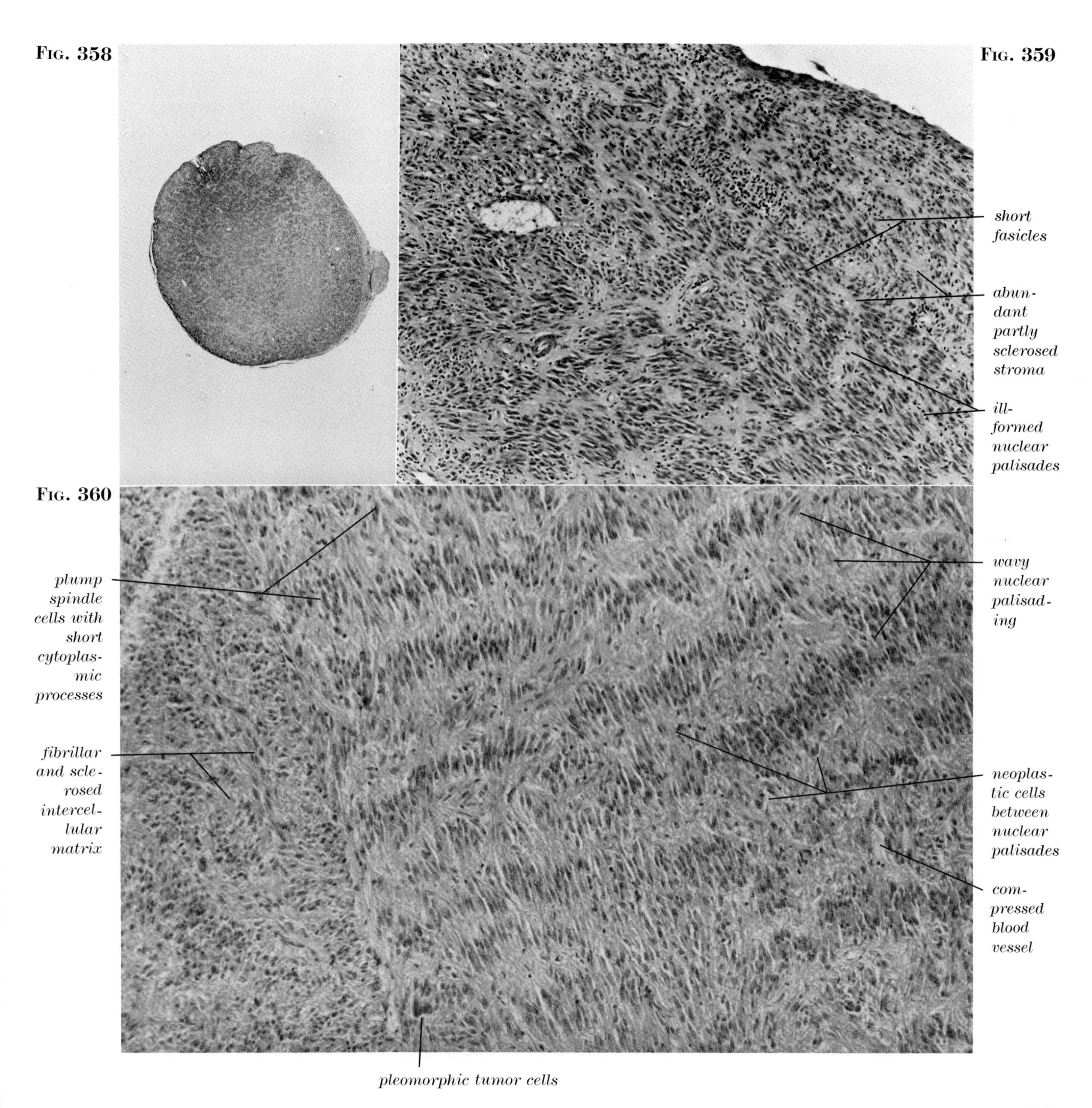

Fig. 361. Leiomyoma of foot (×140) (see Figs. 358, 359, 380, 402, 654, 1010).

Fig. 362. Low grade leiomyosarcoma of retroperitoneum (×140) (see Figs. 72, 273, 341, 347, 360, 366, 381, 404, 588).

Fig. 363. Neurofibroma with schwannian features (×85) (see Figs. 11, 74, 245, 487, 652, 840, 871, 955, 993, 1013).

Fig. 364. Embryonal rhabdomyosarcoma (×85) (see Figs. 10, 20, 78, 207, 372, 378, 379, 426, 427, 443, 446).

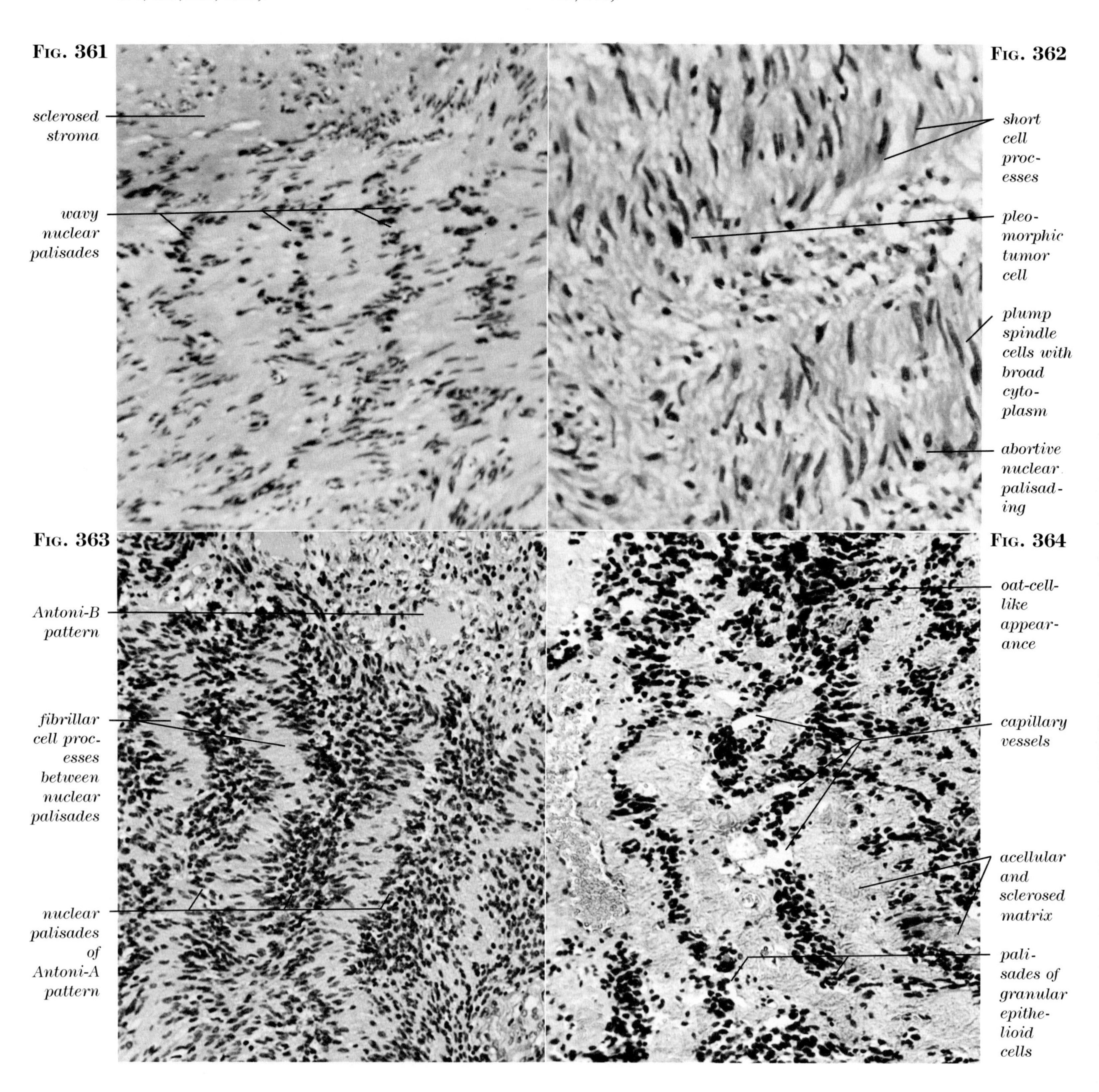

Fig. 365. Plexiform neurofibroma (×85) (see Figs. 11, 74, 363, 487, 527, 627, 842, 942, 1114).

Fig. 366. Low grade cutaneous leiomyosarcoma (×85) (see Figs. 72, 193, 218, 220, 273, 341, 347, 360, 362, 381, 404, 588).

Fig. 367. Aponeurotic fibromatosis of hand (×85) (see Figs. 17, 280, 385, 387, 419, 581, 582, 616, 860).

Fig. 368. Fasciitis (×85) (see Figs. 82, 345, 389, 403, 408, 578, 605, 621, 834, 851, 855, 858, 947, 956, 971, 991).

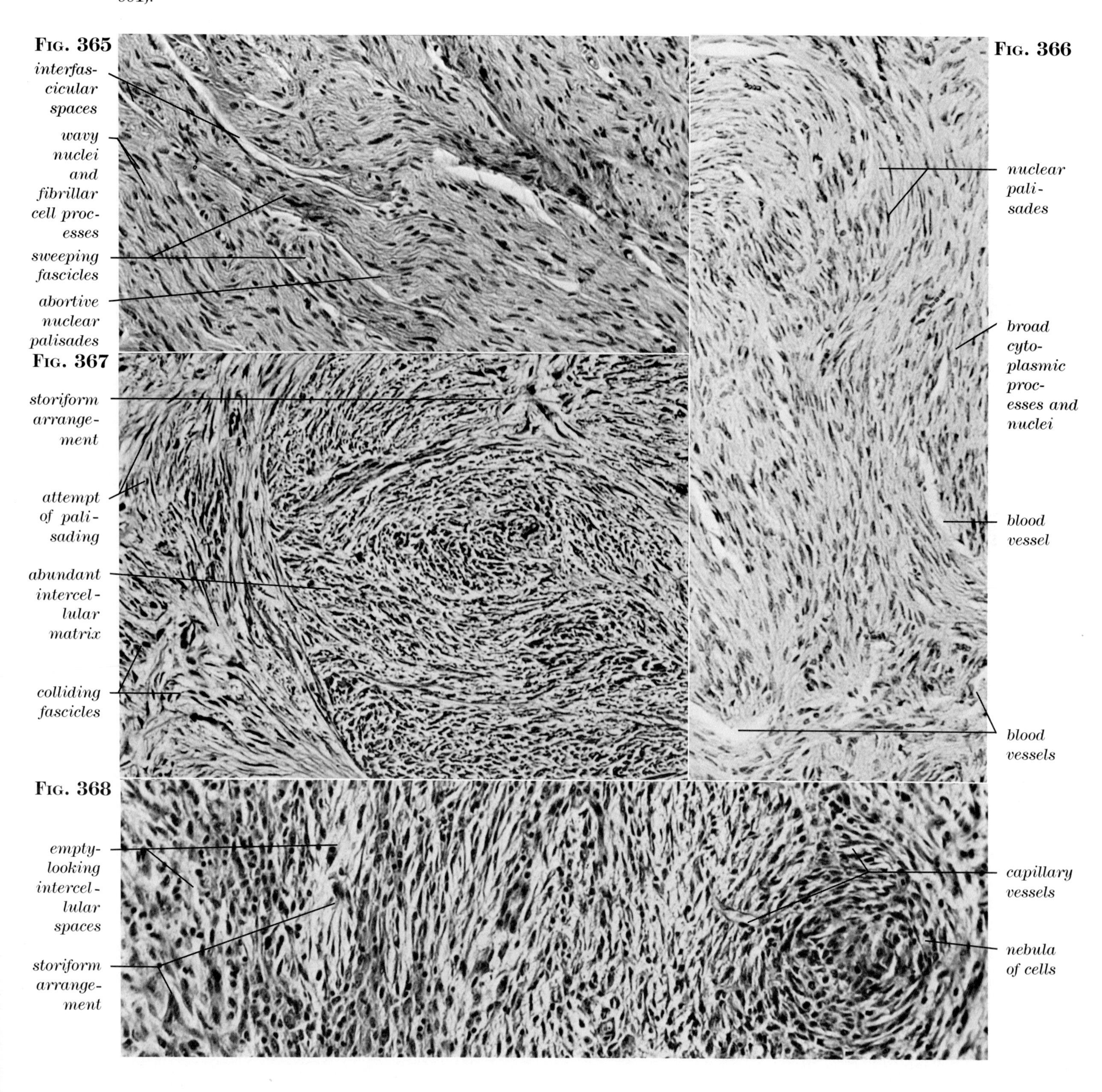

Fig. 369. High grade neurofibrosarcoma (×85) (see Figs. 22, 23, 79, 236, 254, 256, 264, 370, 401, 584, 624, 626, 838, 1116).

Fig. 370. High grade neurofibrosarcoma (×140) (see Figs. 26, 79, 254, 369, 401).

Fig. 371. High grade fibrosarcoma (×85) (see Figs. 33, 91, 240, 297, 374, 393, 601, 608, 625, 647, 660, 665, 671).

Fig. 372. Embryonal rhabdomyosarcoma (×140) (see Figs. 28, 66, 78, 120, 226, 263, 378).

Fig. 373. Malignant fibroblastic fibrous histiocytoma metastatic to lung (×85) (see Figs. 76, 350, 645, 672).

Fig. 374. High grade postirradiation fibrosarcoma (×140) (see Figs. 60, 91, 297, 371, 393, 410, 484, 601, 608).

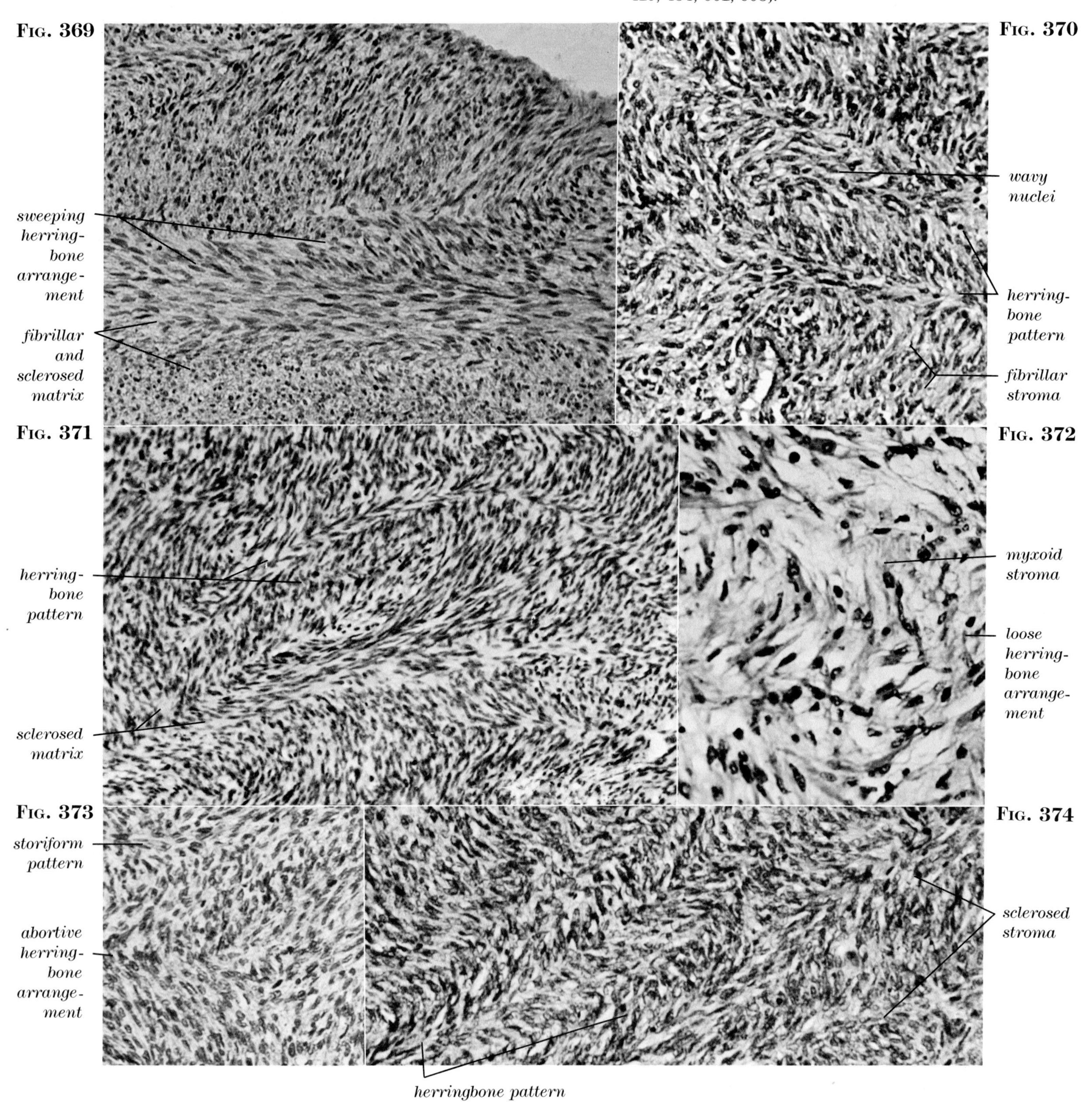

FIG. 375. Malignant thymoma (×140) (see Figs. 337, 546, 617, 966).

FIG. 376. Desmoid tumor, low grade fibrosarcoma (×85) (see Figs. 9, 77, 244, 282, 386, 397, 662).

FIG. 377. Malignant cystosarcoma phyllodes of breast (×140) (see Figs. 47, 351, 657, 663).

FIG. 378. Embryonal rhabdomyosarcoma (×85) (see Figs. 66, 78, 281, 364, 372, 379, 426, 427, 443, 446, 472, 515).

FIG. 379. Embryonal rhabdomyosarcoma (×350) (see Figs. 67, 291, 364, 372, 426, 443, 472, 637).

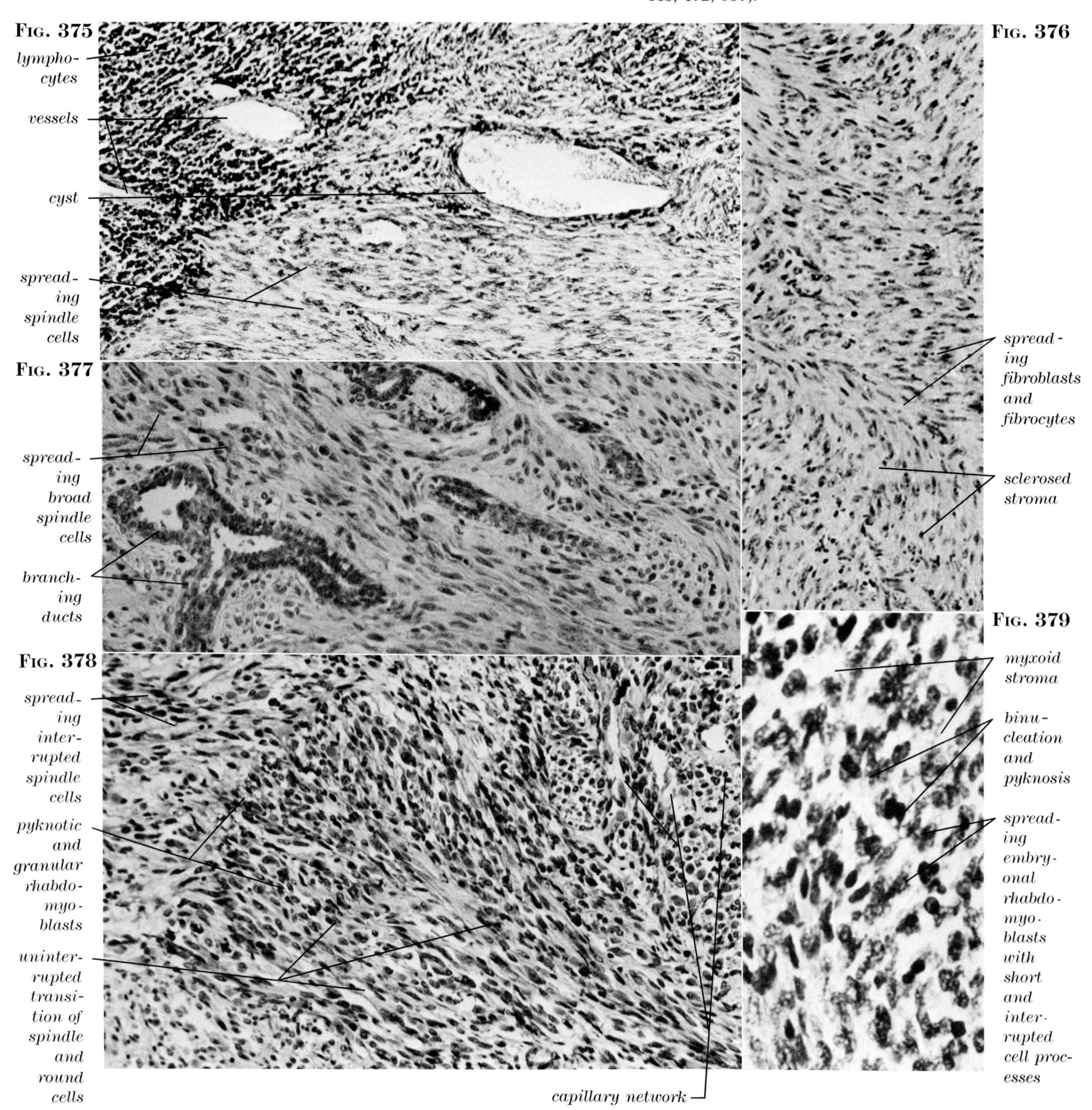

Fig. 380. Cutaneous leiomyoma (×140) (see Figs. 358, 359, 361, 402, 654, 1010).

Fig. 381. High grade leiomyosarcoma (×350) (see Figs. 72, 231, 341, 347, 360, 362, 366, 404).

Fig. 382. High grade malignant schwannoma (×140) (see Figs. 80, 242, 357, 580, 583, 591, 603, 619, 628, 661, 815).

Fig. 383. Adult Wilms' tumor (×350) (see Figs. 45, 835).

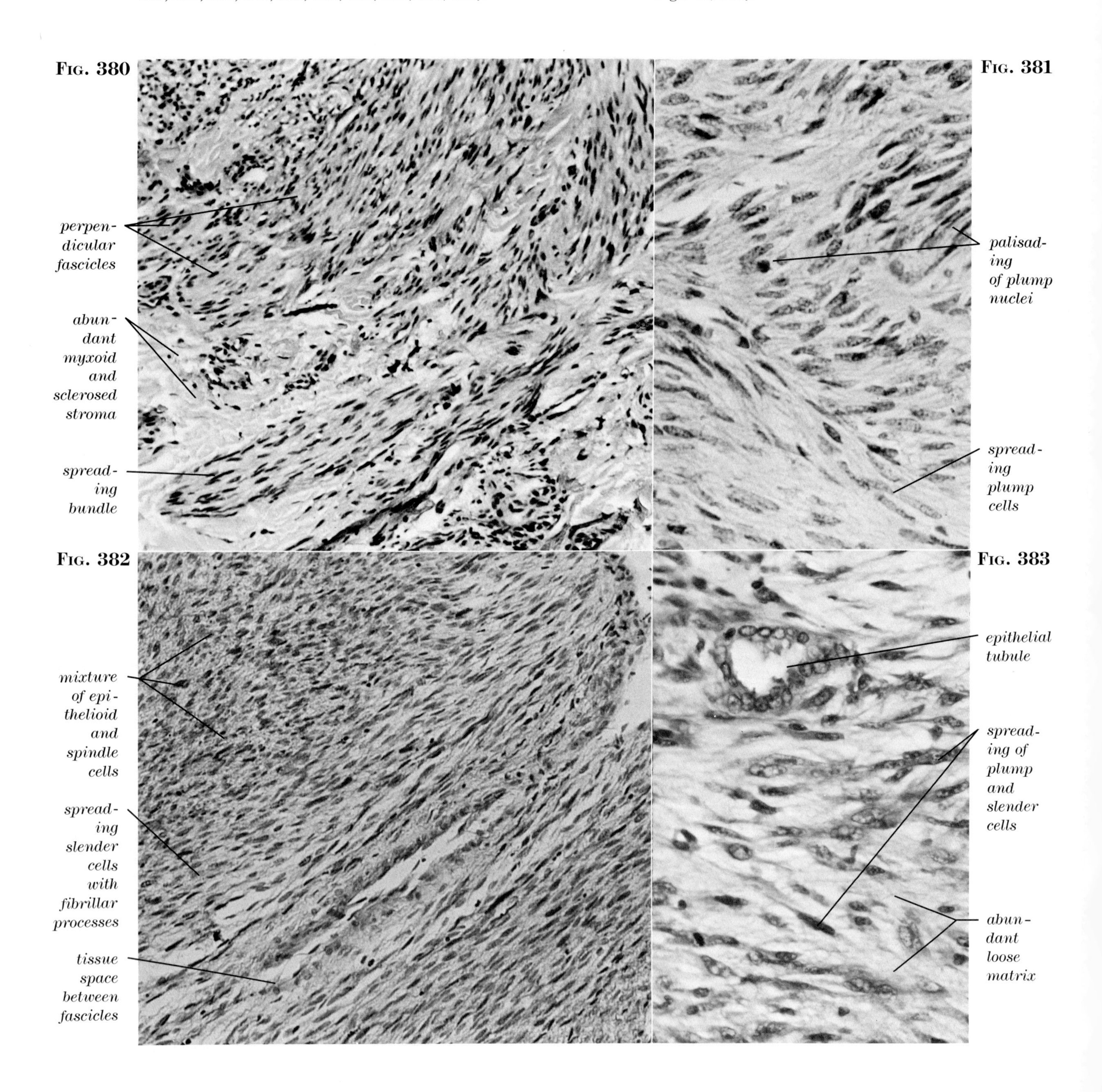

Fig. 384. Myositis ossificans (×140) (see Figs. 145, 620, 913, 1083, 1084, 1085).

Fig. 385. Aponeurotic fibromatosis (×140) (see Figs. 17, 367, 387, 419, 581, 616, 860, 873, 878, 1022).

Fig. 386. Desmoid tumor, low grade fibrosarcoma (×140) (see Figs. 77, 185, 282, 376, 397, 662, 686, 865).

Fig. 387. Aponeurotic fibromatosis (×350) (see Figs. 17, 280, 385, 419, 582, 861, 874, 882, 1025, 1035).

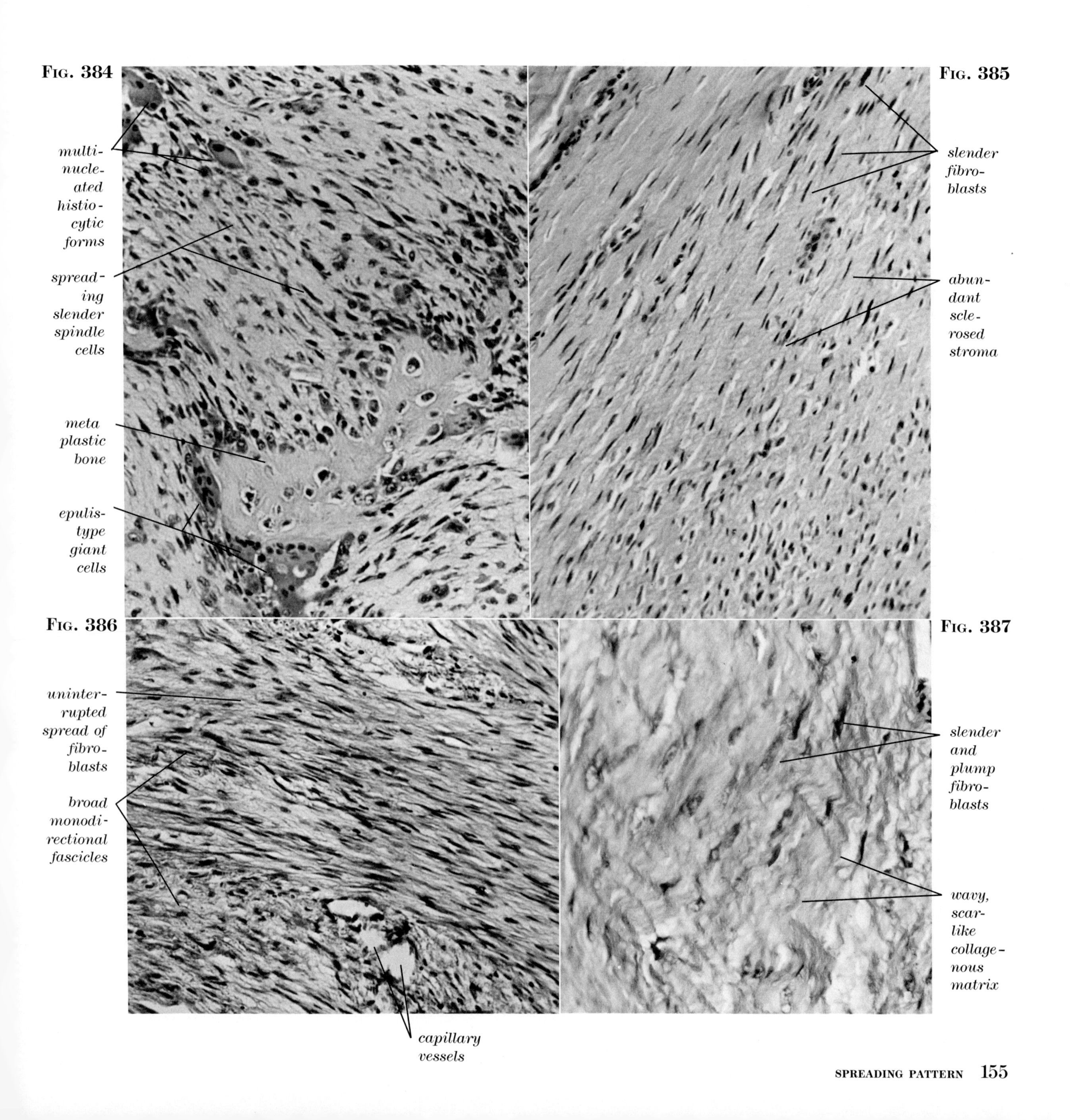

Fig. 388. Kaposi's sarcoma (×140) (see Figs. 63, 629, 655, 658, 977, 986, 1126).

Fig. 390. Mesenchymal chondrosarcoma (×140) (see Figs. 103, 151, 175, 180, 318, 444, 507, 595, 674, 707, 920).

Fig. 389. Fasciitis (×350) (see Figs. 82, 345, 368, 403, 408, 578, 605, 621, 834, 851).

Fig. 391. Monophasic tendosynovial sarcoma, spindle cell type (×350) (see Figs. 39, 85, 179, 392, 396, 398, 399, 467, 491, 492, 542).

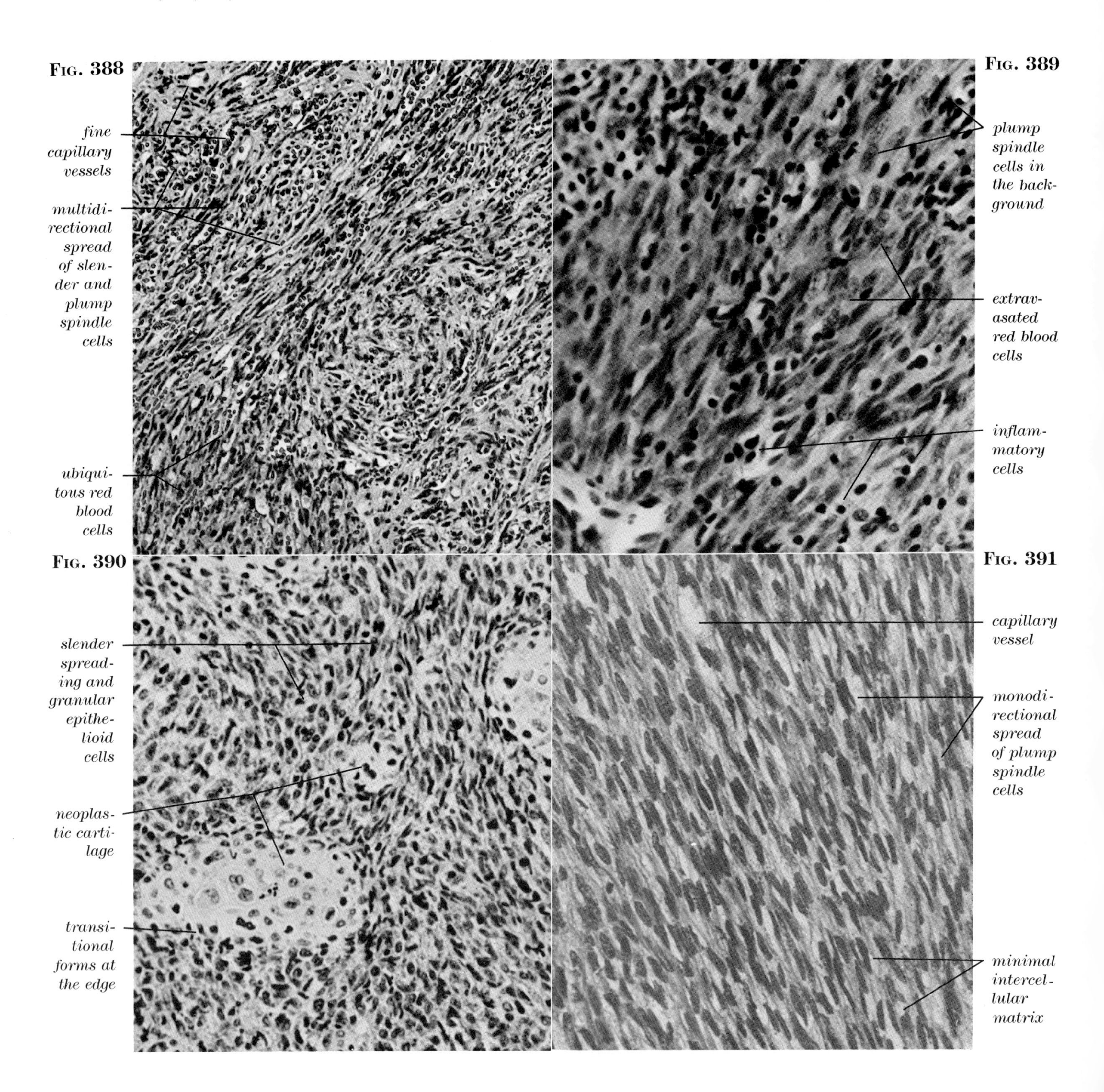

Fig. 392. Monophasic tendosynovial sarcoma, spindle cell type (×140) (see Figs. 64, 85, 179, 190, 315, 320, 326, 391, 396, 398, 399).

Fig. 393. High grade fibrosarcoma (×350) (see Figs. 91, 297, 371, 374, 601, 608, 625, 647, 660, 665, 671).

Fig. 394. Desmoplastic melanoma (×140) (see Figs. 409, 540, 669, 719, 748).

Fig. 395. Poorly differentiated fibroblastic osteosarcoma (×140) (see Figs. 108, 127, 142, 248, 405, 411, 451, 461, 602, 610, 611, 615).

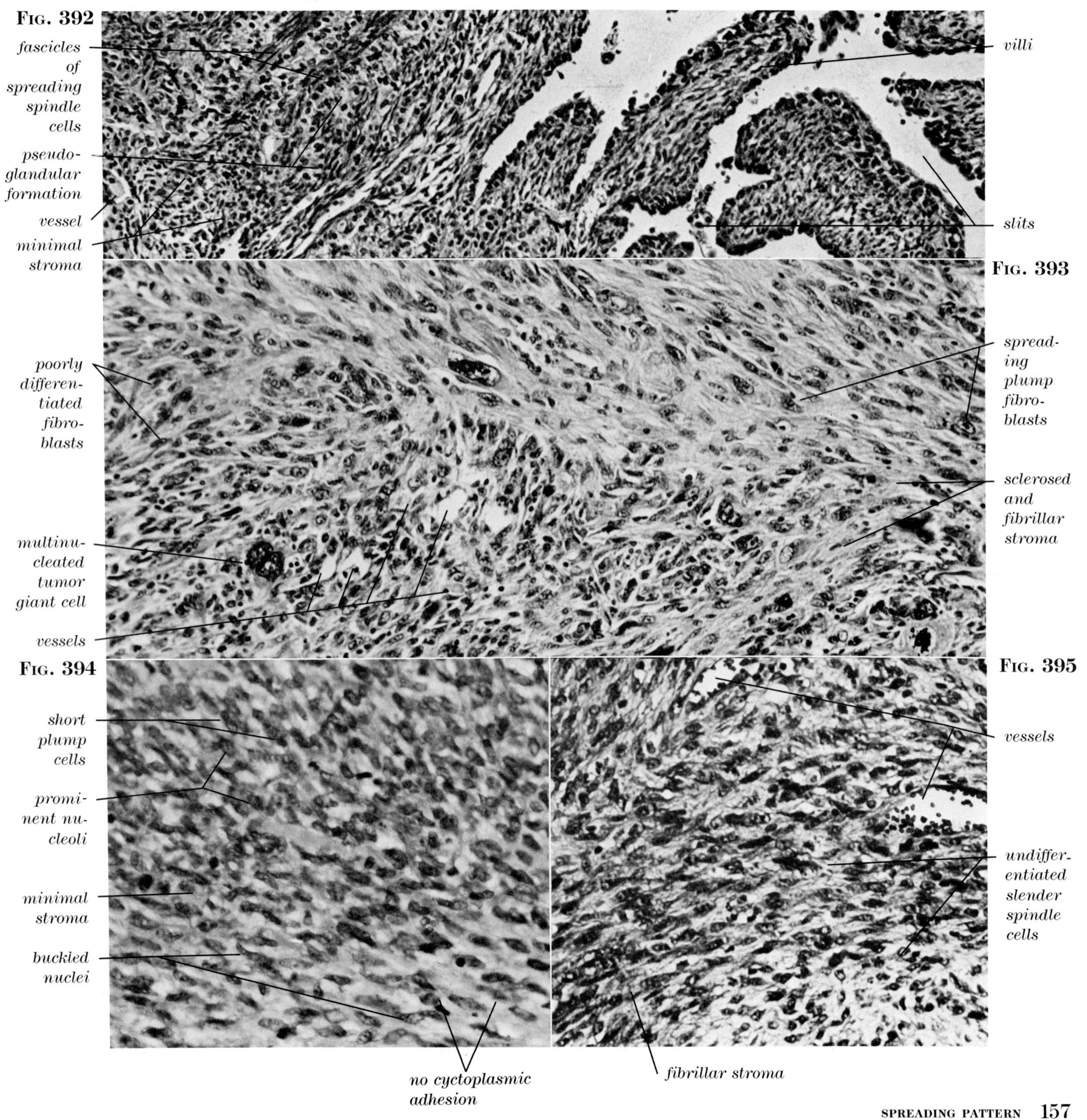

Fig. 396. Monophasic tendosynovial sarcoma, spindle cell type (×350) (see Figs. 39, 64, 85, 179, 190, 315, 391, 392, 398, 399, 467, 491, 492, 542).

Fig. 397. Desmoid tumor, low grade fibrosarcoma (×350) (see Figs. 9, 36, 52, 77, 185, 202, 244, 282, 376, 386, 662, 686, 865, 899, 1118).

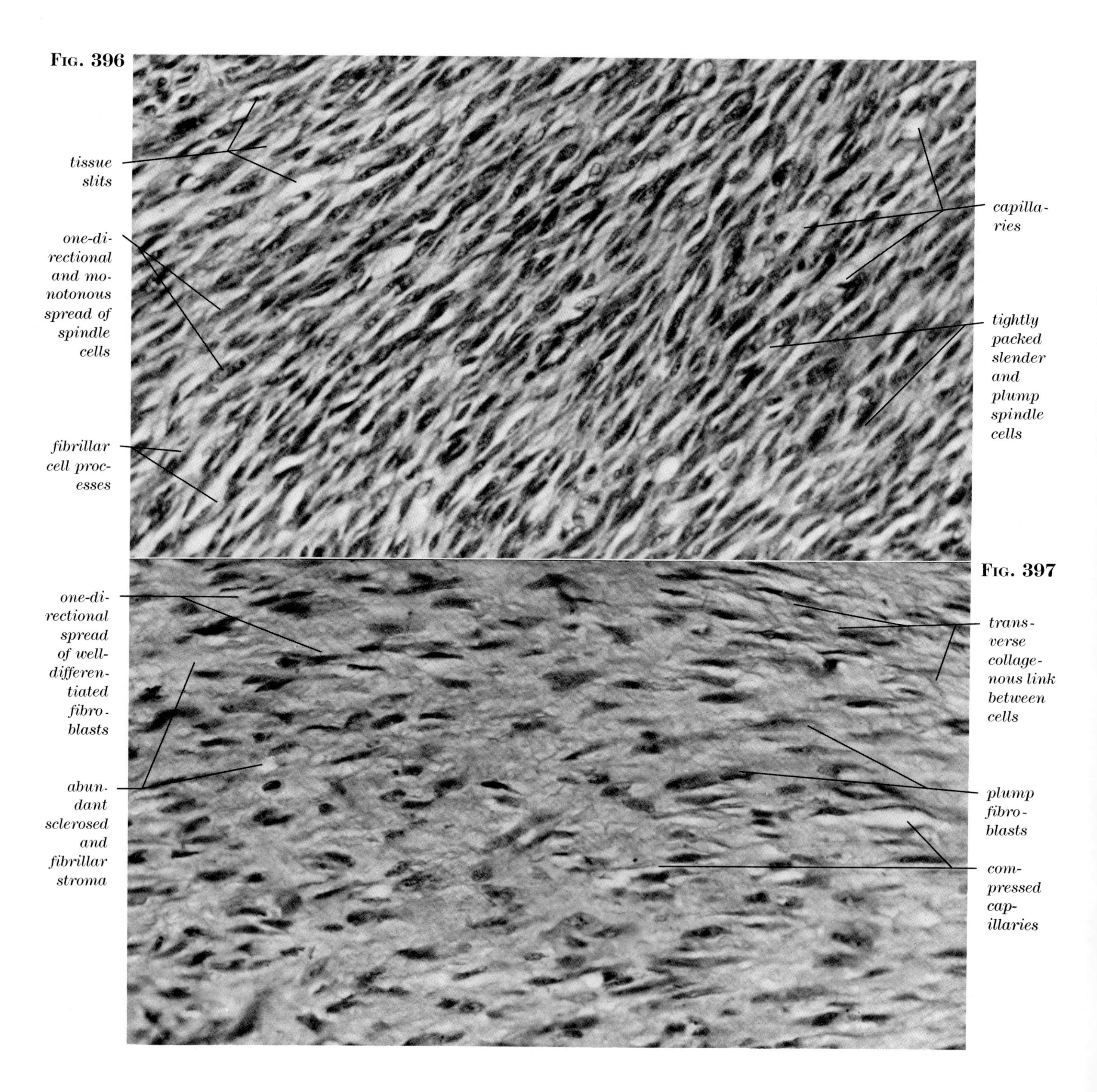

Fig. 398. Monophasic tendosynovial sarcoma, spindle cell type, with sclerosing stroma (×350) (see Figs. 85, 190, 320, 326, 392, 396, 399, 467, 491, 492, 542, 622, 854, 864).

Fig. 399. Monophasic tendosynovial sarcoma, spindle cell type (×140) (see Figs. 85, 392, 396, 467, 491).

Fig. 400. Borderline fibroblastic peripheral nerve tumor, so-called "atypical neurofibroma" (×350) (see Figs. 11, 74, 204, 245, 363, 369, 370, 401, 487, 527, 584).

Fig. 401. High grade neurofibrosarcoma (×350) (see Figs. 34, 256, 369, 370, 584).

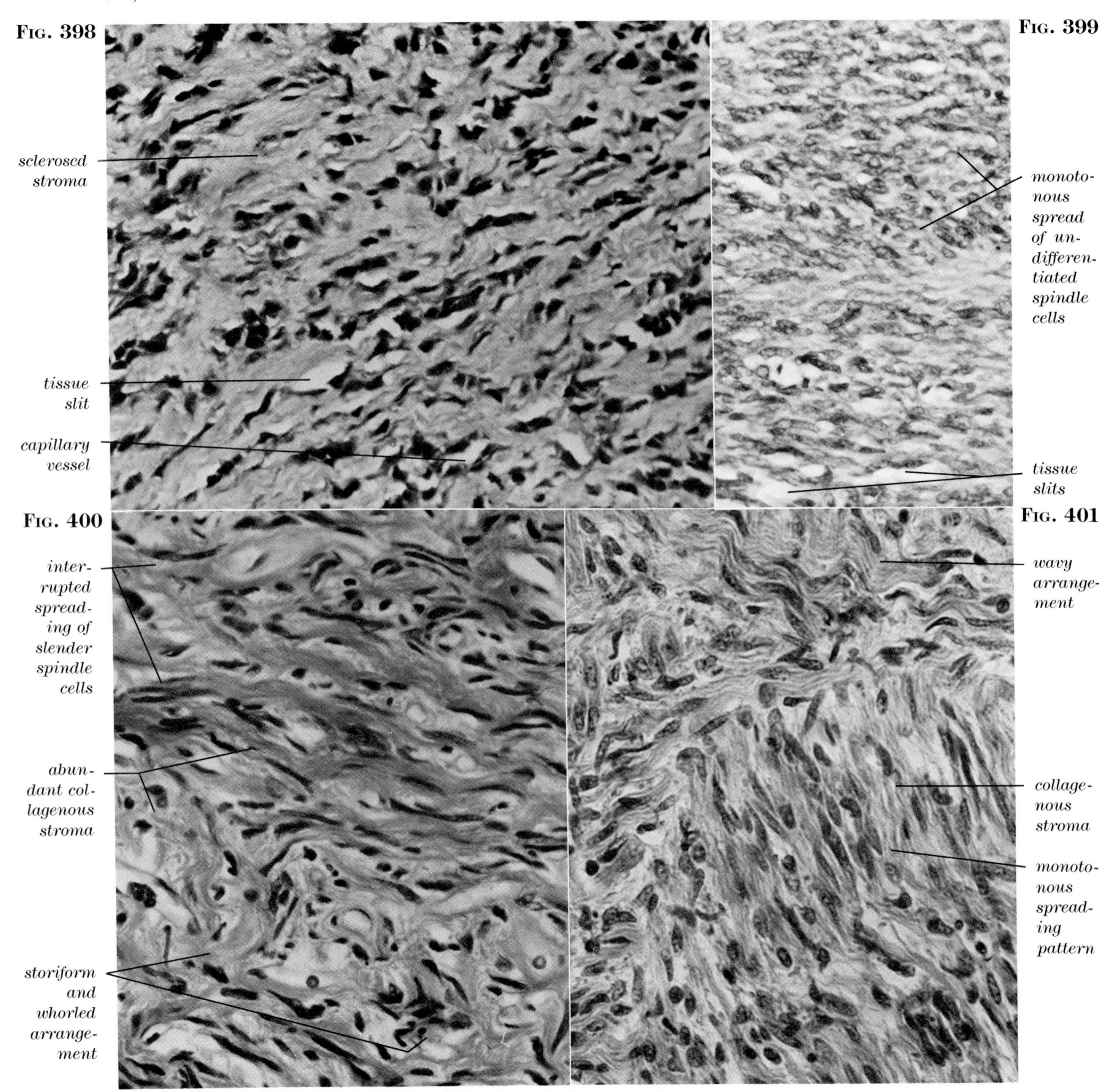

Fig. 402. Peritoneal leiomyomatosis, status 4 years after pregnancy (×140) (see Figs. 50, 358, 359, 361, 380, 654).

Fig. 403. Fasciitis (×350) (see Figs. 82, 345, 368, 389, 408, 578, 605, 621, 834, 851, 855, 858, 947, 956, 971, 991).

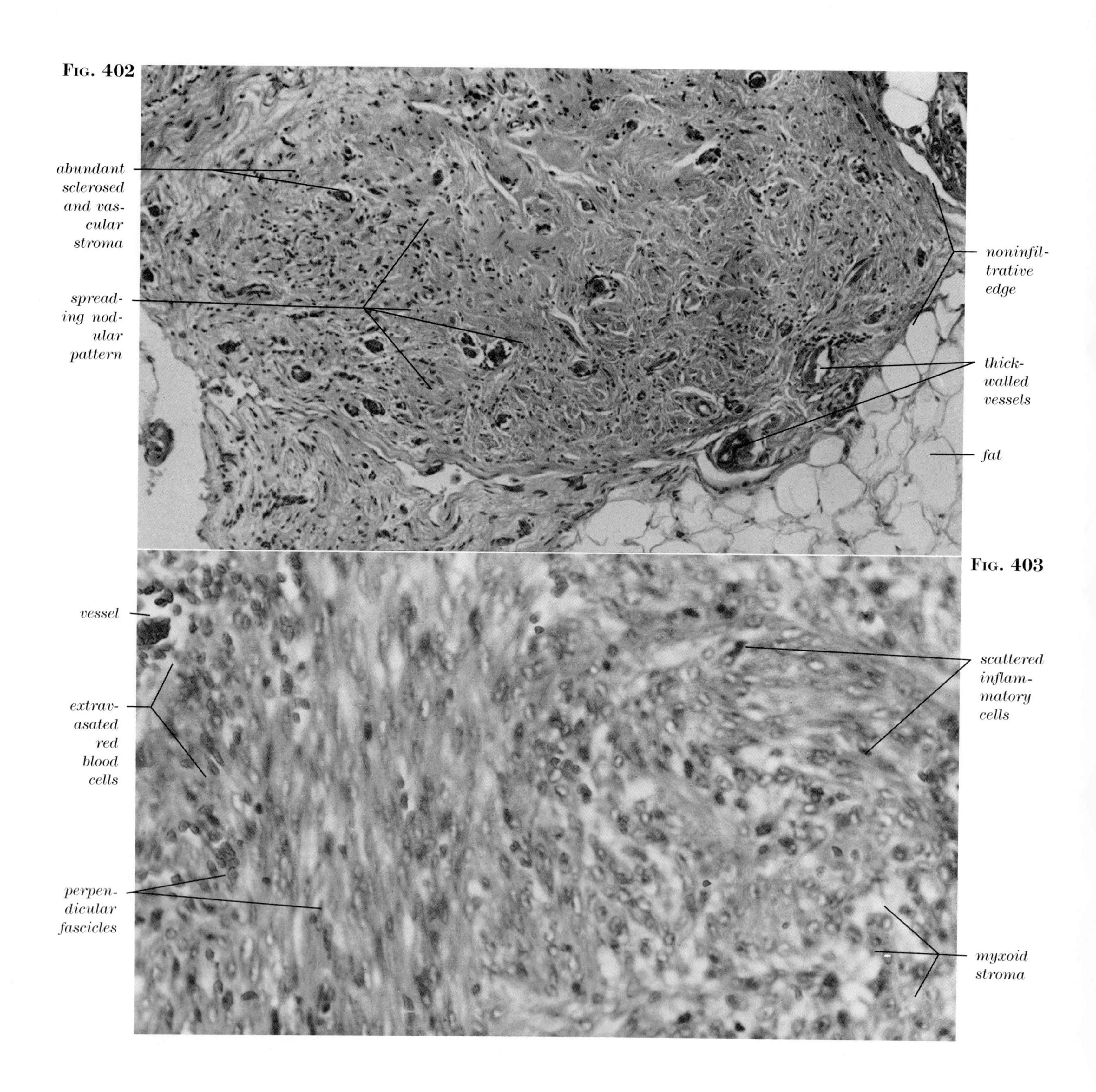

Fig. 404. High grade leiomyosarcoma (×350) (see Figs. 72, 110, 193, 231, 273, 341, 347, 360, 362, 366, 381, 588, 618, 659, 664, 666, 667).

Fig. 405. Osteosarcoma metastatic to lung (×140) (see Figs. 109, 146, 149, 251, 395, 411).

Fig. 406. Malignant fibrous mesothelioma (×350) (see Figs. 189, 266, 268, 614).

Fig. 407. Chondromyxoid fibroma (×350) (see Figs. 94, 113, 114, 116, 128, 191, 506, 772).

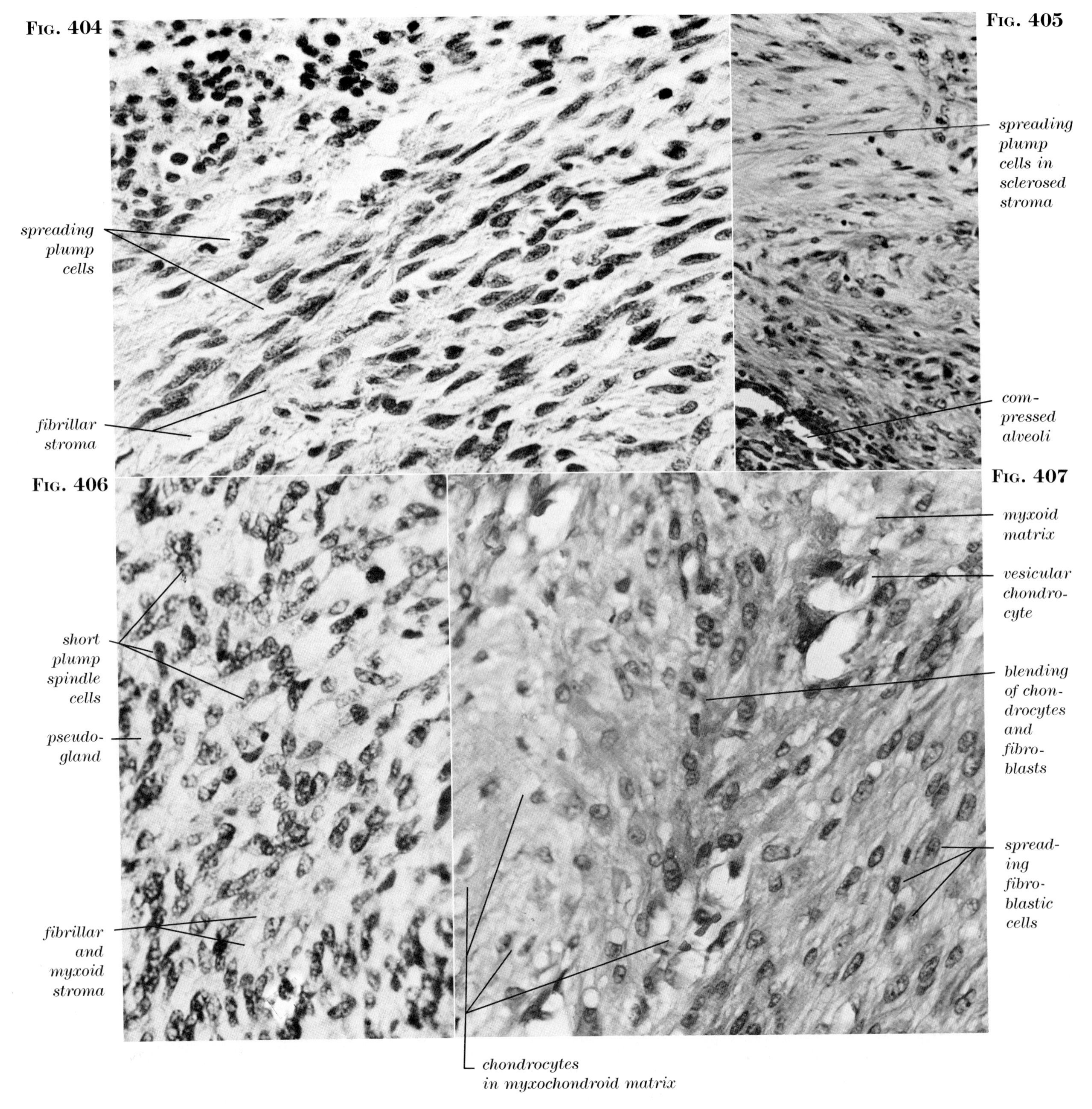

Fig. 408. Fasciitis (×350) (see Figs. 82, 342, 368, 389, 403, 578, 605, 621, 834, 851, 855, 858).

Fig. 409. Desmoplastic melanoma (×350) (see Figs. 394, 540, 669, 719, 748).

Fig. 410. Low grade fibrosarcoma (×350) (see Figs. 91, 131, 260, 376, 386, 397, 662, 686, 832, 900).

Fig. 411. Fibroblastic osteosarcoma (×350) (see Figs. 149, 157, 272, 395, 451).

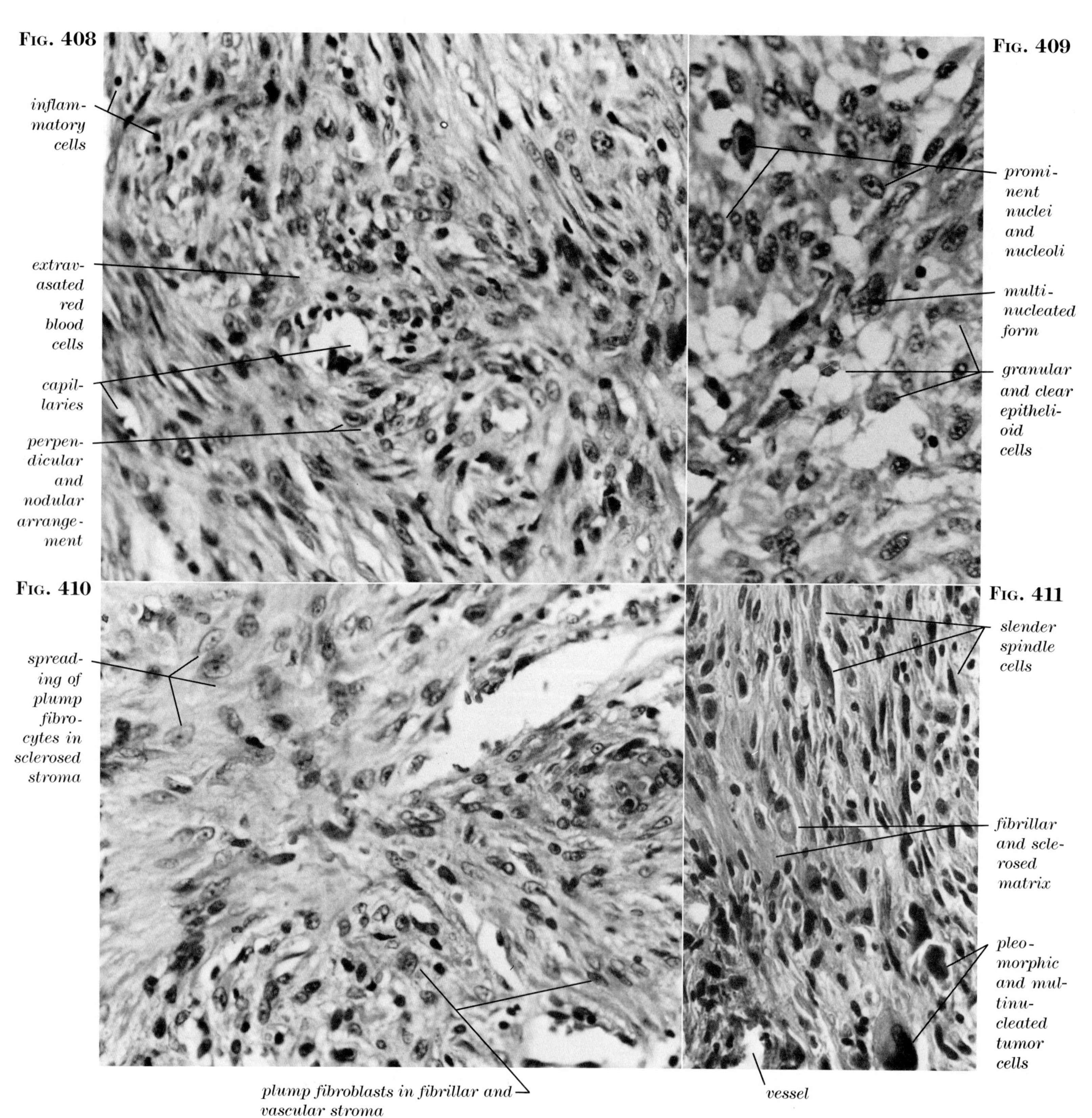

Fig. 412. Lipoblastoma (×85) (see Figs. 413, 437, 996).

Fig. 413. Lipoblastoma (×85) (see Figs. 412, 437, 996).

Fig. 414. Benign cystosarcoma phyllodes of breast (×85) (see Figs. 43, 48, 247, 921, 938).

Fig. 415. Lipoblastic liposarcoma (×85) (see Figs. 55, 86, 199, 448, 450, 543, 687, 753, 934).

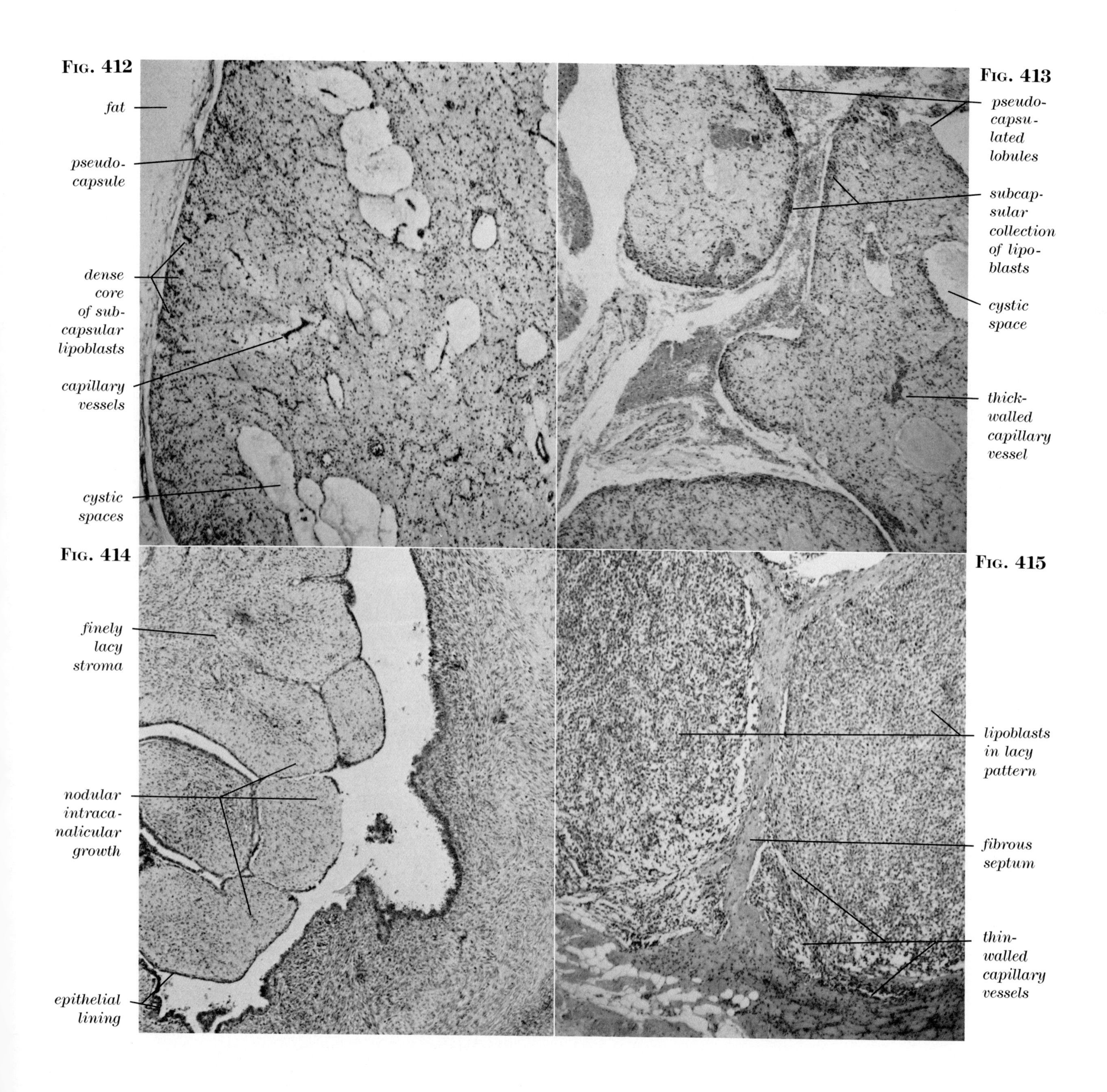

FIG. 416. Embryonal rhabdomyosarcoma, so-called "botryoid or myxoid rhabdomyosarcoma" of nasal cavity (×140) (see Figs. 78, 120, 418, 587, 644, 923, 929).

FIG. 419. Mesenteric fibromatosis infiltrating intestine (×140) (see Figs. 17, 387, 581, 582, 616).

FIG. 417. (Top center) Borderline or atypical, so-called "juvenile cutaneous fibrous histiocytoma," composed of foamy histiocytes and lymphocytes (×140) (see Figs. 346, 422, 459, 722, 782).

FIG. 418. Embryonal rhabdomyosarcoma of bladder (×140) (see Figs. 416, 587, 644, 923, 929, 944, 952, 1033, 1251).

FIG. 420. Leiomyoblastoma (×85) (see Figs. 196, 270, 274, 277, 278, 429, 503, 513, 691, 713, 733).

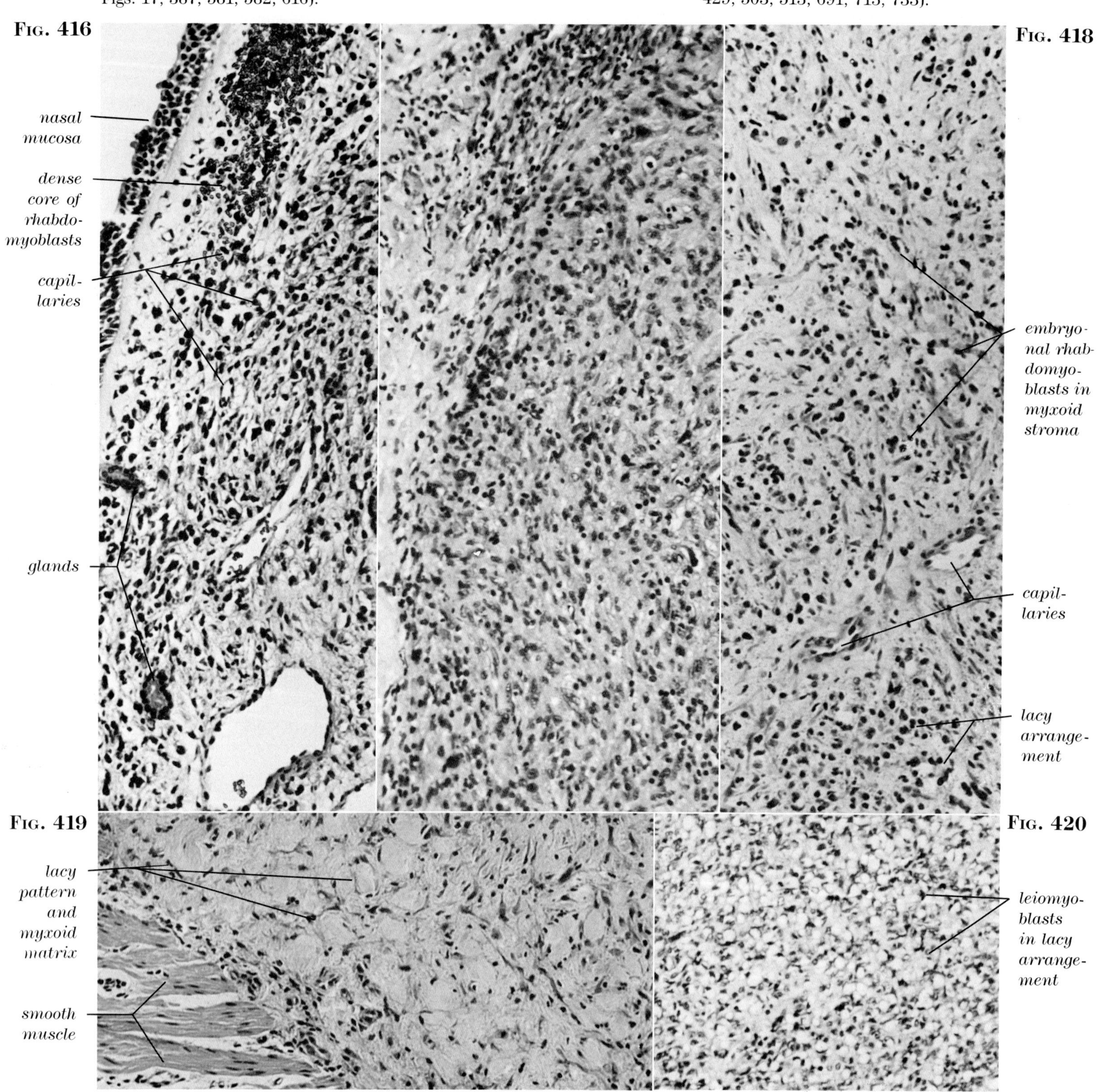

Fig. 421. Benign nevoid, schwannoma (×140) (see Figs. 471, 556).

Fig. 422. Borderline fibrous histiocytoma (×140) (see Figs. 346, 417, 459, 722, 722, 782).

Fig. 423. Malignant lymphoma infiltrating soft tissues (×140) (see Figs. 15, 132, 169, 463, 494, 702, 704, 705, 712, 743).

Fig. 424. Lymphangiosarcoma of leg (×140) (see Figs. 61, 84, 159, 321, 551, 558, 649, 954, 963, 965, 970, 1098).

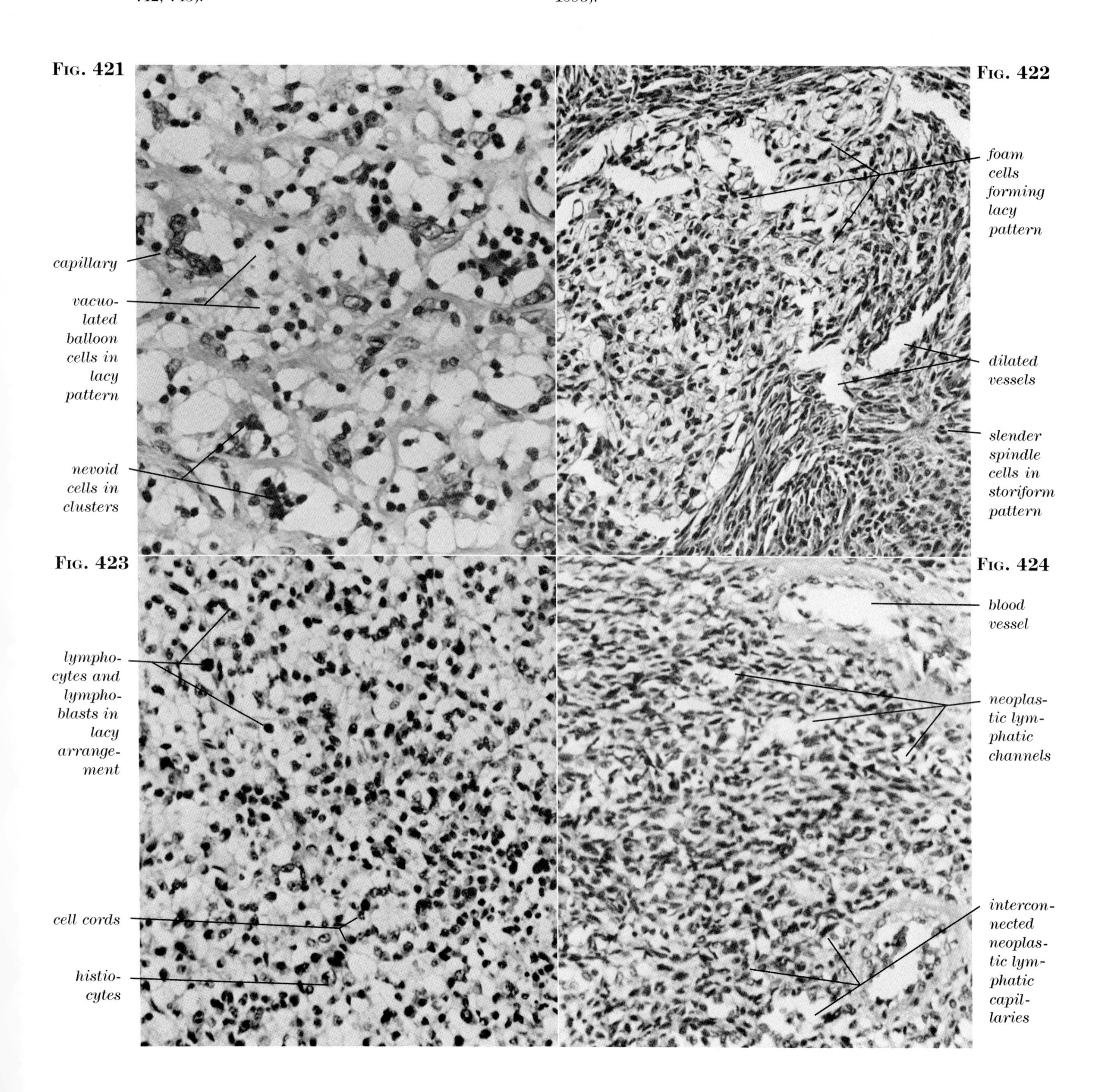

Fig. 425. Fat necrosis (×350) (see Figs. 749, 1042, 1056, 1185).

Fig. 427. Embryonal rhabdomyosarcoma (×350) (see Figs. 372, 378, 379, 416, 418, 426, 443, 587, 644, 923, 929).

Fig. 428. Clear cell carcinoma of ovary (×350) (see Fig. 758).

Fig. 426. Embryonal rhabdomyosarcoma (×350) (see Figs. 78, 263, 281, 364, 372, 378, 379, 427, 443, 446, 472, 515, 637, 689, 694).

Fig. 429. Leiomyoblastoma of stomach (×350) (see Figs. 277, 278, 420, 503, 513, 691, 713, 733, 744, 745, 747).

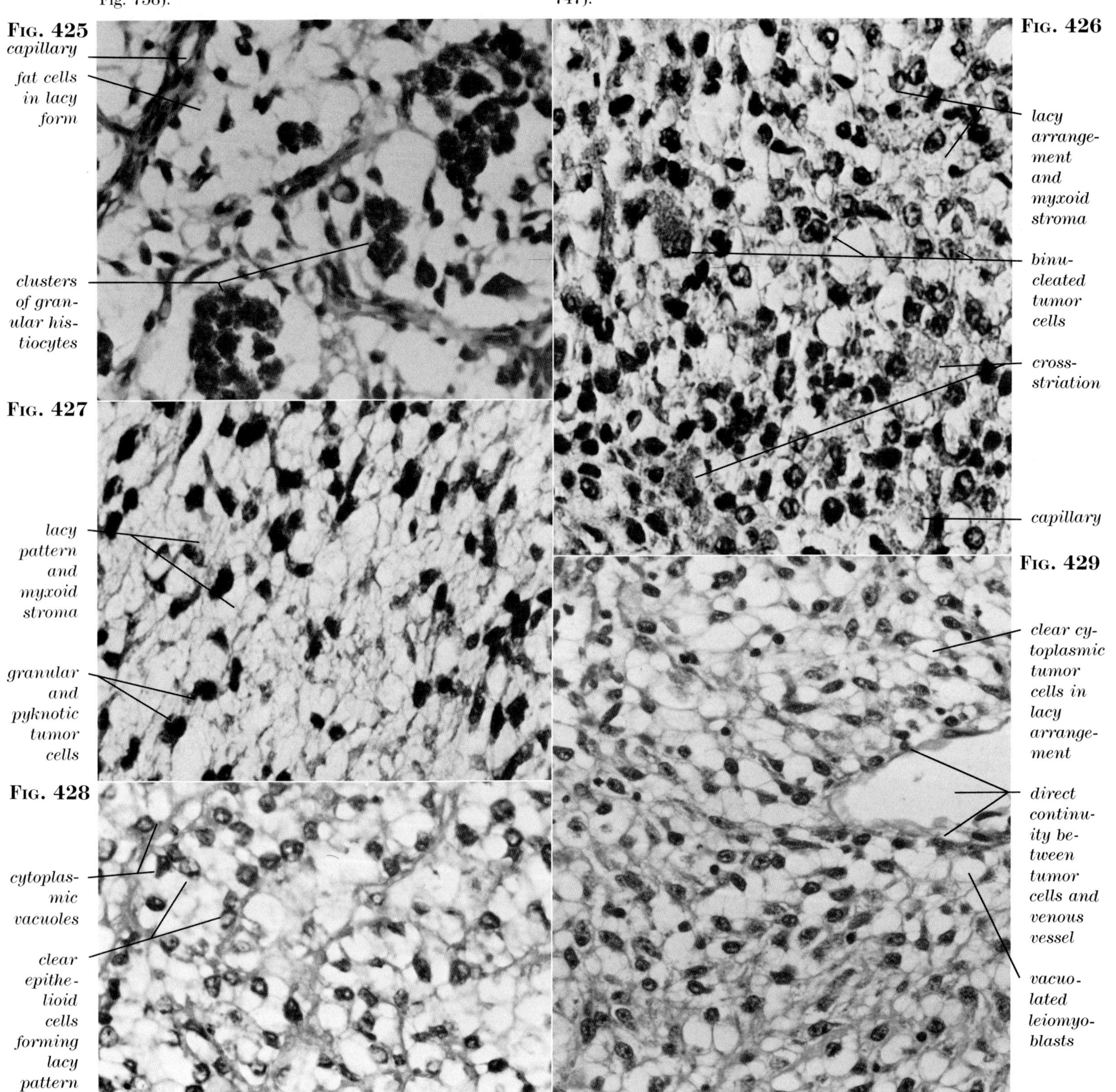

Fig. 430. Proliferative panniculitis (×140) (see Figs. 552, 721, 910).

Fig. 432. Hibernoma (×140) (see Fig. 717, 1162).

Fig. 431. Well-differentiated liposarcoma (×140) (see Figs. 58, 86, 219, 435, 822, 999, 1184, 1233).

Fig. 433. Lipogranuloma (×140) (see Figs. 820, 907, 1043).

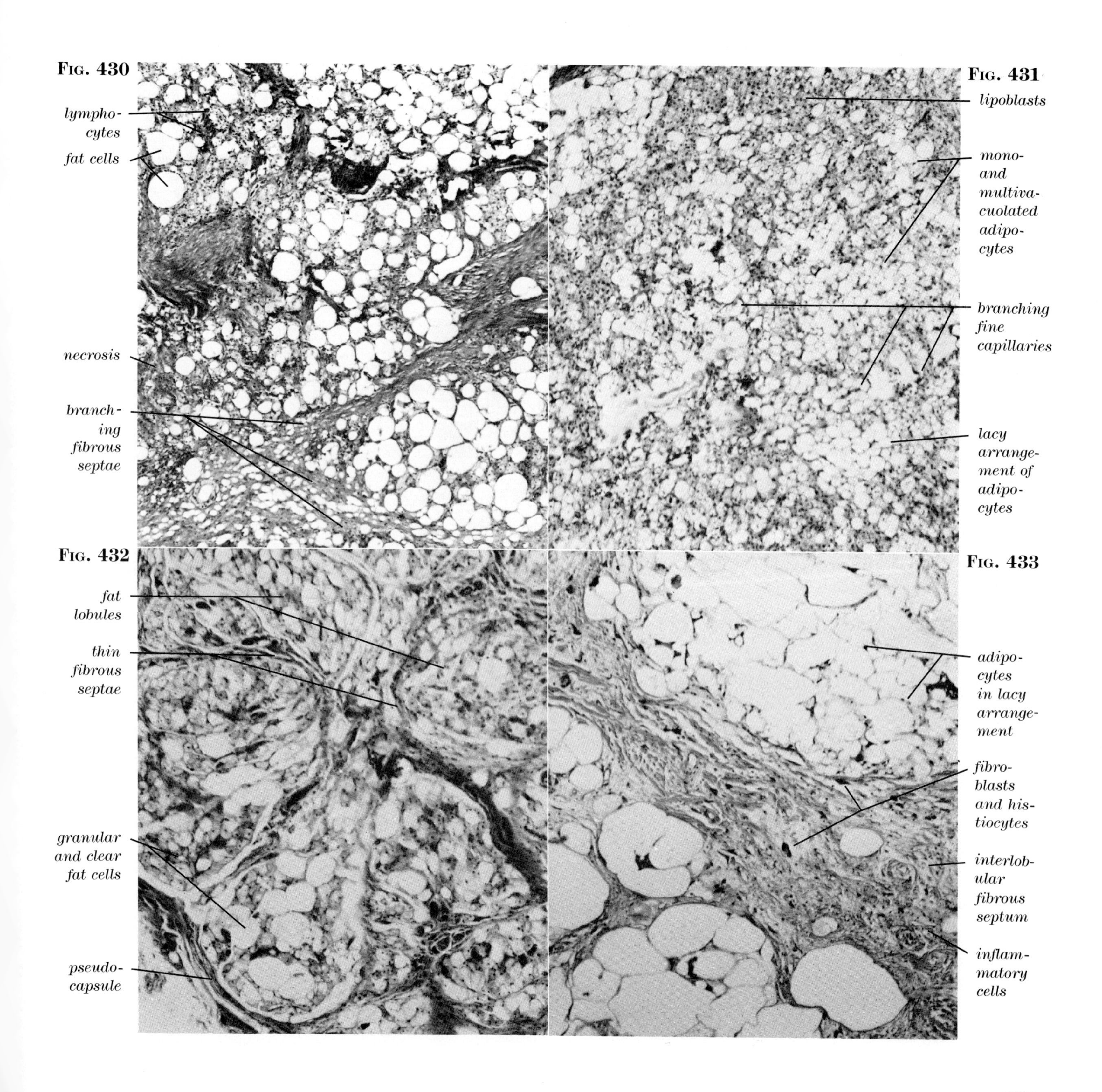

Fig. 434. Poorly differentiated, chondrosarcoma (×350) (see Figs. 95, 124, 126, 137, 317, 407, 441, 445, 638, 1064, 1097, 1102, 1175).

Fig. 435. Well-differentiated liposarcoma (×350) (see Figs. 86, 431, 822, 999, 1184, 1233).

Fig. 436. Pulmonary blastoma (×350).

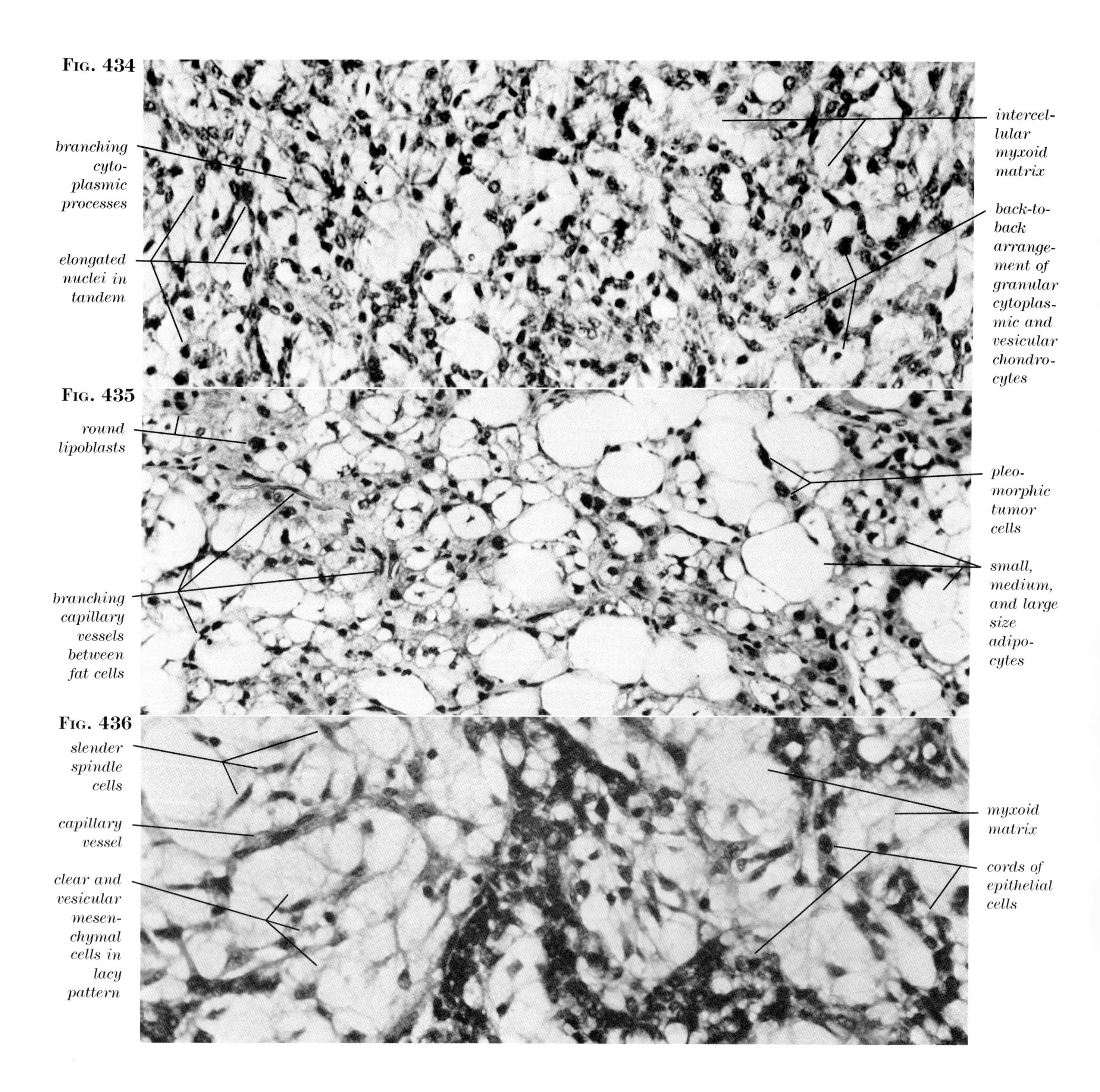

Fig. 437. Lipoblastoma (×140) (see Figs. 412, 413, 996).

Fig. 438. Myxoid liposarcoma (×140) (see Figs. 49, 56, 86, 188, 208, 269, 290, 439).

Fig. 439. Myxoid liposarcoma (×140) (see Figs. 59, 86, 208, 438, 636, 783, 933, 948, 997, 1046, 1113, 1169, 1232).

Fig. 440. Fibroblastic liposarcoma (×140) (see Figs. 86, 343, 641, 668, 837, 848, 926, 998).

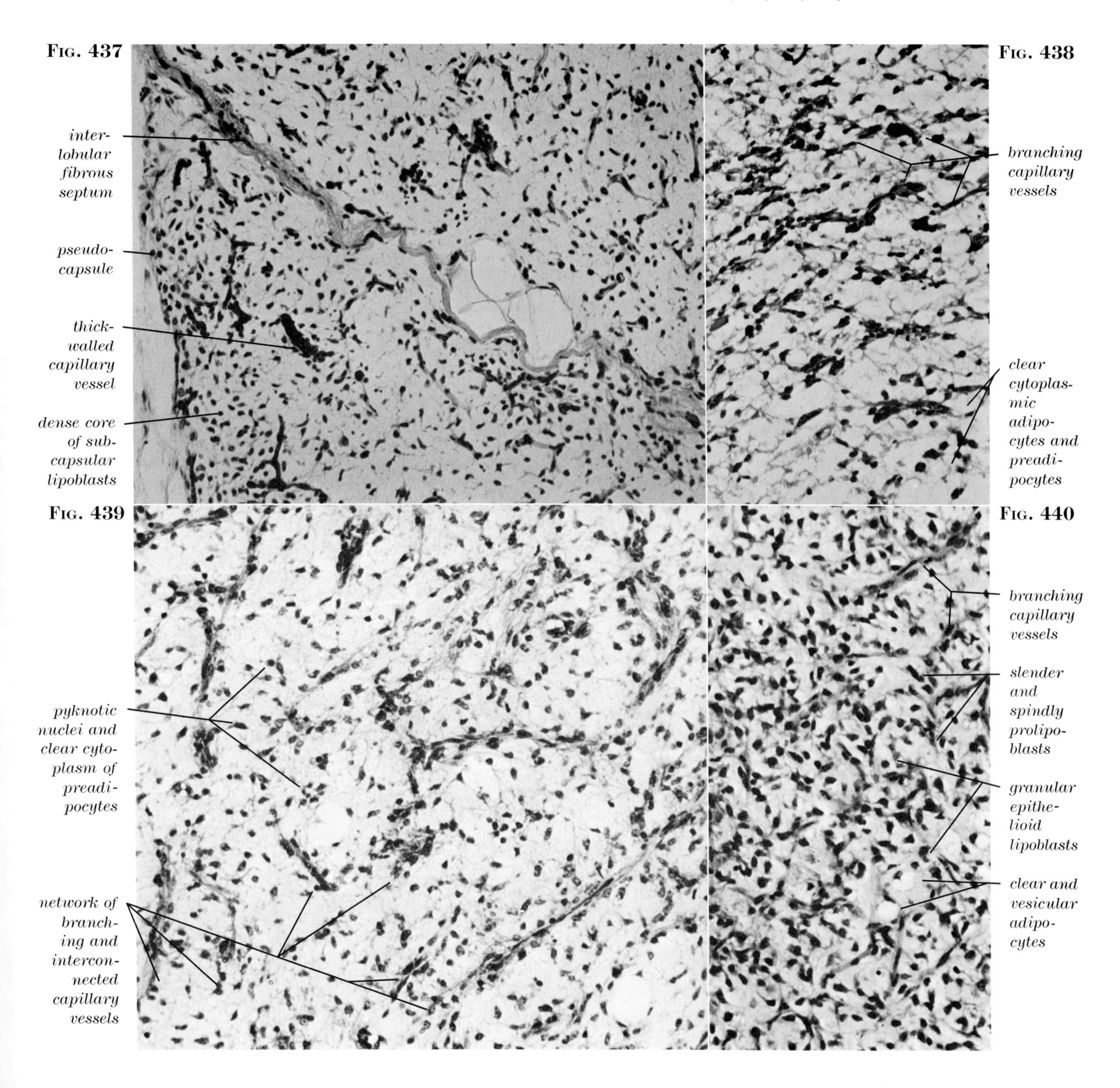

Fig. 441. Poorly differentiated, myxoid chondrosarcoma (×85) (see Figs. 95, 124, 126, 137, 182, 317, 407, 434, 445, 638, 932, 1064, 1097, 1102, 1175).

Fig. 442. Well-differentiated chondrosarcoma (×140) (see Figs. 95, 123, 124, 285, 766).

Fig. 443. Embryonal rhabdomyosarcoma (×140) (see Figs. 379, 426, 427, 446, 472).

Fig. 444. Mesenchymal chondrosarcoma (×140) (see Figs. 103, 151, 175, 180, 318, 390, 507, 595, 674, 707, 920, 1062, 1265).

Fig. 445. Poorly differentiated, chondroblastic chondrosarcoma metastatic to lung (×140) (see Figs. 407, 434, 441, 1064).

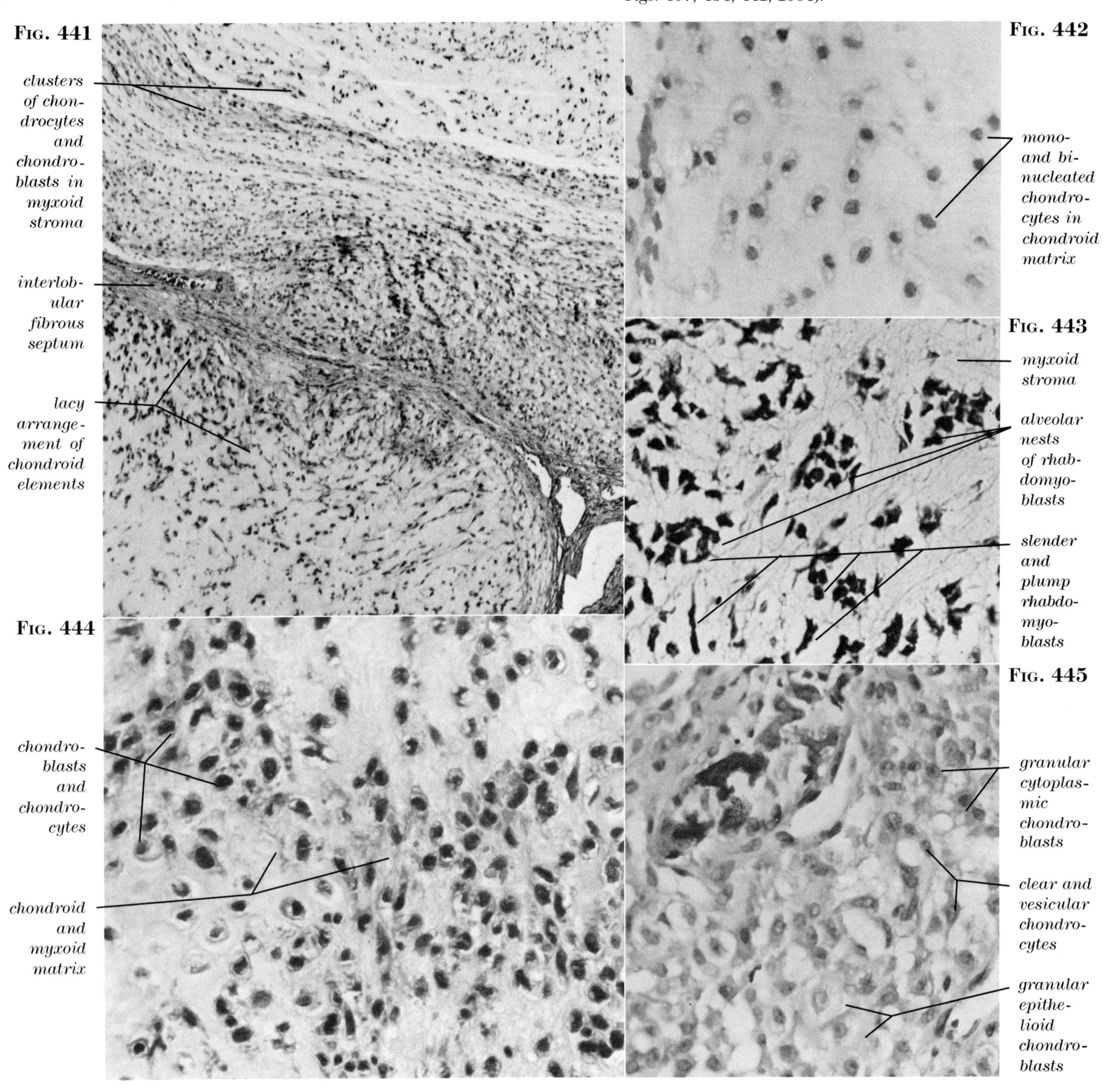

Fig. 446. Embryonal rhabdomyosarcoma (×140) (see Figs. 78, 207, 291, 364, 372, 378, 379, 427, 443, 472, 515, 637, 689, 694, 700, 706, 708, 751).

Fig. 447. Clear cell sarcoma (×140) (see Figs. 62, 85, 322, 492, 526, 759, 761, 896, 927, 1145, 1223, 1244, 1245).

Fig. 448. Lipoblastic liposarcoma (×140) (see Figs. 86, 415, 450, 543, 687, 753, 934, 973, 1231).

Fig. 449. Ewing's sarcoma (×140) (see Figs. 100, 172, 250, 453, 476, 573, 698, 1157, 1230).

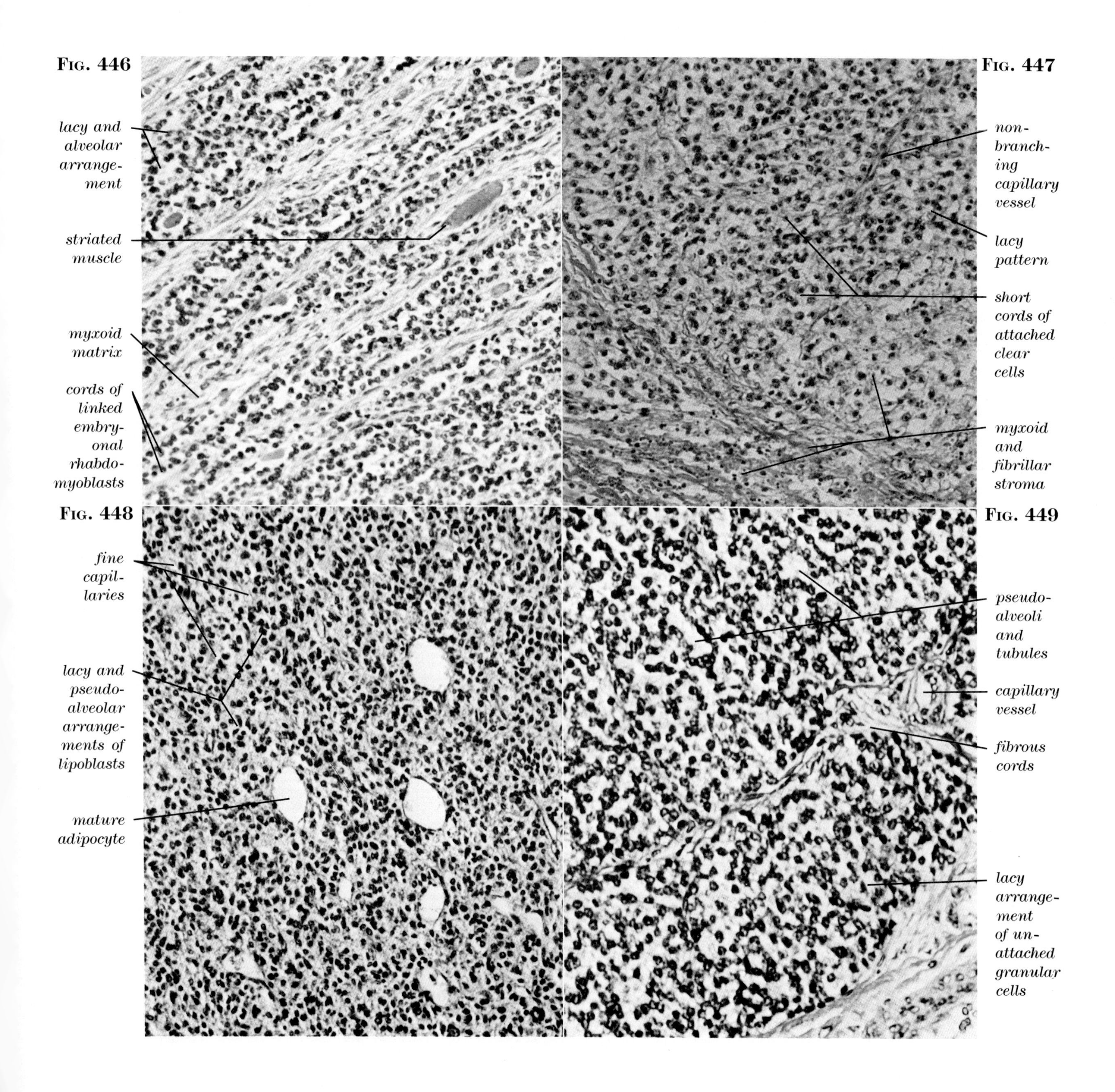

Fig. 450. Lipoblastic liposarcoma (×350) (see Figs. 199, 415, 448, 543, 687, 753, 934, 973, 1231).

Fig. 451. Osteoblastic osteosarcoma (×350) (see Figs. 108, 127, 157, 228, 248, 272, 304, 305, 395, 405, 461, 602, 610, 643, 754, 819, 831, 843, 876).

Fig. 452. Malignant histiocytic fibrous histiocytoma of bone (×350) (see Figs. 300, 534, 592, 600, 711, 756, 775, 776, 784, 791, 816, 872).

Fig. 453. Ewing's sarcoma (×350) (see Figs. 100, 172, 173, 210, 286, 449, 476, 573, 698).

Fig. 454. Plasma cell myeloma (×350) (see Figs. 99, 208A, 692, 741).

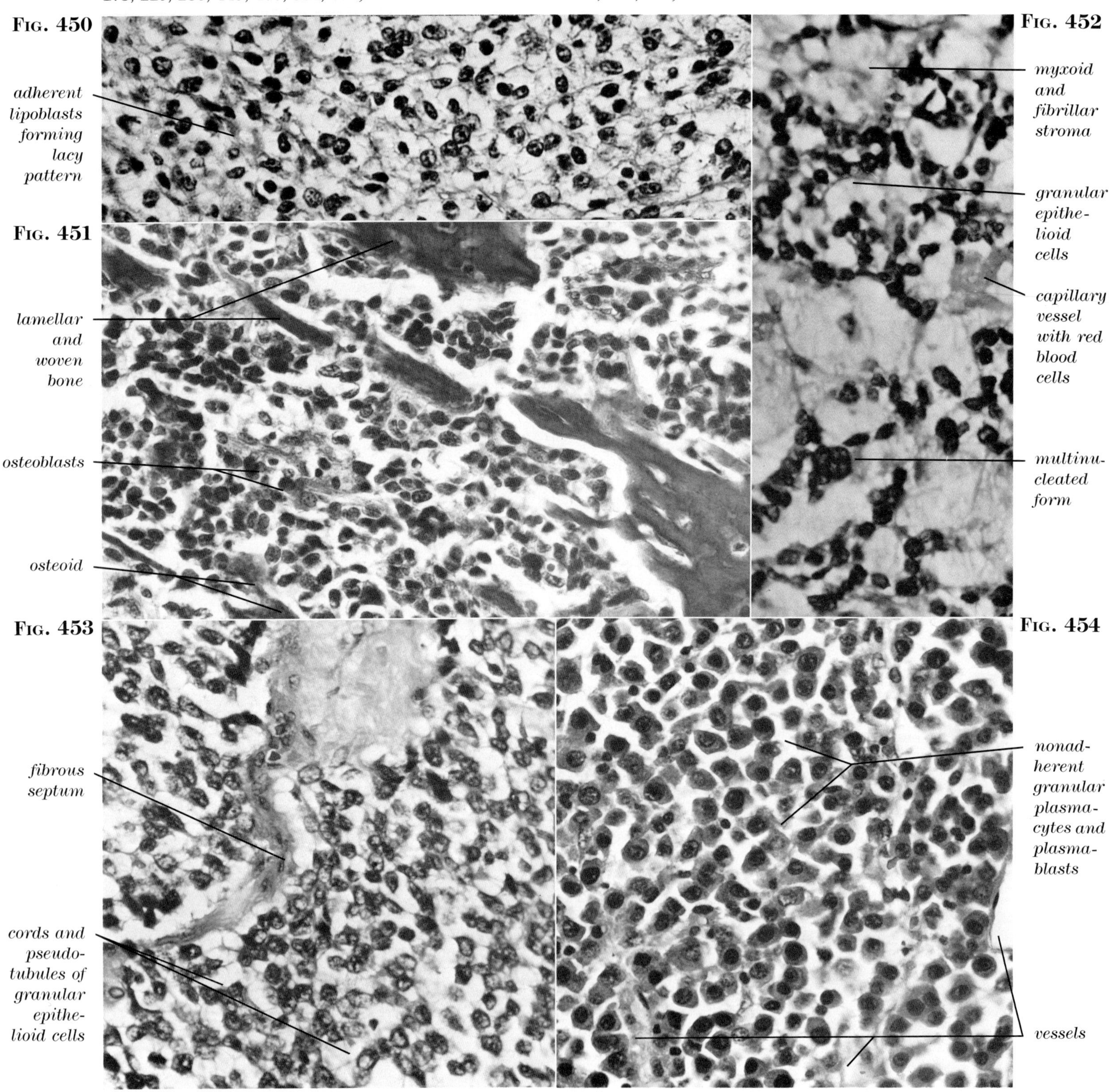

Fig. 455. Well-differentiated, infiltrating lipoma (×85) (see Figs. 24, 73, 198, 243, 1181).

Fig. 456. Chordoid sarcoma (×350) (see Figs. 510, 640, 919).

Fig. 457. Malignant glomus tumor (×350) (see Figs. 482, 550, 728, 980, 1141).

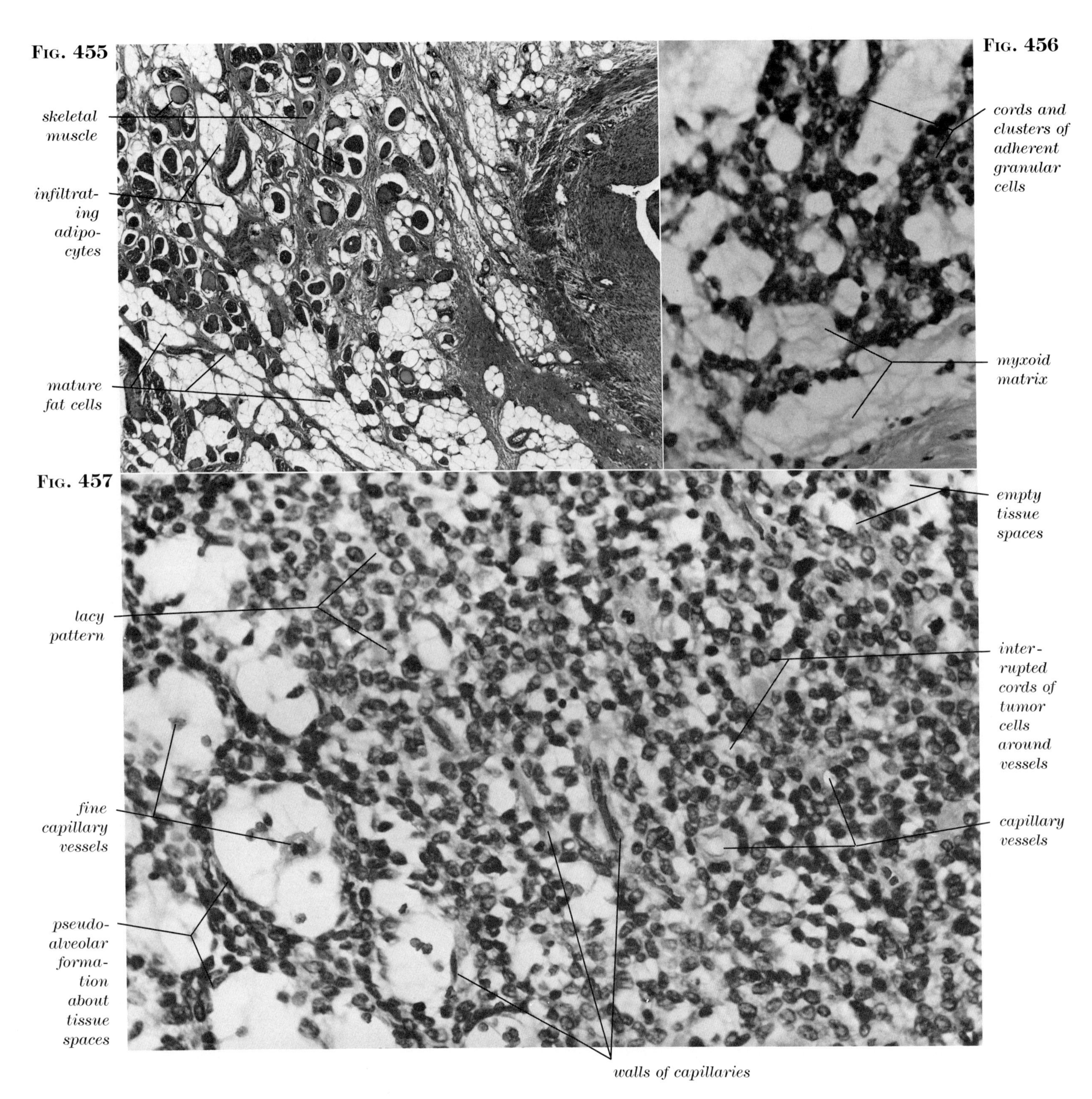

Fig. 458. Pleomorphic liposarcoma (×140) (see Figs. 86, 141, 284, 590, 799, 823, 928, 1019A, 1049).

Fig. 459. Borderline or atypical fibrous histiocytoma (×140) (see Figs. 417, 422, 722, 782).

Fig. 460. High grade malignant pleomorphic fibrous histiocytoma (×350) (see Figs. 88, 111, 130, 344, 577, 579, 598, 679, 780, 788, 789, 792, 798, 805, 818).

Fig. 461. Chondroblastic osteosarcoma (×350) (see Figs. 395, 405, 411, 451, 602).

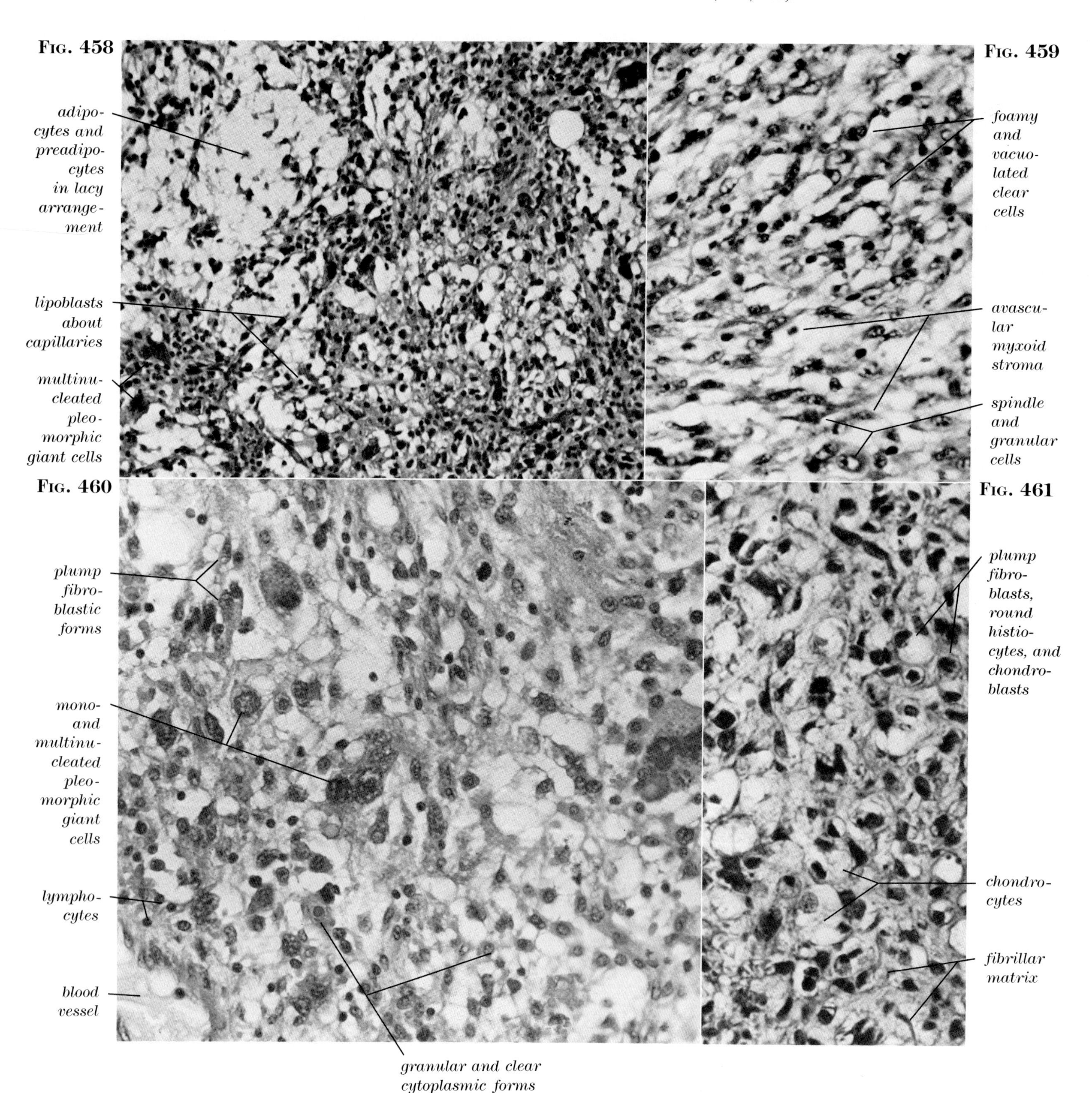

Fig. 462. Epithelioid sarcoma (×140) (see Figs. 85, 181, 258, 464, 468, 491, 676, 830, 866, 895).

Fig. 463. Malignant lymphoma infiltrating muscle (×140) (see Figs. 423, 494, 702, 704, 705, 712, 743).

Fig. 464. Epithelioid sarcoma (×140) (see Figs. 322, 462, 468, 491, 676, 830, 866, 895, 1052, 1059).

Fig. 465. (Bottom center) Stromal sarcoma of uterus (×140) (see Figs. 335, 969).

Fig. 466. Myxopapillary ependymoma (×140) (see Figs. 239, 518).

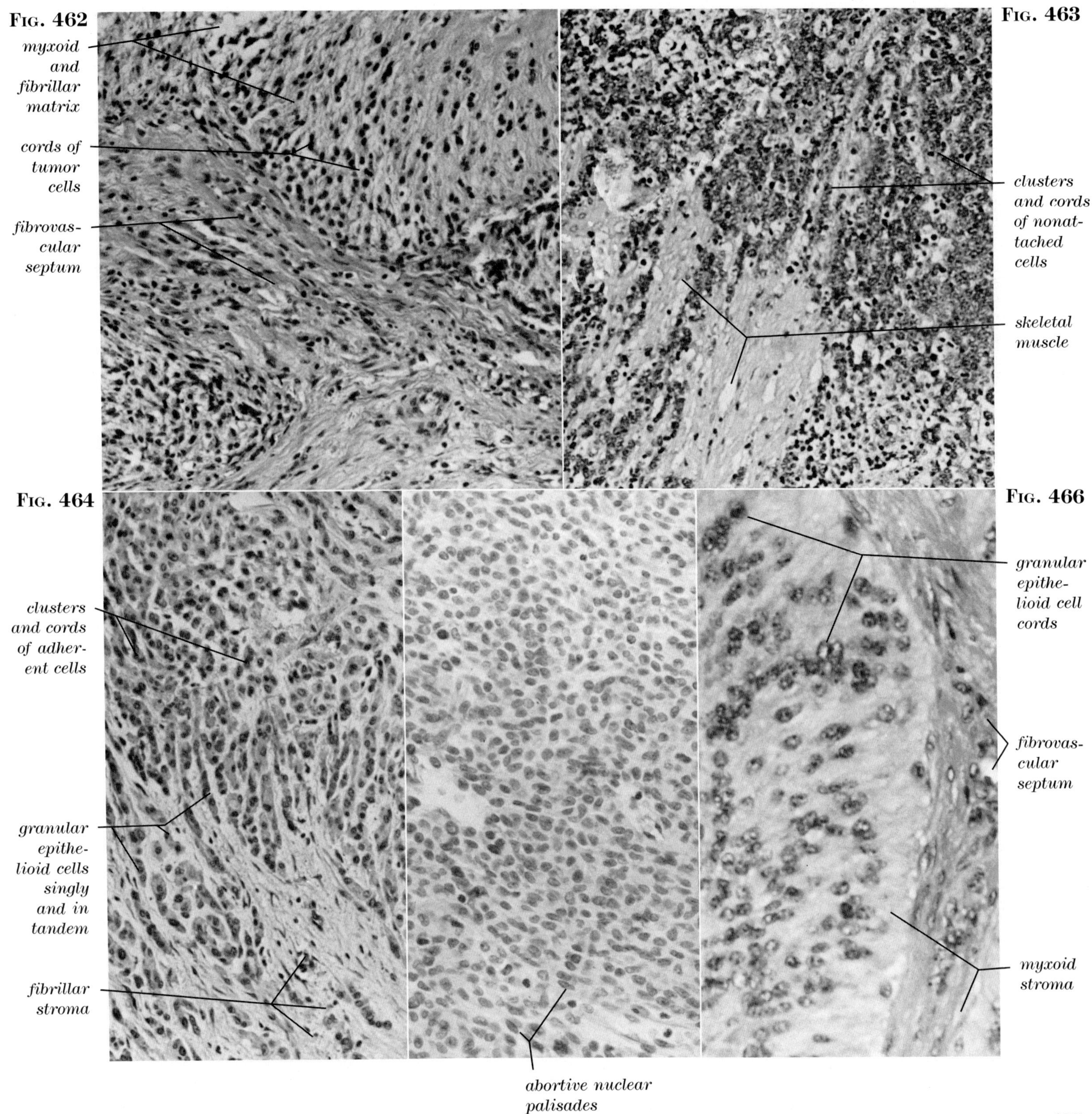

Fig. 467. Monophasic tendosynovial sarcoma, spindle cell type (×140) (see Figs. 392, 396, 398, 399, 491, 492, 542, 622, 642, 646, 648, 650, 675, 683, 684, 854, 864, 870).

Fig. 468. Epithelioid sarcoma (×350) (see Figs. 62, 187, 259, 462, 464, 491, 676, 830, 866, 895, 1052, 1059, 1146, 1170, 1244).

Fig. 469. Metastatic medulloblastoma (×350) (see Figs. 311, 723).

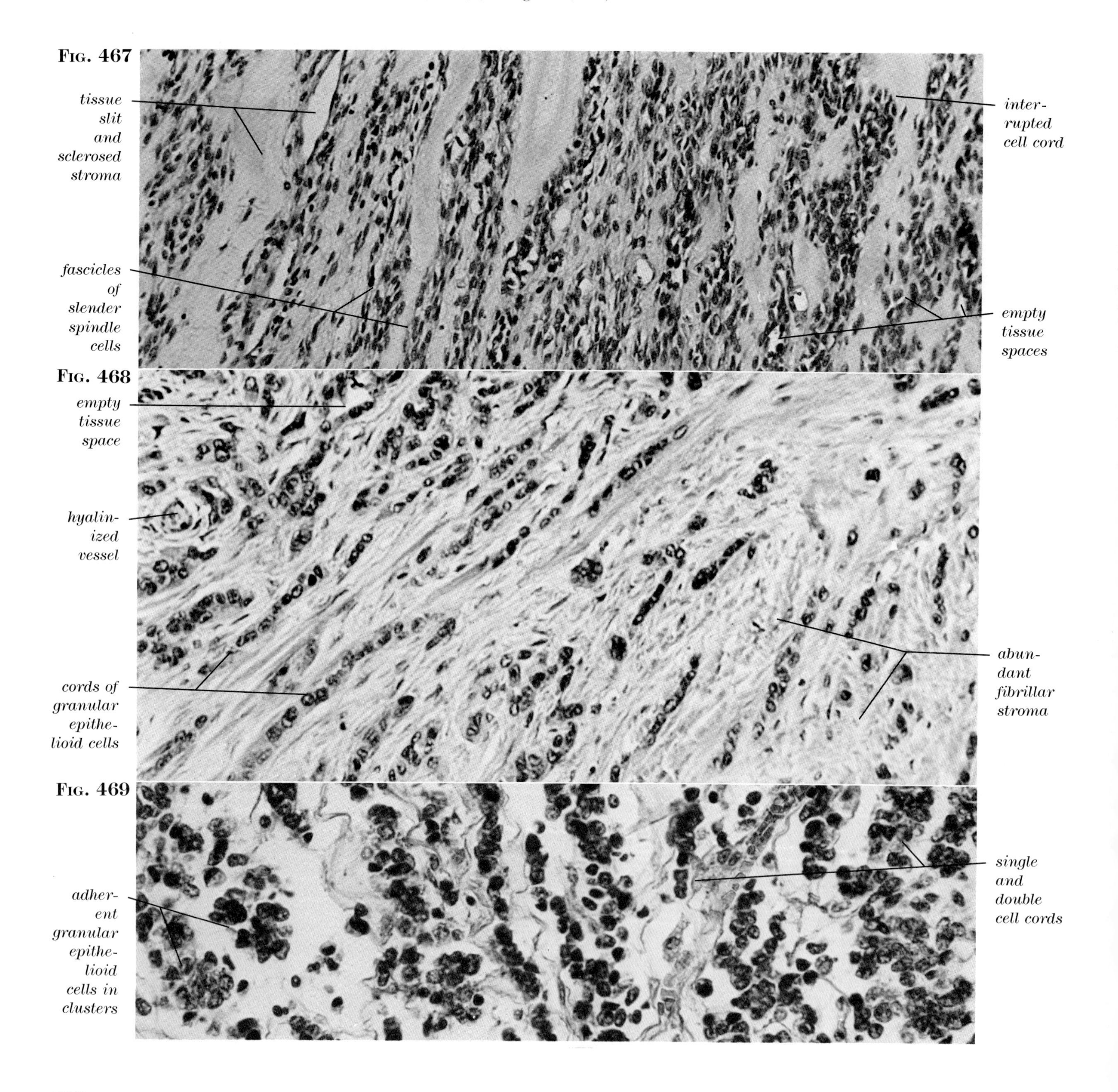

Fig. 470. Low grade hemangiopericytoma (×140) (see Figs. 93, 479, 484, 525).

Fig. 471. (Top center) Benign nevoid schwannoma (×140) (see Figs. 421, 556).

Fig. 472. Embryonal rhabdomyosarcoma (×140) (see Figs. 78, 291, 364, 427, 515).

Fig. 473. Rhabdomyoblastoma (×140) (see Figs. 477, 500, 528, 569, 594, 695, 715, 739, 796, 936).

Fig. 474. Benign histiocytic fibrous histiocytoma (×140) (see Figs. 71, 115, 174, 178, 183, 222, 325, 501, 774, 778).

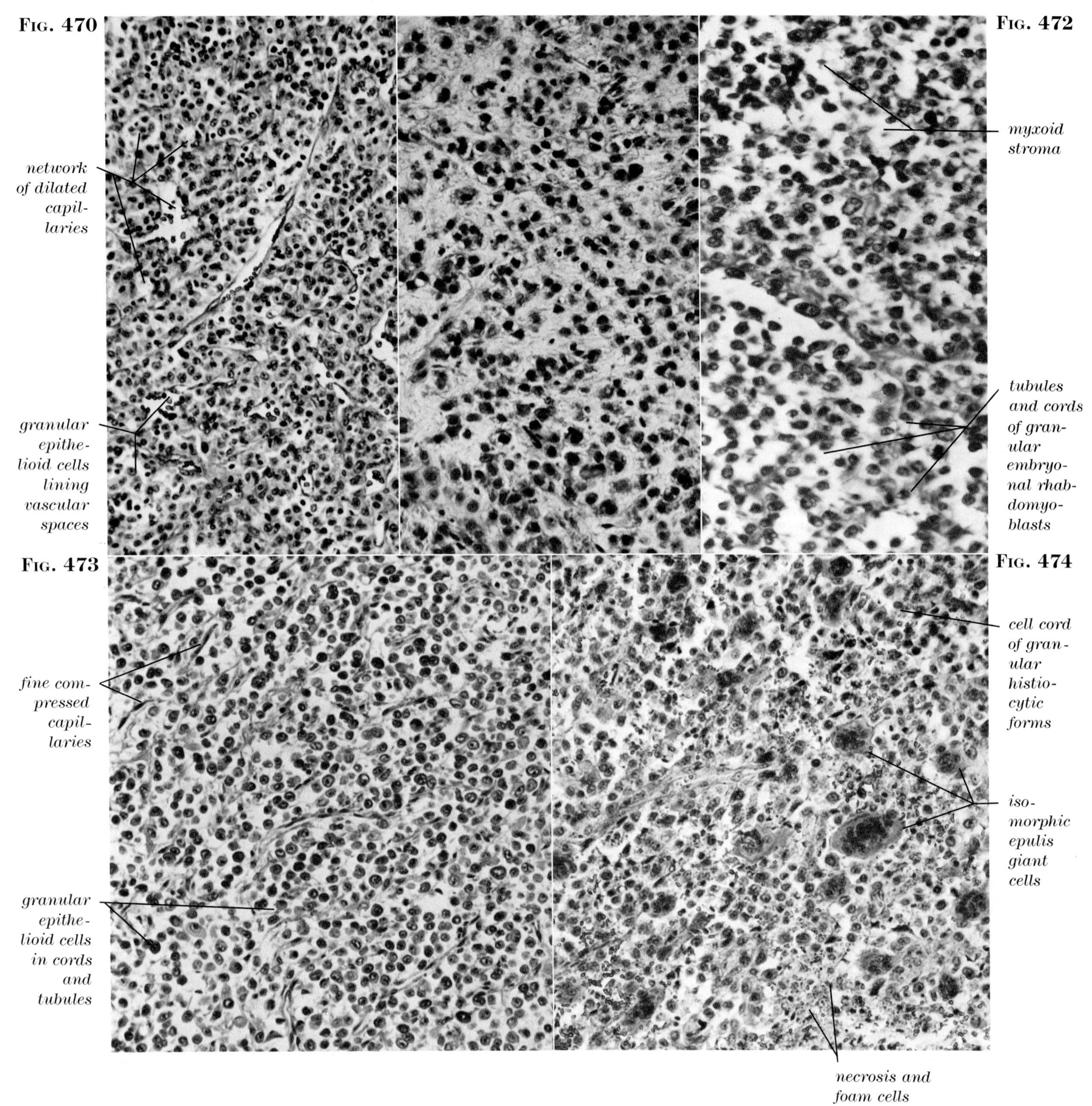

Fig. 475. Benign glomus tumor (×140) (see Figs. 481, 961).

Fig. 476. (Top center) Ewing's sarcoma (×140) (see Figs. 214, 307, 449, 453, 573).

Fig. 477. Rhabdomyoblastoma (×140) (see Figs. 473, 500, 528, 569).

Fig. 478. Malignant neuroepithelioma (×140) (see Figs. 309, 489, 568, 585).

Fig. 479. (Bottom center) Low grade hemangiopericytoma (×140) (see Figs. 93, 470, 484, 525).

Fig. 480. Retinoblastoma (×140).

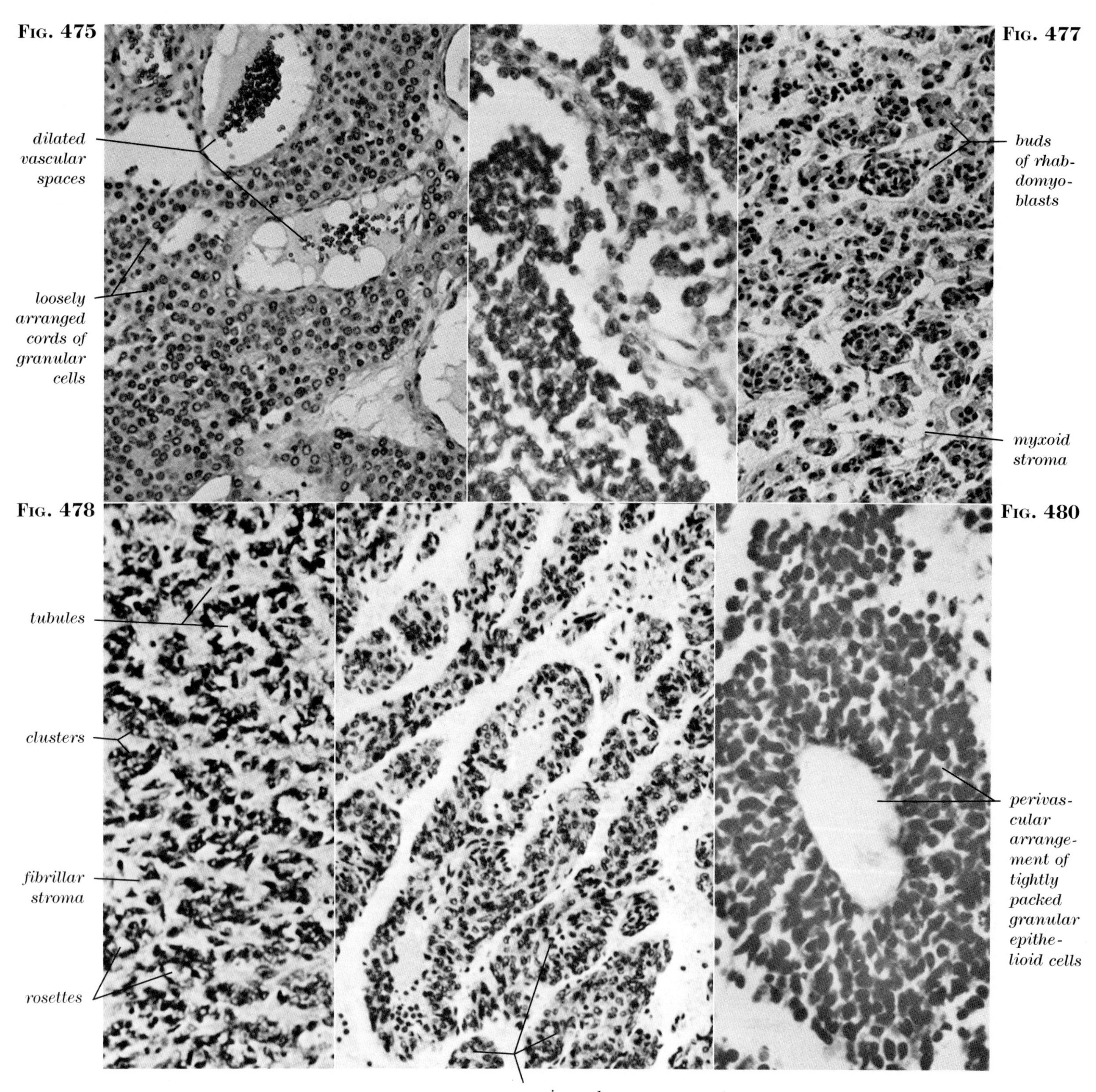

Fig. 481. Benign glomus tumor (×140) (see Figs. 475, 961).

Fig. 482. Malignant glomus tumor (×350) (see Figs. 457, 550, 728, 980, 1141).

Fig. 483. Benign paraganglioma (×85) (see Figs. 276, 544, 714, 790).

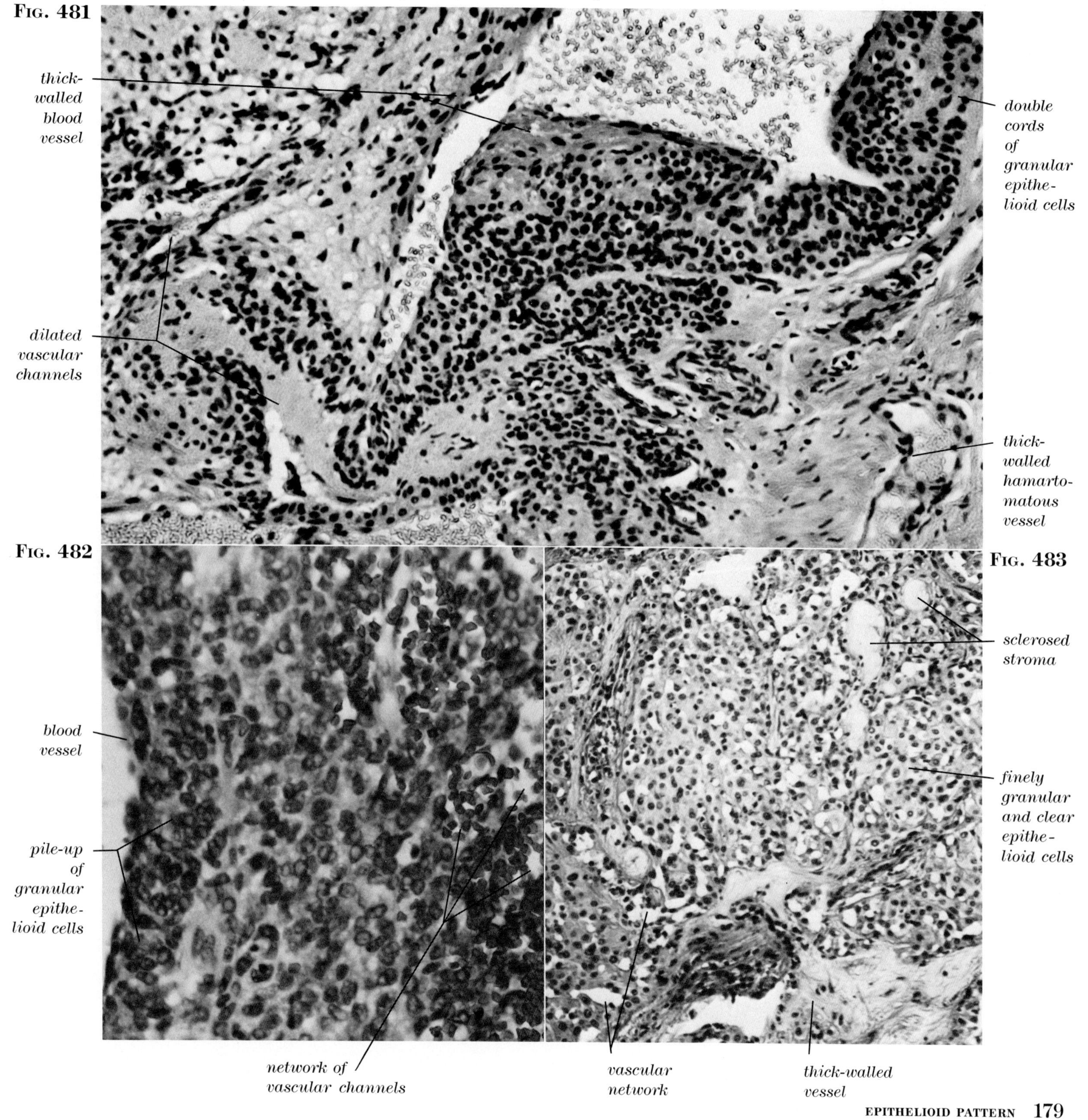

Fig. 484. Post radiation low grade hemangiopericytoma of retroperitoneum (×350) (see Figs. 374, 470, 479, 525, 533, 693, 696, 972, 984, 995, 1203).

Fig. 485. Extracranial meningioma of temporal area (×350) (see Figs. 496, 537, 1096, 1225).

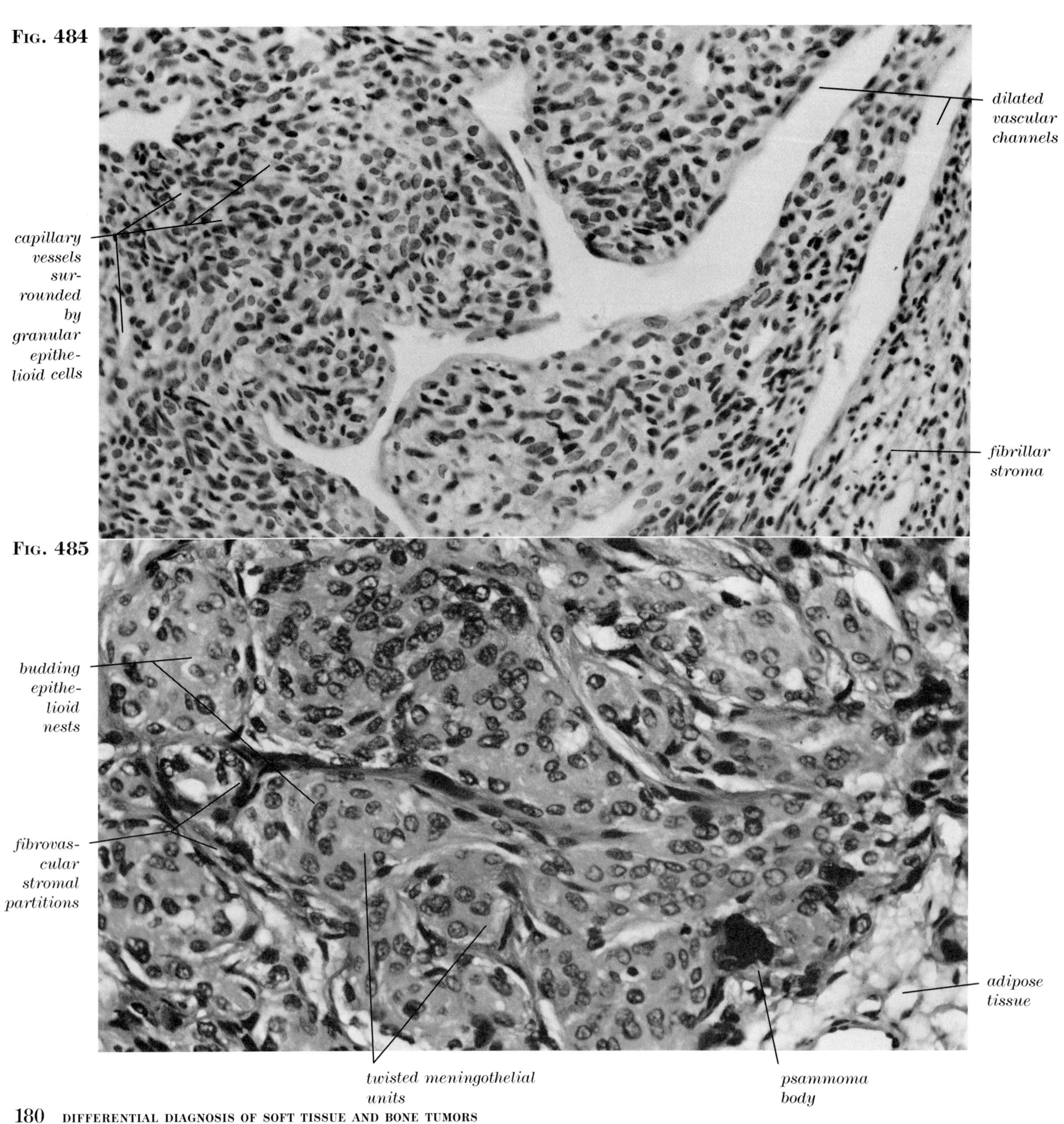

Fig. 486. Ganglioneuroma of mediastinum (×140) (see Figs. 166, 630, 811, 824, 857, 1137, 1191, 1193, 1258).

Fig. 487. Solitary neurofibroma of neck (×140) (see Figs. 11, 74, 363, 652, 840, 871, 955, 993, 1013).

Fig. 488. Granulocytic sarcoma of chest wall (×140) (see Figs. 699, 891, 1194, 1198).

Fig. 489. Malignant neuroepithelioma (×350) (see Figs. 309, 478, 568, 585, 690, 1115, 1238, 1239).

Fig. 490. Borderline or atypical fibrous histiocytoma, so-called "atypical villonodular synovitis," of finger (×350) (see Figs. 71, 174, 501, 778).

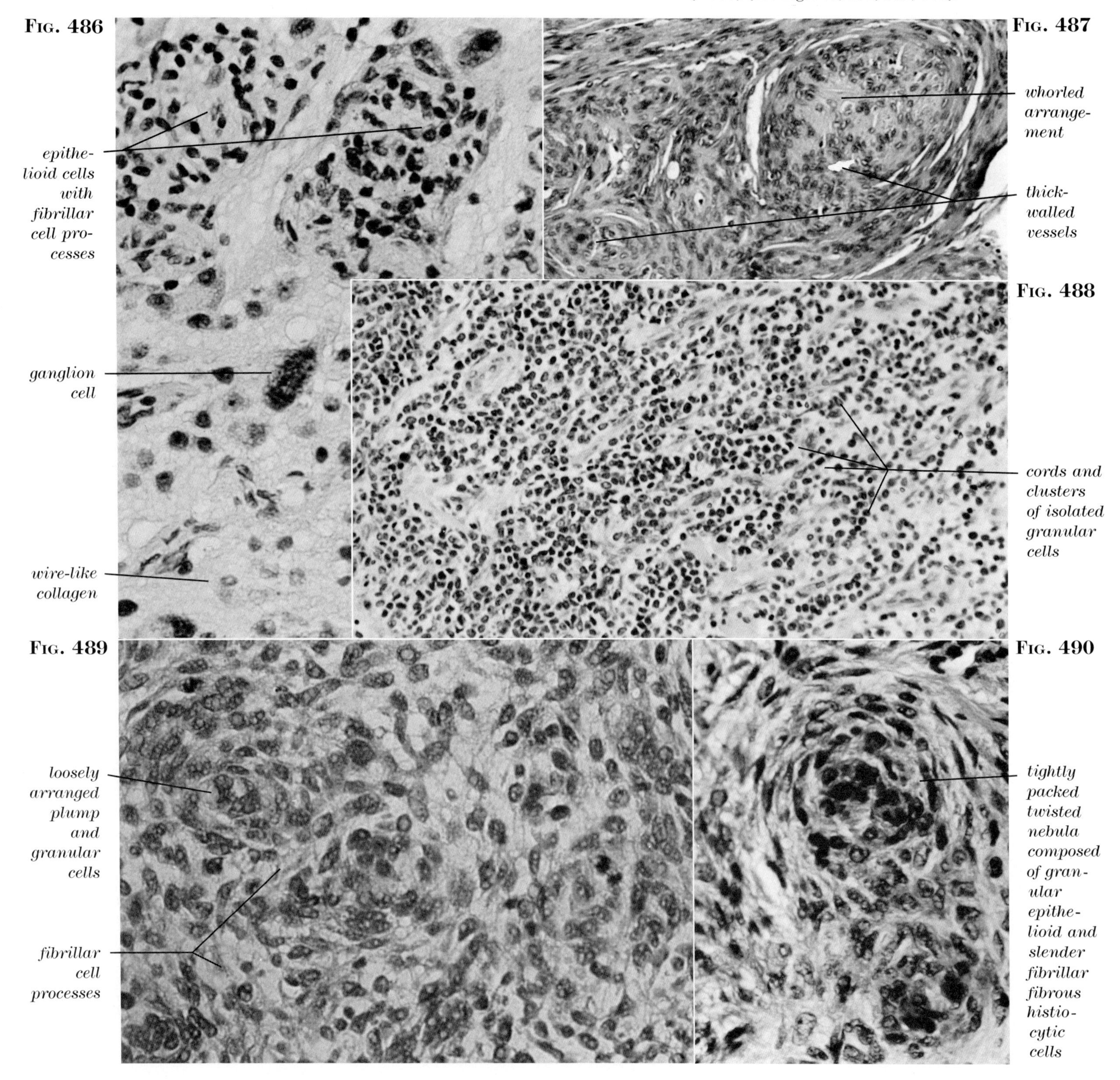

Fig. 491. Tendosynovial sarcoma growing as monophasic spindle cell sarcoma with epithelioid sarcoma component (×350) (see Figs. 85, 399, 462, 464, 467, 468, 492, 542, 622, 642, 646, 648, 650, 675, 676).

Fig. 492. Monophasic tendosynovial sarcoma, spindle cell type, with clear cell sarcoma component (×350) (see Figs. 64, 190, 315, 320, 322, 326, 391, 392, 396, 398, 399, 447, 467, 491, 492, 526).

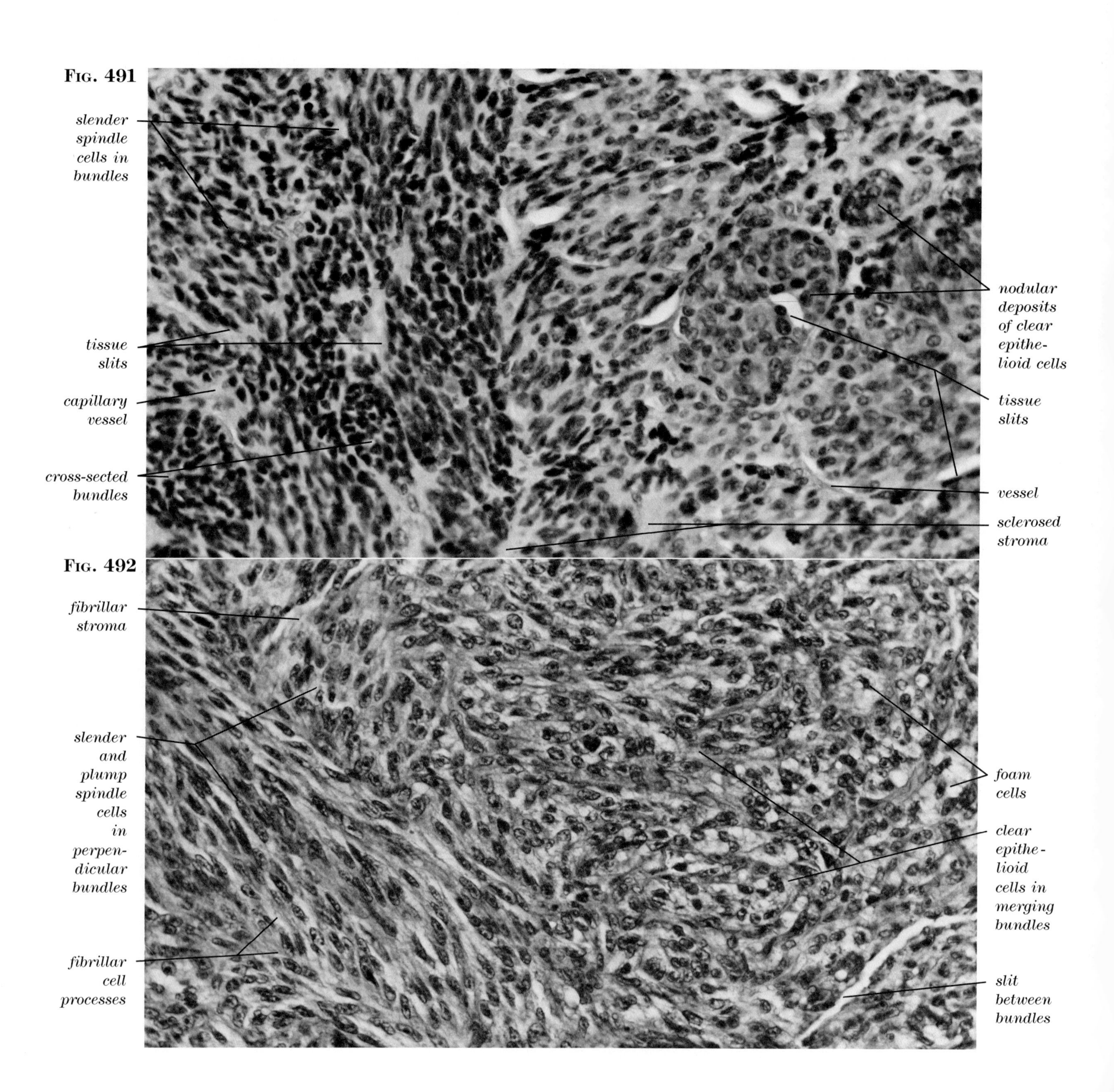

Fig. 493. Adamantinoma of tibia (×350) (see Figs. 171, 634).

Fig. 494. Malignant lymphoma of tibia (×350) (see Figs. 423, 463, 702).

Fig. 495. Meningeal sarcoma (×350) (see Figs. 283, 806, 887).

Fig. 496. Meningioma (×350) (see Figs. 485, 537, 1096).

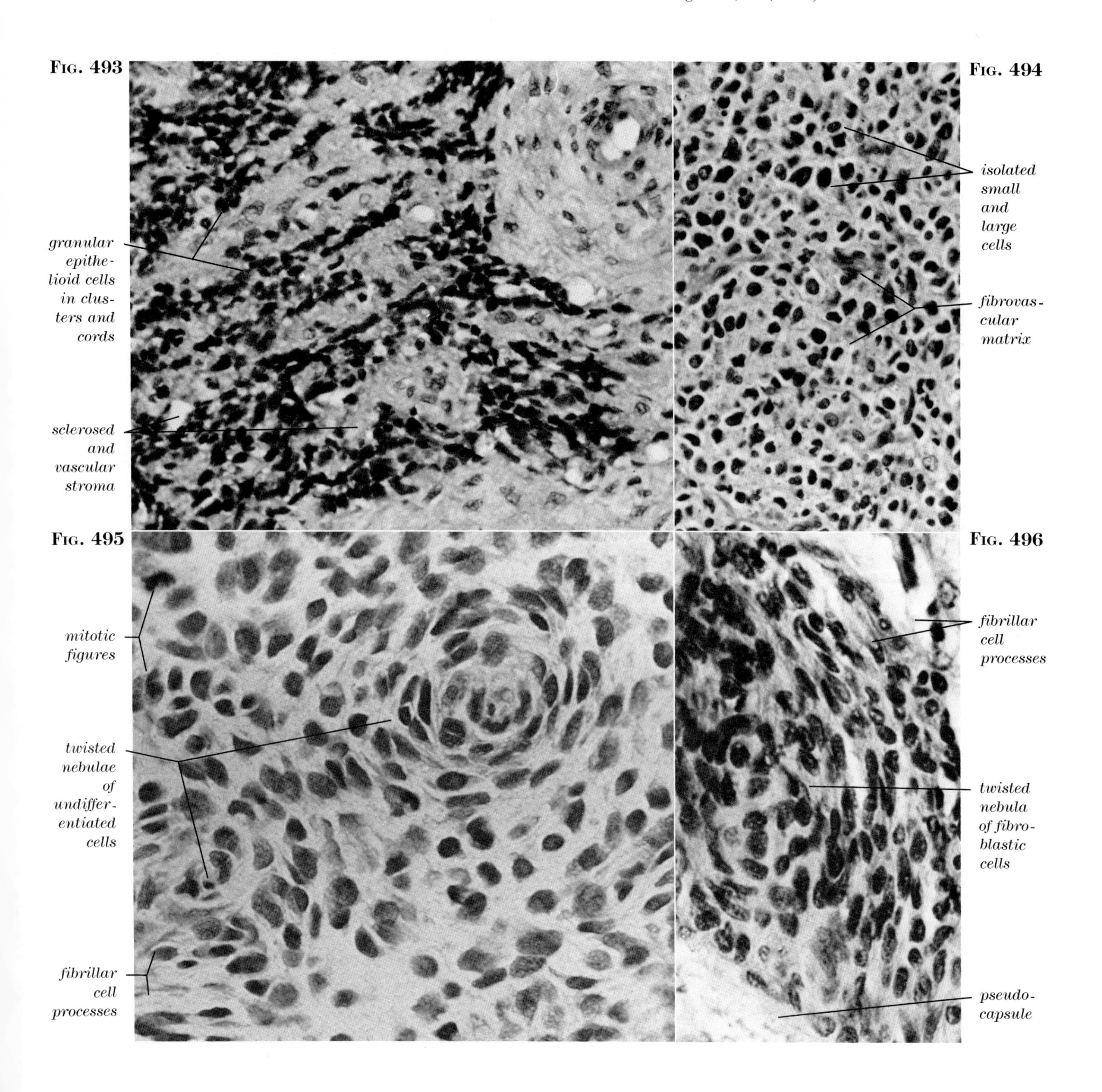

Fig. 497. Rhabdomyoma (×140) (see Figs. 737, 793).

Fig. 498. Benign granular cell tumor of breast (×140) (see Figs. 75, 261, 539, 736, 1160, 1235).

Fig. 499. Malignant granular cell tumor (×140) (see Figs. 740, 1234, 1256).

Fig. 500. Rhabdomyoblastoma (×140) (see Figs. 473, 477, 528, 569, 594, 695, 709, 715, 739, 796).

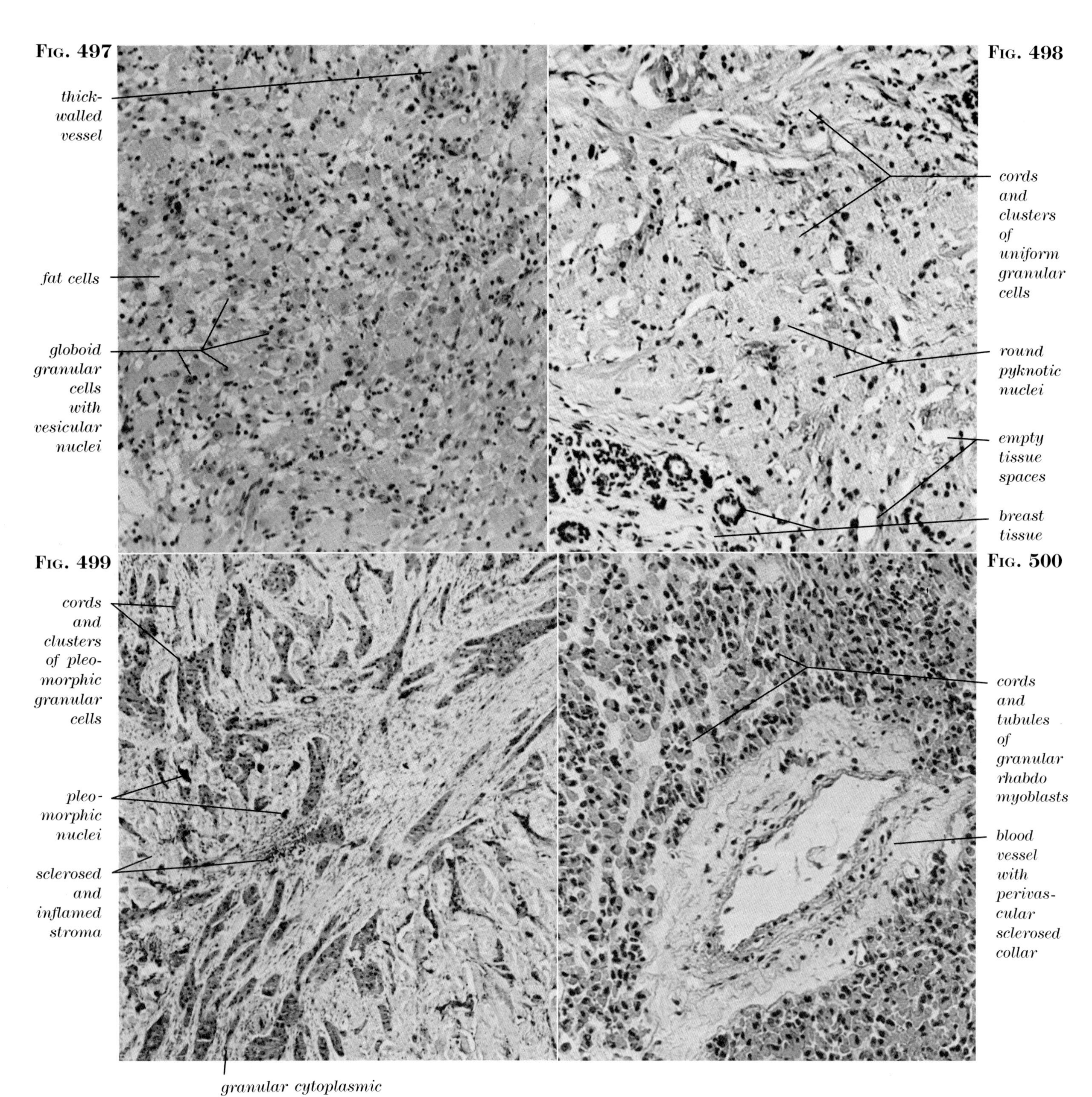

FIG. 501. Borderline histiocytic fibrous histiocytoma of finger (×350) (see Figs. 71, 178, 325, 490, 774, 778).

FIG. 502. Metastatic mesodermal mixed tumor of uterus (×350) (see Figs. 65, 589, 1165, 1217).

FIG. 503. High grade leiomyoblastoma of stomach (×350) (see Figs. 278, 420, 429, 513, 691, 713, 733, 744, 745, 747, 750, 752, 792, 814, 828, 868, 931, 964).

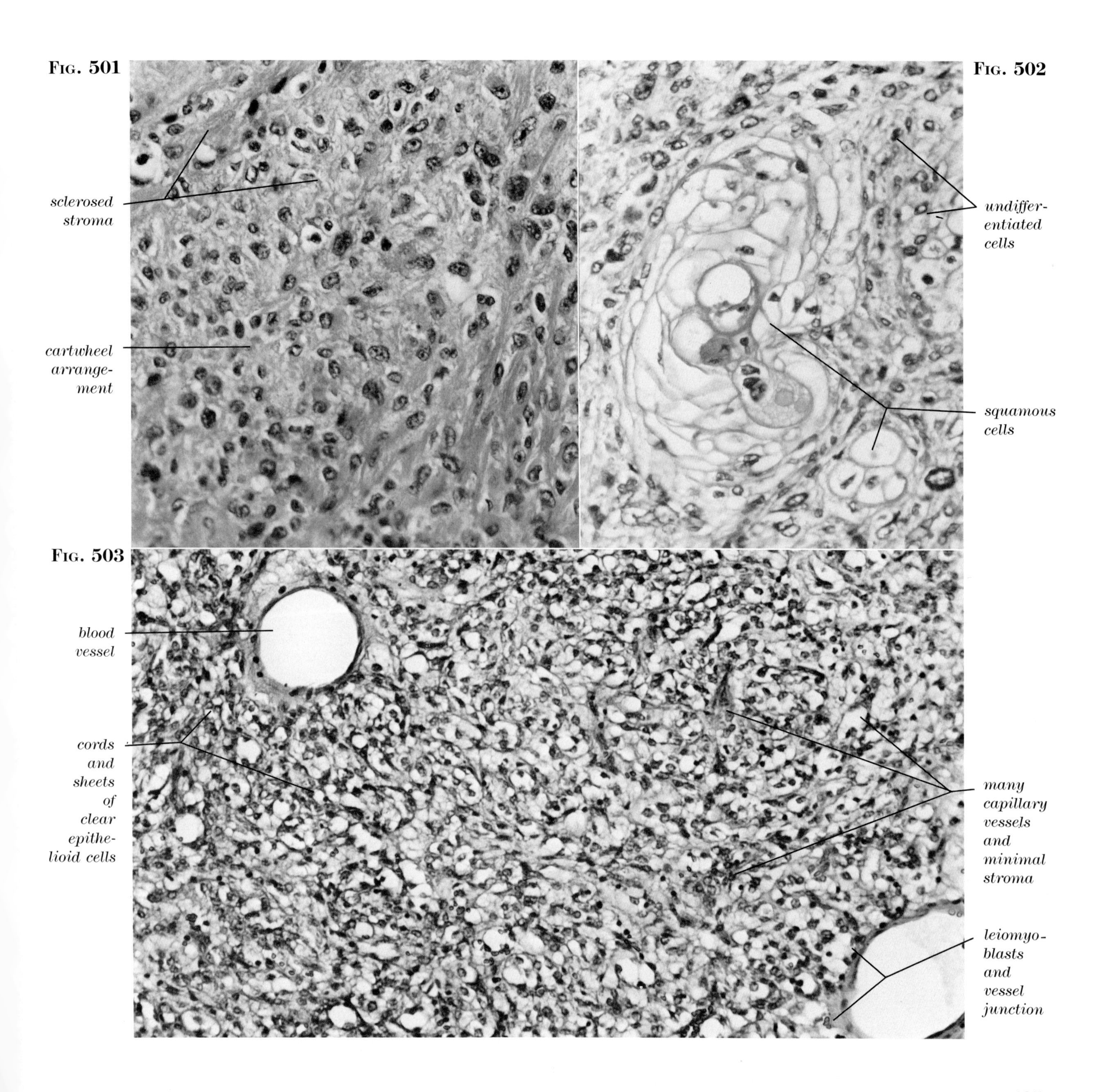

Fig. 504. Medullary carcinoma of thyroid (×350) (see Figs. 532, 916).

Fig. 505. Chordoma (×350) (see Figs. 69, 215, 310, 729, 767, 771, 1174, 1226).

Fig. 506. Atypical chondroblastoma (×350) (see Figs. 116, 128, 1103, 1104, 1111).

Fig. 507. Mesenchymal chondrosarcoma (×350) (see Figs. 103, 390, 444, 595, 674, 707, 920, 1062, 1265).

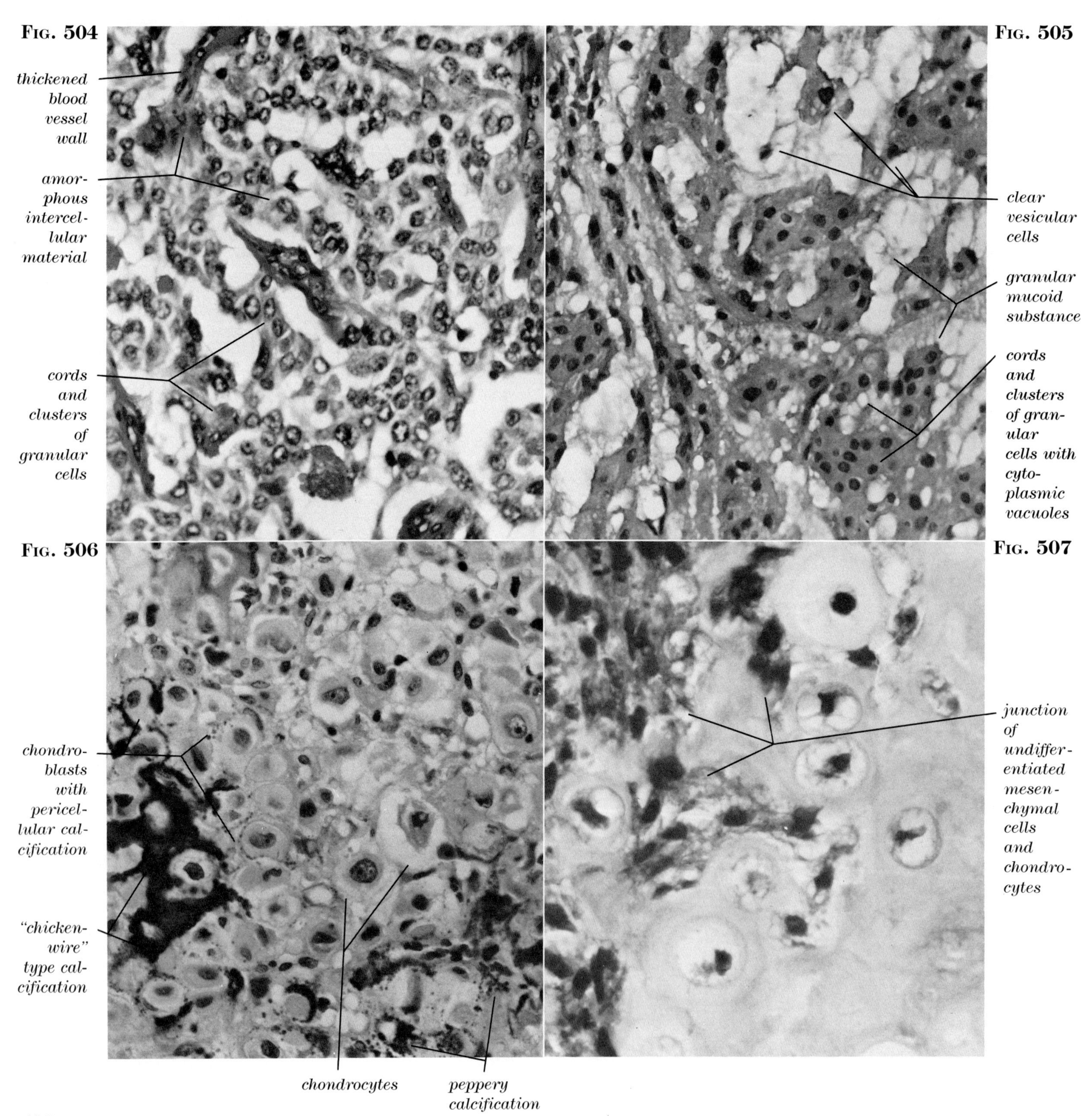

Fig. 508. Intraosseous synovial cyst (×140) (see Figs. 144, 186, 1014).

Fig. 509. Aneurysmal bone cyst (×350) (see Figs. 96, 125, 150, 163, 785, 1015, 1058).

Fig. 510. Chordoid sarcoma (×140) (see Figs. 456, 640, 919, 1176).

Fig. 511. Malignant epithelioid mesothelioma (×140) (see Figs. 265, 267, 531, 869, 1242, 1243).

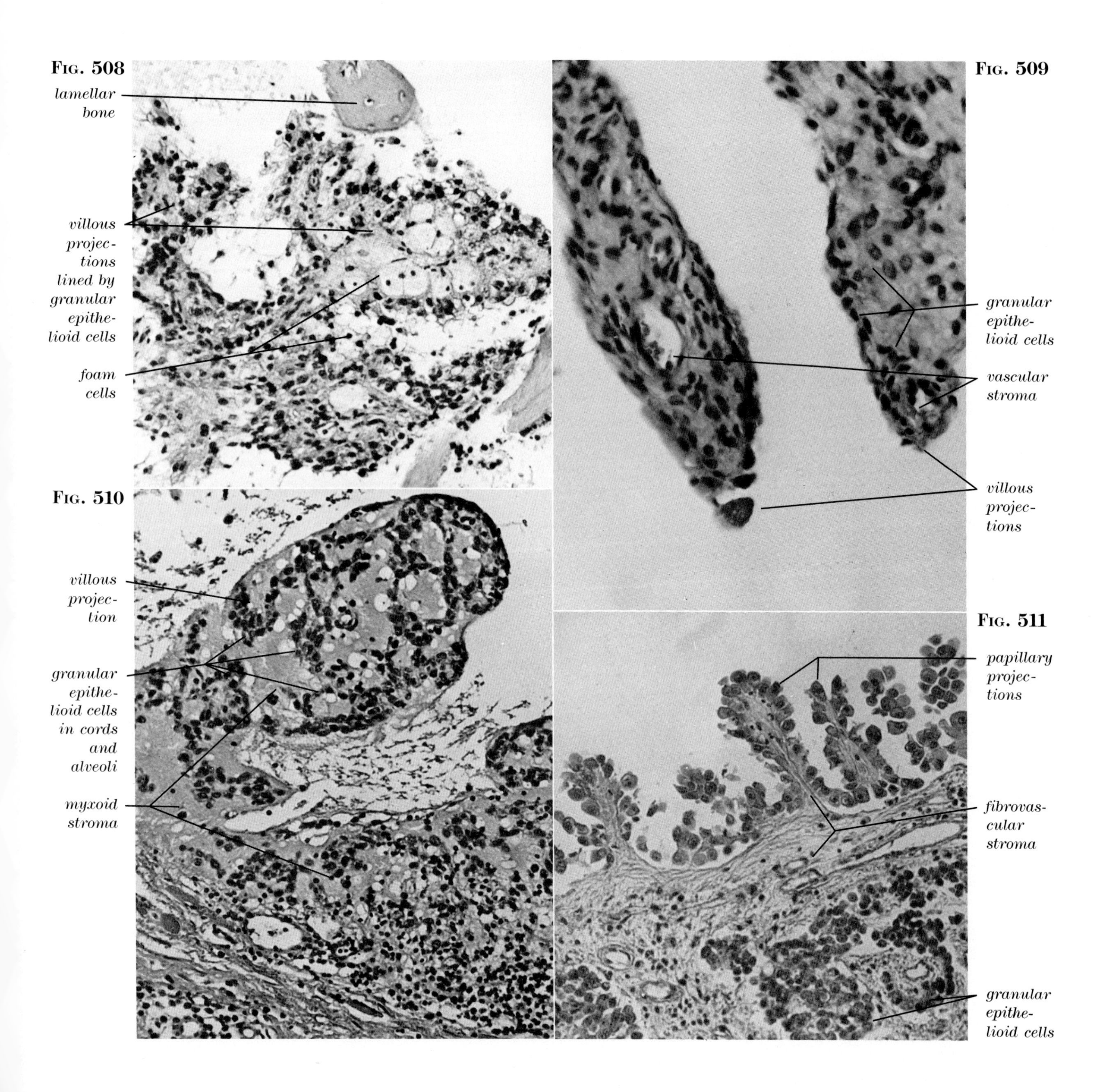

Fig. 512. Malignant thecagranulosa cell tumor of ovary (×350).

Fig. 513. Leiomyoblastoma (×350) (see Figs. 420, 429, 503, 691).

Fig. 514. Biphasic tendosynovial sarcoma (×140) (see Figs. 85, 122, 320, 520, 522, 535, 536, 547, 574, 586, 727, 765, 906, 1099).

Fig. 515. Embryonal rhabdomyosarcoma (×140) (see Figs. 443, 446, 472, 637, 689).

Fig. 516. Benign glandular schwannoma (×140) (see Fig. 517).

Fig. 517. Benign glandular schwannoma (×350) (see Fig. 516).

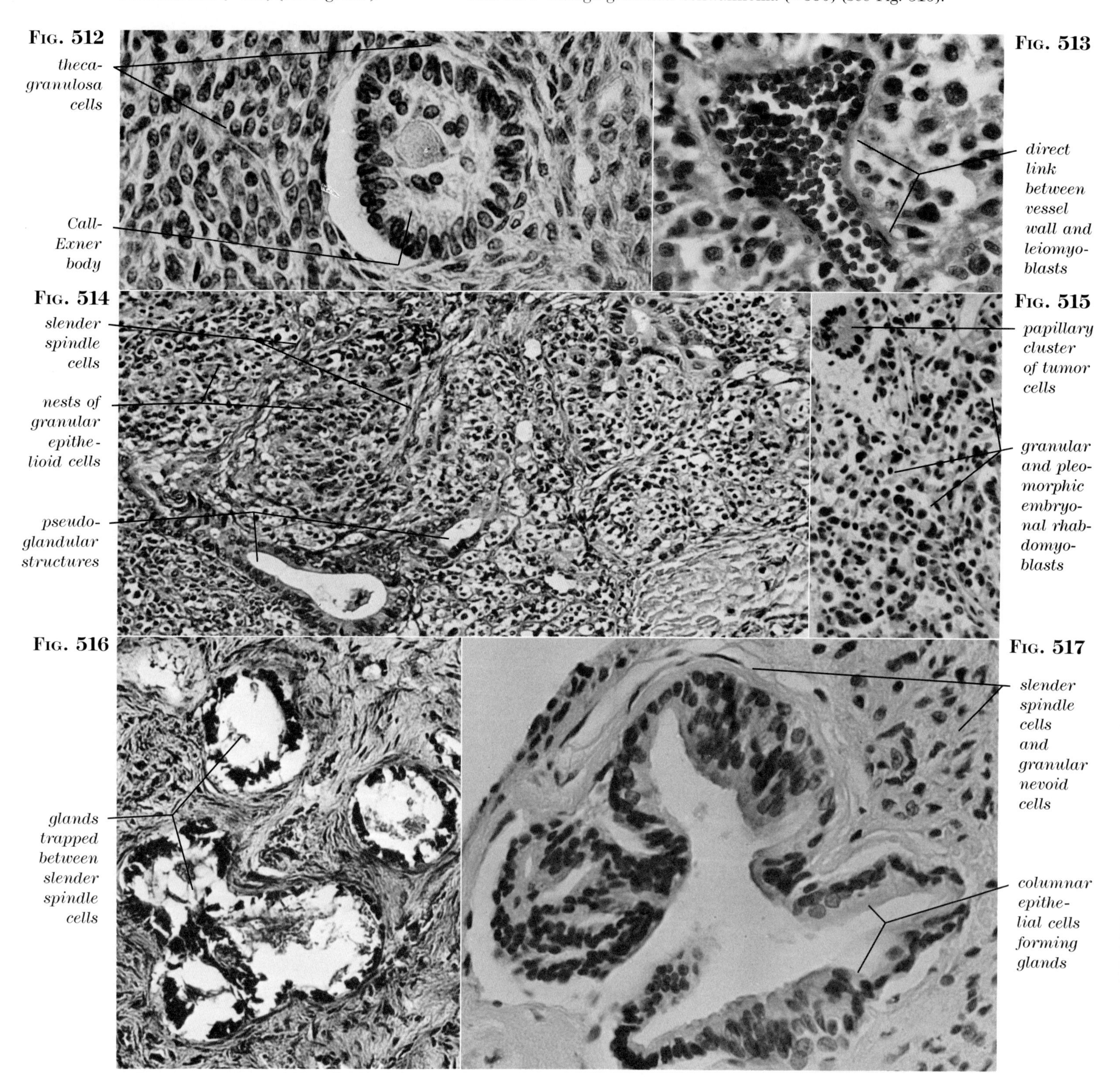

Fig. 518. Myxopapillary ependymoma of pelvis (×350) (see Figs. 239, 466, 1177).

Fig. 519. Malignant undifferentiated peripheral nerve tumor, so-called "pigmented retinal anlage tumor" (×350) (see Figs. 342, 596, 827).

Fig. 520. Biphasic tendosynovial sarcoma (×350) (see Figs. 152, 320, 514, 522, 535, 536, 547, 574, 586, 727, 765, 906).

Fig. 521. Malignant pigmented schwannoma (×350) (see Fig. 701).

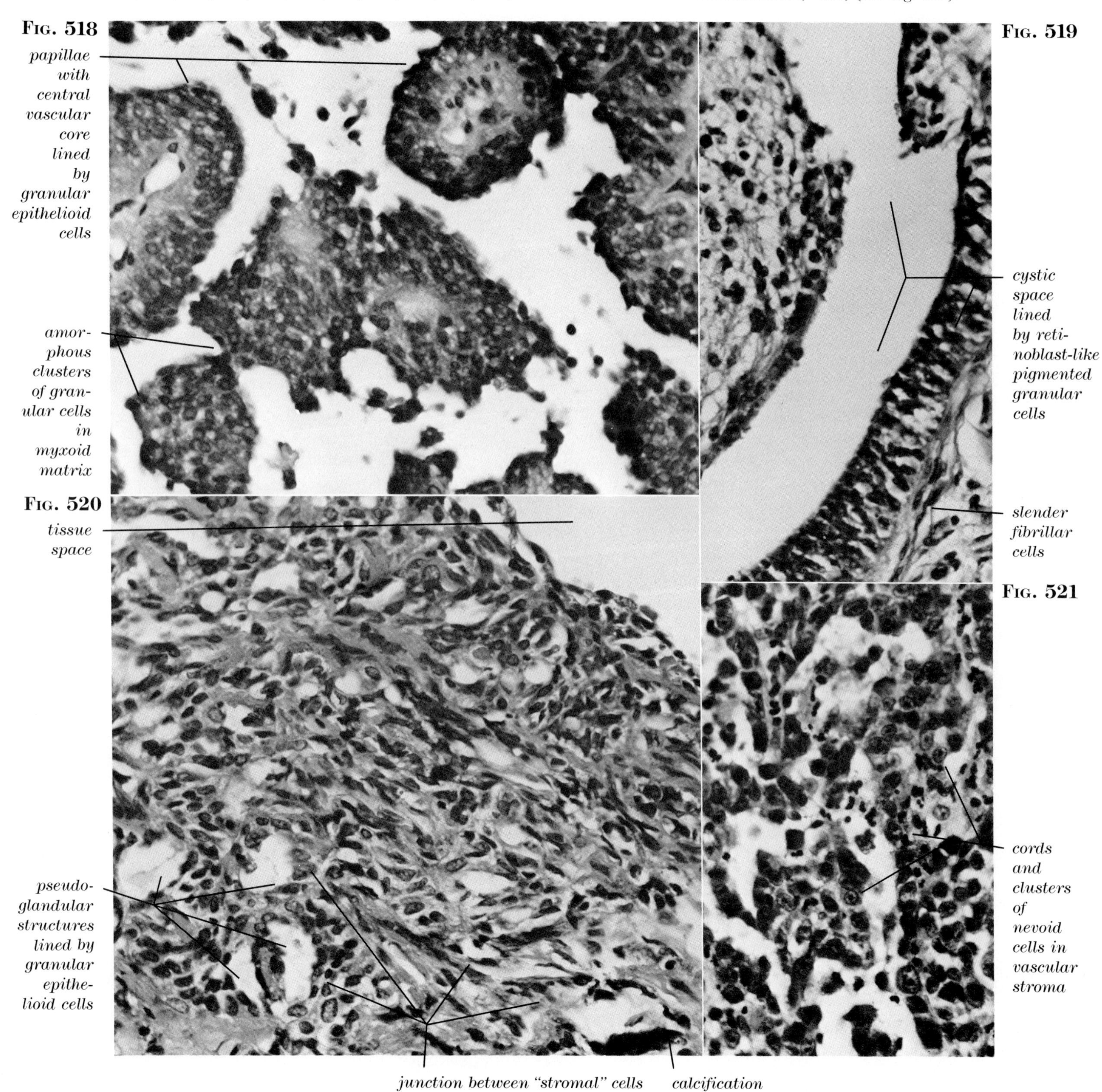

Fig. 522. Biphasic tendosynovial sarcoma (×140) (see Figs. 85, 323, 514, 520, 535, 536, 547, 574, 586, 727).

Fig. 523. High grade hemangiosarcoma (×140) (see Figs. 16, 32, 92, 541, 555, 557, 563, 564, 576, 720).

Fig. 524. Alveolar rhabdomyosarcoma (×140) (see Figs. 78, 206, 529, 530, 567, 570, 571, 804).

Fig. 525. High grade hemangiopericytoma (×140) (see Figs. 93, 470, 479, 533, 693, 696, 972, 984, 995, 1203).

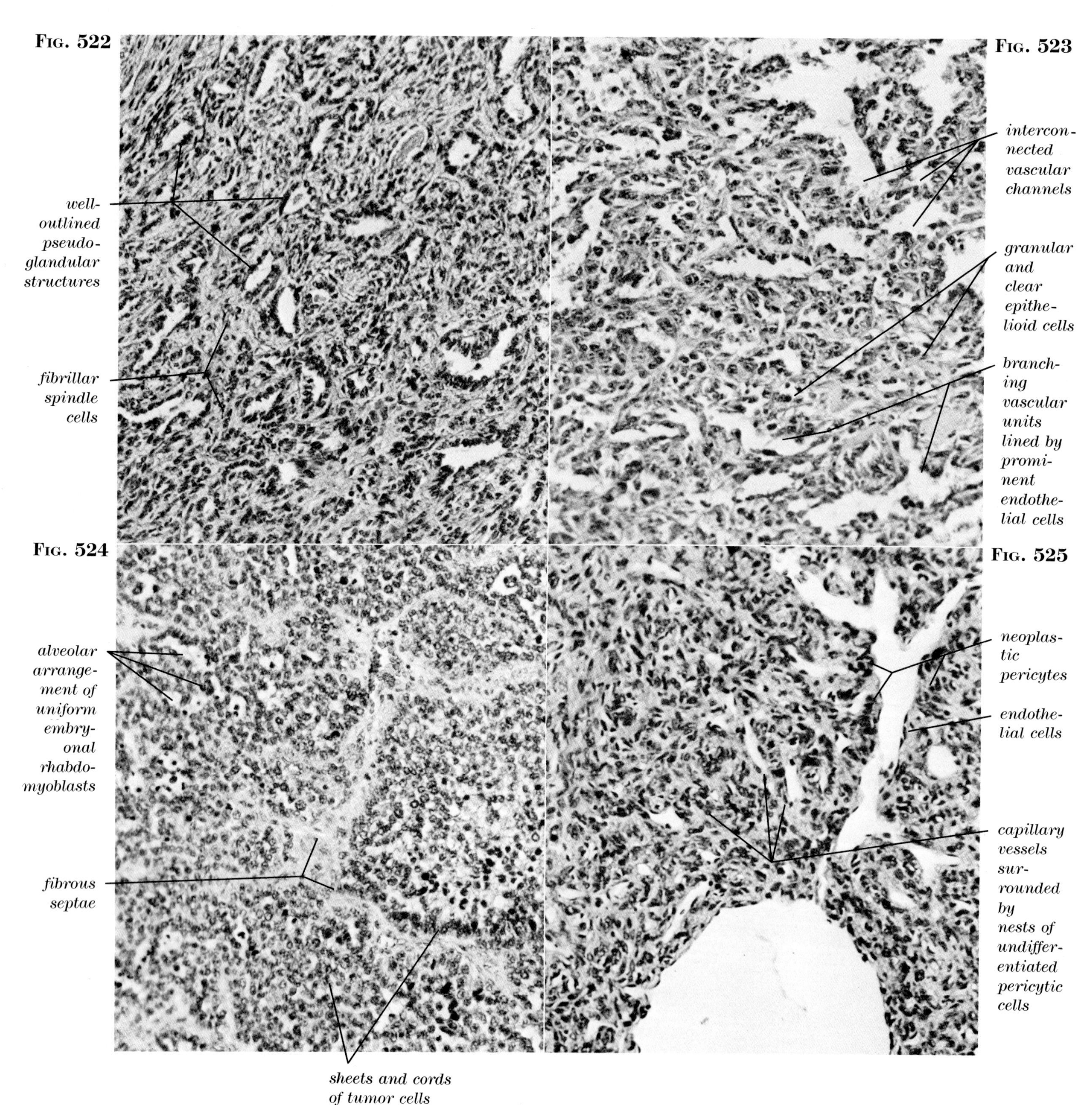

Fig. 526. Clear cell sarcoma (×140) (see Figs. 85, 447, 492, 759, 761, 896, 927, 1145, 1223, 1244, 1245).

Fig. 527. Plexiform neurofibroma (×140) (see Figs. 365, 627, 842, 1114).

Fig. 528. Rhabdomyoblastoma (×140) (see Figs. 477, 500, 569, 594, 695, 715, 739, 796, 936, 1034, 1159, 1260).

Fig. 529. Embryonal rhabdomyosarcoma partly growing as alveolar rhabdomyosarcoma (×140) (see Figs. 78, 524, 530, 567, 570, 571, 637, 689, 751, 804, 949).

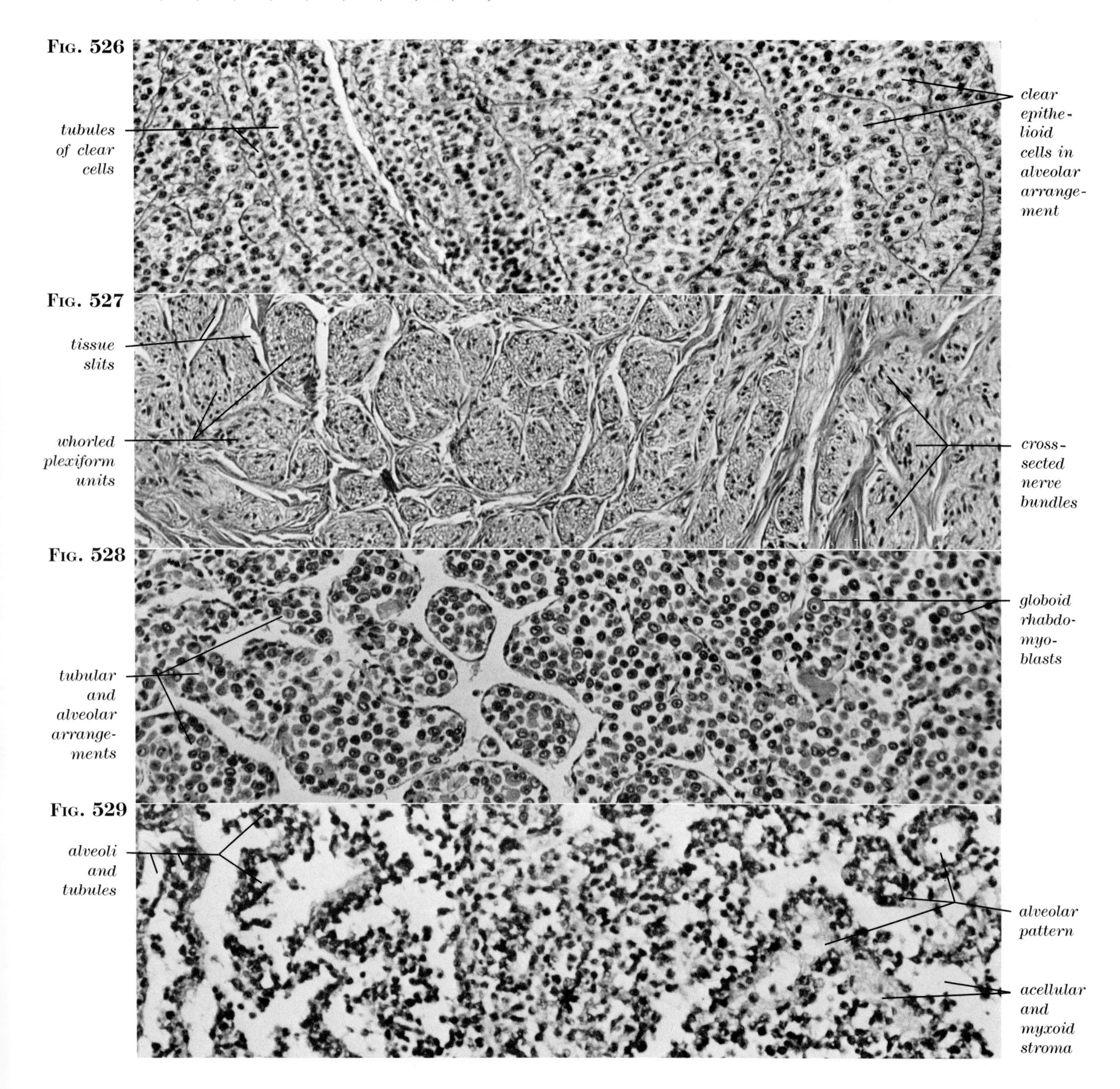

FIG. 530. Alveolar rhabdomyosarcoma (×350) (see Figs. 78, 206, 524, 529, 567, 570, 571, 804).

FIG. 531. Malignant epithelioid mesothelioma (×350) (see Figs. 267, 511, 869).

FIG. 532. Medullary carcinoma of thyroid (×140) (see Figs. 504, 916).

FIG. 533. (Center) Hemangiopericytoma (×140) (see Figs. 470, 479, 484, 525, 693).

FIG. 534. Malignant histiocytic fibrous histiocytoma (×140) (see Figs. 57, 156, 452, 592).

FIG. 535. Biphasic tendosynovial sarcoma (×140) (see Figs. 514, 520, 522, 536).

FIG. 536. Biphasic tendosynovial sarcoma (×140) (see Figs. 85, 514, 522, 547).

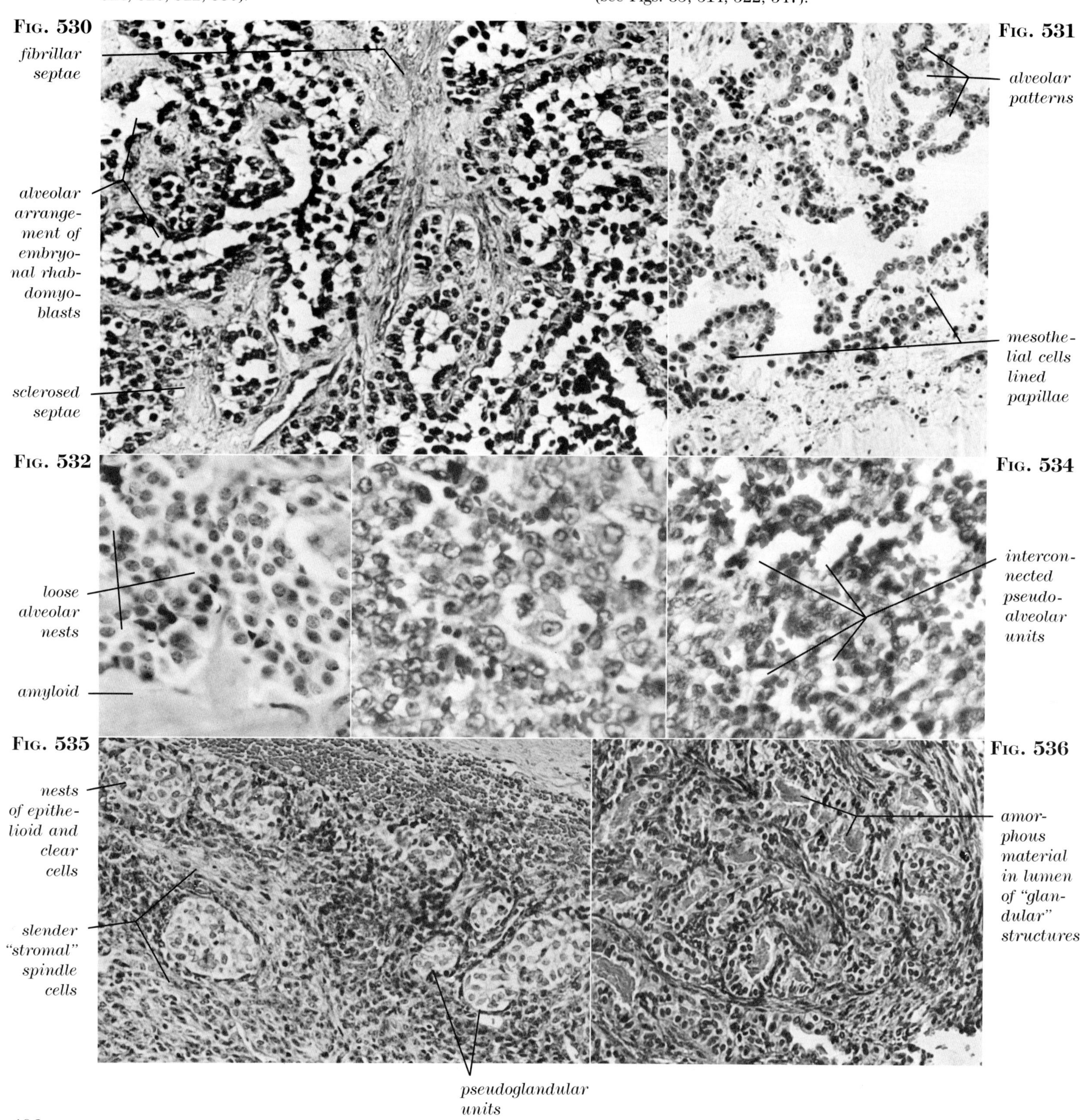

Fig. 537. Extracranial meningioma (×350) (see Figs. 485, 496, 1096, 1225).

Fig. 538. Alveolar soft part sarcoma (×350) (see Figs. 89, 697, 718, 1161, 1199, 1227).

Fig. 539. Benign granular cell tumor (×350) (see Figs. 75, 261, 498, 736, 1160, 1235).

Fig. 540. Metastatic malignant melanoma (×350) (see Figs. 394, 409, 669, 719, 748).

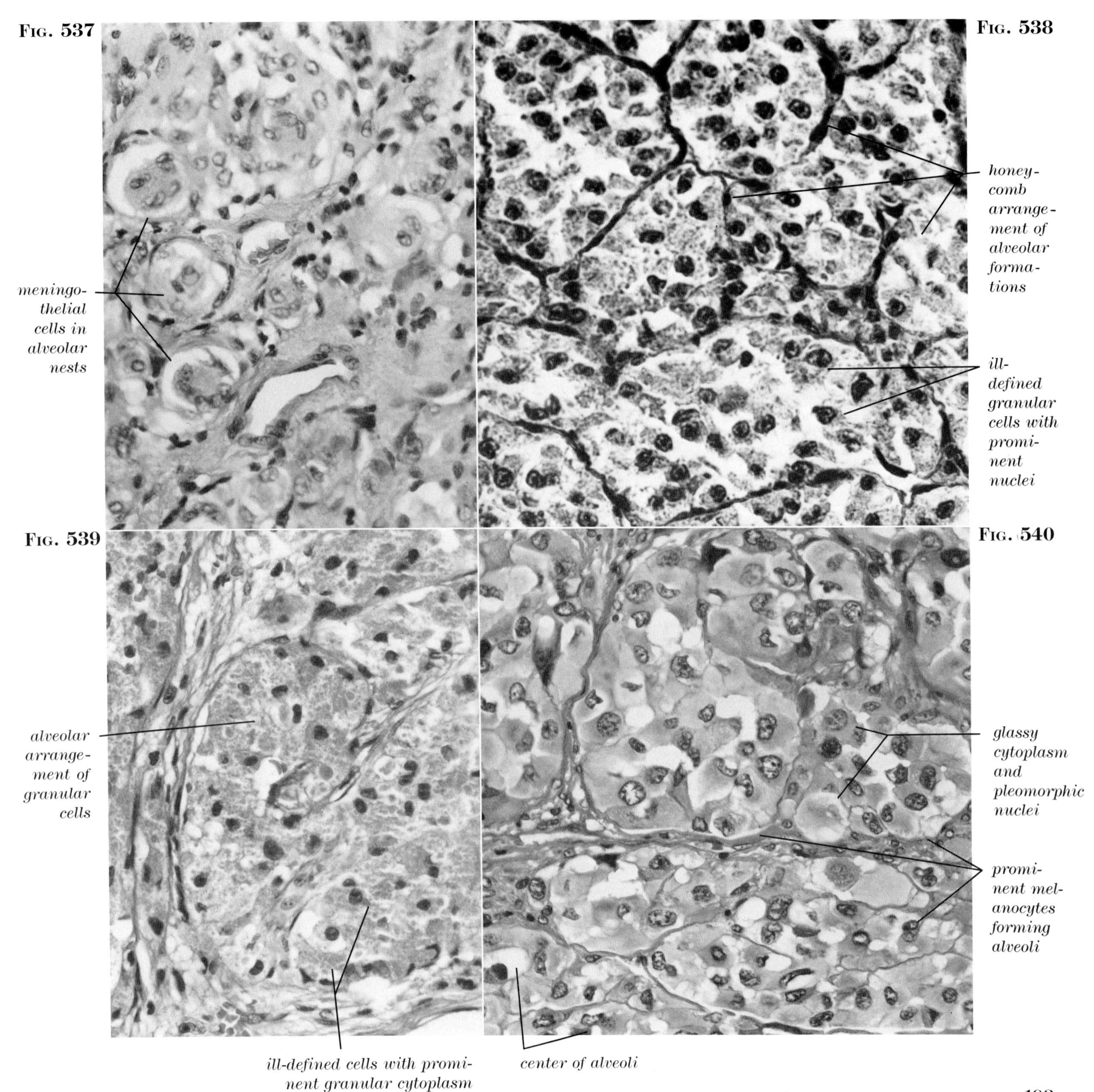

Fig. 541. Low grade hemangiosarcoma (×140) (see Figs. 42, 92, 316, 523, 555, 557, 564, 720, 975, 976, 990, 1005, 1209, 1210).

Fig. 542. High grade hemangiosarcoma (×350) (see Figs. 92, 102, 316, 523, 555, 557, 563, 564, 576, 720, 731, 905, 1210).

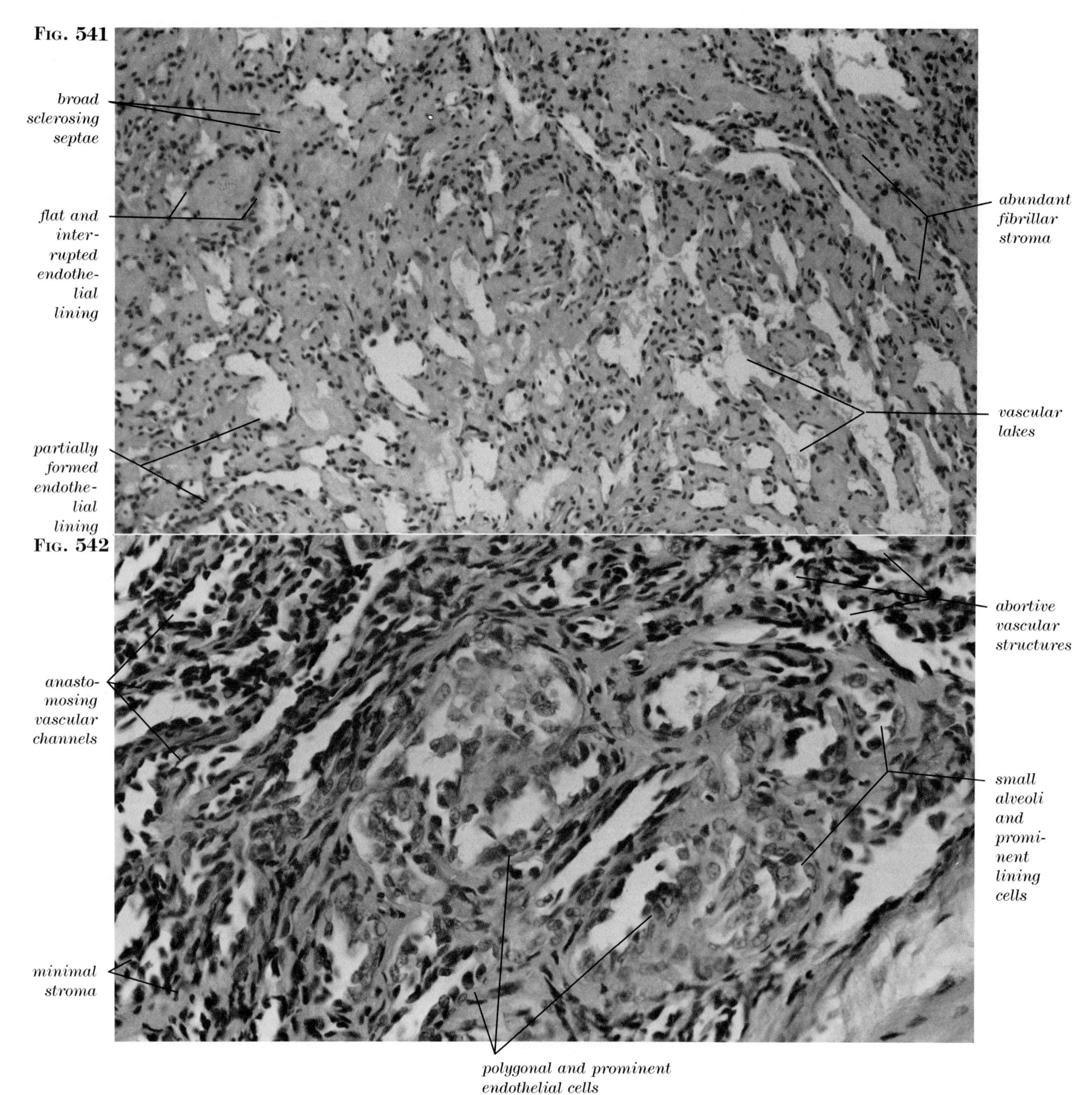

Fig. 543. Lipoblastic liposarcoma (×350) (see Figs. 415, 448, 450, 687).

Fig. 544. (Top center) Malignant paraganglioma composed of ill-defined pleomorphic cells (×350) (see Figs. 483, 714, 790).

Fig. 545. Metastatic clear cell carcinoma of kidney (×350) (see Figs. 651, 716, 764, 768).

Fig. 546. Malignant, epithelioid thymoma (×350) (see Figs. 337, 375, 617).

Fig. 547. (Bottom center) Biphasic tendosynovial sarcoma (×350) (see Figs. 522, 535, 536, 574).

Fig. 548. Poorly differentiated follicular carcinoma of thyroid (×350).

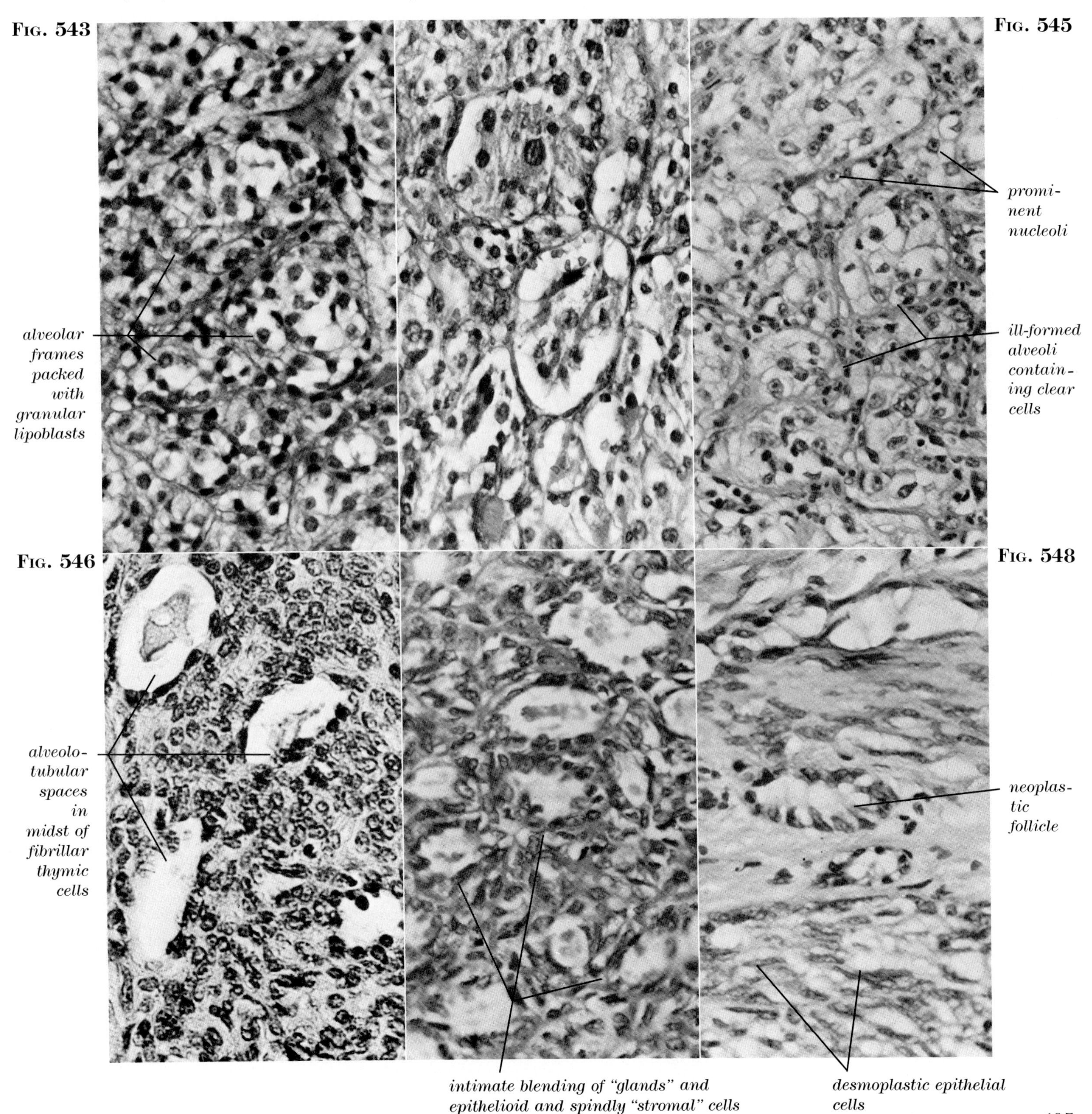

Fig. 549. Capillary hemangioma (×350) (see Figs. 18, 83, 559, 988).

Fig. 550. Malignant glomus tumor (×350) (see Figs. 457, 482, 728, 980, 1141).

Fig. 551. Lymphangiosarcoma of lymphedematous extremity (×350) (see Figs. 41, 61, 159, 321, 424, 558, 649, 954, 963, 965, 970, 1098).

Fig. 552. Proliferative panniculitis (×350) (see Figs. 430, 721, 910).

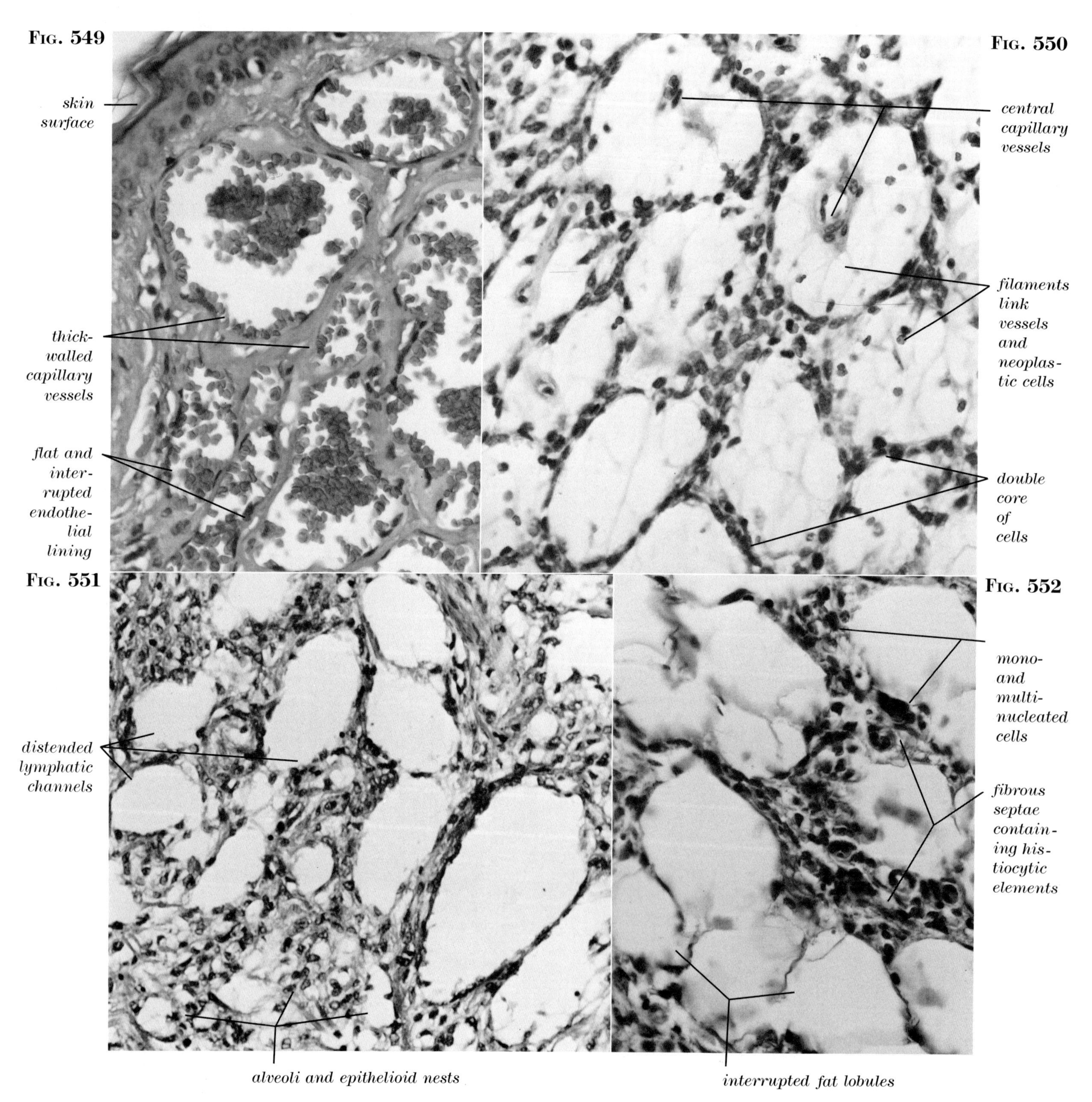

Fig. 553. Borderline vascular neoplasm, so-called "adult hypertrophic hemangioma," resembling glomus tumor (×140) (see Figs. 83, 98, 554).

Fig. 554. Hypertrophic hemangioma (×140) (see Figs. 83, 549, 553, 974, 1221).

Fig. 555. Low grade hemangiosarcoma of breast (×140) (see Figs. 32, 523, 541, 557, 563, 564, 576, 720, 731, 905, 975, 976).

Fig. 556. Benign nevoid schwannoma (×140) (see Figs. 421, 471).

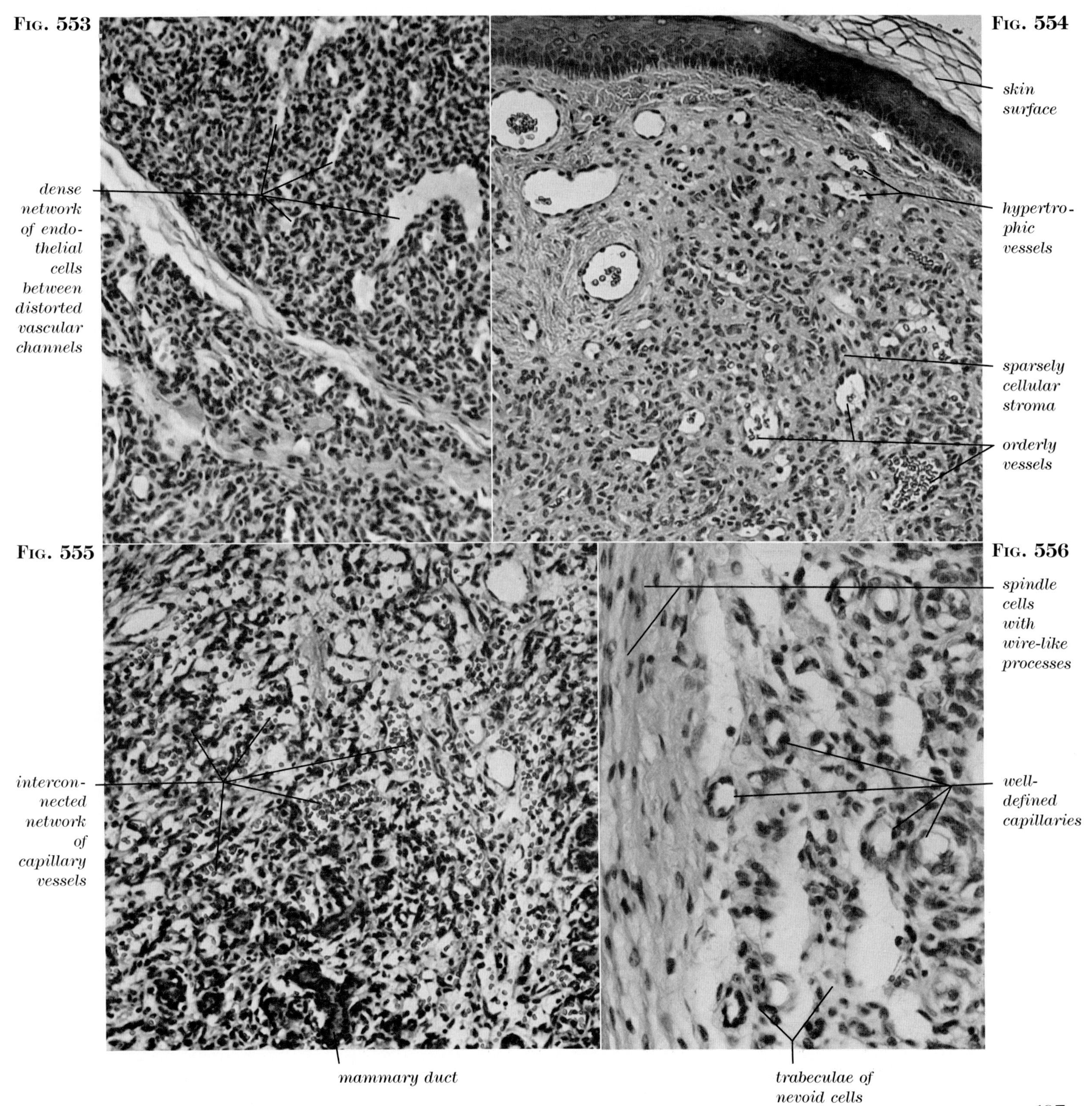

Fig. 557. High grade hemangiosarcoma (×140) (see Figs. 92, 523, 541, 542, 555, 563, 564, 576, 720, 731).

Fig. 558. Lymphangiosarcoma of lymphedematous arm (×140) (see Figs. 35, 41, 154, 321, 424, 551, 649, 954, 963, 965, 970).

Fig. 559. Capillary hemangioma (×140) (see Figs. 18, 83, 549, 988).

Fig. 560. Papillary endothelial hyperplasia (×140) (see Fig. 1004).

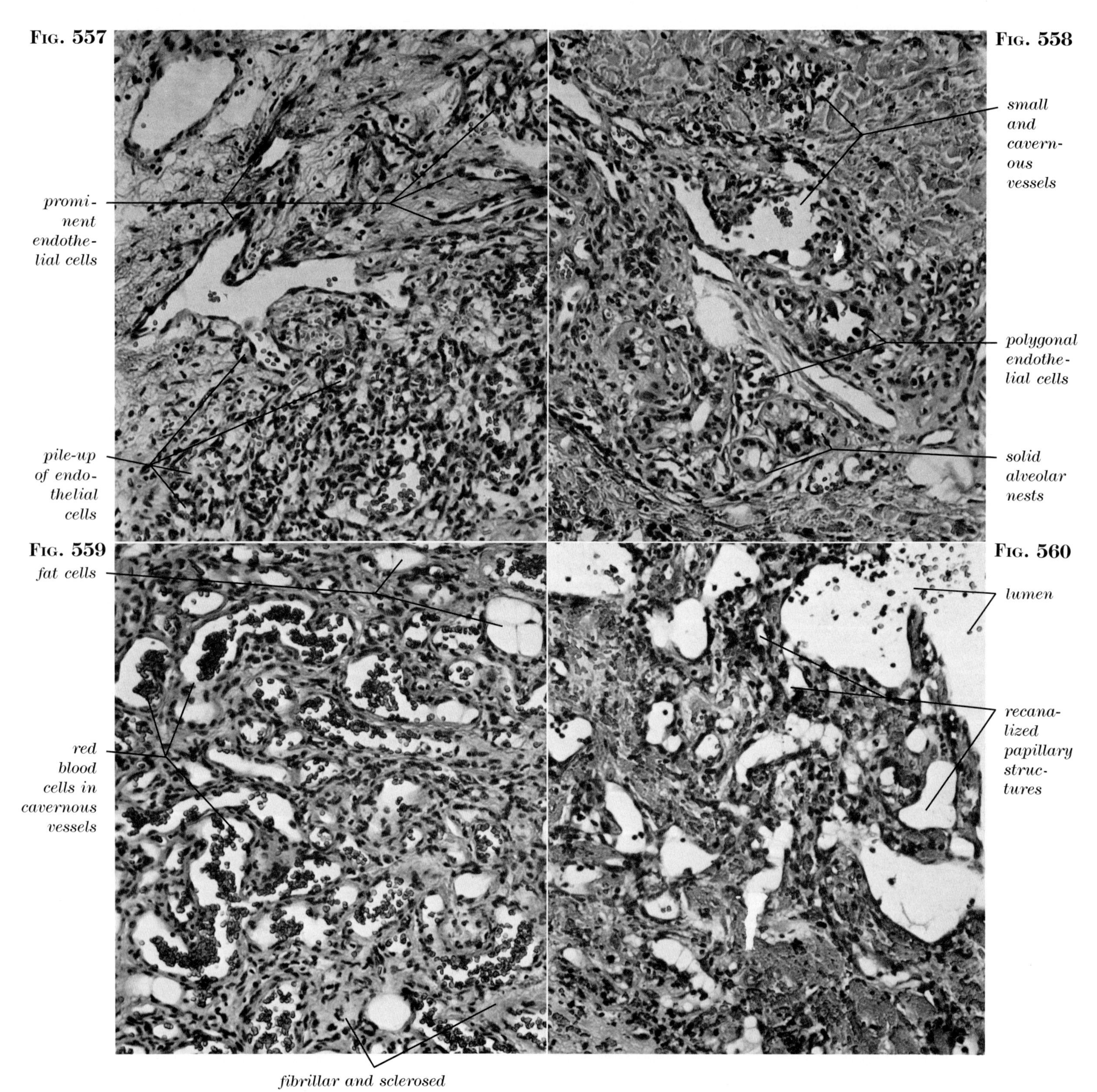

Fig. 561. Primitive neuroectodermal tumor (×140) (see Figs. 565, 572, 982).

Fig. 562. Lymphangiomyoma (×140) (see Fig. 653, 912, 978).

Fig. 563. High grade hemangiosarcoma (×140) (see Figs. 16, 32, 92, 102, 316, 523, 541, 555, 557, 564, 576, 720, 731, 905, 975, 976, 989, 990, 1005).

Fig. 564. High grade hemangiosarcoma (×140) (see Figs. 46, 92, 523, 541, 542, 555, 563, 576, 720, 731, 905, 954, 963, 975, 976, 989, 1098, 1209, 1210).

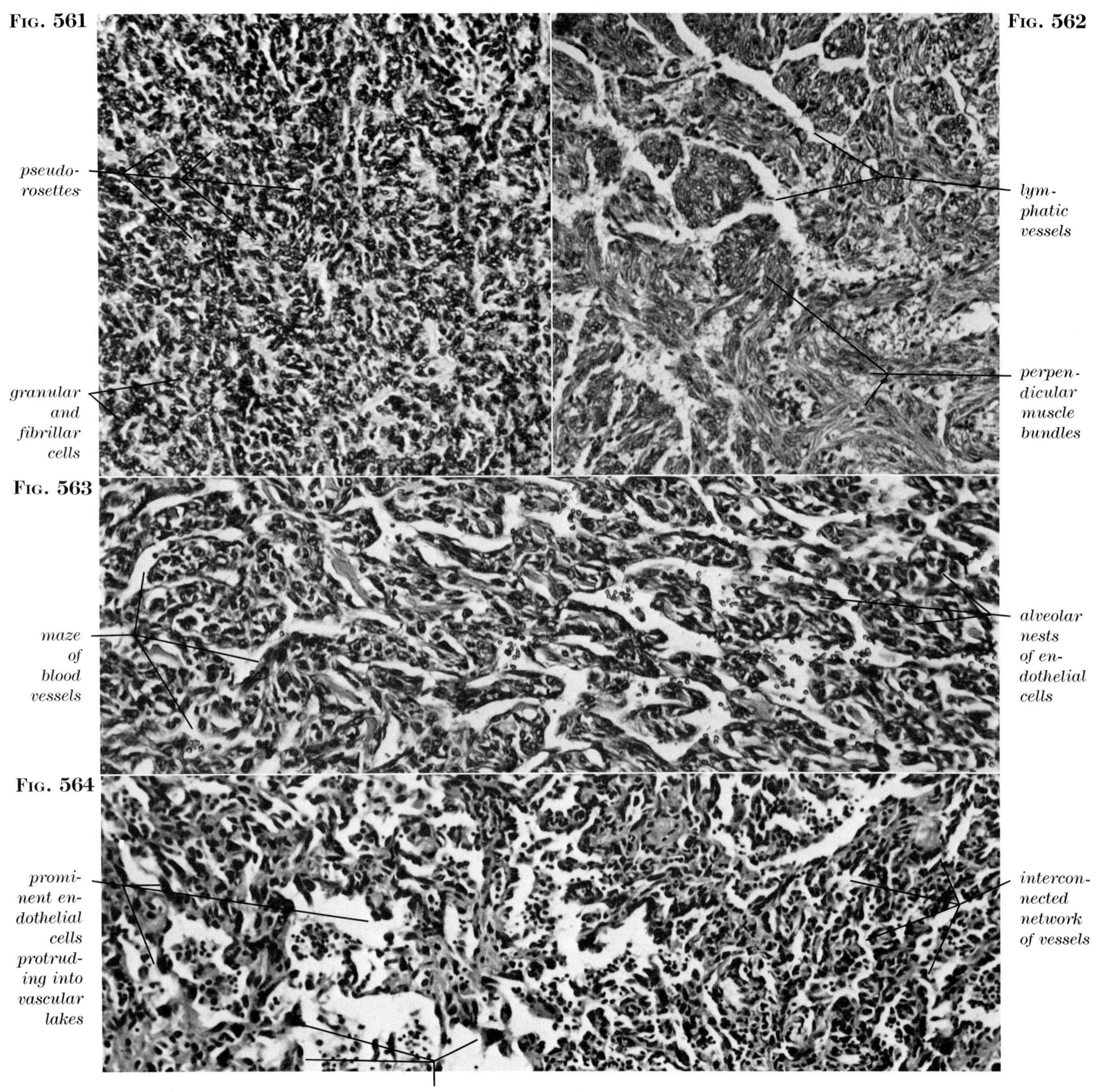

Fig. 565. Primitive neuroectodermal tumor, so-called "peripheral neuroblastoma" (×350) (see Figs. 561, 572, 982).

Fig. 566. (Top center) Seminoma of testis (×350) (see Figs. 14, 757, 1224).

Fig. 567. Alveolar rhabdomyosarcoma (×350) (see Figs. 524, 529, 530, 570).

Fig. 568. Malignant neuroepithelioma of the thigh of an adult (×350) (see Figs. 390, 478, 489, 585).

Fig. 569. Rhabdomyoblastoma (×350) (see Figs. 473, 500, 528, 594, 695, 709, 715, 739, 796, 936, 1034, 1159, 1260).

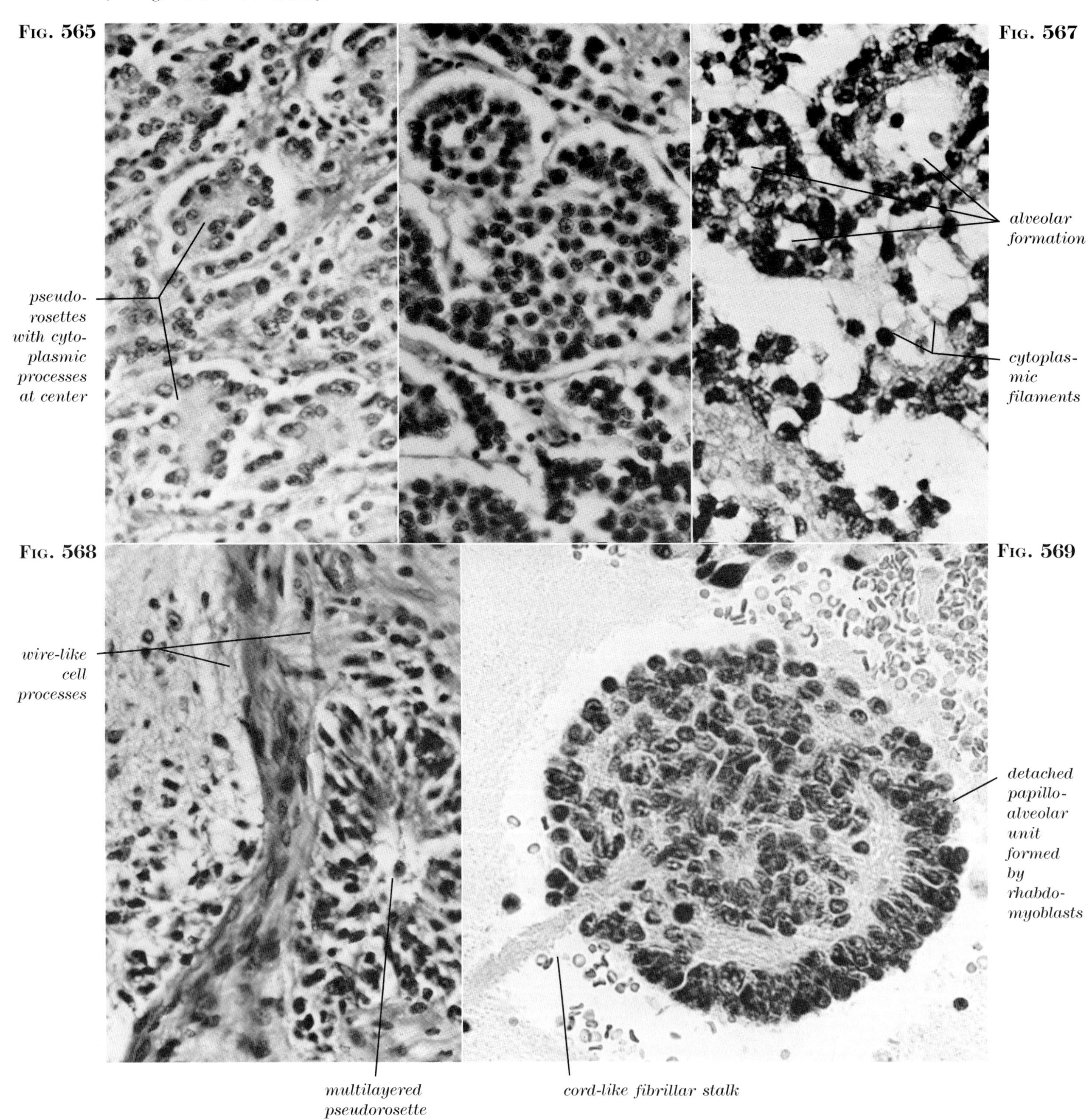

Fig. 570. Alveolar rhabdomyosarcoma (×350) (see Figs. 78, 206, 524, 529, 530, 567, 571, 804, 889, 925, 935, 949, 1129, 1251).

Fig. 571. Alveolar rhabdomyosarcoma (×350) (see Figs. 567, 570, 637, 689, 700, 706, 708, 751, 804, 889, 925, 935, 946, 1138, 1164, 1257).

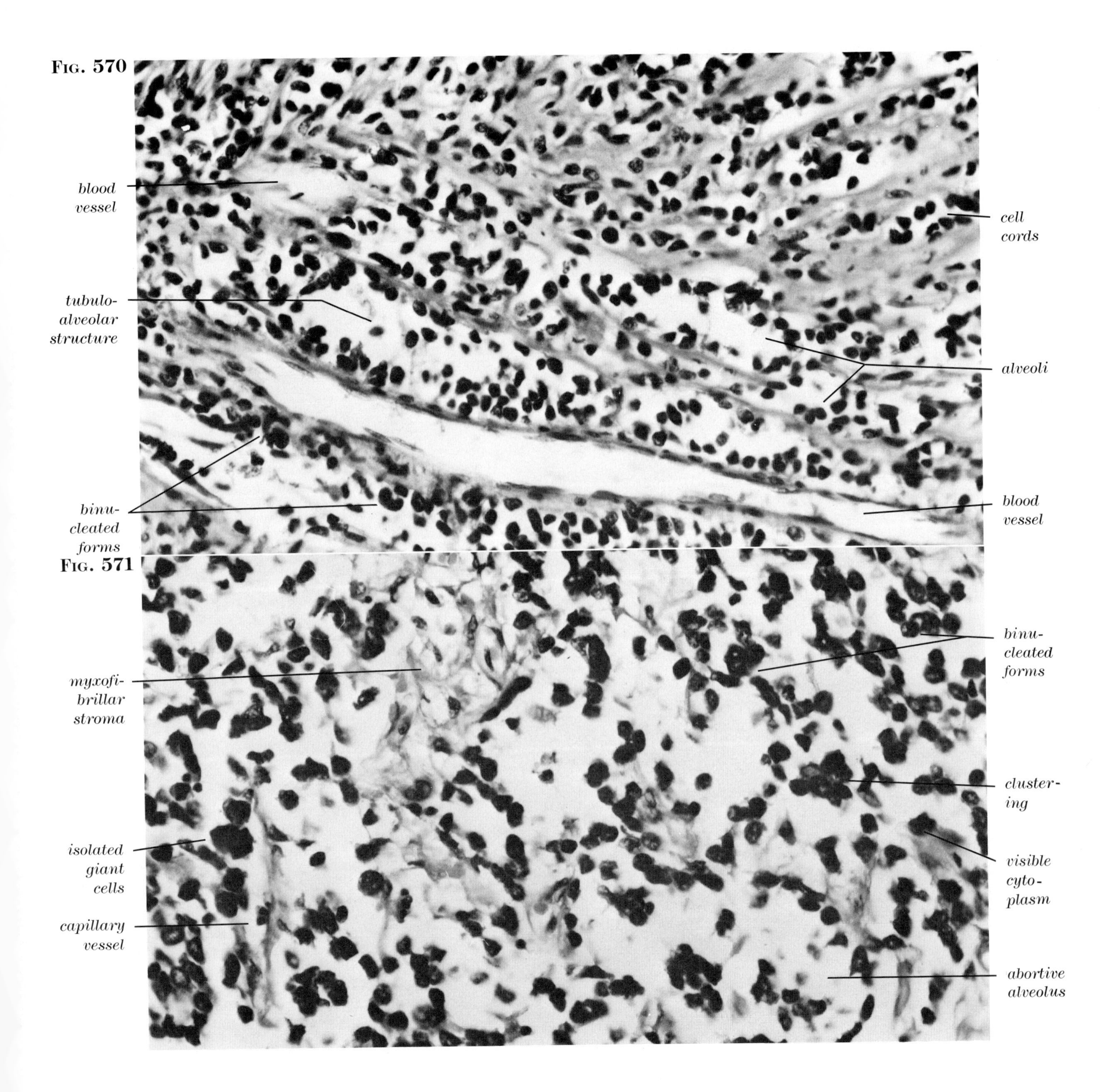

Fig. 572. Melanotic neuroectodermal tumor (×350) (see Figs. 342, 519, 521, 561, 565, 572, 596, 827, 982).

Fig. 573. Ewing's sarcoma (×350) (see Figs. 100, 172, 173, 210, 214, 227, 250, 286, 330, 449, 453, 476, 698, 1157, 1230).

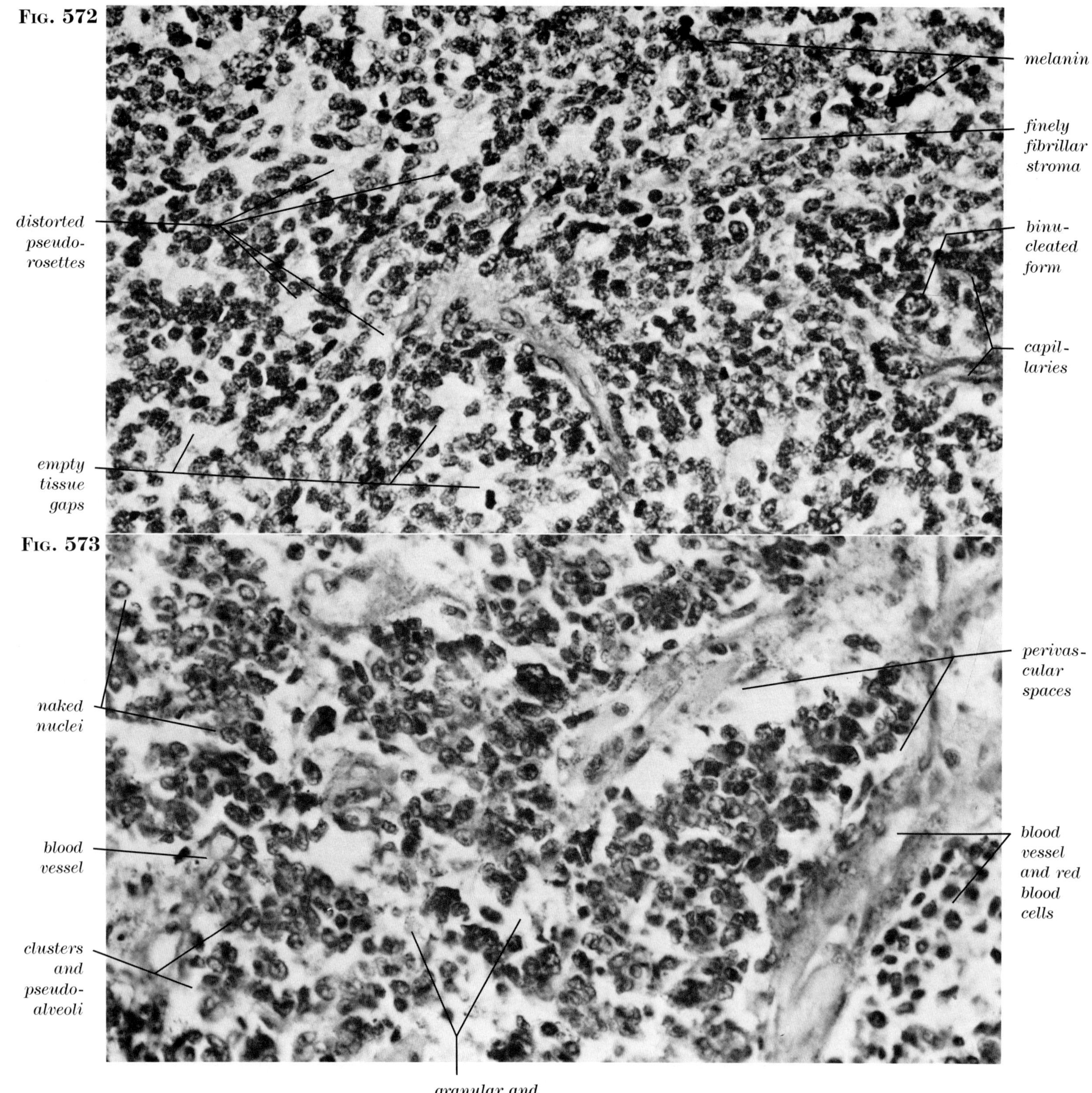

Fig. 574. Biphasic tendosynovial sarcoma (×350) (see Figs. 85, 520, 535, 536, 586).

Fig. 575. Monophasic tendosynovial sarcoma, pseudoglandular type (×350) (see Fig. 85).

Fig. 576. High grade hemangiosarcoma (×350) (see Figs. 92, 102, 316, 523, 541, 542, 557, 563, 564, 576, 720, 731, 905, 975, 976, 989, 990, 1005, 1209, 1210).

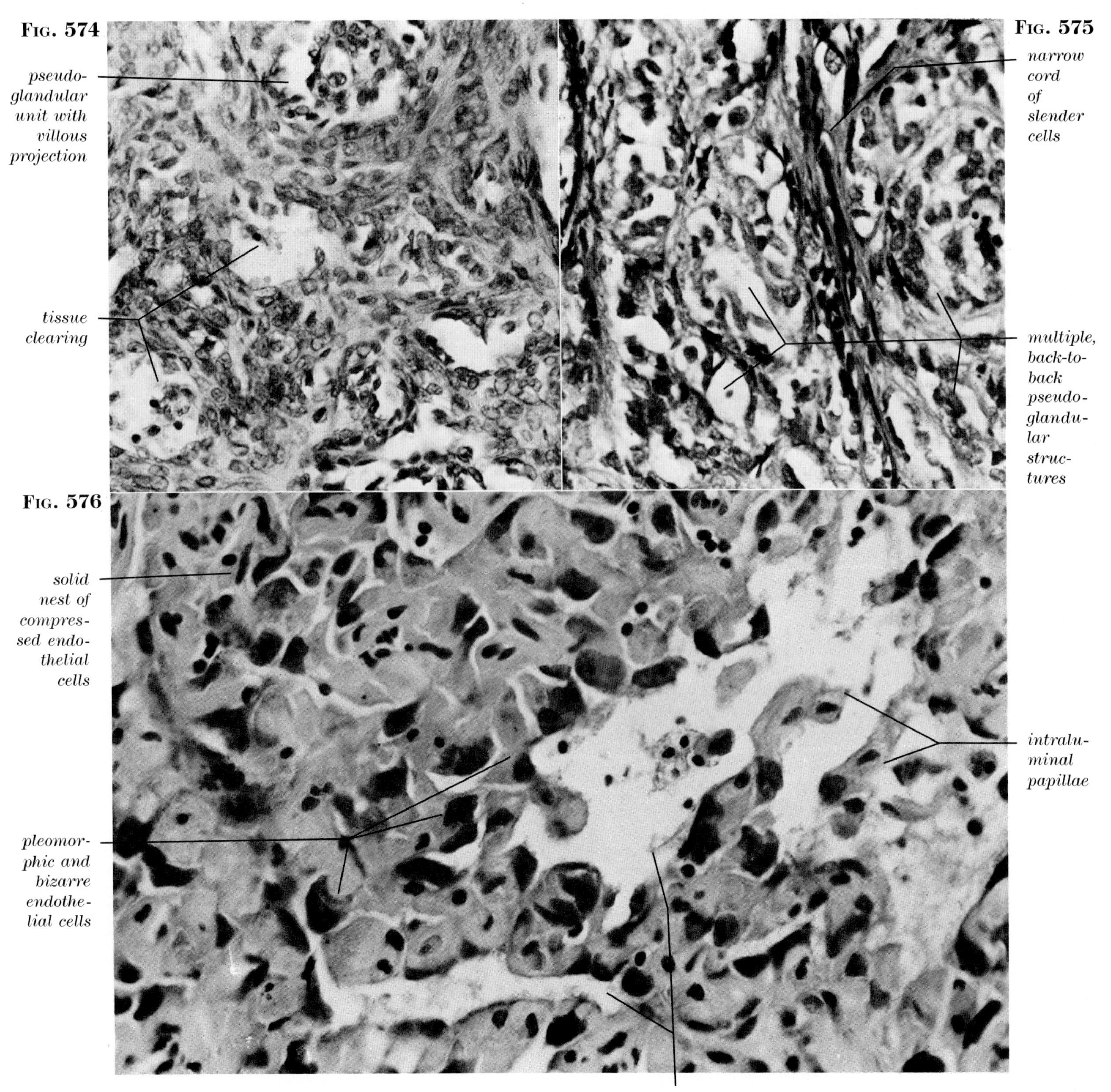

Fig. 577. High grade cutaneous malignant pleomorphic fibrous histiocytoma (×35) (see Figs. 344, 460, 579, 598, 679, 780, 788, 789, 792, 798, 805, 818, 937, 953, 962, 1002).

Fig. 578. Fasciitis (×35) (see Figs. 82, 345, 368, 408, 605, 621, 834, 851, 855, 858, 947, 956, 971, 991, 1020, 1030, 1045).

Fig. 579. High grade metastatic pleomorphic fibrous histiocytoma of tibia (×140) (see Figs. 111, 232, 460, 577, 598).

Fig. 580. High grade malignant schwannoma (×140) (see Figs. 298, 357, 382, 583, 591).

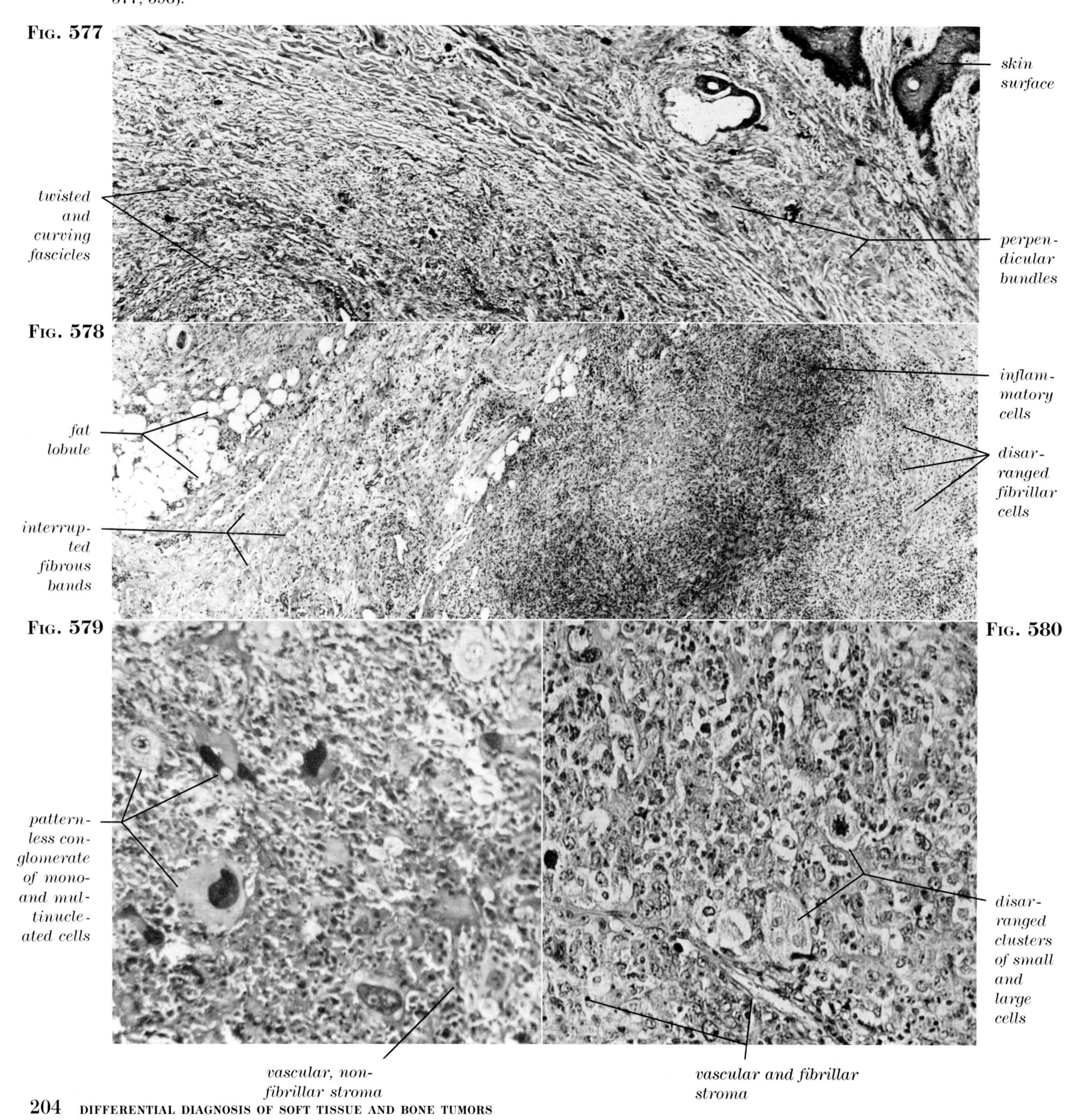

Fig. 581. Fascial fibromatosis (×140) (see Figs. 17, 419, 582, 616, 860).

Fig. 583. High grade malignant schwannoma (×140) (see Figs. 80, 355, 382, 580, 591).

Fig. 582. Subcutaneous fibromatosis (×140) (see Figs. 17, 280, 367, 385, 387, 419, 581, 616, 860, 861, 873, 874, 878).

Fig. 584. High grade neurofibrosarcoma (×350) (see Figs. 79, 264, 369, 370, 401, 624, 626, 838, 1116, 1229, 1263, 1266).

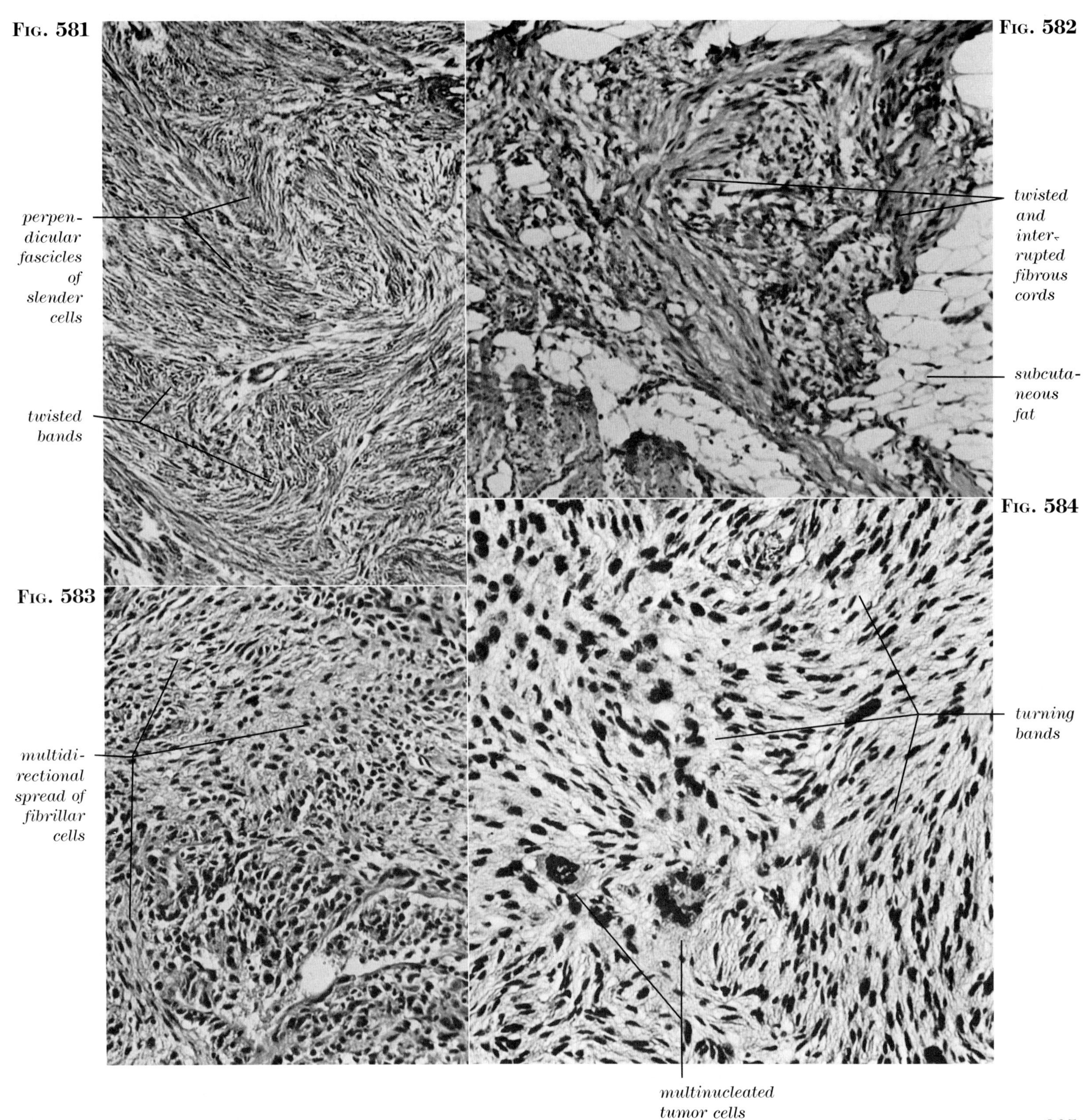

Fig. 585. Malignant neuroepithelioma of the retroperitoneum (×140) (see Figs. 478, 489, 519, 568, 572, 690, 1115, 1238, 1239).

Fig. 586. Biphasic tendosynovial sarcoma metastatic to bone (×140) (see Figs. 536, 547, 574, 727, 765, 906, 1099, 1100, 1173, 1202).

Fig. 587. Myxoid rhabdomyosarcoma, botryoid sarcoma, of bladder (×140) (see Figs. 416, 418, 644, 923, 929, 944, 952, 1033, 1251).

Fig. 588. High grade leiomyosarcoma of retroperitoneum (×140) (see Figs. 362, 366, 381, 404, 618, 659, 664, 666, 667, 680, 681, 685, 746).

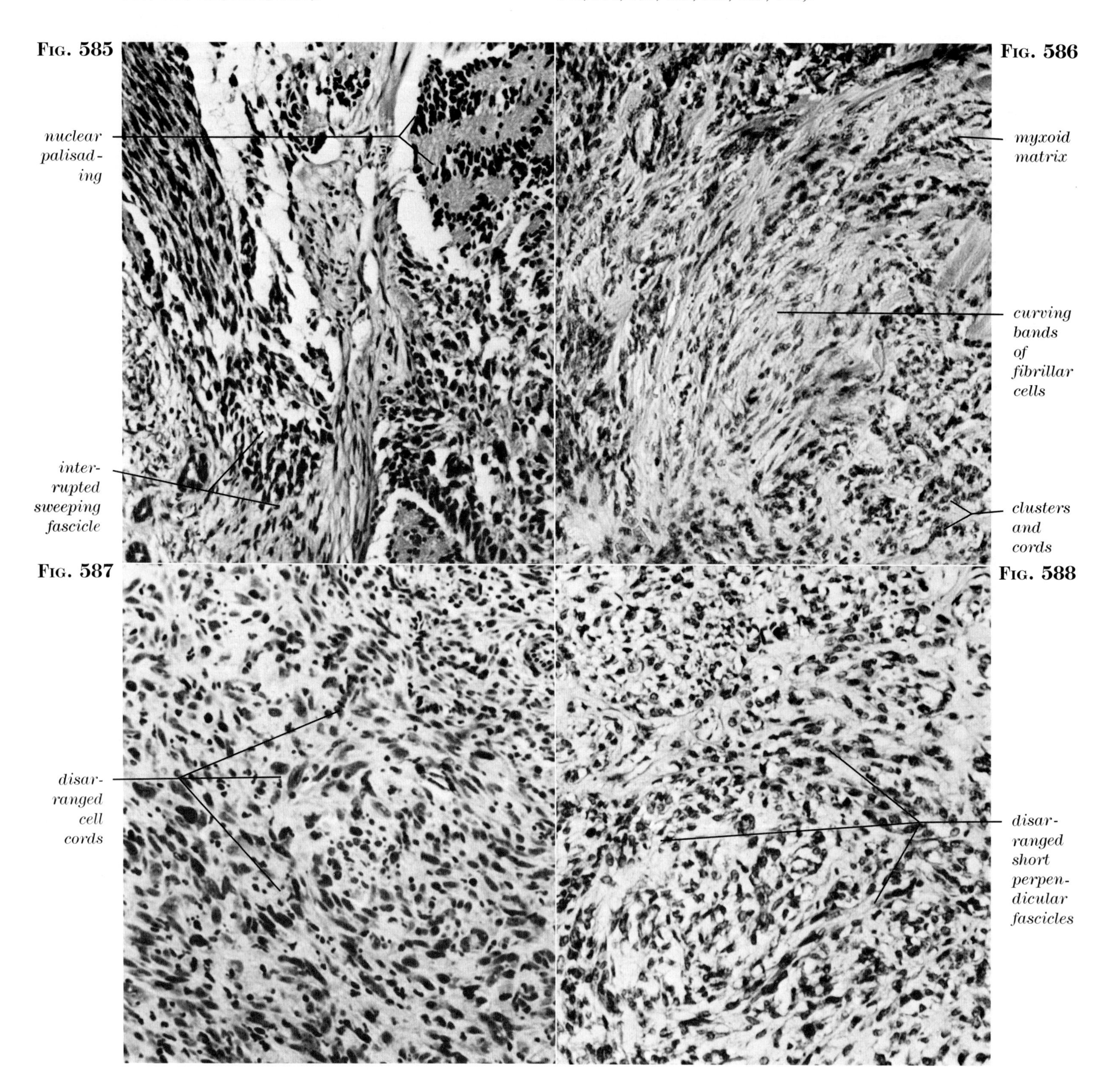

Fig. 589. Mesodermal mixed tumor of uterus (×140) (see Figs. 65, 763, 969, 1165, 1217).

Fig. 590. Pleomorphic liposarcoma (×140) (see Figs. 294, 458, 799, 823, 928, 1019A, 1049).

Fig. 591. High grade malignant schwannoma (×140) (see Figs. 357, 382, 580, 583, 603, 619, 628, 661, 815, 836).

Fig. 592. High grade malignant histiocytic fibrous histiocytoma (×140) (see Figs. 234, 452, 534, 600, 711, 756, 775, 776, 784, 791, 816, 872).

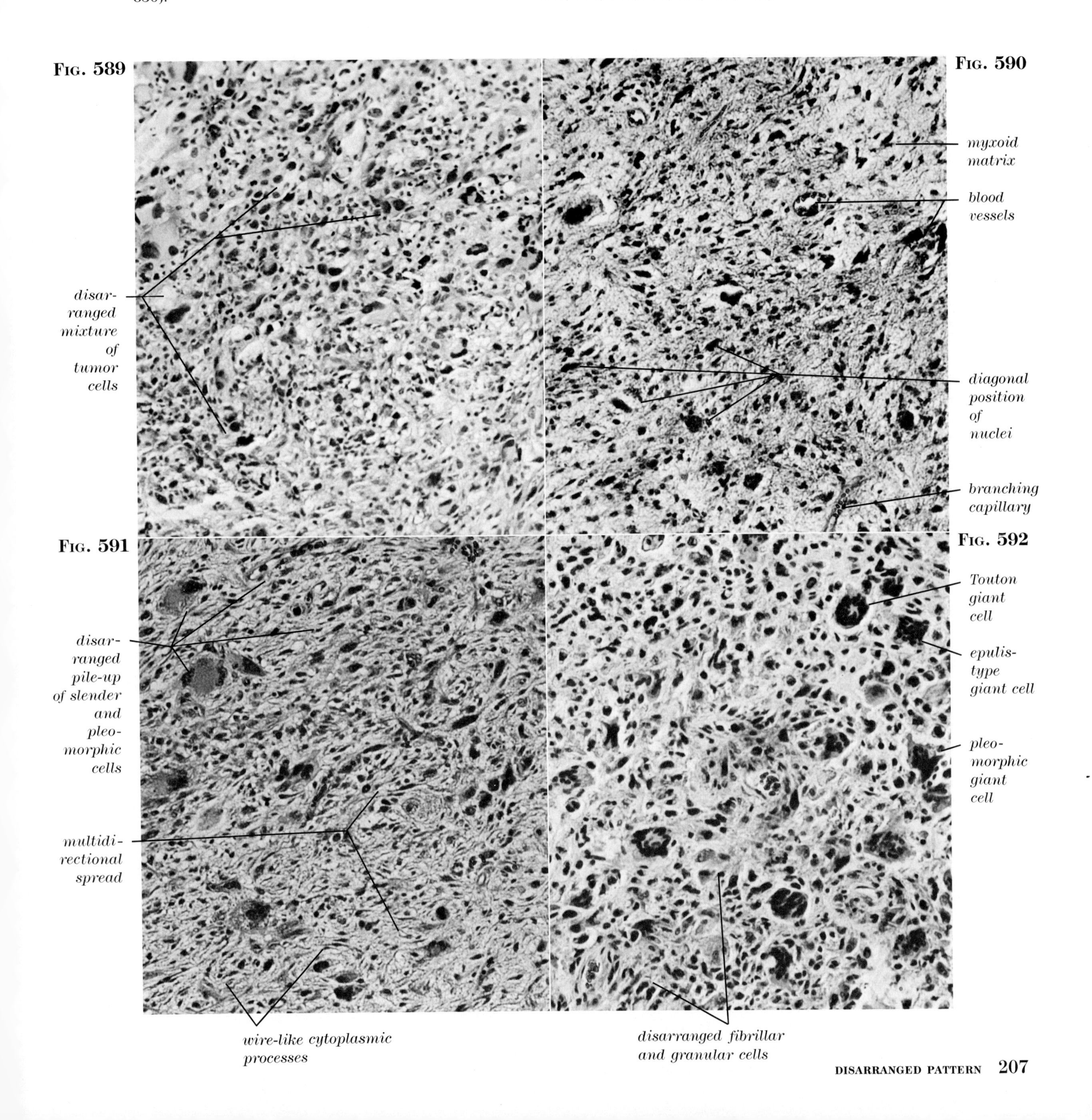

FIG. 593. Malignant Triton tumor (×140) (see Figs. 606, 732, 812, 813, 1259).

FIG. 594. Rhabdomyoblastoma (×140) (see Figs. 500, 528, 569, 695, 709, 715, 739, 796, 936).

FIG. 595. Mesenchymal chondrosarcoma (×140) (see Figs. 103, 151, 175, 180, 318, 390, 444, 507, 674, 707, 920, 1062, 1265).

FIG. 596. Malignant undifferentiated peripheral nerve tumor (×140) (see Figs. 342, 519, 572, 827, 982).

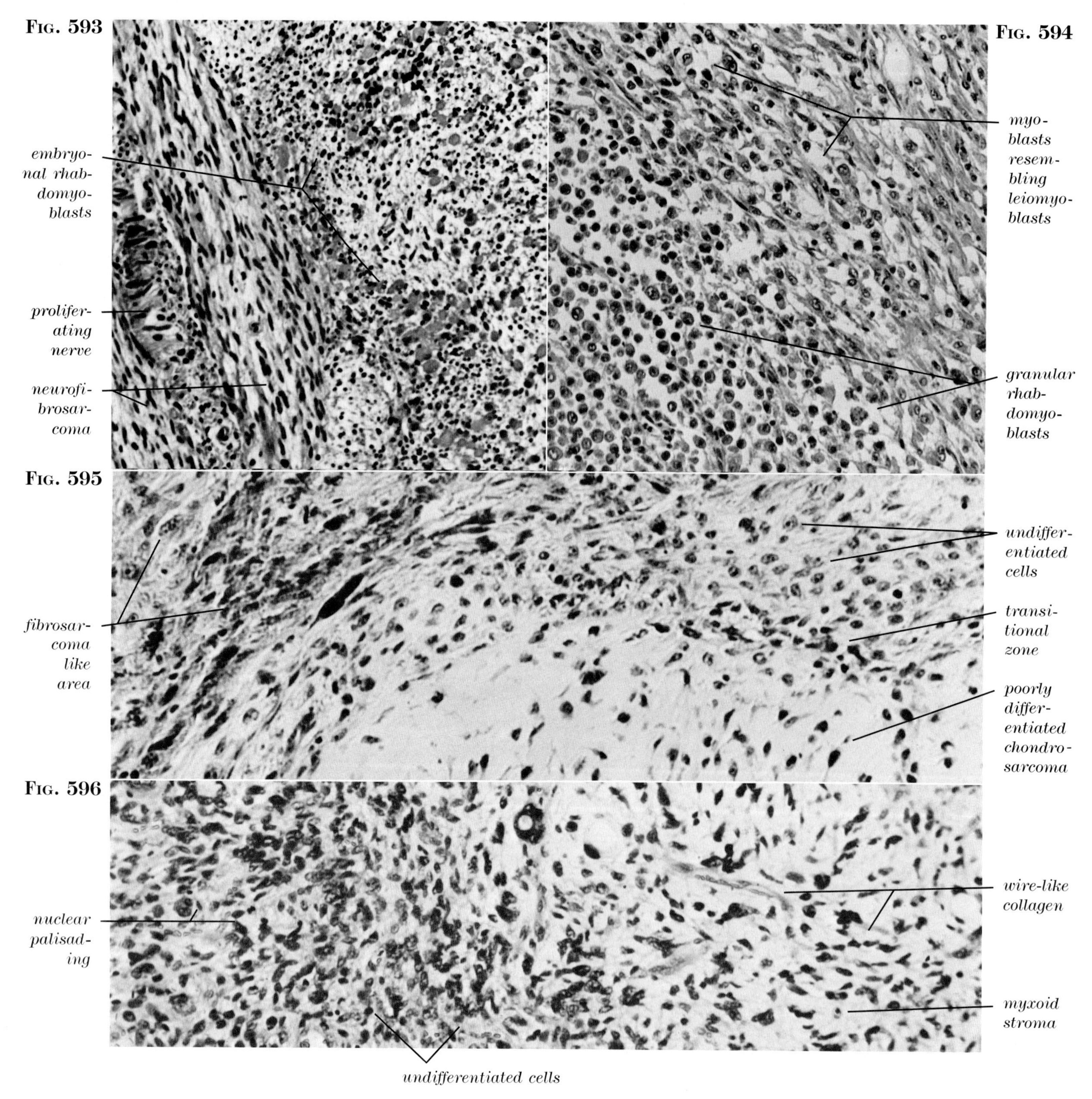

Fig. 597. Proliferative myositis (×140) (see Figs. 597, 738, 863, 951).

Fig. 598. High grade malignant pleomorphic fibrous histiocytoma (×140) (see Figs. 460, 577, 579, 679, 780, 788, 789, 792, 798, 805, 818, 937, 953, 962, 1002, 1003, 1008, 1032).

Fig. 599. Pleomorphic rhabdomyosarcoma (×140) (see Figs. 29, 87, 121, 604, 609, 612, 800, 802, 804, 810, 813, 1130, 1166).

Fig. 600. High grade malignant histiocytic fibrous histiocytoma (×140) (see Figs. 452, 534, 592, 711, 756, 775, 776, 784, 791, 816, 872, 892, 893, 930, 1018B).

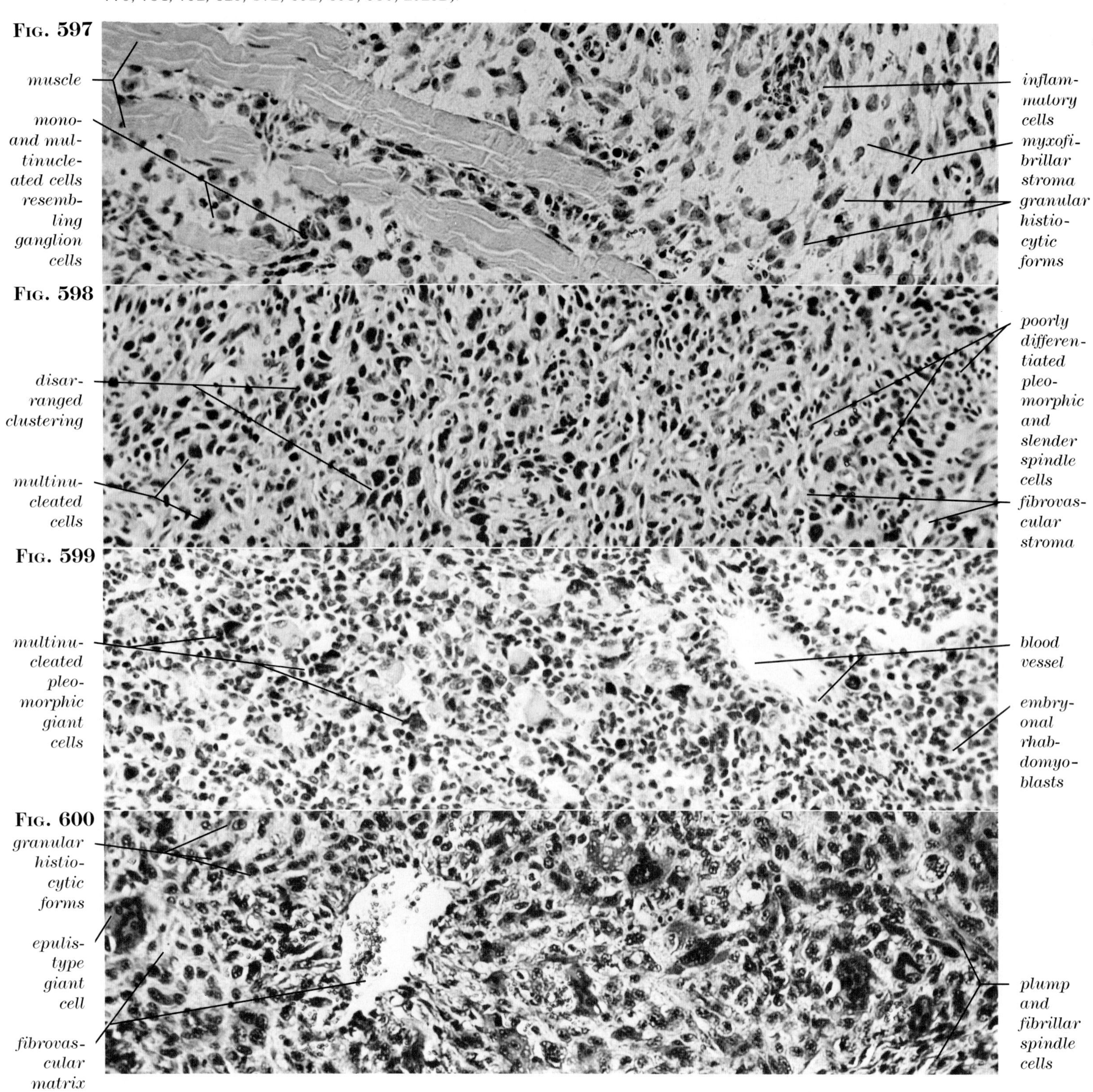

Fig. 601. Pleomorphic fibrosarcoma (×140) (see Figs. 91, 371, 374, 393, 608, 625, 647, 660, 665, 671, 809, 825).

Fig. 602. Fibrosarcomatous osteosarcoma (×140) (see Figs. 411, 451, 461, 610, 611, 615, 643, 677, 730, 754, 769, 777, 801, 807, 817, 819).

Fig. 603. High grade malignant schwannoma (×140) (see Figs. 382, 580, 583, 591, 619, 628, 661, 815, 836, 839, 845).

Fig. 604. Pleomorphic rhabdomyosarcoma (×140) (see Figs. 121, 292, 599, 609, 612, 800, 802, 804, 810, 813).

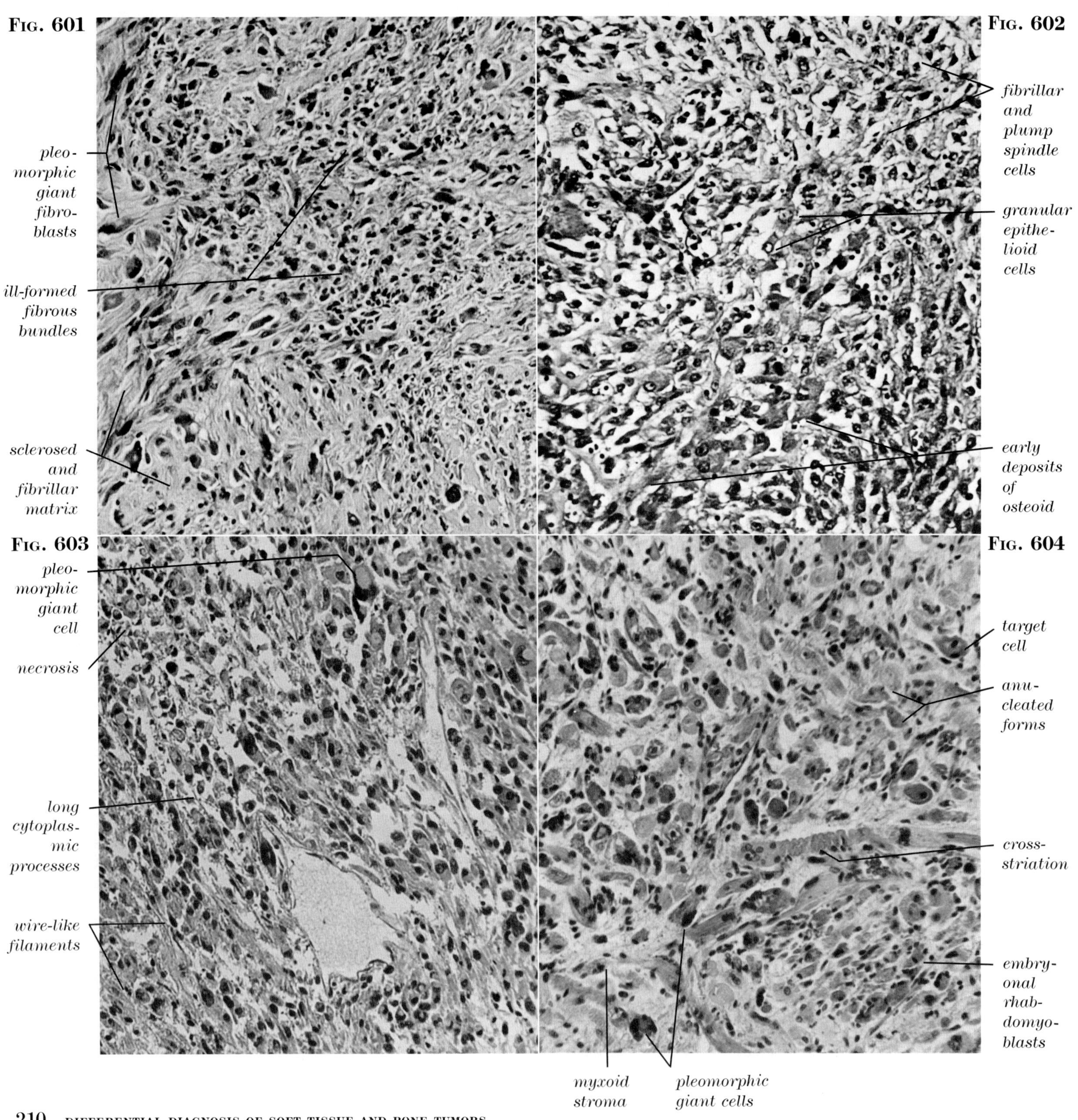

Fig. 605. Fasciitis (×85) (see Figs. 82, 368, 408, 578, 621, 834, 851, 855, 858, 947).

Fig. 606. Malignant Triton tumor (×350) (see Figs. 593, 732, 812, 813, 1259).

Fig. 607. Xanthogranuloma, so-called "juvenile cutaneous xanthogranuloma" (×350) (see Figs. 734, 779, 1031, 1127, 1201).

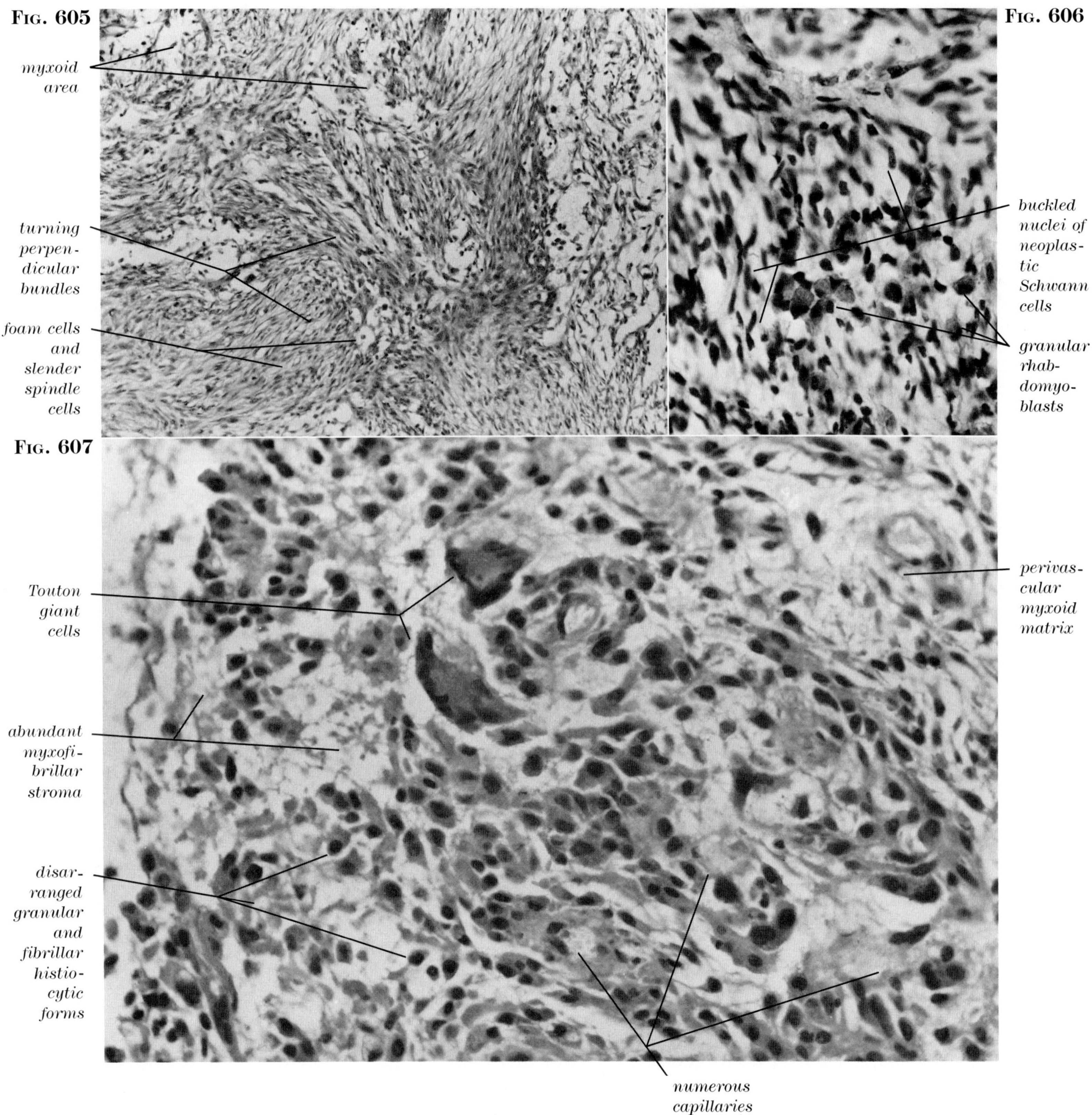

Fig. 608. High grade intramedullary fibrosarcoma (×140) (see Figs. 117, 297, 393, 601, 625, 647, 660, 665, 671, 809, 825).

Fig. 609. Pleomorphic rhabdomyosarcoma (×140) (see Figs. 87, 295, 599, 604, 612, 800, 802, 804, 810, 813).

Fig. 610. Osteosarcoma (×140) (see Figs. 90, 395, 411, 461, 611, 615, 643, 677, 730, 754, 817).

Fig. 611. Osteosarcoma (×140) (see Figs. 108, 109, 405, 451, 615, 769, 777, 801, 807, 819, 831, 843).

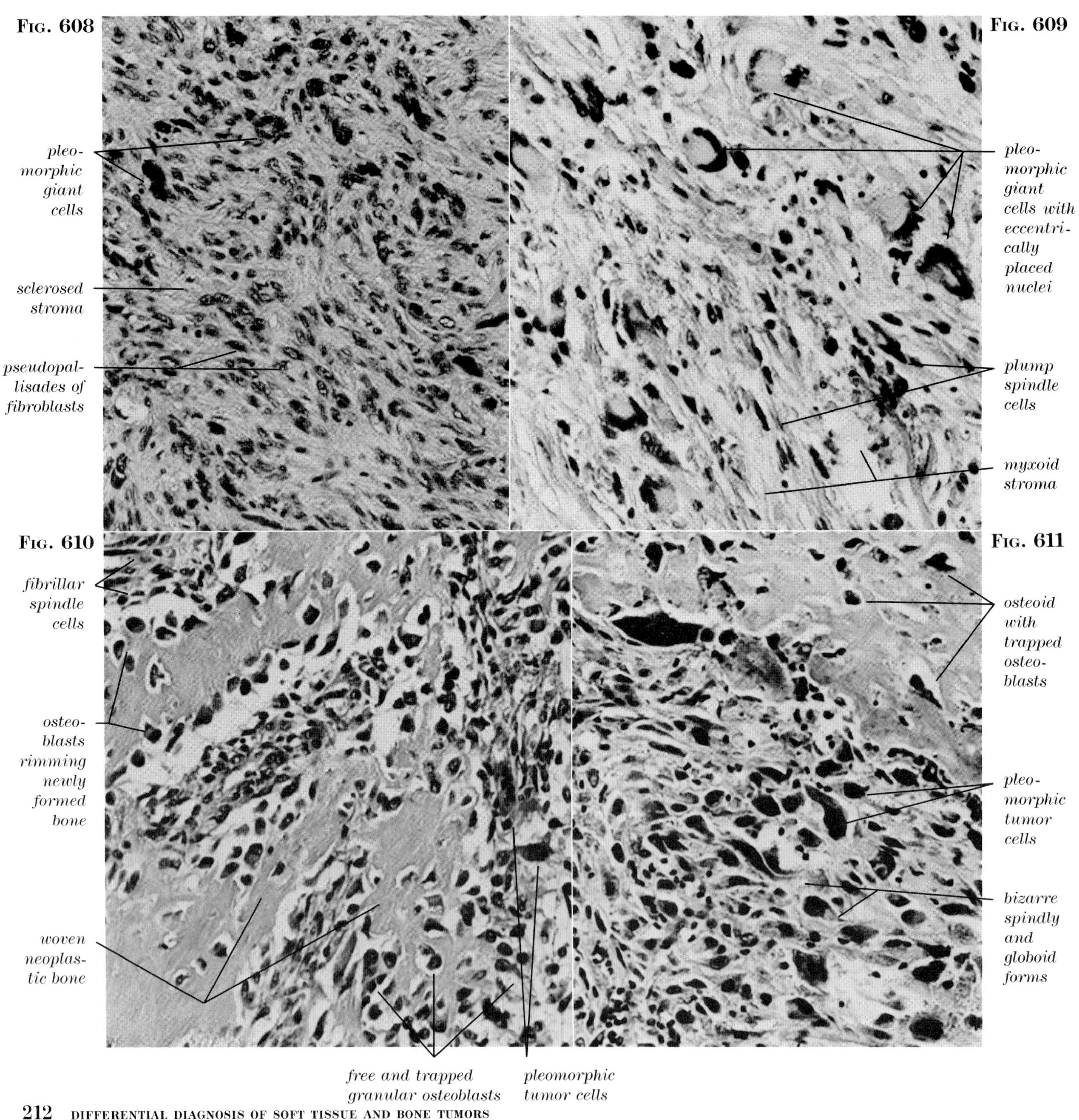

Fig. 612. Pleomorphic rhabdomyosarcoma (×350) (see Figs. 313, 599, 604, 609, 800, 802, 804, 810, 813, 1130, 1166, 1167, 1216, 1250, 1255, 1273).

Fig. 613. Pleomorphic lipoma (×140) (see Figs. 73, 821, 909, 957).

Fig. 614. Low grade intrascrotal malignant fibrous mesothelioma (×350) (see Figs. 268, 406, 678).

Fig. 615. Osteosarcoma (×350) (see Figs. 461, 602, 610, 611, 643, 677, 730, 754, 769, 777, 801, 807, 817, 819, 831, 843, 876, 897, 898, 958).

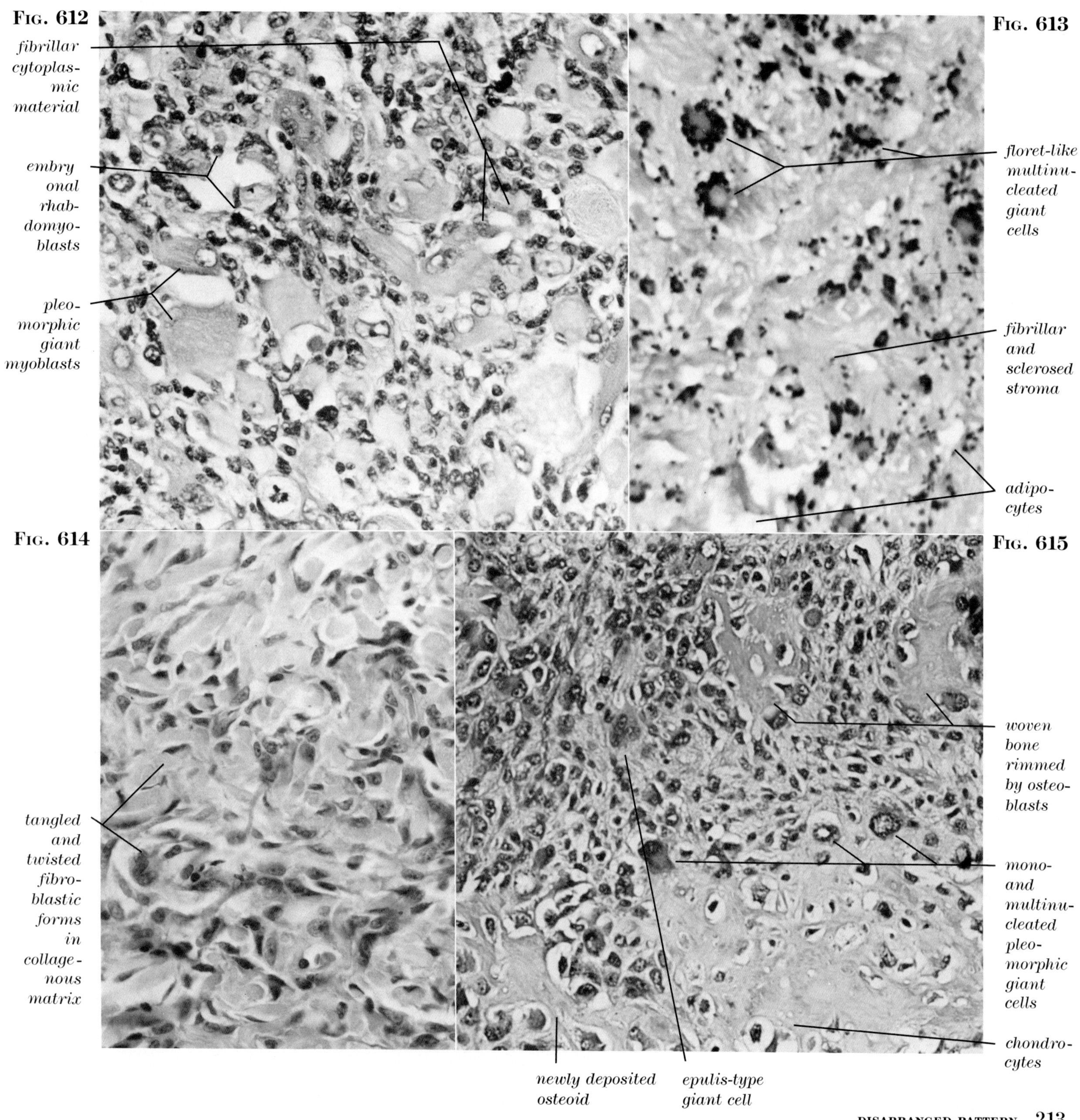

"The cell is the most important invention in nature and must be the continuous wonder of any thoughtful person"

SIR RUDOLPH PETERS (1889–1982)

11. Cell Morphology

	NON-NEOPLASTIC LESIONS	BENIGN NEOPLASMS	MALIGNANT NEOPLASMS
1. **Slender Spindle Cells.** See Tables 51 to 53. See Figures:	616, 620, 621, 623, 633	627, 630–632, 635, 652–654	617–619, 622, 624–626, 628, 629, 634, 636–651, 655–657
2. **Plump Spindle Cells.** See Tables 51 to 53. See Figures:		679	658–678, 680–686
3. **Granular Epithelioid Cells.** See Tables 51 to 53. See Figures:	703, 721, 734, 739	711, 717, 722, 725, 728, 736, 737	687–702, 704–710, 712–716, 718–720, 723, 724, 726, 727, 729–733, 735, 738, 740
4. **Clear Epithelioid Cells.** See Tables 51 to 53. See Figures:	749, 770	762, 772	741–748, 750–761, 763–769, 771, 773
5. **Isomorphic Giant Cells.** See Tables 51 to 53. See Figures:	779, 786, 787	774, 776, 778, 780–782, 785	775, 777, 783, 784, 788
6. **Pleomorphic Giant Cells.** See Tables 51 to 53. See Figures:	820	789, 793, 794, 805, 811, 821, 824	790–792, 795–804, 806–810, 812–819, 822, 823, 825–829

THE morphology of tumor cells is influenced by a number of factors—genetic coding, phase of growth, degree of differentiation, and local tissue conditions—and is expressed in the appearance of nuclei and cytoplasm. Rindfleisch believed, more than 100 years ago, that spindle cells are converted into round cells and vice versa (see Figure 5). Soft tissue and bone tumors are composed of an array of diverse cells, and accurate identification of the predominant cell type occasionally can be a formidable task. Despite such difficulties, it is a task the pathologist must fulfill.

The accurate identification of the predominant cell type is one of the major goals of tumor diagnosis. Often, sections stained with hematoxylin and eosin do not allow accurate assessment of the nature of cellular elements, and the pathologist may carry out special histochemical, immunocytochemical, and ultrastructural studies prior to final conclusion (see Chapter 13). Such studies are almost a prerequisite in epithelioid or so-called "round cell" soft tissue and bone tumors to accurate diagnosis. Although the specific histogenetic identity of the cellular elements of tumors is important, for diagnostic purposes it is less important than the recognition of the shape and size of the cells and nuclei. It must be remembered, however, that nuclear or cytoplasmic shape as well as size can be misleading without attention to growth pattern (see Chapter 10).

Each soft tissue and bone tumor can contain one or more of the six cell types: slender spindle, plump spindle, granular epithelioid, clear epithelioid, isomorphic, giant and pleomorphic giant.

The *slender spindle cells* have fibrillar, almost filamentous cytoplasm and elongated, buckled, or wavy nuclei that often bulge out from the cytoplasm, giving the impression of naked nuclei or nuclei that are adjacent to cytoplasmic processes.

Plump spindle cells possess cigar-shaped granular or vesicular nuclei. The nuclei are blunt-ended and deposited at the mid-point in broad, moderately elongated, and well-defined cytoplasm.

Granular epithelioid cells are fairly uniform round cells with well-defined, chromatin-rich nuclei and granular cytoplasm.

Clear epithelioid cells contain vacuolated or empty-looking cytoplasm and well-demarcated finely granular nuclei with prominent nucleoli.

Isomorphic giant cells can be mononucleated or multinucleated. The nuclei are uniform and round and may be placed anywhere in the finely granular, well-outlined, and globoid cytoplasm.

Pleomorphic giant cells are bizarre and distorted cells and exhibit marked anisocytosis and anisokaryosis. The nuclei may be multiple, and the cytoplasm may expand to several hundreds of microns.

TABLE 51. *Cell Morphology of Non-neoplastic Lesions**

	SLENDER SPINDLE CELLS	PLUMP SPINDLE CELLS	GRANULAR EPITHELIOID CELLS	CLEAR EPITHELIOID CELLS	ISOMORPHIC GIANT CELLS	PLEOMORPHIC GIANT CELLS
● Aneurysomal bone cyst	X		X		X	
● Giant cell granuloma			X		X	
● Hyperparathyroidism			X		X	
○ Xanthogranuloma	X			X	X	
○ Fasciitis	X	X				
○ Keloid		X				
○ Fibromatosis	X	X				
● Fibrous dysplasia	X	X				
○ Elastofibroma		X				
○ Collagenoma		X				
○ Tendosynovitis	X		X		X	
◑ Tendosynovial cyst	X		X	X		
○ Tendosynovial chondromatosis			X	X		
○ Gouty arthritis	X			X		
◑ Fat necrosis			X	X		X
◑ Lipogranuloma			X	X	X	
○ Proliferative panniculitis	X		X	X		
○ Proliferative myositis	X		X		X	X
○ Myositis ossificans	X	X	X		X	
○ Pyogenic granuloma	X		X			
○ Angiofollicular lymphoid hyperplasia	X		X			
◑ Arteriovenous malformation	X	X	X			
○ Traumatic neuroma	X					
● Callus	X	X	X	X	X	
◑ Eosinophilic granuloma			X		X	

*○ Primary soft tissue tumor. ● Primary bone tumor. ◑ Can be primary soft tissue or bone tumor.

TABLE 52. *Cell Morphology of Benign Neoplasms**

	SLENDER SPINDLE CELLS	PLUMP SPINDLE CELLS	GRANULAR EPITHELIOID CELLS	CLEAR EPITHELIOID CELLS	ISOMORPHIC GIANT CELLS	PLEOMORPHIC GIANT CELLS
◑ Benign fibroblastic fibrous histiocytoma	X					
◑ Benign histiocytic fibrous histiocytoma			X		X	
◑ Benign pleomorphic fibrous histiocytoma			X	X	X	
● Nonossifying fibroma	X			X	X	
○ Fibroma		X				
● Desmoplastic fibroma		X				
◑ Well differentiated lipoma				X		
◑ Myxoid lipoma	X			X		
○ Fibroblastic lipoma	X					
○ Lipoblastoma			X	X		
○ Pleomorphic lipoma	X		X	X		X
○ Angiomyolipoma		X		X		
○ Hibernoma			X			
◑ Leiomyoma		X				
○ Rhabdomyoma			X			X
◑ Hemangioma	X		X			
◑ Benign glomus tumor			X			
◑ Neurofibroma	X	X				
○ Plexiform neurofibroma	X					
◑ Benign schwannoma	X					
○ Benign nevoid schwannoma	X		X			
◑ Benign undifferentiated peripheral nerve tumor	X		X			
◑ Ganglioneuroma	X				X	X
○ Benign paraganglioma	X		X			
○ Benign fibrous mesothelioma		X				
● Chondroblastoma			X		X	
● Chondromyxoid fibroma	X		X	X	X	
◑ Chondroma				X		
● Osteochondroma				X		
● Osteoblastoma	X		X		X	
● Osteoid osteoma		X	X		X	
○ Benign granular cell tumor			X			
○ Meningioma	X	X	X			
○ Benign cystosarcoma phyllodes	X	X				

*○ Primary soft tissue tumor. ● Primary bone tumor. ◑ Can be primary soft tissue or bone tumor.

TABLE 53. *Cell Morphology of Malignant Neoplasms**

	SLENDER SPINDLE CELLS	PLUMP SPINDLE CELLS	GRANULAR EPITHELIOID CELLS	CLEAR EPITHELIOID CELLS	ISOMORPHIC GIANT CELLS	PLEOMORPHIC GIANT CELLS
◑ Malignant fibroblastic fibrous histiocytoma	X					
◑ Malignant histiocytic fibrous histiocytoma	X		X		X	
◑ Malignant pleomorphic fibrous histiocytoma	X	X	X		X	X
◑ Desmoid tumor		X				
◑ Fibroblastic fibrosarcoma		X				
◑ Pleomorphic fibrosarcoma		X				X
○ Biphasic tendosynovial sarcoma	X		X			
○ Monophasic tendosynovial sarcoma, spindle cell type	X	X				
○ Monophasic tendosynovial sarcoma, pseudoglandular type			X			
○ Epithelioid sarcoma			X			
○ Clear cell sarcoma				X		
○ Chordoid sarcoma			X			
◑ Well differentiated liposarcoma	X			X		
◑ Myxoid liposarcoma	X			X		
○ Lipoblastic liposarcoma			X			
○ Fibroblastic liposarcoma	X					
○ Pleomorphic liposarcoma	X	X	X	X		X
◑ Leiomyoblastoma				X		
◑ Leiomyosarcoma		X				X
○ Embryonal rhabdomyosarcoma	X		X[†]			
○ Alveolar rhabdomyosarcoma			X[†]			
○ Myxoid rhabdomyosarcoma	X		X			
○ Rhabdomyoblastoma			X		X	
○ Pleomorphic rhabdomyosarcoma	X	X	X			X
◑ Hemangiopericytoma	X		X			
◑ Hemangiosarcoma			X			X
○ Lymphangiosarcoma	X		X			X
◑ Neurofibrosarcoma	X	X				
◑ Malignant schwannoma	X		X			
◑ Malignant undifferentiated peripheral nerve tumor			X			
○ Malignant epithelioid mesothelioma			X			X
○ Malignant fibrous mesothelioma		X				X
◑ Chondrosarcoma			X	X[†]		
◑ Mesenchymal chondrosarcoma	X		X	X		
◑ Osteosarcoma		X	X		X	X
◑ Parosteal osteosarcoma	X	X	X			
● Paget's sarcoma		X	X		X	X
◑ Granulocytic sarcoma			X			
● Plasma cell myeloma			X			
○ Malignant granular cell tumor			X			X
○ Alveolar soft part sarcoma			X			X
○ Kaposi's sarcoma	X	X				
◑ Ewing's sarcoma			X			
● Chordoma			X	X		
○ Myxopapillary ependymoma			X	X		

*○ Primary soft tissue tumor. ● Primary bone tumor. ◑ Can be primary soft tissue or bone tumor.
[†]Many cells are binucleated.

Fig. 616. Mesenteric fibromatosis (×570) (see Figs. 280, 419, 860, 861, 873, 874, 878, 1022).

Fig. 617. Malignant thymoma (×570) (see Figs. 337, 375, 546, 966).

Fig. 618. Low grade cutaneous leiomyosarcoma (×570) (see Figs. 72, 193, 273, 341, 659, 664, 666, 667).

Fig. 619. Low grade malignant schwannoma (×570) (see Figs. 80, 200, 238, 355, 628, 661, 815, 836).

Fig. 620. Myositis ossificans (×570) (see Figs. 145, 384, 913, 1083–1085).

Fig. 621. Fasciitis (×350) (see Figs. 82, 345, 605, 834, 851, 855, 858, 947, 956, 971, 991, 1020).

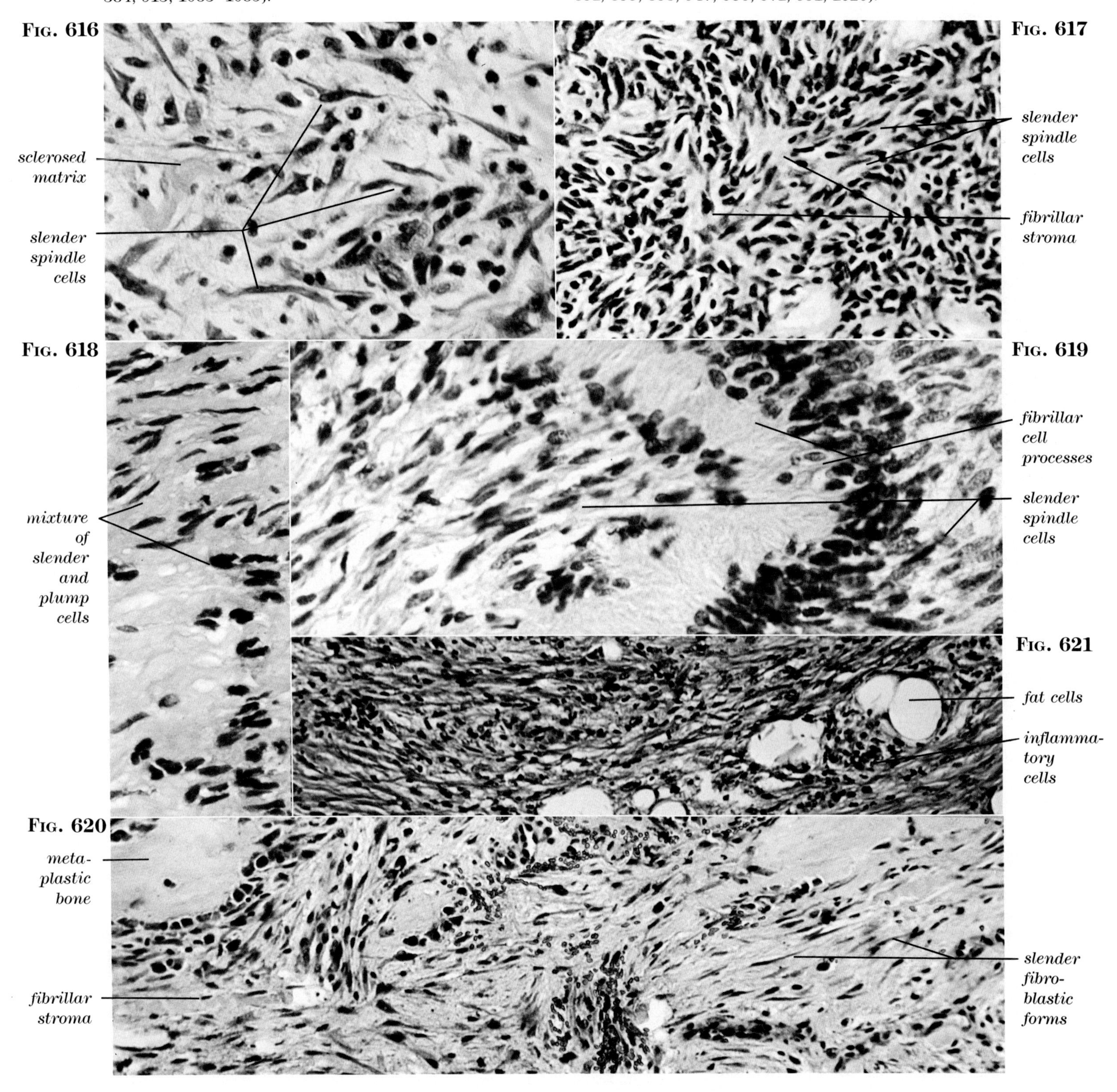

Fig. 622. Monophasic tendosynovial sarcoma, spindle cell type (×140) (see Figs. 467, 491, 492, 542, 642, 646, 648, 650, 675, 683, 684).

Fig. 623. (Top center) Ovarian stroma (×350) (see Fig. 348).

Fig. 624. Low grade neurofibrosarcoma (×350) (see Figs. 79, 236, 369, 370, 626, 838, 1116, 1229, 1263).

Fig. 625. Fibrosarcoma (×570) (see Figs. 91, 393, 601, 608, 647).

Fig. 626. (Bottom center) High grade neurofibrosarcoma (×570) (see Figs. 53, 401, 584, 624, 838).

Fig. 627. Plexiform neurofibroma (×350) (see Figs. 365, 527, 842, 1114).

Fig. 628. Low grade malignant schwannoma (×350) (see Figs. 242, 357, 382, 619, 661, 815, 836, 839).

Fig. 629. Kaposi's sarcoma (×570) (see Figs. 63, 388, 655, 658, 977).

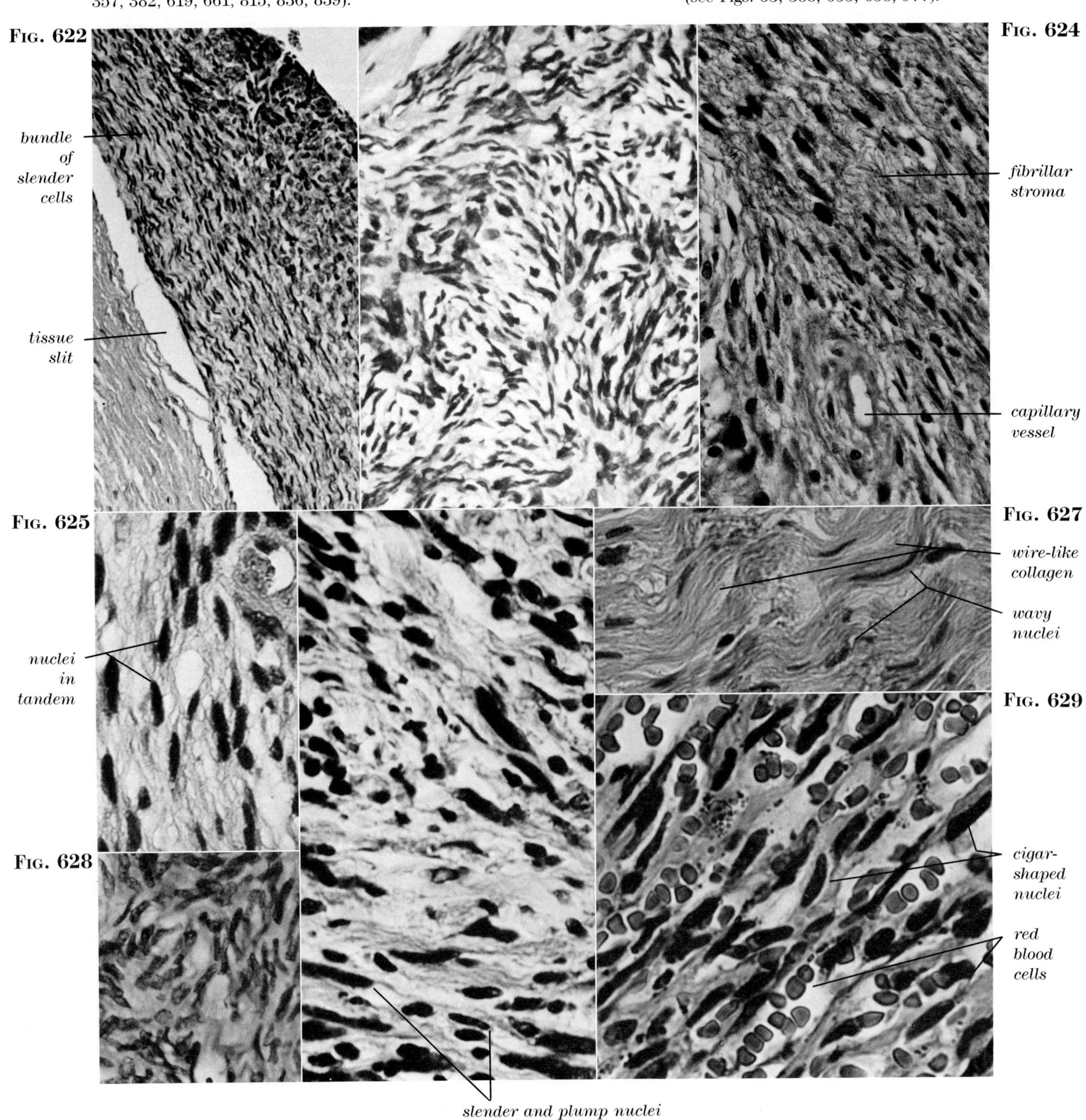

Fig. 630. Ganglioneuroma (×570) (see Figs. 486, 811, 824, 857).

Fig. 631. Nonossifying fibroma (×350) (see Figs. 112, 177, 852).

Fig. 632. Giant cell granuloma (×350) (see Figs. 119, 787, 1000, 1069).

Fig. 633. Desmoplastic fibroma (×350) (see Fig. 106).

Fig. 634. Adamantinoma of tibia (×50) (see Figs. 171, 493).

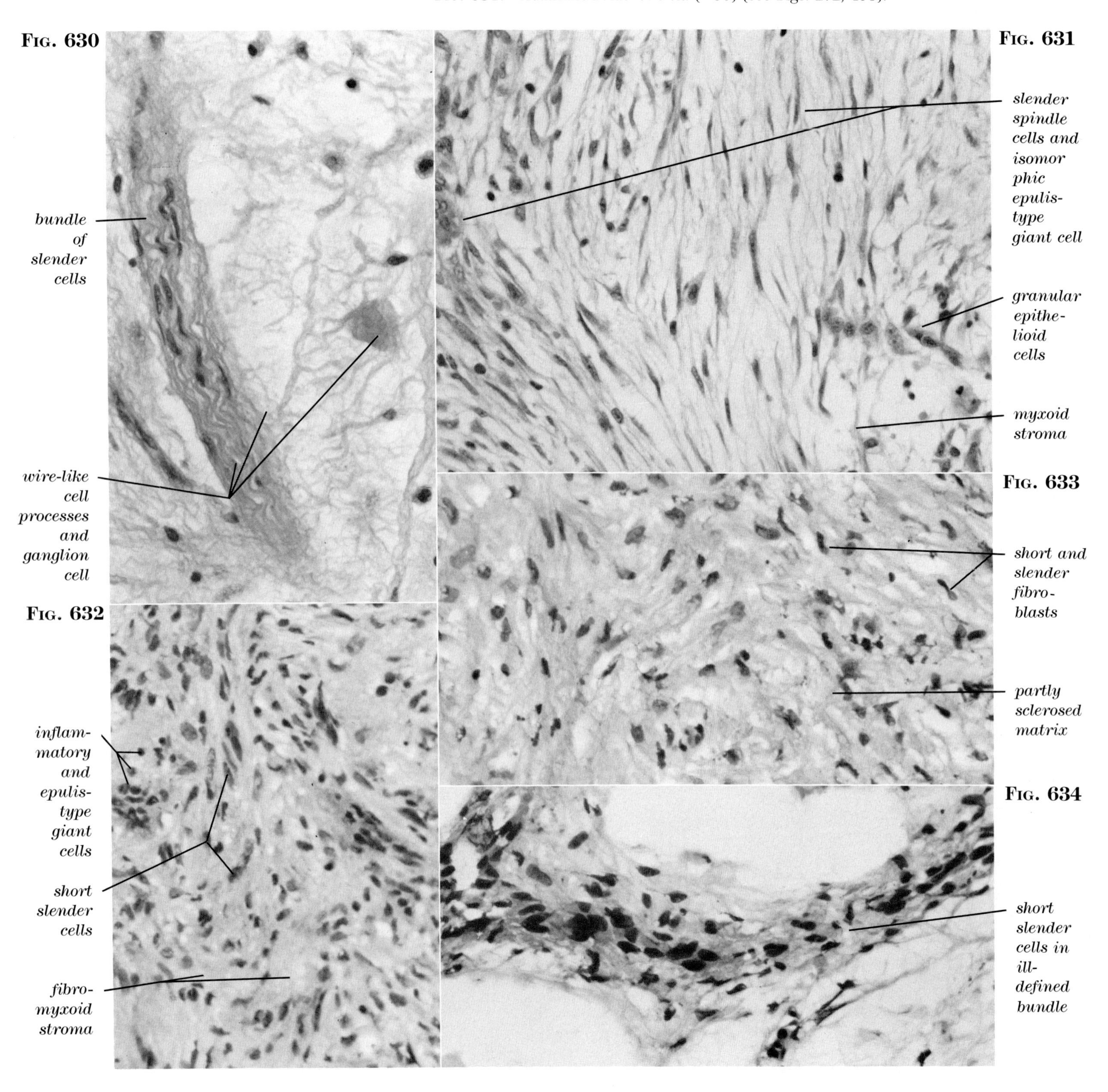

Fig. 635. Myxoid lipoma, so-called "myxoma" (×570) (see Figs. 73, 939, 941, 950, 1168, 1187).

Fig. 636. (Top center) Myxoid liposarcoma (×570) (see Figs. 49, 86, 188, 438, 439, 783, 933, 948, 997).

Fig. 637. Embryonal rhabdomyosarcoma (×350) (see Figs. 78, 207, 263, 364, 689, 694, 700, 706).

Fig. 638. Myxoid chondrosarcoma (×570) (see Figs. 124, 182, 308, 329, 932, 1175).

Fig. 639. Basal cell carcinoma, morphea type, infiltrating fat (×350).

Fig. 640. Chordoid sarcoma (×350) (see Figs. 456, 510, 919, 1176).

Fig. 641. Fibroblastic liposarcoma (×570) (see Figs. 86, 343, 440, 668).

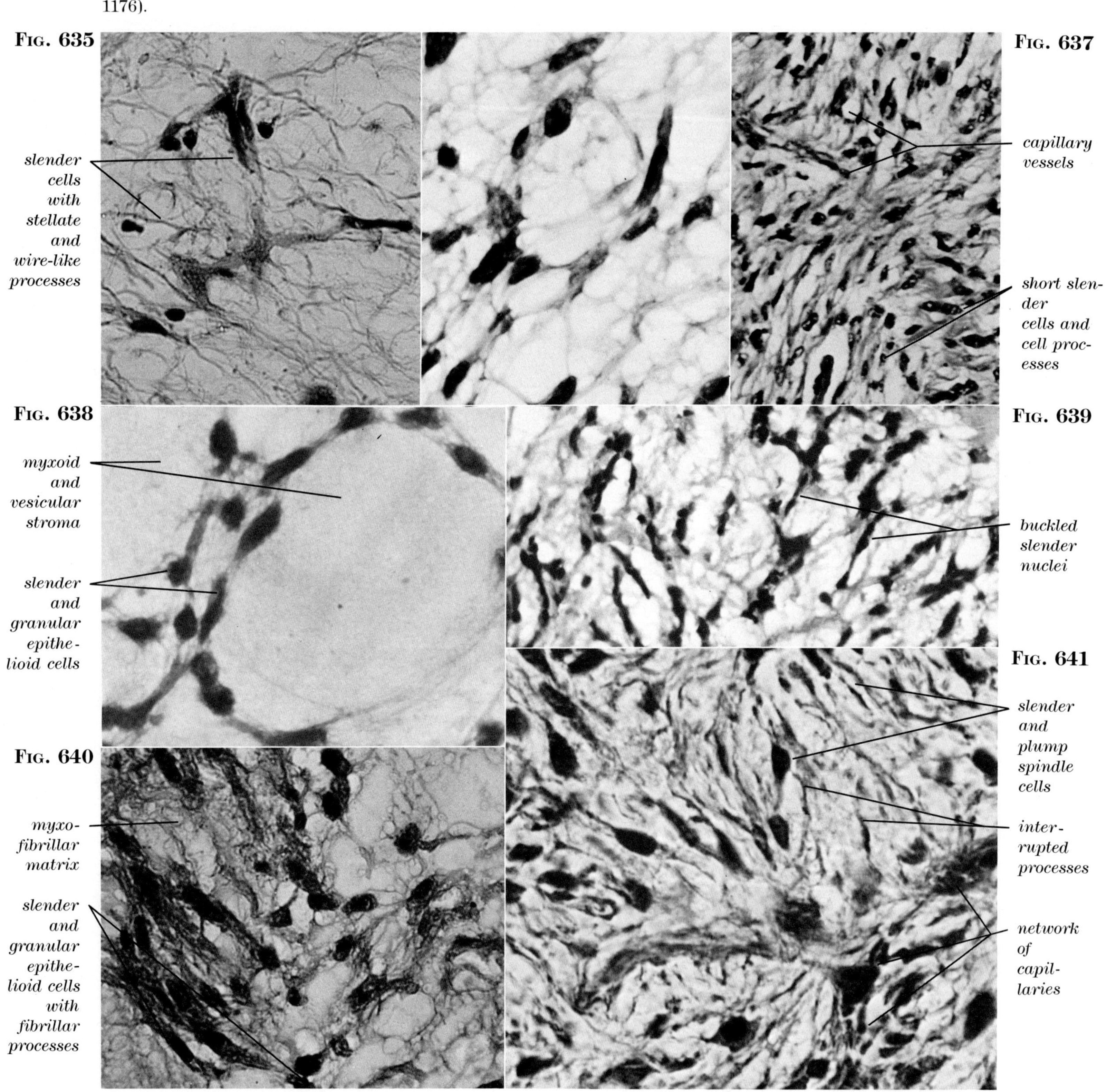

Fig. 642. Monophasic tendosynovial sarcoma, spindle cell type (×140) (see Figs. 542, 622, 646, 648).

Fig. 645. Malignant fibroblastic fibrous histiocytoma (×350) (see Figs. 76, 339, 340, 672).

Fig. 643. (Top center) Osteosarcoma (×140) (see Figs. 90, 142, 248, 395, 405, 677).

Fig. 646. (Bottom center) Monophasic tendosynovial sarcoma, spindle cell type (×350) (see Figs. 542, 622, 642, 648).

Fig. 644. Embryonal rhabdomyosarcoma, myxoid type (×350) (see Figs. 120, 416, 418, 587, 923).

Fig. 647. High grade fibrosarcoma (×350) (see Figs. 601, 608, 625, 660, 665).

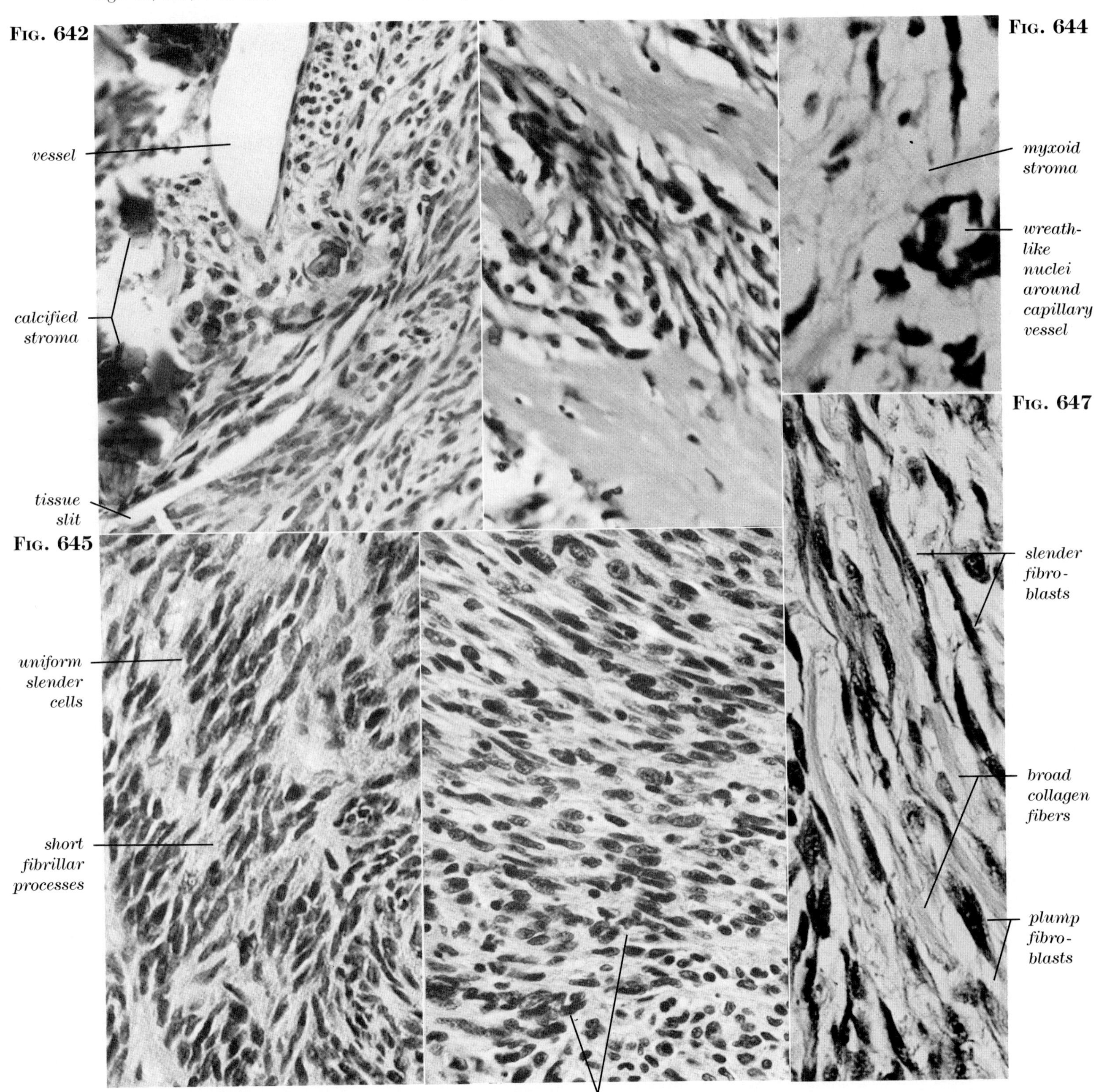

Fig. 648. Tendosynovial sarcoma invading bone (×350) (see Figs. 39, 62, 64, 85, 179, 190, 315, 320, 326, 622, 642, 646, 650).

Fig. 649. Lymphangiosarcoma (×350) (see Figs. 61, 84, 159, 321, 424, 551, 558, 954, 963, 965, 970, 1098).

Fig. 650. Monophasic tendosynovial sarcoma, spindle cell type (×350) (see Figs. 642, 646, 648, 650, 675, 683, 684, 854, 864, 870, 894, 902, 903).

Fig. 651. Desmoplastic clear cell carcinoma of kidney (×350) (see Figs. 545, 716, 764, 768, 1018A, 1270).

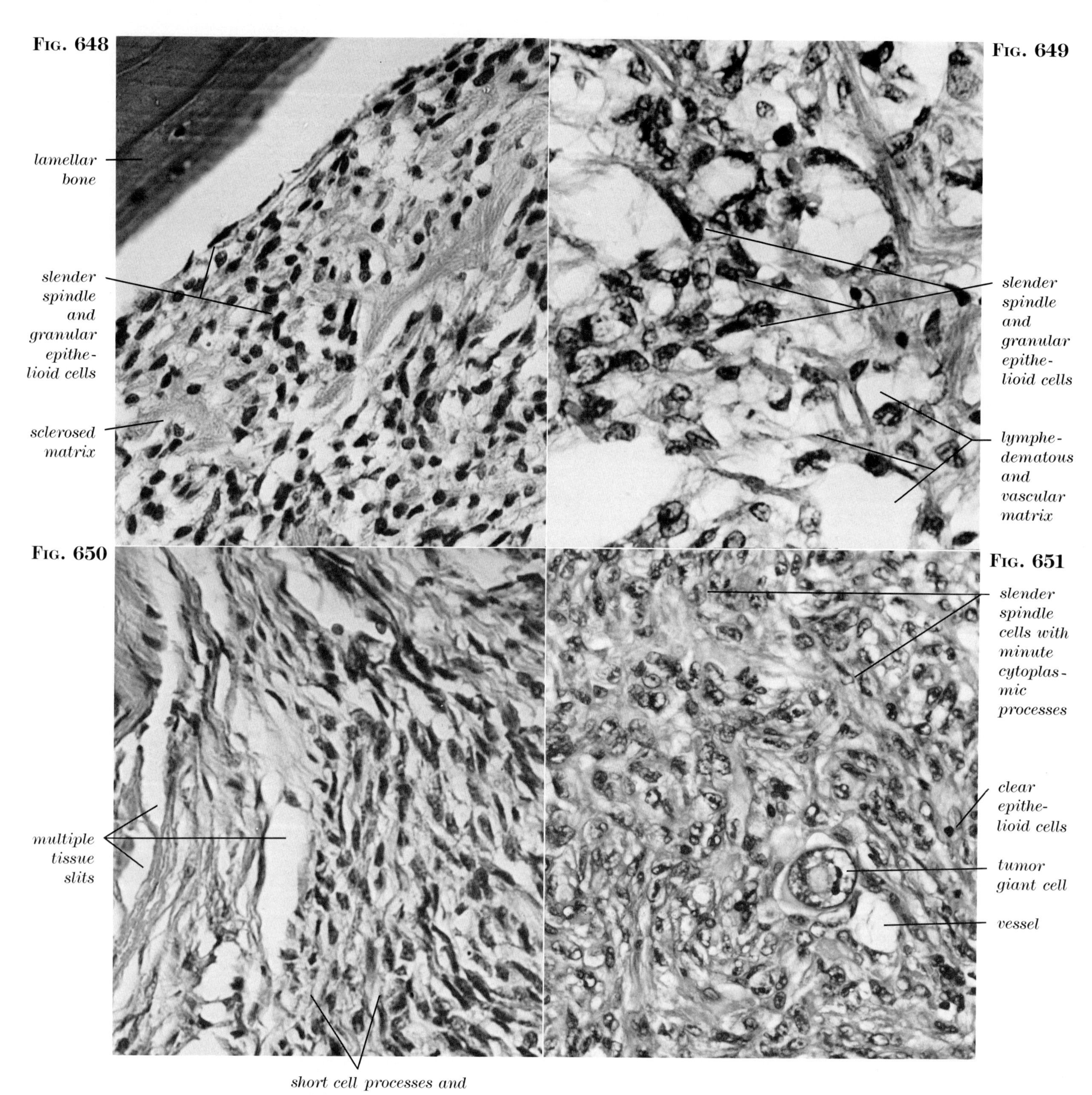

FIG. 652. Neurofibroma (×350) (see Figs. 74, 363, 487, 840, 871, 955, 993, 1013, 1142, 1143).

FIG. 653. Angiomyoma (×350) (see Figs. 912, 985, 1254).

FIG. 654. Peritoneal leiomyomatosis (×350) (see Figs. 50, 358, 359, 361, 402).

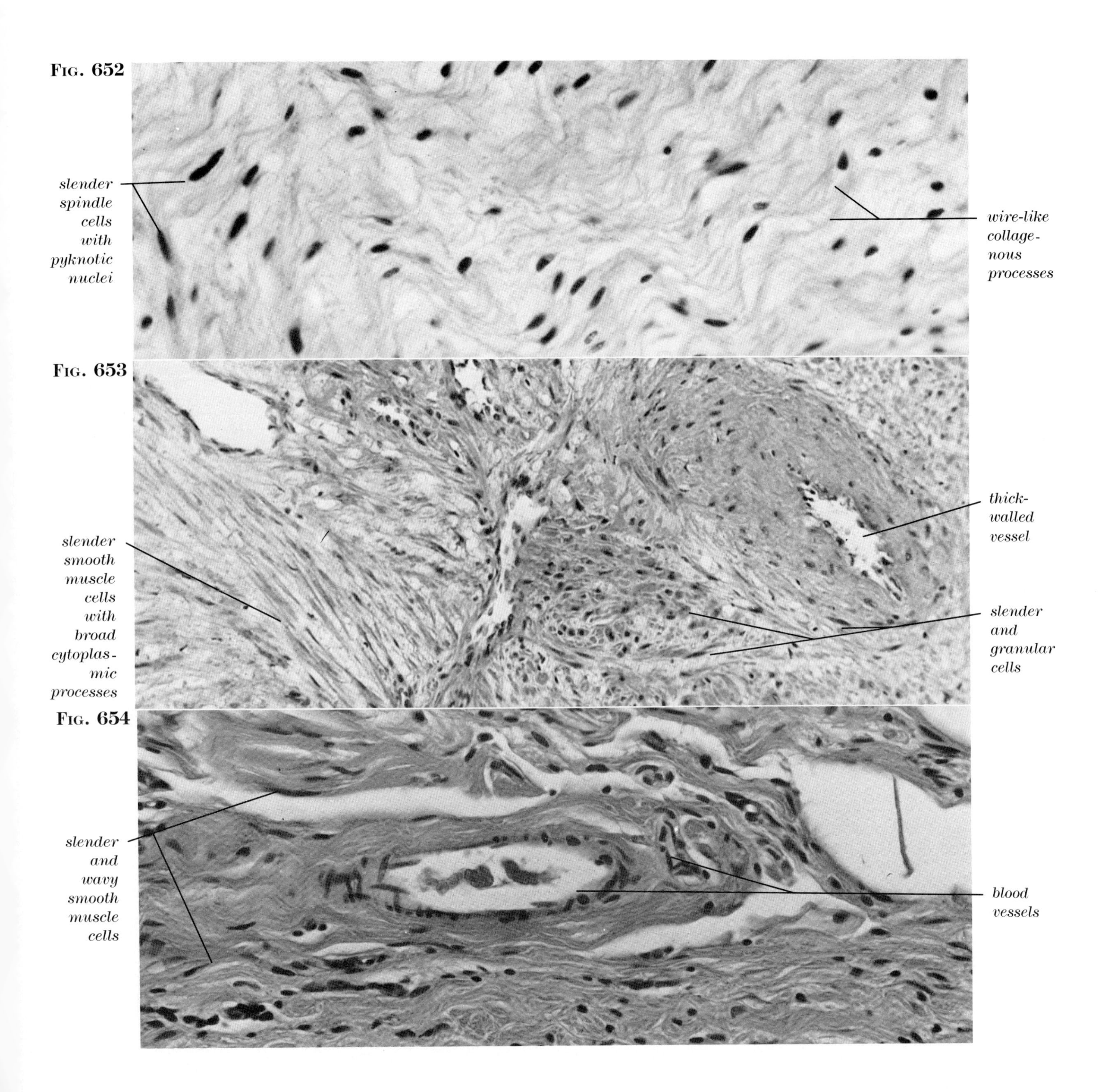

Fig. 655. Kaposi's sarcoma associated with acquired immunodeficiency (×350) (see Figs. 63, 388, 629, 658, 977, 986, 1126).

Fig. 656. Parosteal osteosarcoma (×350) (see Figs. 109, 129, 217, 249, 287, 353, 354, 1082, 1154).

Fig. 657. Malignant cystosarcoma phyllodes of breast (×350) (see Figs. 47, 351, 377, 663).

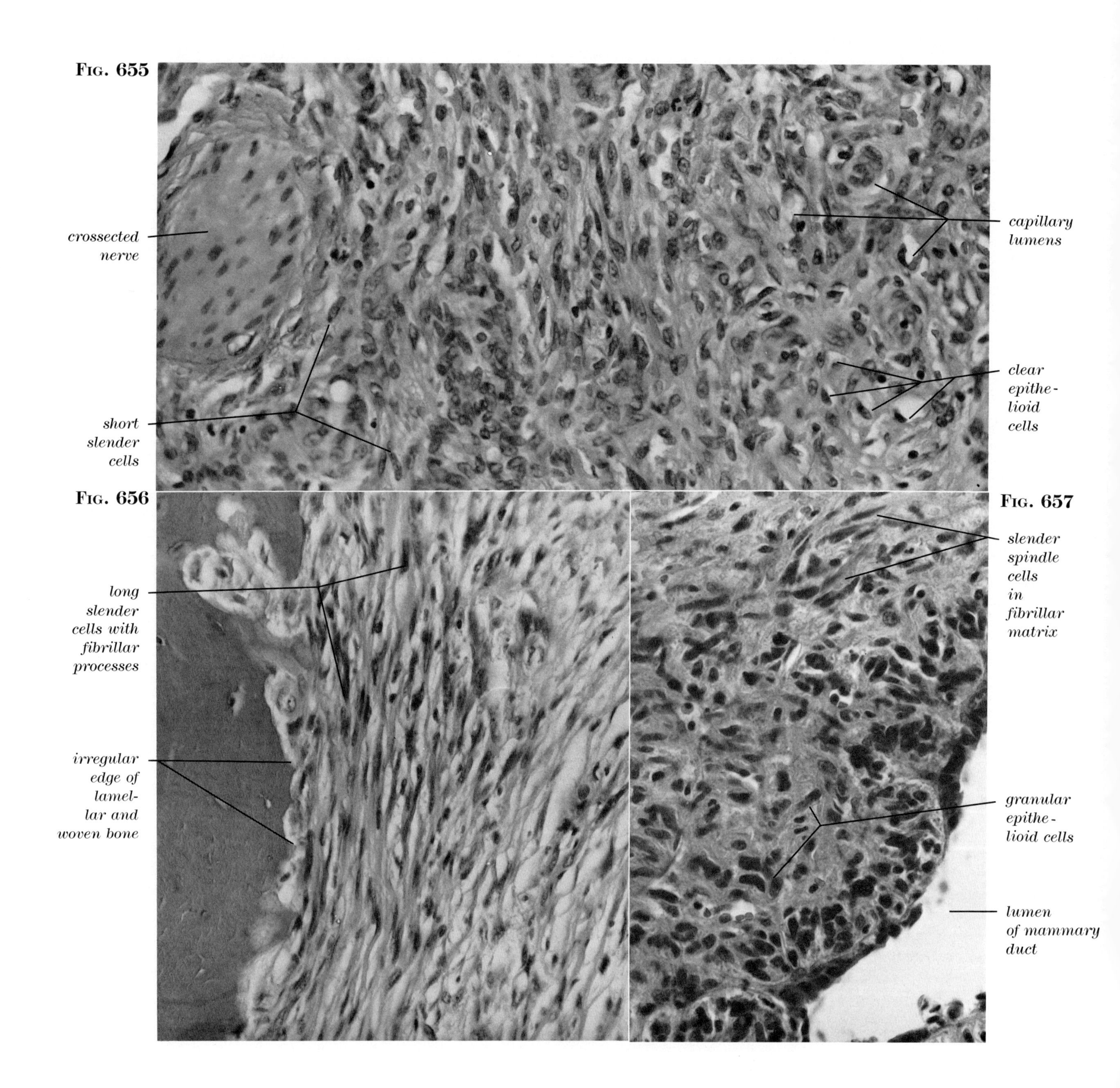

Fig. 658. Kaposi's sarcoma (×570) (see Figs. 63, 388, 629, 655, 977, 986, 1126).

Fig. 660. High grade fibrosarcoma (×570) (see Figs. 91, 240, 297, 371, 374, 393, 601, 608, 625, 647, 665, 671, 809, 825).

Fig. 662. Desmoid tumor, low grade fibrosarcoma (×570) (see Figs. 77, 185, 202, 209, 244, 282, 376, 386, 397, 686).

Fig. 659. High grade leiomyosarcoma (×570) (see Figs. 347, 360, 362, 618, 664, 666, 667, 670, 680).

Fig. 661. High grade malignant schwannoma (×570) (see Figs. 51, 80, 200, 238, 242, 355, 357, 382, 580, 583, 619, 628, 815).

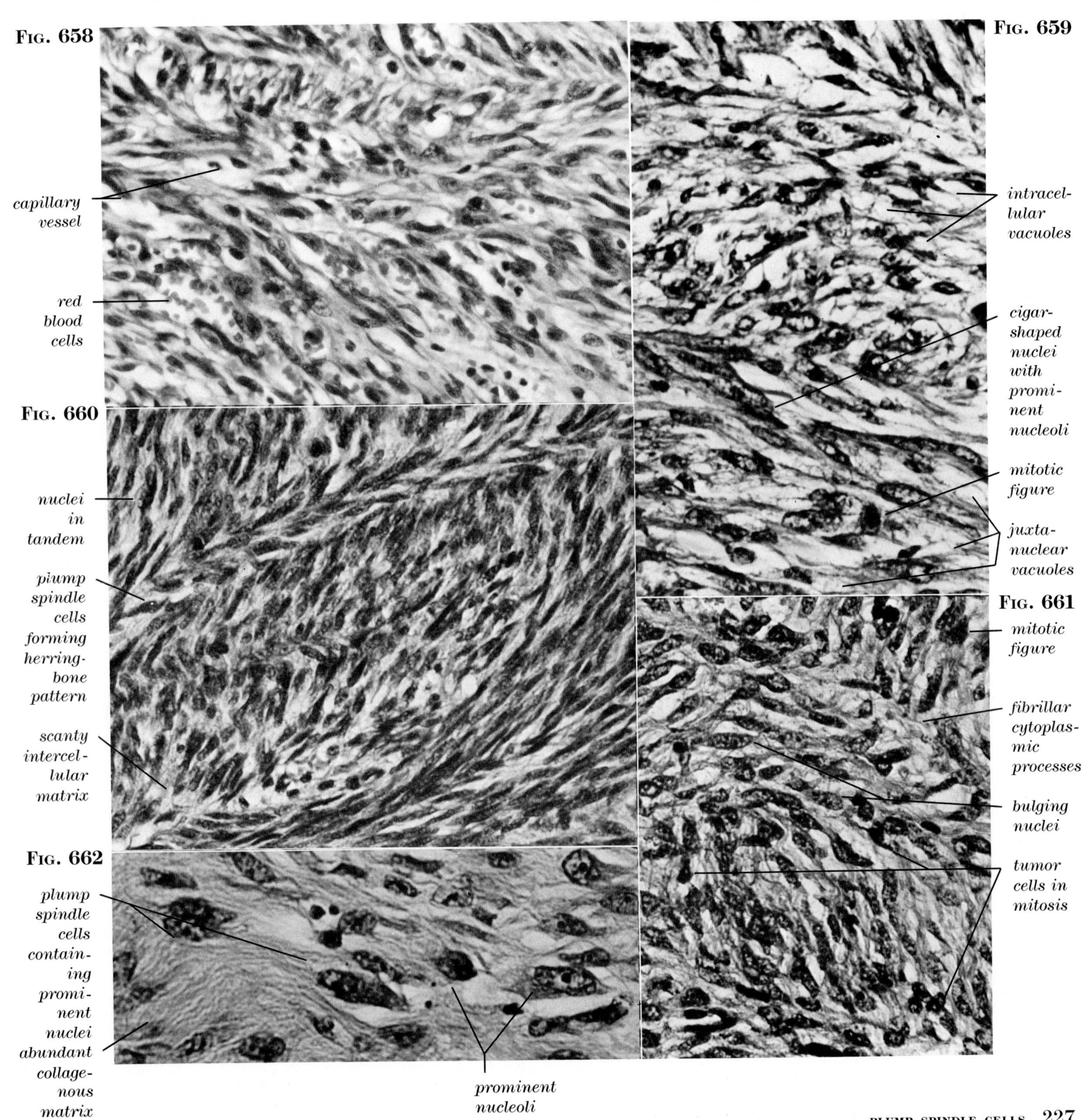

Fig. 663. Malignant cystosarcoma phyllodes of breast (×570) (see Figs. 47, 351, 377, 657, 663).

Fig. 664. High grade leiomyosarcoma of foot (×570) (see Figs. 72, 110, 193, 218, 220, 231, 273, 341, 347, 360, 362, 366, 618, 659, 666).

Fig. 665. High grade fibrosarcoma (×570) (see Figs. 91, 117, 240, 371, 374, 393, 601, 608, 625, 647, 660, 671, 809).

Fig. 666. High grade leiomyosarcoma of retroperitoneum (×570) (see Figs. 618, 659, 664, 667, 670, 680, 681, 685, 746, 803, 808).

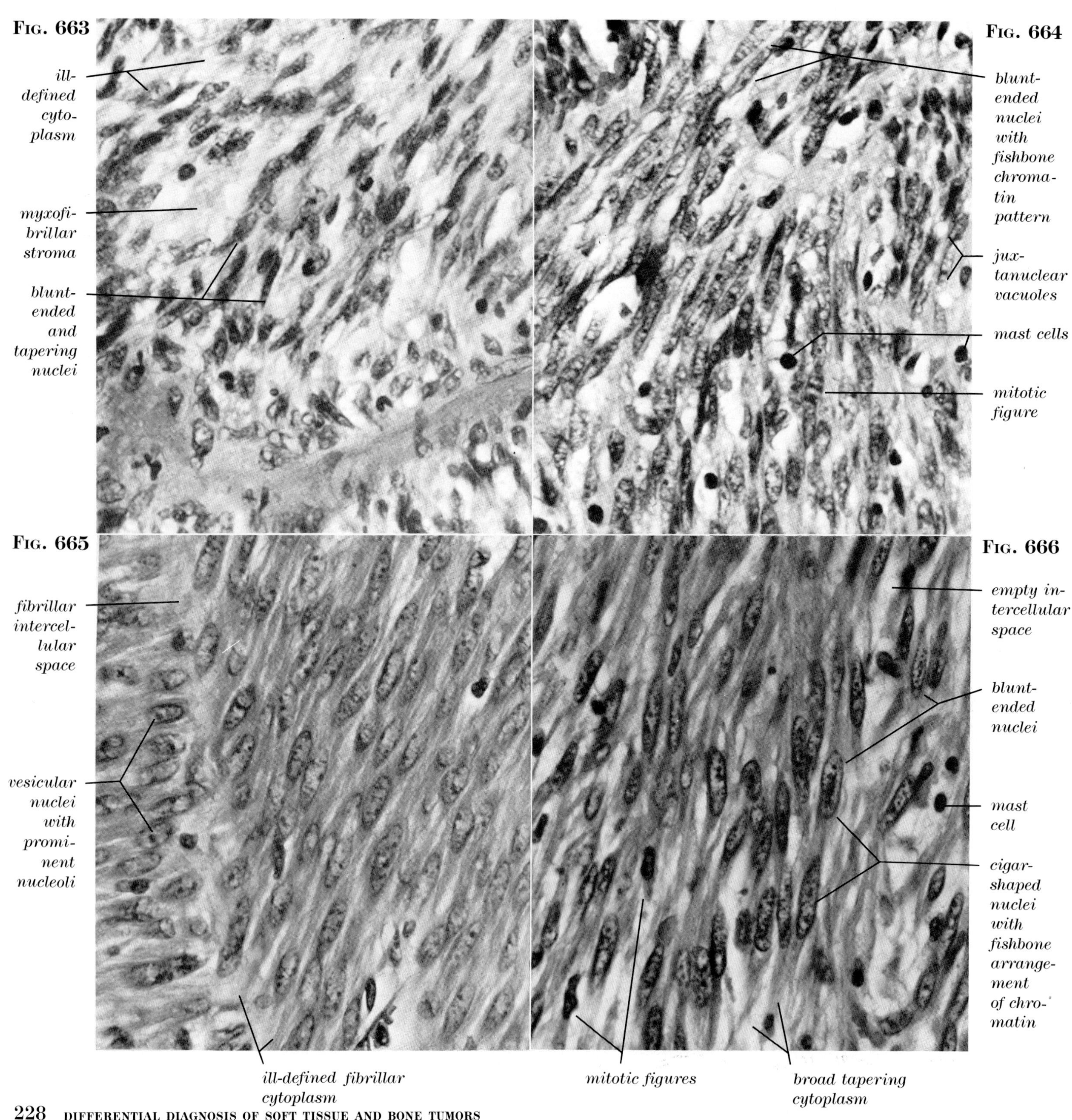

FIG. 667. Low grade leiomyosarcoma of broad ligament (×570) (see Figs. 659, 664, 666, 670, 680, 681, 685, 746, 803, 808, 841, 844, 849, 856, 875).

FIG. 669. Desmoplastic malignant melanoma (×570) (see Figs. 409, 540, 719, 748, 797, 826, 1188, 1189, 1190, 1219, 1262).

FIG. 671. High grade fibrosarcoma (×350) (see Figs. 91, 117, 240, 371, 625, 647, 660, 665, 809, 825, 832, 901, 1117, 1205).

FIG. 668. Fibroblastic liposarcoma (×570) (see Figs. 440, 641, 837, 848).

FIG. 670. Low grade leiomyosarcoma (×570) (see Figs. 664, 666, 667, 680, 681, 685, 746, 803, 808).

FIG. 672. Malignant fibroblastic fibrous histiocytoma (×350) (see Figs. 340, 350, 645, 710).

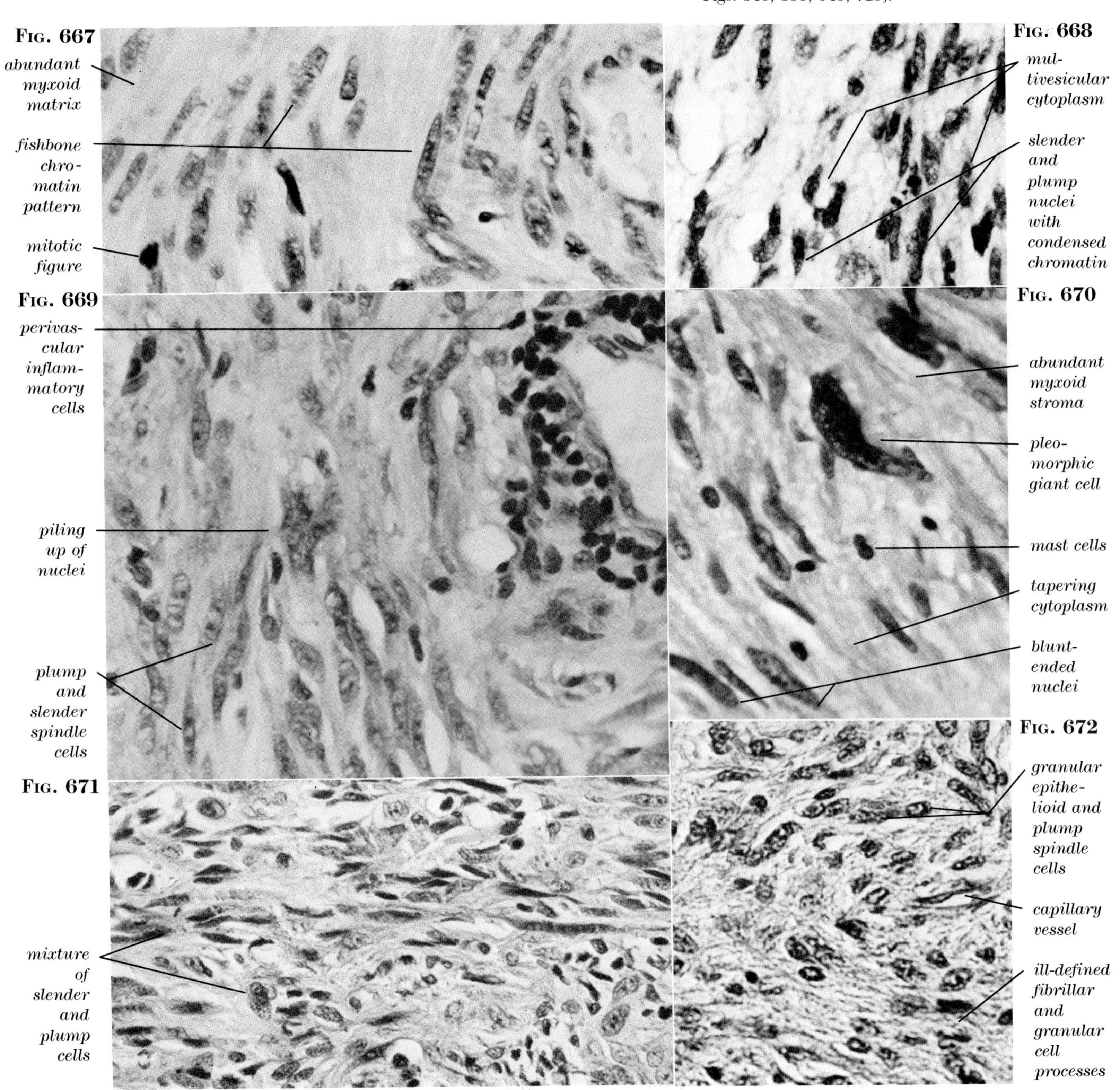

Fig. 673. Desmoplastic epidermoid carcinoma (×570) (see Figs. 338).

Fig. 675. Tendosynovial sarcoma (×570) (see Figs. 39, 62, 64, 85, 391, 462, 514, 522, 526, 542, 646, 648, 650, 683).

Fig. 677. Osteosarcoma, telangiectatic type (×570) (see Figs. 411, 451, 461, 602, 610, 611, 615, 643, 730, 754, 769, 777, 801, 807, 817, 819).

Fig. 674. Mesenchymal chondrosarcoma (×570) (see Figs. 103, 151, 318, 390, 707, 920, 1062, 1265).

Fig. 676. Epithelioid sarcoma (×570) (see Figs. 85, 181, 258, 462, 830, 866, 895, 1052, 1059, 1146, 1170, 1244, 1245).

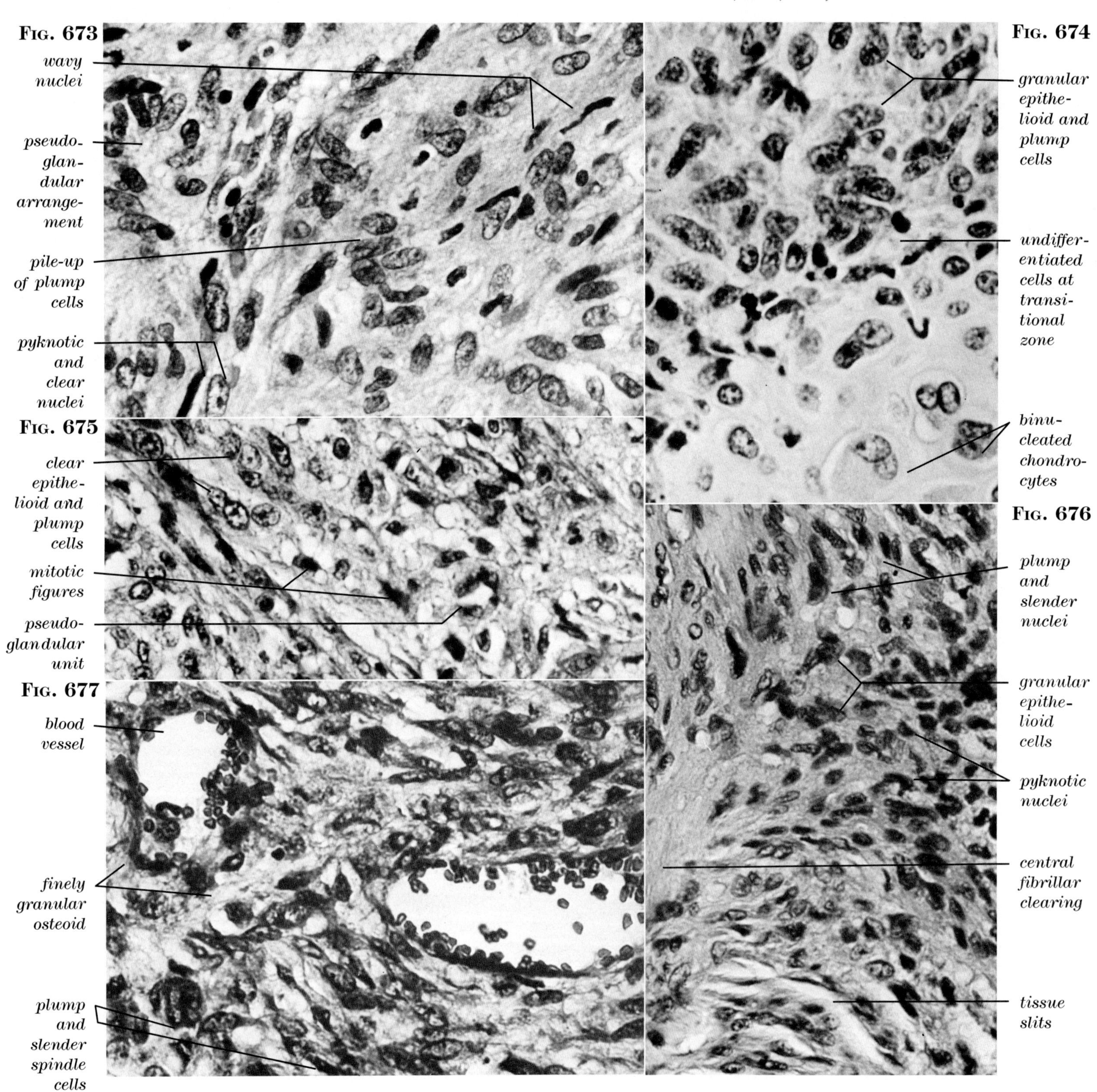

Fig. 678. Malignant fibrous mesothelioma (×570) (see Figs. 189, 266, 268, 406, 614, 869, 880, 881, 1240, 1241).

Fig. 679. Borderline fibrous histiocytoma (×350) (see Figs. 344, 460, 577, 780).

Fig. 680. Low grade leiomyosarcoma (×350) (see Figs. 366, 667, 670, 681, 685).

Fig. 681. High grade leiomyosarcoma (×570) (see Figs. 72, 110, 193, 273, 341, 347, 360, 362, 588, 667, 670, 681, 685, 746, 803, 808, 841, 844, 849, 856).

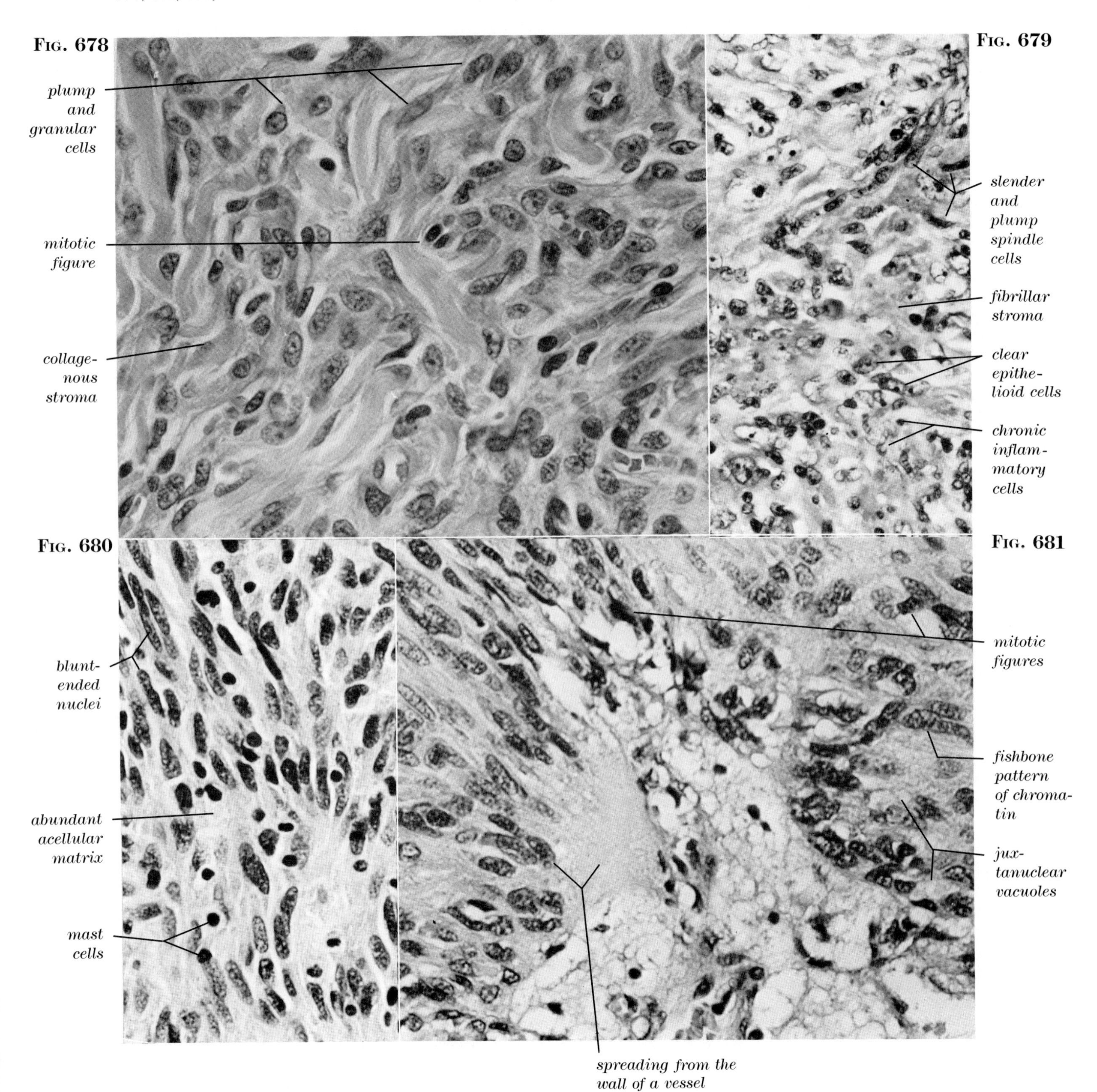

Fig. 682. Desmoplastic mammary carcinoma (×570). (see Figs. 160, 1269).

Fig. 683. Monophasic tendosynovial sarcoma, spindle cell type (×570) (see Figs. 391, 392, 396, 398, 399, 467, 491, 492, 542, 622, 642, 646, 648, 650, 684, 854, 864, 870, 894, 902, 903, 1144, 1145).

Fig. 684. Tendosynovial sarcoma growing as monophasic spindle cell sarcoma with clear cell sarcoma component (×570) (see Figs. 391, 392, 399, 447, 467, 491, 492, 526, 542, 622, 642, 646, 648, 650, 675, 683, 854).

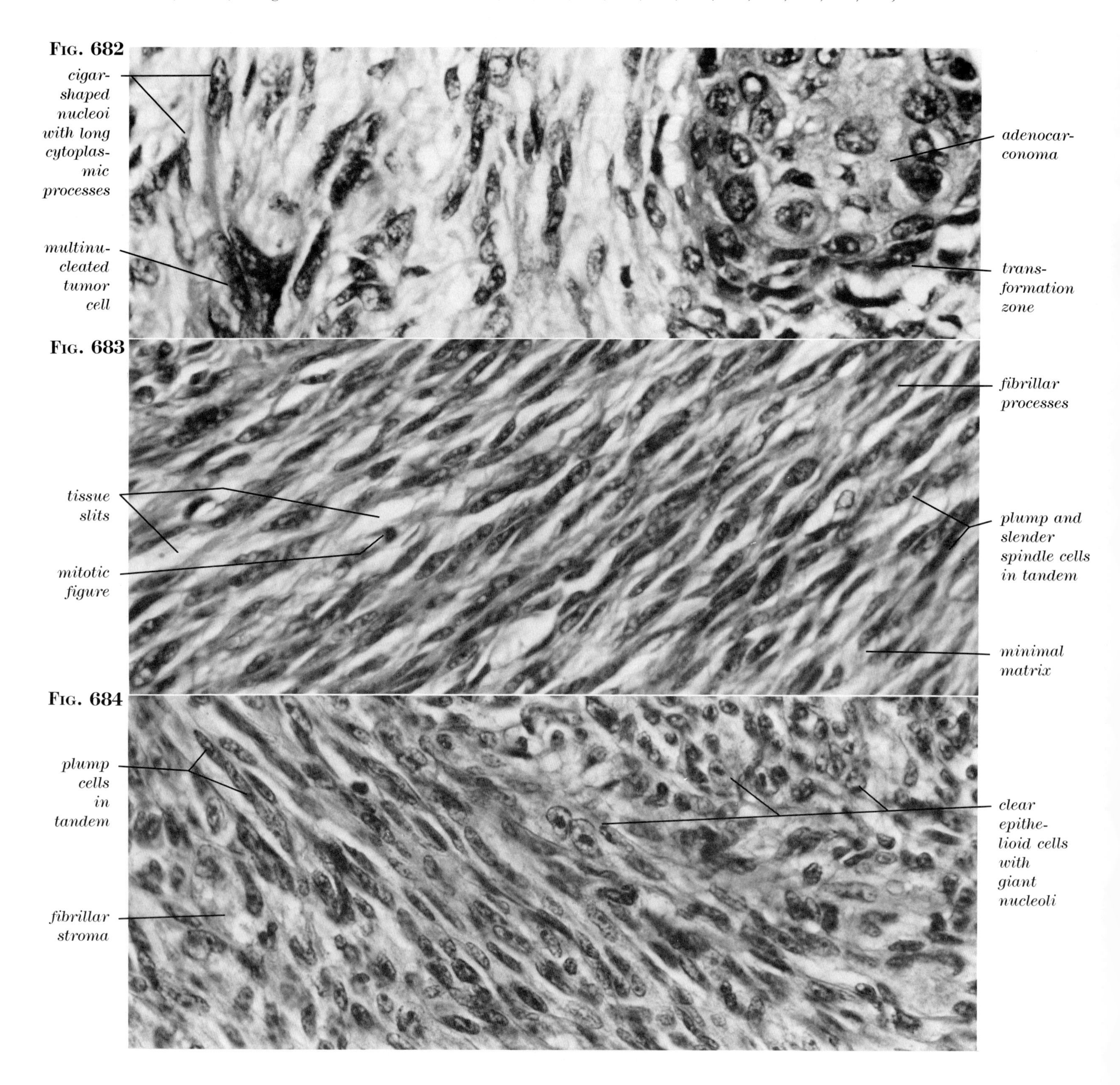

Fig. 685. Low grade leiomyosarcoma arising in teratoma of testis (×350) (see Figs. 670, 680, 681, 746).

Fig. 686. Desmoid tumor of pelvis (×350) (see Figs. 9, 36, 52, 77, 185, 202, 209, 244, 282, 376, 386, 397, 662, 865, 899, 1118).

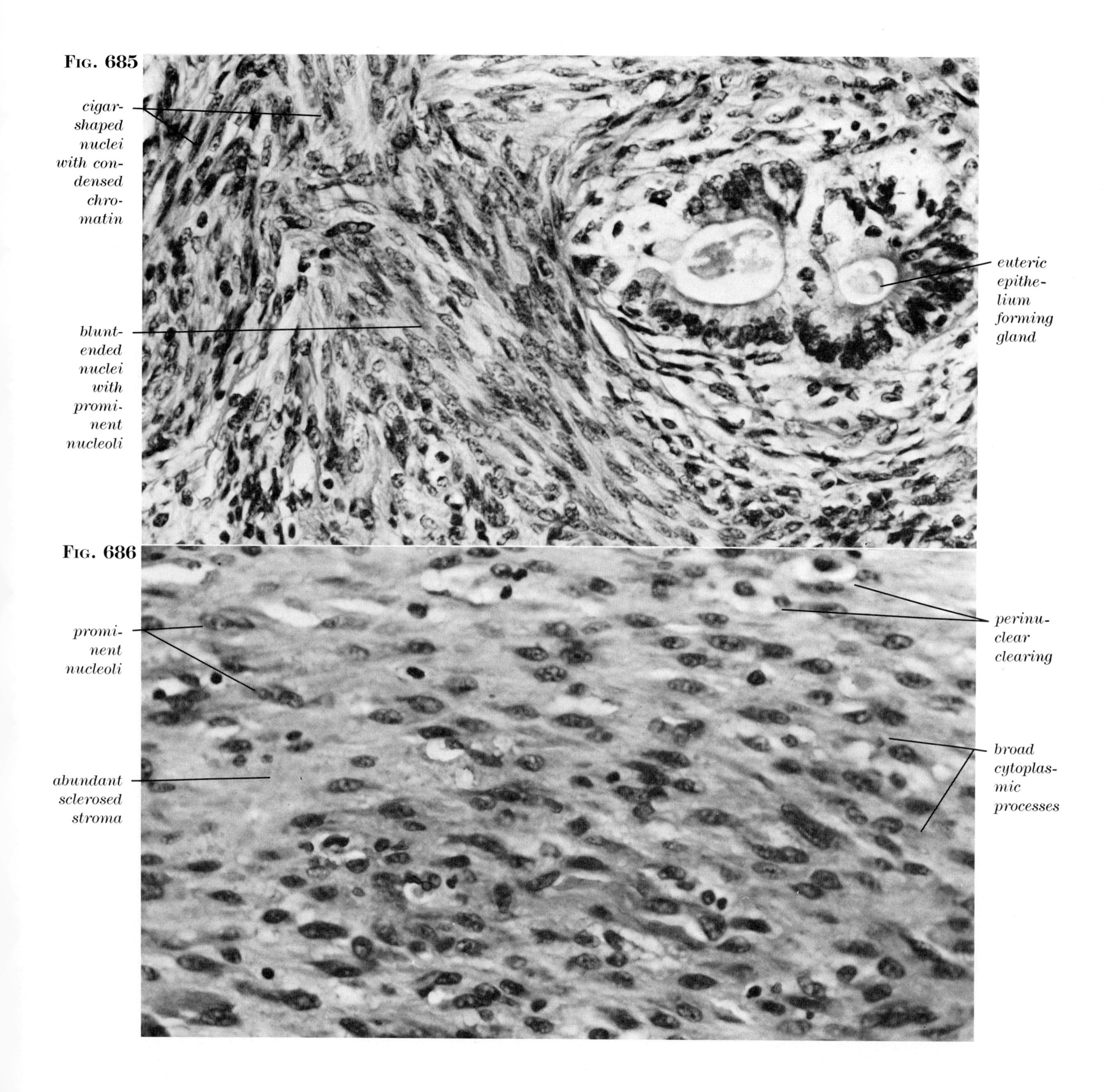

Fig. 687. Lipoblastic liposarcoma (×350) (see Figs. 55, 415, 448, 450, 543, 687, 753, 934, 973, 1231).

Fig. 688. Seminoma (×350) (see Figs. 14, 279, 349, 566, 757, 922, 1224).

Fig. 689. Embryonal rhabdomyosarcoma (×350) (see Figs. 372, 378, 379, 637, 694, 700, 706, 708, 751, 889, 925, 935).

Fig. 690. Malignant neuroepithelioma (×350) (see Figs. 309, 478, 489, 568, 585, 1115, 1238, 1239).

Fig. 691. Leiomyoblastoma of stomach (×350) (see Figs. 196, 270, 420, 429, 713, 733, 744, 745, 747, 750, 752, 795, 814, 828).

Fig. 692. Plasma cell myeloma (×350) (see Figs. 99, 208A, 454, 726, 741).

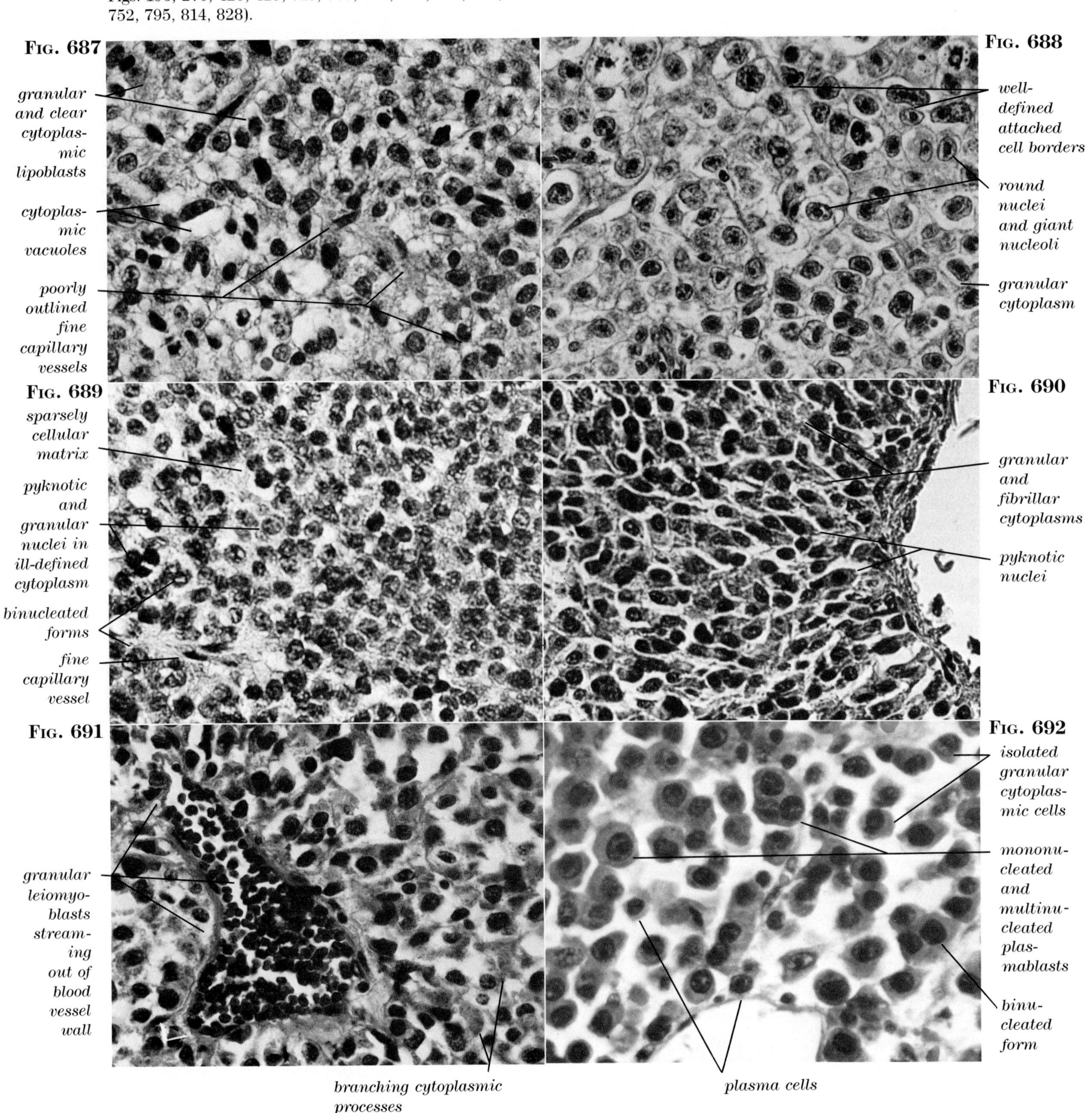

Fig. 693. High grade hemangiopericytoma (×350) (see Figs. 93, 470, 479, 484, 696, 972, 984, 995, 1203).

Fig. 695. Rhabdomyoblastoma (×350) (see Figs. 473, 477, 500, 528, 569, 594, 709, 715, 739, 796, 936, 1034, 1159, 1260).

Fig. 697. Alveolar rhabdomyosarcoma (×350) (see Figs. 78, 206, 524, 529, 530, 567, 570, 571, 804).

Fig. 694. Embryonal rhabdomyosarcoma (×350) (see Figs. 379, 426, 427, 637, 689, 700, 706, 708, 751, 889, 925, 935, 946, 949).

Fig. 696. Low grade hemangiopericytoma (×350) (see Figs. 374, 470, 479, 484, 525, 533, 693, 972, 984).

Fig. 698. Ewing's sarcoma (×350) (see Figs. 100, 172, 250, 449, 453, 476, 573, 1157, 1230).

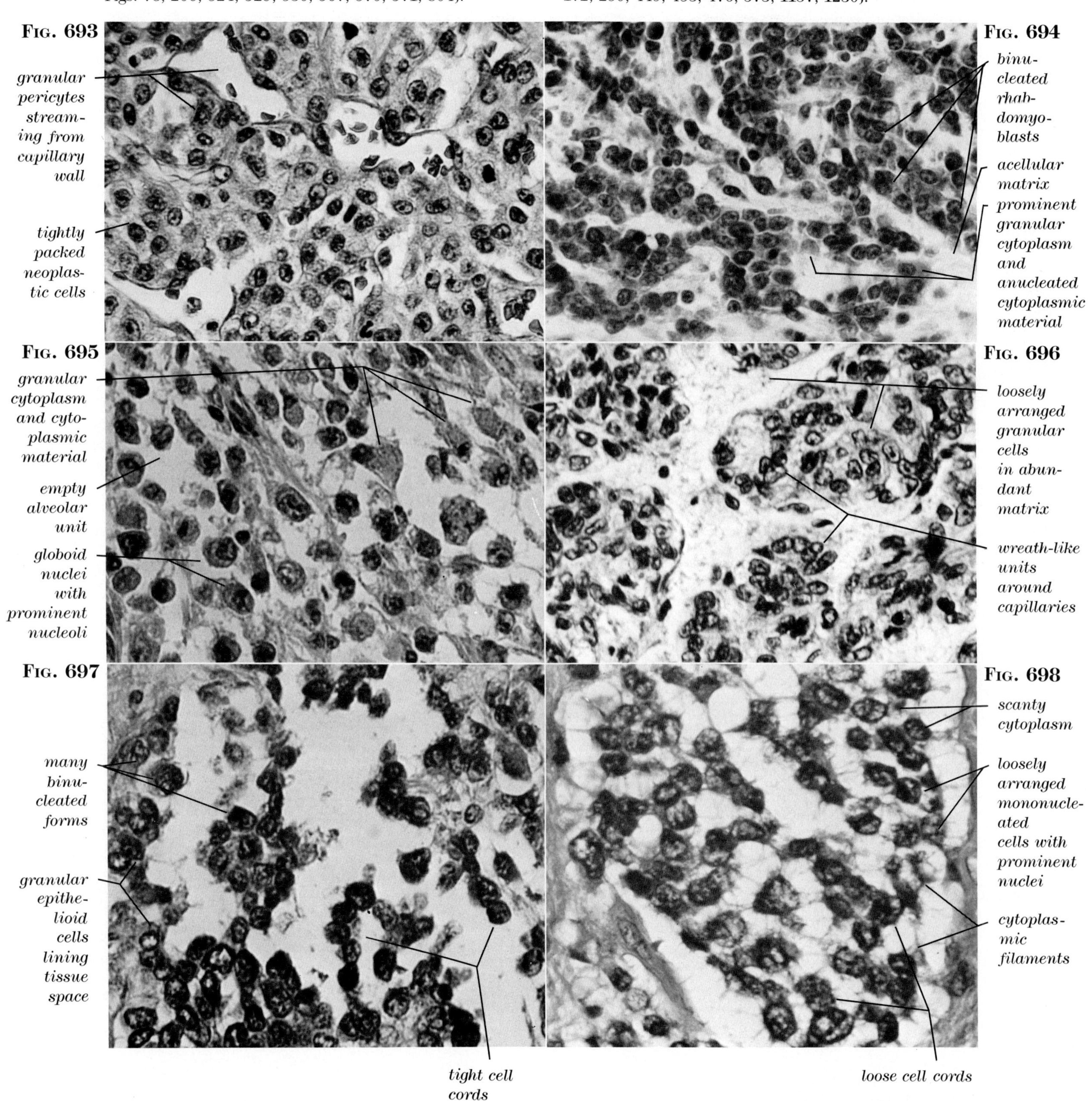

Fig. 699. Granulocytic sarcoma (×350) (see Figs. 488, 891, 1037, 1194, 1198).

Fig. 701. Malignant pigmented schwannoma (×350) (see Figs. 80, 521, 735, 967, 1192).

Fig. 703. Mast cells in smooth muscle (Giemsa's stain, ×350).

Fig. 700. Embryonal rhabdomyosarcoma (×350) (see Figs. 78, 263, 364, 443, 637, 689, 694, 706, 708, 751, 889, 925, 935, 943, 946).

Fig. 702. Malignant lymphoma, mixed cell type (×350) (see Figs. 15, 132, 169, 423, 463, 494, 704, 705, 712, 743, 888).

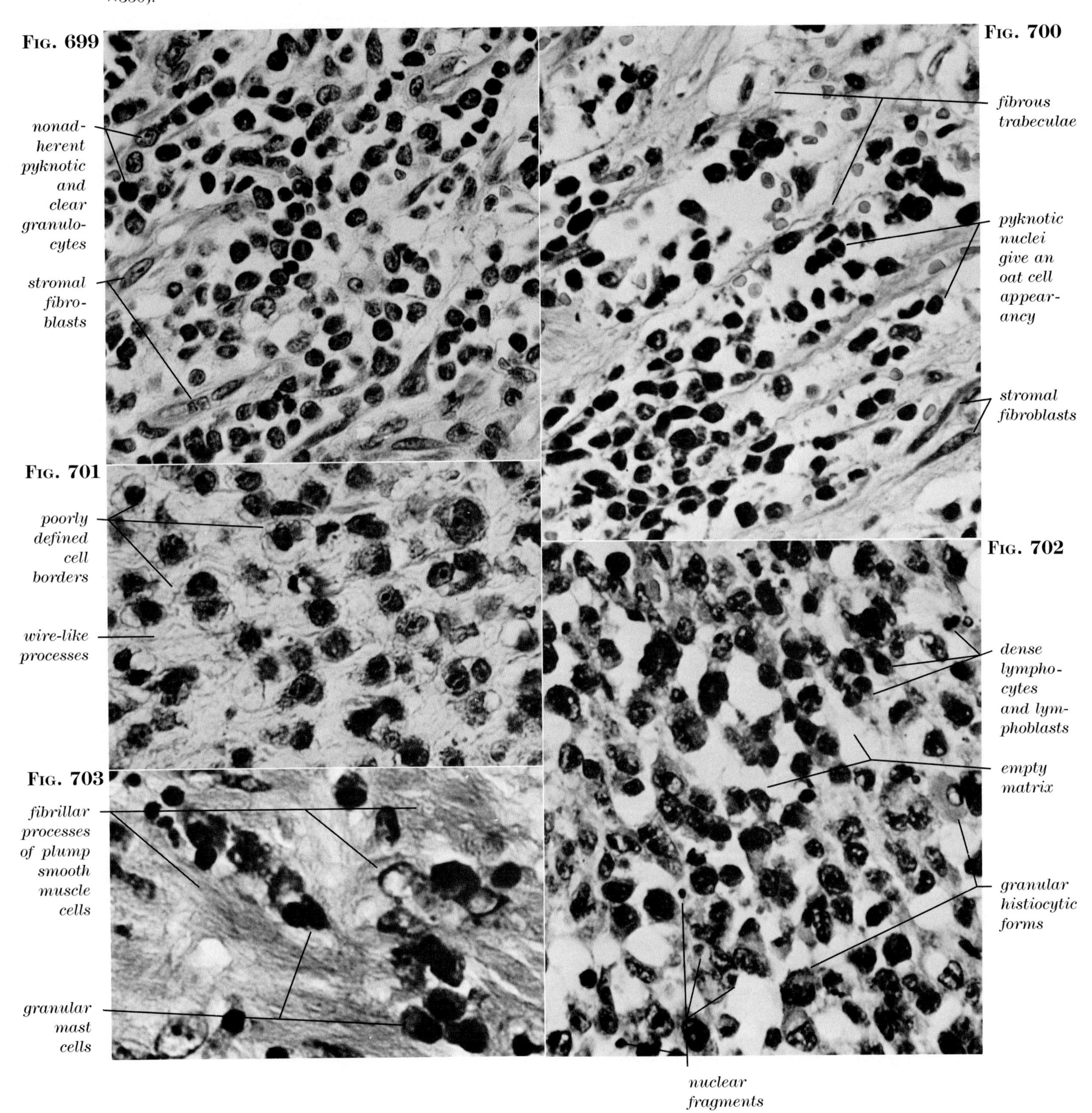

Fig. 704. Malignant lymphoma infiltrating bone (×140) (see Figs. 15, 132, 169, 423, 463, 494, 702, 705, 712, 742, 743, 888, 1024).

Fig. 705. Primary intraskeletal lymphoma (×350) (see Figs. 423, 463, 494, 702, 704, 743, 885, 888).

Fig. 706. Embryonal rhabdomyosarcoma invading bone (×350) (see Figs. 637, 689, 694, 700, 708, 751, 889, 925, 935, 943, 946, 949).

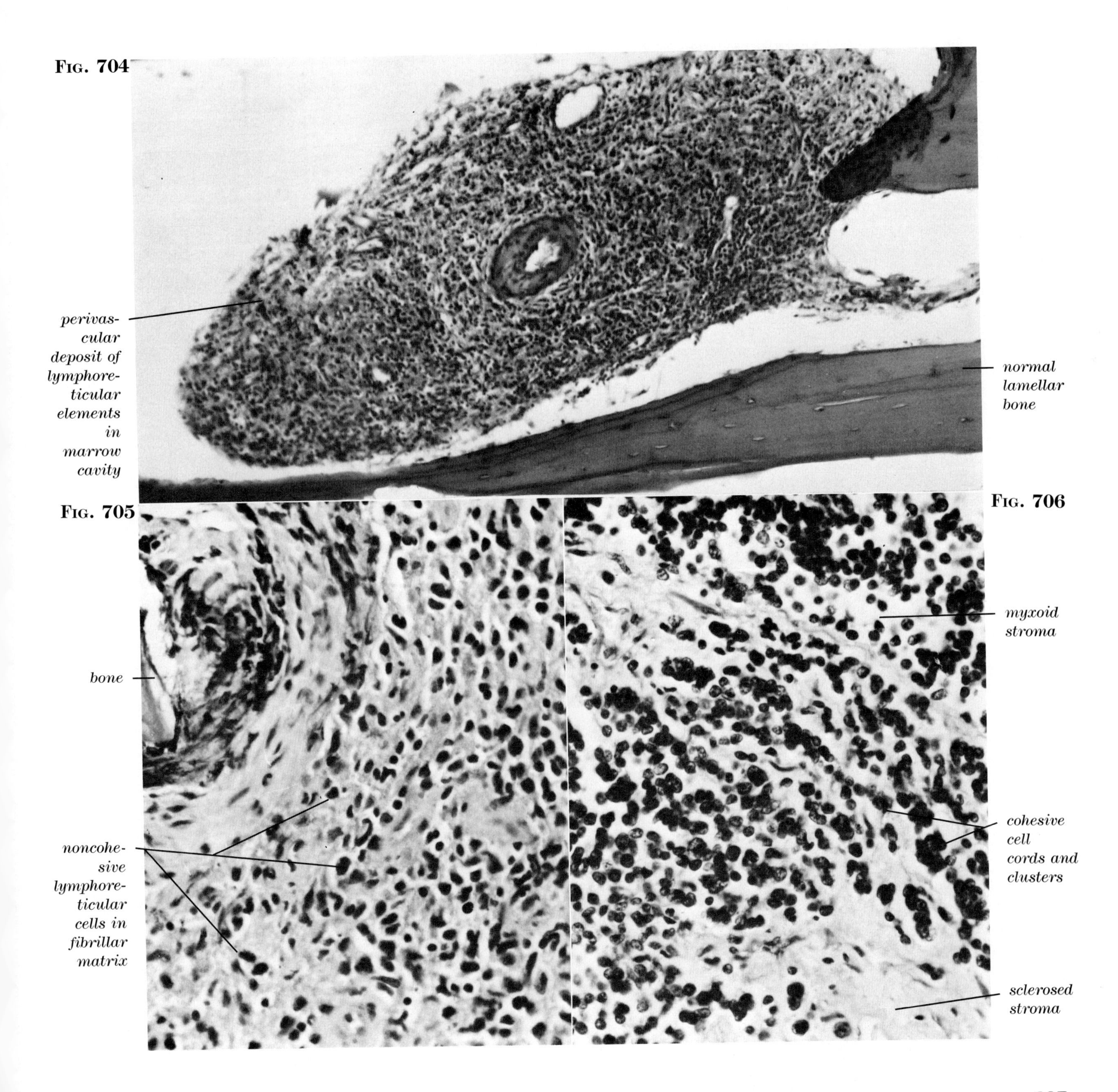

Fig. 707. Mesenchymal chondrosarcoma (×350) (see Figs. 103, 151, 175, 180, 318, 390, 444, 507, 595, 674, 707, 920, 1062, 1265).

Fig. 708. Embryonal rhabdomyosarcoma, epithelioid type (×350) (see Figs. 78, 364, 372, 378, 379, 515, 637, 689, 694, 700, 706, 751).

Fig. 709. Embryonal rhabdomyosarcoma, rhabdomyoblastic type (×350) (see Figs. 473, 477, 500, 528, 569, 594, 695, 715, 739, 796, 936, 1034).

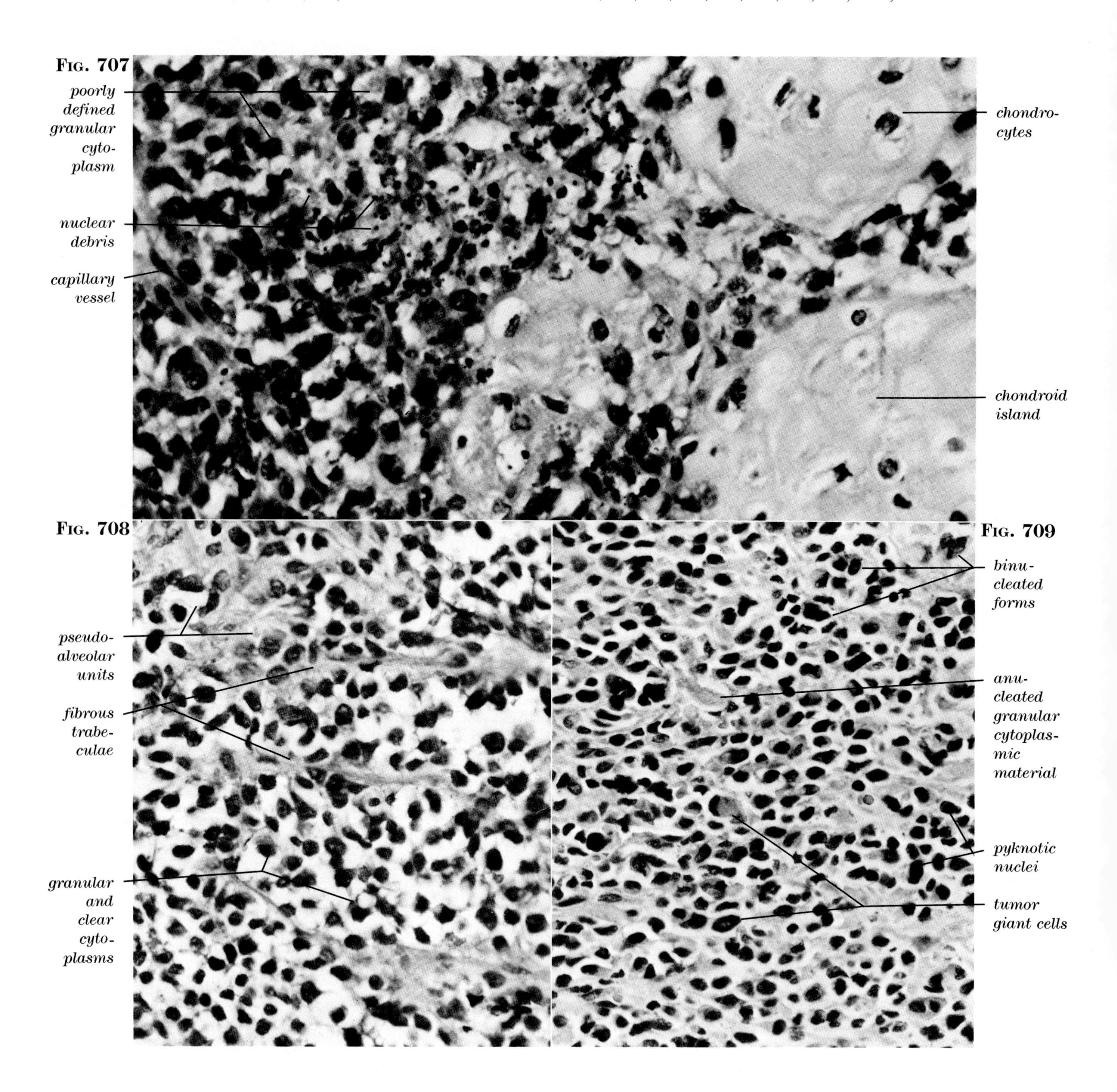

Fig. 710. Malignant fibroblastic fibrous histiocytoma (×350) (see Figs. 339, 340, 350, 373, 645, 672, 710, 760, 833, 1026, 1133–1135, 1147, 1178, 1200).

Fig. 711. Borderline histiocytic fibrous histiocytoma, so-called "atypical giant cell tumor" (×350) (see Figs. 452, 534, 592, 756, 775, 776, 784, 791, 816).

Fig. 712. Malignant lymphoma, histiocytic type in soft tissues (×350) (see Figs. 702, 704, 705, 743).

Fig. 713. (Bottom center) High grade leiomyoblastoma (×350) (see Figs. 503, 513, 691, 733).

Fig. 714. Metastatic paraganglioma (×350) (see Figs. 276, 483, 544, 790).

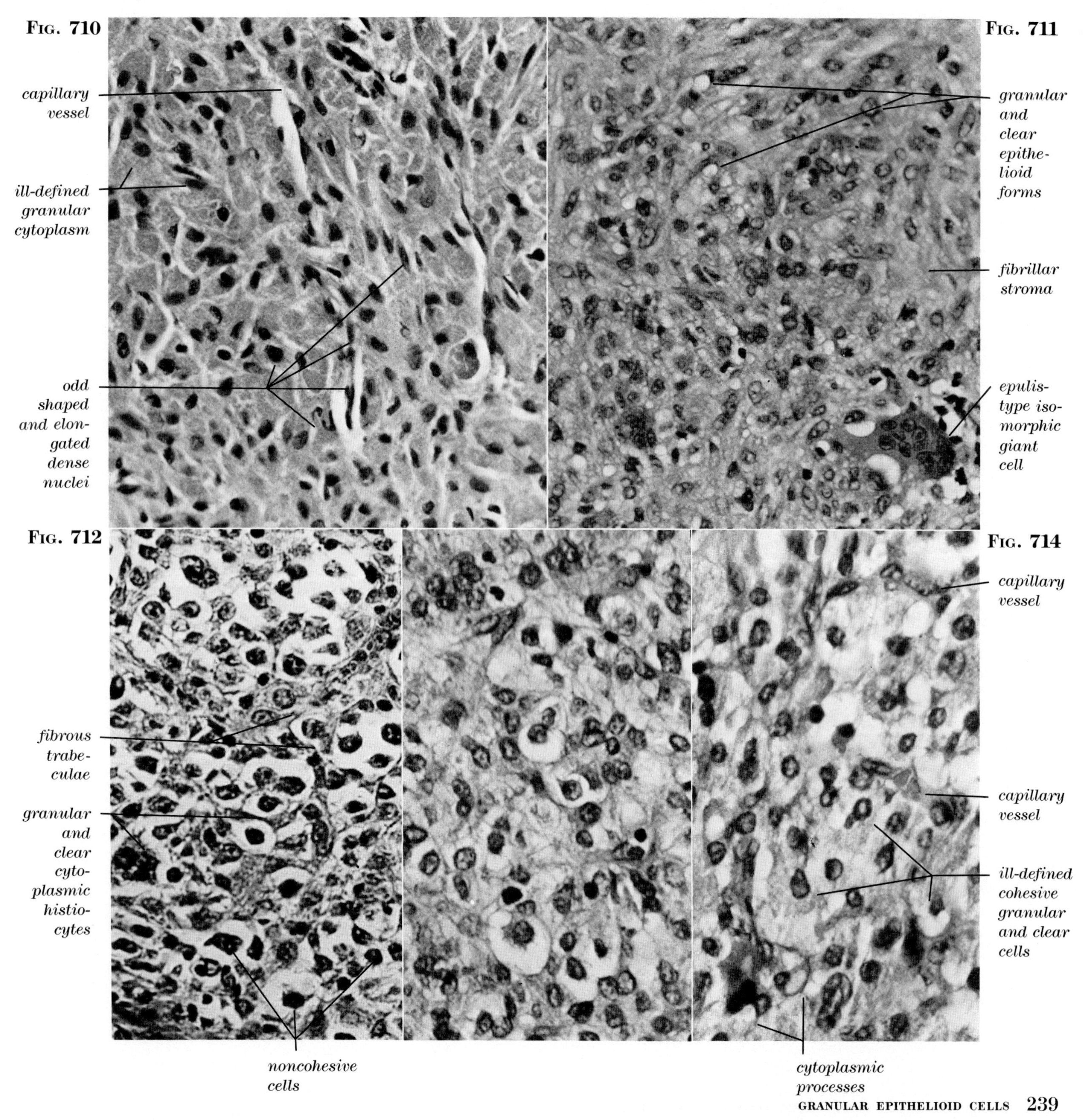

Fig. 715. Rhabdomyoblastoma (×350) (see Figs. 528, 569, 594, 695, 709, 715, 739, 796, 936, 1034).

Fig. 716. Oncocitic renal cell carcinoma (×350) (see Figs. 545, 651, 764, 768, 1018A, 1270).

Fig. 717. Hibernoma (×350) (see Figs. 432, 1162).

Fig. 718. Alveolar soft part sarcoma (×350) (see Figs. 89, 538, 1161, 1199, 1227).

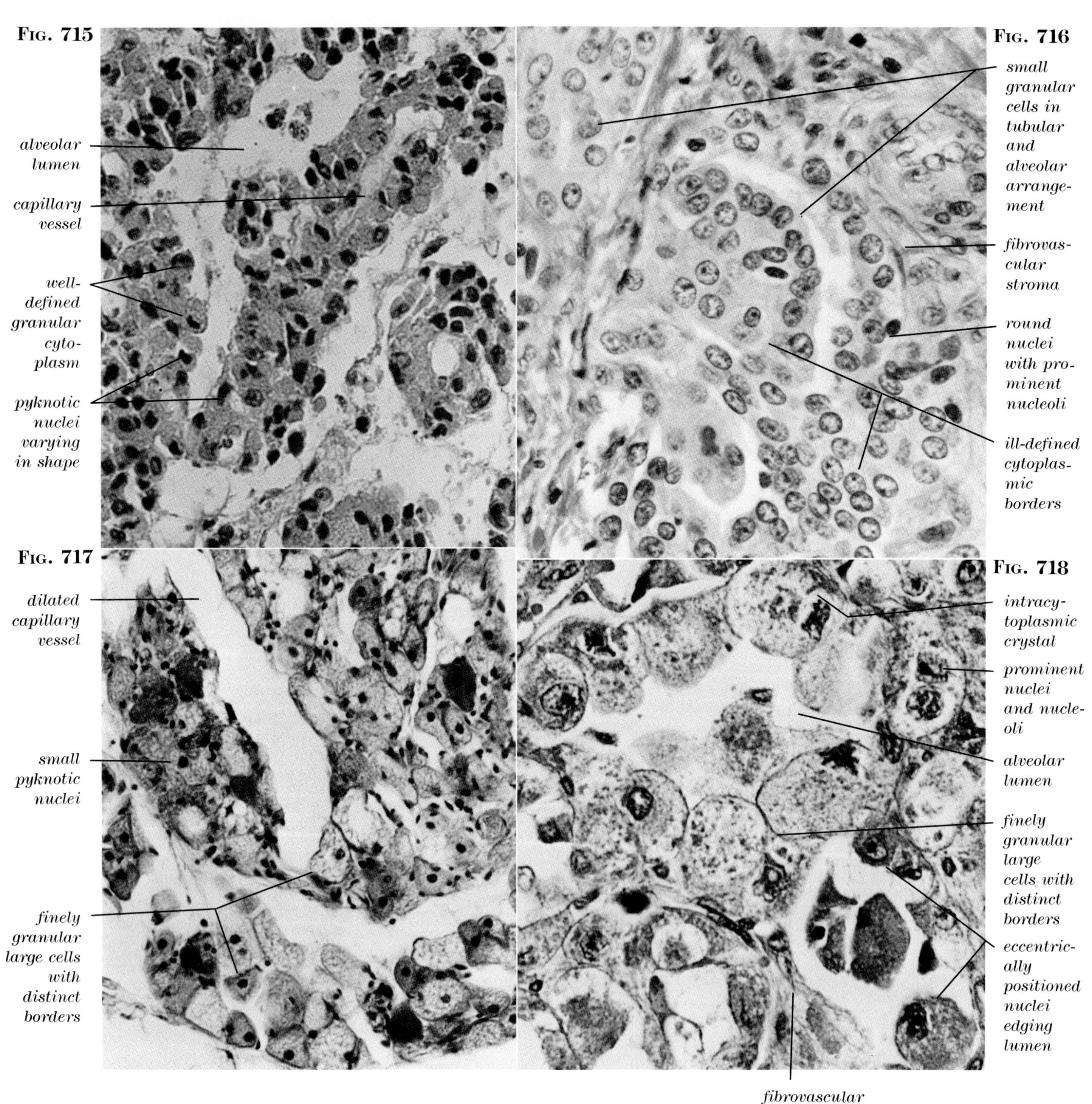

Fig. 719. Malignant melanoma (×350) (see Figs. 540, 669, 719, 748, 797, 826, 1188, 1189, 1190, 1219, 1262).

Fig. 720. High grade hemangiosarcoma (×350) (see Figs. 92, 102, 316, 523, 541, 542, 555, 557, 563, 564, 576, 720, 731).

Fig. 721. Proliferative panniculitis (×140) (see Figs. 430, 552, 910).

Fig. 722. Benign pleomorphic fibrous histiocytoma, so-called "xanthoma of femur" (×570) (see Figs. 135, 346, 417, 422, 459, 782).

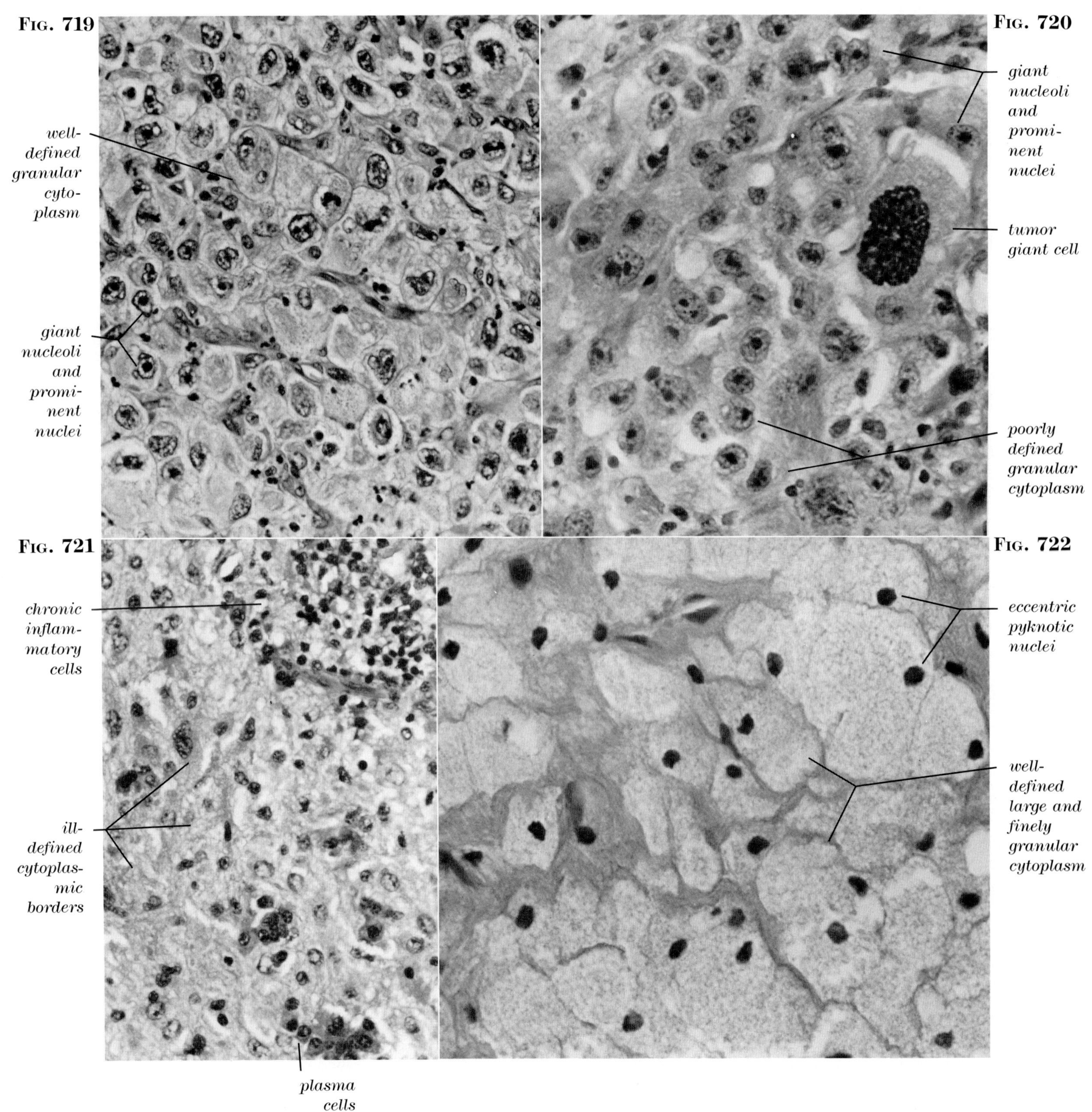

Fig. 723. Medulloblastoma (×570) (see Figs. 311, 469).

Fig. 724. Extraadrenal neuroblastoma (×570) (see Fig. 1218).

Fig. 725. Pinealoma (×570)

Fig. 726. Extraskeletal plasmacytoma (×570) (see Figs. 133, 192).

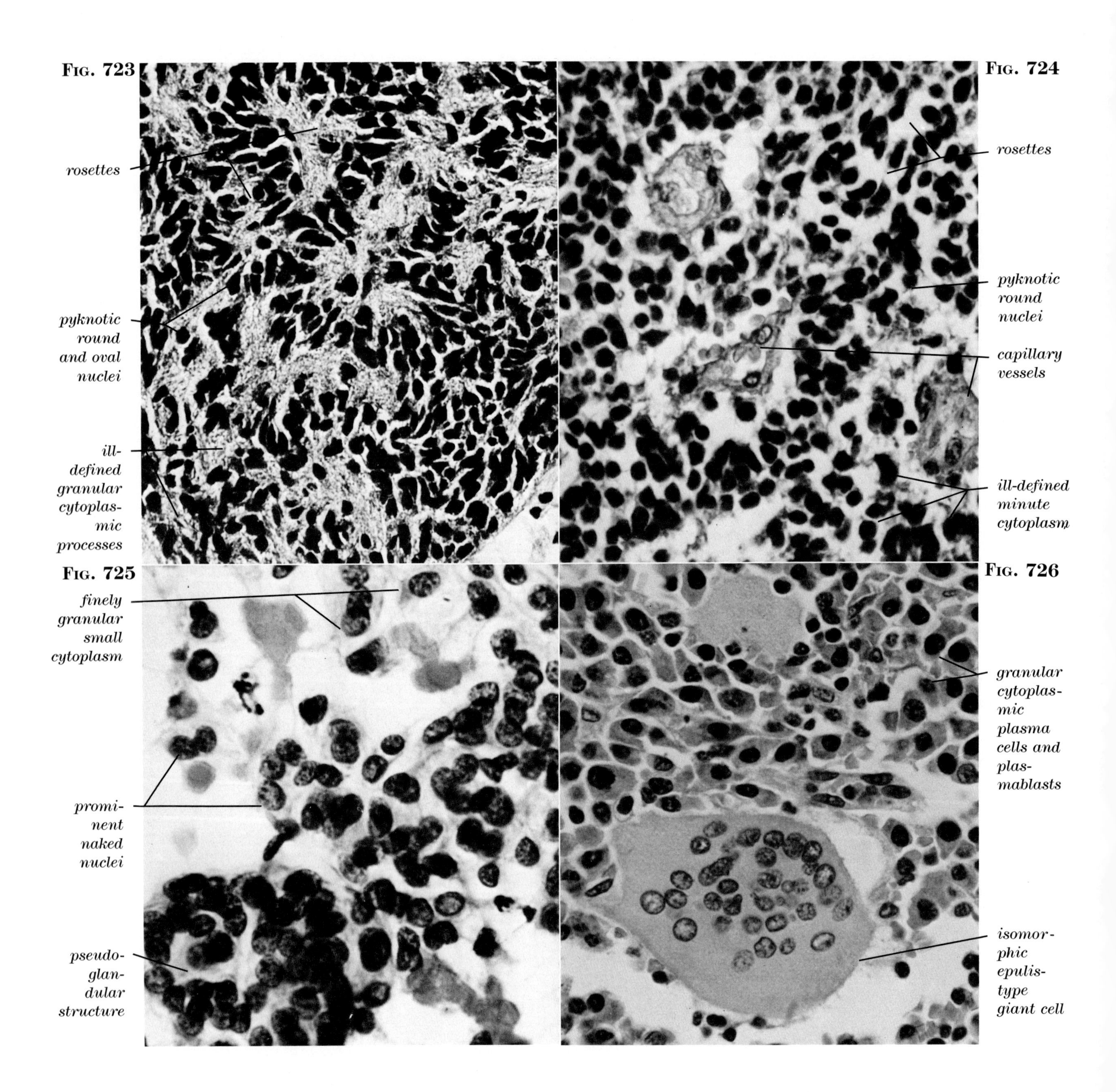

Fig. 727. Biphasic tendosynovial sarcoma (×350) (see Figs. 85, 514, 520, 522, 765, 906, 1099, 1100).

Fig. 729. Chordoma (×350) (see Figs. 69, 215, 310, 505, 767, 771, 1174, 1226).

Fig. 731. High grade hemangiosarcoma (×350) (see Figs. 555, 557, 563, 564, 576, 720, 905).

Fig. 728. Borderline or atypical glomus tumor (×350) (see Figs. 457, 482, 550, 980, 1141).

Fig. 730. Osteosarcoma, osteoblastic type (×570) (see Figs. 108, 602, 610, 643, 677, 730, 754, 769, 777, 801, 807, 817, 819, 831).

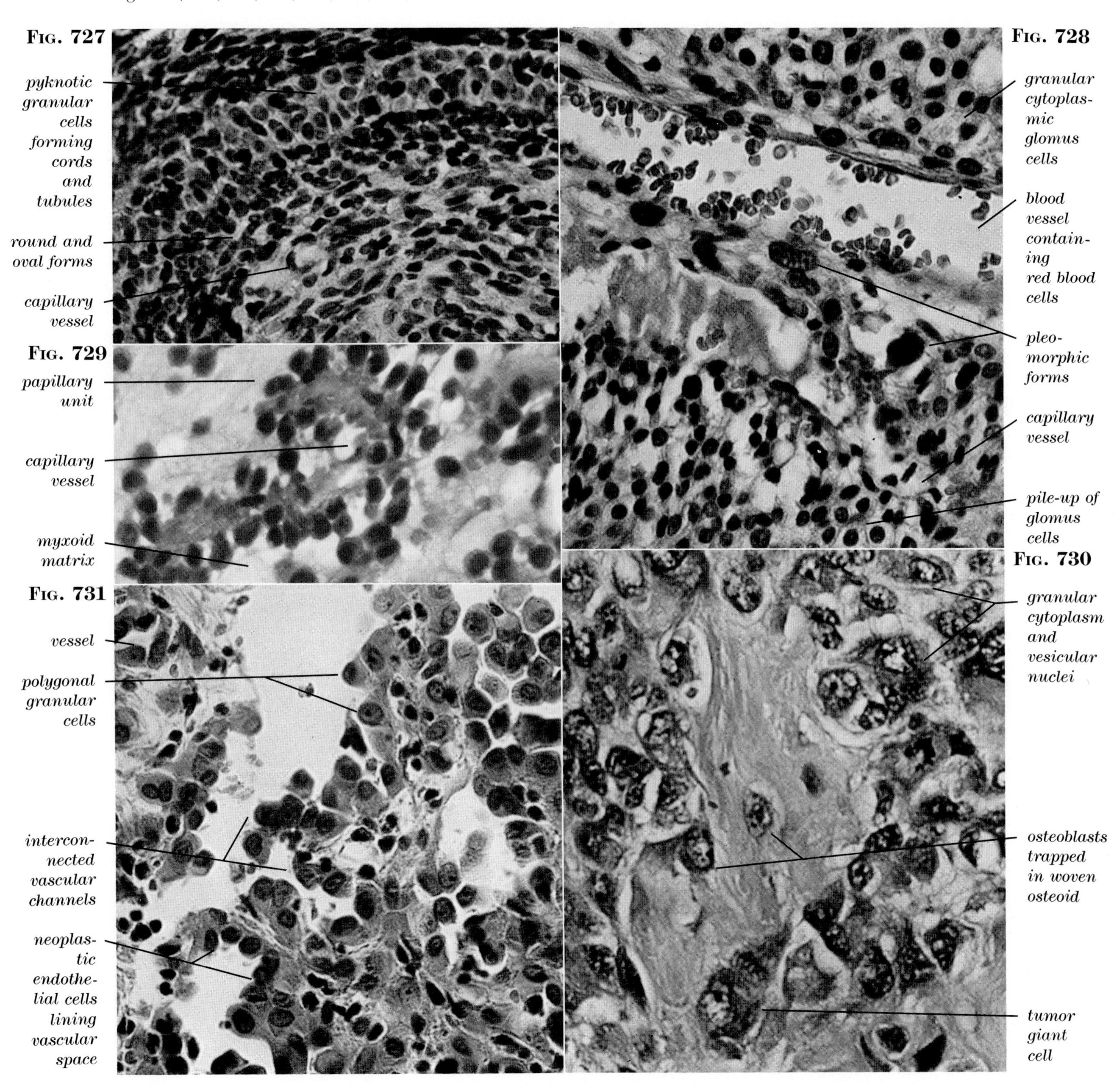

Fig. 732. Malignant Triton tumor (×570) (see Figs. 593, 606, 812, 813, 1259).

Fig. 733. Low grade leiomyoblastoma (×350) (see Figs. 196, 270, 274, 420, 429, 503, 513, 691, 713, 744, 745, 747, 750, 752).

Fig. 734. Xanthogranuloma (×350) (see Figs. 607, 779, 1031, 1127, 1201).

Fig. 735. Malignant nevoid schwannoma (×570) (see Figs. 521, 701, 967, 1192).

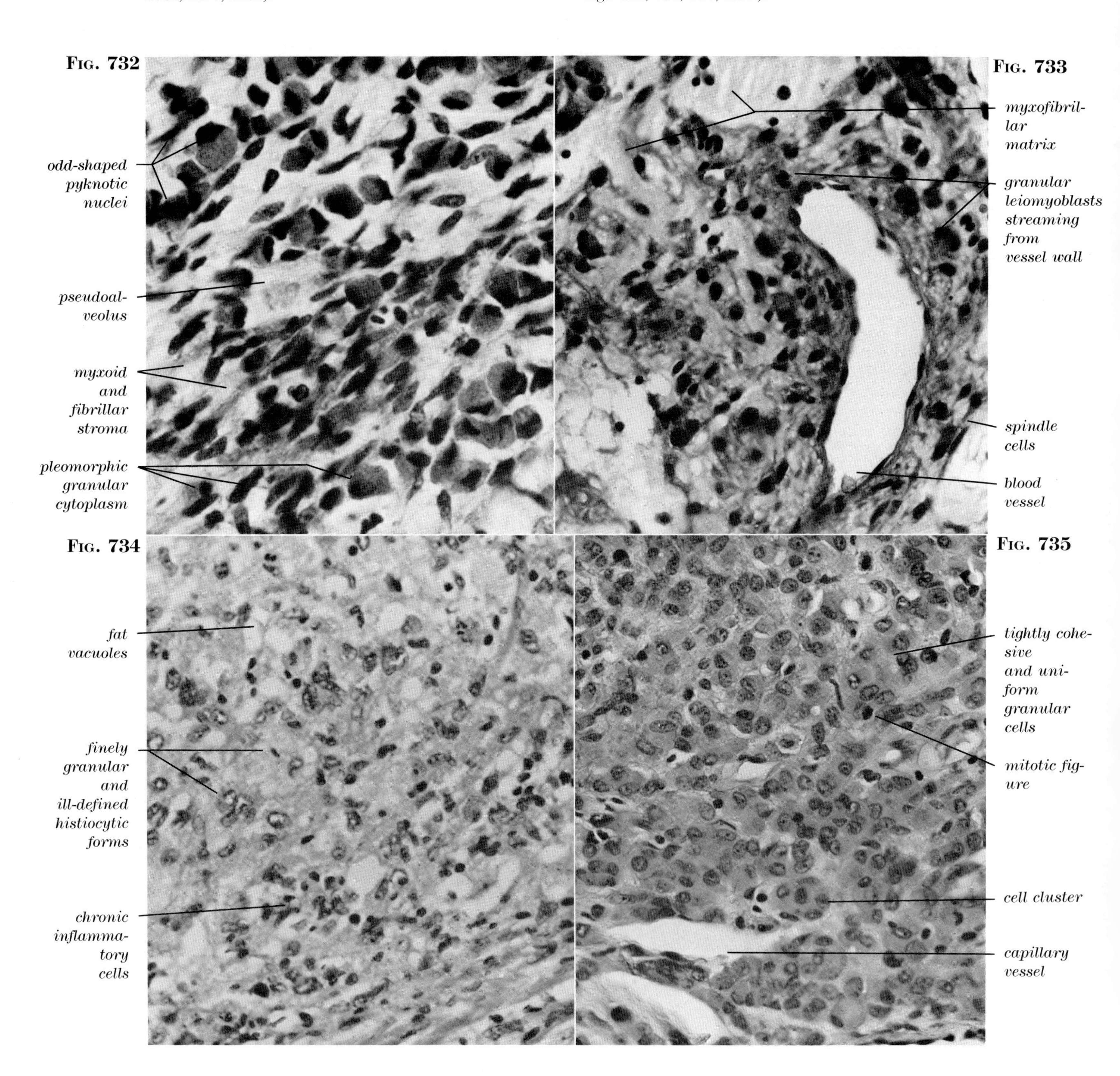

Fig. 736. Benign granular cell tumor (×570) (see Figs. 75, 261, 498, 539, 1160, 1235).

Fig. 738. Proliferative myositis (×570) (see Figs. 597, 863, 951).

Fig. 737. Rhabdomyoma (×570) (see Figs. 497, 793).

Fig. 739. Rhabdomyoblastoma (×570) (see Figs. 594, 695, 709, 715, 739, 796, 936, 1034, 1159, 1260).

Fig. 740. Malignant granular cell tumor (×570) (see Figs. 499, 1234, 1256).

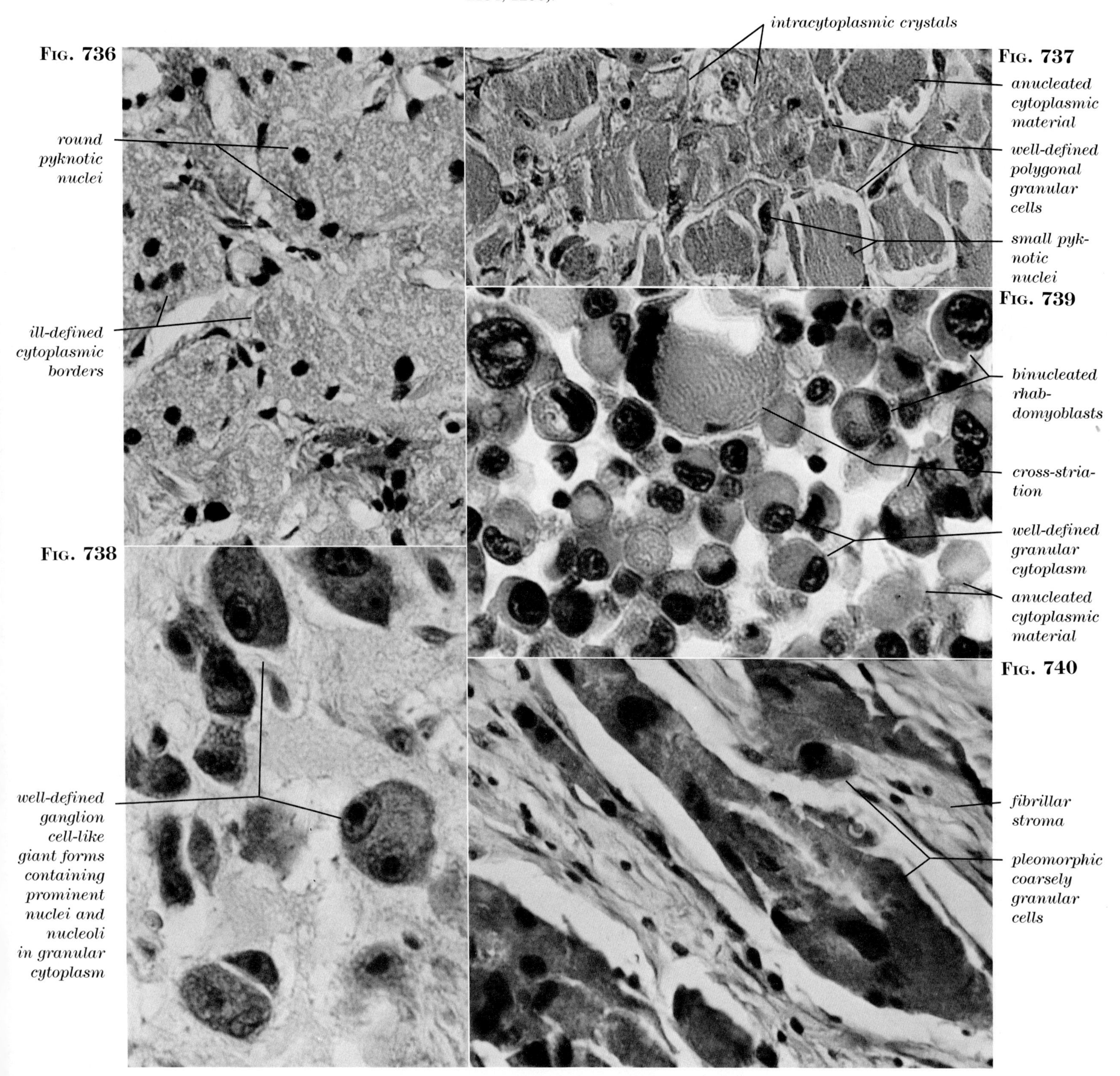

Fig. 741. Plasma cell myeloma (×350) (see Figs. 99, 454, 692).

Fig. 742. (Top center) Hodgkin's disease (×350) (see Figs. 27, 885, 1024).

Fig. 743. Malignant lymphoma (×350) (see Figs. 463, 494, 702, 704, 705, 888).

Fig. 744. Low grade leiomyoblastoma of stomach (×350) (see Figs. 691, 713, 733, 745).

Fig. 745. (Bottom center) Low grade leiomyoblastoma of vagina (×350) (see Figs. 713, 733, 744, 747).

Fig. 746. Leiomyosarcoma with leiomyoblastic elements (×350) (see Figs. 680, 681, 685, 808).

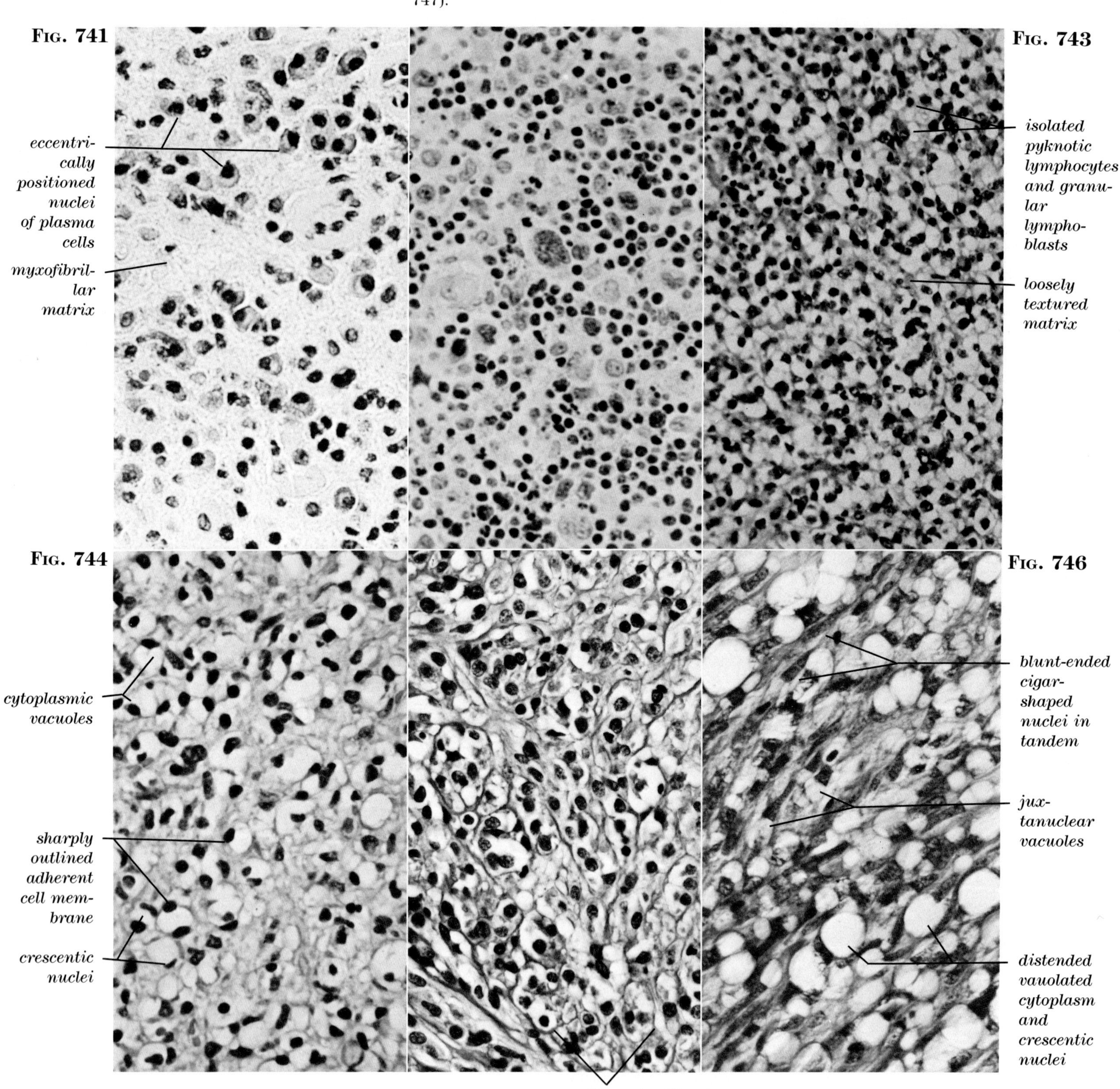

Fig. 747. Low grade leiomyoblastoma (×350) (see Figs. 196, 270, 274, 277, 278, 420, 429, 503, 513, 691, 733, 750).

Fig. 748. Balloon cell melanoma (×350) (see Figs. 719, 797, 826, 1188, 1189, 1190, 1219, 1262).

Fig. 749. Fat necrosis (×350) (see Figs. 425, 1042, 1056, 1185).

Fig. 750. High grade leiomyoblastoma (×350) (see Figs. 733, 744, 745, 747, 752, 795, 814, 828, 868, 931, 964, 1011).

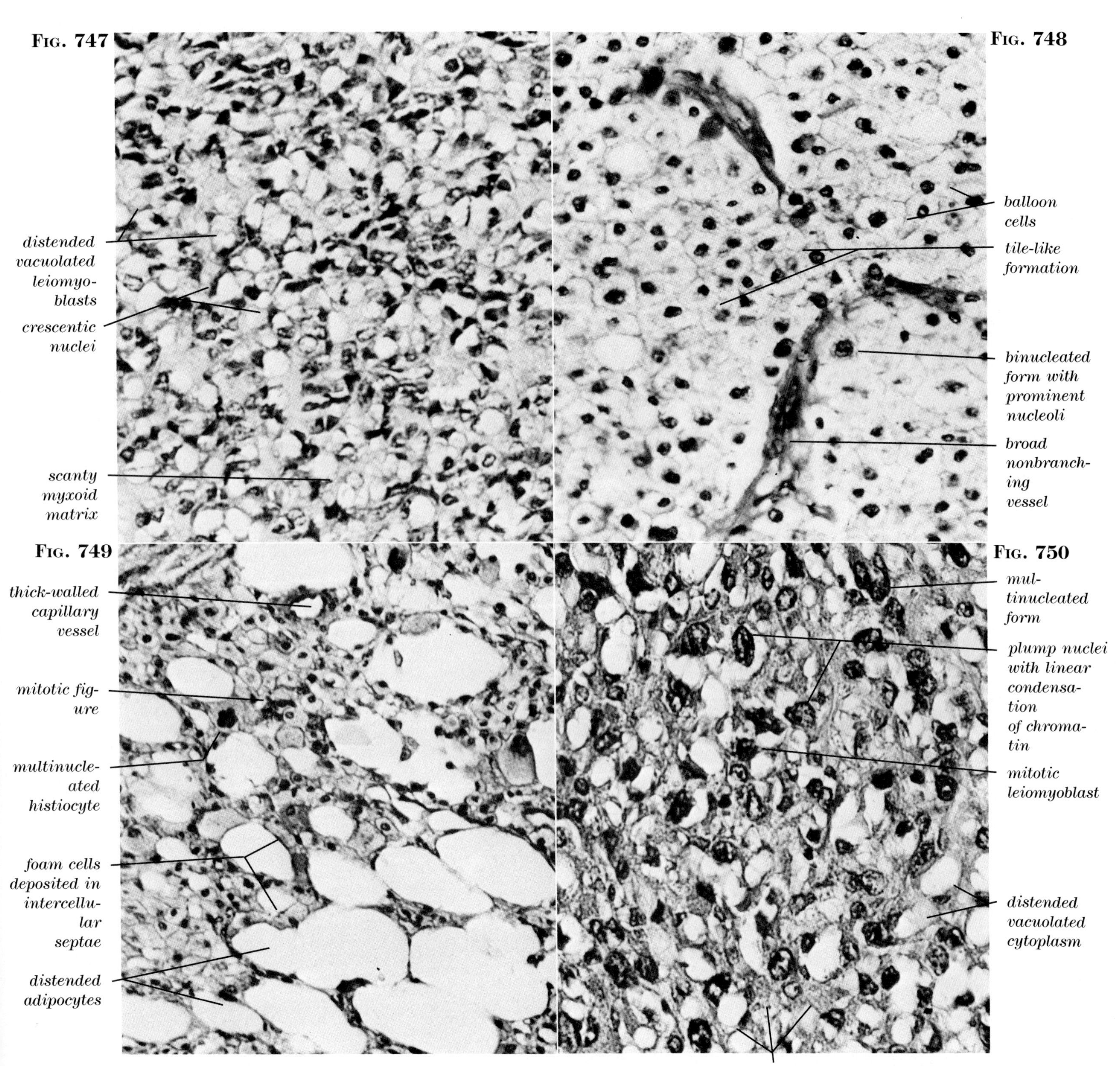

FIG. 751. Embryonal rhabdomyosarcoma (×570) (see Figs. 515, 637, 689, 694, 700, 706, 708, 889, 925, 935, 943, 946).

FIG. 752. Leiomyoblastoma (×570) (see Figs. 274, 503, 744, 745, 747, 750, 795, 814, 828, 868, 931).

FIG. 753. Lipoblastic liposarcoma (×570) (see Figs. 199, 415, 448, 450, 543, 687, 934, 973, 1231).

FIG. 754. Osteosarcoma with chondroblastic elements (×570) (see Figs. 90, 108, 142, 248, 328, 395, 615, 643, 677, 730, 769, 777, 801, 807, 817, 819, 831).

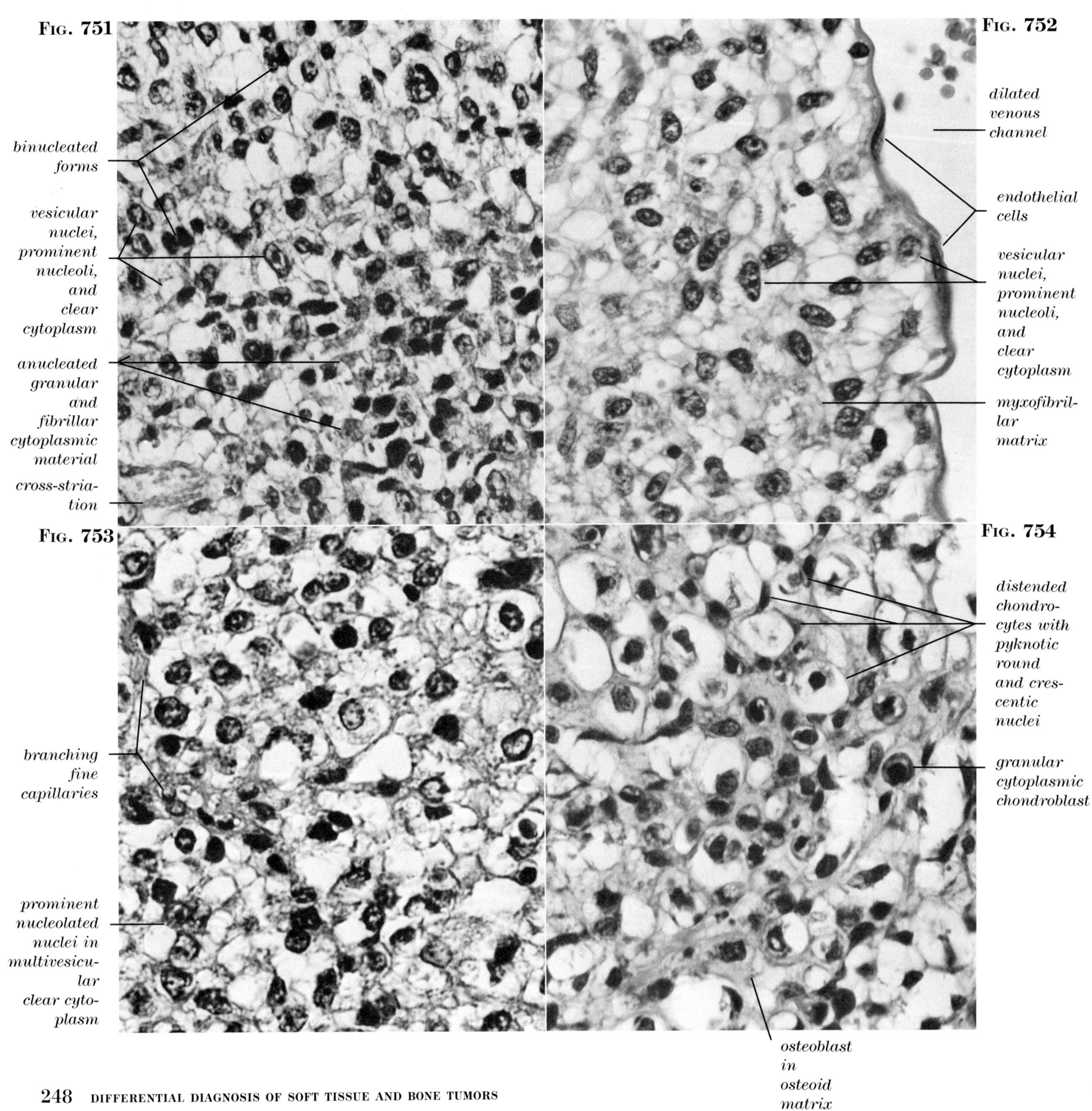

Fig. 755. Embryonal carcinoma of testis (×350).

Fig. 757. Seminoma (×350) (see Figs. 14, 566, 1224).

Fig. 759. Clear cell sarcoma (×350) (see Figs. 62, 322, 447, 492, 526, 761, 896, 927, 1145, 1223).

Fig. 756. Malignant histiocytic fibrous histiocytoma (×350) (see Figs. 57, 156, 161, 300, 452, 534, 592, 600, 711, 775, 776, 784, 791, 816, 872, 892).

Fig. 758. Clear cell carcinoma of ovary (×350) (see Fig. 428).

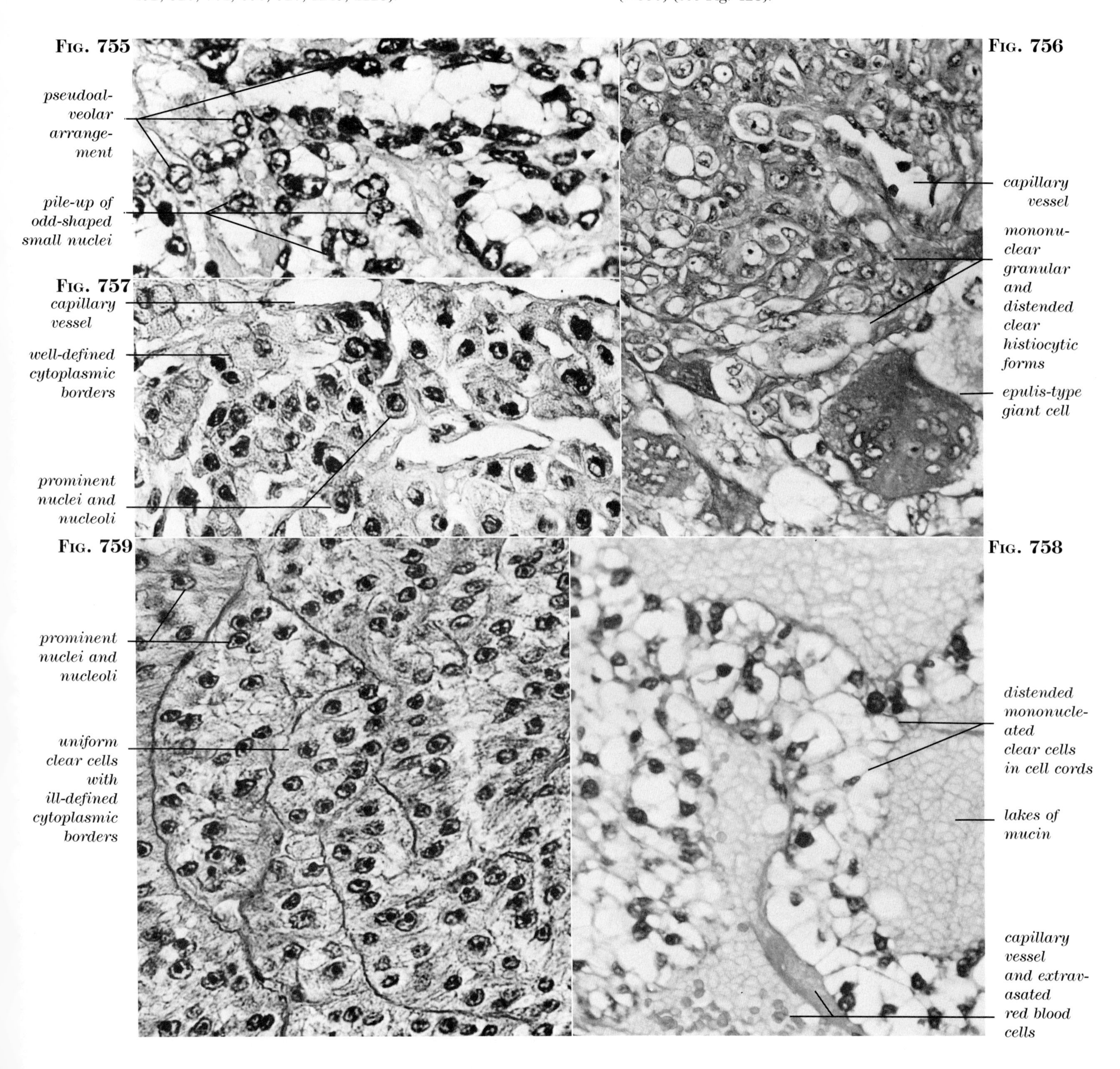

Fig. 760. Metastic malignant fibroblastic fibrous histiocytoma (×350) (see Figs. 645, 672, 710, 833).

Fig. 761. (Top center) Clear cell sarcoma (×570) (see Figs. 85, 526, 759, 896, 927).

Fig. 762. Inflamed nasal polyp (×350).

Fig. 763. Metastatic mesodermal mixed tumor (×350). (see Figs. 65, 589, 969, 1165, 1217).

Fig. 764. Metastatic clear cell carcinoma of kidney (×350) (see Figs. 545, 651, 716, 768, 1081A, 1270).

Fig. 765. Biphasic tendosynovial sarcoma (×350) (see Figs. 547, 574, 586, 727, 906, 1099, 1100, 1173, 1202).

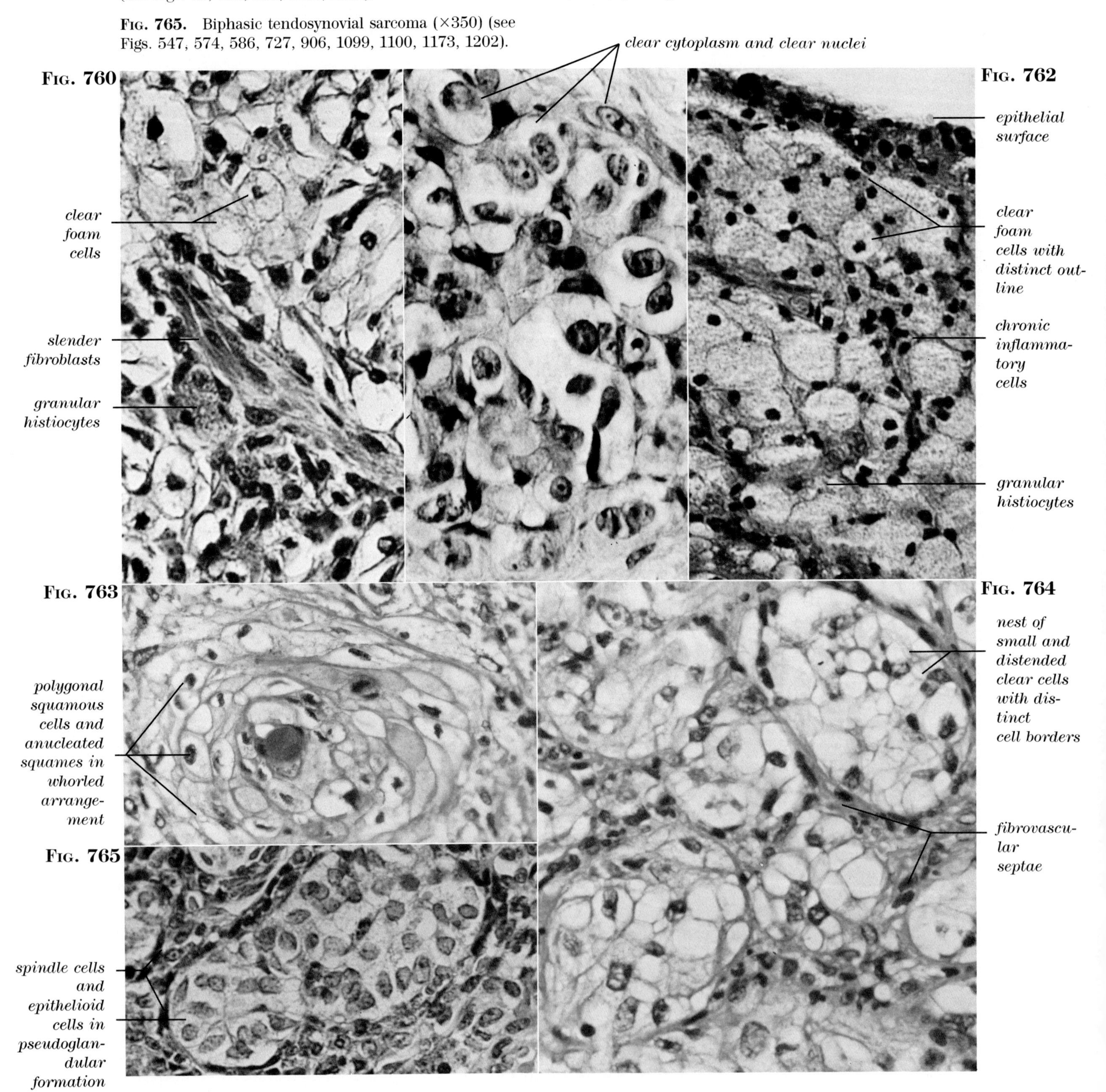

Fig. 766. Well-differentiated chondrosarcoma (×570) (see Figs. 81, 95, 123, 124, 136, 147, 216, 223, 285, 442, 773, 1061, 1065, 1093, 1101, 1110, 1222, 1275).

Fig. 767. Chordoma (see Figs. 310, 505, 729, 771, 1174).

Fig. 768. Clear cell carcinoma of kidney metastatic to bone (×350) (see Figs. 764, 1018A, 1270).

Fig. 769. Osteosarcoma with chondrosarcoma component (×570) (see Figs. 90, 108, 127, 248, 395, 643, 677, 730, 754, 777, 801, 807, 817, 819, 831, 843, 876, 897, 898).

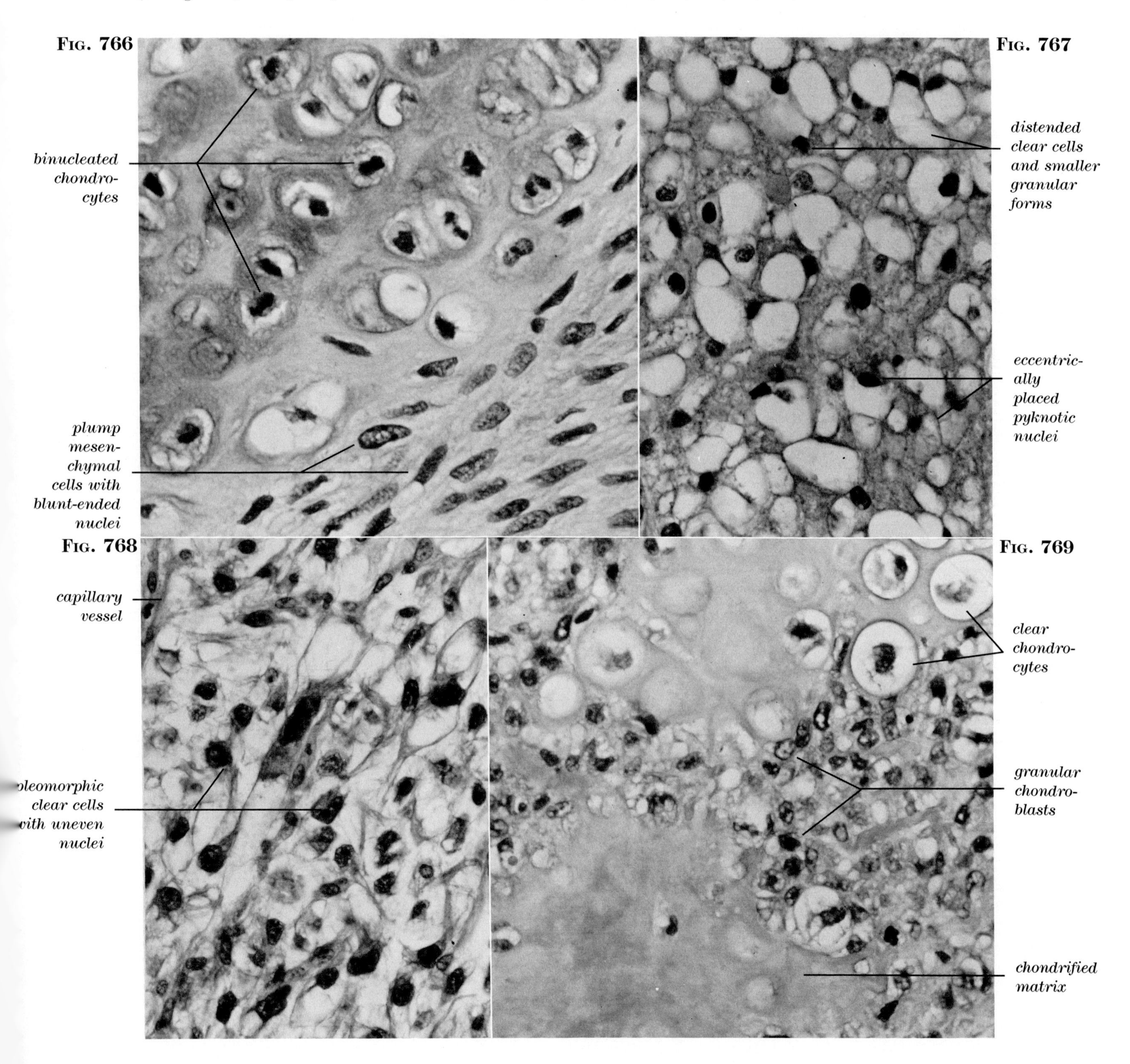

Fig. 770. Decidua in pelvic lymph node (×570) (see Fig. 924).

Fig. 771. Chordoma (×570) (see Figs. 69, 215, 310, 505, 729, 767, 1174, 1226).

Fig. 772. Osteochondroma (×570) (see Figs. 113, 155, 167, 184, 302, 303, 1063, 1066, 1090, 1152).

Fig. 773. Well-differentiated chondrosarcoma (×570) (see Figs. 95, 327, 442, 766, 1061, 1065, 1093, 1101, 1110, 1222, 1275).

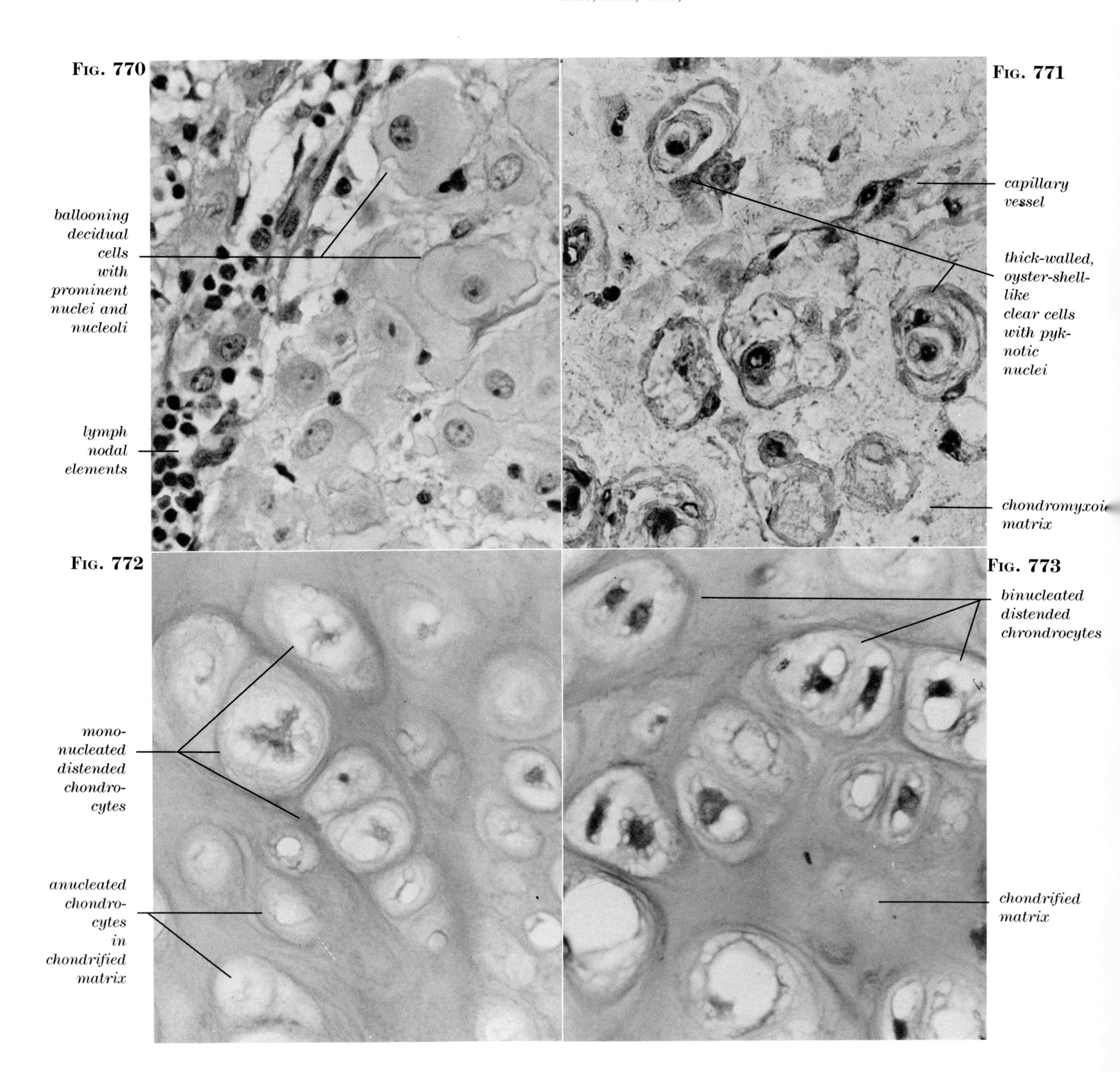

Fig. 774. Benign histiocytic fibrous histiocytoma (×570) (see Figs. 490, 501, 778).

Fig. 776. Borderline or atypical histiocytic fibrous histiocytoma (×350) (see Figs. 711, 756, 775, 784).

Fig. 778. Benign histiocytic fibrous histiocytoma (×140) (see Figs. 71, 174, 178, 490, 774).

Fig. 775. Malignant histiocytic fibrous histiocytoma (×140) (see Figs. 57, 156, 161, 300, 452, 534, 592, 600, 711, 756, 776, 784, 791, 816, 872, 892, 893, 930, 1018B, 1048).

Fig. 777. Fibrous histiocytic osteosarcoma (×140) (see Figs. 90, 108, 109, 142, 146, 248, 251, 395, 405, 602, 643, 769, 801, 807, 817, 819, 831, 843, 876, 897, 898, 958, 968).

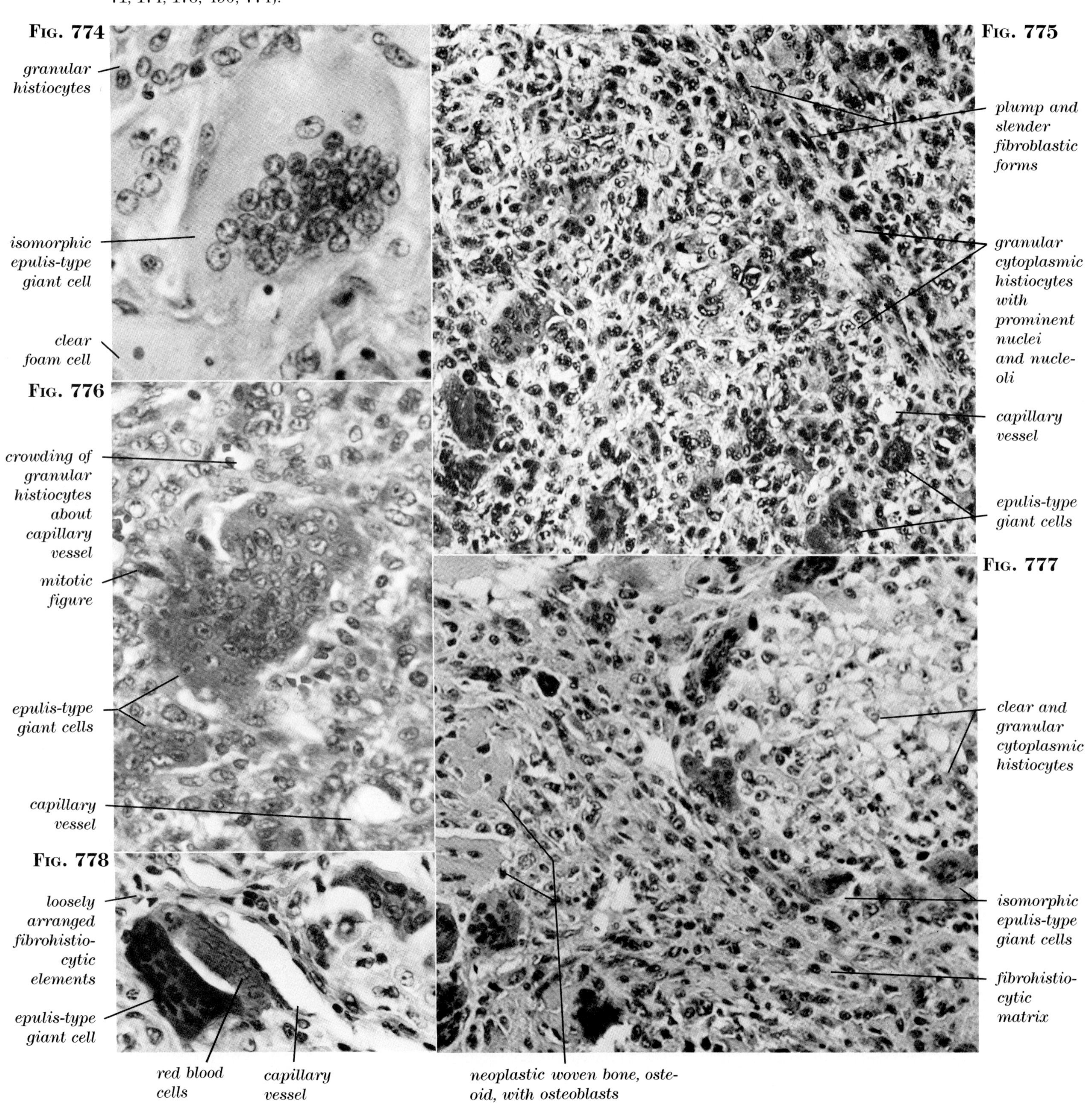

Fig. 779. Xanthogranuloma (×350) (see Figs. 607, 734, 1031, 1127, 1201).

Fig. 780. Atypical cutaneous fibrous histiocytoma (×350) (see Figs. 577, 579, 598, 679, 788).

Fig. 781. Pleomorphic lipoma (×350) (see Figs. 613, 821, 909, 957, 1182).

Fig. 782. Benign pleomorphic fibrous histiocytoma (×350) (see Figs. 135, 346, 417, 422, 459, 722).

Fig. 783. Myxoid liposarcoma (×350) (see Figs. 86, 208, 290, 636, 933).

Fig. 784. Malignant histiocytic fibrous histiocytoma (×350) (see Figs. 452, 534, 592, 600, 711, 756, 775, 776, 791, 816, 872, 892, 893).

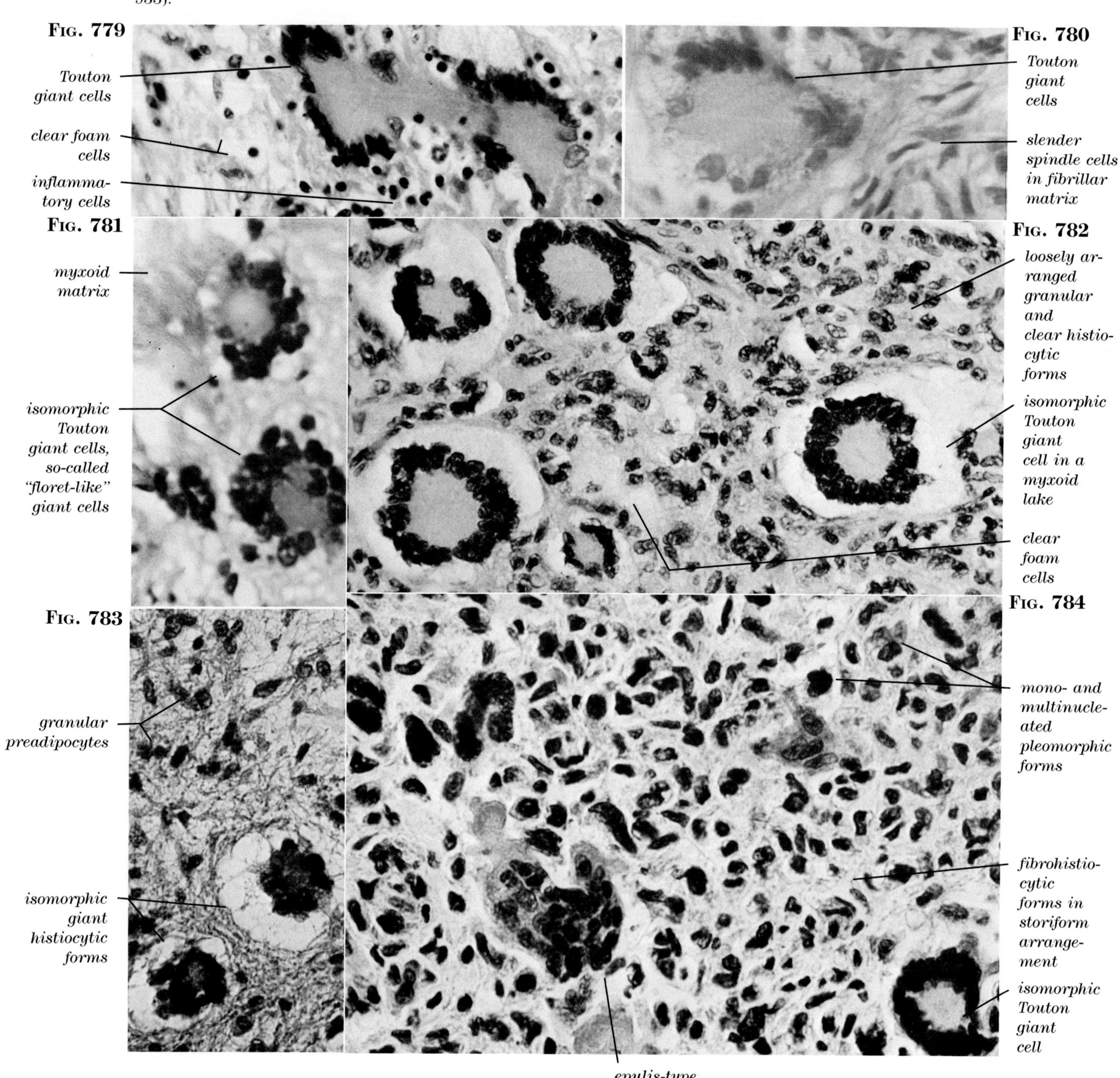

Fig. 785. Aneurysmal bone cyst (×350) (see Figs. 96, 125, 150, 163, 509, 1015, 1058).

Fig. 786. Tendosynovitis (×350) (see Figs. 70, 867, 1023, 1029).

Fig. 787. Giant cell granuloma (×350) (see Figs. 119, 632, 1000, 1069).

Fig. 788. Borderline pleomorphic fibrous histiocytoma (×350) (see Figs. 598, 679, 780, 789).

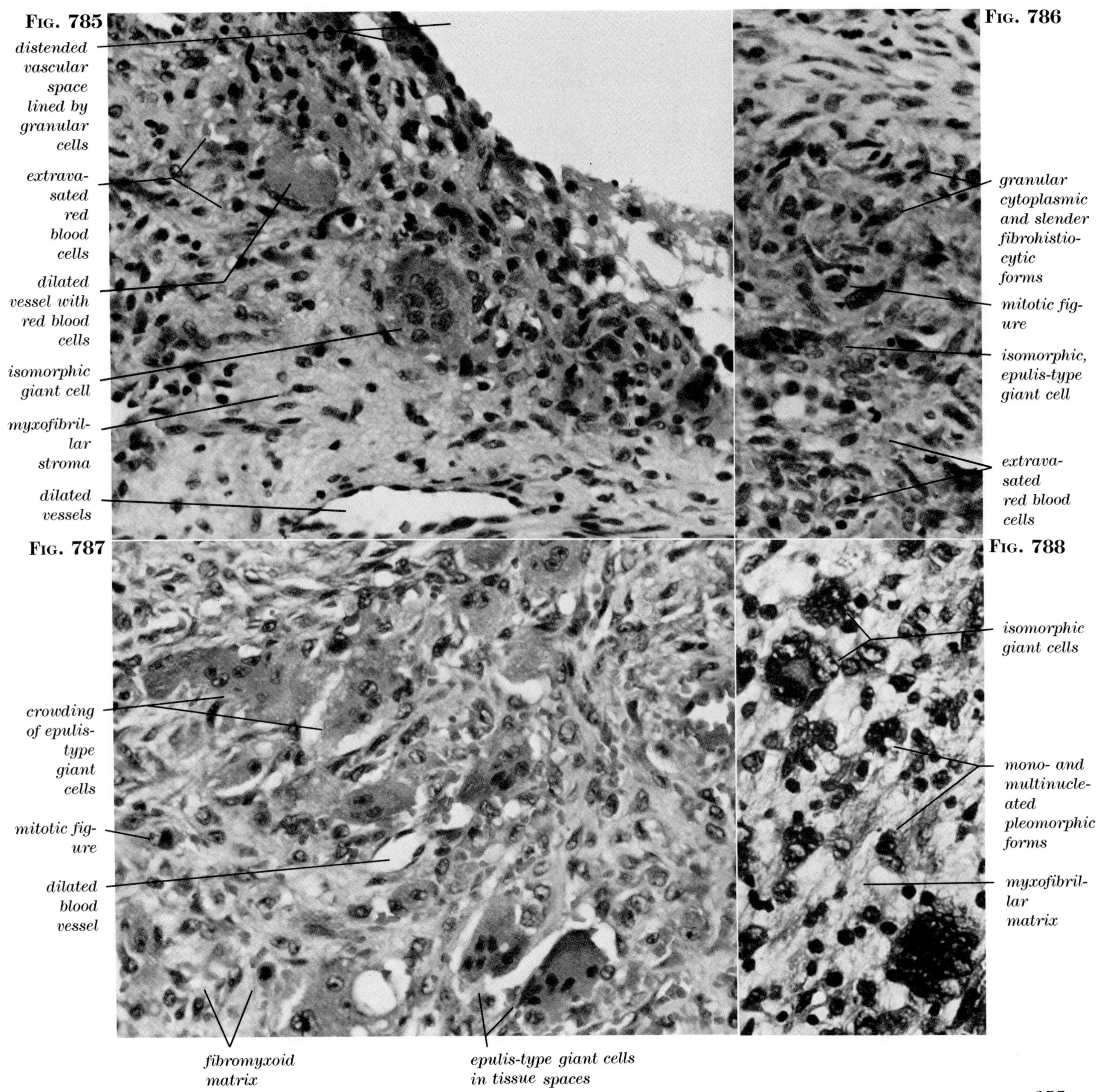

Fig. 789. Borderline or atypical pleomorphic fibrous histiocytoma (×350) (see Figs. 679, 780, 788, 792, 798, 805, 818).

Fig. 790. Malignant paraganglioma (×350) (see Figs. 276, 483, 544, 714, 794, 853, 1196, 1220).

Fig. 791. Malignant histiocytic fibrous histiocytoma, high grade (×350) (see Figs. 775, 776, 784, 816, 872, 892, 893, 930, 1018B, 1048, 1149).

Fig. 792. Malignant pleomorphic fibrous histiocytoma, high grade (×350) (see Figs. 88, 111, 232, 237, 344, 598, 789, 798, 805, 818, 937, 952, 962, 1002, 1003, 1008).

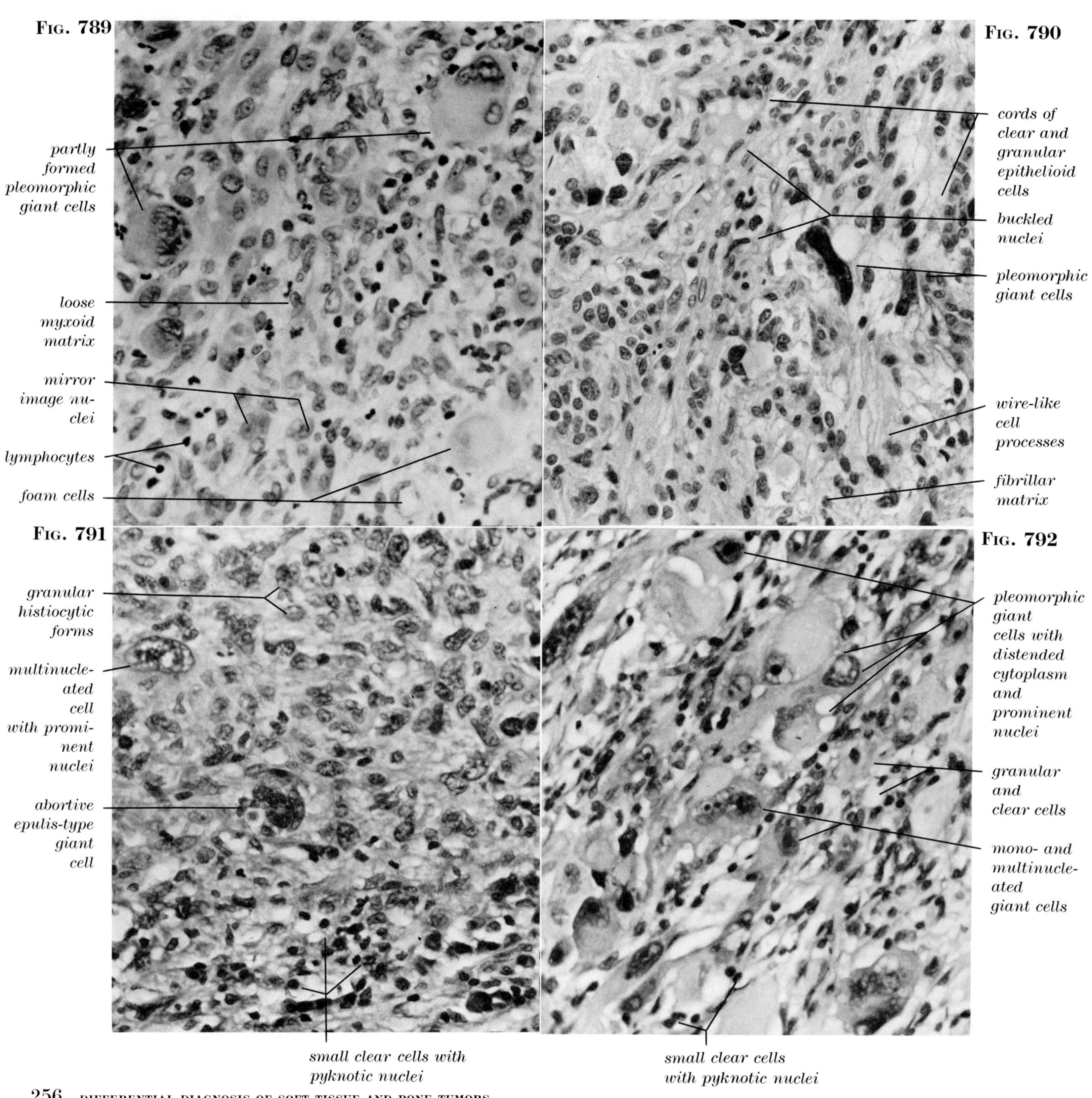

Fig. 793. Rhabdomyoma (×350) (see Figs. 497, 737).

Fig. 794. (Top center) Borderline or atypical paraganglioma of retroperitoneum (×350) (see Figs. 544, 714, 790, 853).

Fig. 795. High grade leiomyoblastoma (×350) (see Figs. 745, 747, 750, 752, 814).

Fig. 796. Rhabdomyoblastoma (×350) (see Figs. 695, 709, 715, 739, 936).

Fig. 797. Malignant melanoma (×350) (see Figs. 669, 719, 748, 826, 1188, 1189, 1190, 1219, 1262).

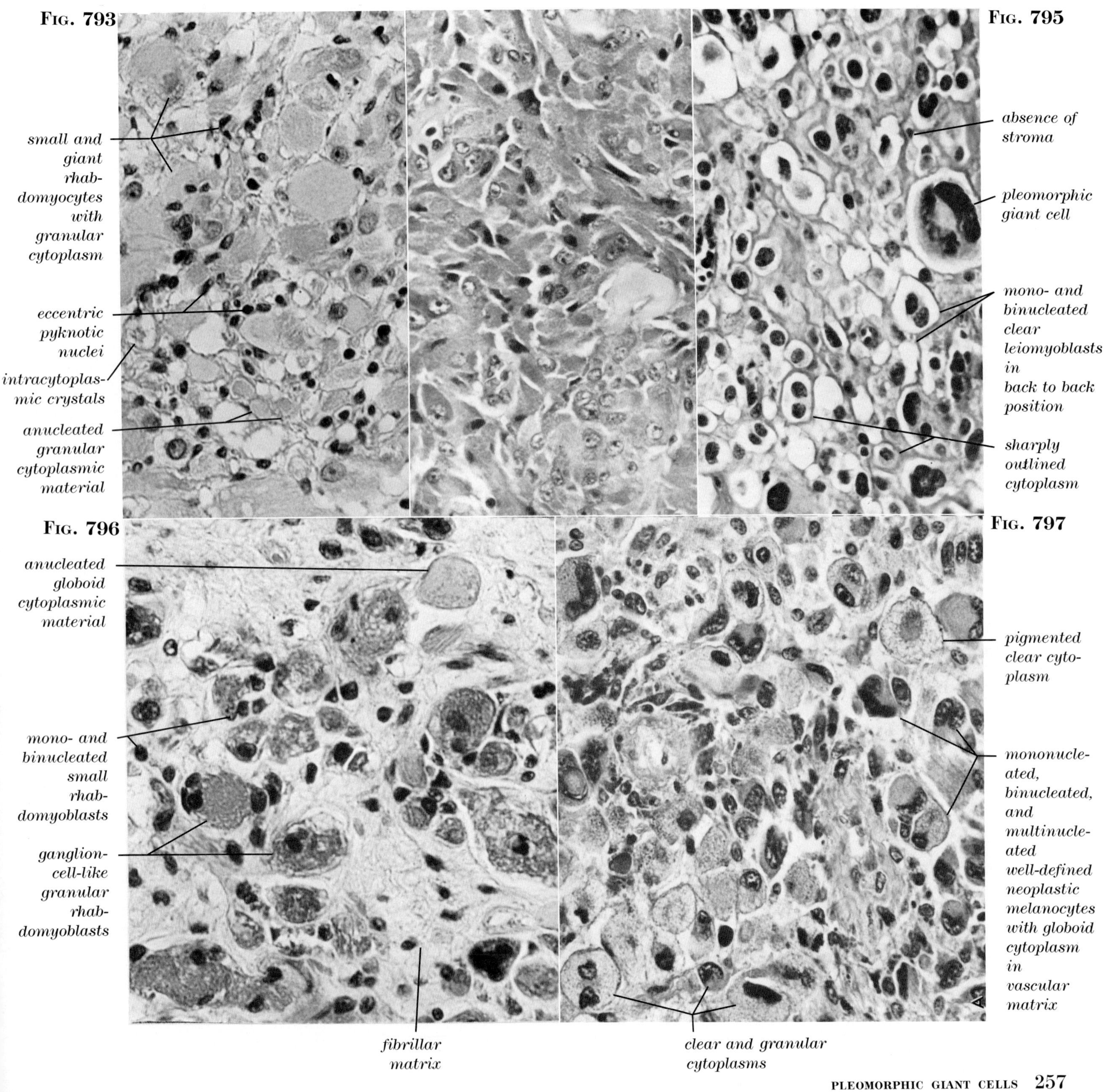

Fig. 798. High grade malignant pleomorphic fibrous histiocytoma (×350) (see Figs. 88, 111, 344, 460, 577, 579, 598, 679, 780, 788, 789, 792, 805, 818).

Fig. 799. Pleomorphic liposarcoma (×350) (see Figs. 86, 141, 284, 458, 590, 823, 928, 1019A, 1049).

Fig. 800. Pleomorphic rhabdomyosarcoma (×350) (see Figs. 87, 121, 292, 599, 604, 609, 612, 802, 804, 810, 813, 1130, 1166, 1167).

Fig. 801. Osteosarcoma (×350) (see Figs. 90, 108, 109, 221, 312, 602, 643, 677, 730, 754, 769, 777, 807, 817, 819, 831, 843, 876, 898, 958, 968, 1001).

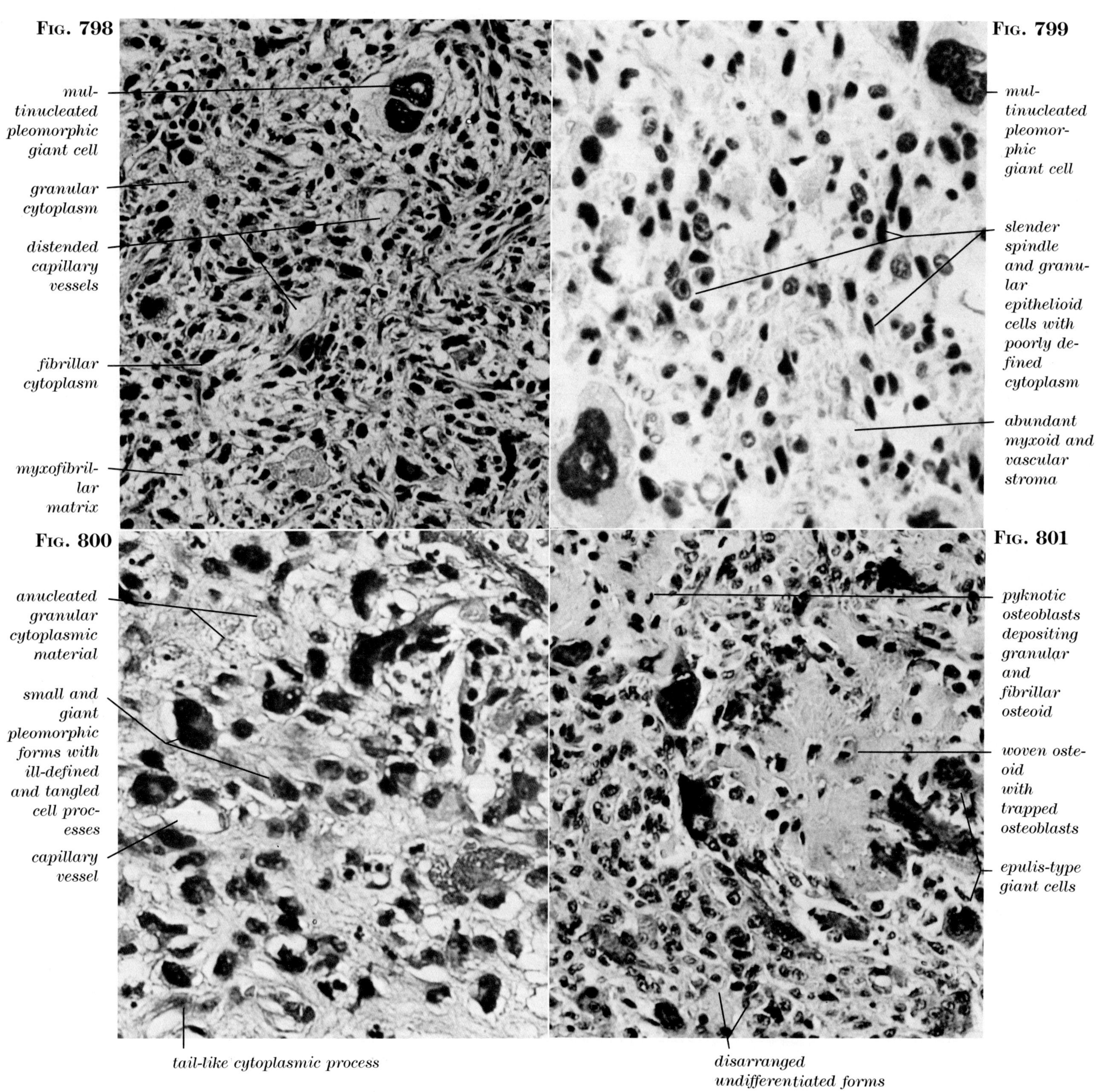

Fig. 802. Pleomorphic rhabdomyosarcoma (×350) (see Figs. 87, 121, 292, 295, 612, 800, 802, 804, 810, 813, 1130, 1166, 1167, 1216, 1250).

Fig. 803. Pleomorphic leiomyosarcoma (×350) (see Figs. 72, 110, 193, 273, 341, 588, 618, 659, 664, 666, 667, 670, 680, 681, 685, 746, 808, 841).

Fig. 804. Alveolar rhabdomyosarcoma metastatic to lung as pleomorphic rhabdomyosarcoma (×350) (see Figs. 78, 87, 121, 206, 524, 529, 530, 567, 800, 802, 810, 813, 1130, 1166, 1167, 1216, 1250, 1255, 1273).

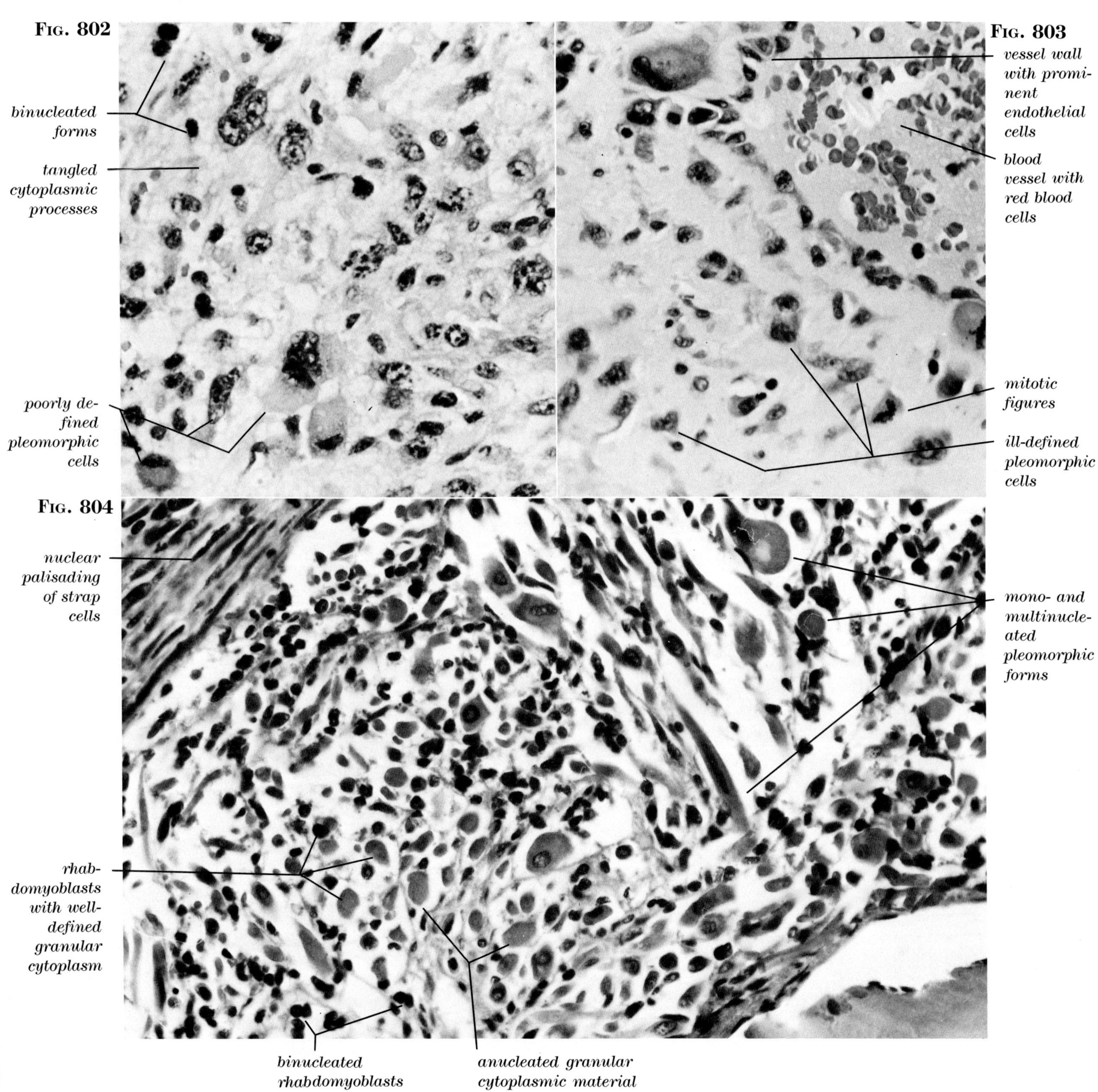

Fig. 805. Borderline pleomorphic fibrous histiocytoma (×350) (see Figs. 88, 111, 140, 288, 344, 460, 577, 579, 598, 679, 780, 788, 818).

Fig. 807. Osteosarcoma (×350) (see Figs. 90, 108, 221, 306, 602, 610, 611, 615, 643, 677, 730, 754, 769, 801, 817).

Fig. 809. High grade fibrosarcoma (×350) (see Figs. 91, 117, 297, 371, 601, 608, 625, 647, 660, 665, 671, 825, 832).

Fig. 806. Meningeal sarcoma (×350) (see Figs. 283, 495, 887).

Fig. 808. High grade leiomyosarcoma (×350) (see Figs. 618, 659, 664, 666, 667, 670, 680, 681, 685, 746, 803).

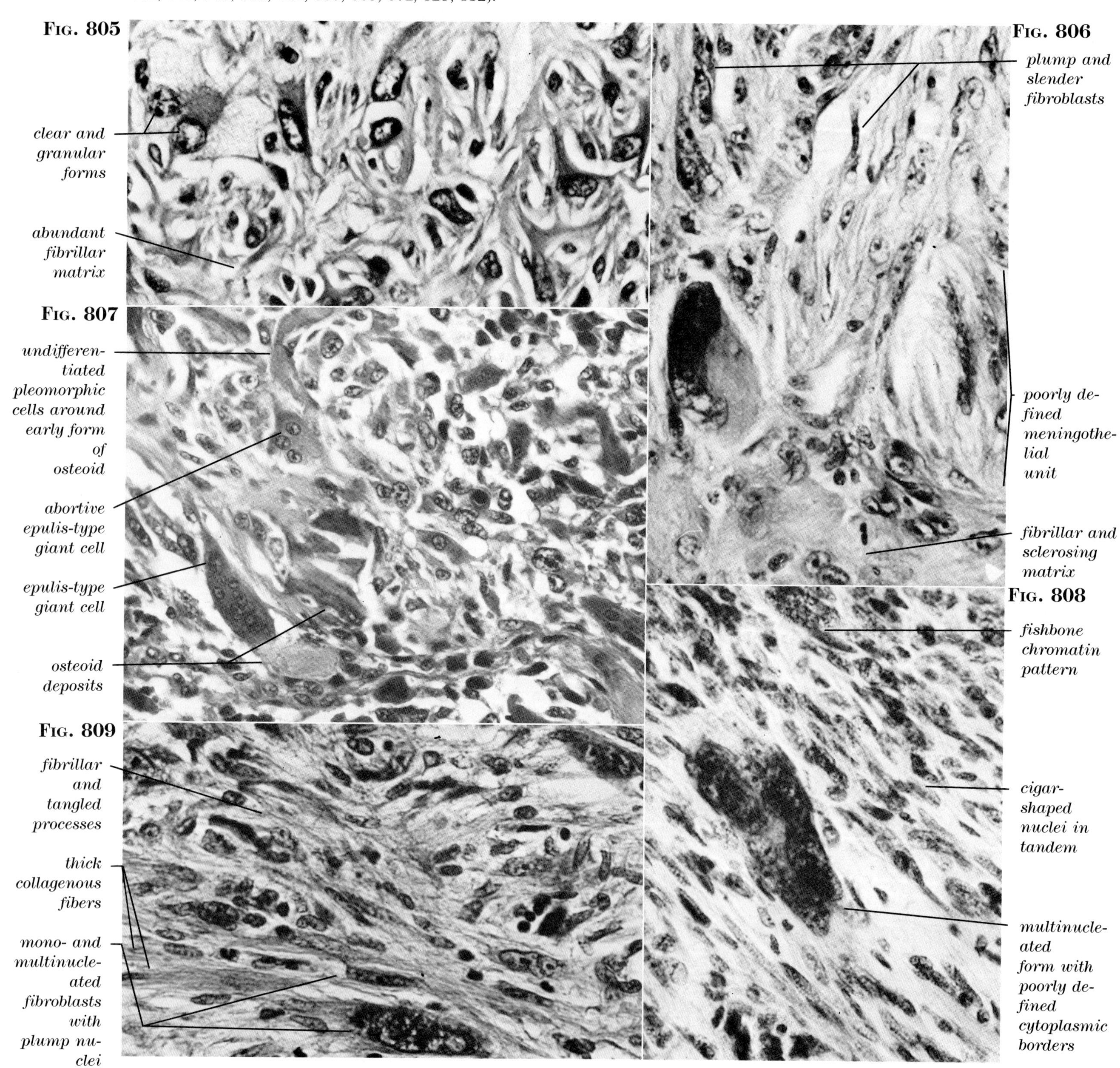

Fig. 810. Pleomorphic rhabdomyosarcoma (×350) (see Figs. 87, 121, 292, 313, 599, 604, 609, 612, 800, 813).

Fig. 811. Ganglioneuroma (×350) (see Figs. 166, 486, 630, 824, 857, 1137, 1191, 1193, 1258).

Fig. 812. Malignant Triton tumor (×350) (see Figs. 593, 606, 732, 813, 1259).

Fig. 813. Pleomorphic rhabdomyosarcoma in malignant Triton tumor (×350) (see Figs. 593, 612, 800, 802, 804, 810, 813, 1130, 1166, 1167, 1216).

Fig. 814. High grade leiomyoblastoma (×350) (see Figs. 691, 713, 733, 744, 745, 747, 750, 752, 795, 828).

Fig. 815. Malignant schwannoma (×350) (see Figs. 80, 200, 246, 603, 619, 628, 661, 836, 839).

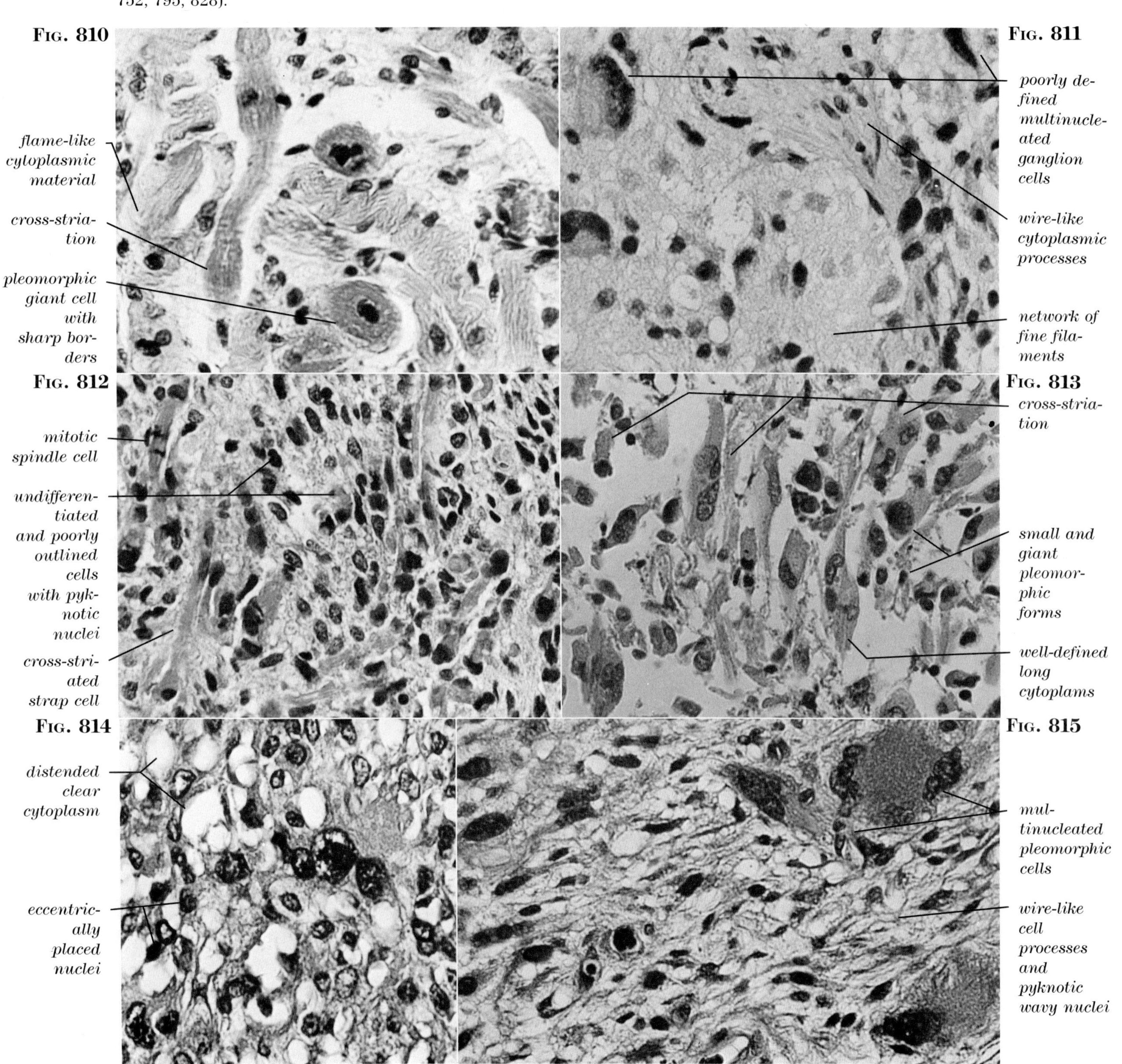

Fig. 816. Malignant histiocytic fibrous histiocytoma, low grade (×350) (see Figs. 775, 776, 784, 791, 872).

Fig. 817. Telangiectatic osteosarcoma (×350) (see Figs. 730, 754, 777, 807, 819).

Fig. 818. High grade malignant pleomorphic fibrous histiocytoma (×570) (see Figs. 88, 111, 140, 237, 344, 460, 598, 679, 789, 792, 798, 937).

Fig. 819. Osteosarcoma (×570) (see Figs. 90, 108, 109, 328, 395, 405, 411, 451, 461, 602, 610, 611, 615, 817, 831).

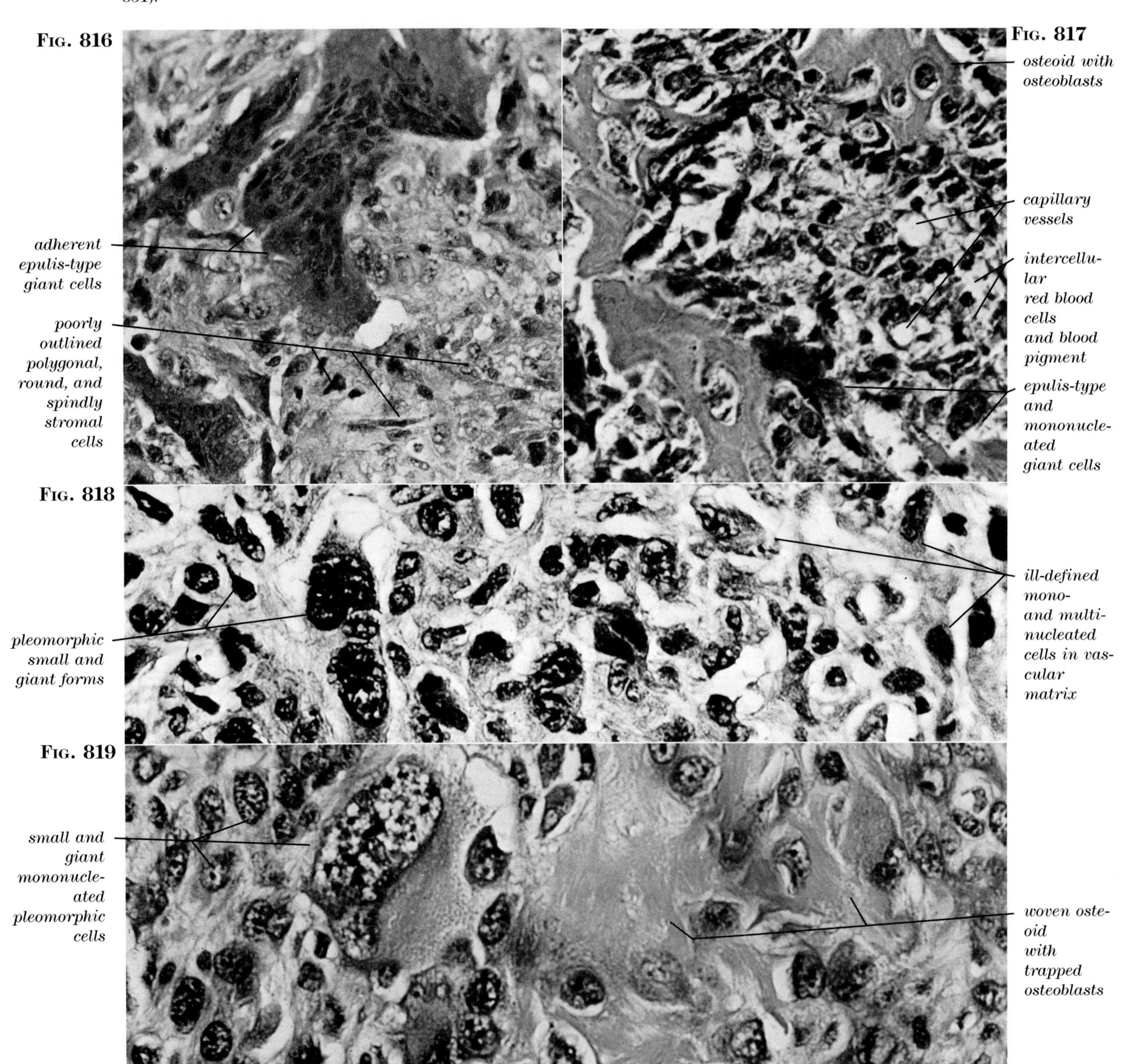

Fig. 820. Lipogranuloma (×350) (see Figs. 433, 907, 1043).

Fig. 821. Pleomorphic lipoma (×350) (see Figs. 73, 613, 781, 909, 957, 1182, 1183).

Fig. 822. Well-differentiated liposarcoma (×350) (see Figs. 86, 431, 435, 999, 1184, 1233).

Fig. 823. Pleomorphic liposarcoma (×350) (see Figs. 86, 141, 284, 294, 458, 590, 799, 928, 1019*A*, 1049).

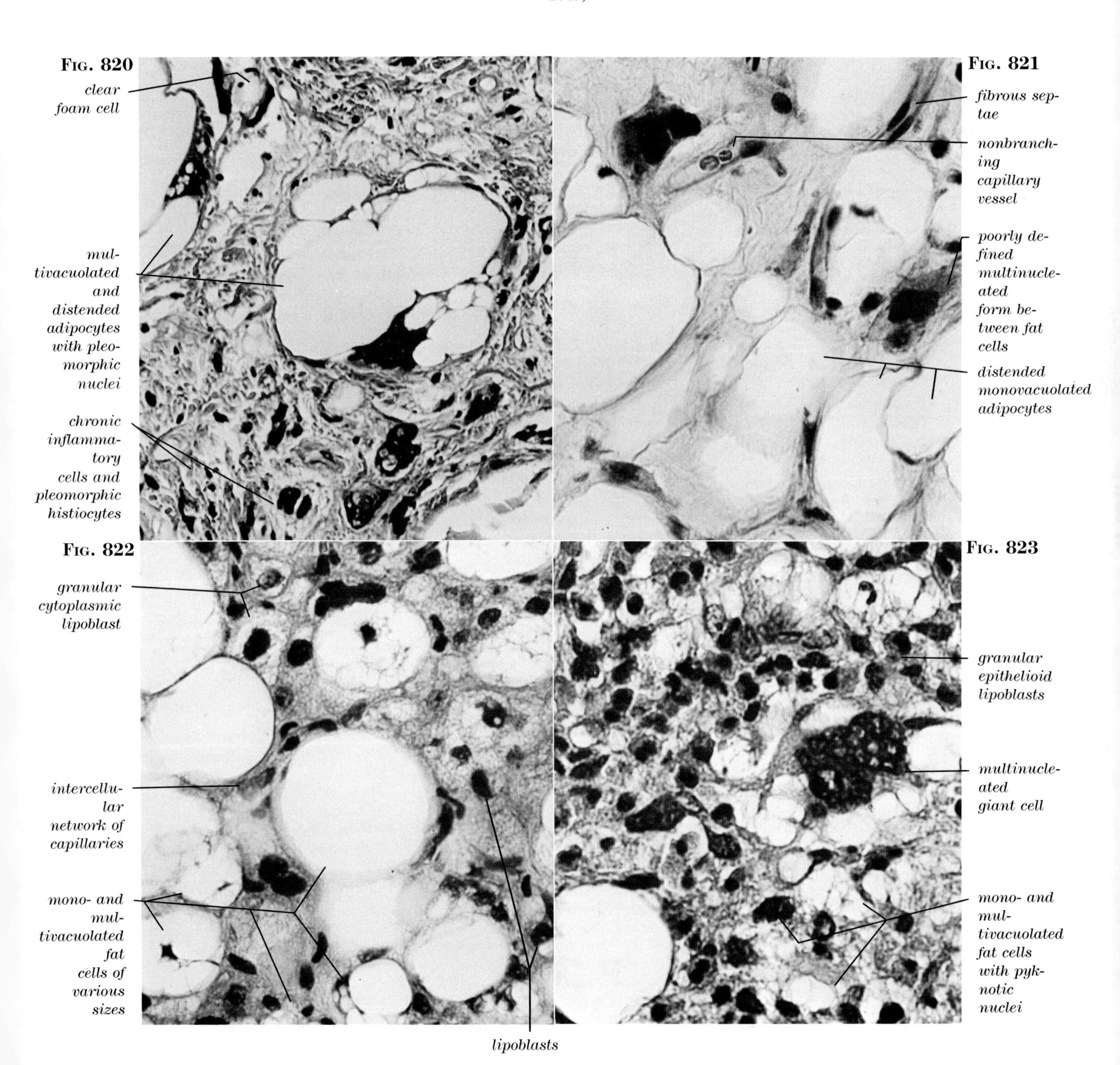

Fig. 824. Ganglioneuroma of gall bladder (×570) (see Figs. 630, 811, 857, 1137).

Fig. 825. (Top center) High grade fibrosarcoma (×570) (see Figs. 625, 647, 665, 671, 832).

Fig. 826. Metastatic malignant melanoma (×570) (see Figs. 394, 409, 540, 669, 719, 748, 797, 1188, 1190, 1219, 1262).

Fig. 827. Malignant undifferentiated peripheral nerve tumor (×570) (see Figs. 342, 519, 572, 596).

Fig. 828. Leiomyoblastoma (×140) (see Figs. 752, 795, 814, 868, 931).

Fig. 829. Metastatic choriocarcinoma of ovary (×350) (see Figs. 195, 1055).

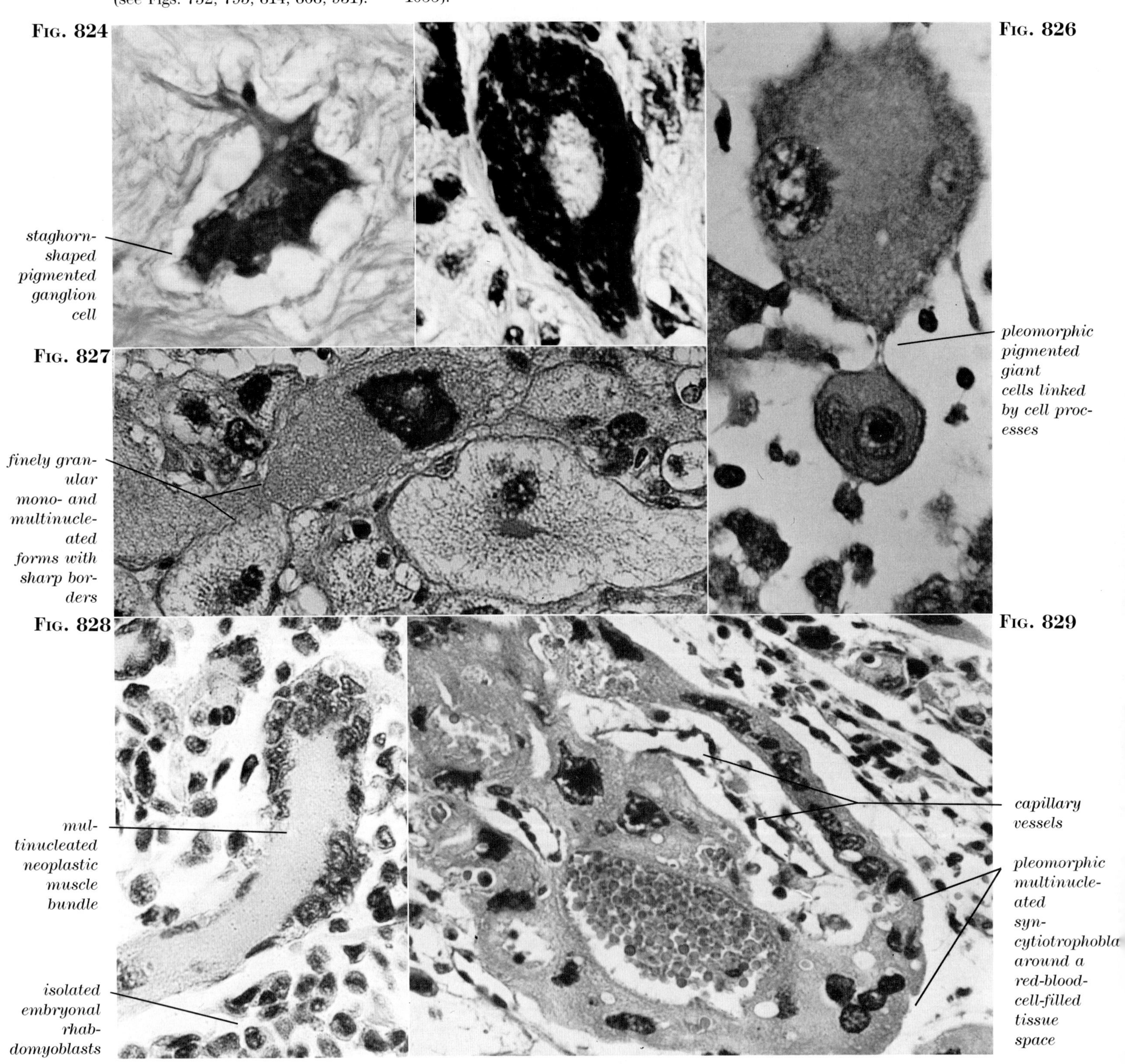

"Varieties of sarcoma are dependent upon the form and arrangement of the fibers"

CARL ROKITANSKY (1804–1878)

12. Appearance of Stroma

	NON-NEOPLASTIC LESIONS	BENIGN NEOPLASMS	MALIGNANT NEOPLASMS
1. **Fibrillar Stroma.** See Tables 54 to 56.			
See Figures:	834, 851, 855	840, 842, 846, 847, 850, 852, 857	830–833, 835–839, 841, 843–845, 848, 849, 853, 854, 856
2. **Sclerosed Stroma.** See Tables 54 to 56.			
See Figures:	858, 860, 861, 863, 867, 873, 874, 877, 878, 882, 907, 910, 911, 913, 914	859, 862, 871, 879, 880, 883, 884, 886, 892, 908, 909, 912, 915, 917	864–866, 868–870, 872, 875, 876, 881, 885, 887–891, 893–906, 916, 918
3. **Myxoid Stroma.** See Tables 54 to 56.			
See Figures:	924, 947, 951	921, 938–942, 950	919, 920, 922, 923, 925–937, 943–946, 948, 949, 952, 953
4. **Vascular Stroma.** See Tables 54 to 56.			
See Figures:	956, 971, 979, 981, 991, 1000, 1006, 1009, 1014–1016	955, 957, 959–962, 974, 976, 978, 980, 983, 985, 987, 988, 993, 996, 1003–1005, 1010, 1013	954, 958, 963–970, 972, 973, 975, 977, 982, 984, 986, 989, 990, 992, 994, 995, 997–999, 1001, 1002, 1007, 1008, 1011, 1012, 1017–1019

	NON-NEOPLASTIC LESIONS	BENIGN NEOPLASMS	MALIGNANT NEOPLASMS
5. **Inflamed Stroma.** See Tables 54 to 56. See Figures:	1020, 1022, 1023, 1025, 1028–1031, 1035, 1036, 1038–1040, 1042, 1043, 1045	1021, 1026, 1027, 1041, 1044, 1048	1024, 1032–1034, 1037, 1046, 1047
6. **Necrotic Stroma.** See Tables 54 to 56. See Figures:	1050, 1054, 1056–1058		1049, 1051–1053, 1055, 1059
7. **Chondrified Stroma.** See Tables 54 to 56. See Figures:	1060	1063	1061, 1062, 1064, 1065
8. **Ossified Stroma.** See Tables 54 to 57. See Figures:	1067, 1069, 1080, 1081, 1083–1085	1066, 1089, 1090, 1092	1068, 1070–1079, 1082, 1086–1088, 1091, 1093
9. **Calcified Stroma.** See Tables 54 to 56. See Figures:	1094, 1095	1096, 1104, 1109, 1111	1097–1103, 1105–1108, 1110, 1112

A significant portion of soft tissue and bone tumors is made up of extracellular matrix or so-called "biomatrix" (the stroma). It is thought that biomatrix is species specific and organ specific. Stroma is composed of a meshwork of collagen, elastin, proteoglycans, and glycoproteins. As they grow, tumors may change the matrix by overproduction or depletion of stromal components. Some workers proposed a classification scheme for soft tissue tumors by studying the stromal component analogous to that found in certain bone tumors, for example, osteo- and chondrosarcoma. Although such attempts are commendable, collagen is an ubiquitous product of connective tissue cells and even many epithelial cells.

It is hard to accept that collagen, laminin, and fibronectin could serve as useful markers for the precise diagnosis of soft tissue and bone tumors. Despite the limited role of stromal or interstitial collagen in typing of connective tissue tumors, it is well established that the matrix of poorly differentiated tumors, for example, fibrous histiocytomas, contain more collagen type III (demonstrable with reticulin stain) than well-differentiated tumors. On the other hand, well-differentiated tumors, for example, fibrosarcomas, contain more collagen type I (demonstrable with trichrome stain) than poorly differentiated tumors. It is possible that newly refined cyto- and immunochemical techniques will contribute to a better understanding of the structural and functional role of tumor matrices.

The most common components of the matrix can be demonstrated by various histologic techniques, ranging from simple special stains to monoclonal antibodies, but in most instances good quality hematoxylin and eosin stain is completely adequate and satisfactory for the demonstration of stromal components.

Fibrillar stroma can be demonstrated by reticulin stain that stains the finest polymerized collagen type III fibers.

Sclerosed stroma stains strongly with trichrome stain and is a useful marker for the highly polymerized collagen type I. The microarchitecture of mature collagen type I can also be demonstrated by the use of polarized light.

Myxoid stroma is the result of overproduction of mucopolysaccharide substances by the tumor cells and the stromal cells. Good quality hematoxylin and eosin stain is sufficient for the demonstration of myxoid stroma, but in case of doubt, a simple alcian blue stain may enhance the bluish color of the connective tissue mucin which, partly depolymerized by hyaluronidase, results in a negative or pale blue stain.

The *vascularity of the stroma* may vary from tumor to tumor and has a limited role in classification of tumors but is important in the grading of sarcomas (see Chapter 14).

Inflammatory and chronic inflammatory elements are not uncommon findings in soft tissue and bone tumors for example, fibrous histiocytomas and muscle tumors, but are observed most commonly in reactive, non-neoplastic lesions.

Necrosis in tumors is an alarming sign and in sarcomas almost invariably a sign of high grade malignancy.

The deposit of *chondroid* or *bony matrix* may be nothing other than the metaplastic transformation of stromal elements or the body's attempt to heal itself, and is best expressed in the formation of callus. The formation of a callus begins with (1) the appearance of ominous-looking preosteoblasts and prechondroblasts, which gives the callus a resemblance to fasciitis; (2) the beginning of ossification is signaled by the appearance of primitive osteoid or cartilage resembling that found in osteosarcoma and in chondrosarcoma; and (3) the deposition of calcium within the osteoid is tied to the transformation of woven bone to lamellar bone. The osteoid, or woven, immature bone is a specialized form of collagen produced by osteoblasts, and the collagen is deposited in a tangled or criss-cross pattern. Lamellar, mature bone shows an orderly, uninterrupted, almost parallel positioning of seemingly acellular mature collagen (Table 57). The deposit of calcium is a prerequisite for ossification. However, calcium may be deposited at sites of chronic irritation, vascular occlusion, and tissue necrosis.

Stromal calcification can seldom be used as a diagnostic criterion. The only exception is perhaps the lacy or trabecular calcification commonly seen in chondroblastoma.

Table 54. *Appearance of Stroma of Non-neoplastic Lesions**

	FIBRILLAR	SCLEROSED	MYXOID	VASCULAR	INFLAMED	NECROTIC	CHONDRIFIED	OSSIFIED	CALCIFIED
● Aneurysmal bone cyst[†]	X		X	X				X	
● Giant cell granuloma			X	X		X			X
● Hyperparathyroidism	X			X					
○ Xanthogranuloma	X			X	X				
○ Fasciitis	X		X	X	X			X[‡]	X[§]
○ Keloid		X							
○ Fibromatosis	X	X					X[¶]	X[‡]	X[§]
● Fibrous dysplasia	X	X						X	X
○ Elastofibroma	X[#]		X			X			
○ Collagenoma		X							X
○ Tendosynovitis[†]	X		X	X	X				
◑ Tendosynovial cyst[†]	X		X		X				X
○ Tendosynovial chondromatosis		X	X				X		X[§]
○ Gouty arthritis	X			X	X	X			X
◑ Fat necrosis	X		X		X	X			X[§]
◑ Lipogranuloma			X		X	X			X[§]
○ Proliferative panniculitis	X		X		X		X[¶]	X[‡]	
○ Proliferative myositis	X		X						
○ Myositis ossificans	X				X	X		X	X
○ Pyogenic granuloma[†]	X			X	X				
○ Angifollicular lymphoid hyperplasia	X			X	X**				
◑ Arteriovenous malformation	X	X		X					X[§]
○ Traumatic neuroma	X								
● Callus	X	X					X	X	X[§]
◑ Eosinophilic granuloma				X	X**	X			

*○ Primary soft tissue tumor. ● Primary bone tumor. ◑ Can be primary soft tissue or bone tumor.

[†]Can be pigmented.

[‡]Can be ossified.

[§]Can be calcified.

[¶]Can be chondrified.

[#]Elastic fibers.

**Eosinophiles.

Table 55. *Appearance of Stroma of Benign Neoplasms*[*]

	FIBRILLAR	SCLEROSED	MYXOID	VASCULAR	INFLAMED	NECROTIC	CHONDRIFIED	OSSIFIED	CALCIFIED
◑ Benign fibroblastic fibrous histiocytoma[†]	X			X					
◑ Benign histiocytic fibrous histiocytoma[†]	X			X	X[‡]				
◑ Benign pleomorphic fibrous histiocytoma	X		X	X	X[‡]				
● Nonossifying fibroma	X		X		X				
○ Fibroma		X	X[§]						X
● Desmoplastic fibroma		X							
◑ Well-differentiated lipoma			X						
◑ Myxoid lipoma			X						
○ Fibroblastic lipoma	X	X	X						
○ Lipoblastoma			X	X					
○ Pleomorphic lipoma	X	X	X						
○ Angiomyolipoma	X	X		X[¶]					
○ Hibernoma			X	X					
◑ Leiomyoma	X	X							
○ Rhabdomyoma			X						X[#]
◑ Hemangioma[†]	X			X					
◑ Benign glomus tumor	X			X					
◑ Neurofibroma	X	X	X[§]	X[¶]					
○ Plexiform neurofibroma	X								
◑ Benign schwannoma[†]	X		X						
○ Benign nevoid schwannoma[†]	X		X						
◑ Benign undifferentiated peripheral nerve tumor	X		X	X					
◑ Ganglioneuroma	X	X	X	X					X[#]
○ Benign paraganglioma	X		X	X					X[#]
○ Benign fibrous mesothelioma	X	X			X				X[#]
● Chondroblastoma			X	X		X	X		X
● Chondromyxoid fibroma	X		X	X			X		
◑ Chondroma			X				X		
● Osteochondroma							X	X	X
● Osteoblastoma	X			X				X	
● Osteoid osteoma	X	X						X	
○ Benign granular cell tumor	X	X							X
○ Benign cystosarcoma phyllodes	X		X						

[*]○ Primary soft tissue tumor. ● Primary bone tumor. ◑ Can be primary soft tissue or bone tumor.
[†]Can be pigmented.
[‡]Foam cells.
[§]Can be myxoid.
[¶]Hamartomatous vessels.
[#]Can be calcified.

Table 56. *Appearance of Stroma of Malignant Neoplasms**

	FIBRILLAR	SCLEROSED	MYXOID	VASCULAR	INFLAMED	NECROTIC	CHONDRIFIED	OSSIFIED	CALCIFIED
◑ Malignant fibroblastic fibrous histiocytoma	X		X						
◑ Malignant histiocytic fibrous histiocytoma	X			X	X				
◑ Malignant pleomorphic fibrous histiocytoma	X		X	X	X	X			
◑ Desmoid tumor		X							X†
◑ Fibroblastic fibrosarcoma	X	X							
◑ Pleomorphic fibrosarcoma		X				X			
○ Biphasic tendosynovial sarcoma	X		X	X					X
○ Monophasic tendosynovial sarcoma, spindle cell type	X	X	X	X					X†
○ Monophasic tendosynovial sarcoma, pseudoglandular type	X			X					X†
○ Epithelioid sarcoma	X	X				X			
○ Clear cell sarcoma	X		X	X					
○ Chordoid sarcoma			X	X		X	X		
◑ Well-differentiated liposarcoma			X	X					
◑ Myxoid liposarcoma			X	X			X#		
○ Lipoblastic liposarcoma			X	X					
○ Fibroblastic liposarcoma	X			X					
○ Pleomorphic liposarcoma	X		X	X		X			
◑ Leiomyoblastoma			X	X		X			
◑ Leiomyosarcoma	X	X		X		X			
○ Embryonal rhabdomyosarcoma			X	X		X			
○ Alveolar rhabdomyosarcoma	X		X	X					
○ Myxoid rhabdomyosarcoma	X		X	X					
○ Rhabdomyoblastoma				X					
○ Pleomorphic rhabdomyosarcoma	X			X		X			
◑ Hemangiopericytoma	X			X					
◑ Hemangiosarcoma				X		X			
○ Lymphangiosarcoma	X			X		X			X†
◑ Neurofibrosarcoma	X	X				X			
◑ Malignant schwannoma‡	X		X	X		X			
◑ Malignant undifferentiated peripheral nerve tumor	X		X			X			
○ Malignant epithelioid mesothelioma	X			X	X	X			X
○ Malignant fibrous mesothelioma	X	X							X†
◑ Chondrosarcoma			X				X	X	X
◑ Mesenchymal chondrosarcoma	X		X				X		
◑ Osteosarcoma		X				X	X	X	
◑ Parosteal osteosarcoma	X	X	X					X	
● Paget's sarcoma	X	X		X	X	X		X	X
◑ Granulocytic sarcoma			X						
● Plasma cell myeloma			X	X	X	X			
○ Malignant granular cell tumor	X					X			
○ Alveolar soft part sarcoma	X		X			X			X†
○ Kaposi's sarcoma	X		X	X					
◑ Ewing's sarcoma			X	X		X			
● Chordoma		X	X				X		X
○ Myxopapillary ependymoma			X	X		X			X

*○ Primary soft tissue tumor. ● Primary bone tumor. ◑ Can be primary soft tissue or bone tumor.

†Can be calcified.

‡Can be pigmented.

#Can be chondrified.

Table 57A. *Comparison of Type of Ossified Matrix*

	WOVEN OR IMMATURE BONE	LAMELLAR OR MATURE BONE
Fetal bone	+	−
Adult bone	−	+
Early callus	+	−
Late callus	−	+
Early myositis ossificans	+	−
Late myositis ossificans	−	+
Early osteoid metaplasia	+	−
Late osteoid metaplasia	−	+
Osteogenesis imperfecta	+	−
Fibrous dysplasia	+	−
Paget's disease	+	−
Osteoid osteoma	+	−
Osteoblastoma	+	−
Osteoma	−	+
Osteosarcoma	+	−
Parosteal osteosarcoma	−+	+

Table 57B. *Soft Tissue and Bone Tumors with Chondrified or Ossified Matrix*

	CHONDRIFIED	OSSIFIED
Non-Neoplastic Lesions		
Fasciitis		+**
Fibromatosis	+*	+**
Fibrous dysplasia		+
Tendosynovial chondromatosis	+	
Proliferative panniculitis	+*	+**
Myositis ossificans		+
Callus	+	+
Benign Neoplasms		
Ossifying fibroma		+
Chondroblastoma	+	
Chondromyxoid fibroma	+	
Chondroma	+	
Osteochondroma	+	+
Osteoblastoma		+
Osteoid osteoma		+
Maligant Neoplasms		
Chordoid sarcoma	+*	
Myxoid liposarcoma	+*	
Chondrosarcoma	+	+**
Mesenchymal chondrosarcoma	+	
Osteosarcoma	+*	+
Parosteal osteosarcoma		+
*Paget's sarcoma	+*	+**
Chordoma	+*	

*Can be chondrified.
**Can be ossified.

Fig. 830. Epithelioid sarcoma with areas of clear cell sarcoma (×140) (see Figs. 85, 181, 258, 462, 676).

Fig. 831. (Top center) Osteosarcoma. Undifferentiated mesenchymal cells in fibrovascular stroma. (×140) (see Figs. 108, 395, 643, 843).

Fig. 832. High grade fibroblastic fibrosarcoma (×140) (see Figs. 91, 131, 260, 410, 625, 647, 900).

Fig. 833. Malignant fibroblastic fibrous histiocytoma (×140) (see Figs. 76, 339, 340, 645, 672, 1026).

Fig. 834. (Bottom center) Fasciitis. Note feathery spreading (×140) (see Figs. 82, 345, 368, 403, 621, 851).

Fig. 835. Wilms' tumor (×140) (see Figs. 45, 383).

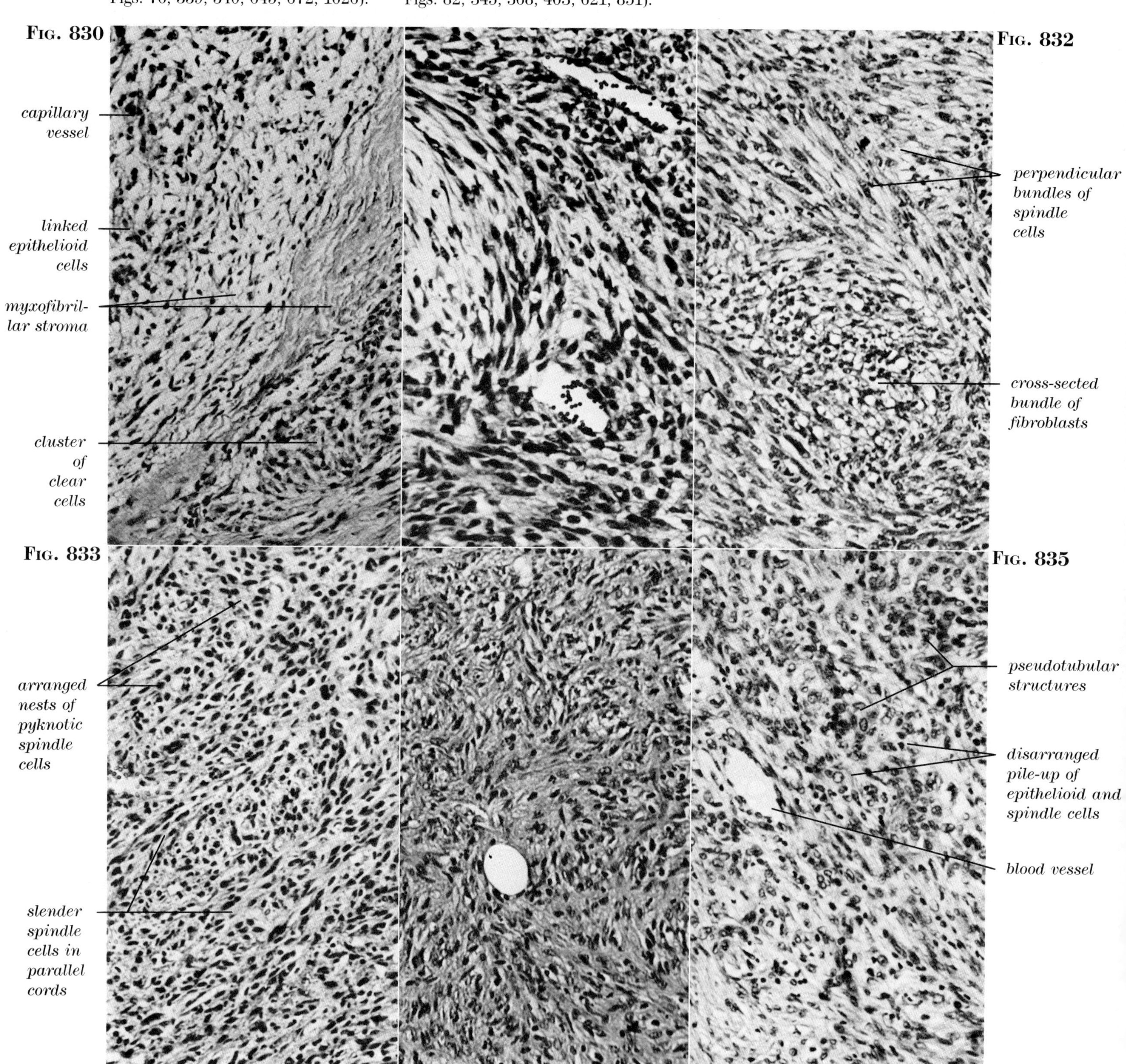

Fig. 836. Low grade malignant schwannoma (×140) (see Figs. 21, 38, 51, 80, 355, 357, 619, 839, 845).

Fig. 837. Fibroblastic liposarcoma (×140) (see Figs. 86, 343, 440, 641, 668, 848, 926, 998).

Fig. 838. High grade neurofibrosarcoma (×140) (see Figs. 79, 369, 370, 401, 584, 624, 626, 1116, 1229, 1263, 1266).

Fig. 839. Low grade malignant schwannoma (×140) (see Figs. 580, 583, 591, 603, 619, 628, 661, 836, 845, 1207, 1208).

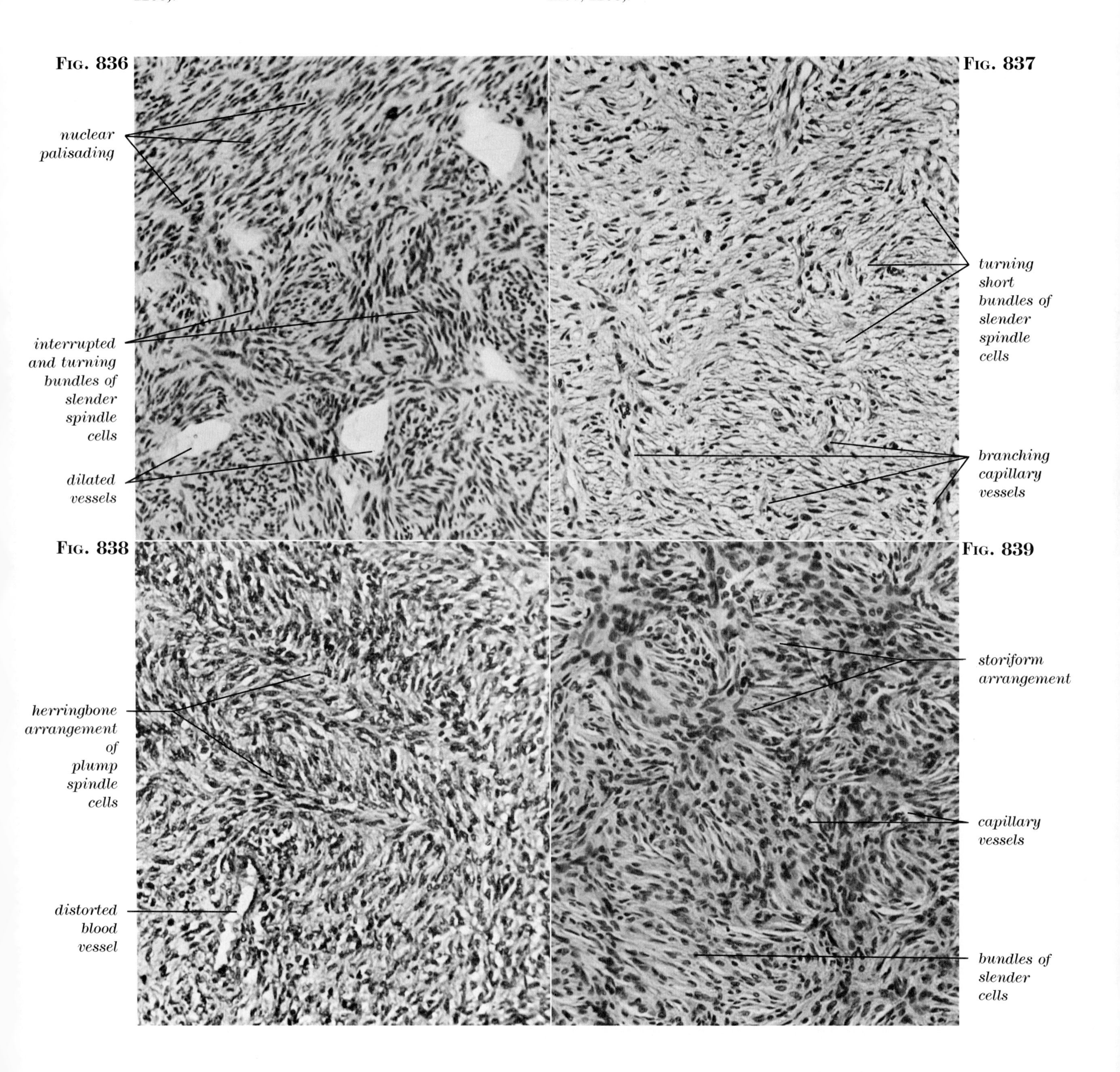

Fig. 840. Neurofibroma (×140) (see Figs. 11, 74, 245, 363, 652, 871).

Fig. 841. (Top center) Low grade leiomyosarcoma. Abundant intercellular stoma and mast cell (*arrow*) (×350) (see Figs. 72, 341, 618, 844, 849, 856).

Fig. 842. Plexiform neurofibroma (×140) (see Figs. 365, 527, 627, 1114).

Fig. 843. Osteosarcoma (×140) (see Figs. 405, 677, 831, 876, 897, 898, 958, 968, 1001, 1068, 1070–1079).

Fig. 844. (Bottom center) High grade leiomyosarcoma. Mitotic figures (*arrows*) (see Figs. 347, 659, 841, 849, 856, 875).

Fig. 845. High grade malignant schwannoma (×140) (see Figs. 619, 628, 661, 836, 839, 1207, 1208).

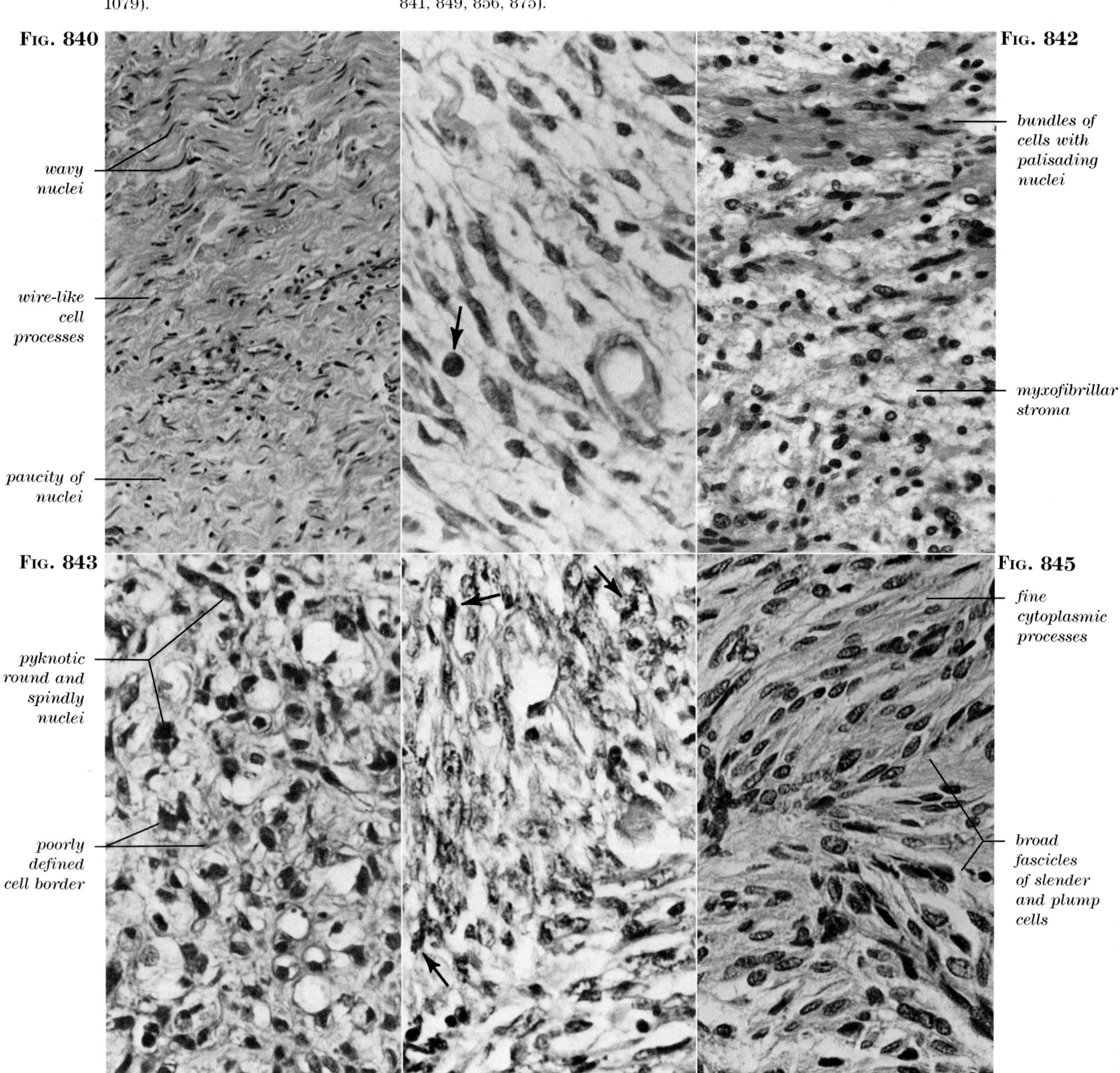

Fig. 846. Benign cystic schwannoma (×140) (see Figs. 257, 356, 850, 1228).

Fig. 847. Fibroblastic lipoma (×140) (see Figs. 73, 908, 940).

Fig. 848. Fibroblastic liposarcoma (×350) (see Figs. 30, 86, 343, 440, 641, 668, 837, 926, 998).

Fig. 849. Low grade leiomyosarcoma (×350) (see Figs. 72, 231, 341, 366, 404, 588, 803, 841, 844, 856, 875).

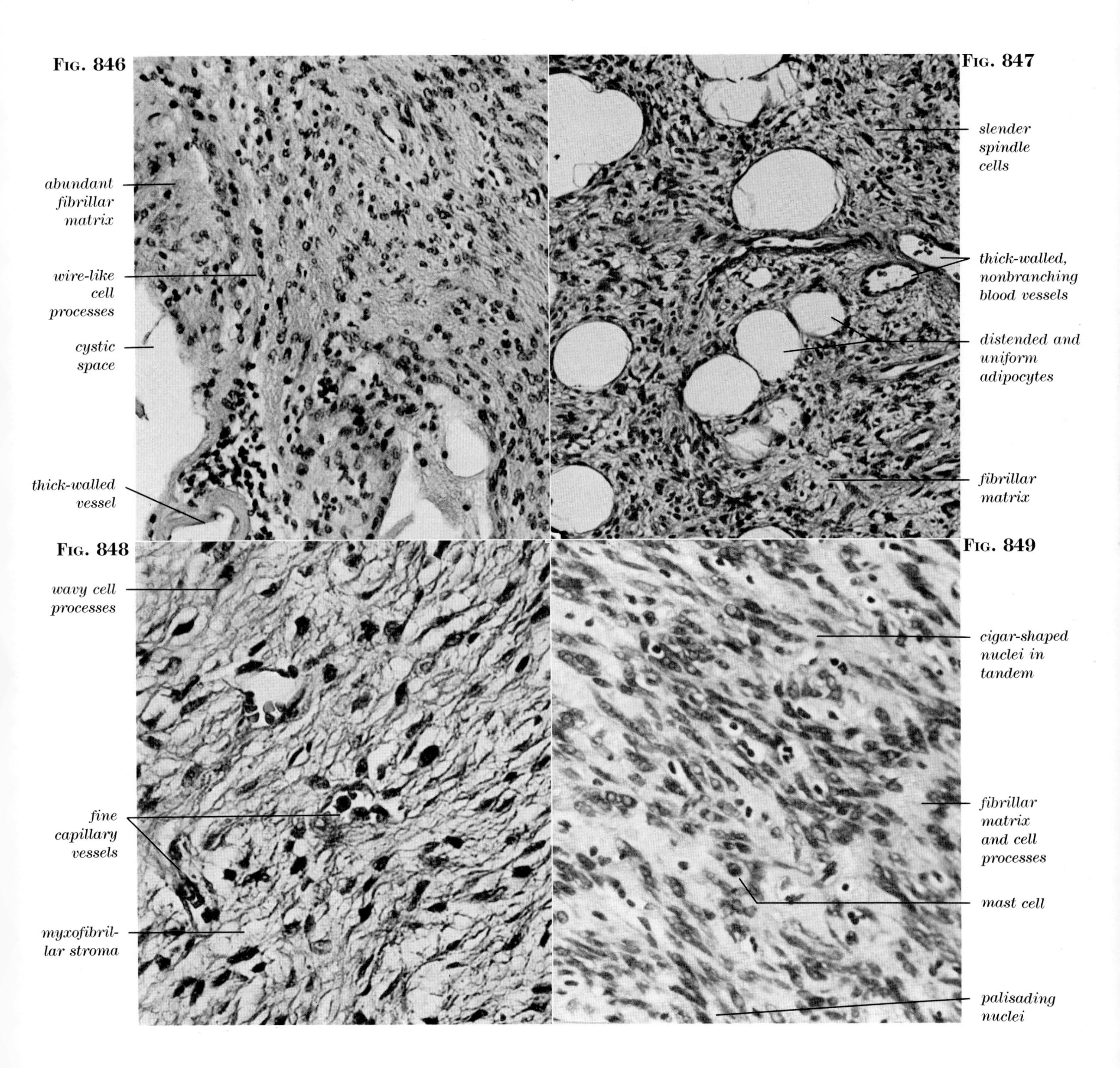

Fig. 850. Benign schwannoma (×350) (see Figs. 257, 356, 846).

Fig. 851. Fasciitis (×350) (see Figs. 82, 345, 368, 389, 403, 408, 578, 605, 621, 834, 855, 858, 947, 956, 971, 991, 1020, 1030, 1045).

Fig. 852. Nonossifying fibroma (×350) (see Figs. 112, 177, 631).

Fig. 853. Malignant paraganglioma (×350) (see Figs. 483, 544, 714, 790, 794, 1196).

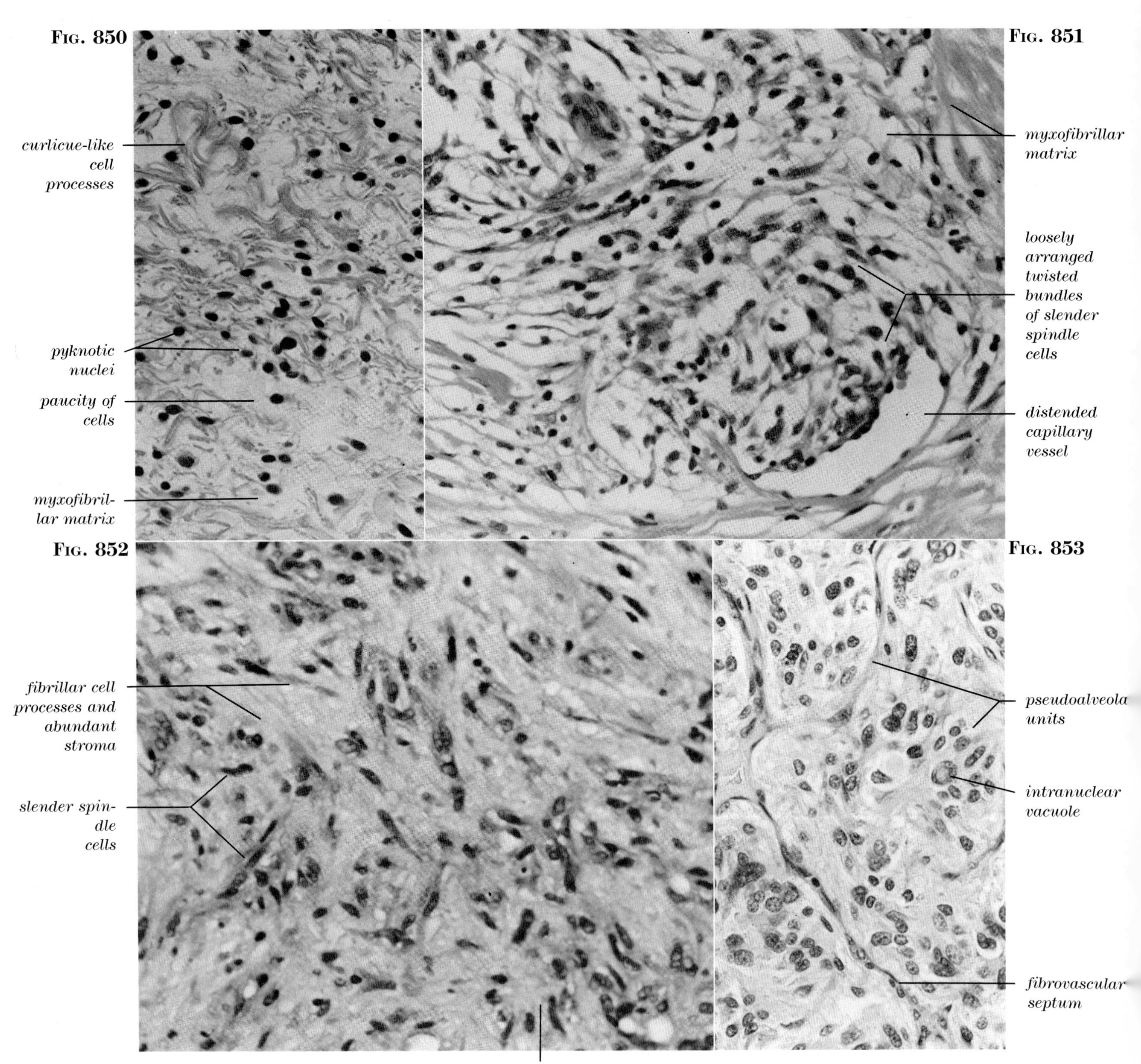

Fig. 854. Monophasic tendosynovial sarcoma, spindle cell type (×350) (see Figs. 85, 315, 391, 622, 642, 864).

Fig. 855. Fasciitis (×350) (see Figs. 578, 605, 621, 834, 851, 858, 947, 956, 971, 991).

Fig. 856. Low grade leiomyosarcoma (×350) (see Figs. 360, 664, 841, 844, 849, 875).

Fig. 857. Ganglioneuroma (×350) (see Figs. 166, 486, 630, 811, 824, 1137, 1191, 1193, 1258).

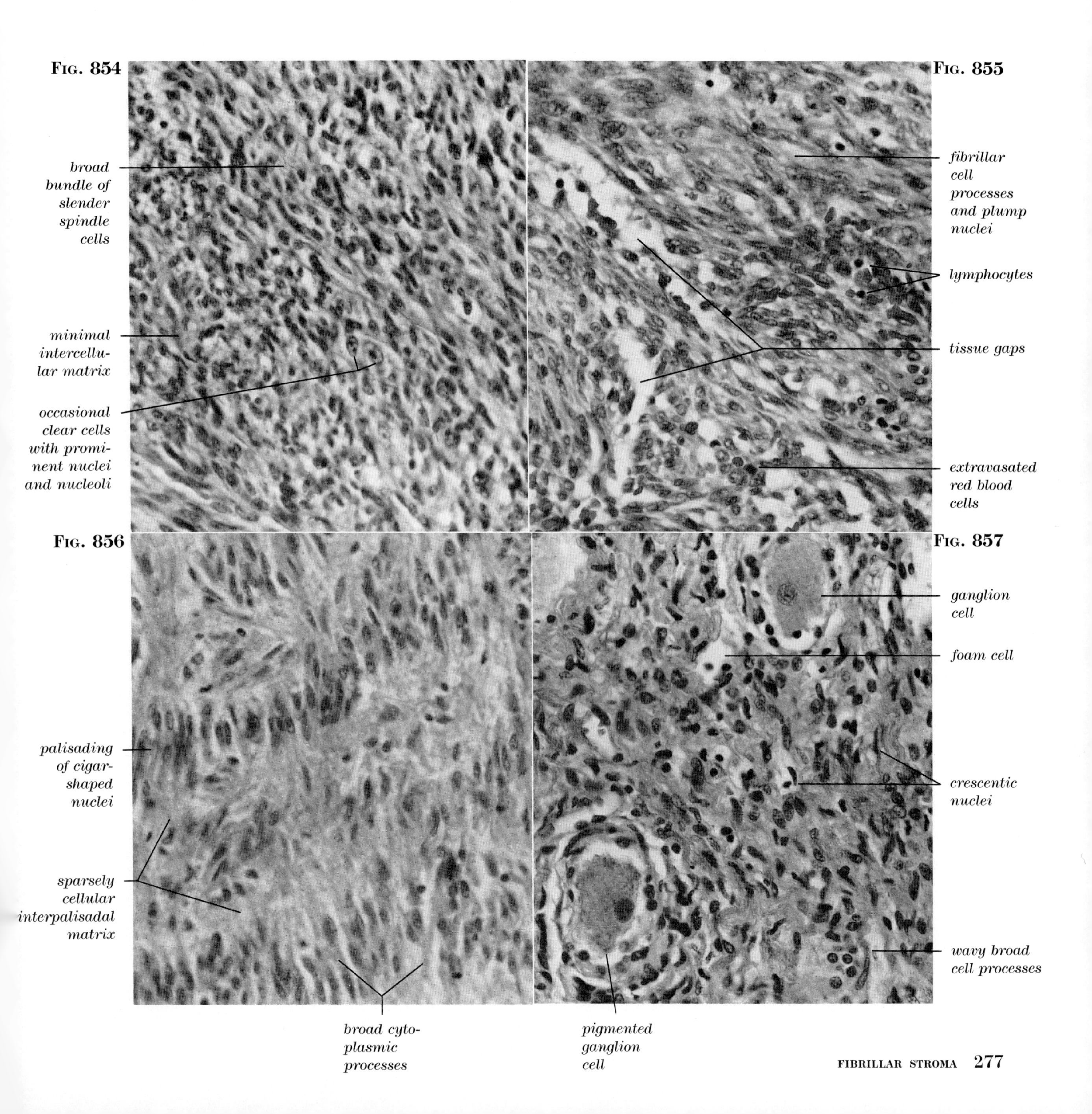

Fig. 858. Fasciitis (×140) (see Figs. 605, 621, 834, 851, 855, 947).

Fig. 859. (Top center) Fibroma, so-called "dermatofibroma," sclerosed cartwheel formation (*arrows*) (see Figs. 336, 884, 886, 904, 915, 1204).

Fig. 860. Palmar fibromatosis (×140) (see Figs. 17, 280, 367, 385, 387, 861).

Fig. 861. Mesenteric fibromatosis (×140) (see Figs. 419, 581, 582, 616, 860, 873, 874, 878, 882, 1022, 1025, 1035, 1036).

Fig. 862. Angiofibroma (×140) (see Figs. 653, 912, 985).

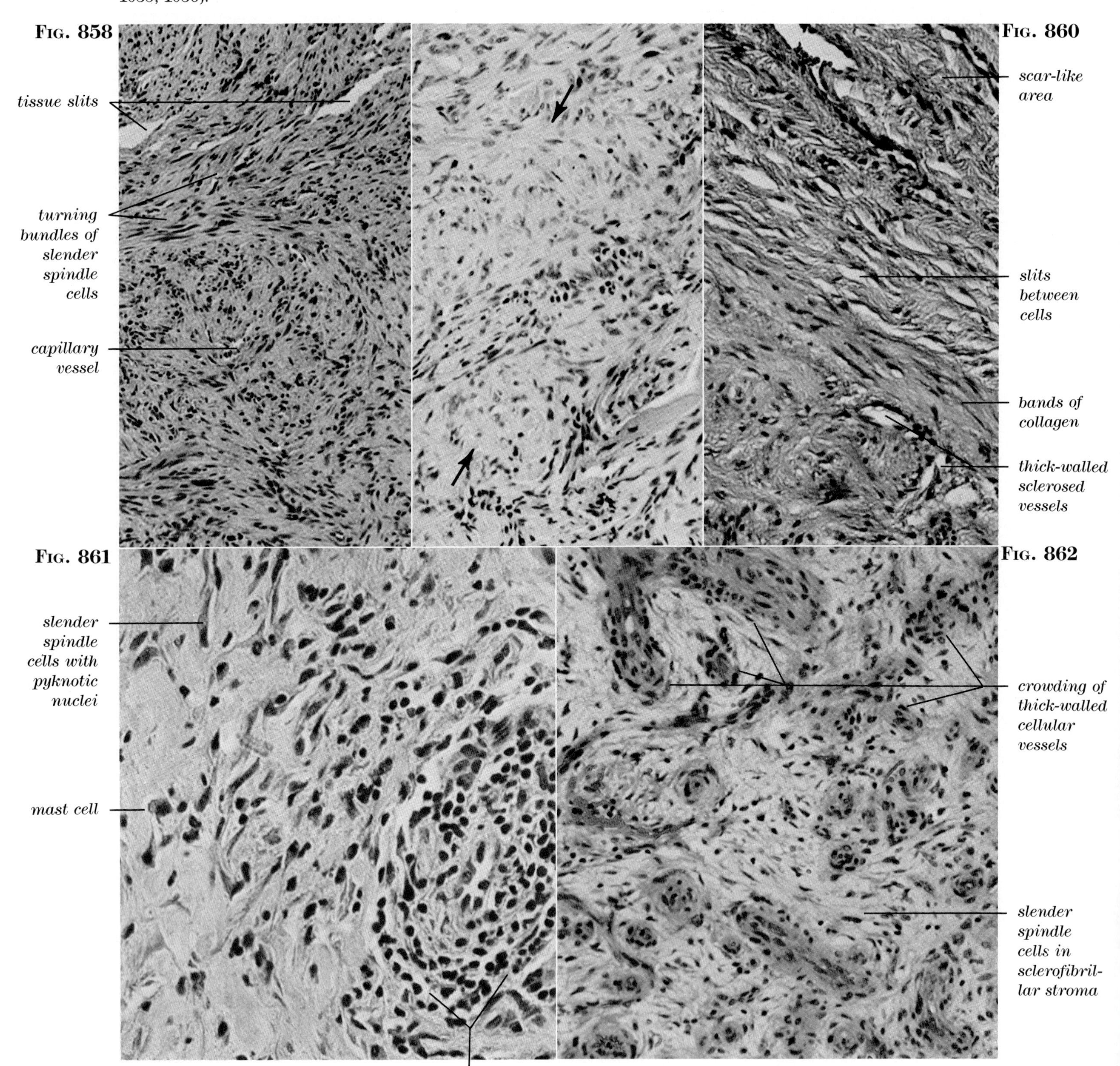

Fig. 863. Proliferative myositis (×140) (see Figs. 597, 738, 951).

Fig. 865. Desmoid tumor, low grade fibrosarcoma (×140) (see Figs. 77, 185, 244, 376, 662, 686, 899, 1118).

Fig. 867. Tendosynovitis (×140) (see Figs. 70, 786, 1023, 1029, 1136).

Fig. 864. Monophasic tendosynovial sarcoma, spindle cell type (×140) (see Figs. 392, 646, 648, 854, 870, 894, 902, 903, 1144, 1145).

Fig. 866. Epithelioid sarcoma (×140) (see Figs. 62, 85, 181, 258, 462, 464, 468, 676, 830, 895, 1052).

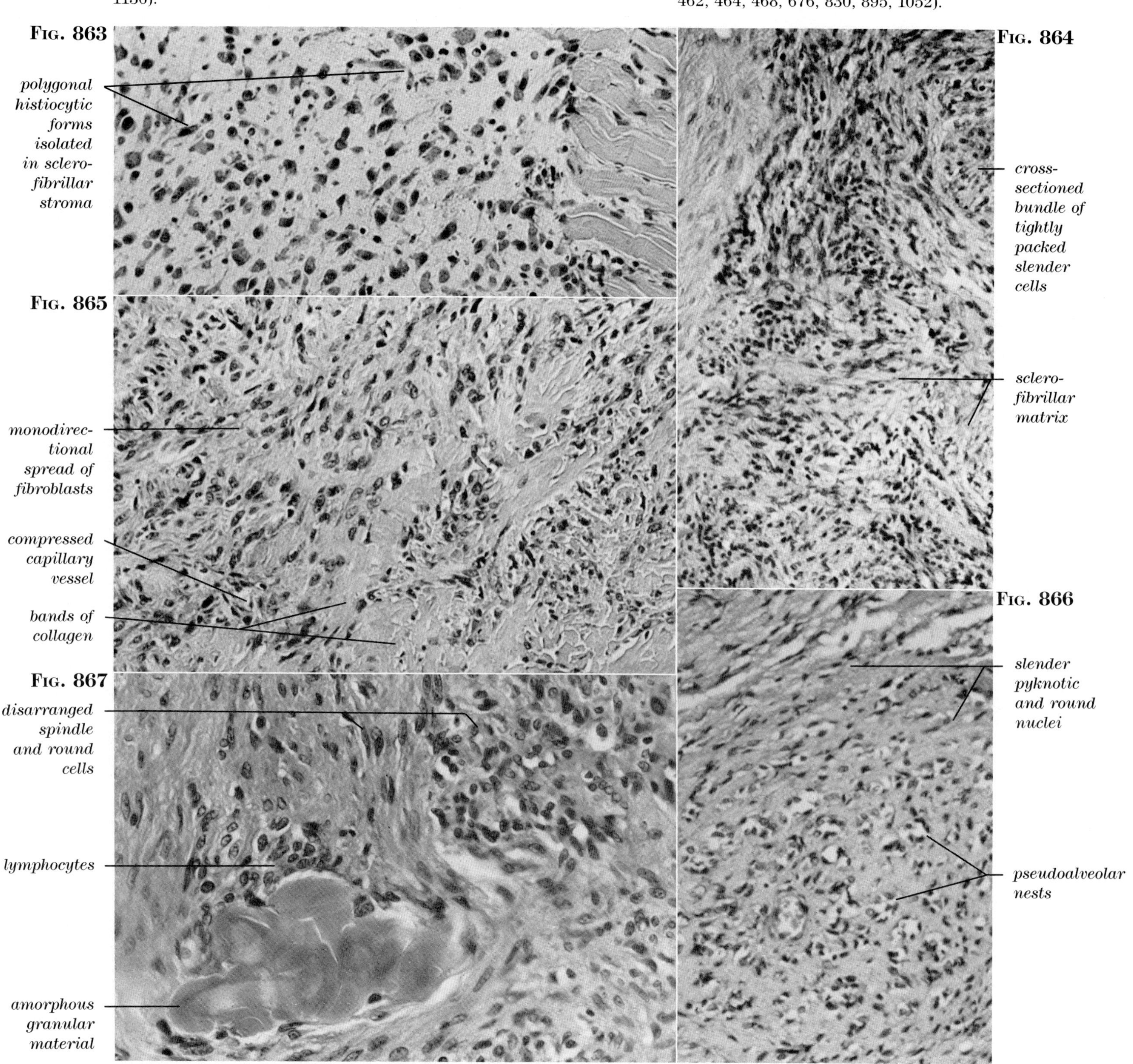

Fig. 868. Low grade leiomyoblastoma (×350) (see Figs. 420, 429, 691, 713, 931, 964, 1011, 1017B).

Fig. 869. Malignant fibroepithelioid mesothelioma (×140) (see Figs. 265, 266, 267, 268, 406, 511, 531, 880, 881, 1240, 1241, 1242).

Fig. 870. Monophasic tendosynovial sarcoma, spindle cell type (×140) (see Figs. 542, 622, 642, 646, 648, 650, 675, 854, 864, 894, 902, 903, 1144, 1145).

Fig. 871. Neurofibroma (×140) (see Figs. 253, 487, 652, 840, 955, 993).

Fig. 872. Low grade malignant histiocytic fibrous histiocytoma (×350) (see Figs. 57, 156, 452, 711, 756, 775, 776, 784, 791, 816, 892, 893, 930, 1018B, 1048).

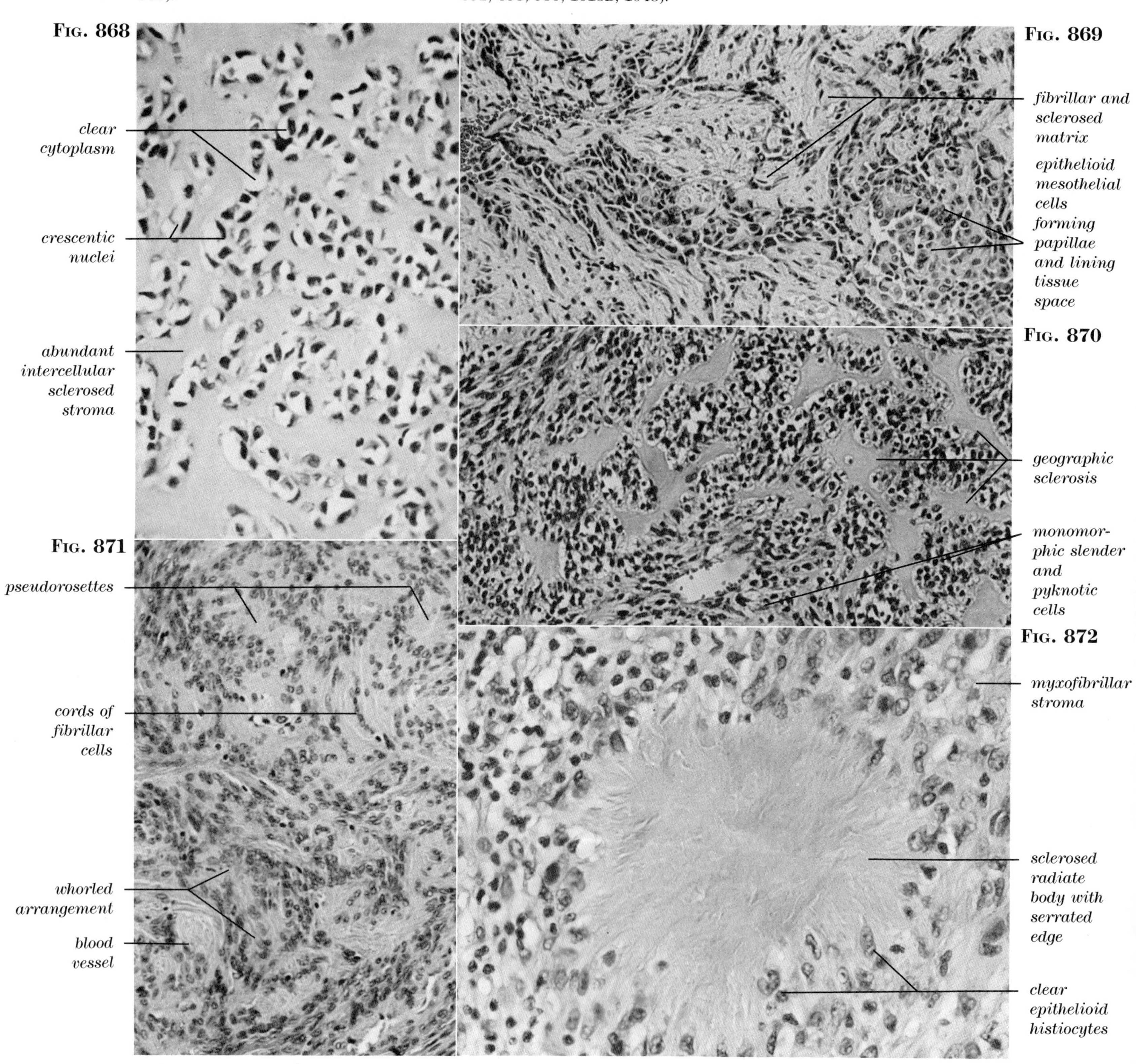

Fig. 873. Fibromatosis (×350) (see Figs. 582, 616, 860, 861, 874, 878, 882, 1022, 1025, 1035, 1036).

Fig. 874. Post-radiation fibromatosis (×350) (see Figs. 280, 367, 385, 387, 419, 581, 616, 860, 861, 878, 882).

Fig. 875. Low grade leiomyosarcoma (×140) (see Figs. 841, 844, 849, 856, 890, 994, 1007, 1012, 1017A, 1051).

Fig. 876. Sclerosing osteosarcoma (×140) (see Figs. 615, 819, 831, 843, 897, 898, 958, 968, 1001, 1068, 1070–1079).

Fig. 877. Keloid (×140) (see Fig. 1153).

Fig. 878. Aponeurotic fibromatosis (×350) (see Figs. 17, 280, 385, 616, 860, 861, 873, 882, 1022, 1025, 1035, 1036, 1050).

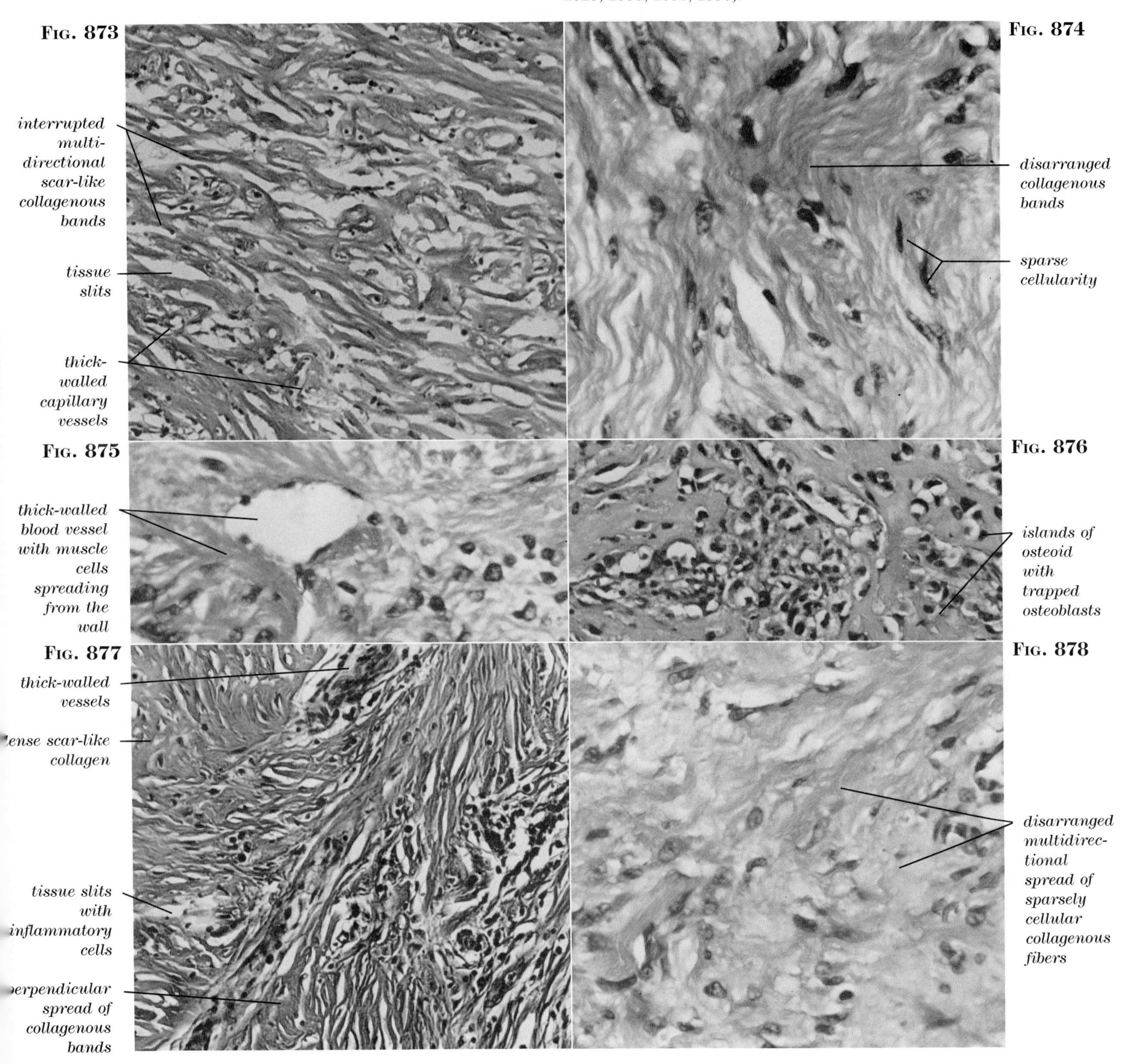

Fig. 879. Benign fibrous mesothelioma (×350) (see Figs. 869, 880).

Fig. 880. (Top center) Borderline fibrous mesothelioma showing increased cellularity (×350) (see Figs. 614, 678, 869, 881).

Fig. 881. Malignant fibrous mesothelioma (×350) (see Figs. 189, 266, 268, 678, 869, 880, 1240, 1241).

Fig. 882. Plantar fibromatosis (×140) (see Figs. 17, 280, 367, 616, 860, 861, 873, 874, 878, 1022, 1025, 1035, 1036, 1050, 1081, 1095).

Fig. 883. Stroma of fibroepithelial papilloma of skin (×140).

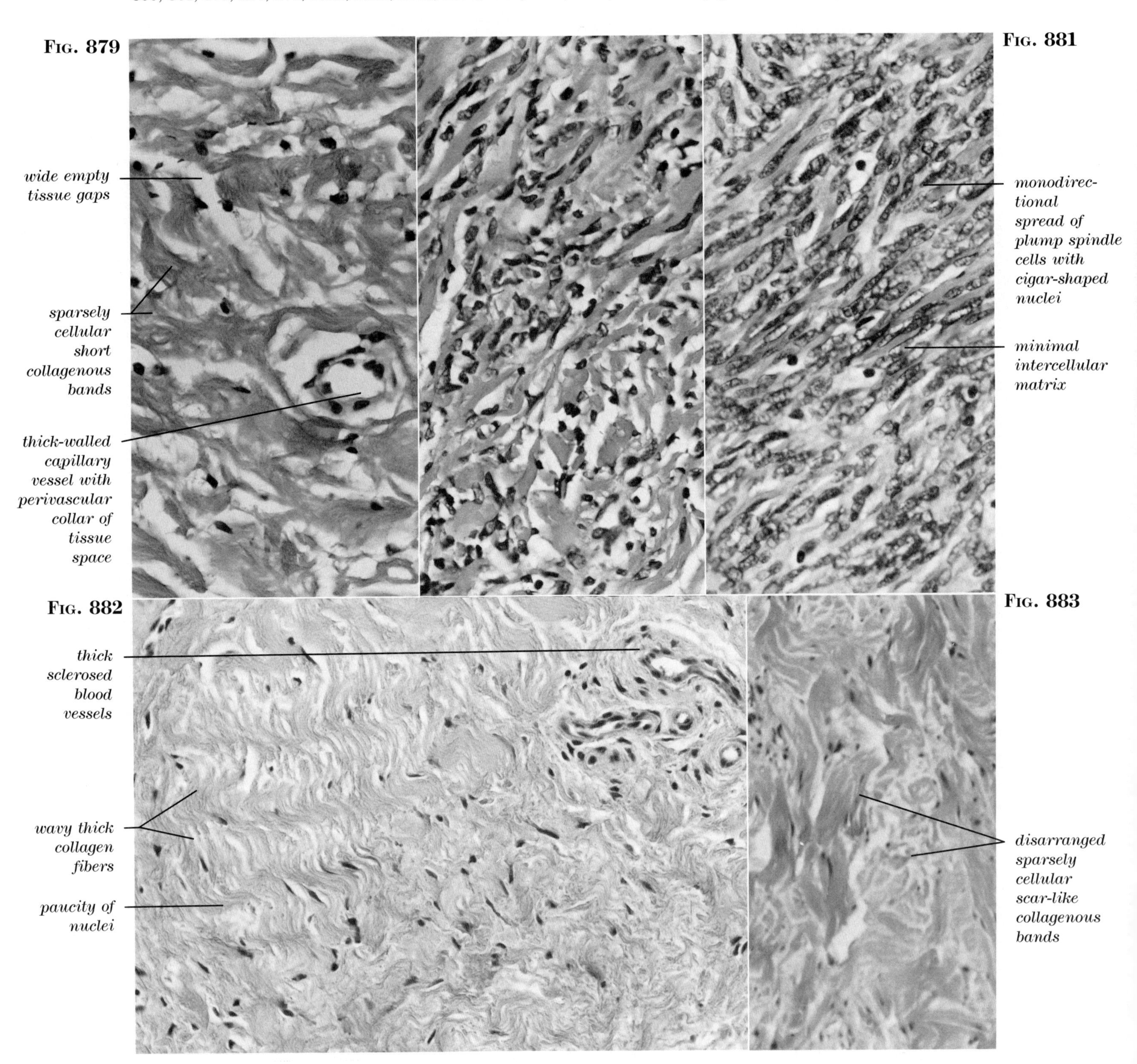

Fig. 884. Fibroma of tendosynovium (×140) (see Figs. 336, 859, 886, 904, 915, 917, 1027, 1109, 1204).

Fig. 885. Mediastinal Hodgkin's disease (×350) (see Figs. 27, 742, 743, 888, 1024).

Fig. 886. Fibroma, so-called "dermatofibroma" (×350) (see Figs. 336, 859, 884, 904, 915, 917, 1027, 1109, 1204).

Fig. 887. Low grade meningeal sarcoma (×350) (see Figs. 283, 495, 806).

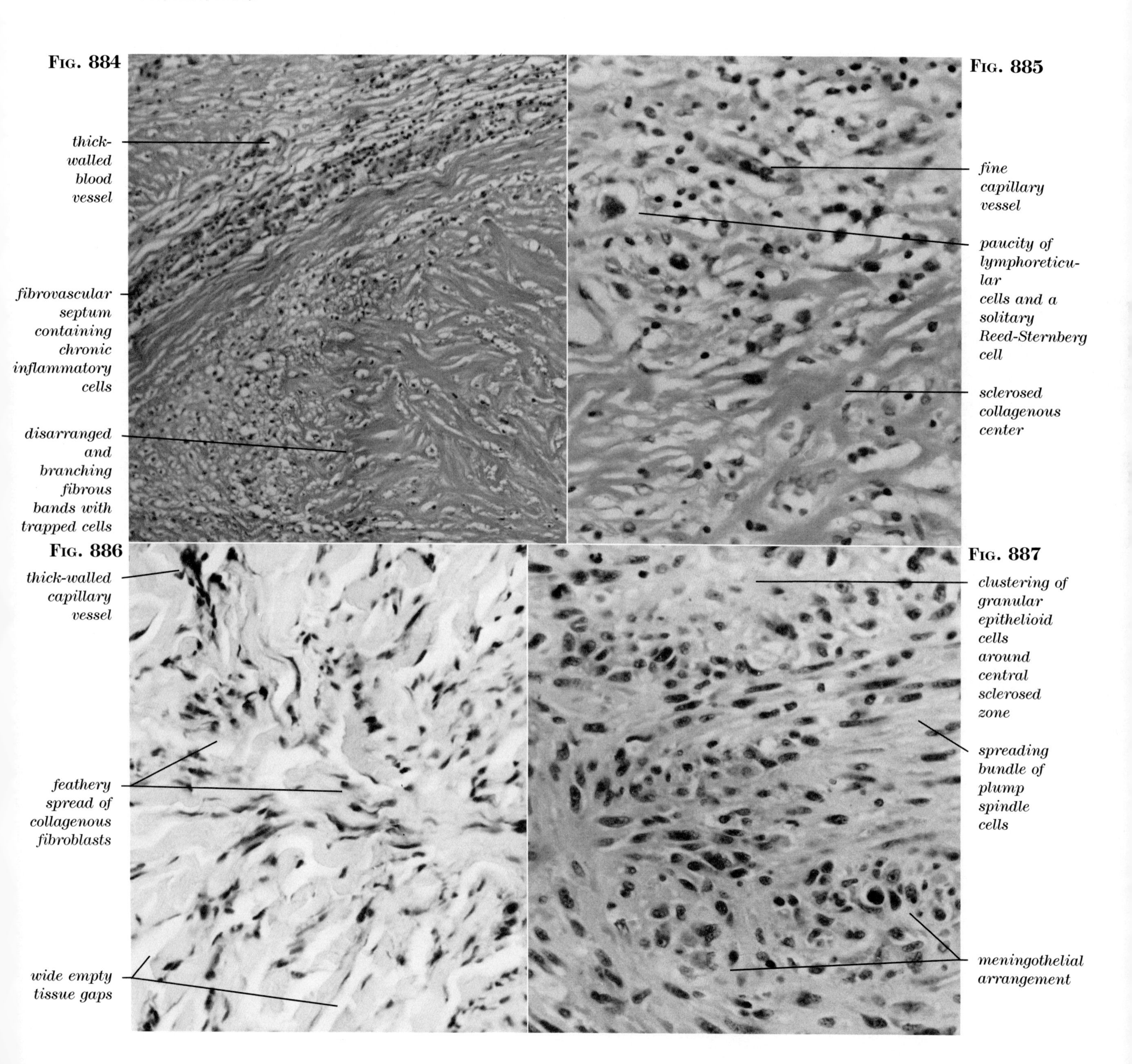

Fig. 888. Non-Hodgkin's lymphoma infiltrating sclerosed tissue (×140) (see Figs. 15, 132, 169, 423, 463, 494, 702, 704, 705, 712, 743).

Fig. 889. Embryonal rhabdomyosarcoma (×350) (see Figs. 78, 364, 637, 925, 935, 943, 946, 949, 1129, 1138, 1164).

Fig. 890. Low grade vascular leiomyosarcoma (×140) (see Figs. 664, 666, 676, 841, 844, 849, 856, 875, 994, 1007, 1012).

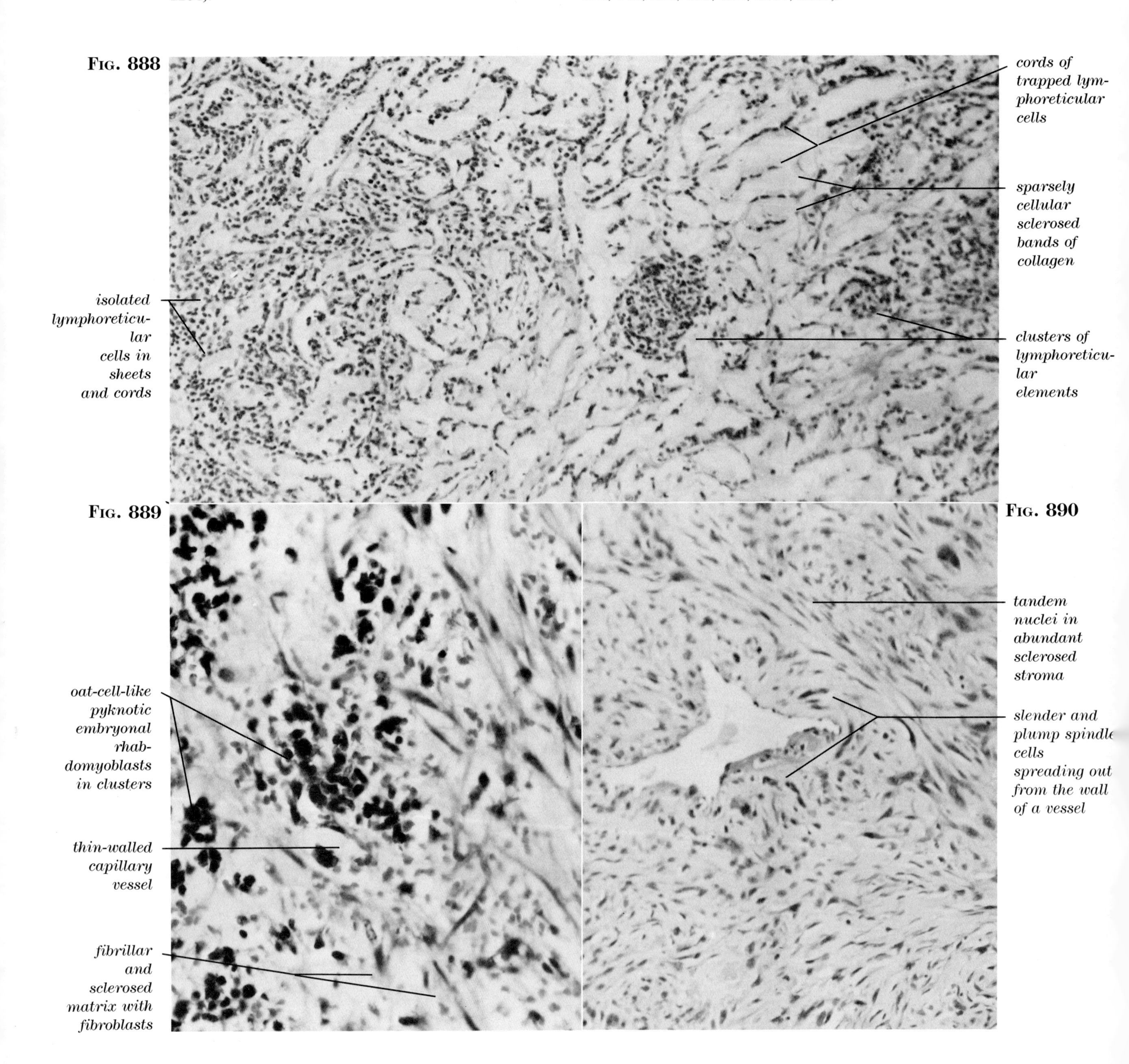

Fig. 891. Granulocytic sarcoma (×350) (see Figs. 488, 699, 1194, 1198).

Fig. 892. Borderline histiocytic fibrous histiocytoma (×350) (see Figs. 534, 775, 776, 872, 893, 930, 1018B, 1048, 1149).

Fig. 893. Low grade malignant histiocytic fibrous histiocytoma (×350) (see Figs. 161, 300, 600, 711, 791, 816, 872, 892, 930, 1018B, 1048, 1149, 1172).

Fig. 894. Monophasic tendosynovial sarcoma, spindle cell type (×140) (see Figs. 85, 326, 542, 684, 854, 864, 870, 902, 903, 1144, 1145).

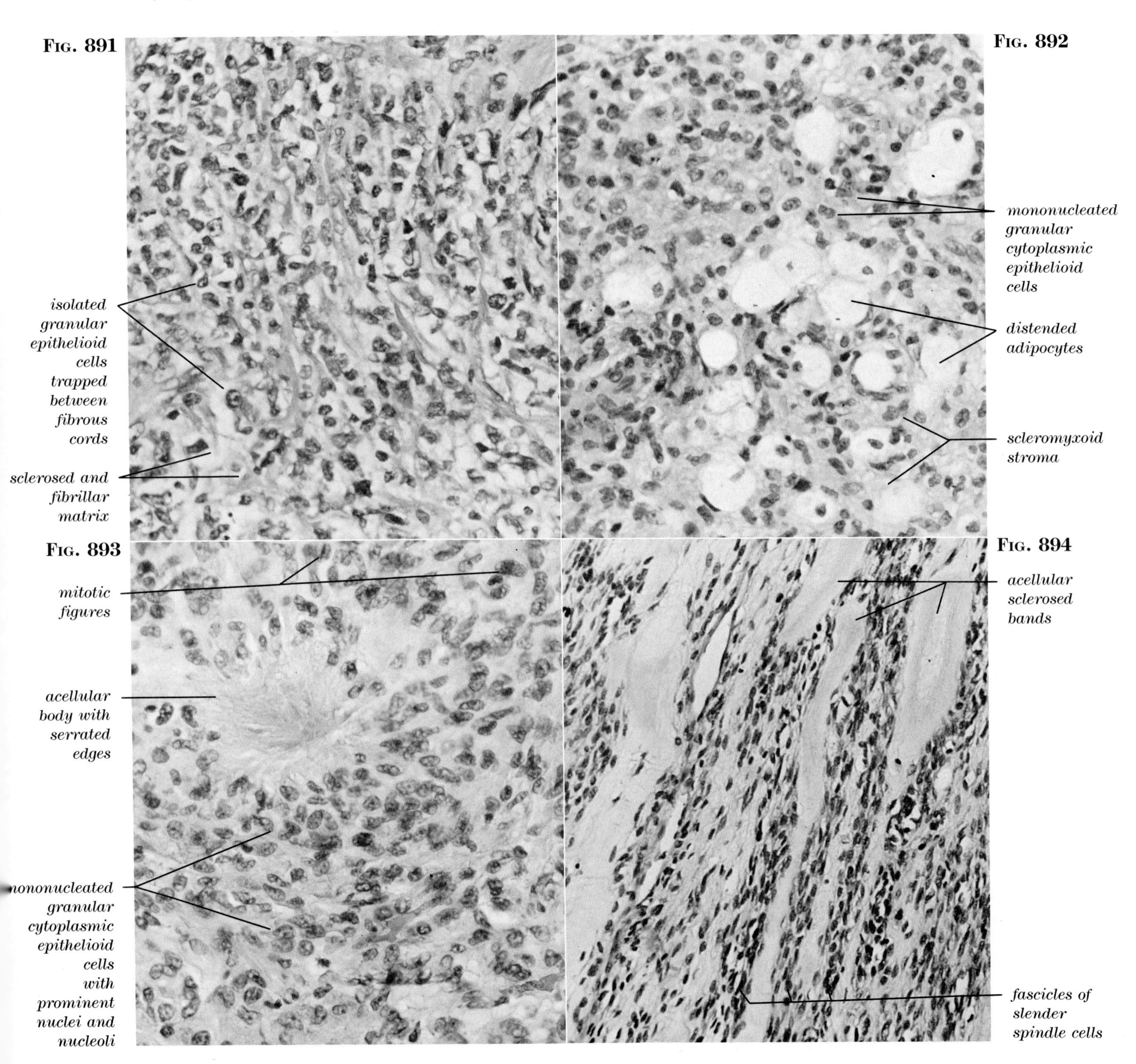

FIG. 895. Epithelioid sarcoma (×140) (see Figs. 62, 85, 181, 258, 468, 491, 676, 830, 866, 1052, 1059).

FIG. 896. Clear cell sarcoma (×140) (see Figs. 62, 85, 322, 447, 492, 759, 761, 927, 1145, 1223, 1244, 1245).

FIG. 897. Sclerosing osteosarcoma (×140) (see Figs. 90, 108, 142, 248, 395, 643, 817, 831, 843, 876, 898, 958, 968, 1001, 1068, 1070–1079).

FIG. 898. Treated osteosarcoma (×350) (see Figs. 90, 108, 228, 328, 615, 819, 831, 843, 876, 897, 958, 968, 1070–1079, 1086, 1087, 1088).

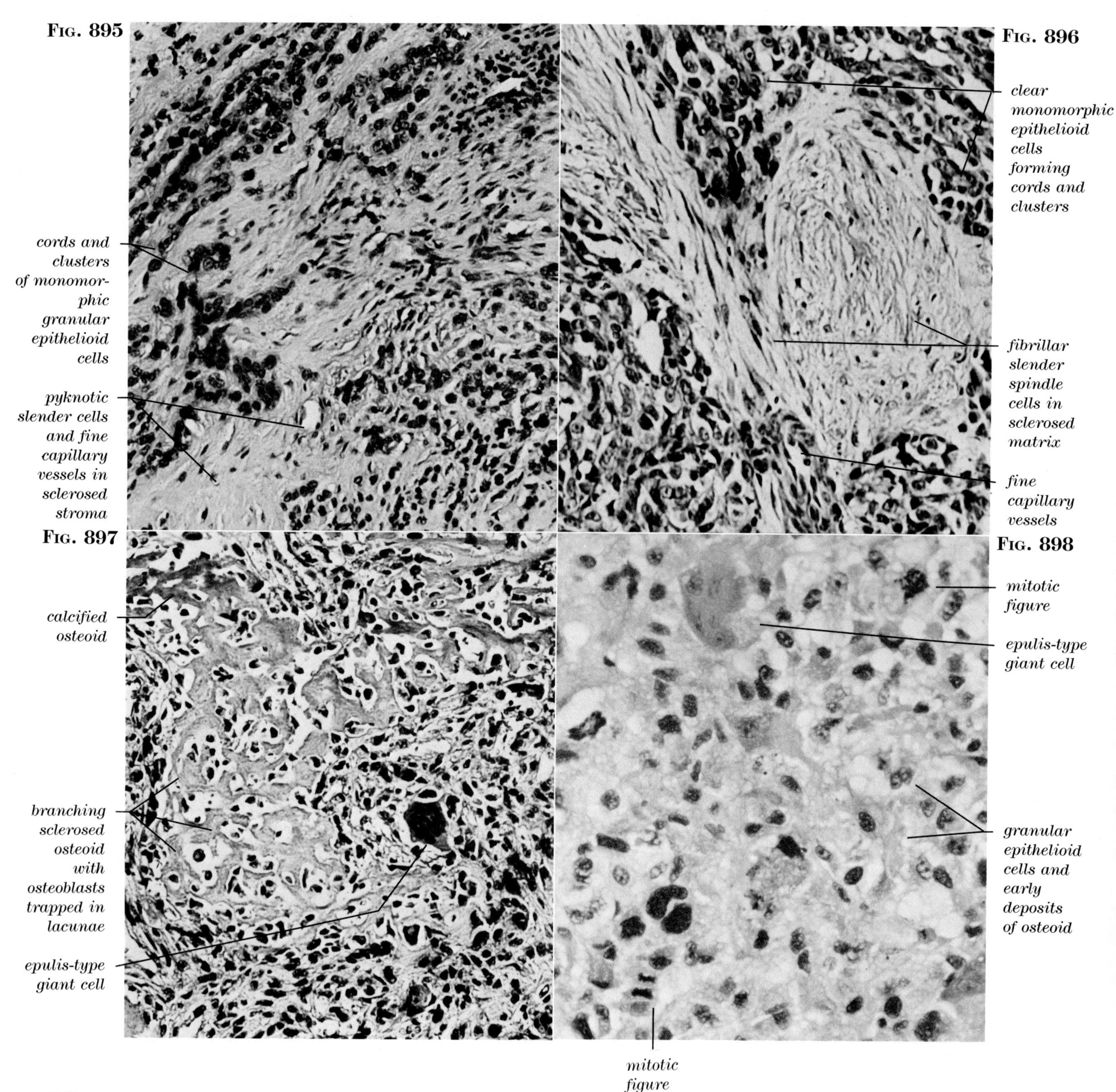

Fig. 899. Desmoid tumor, low grade fibrosarcoma (×350) (see Figs. 77, 185, 244, 282, 386, 686, 865, 1118).

Fig. 900. Fibrosarcoma, so-called juvenile fibrosarcoma (×350) (see Figs. 91, 131, 170, 260, 410, 832).

Fig. 901. High grade pleomorphic fibrosarcoma (×350) (see Figs. 91, 117, 297, 601, 608, 625, 647, 660, 665, 671, 809, 825, 832, 1117, 1205).

Fig. 902. Tendosynovial sarcoma, monoplastic spindle cell type (×350) (see Figs. 64, 85, 190, 326, 542, 684, 854, 864, 870, 894, 903, 1144, 1145).

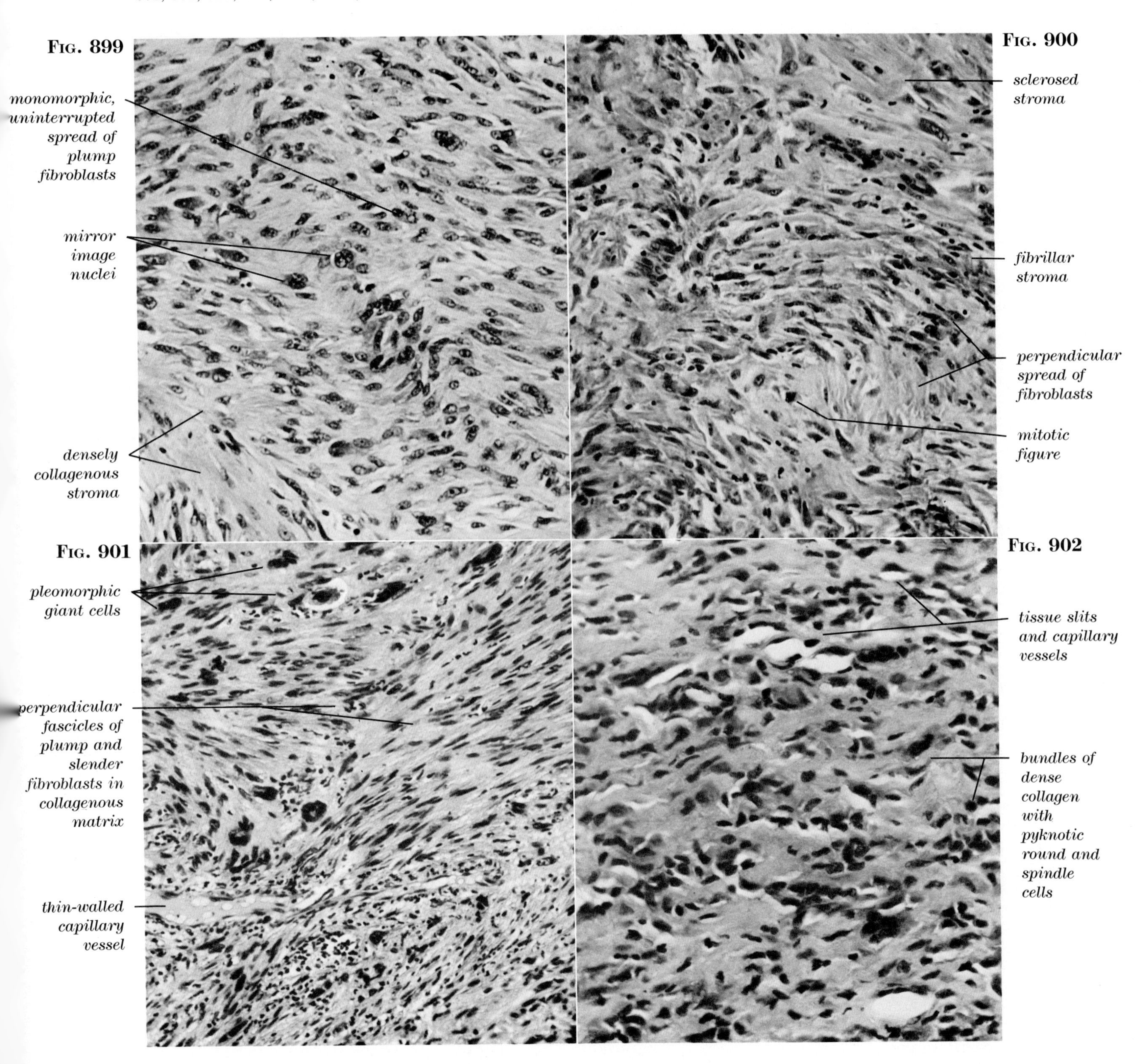

Fig. 903. Monophasic tendosynovial sarcoma, spindle cell type (×350) (see Figs. 85, 190, 320, 492, 684, 854, 864, 870, 894, 902, 1144, 1145).

Fig. 904. Tendosynovial fibroma (×350) (see Figs. 336, 859, 884, 886, 915, 917, 1027, 1109, 1204).

Fig. 905. Low grade hemangiosarcoma (×350) (see Figs. 92, 523, 720, 975, 976, 989, 990, 1005, 1209, 1210).

Fig. 906. Biphasic tendosynovial sarcoma (×350) (see Figs. 85, 514, 520, 727, 765, 1099, 1100, 1173, 1202).

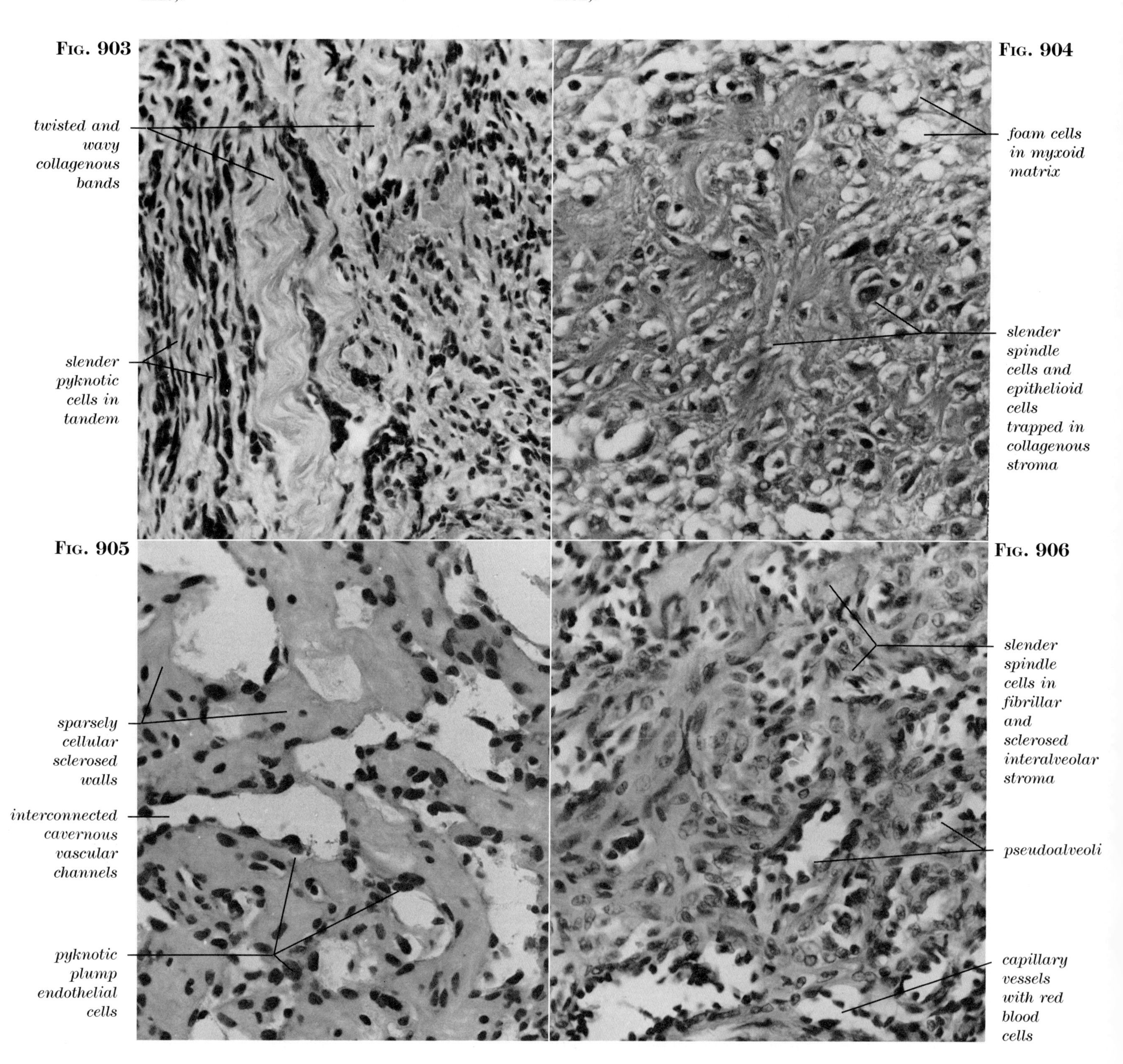

Fig. 907. Lipogranuloma (×350) (see Figs. 433, 820, 1043).

Fig. 908. Fibroblastic lipoma (×350) (see Figs. 73, 847, 940).

Fig. 909. Pleomorphic lipoma (×350) (see Figs. 73, 613, 781, 821, 957, 1182, 1183).

Fig. 910. Proliferative panniculitis (×350) (see Figs. 430, 552, 721).

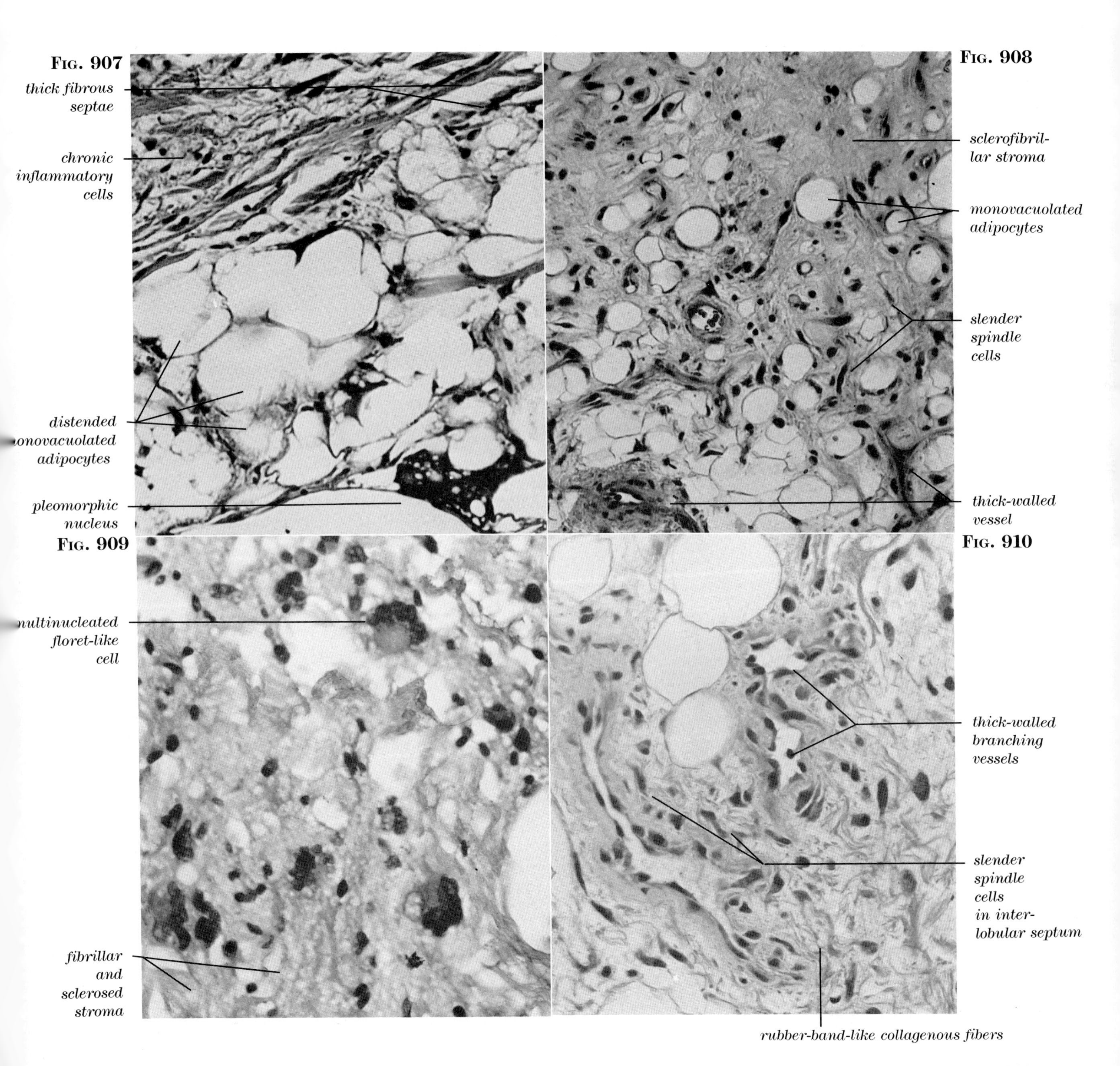

Fig. 911. Arteriovenous malformation in muscle (×85) (see Figs. 139, 197, 1006, 1009).

Fig. 912. Angiomyoma (×85) (see Figs. 653, 985, 1254).

Fig. 913. Myositis ossificans (×85) (see Figs. 145, 384, 620, 1083, 1084, 1085).

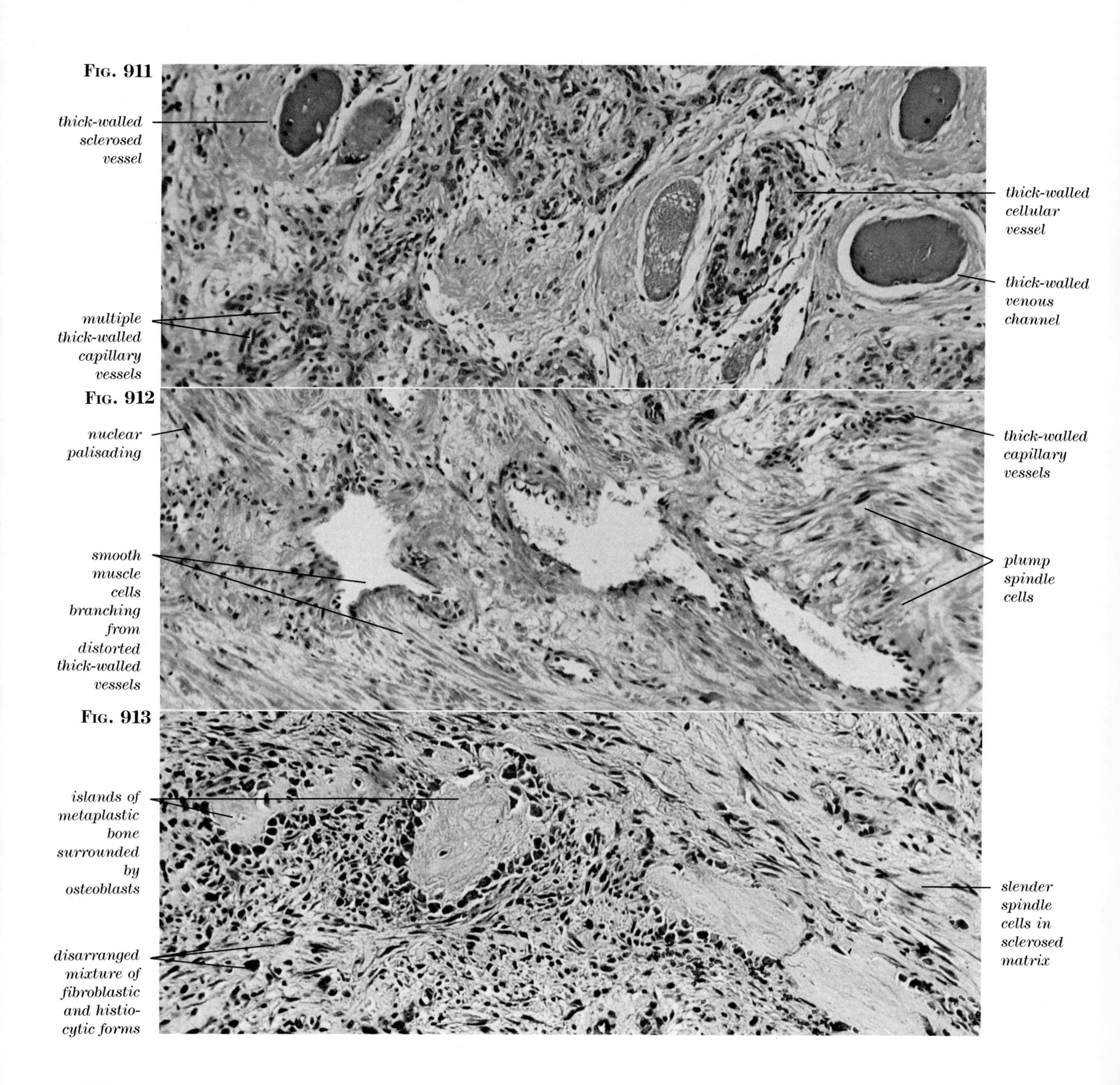

Fig. 914. Elastofibroma (×350) (see Figs. 1123, 1140, 1206).

Fig. 916. Medullary carcinoma of thyroid (×350) (see Figs. 504, 532).

Fig. 918. Paget's sarcoma of bone (×140) (see Figs. 107, 134, 1103, 1128).

Fig. 915. Tendosynovial fibroma, so-called "aponeurotic fibroma" (×350) (see Figs. 336, 859, 884, 886, 904, 917, 1027, 1109, 1204).

Fig. 917. Fascial fibroma (×140) (see Figs. 859, 884, 886, 904, 915, 1027, 1109).

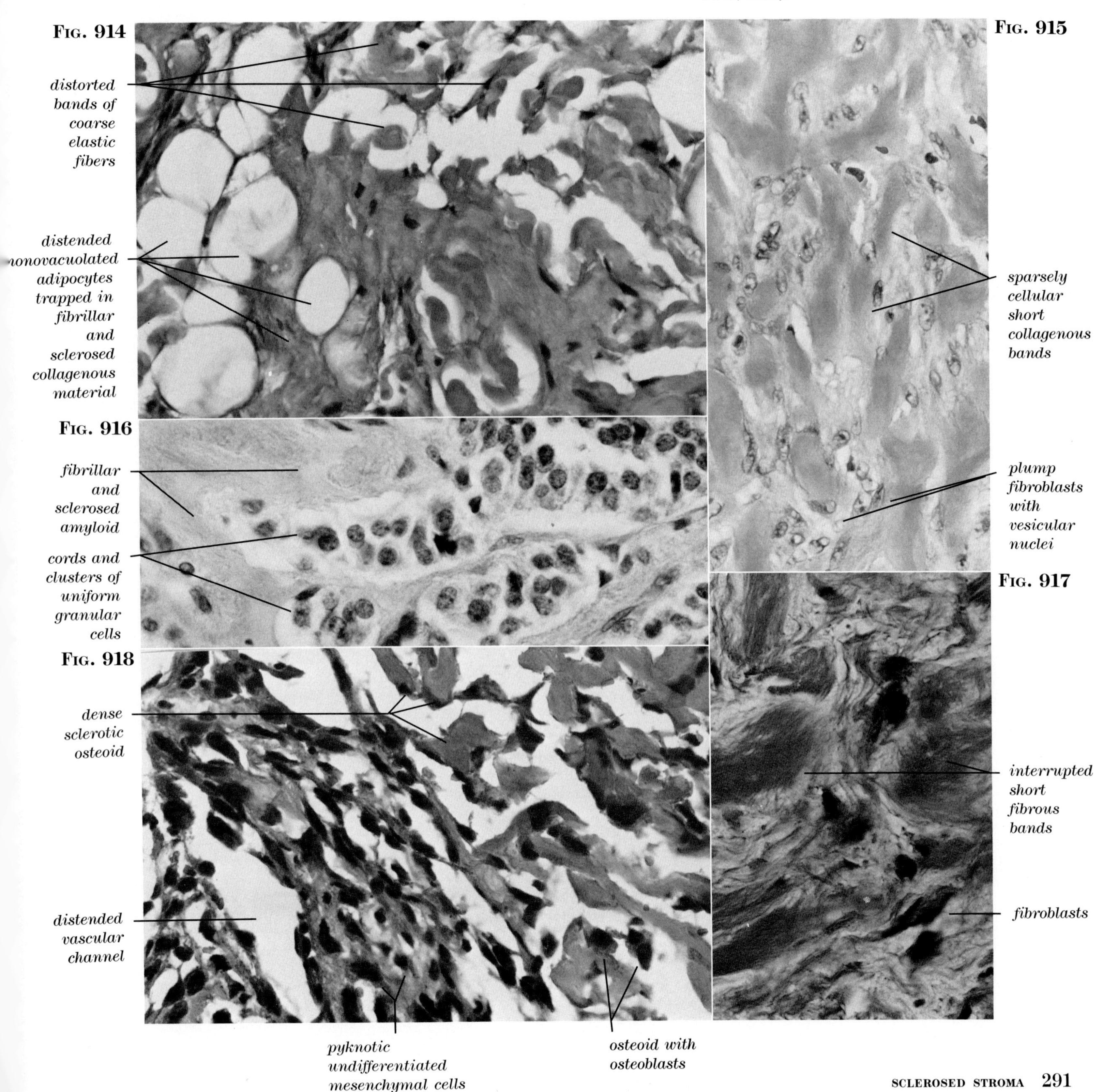

FIG. 919. Chordoid sarcoma (×140) (see Figs. 456, 510, 640, 1176).

FIG. 920. Mesenchymal chondrosarcoma (×140) (see Figs. 103, 390, 444, 674, 1062, 1265).

FIG. 921. Benign cystosarcoma phyllodes of breast (×140) (see Figs. 43, 48, 247, 414, 938).

FIG. 922. Teratoma with "immature" sarcomatous stroma (×140) (see Figs. 8, 235, 279, 349, 688, 1250).

FIG. 923. Myxoid rhabdomyosarcoma, so-called "botryoid sarcoma" (×35) (see Figs. 78, 120, 416, 418, 587, 644, 929, 944, 952, 1033, 1251).

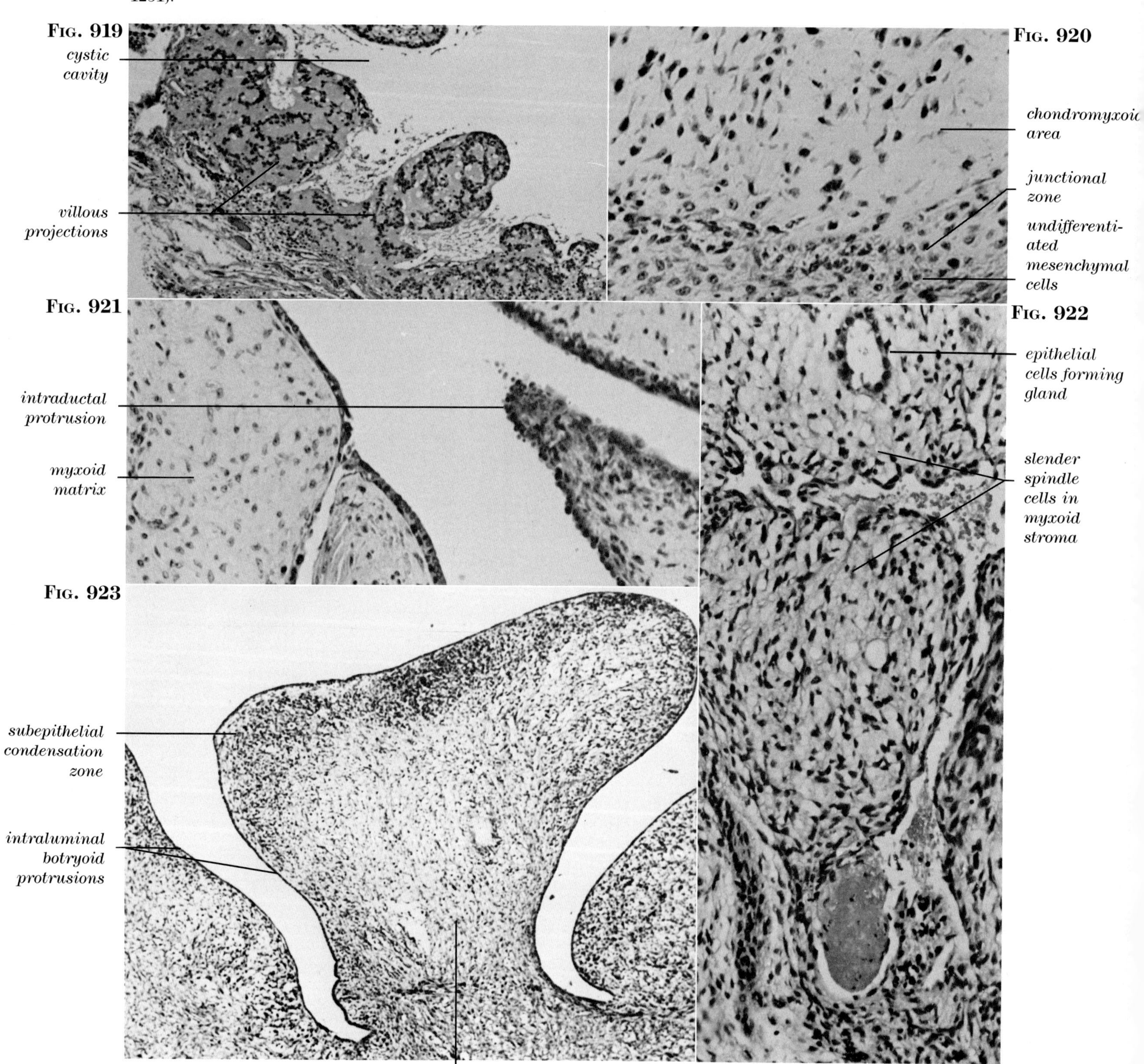

Fig. 924. Decidua (×140) (see Fig. 770).

Fig. 925. (Top center) Embryonal rhabdomyosarcoma, notice multiple thin-walled capillaries. (×140) (see Figs. 372, 689, 889, 935, 943, 946).

Fig. 926. Low grade fibroblastic liposarcoma (×140) (see Figs. 440, 668, 837, 848, 998).

Fig. 927. Clear cell sarcoma (×140) (see Figs. 492, 526, 759, 761, 896, 1145).

Fig. 928. Pleomorphic liposarcoma (×140) (see Figs. 86, 284, 458, 590, 799, 1019*a*).

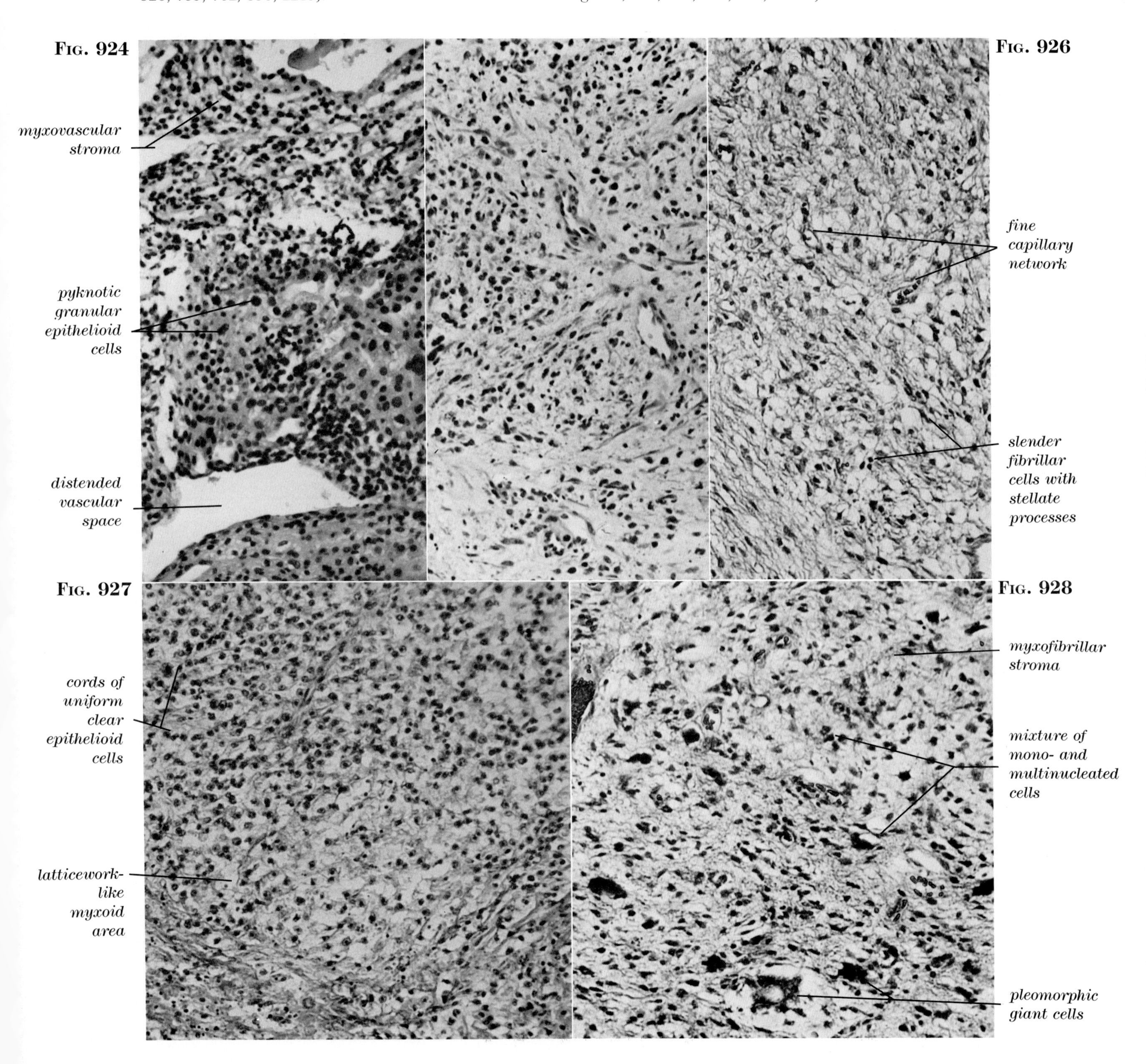

Fig. 929. Myxoid rhabdomyosarcoma (×350) (see Figs. 587, 644, 923, 944, 952).

Fig. 930. (Top center) Low grade malignant histiocytic fibrous histiocytoma (×350) (see Figs. 791, 816, 872, 892, 893, 1018*b*).

Fig. 931. Low grade leiomyoblastoma (×350) (see Figs. 503, 733, 744, 868, 964, 1011).

Fig. 932. Myxoid chondrosarcoma (×350) (see Figs. 81, 95, 124, 308, 329, 638, 1175).

Fig. 933. Myxoid liposarcoma (×350) (see Figs. 86, 438, 439, 636, 783, 945, 948).

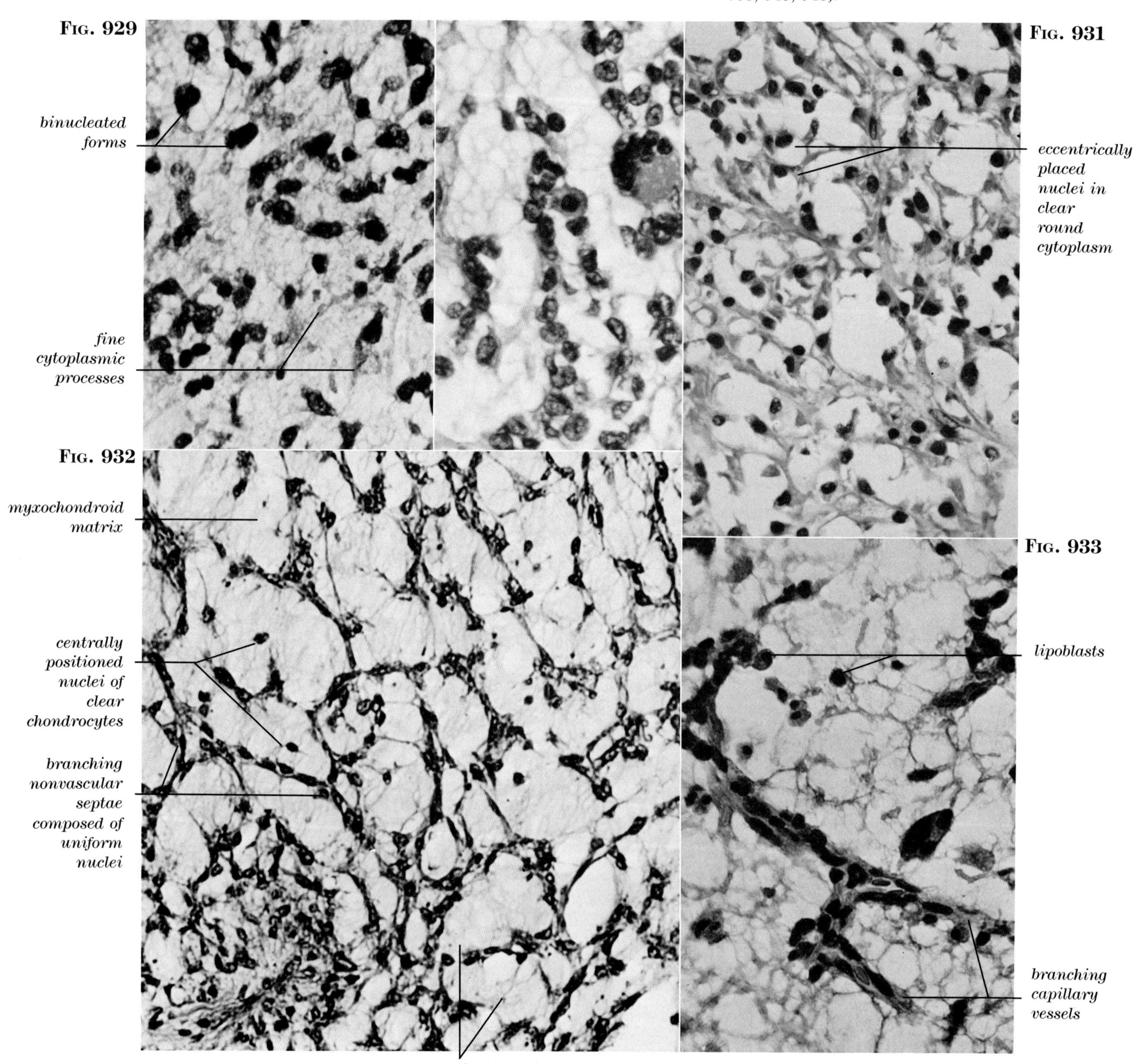

Fig. 934. Lipoblastic liposarcoma (×350) (see Figs. 86, 199, 415, 448, 687, 753, 945, 973, 1231).

Fig. 935. Embryonal rhabdomyosarcoma (×350) (see Figs. 78, 226, 263, 364, 379, 700, 889, 925, 943, 946, 949).

Fig. 936. Rhabdomyoblastoma (×350) (see Figs. 473, 477, 500, 528, 569, 594, 695, 715, 1034, 1159, 1260).

Fig. 937. Low grade malignant pleomorphic fibrous histiocytoma (×350) (see Figs. 25, 88, 111, 130, 344, 679, 953, 962, 1002, 1003, 1008, 1032, 1047).

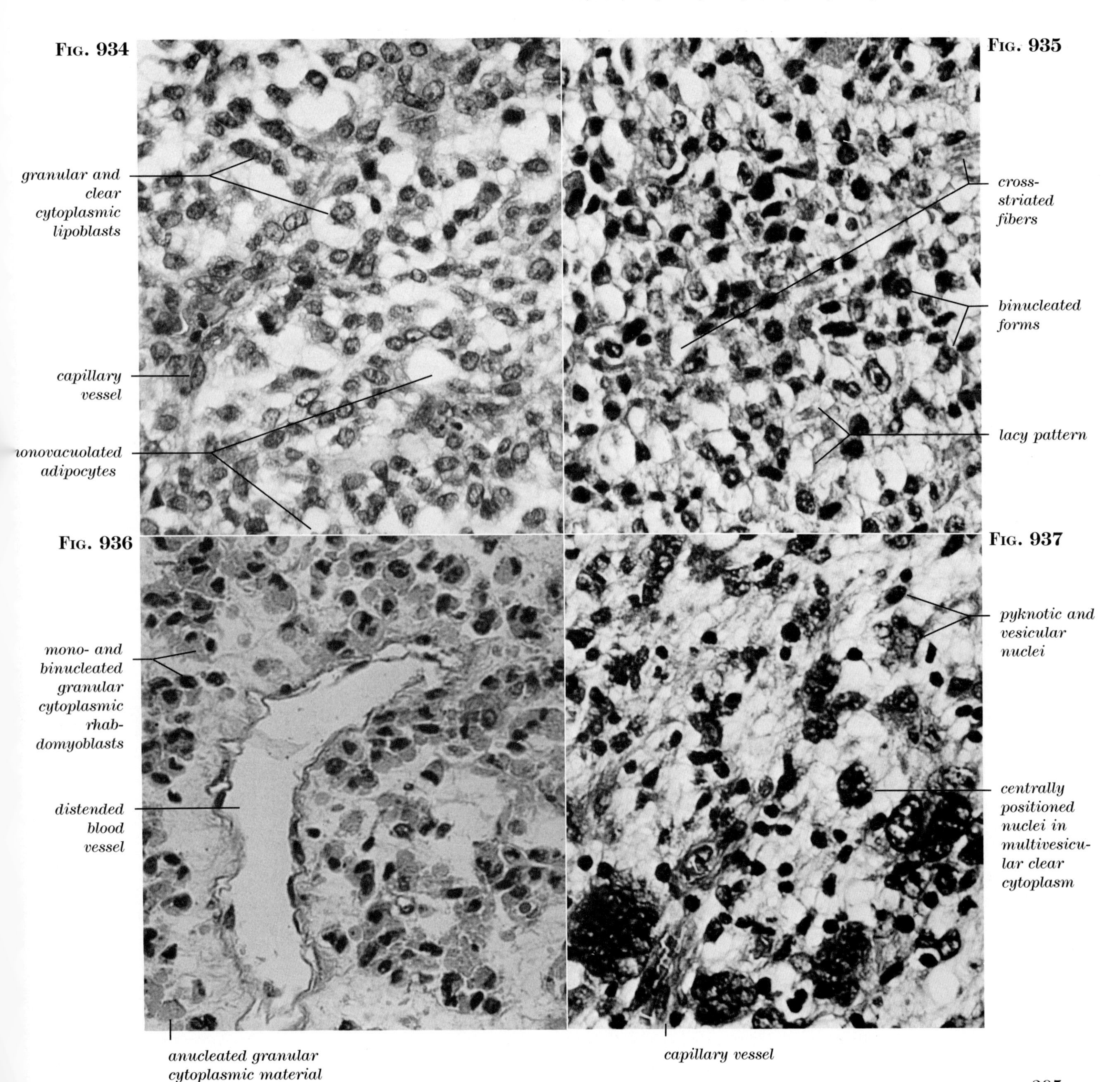

Fig. 938. Benign cystosarcoma phyllodes of breast (×35) (see Figs. 43, 48, 247, 414, 921).

Fig. 940. Fibroblastic lipoma (×140) (see Figs. 73, 847, 908).

Fig. 942. Neurofibroma, so-called myxoid neurofibroma (×140) (see Figs. 11, 74, 245, 363, 840, 842, 1114).

Fig. 939. Myxoid lipoma, so-called "myxoma" infiltrating muscle (×140) (see Figs. 73, 635, 941, 950, 1168, 1187).

Fig. 941. Myxoid lipoma, so-called "myxoma" (×140) (see Figs. 73, 635, 939, 950, 1168, 1187).

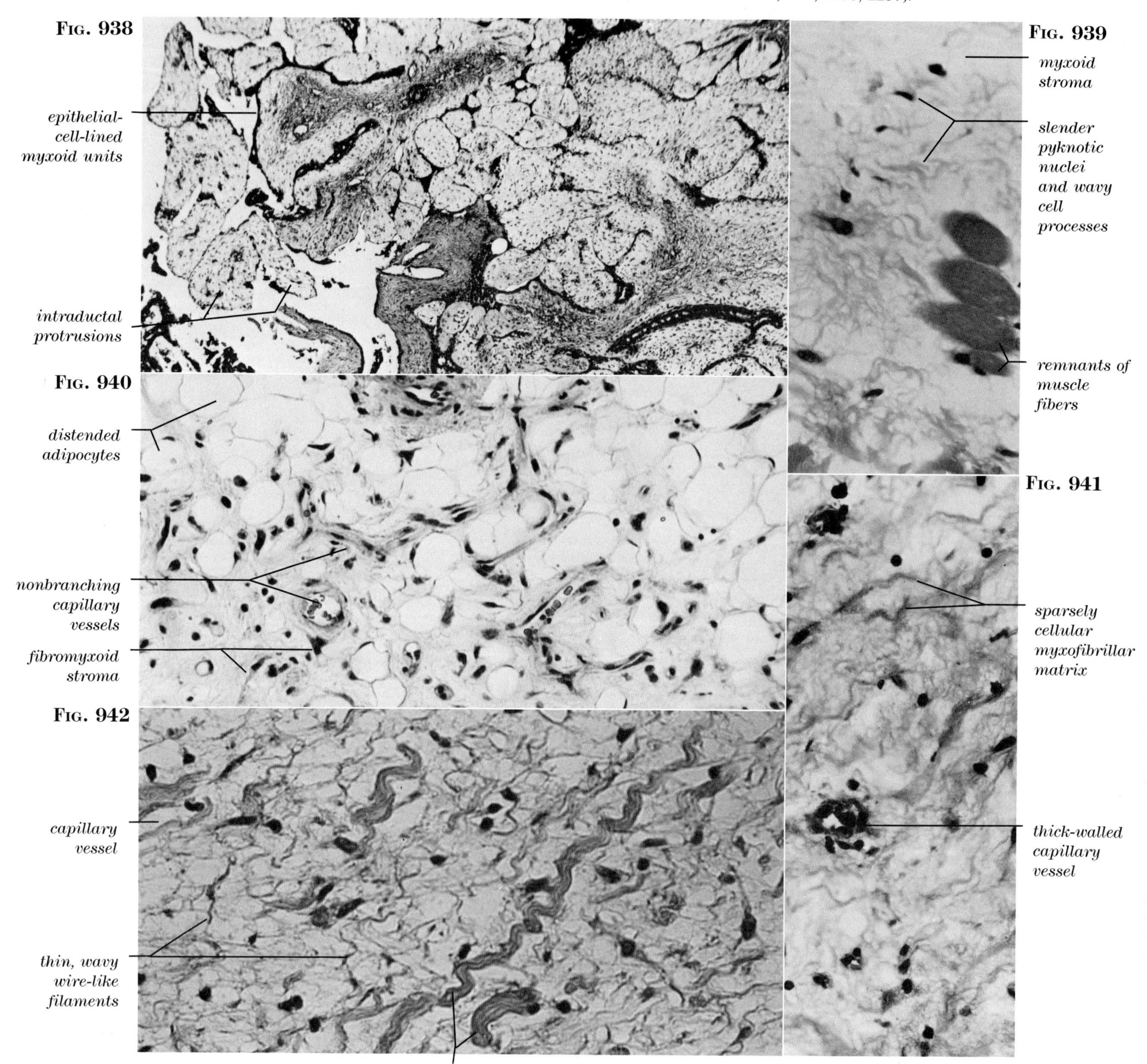

Fig. 943. Embryonal rhabdomyosarcoma, spreading type (×350) (see Figs. 78, 379, 427, 446, 706, 889, 925, 935, 946, 949).

Fig. 944. Myxoid embryonal rhabdomyosarcoma (×350) (see Figs. 78, 120, 416, 418, 587, 644, 923, 929, 952, 1033, 1251).

Fig. 945. Myxoid liposarcoma containing lipoblasts (× 350) (see Figs. 86, 118, 269, 438, 439, 636, 783, 933, 948, 997, 1046, 1113, 1169, 1232).

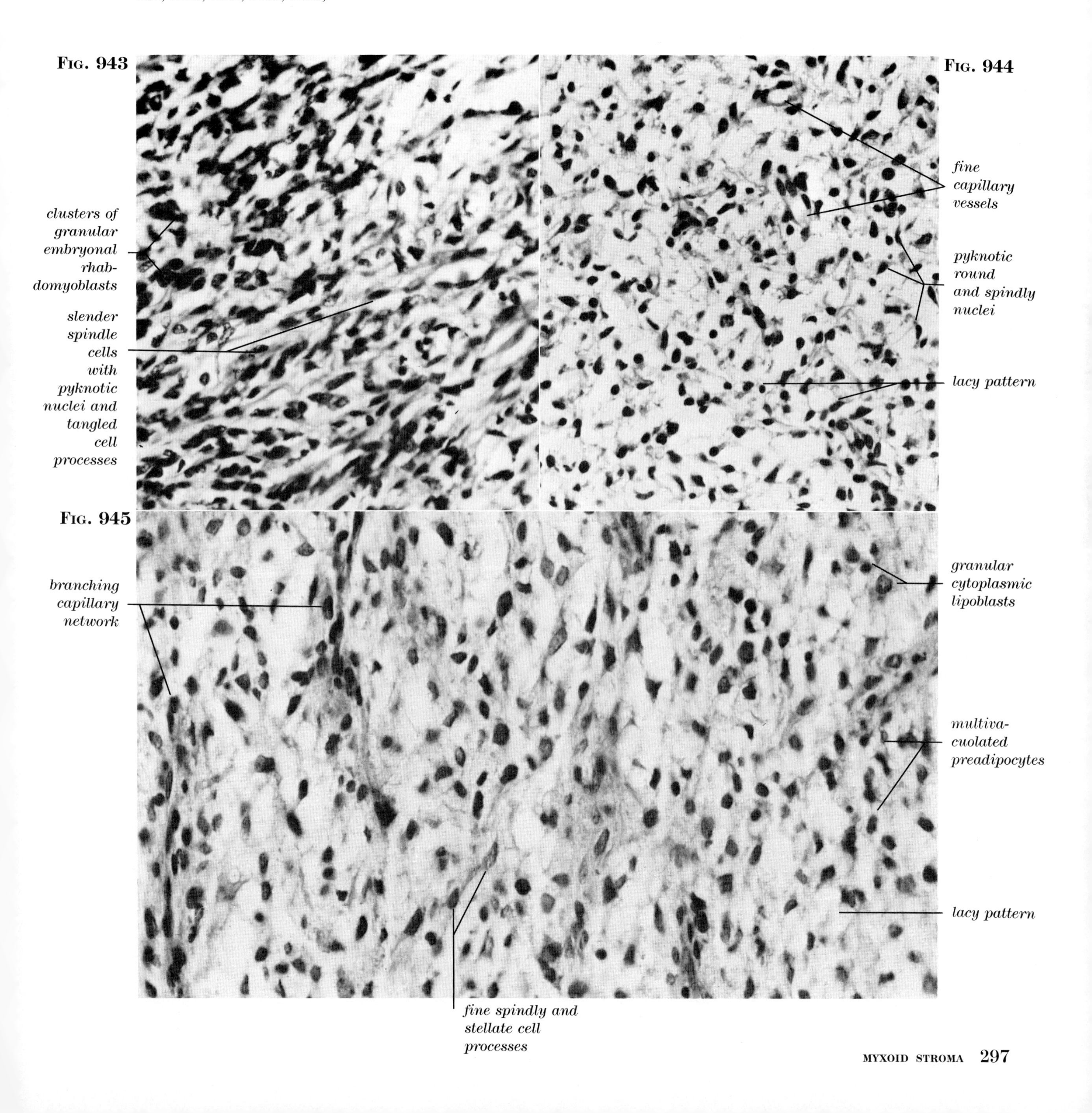

Fig. 946. Embryonal rhabdomyosarcoma (×350) (see Figs. 751, 889, 925, 935, 943, 949, 1129, 1138, 1164, 1171, 1215, 1257).

Fig. 947. Fasciitis (×350) (see Figs. 82, 345, 621, 834, 858, 956, 971, 991, 1020, 1030, 1045).

Fig. 948. Myxoid liposarcoma (×350) (see Figs. 86, 188, 269, 438, 439, 636, 783, 933, 945, 997, 1046, 1113, 1169, 1232).

Fig. 949. Embryonal rhabdomyosarcoma (×350) (see Figs. 78, 226, 263, 291, 364, 372, 637, 689, 751, 889, 925, 935, 943, 946, 1129).

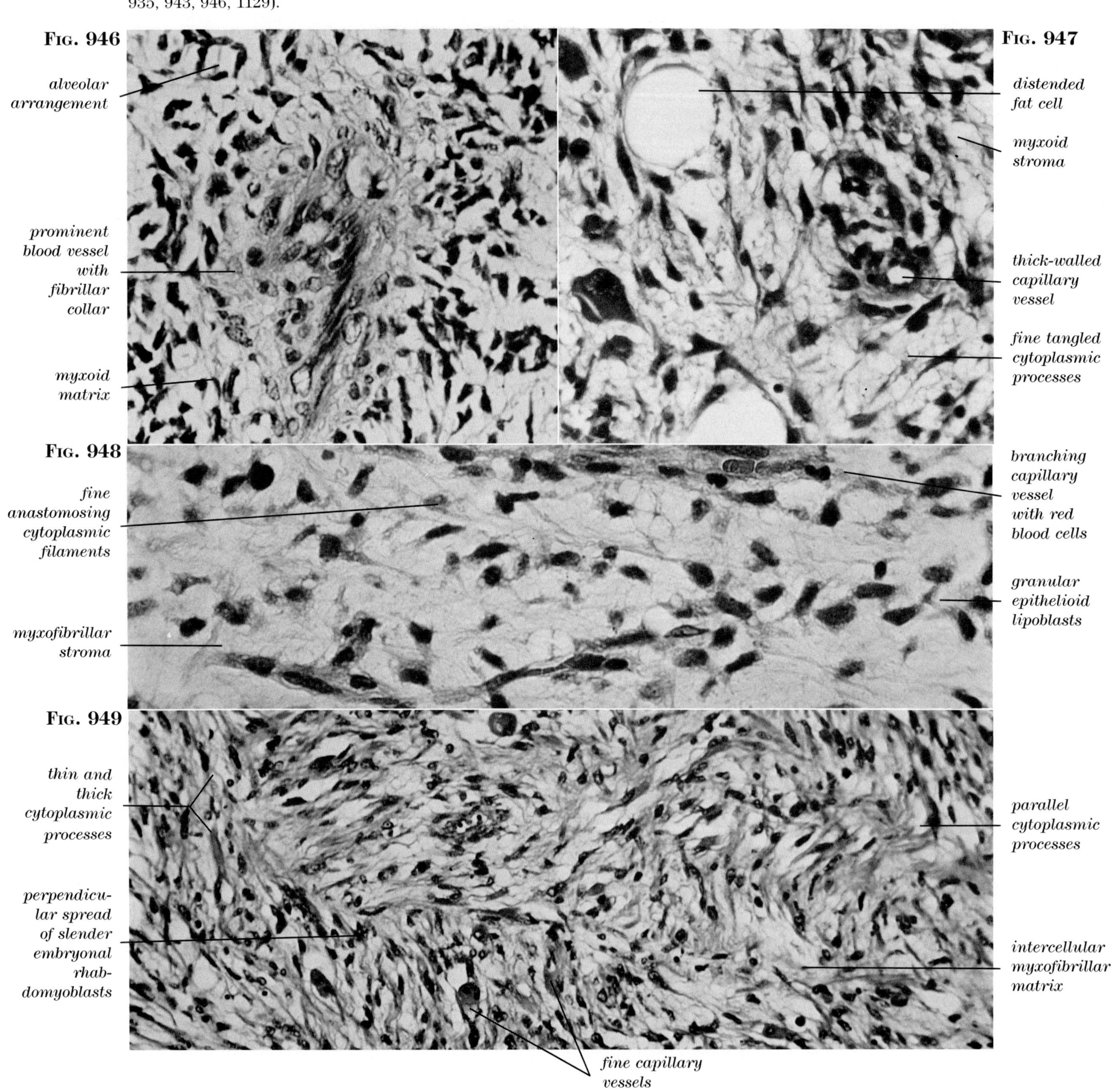

Fig. 950. Myxoid lipoma, so-called "myxoma" (×350) (see Figs. 73, 635, 939, 941, 1168, 1187).

Fig. 951. Proliferative myositis (×140) (see Figs. 597, 738, 863).

Fig. 952. Myxoid rhabdomyosarcoma (×350) (see Figs. 644, 923, 929, 944, 1033).

Fig. 953. Low grade malignant pleomorphic fibrous histiocytoma (×350) (see Figs. 460, 577, 780, 788, 937, 962).

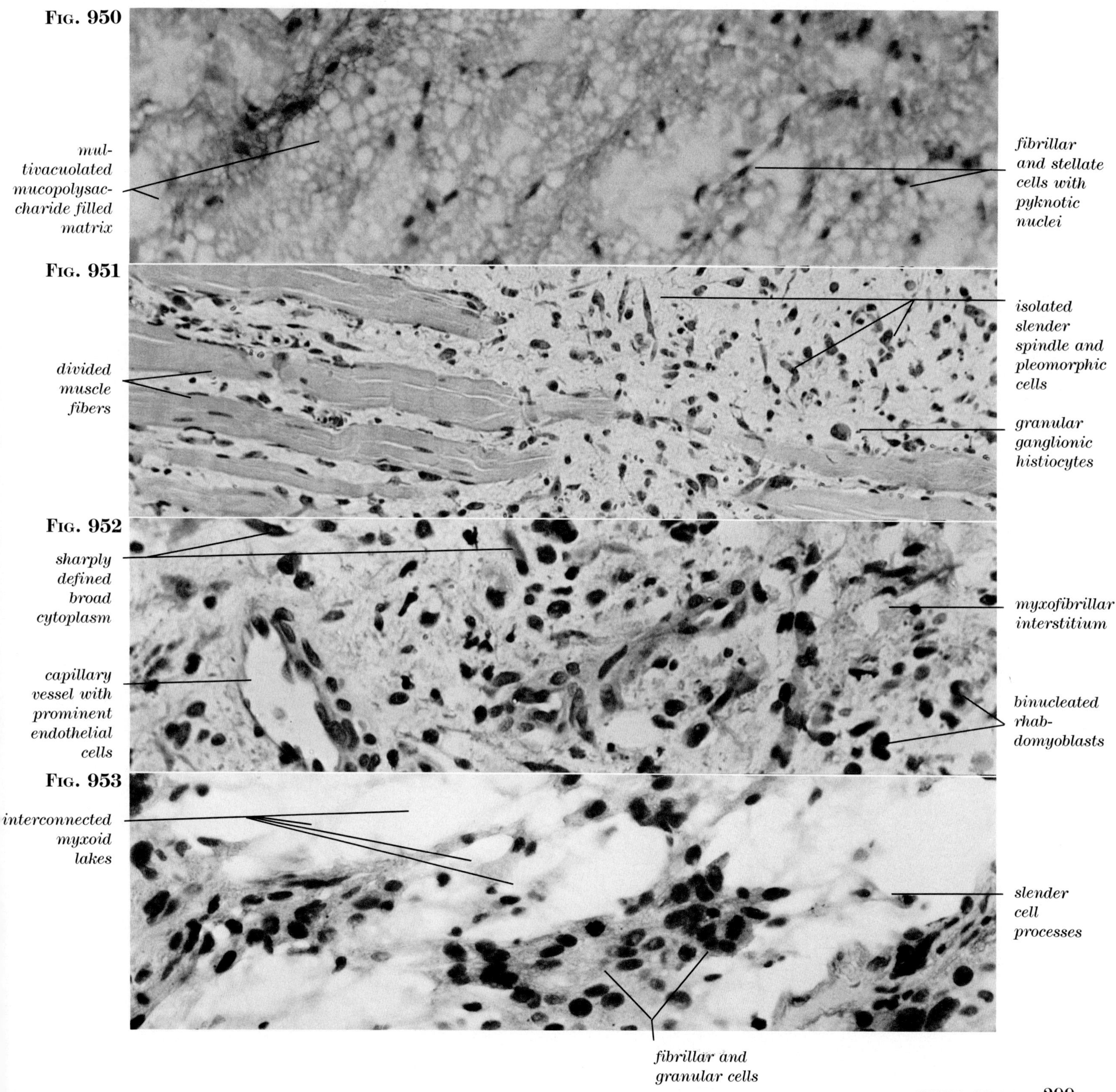

Fig. 954. Lymphangiosarcoma (×140) (see Figs. 84, 321, 424, 649, 963, 965, 970, 1098).

Fig. 957. Pleomorphic lipoma (×140) (see Figs. 73, 613, 781, 821, 909, 1182).

Fig. 955. Neurofibroma, so-called "angioneuroma" or "ancient neuroma" (×140) (see Figs. 652, 840, 871, 993, 1013).

Fig. 956. Fasciitis (×140) (see Figs. 834, 851, 855, 858, 947, 971, 991, 1020, 1030, 1045).

Fig. 958. Telangiectatic osteosarcoma (×140) (see Figs. 90, 108, 109, 831, 843, 876, 897, 898, 968, 1001, 1068, 1070–1079, 1086).

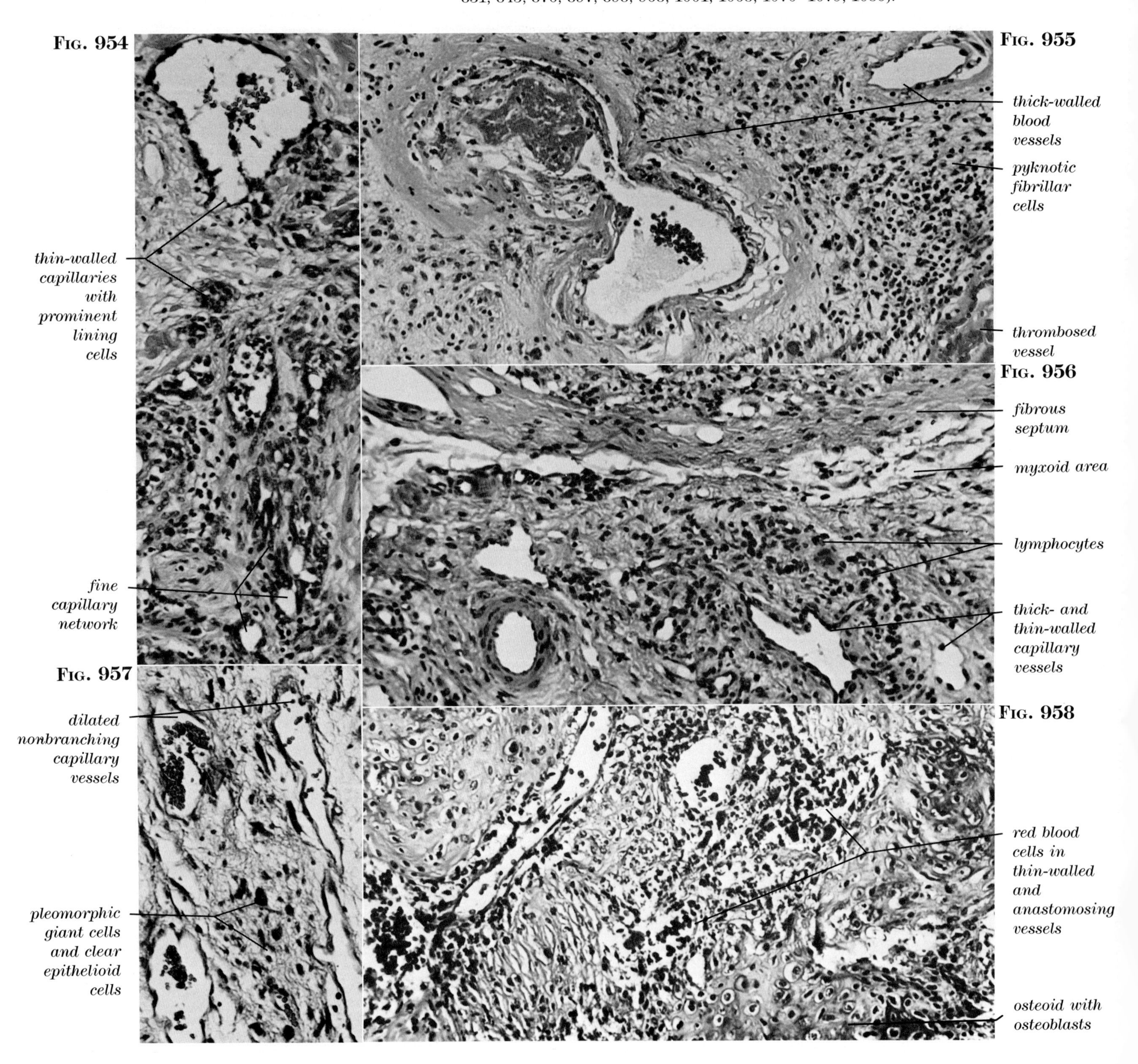

Fig. 959. Angiomyolipoma (×85) (see Figs. 653, 912).

Fig. 961. Glomus tumor (×140) (see Figs. 475, 481).

Fig. 963. Lymphangiosarcoma associated with lymphedema (×140) (see Figs. 321, 551, 649, 954, 965).

Fig. 960. Benign fibroblastic fibrous histiocytoma, so-called "sclerosing hemangioma" (×140) (see Figs. 333, 334, 352, 1021, 1044).

Fig. 962. Borderline cutaneous fibrous histiocytoma (×140) (see Figs. 579, 788, 789, 937, 953, 1002).

Fig. 964. Low grade leiomyoblastoma (×140) (see Figs. 513, 745, 868, 931, 1011, 1071*B*).

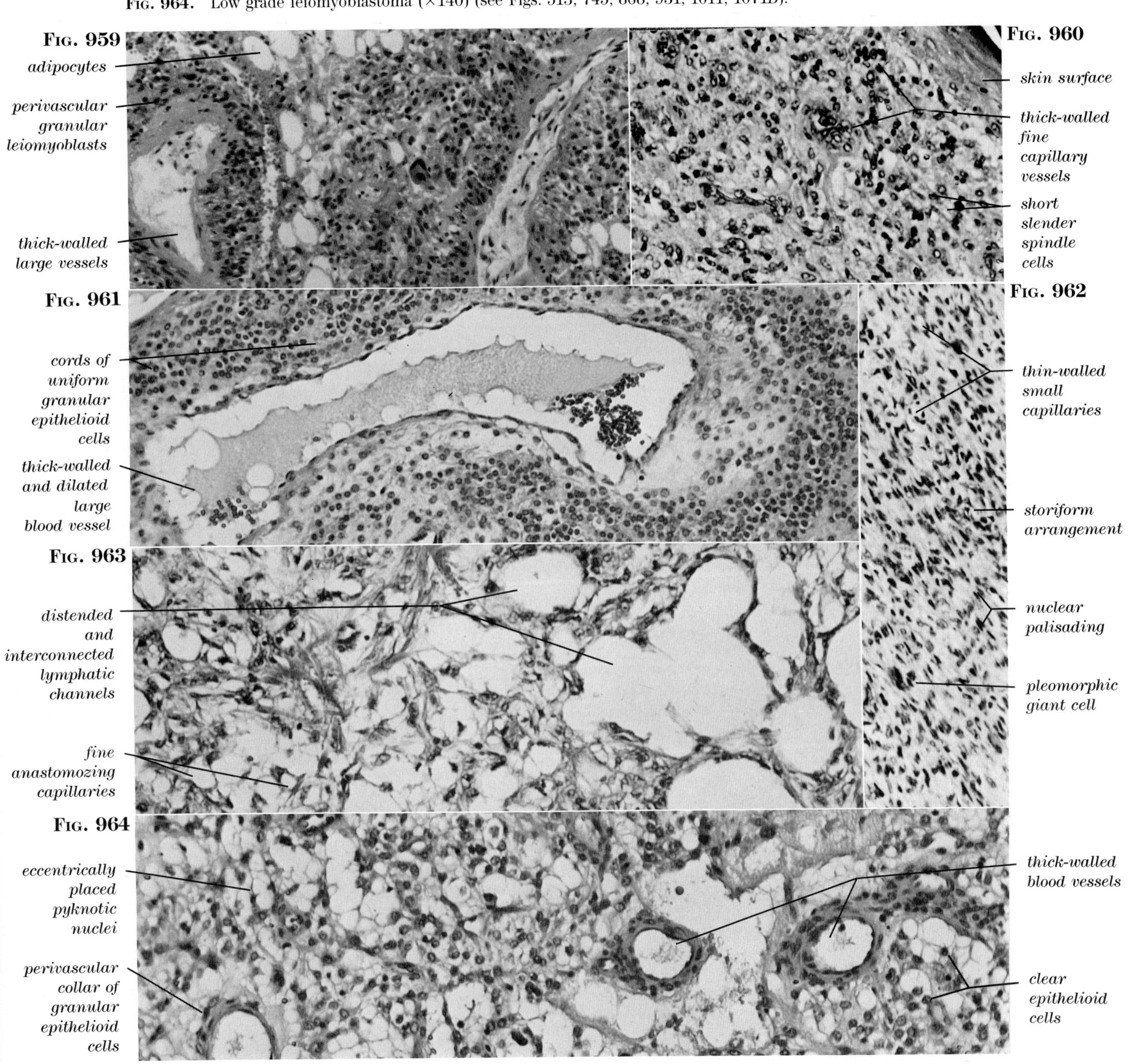

Fig. 965. Lymphangiosarcoma (×140) (see Figs. 84, 159, 551, 558, 649, 954, 963, 970).

Fig. 966. Malignant thymoma (×350) (see Figs. 337, 375, 546, 617).

Fig. 967. Malignant nevoid schwannoma (×140) (see Fig. 735).

Fig. 968. (Bottom center) Osteosarcoma with fine vascular matrix (see Figs. 843, 876, 897, 958, 1001).

Fig. 969. Stromal sarcoma of uterus (×140) (see Figs. 465, 589, 763, 1165, 1217).

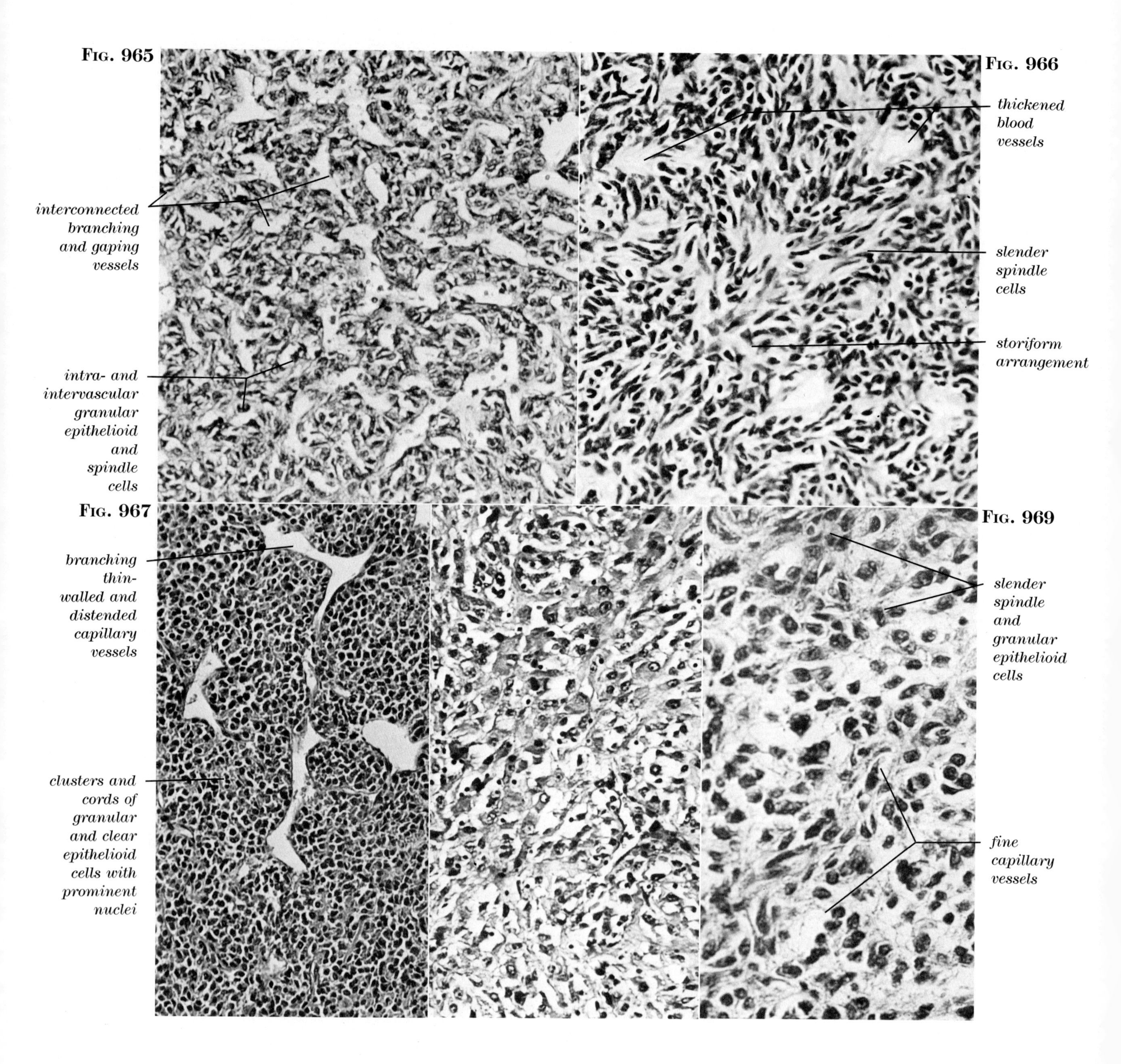

Fig. 970. Lymphangiosarcoma (×140) (see Figs. 84, 424, 558, 649, 954, 963, 965, 1098).

Fig. 971. Fasciitis (×350) (see Figs. 82, 345, 368, 605, 621, 851, 855, 858, 956, 991, 1020, 1030, 1045).

Fig. 972. Low grade hemangiopericytoma (×350) (see Figs. 93, 470, 479, 693, 984, 995, 1203).

Fig. 973. Lipoblastic liposarcoma (×350) (see Figs. 86, 415, 448, 450, 543, 687, 753, 934, 1231).

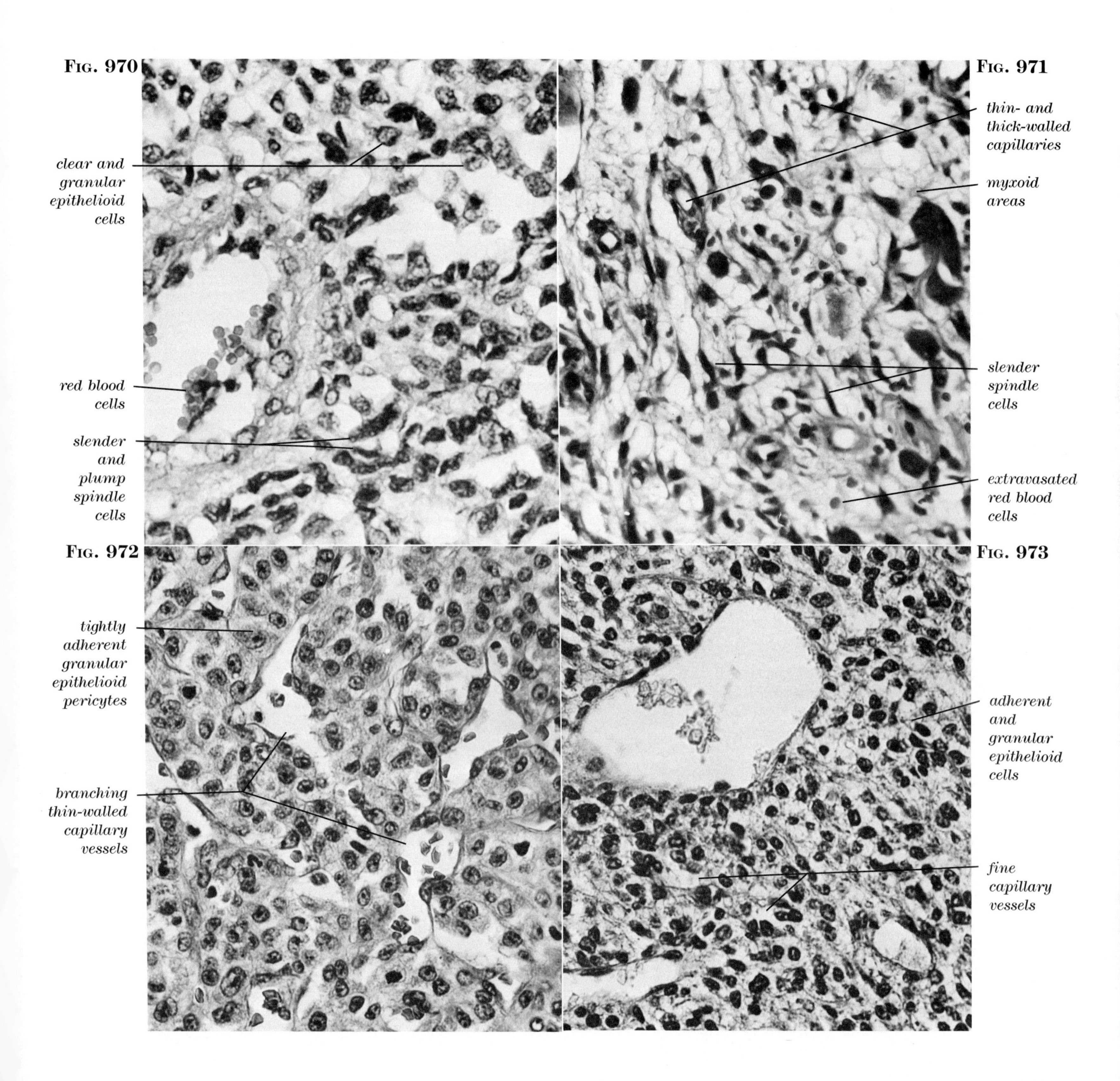

Fig. 974. Hypertrophic hemangioma of skin (×140) (see Figs. 553, 554, 1221).

Fig. 975. Hemangiosarcoma of skin (×140) (see Figs. 92, 102, 316, 523, 576, 720, 731, 905, 976, 989, 990, 1005, 1209, 1210).

Fig. 976. Borderline vascular neoplasm of breast (×140) (see Figs. 731, 905, 975, 989, 990, 1005, 1209, 1210).

Fig. 977. Kaposi's sarcoma (×140) (see Figs. 63, 388, 629, 655, 658, 986, 1126).

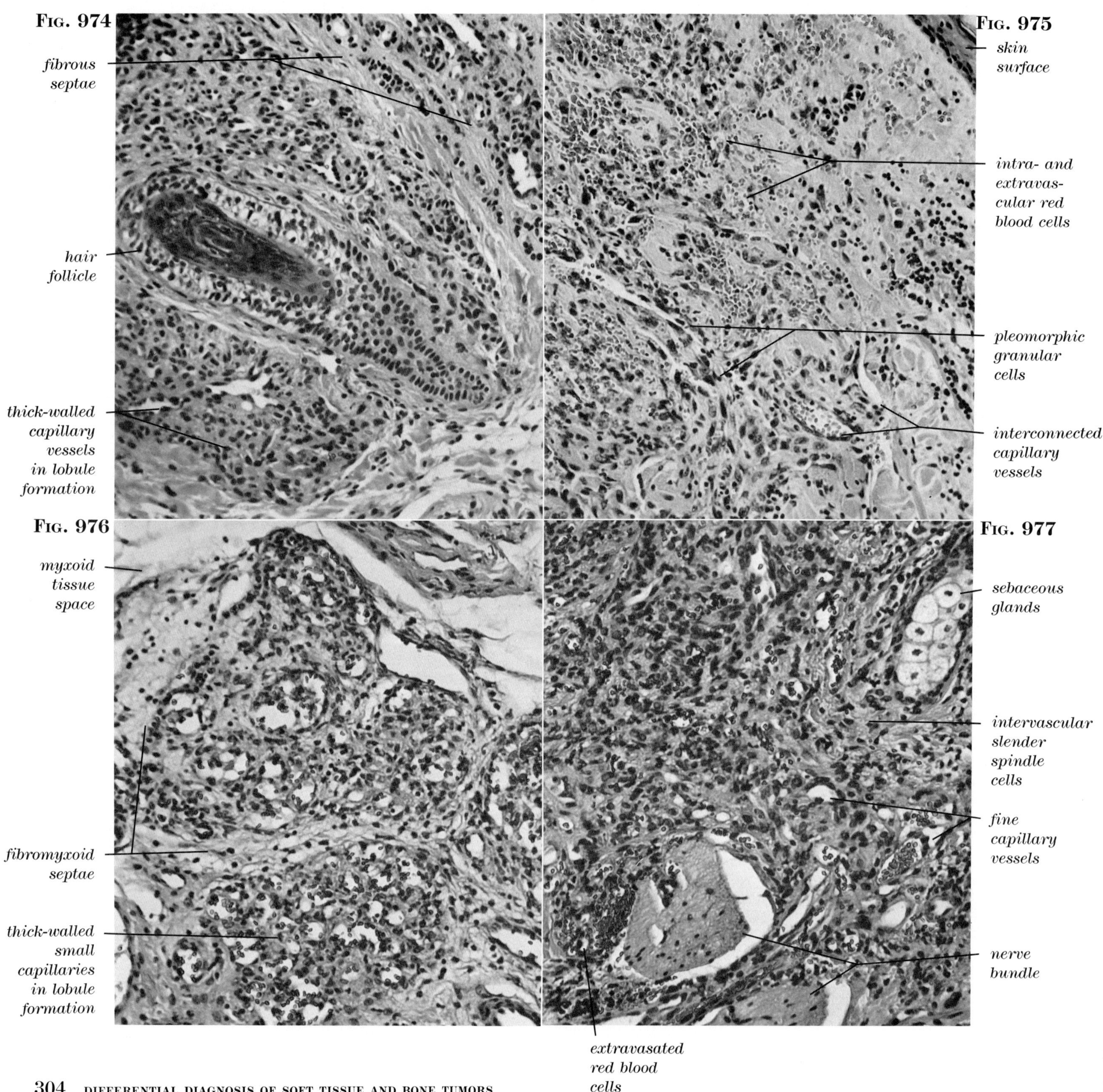

Fig. 978. Lymphangiomyoma (×140) (see Fig. 562).

Fig. 979. Angiofollicular lymphoid hyperplasia (×140) (see Figs. 12, 13, 1040).

Fig. 980. Borderline or atypical glomus tumor (×140) (see Figs. 457, 482, 550, 728, 961, 1141).

Fig. 981. Pyogenic granuloma (×350) (see Figs. 331, 332).

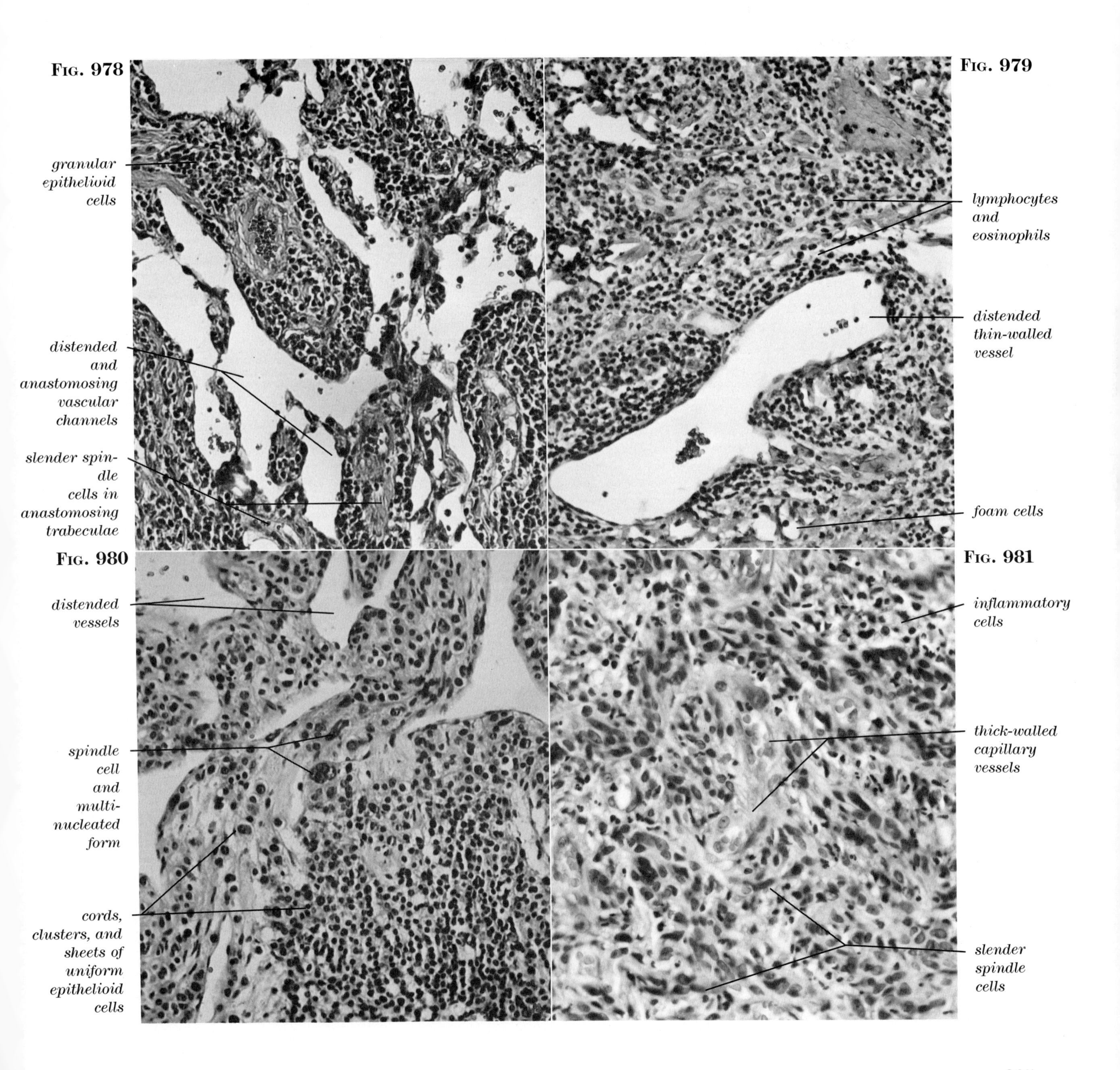

Fig. 982. Primitive neuroectodermal tumor (×350) (see Figs. 561, 565, 572).

Fig. 984. High grade hemangiopericytoma (×350) (see Figs. 93, 374, 470, 479, 484, 525, 533, 696, 972, 995, 1203).

Fig. 986. Kaposi's sarcoma (×350) (see Figs. 63, 388, 629, 655, 658, 977, 1126).

Fig. 983. Angiolipoma (×140) (see Fig. 959).

Fig. 985. Angiomyoma (×140) (see Fig. 653, 912, 959, 1254).

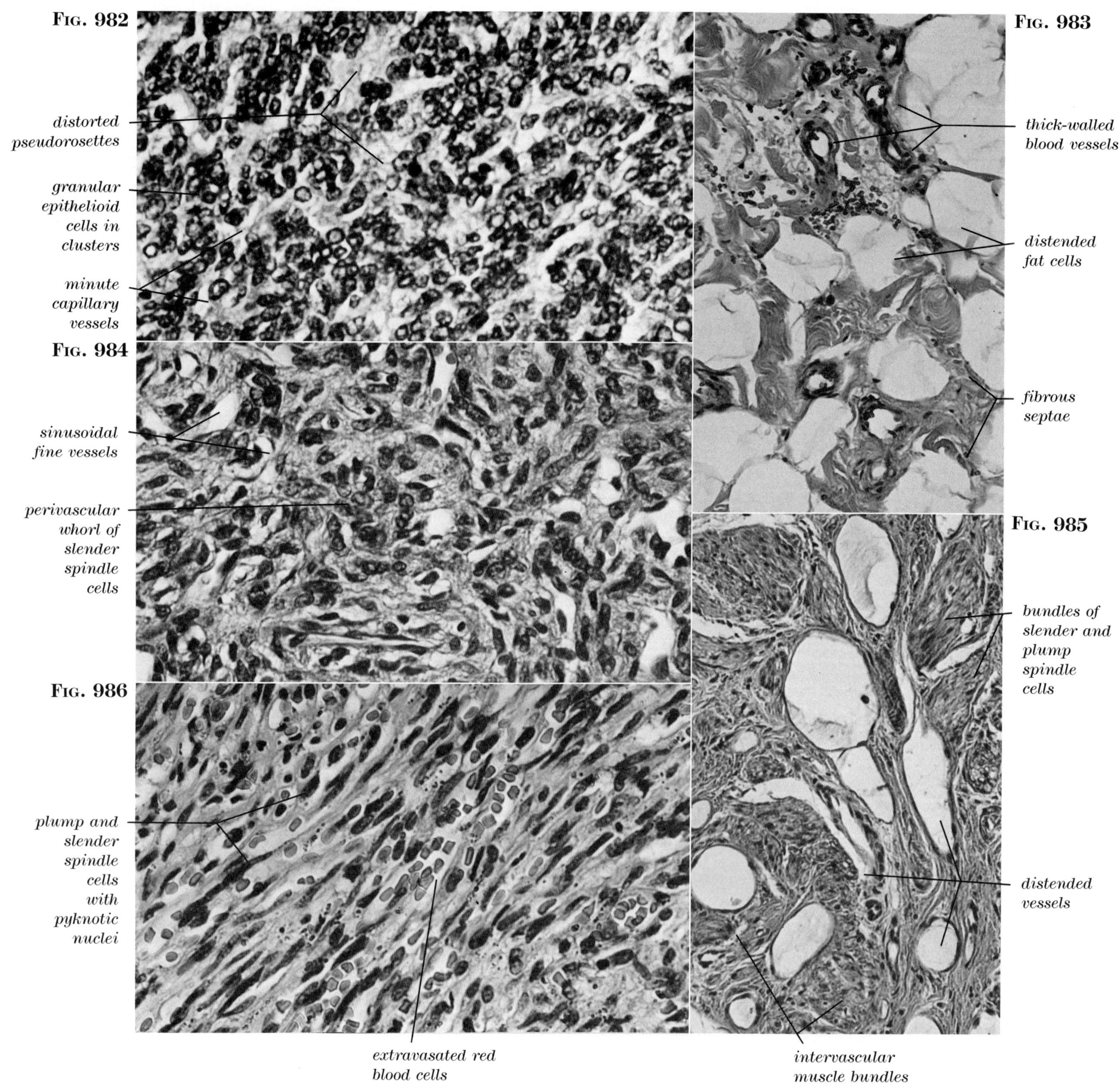

Fig. 987. Borderline vascular neoplasm, so-called "cerebellar hemangioblastoma" (×350).

Fig. 988. Hemangioma (×350) (see Figs. 18, 83, 549, 553, 554, 559, 974, 1221).

Fig. 989. Low grade hemangiosarcoma (×350) (see Figs. 92, 102, 523, 541, 542, 905, 975, 976, 990, 1005, 1209, 1210).

Fig. 990. High grade hemangiosarcoma (×350) (see Figs. 92, 564, 731, 976, 989, 1005).

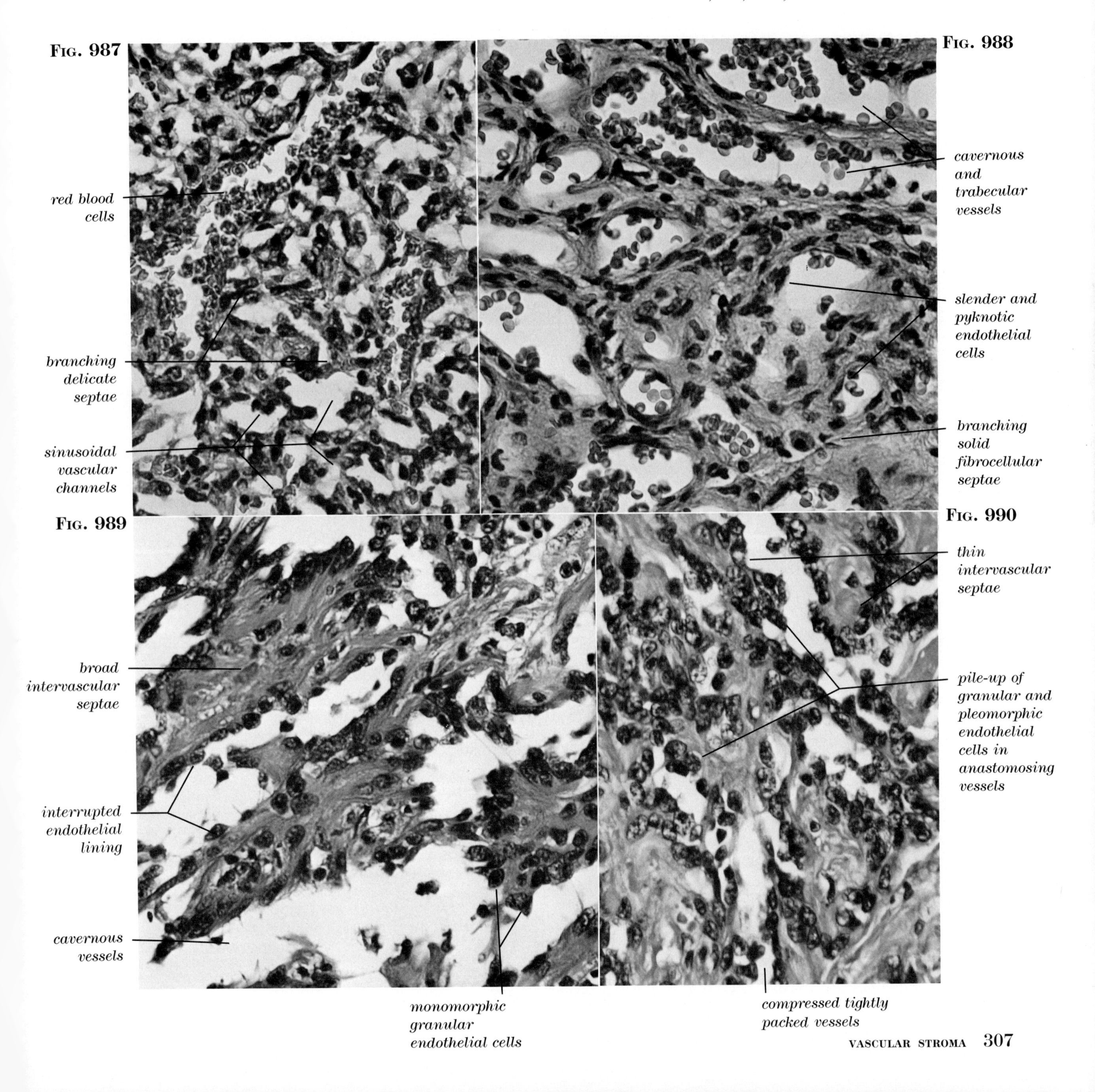

FIG. 991. Fasciitis (×140) (see Figs. 855, 858, 947, 956, 971, 1020).

FIG. 992. (Top center) Metastatic glioblastoma (×140) (see Fig. 1261).

FIG. 993. Neurofibroma (×140) (see Figs. 840, 871, 955, 1013, 1142).

FIG. 994. Low grade leiomyosarcoma (×140) (see Figs. 72, 110, 193, 273, 341, 618, 841, 844, 849, 856, 875, 890, 1007, 1012, 1017A).

FIG. 995. High grade hemangiopericytoma (×140) (see Figs. 93, 374, 470, 479, 484, 525, 696, 972, 984, 1203).

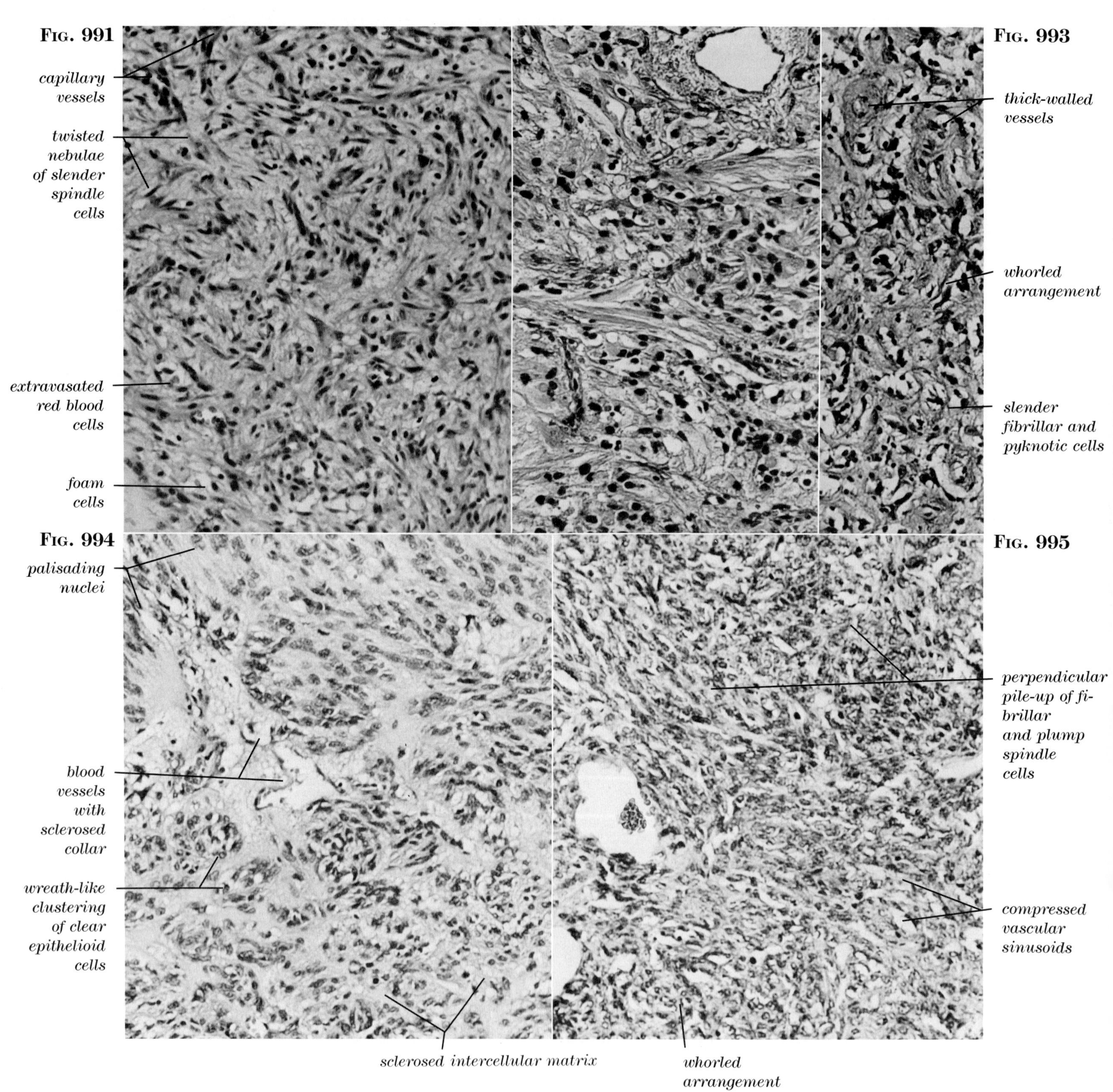

Fig. 996. Lipoblastoma (×140) (see Figs. 412, 413, 437).

Fig. 997. Myxoid liposarcoma (×140) (see Figs. 439, 783, 933, 945, 948, 1046, 1113, 1169, 1232).

Fig. 998. Fibroblastic liposarcoma (×140) (see Figs. 343, 440, 641, 668, 837, 848, 926).

Fig. 999. Well-differentiated liposarcoma (×140) (see Figs. 86, 431, 435, 822, 1184, 1233).

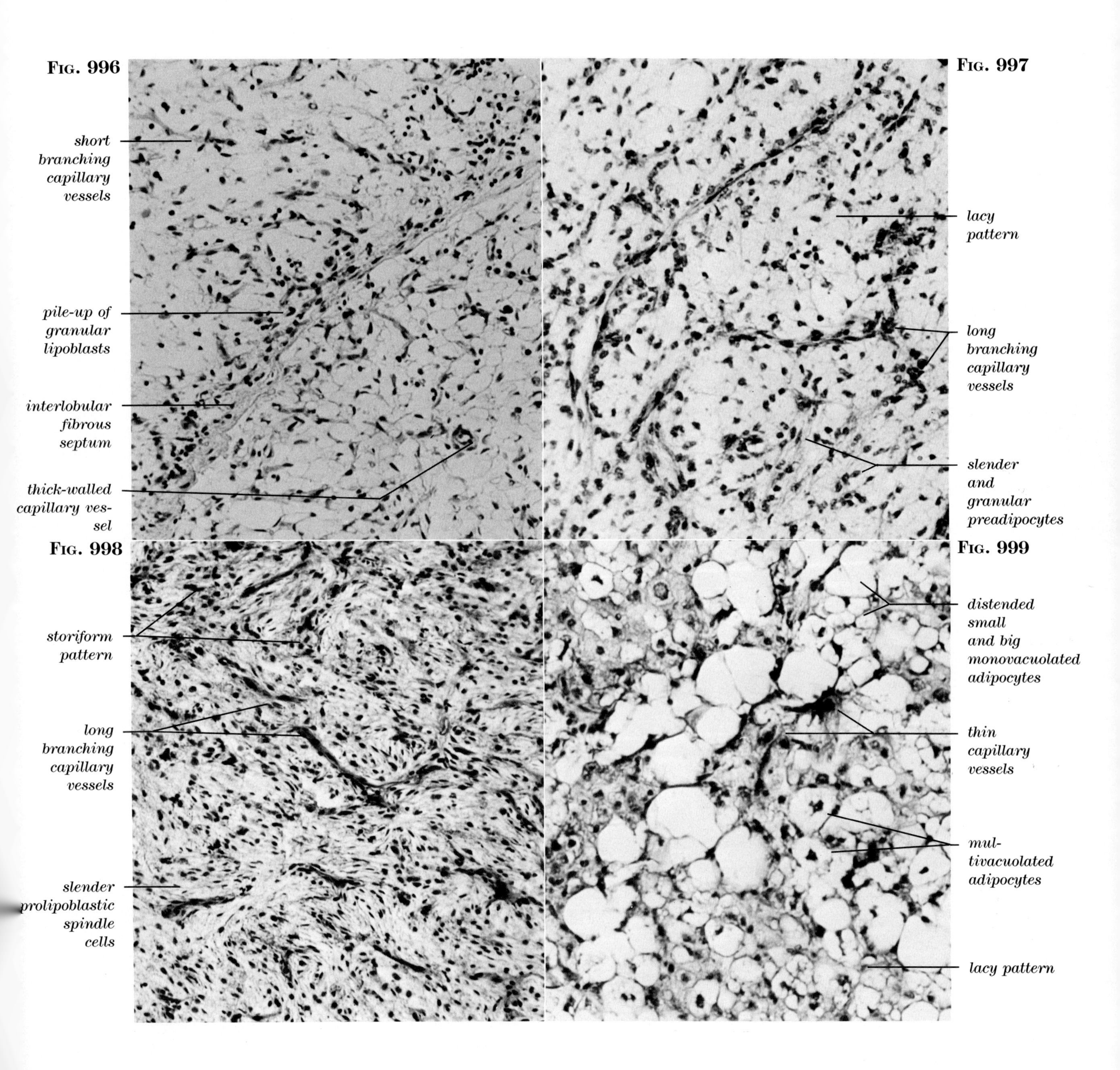

Fig. 1000. Giant cell granuloma of mandible (×350) (see Figs. 119, 632, 787, 1069).

Fig. 1001. Telangiectatic osteosarcoma (×140) (see Figs. 108, 615, 819, 831, 843, 876, 897, 898, 958, 968, 1068, 1070–1079, 1086).

Fig. 1002. Low grade malignant pleomorphic fibrous histiocytoma (×140) (see Figs. 805, 818, 937, 953, 962, 1003).

Fig. 1003. Borderline histiocytic fibrous histiocytoma (×140) (see Figs. 452, 534, 592, 600, 711, 756, 775, 776, 784, 791, 816, 872, 892, 893, 930, 1018B, 1048).

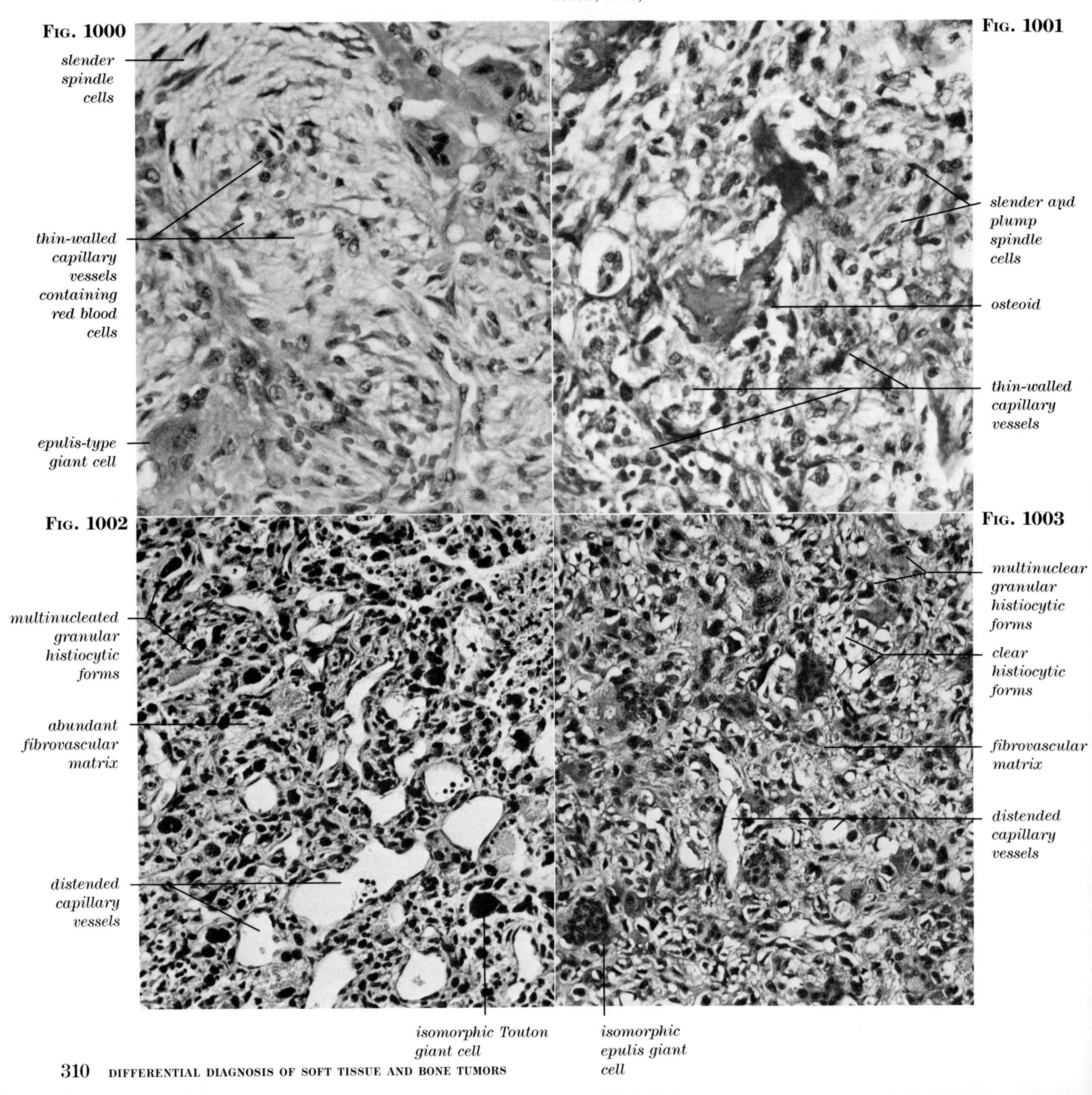

Fig. 1004. Papillary endothelial hyperplasia (×140) (see Fig. 560).

Fig. 1005. Borderline or atypical vascular neoplasm of breast (×140) (see Figs. 905, 975, 976, 989, 990, 1209, 1210).

Fig. 1006. Arteriovenous malformation (×140) (see Figs. 139, 197, 911, 1009).

Fig. 1007. High grade leiomyosarcoma of small intestine (×350) (see Figs. 856, 875, 890, 994, 1012, 1017*A*, 1051, 1131, 1132).

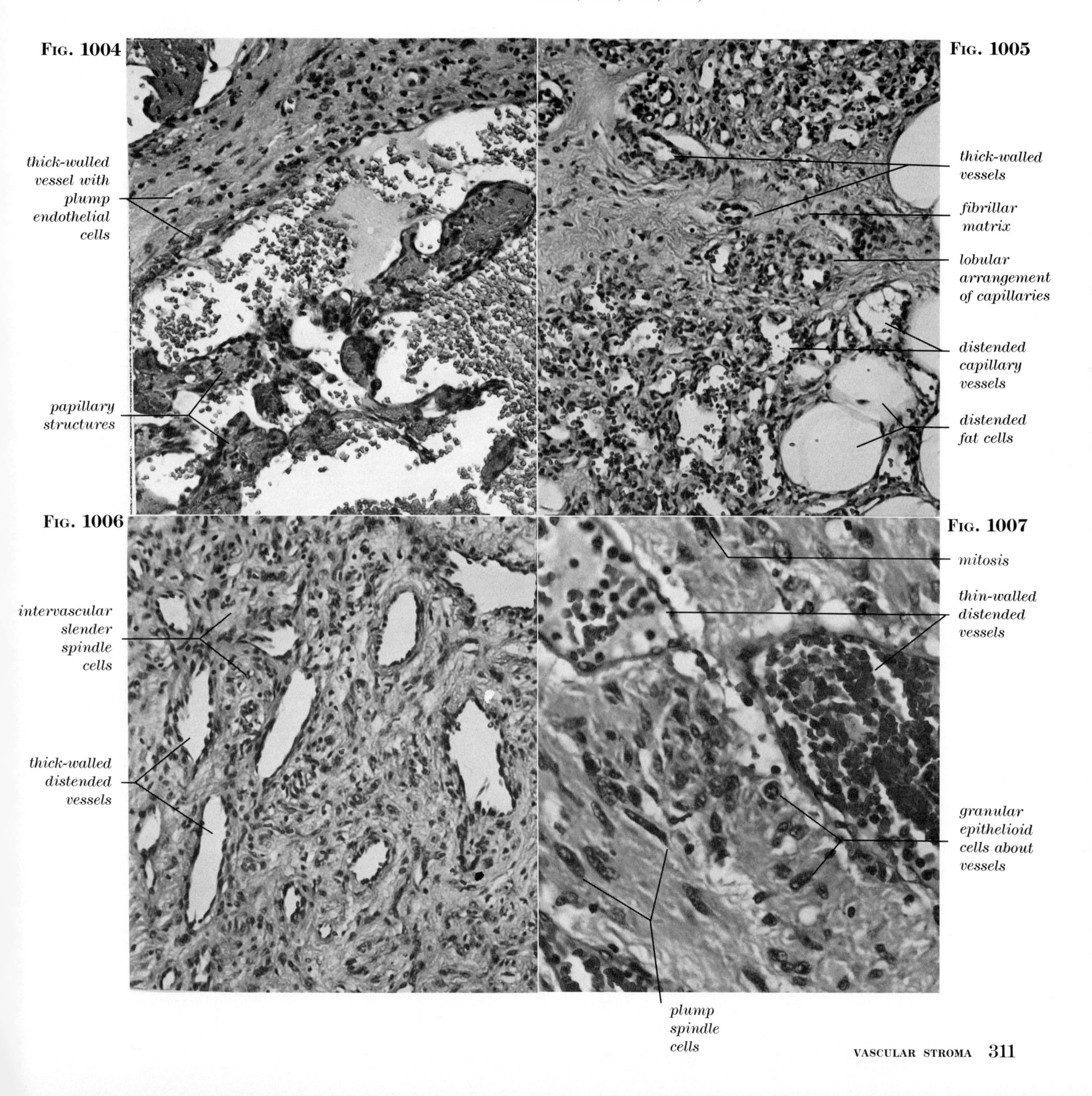

Fig. 1008. Borderline pleomorphic fibrous histiocytoma (×350) (see Figs. 953, 962, 1002, 1003, 1032, 1047).

Fig. 1009. Arteriovenous malformation (×140) (see Figs. 139, 197, 911, 1006).

Fig. 1010. Vascular leiomyoma (×350) (see Figs. 358, 359, 361, 380).

Fig. 1011. Low grade leiomyoblastoma of stomach (×140) (see Figs. 513, 691, 713, 733, 814, 828, 868, 931, 964, 1017*B*, 1053).

Fig. 1012. Low grade leiomyosarcoma of stomach (×140) (see Figs. 404, 666, 841, 844, 849, 856, 875, 890, 994, 1007, 1017*A*).

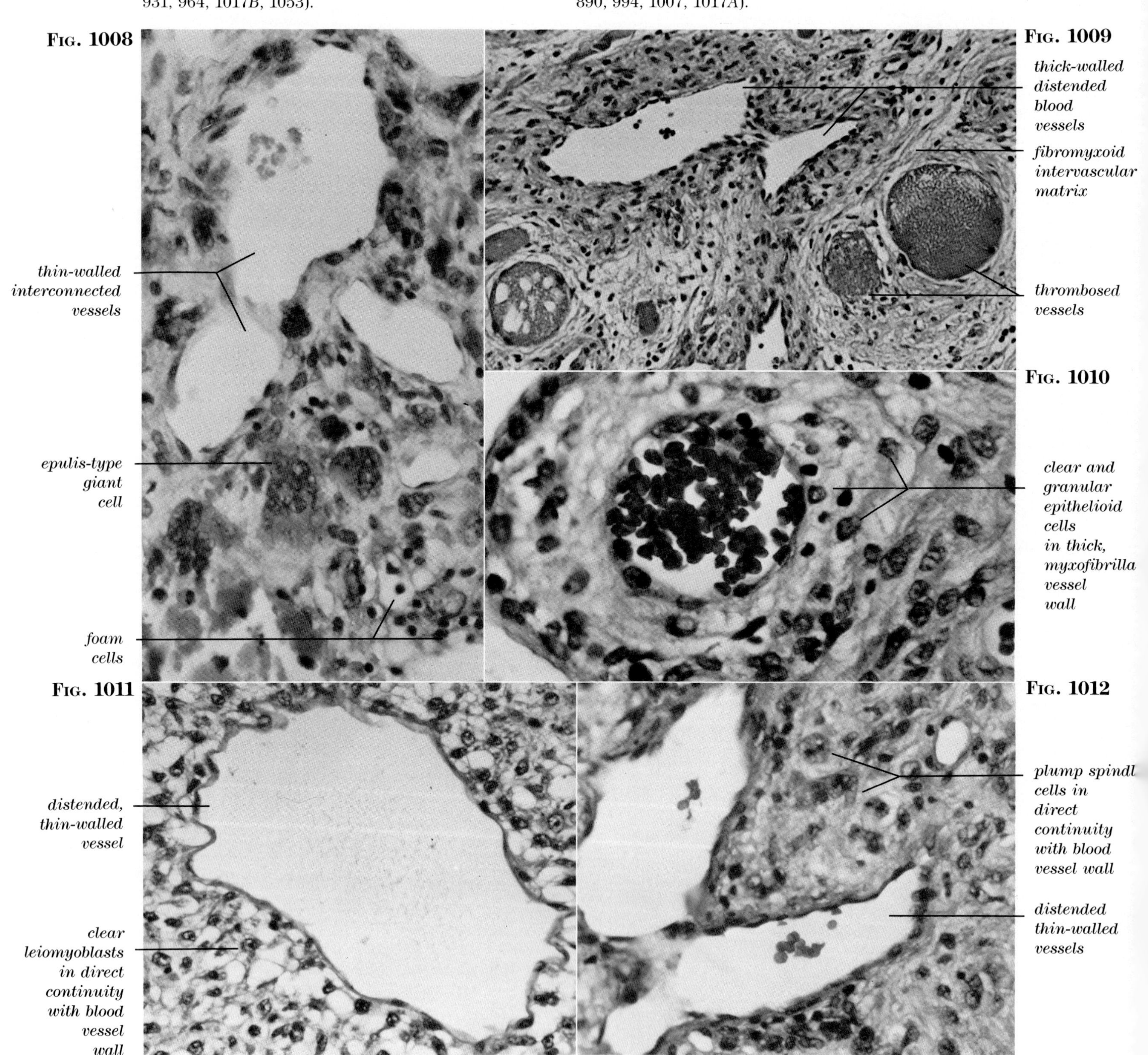

Fig. 1013. Neurofibroma (×350) (see Figs. 840, 871, 955, 993, 1142, 1143).

Fig. 1014. Tendosynovial cyst (×350) (see Figs. 70, 144, 186, 508, 786, 867, 1023, 1029, 1136).

Fig. 1015. Aneurysmal bone cyst (×350) (see Figs. 96, 125, 150, 163, 509, 785, 1058).

Fig. 1016. Cystic hygroma (×140) (see Fig. 19).

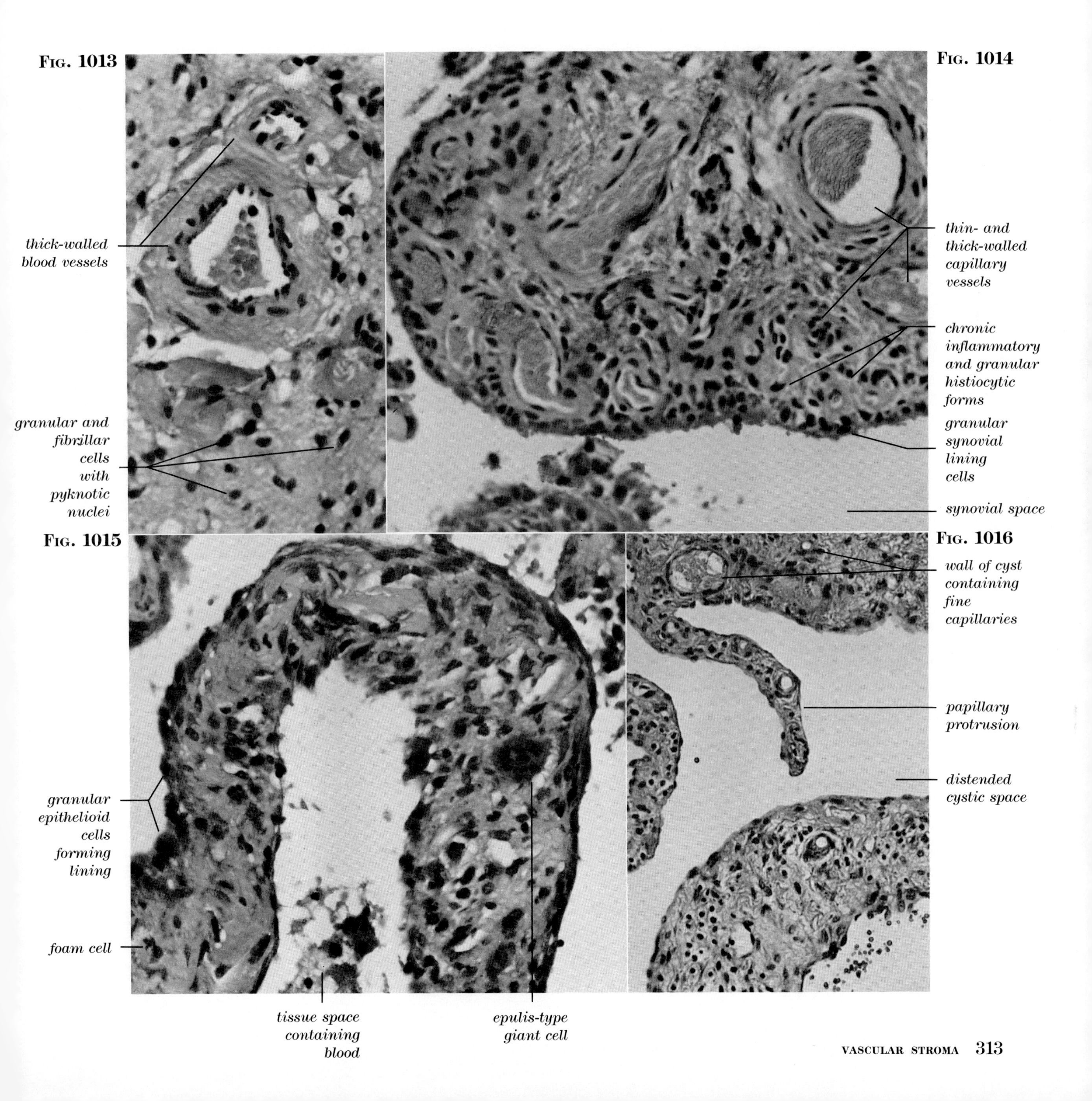

Fig. 1017A. High grade vascular leiomyosarcoma (×350) (see Figs. 72, 110, 341, 347, 618, 659, 841, 844, 849, 875, 890, 994, 1007, 1012, 1051, 1131).

Fig. 1017B. High grade vascular leiomyoblastoma (×350) (see Figs. 420, 429, 503, 513, 691, 713, 733, 744, 745, 747, 750, 752, 795, 814, 828, 868, 931, 964, 1011, 1053).

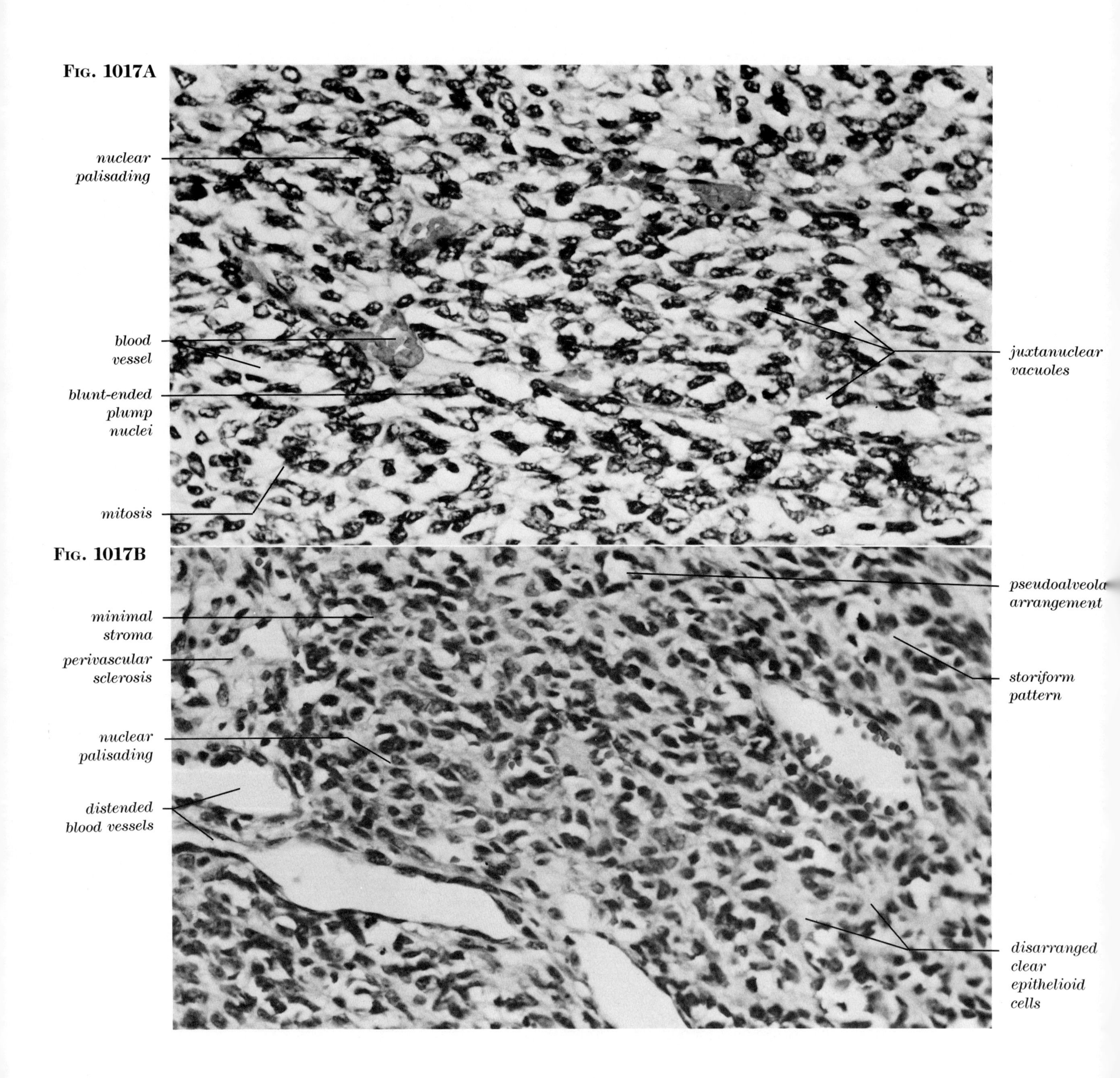

Fig. 1018A. Renal cell carcinoma metastatic to patella (×350) (see Figs. 545, 651, 716, 764, 768, 1270).

Fig. 1018B. Malignant histiocytic fibrous histiocytoma (×350) (see Figs. 161, 452, 534, 592, 872, 892, 893, 930, 1003, 1048).

Fig. 1019A. A sparsely cellular area of pleomorphic liposarcoma (×350) (see Figs. 86, 284, 458, 590, 799, 823, 928, 1049).

Fig. 1019B. Merkel cell tumor (×350) (see Figs. 1112, 1236, 1237).

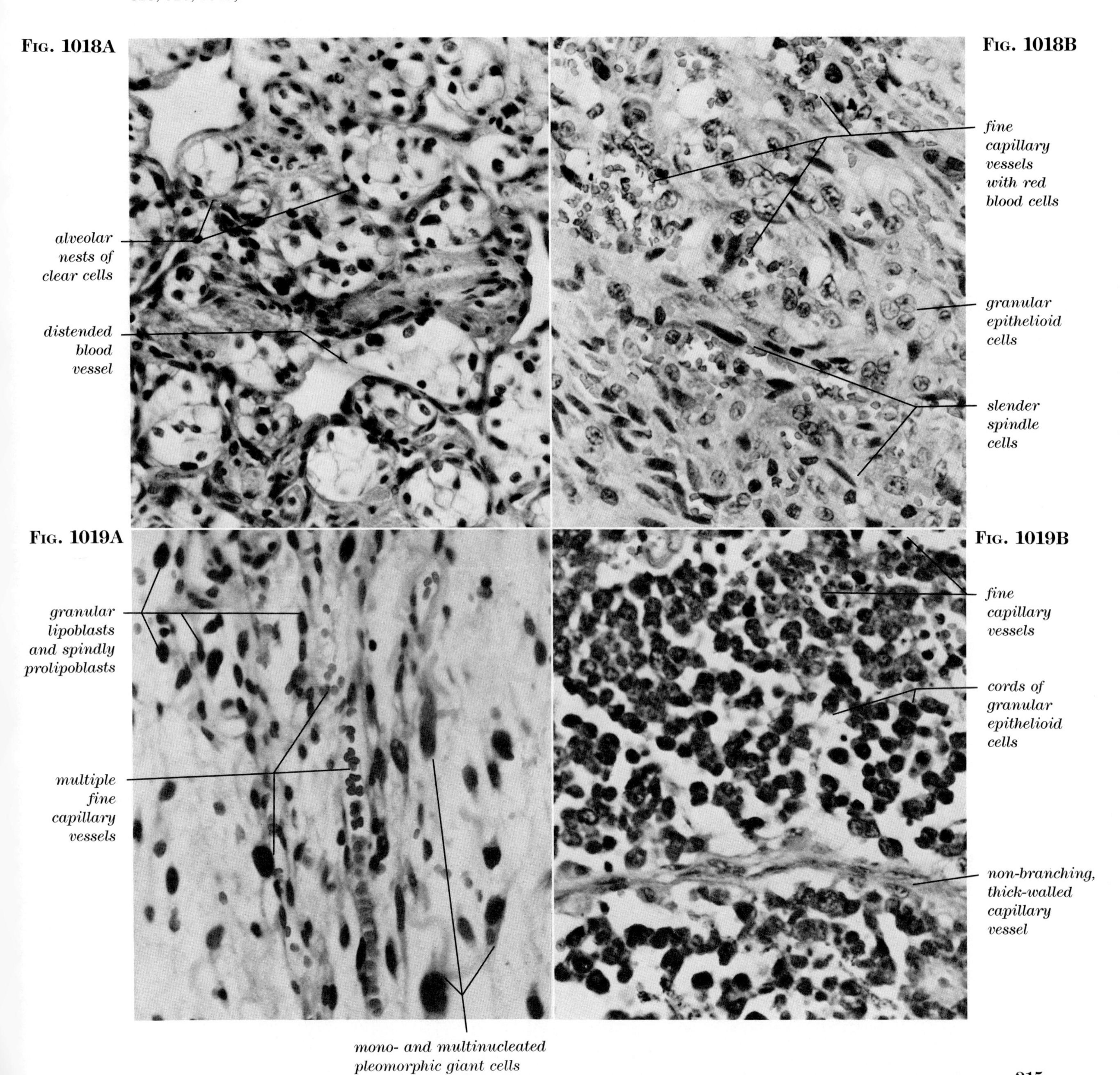

FIG. 1020. Fasciitis (×140) (see Figs. 82, 345, 368, 389, 621, 834, 851, 855, 858, 947, 956, 971, 991, 1030).

FIG. 1021. Benign fibroblastic fibrous histiocytoma, so-called "sclerosing hemangioma" (×140) (see Figs. 333, 334, 352, 960, 1044).

FIG. 1022. Mesenteric fibromatosis (×140) (see Figs. 367, 385, 387, 419, 581, 582, 616, 860, 861, 873, 874, 878, 882, 1025).

FIG. 1023. Tendosynovitis, so-called "villonodular synovitis" (×140) (see Figs. 70, 786, 867, 1029, 1136).

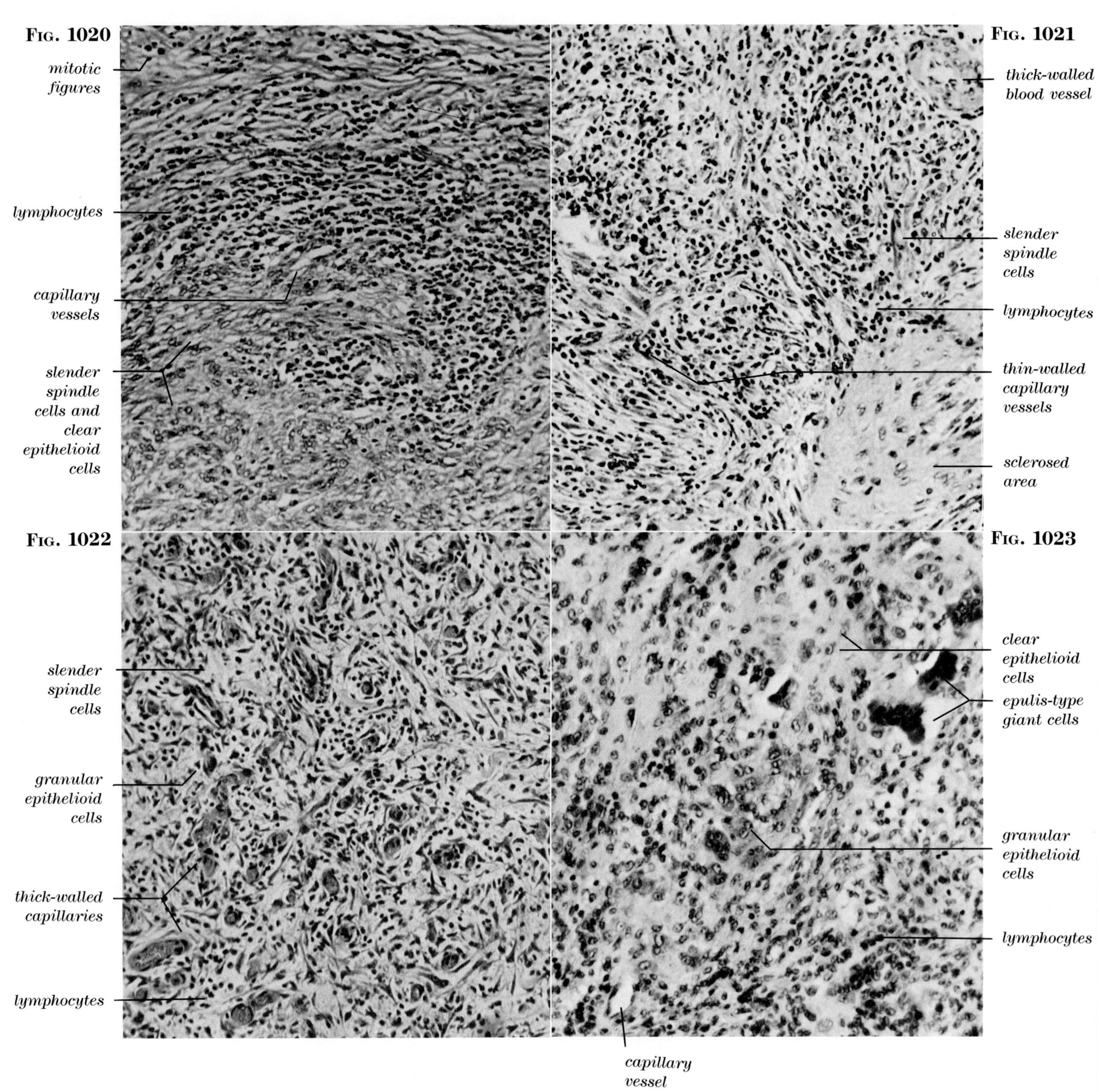

Fig. 1024. Hodgkin's disease (×140) (see Figs. 27, 742, 885).

Fig. 1025. (Top center) Mediastinal fibromatosis with distended capillary vessels (×140) (see Figs. 874, 878, 882, 1035).

Fig. 1026. Borderline fibroblastic fibrous histiocytoma (×350) (see Figs. 76, 373, 710, 760, 833, 1133).

Fig. 1027. Aponeurotic fibroma (×350) (see Figs. 336, 859, 884, 886, 904, 915, 917, 1109, 1204).

Fig. 1028. Eosinophilic granuloma (×350) (see Figs. 118).

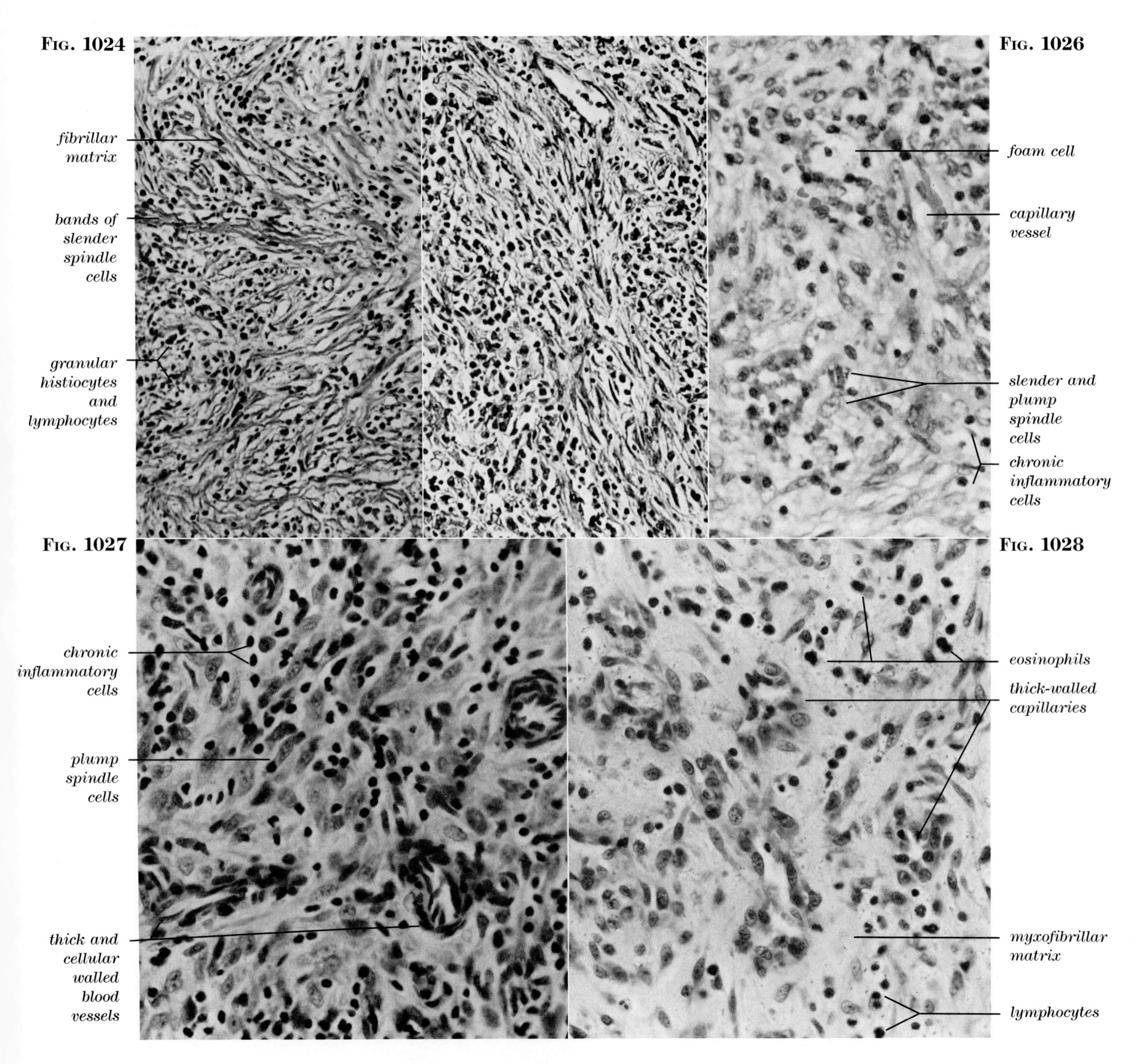

Fig. 1029. Tendosynovitis (×140) (see Figs. 70, 786, 867, 1023, 1136).

Fig. 1030. Fasciitis (×140) (see Figs. 82, 605, 956, 971, 991, 1045).

Fig. 1031. Xanthogranuloma, so-called "juvenile cutaneous xanthogranuloma" (×140) (see Figs. 607, 734, 779, 1127, 1201).

Fig. 1032. High grade malignant pleomorphic fibrous histiocytoma (×140) (see Figs. 937, 953, 962, 1002, 1008, 1047).

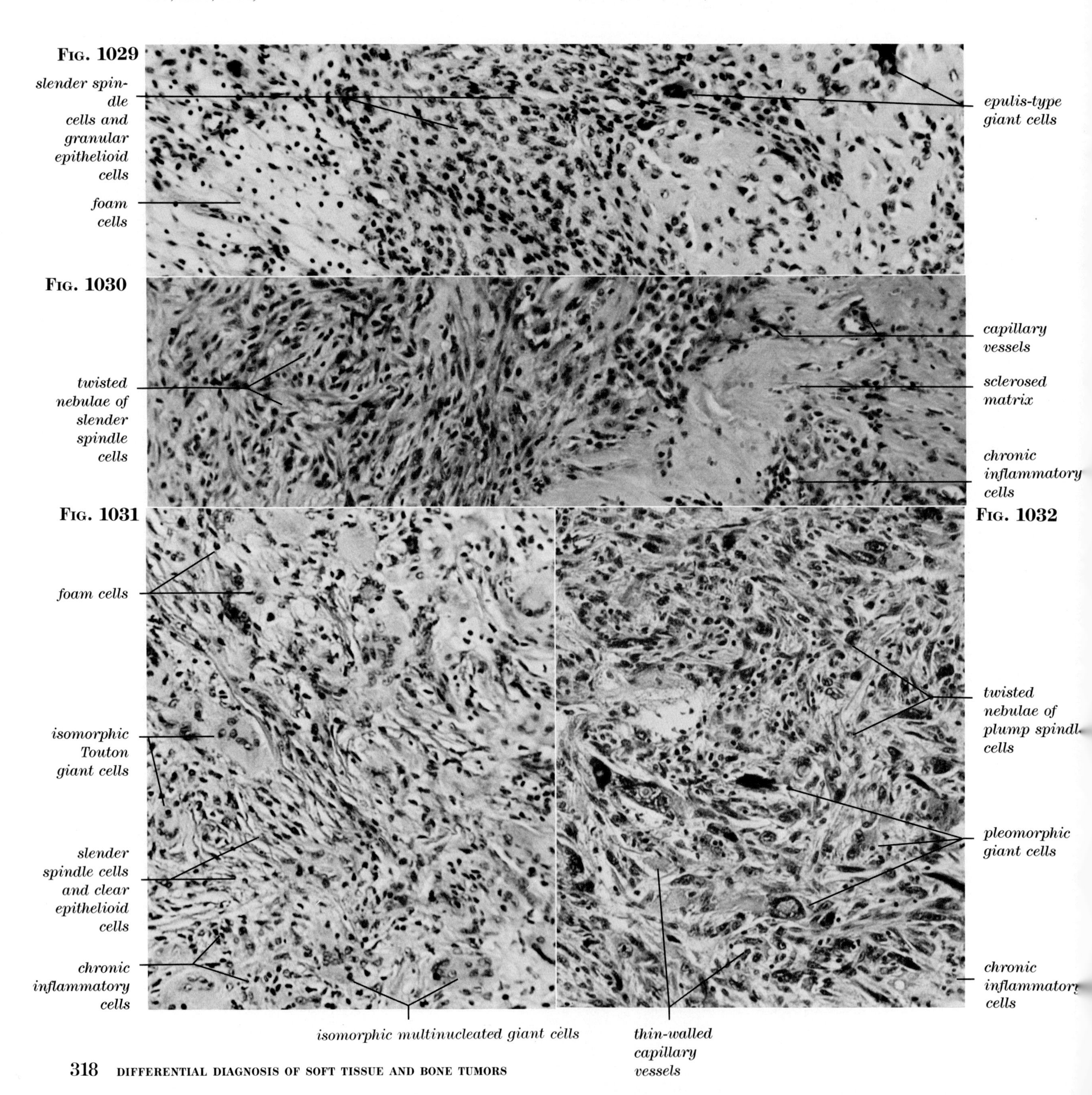

Fig. 1033. Myxoid embryonal rhabdomyosarcoma (×140) (see Figs. 923, 929, 944, 952, 1251).

Fig. 1034. Rhabdomyoblastoma (×140) (see Figs. 569, 594, 709, 796, 936, 1159).

Fig. 1035. Mesenteric fibromatosis (×140) (see Figs. 860, 861, 873, 874, 878, 882, 1022, 1025, 1036, 1050, 1081, 1095).

Fig. 1036. Retroperitoneal fibromatosis (×140) (see Figs. 17, 280, 367, 616, 860, 861, 873, 874, 878, 882, 1022, 1025, 1035, 1050).

Fig. 1037. Leukemic infiltrate in connective tissue (×140) (see Fig. 888).

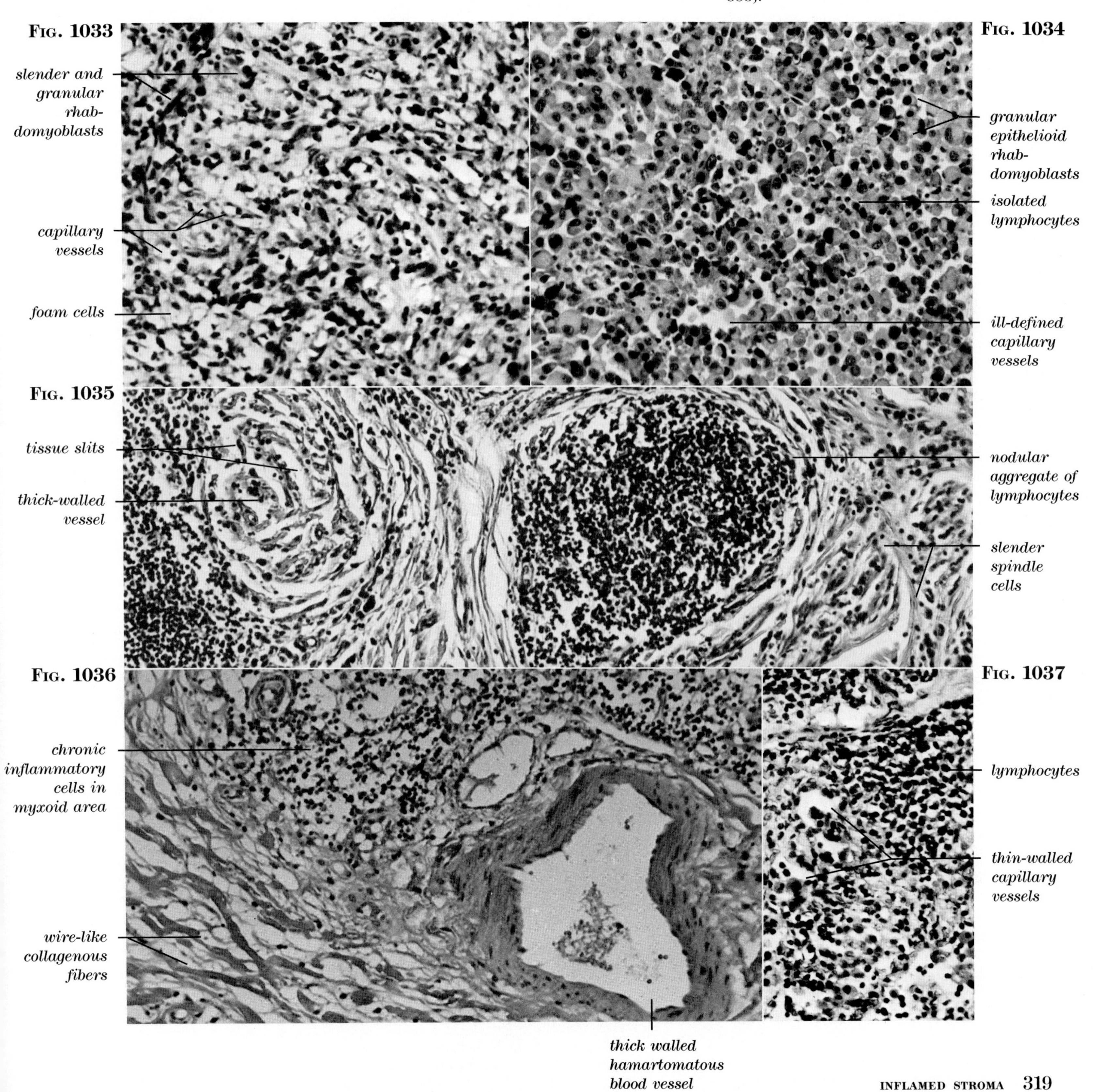

Fig. 1038. Dystrophy of muscle (×85) (see Figs. 1039, 1094, 1186).

Fig. 1039. Myositis (×85) (see Figs. 1038, 1094, 1186).

Fig. 1040. Angiofollicular lymphoid hyperplasia (×85) (see Figs. 12, 13, 979).

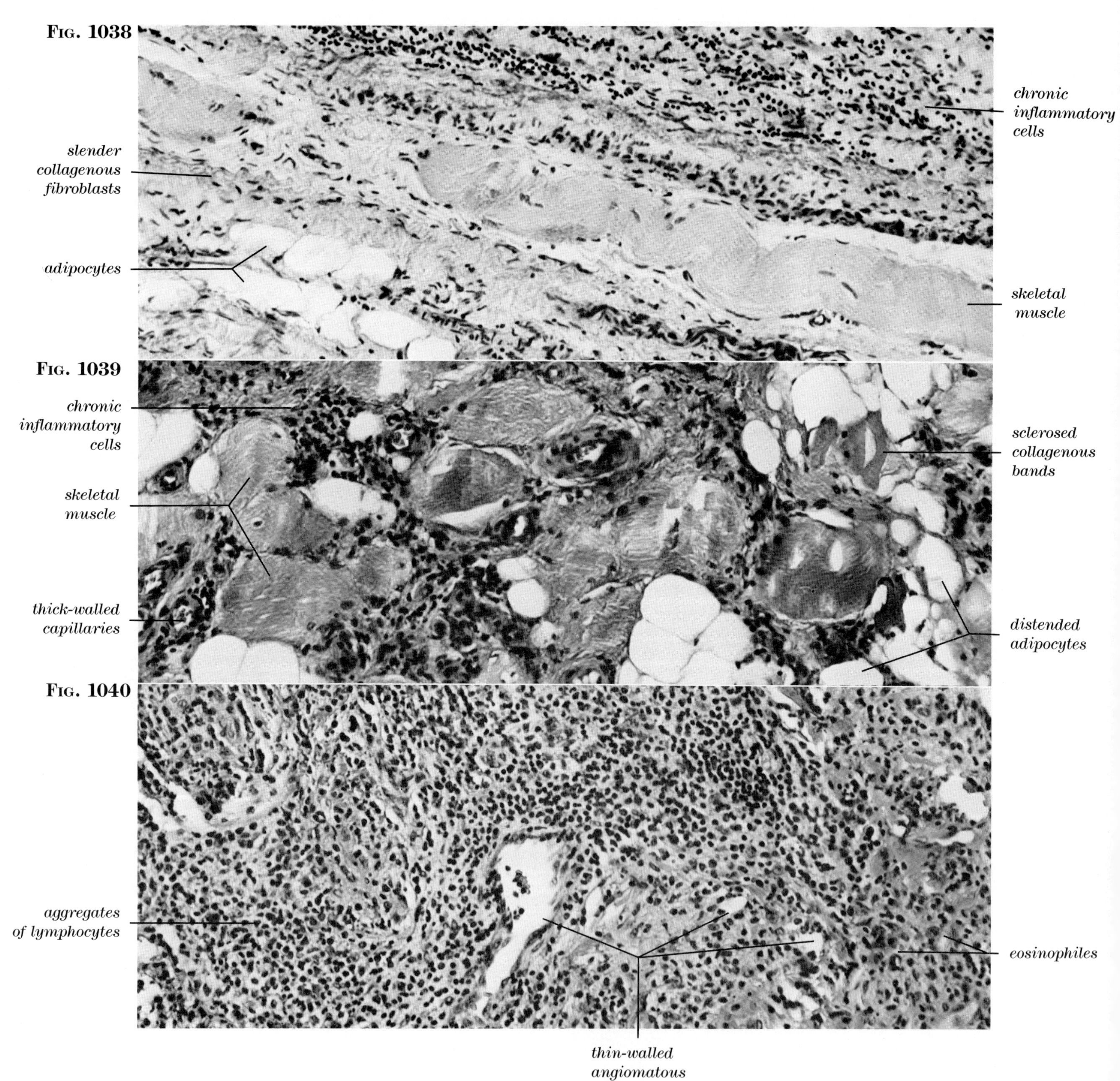

FIG. 1041. Myelolipoma (×140) (see Fig. 959).

FIG. 1042. Fat necrosis (×140) (see Figs. 425, 749, 1056, 1185).

FIG. 1043. Lipogranuloma (×140) (see Figs. 433, 820, 907).

FIG. 1044. Benign fibroblastic fibrous histiocytoma, so-called "sclerosing hemangioma" (×350) (see Figs. 333, 334, 352, 960, 1021).

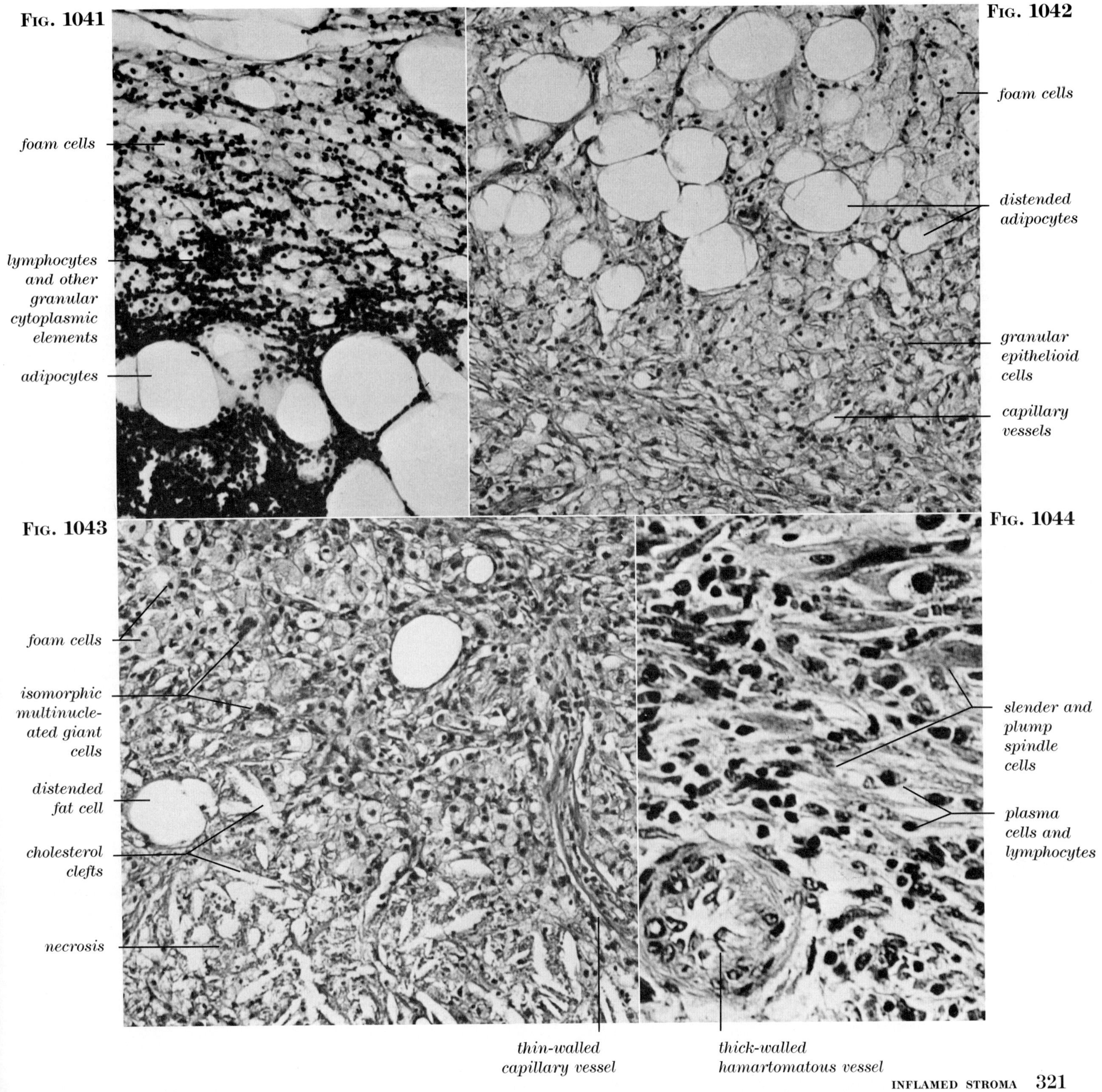

Fig. 1045. Fasciitis (×140) (see Figs. 345, 368, 389, 403, 408, 578, 605, 621, 834, 851, 855, 858, 947, 956, 971, 991, 1020, 1030, 1080).

Fig. 1046. Inflamed myxoid liposarcoma (×140) (see Figs. 783, 933, 948, 997, 1113, 1169).

Fig. 1047. High grade malignant pleomorphic fibrous histiocytoma (×85) (see Figs. 962, 1002, 1003, 1008, 1032, 1148).

Fig. 1048. Borderline histiocytic fibrous histiocytoma of tendosynovium (×350) (see Figs. 452, 534, 592, 600, 711, 756, 775, 776, 784, 791, 816, 872, 892, 893, 930, 1018*B*, 1149).

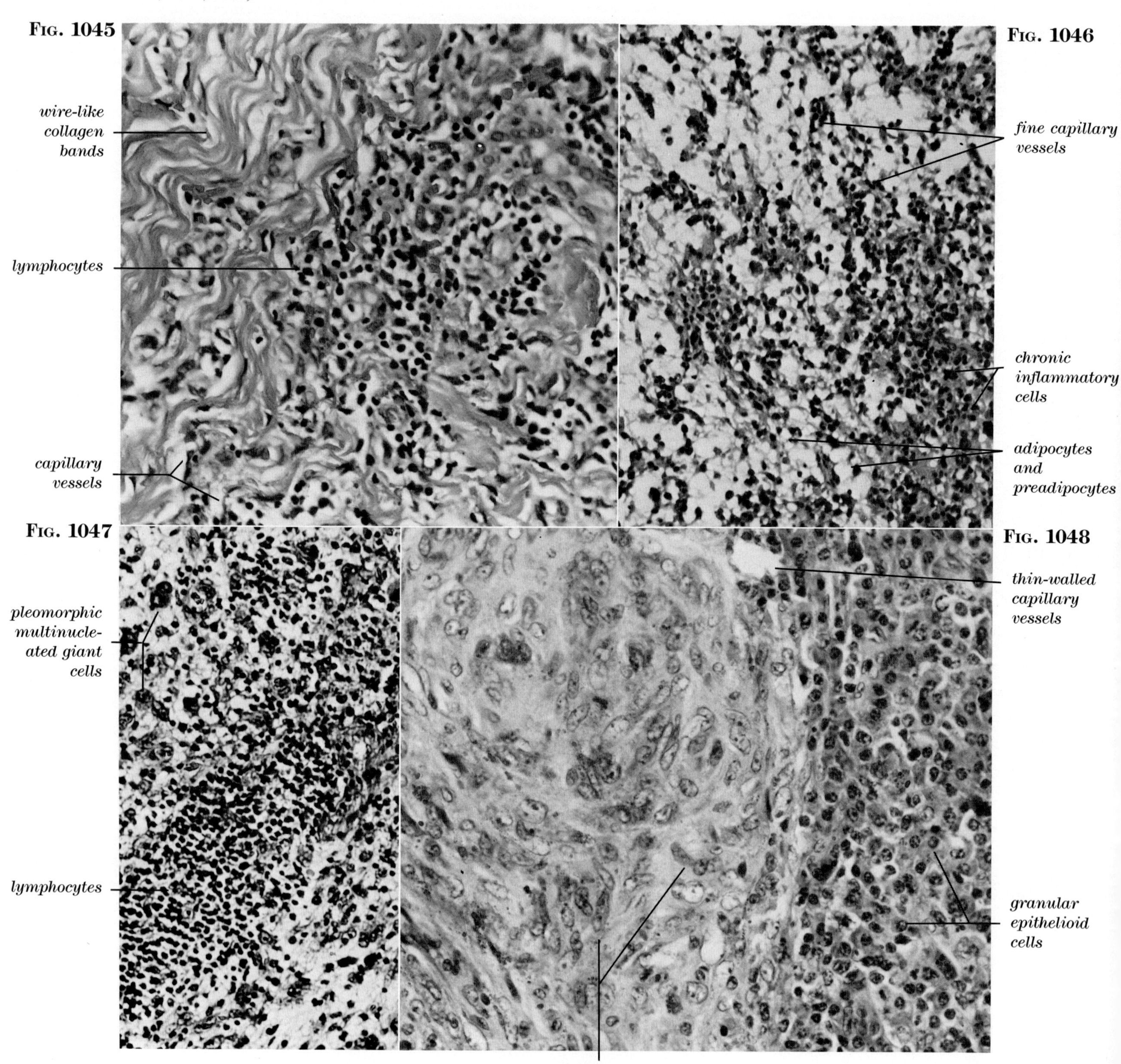

Fig. 1049. Pleomorphic liposarcoma (×140) (see Figs. 86, 294, 590, 799, 928, 1019A).

Fig. 1050. Mesenteric fibromatosis (×140) (see Figs. 873, 874, 878, 882, 1022, 1081).

Fig. 1051. High grade leiomyosarcoma (×140) (see Figs. 841, 844, 849, 875, 890, 994, 1007, 1012, 1017A., 1131, 1132).

Fig. 1052. Epithelioid sarcoma (×140) (see Figs. 676, 830, 866, 895, 1059, 1146, 1170, 1244).

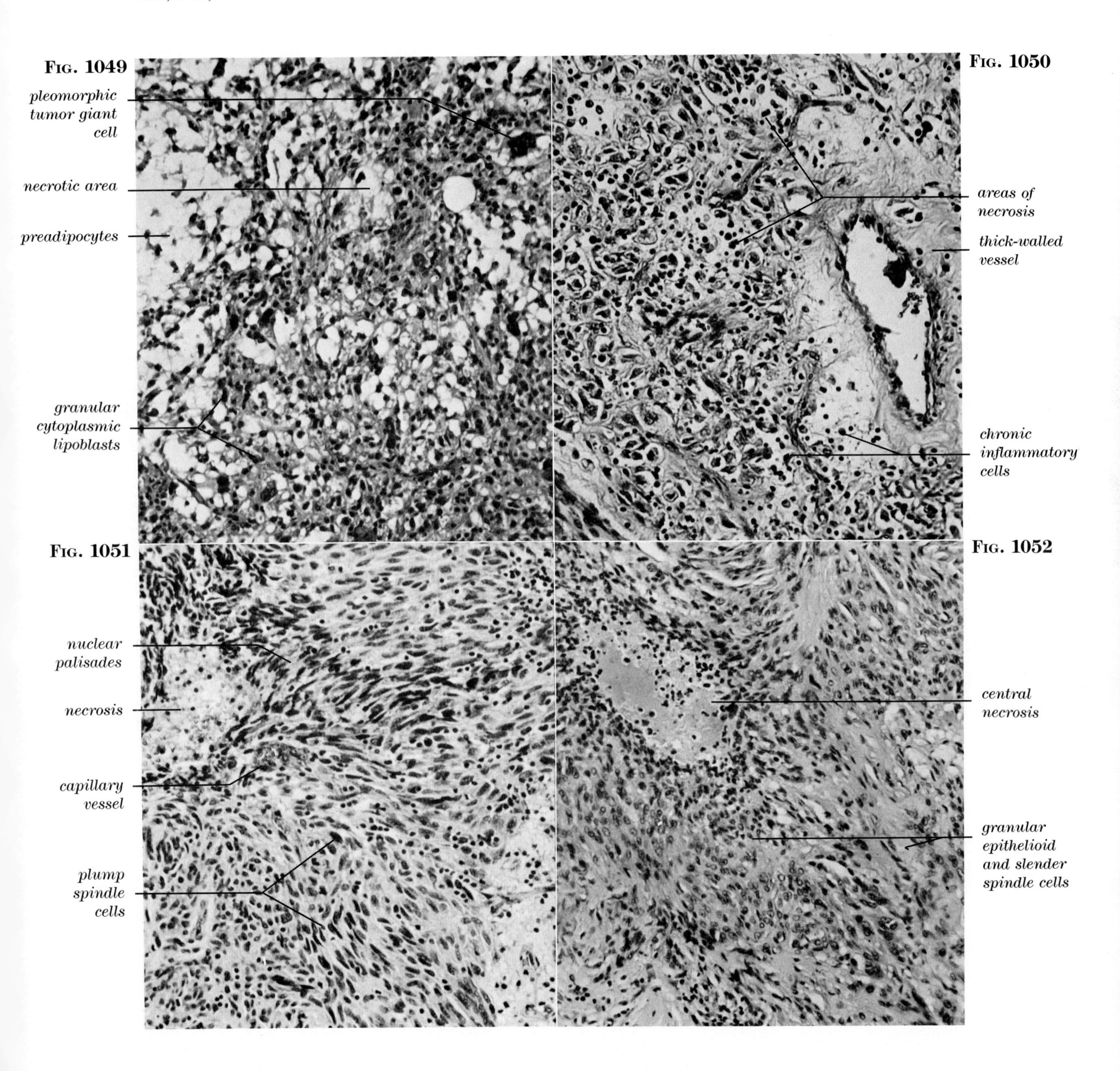

Fig. 1053. Leiomyoblastoma (×350) (see Figs. 868, 931, 964, 1011, 1017*B*, 1139).

Fig. 1054. Cysticercus cellulosae in subcutaneous tissue (×10).

Fig. 1055. Choriocarcinoma metastatic to retroperitoneum (×140) (see Figs. 195, 829).

Fig. 1056. Fat necrosis (×140) (see Figs. 425, 749, 1042, 1185).

Fig. 1057. Gouty arthritis (×140) (see Figs. 1197).

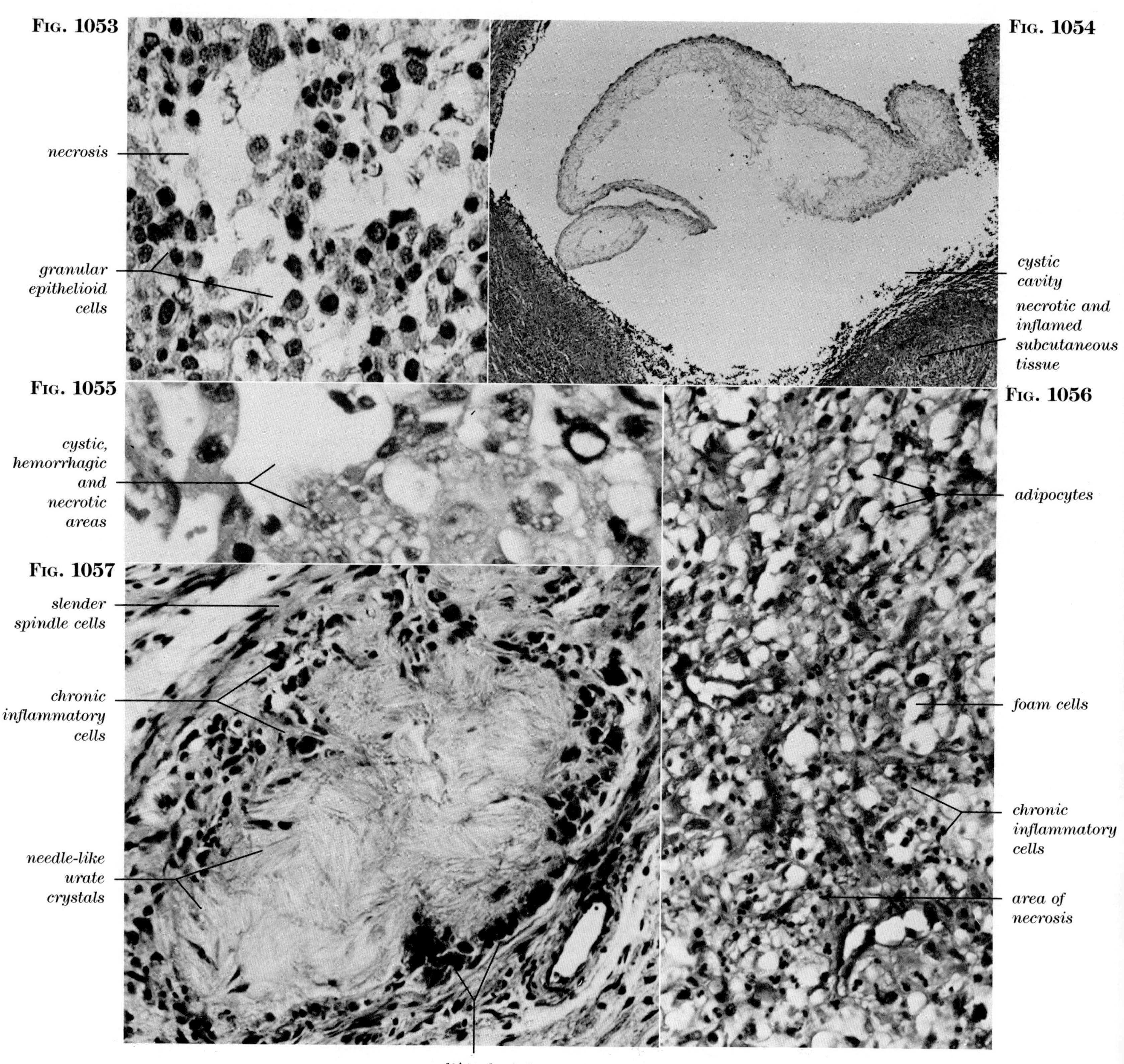

Fig. 1058. Aneurysmal bone cyst (×350) (see Figs. 96, 125, 150, 163, 509, 785, 1015, 1058).

Fig. 1059. Epithelioid sarcoma (×350) (see Figs. 85, 181, 187, 258, 259, 322, 326, 462, 464, 468, 491, 676, 830, 866, 895, 1052, 1146, 1170, 1244, 1245).

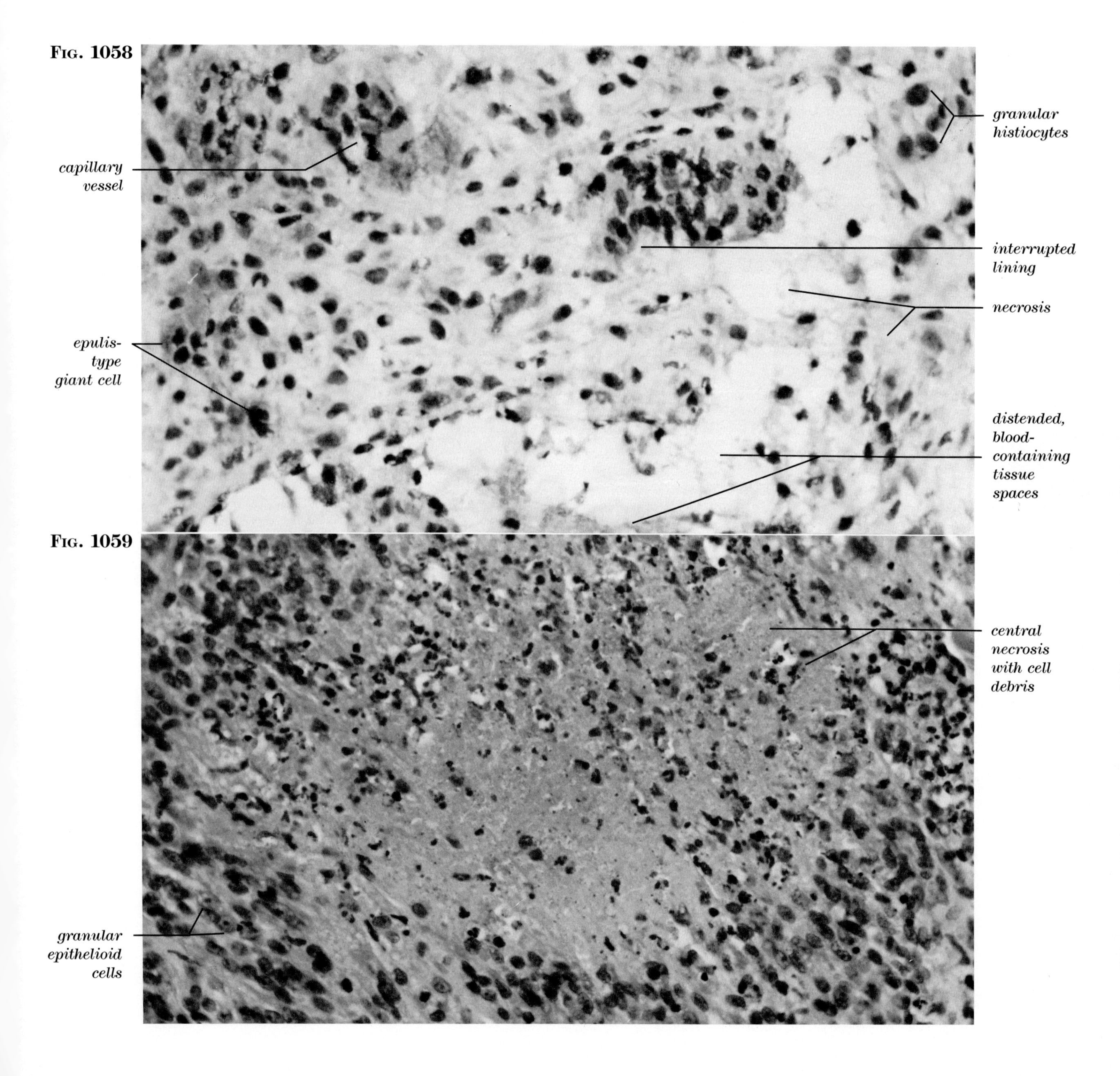

Fig. 1060. Chondroid metaplasia in fibrous tissue (×140) (see Fig. 1067).

Fig. 1062. Mesenchymal chondrosarcoma (×140) (see Figs. 103, 151, 175, 180, 318, 390, 444, 507, 595, 674, 707, 920, 1265).

Fig. 1061. Well-differentiated chondrosarcoma (×350) (see Figs. 95, 285, 442, 766, 1065, 1093).

Fig. 1063. Osteochondroma (×10) (see Figs. 113, 155, 167, 184, 302, 303, 772, 1066, 1090, 1152).

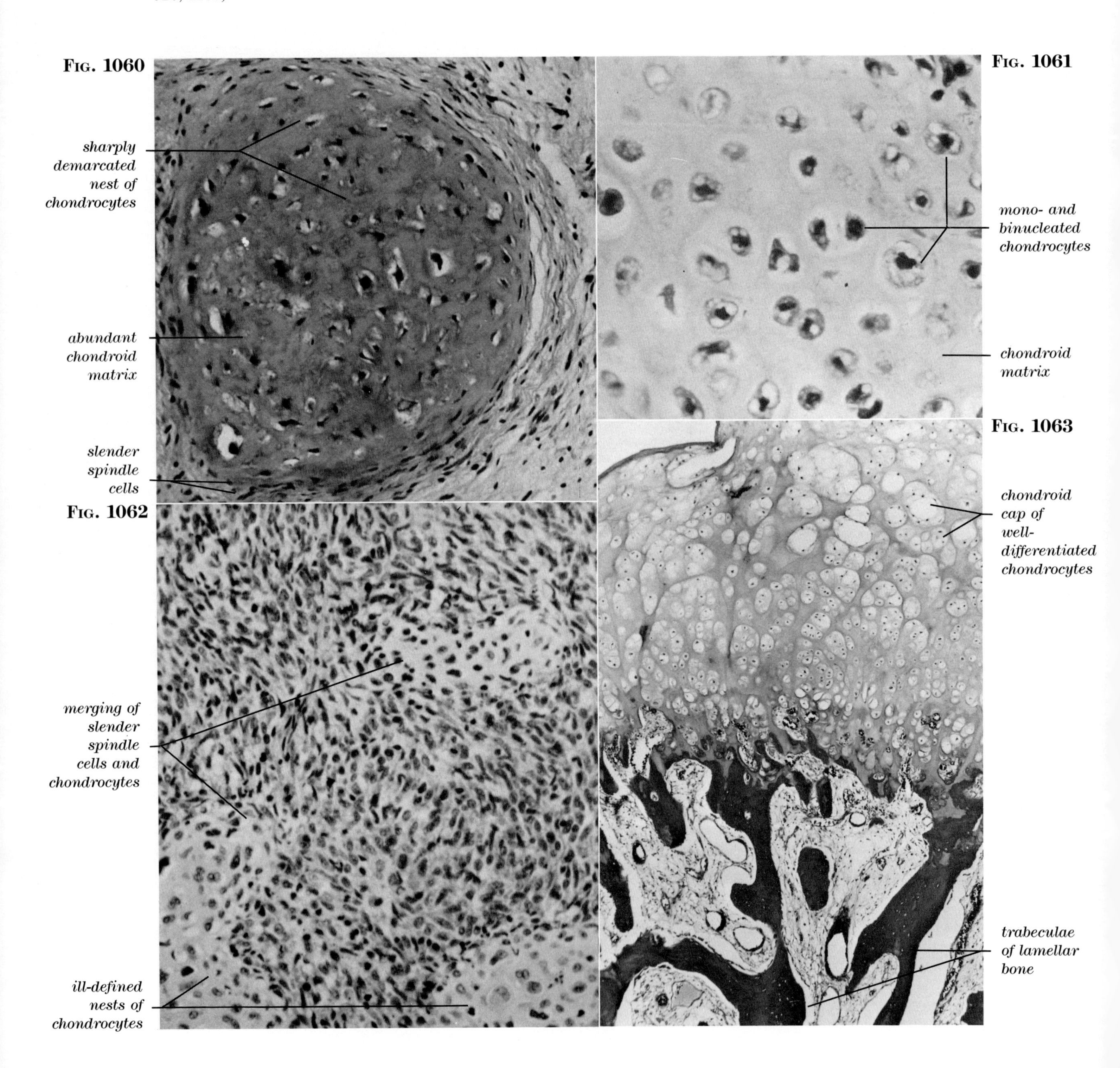

Fig. 1064. Poorly differentiated chondrosarcoma (×350) (see Figs. 95, 124, 126, 137, 317, 407, 434, 441, 445, 1097, 1102).

Fig. 1065. Well-differentiated chondrosarcoma (×350) (see Figs. 81, 95, 123, 124, 136, 147, 216, 223, 285, 327, 442, 766, 773, 1061, 1093, 1101, 1110, 1222, 1275).

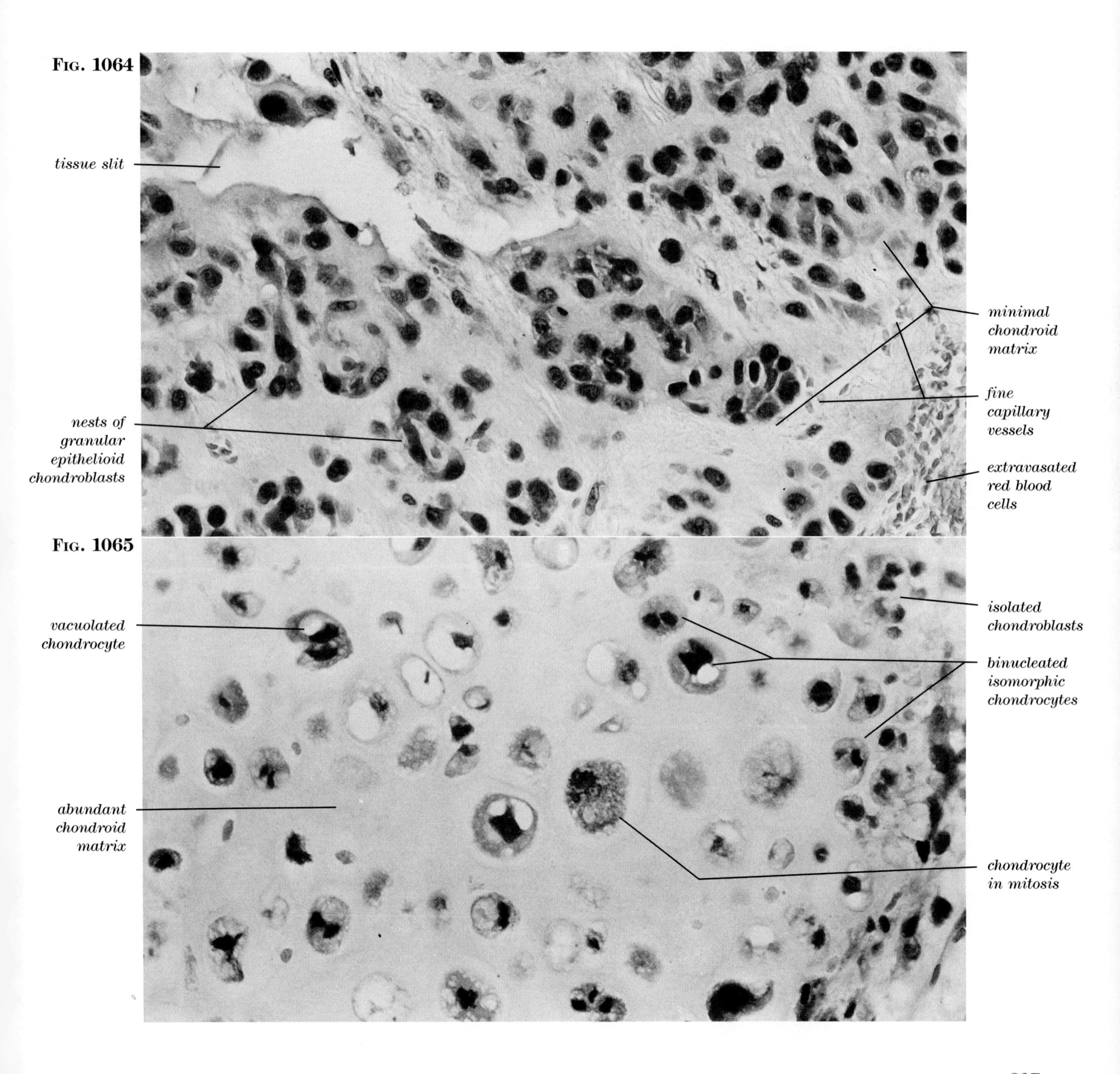

Fig. 1066. Osteochondroma (×35) (see Figs. 113, 155, 167, 184, 302, 303, 772, 1063, 1090, 1152).

Fig. 1067. Osteoid metaplasia (×35) (see Figs. 1060).

Fig. 1068. Osteoblastic osteosarcoma (×85) (see Figs. 109, 395, 405, 411, 451, 643, 677, 968, 1001, 1070–1079, 1086).

Fig. 1069. Giant cell granuloma (×140) (see Figs. 119, 632, 787, 1000).

Fig. 1070. Sclerosing osteosarcoma (×140) (see Figs. 108, 1001, 1071–1079, 1086, 1087, 1088, 1091, 1105–1108).

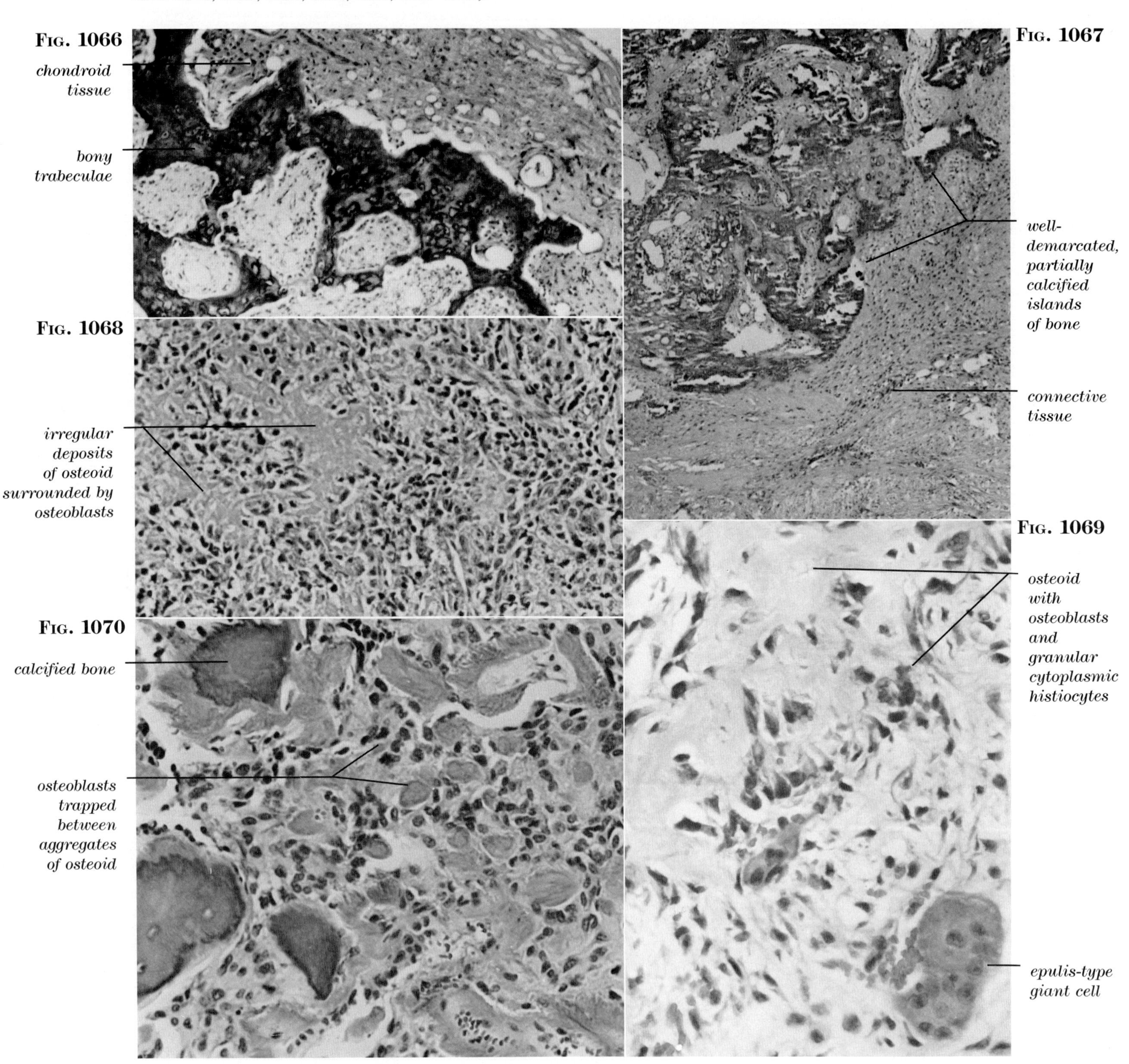

Fig. 1071. Osteosarcoma extending between trabeculae of benign bone (×140) (see Figs. 1001, 1072–1079, 1086).

Fig. 1072. Osteosarcoma (×85) (see Figs. 1001, 1073–1079, 1086).

Fig. 1073. Fibrosarcomatous osteosarcoma (×140) (see Figs. 1001, 1074–1079, 1086).

Fig. 1074. Osteosarcoma (×140) (see Figs. 1001, 1075–1079, 1086, 1087).

Fig. 1075. Telangiectatic osteosarcoma (×140) (see Figs. 1070–1079, 1086, 1087).

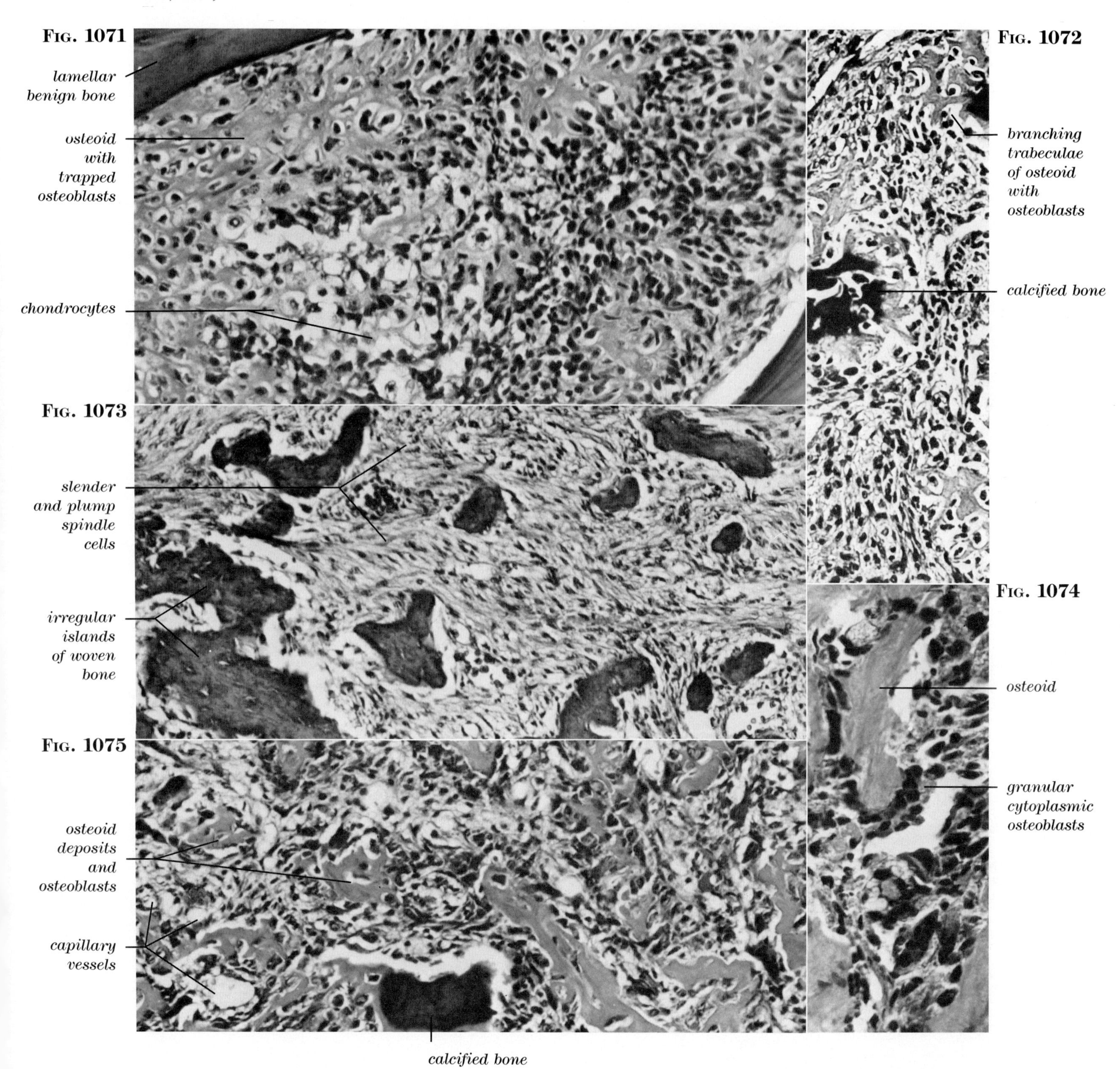

Fig. 1076. Fibrous histiocytic osteosarcoma (×350) (see Figs. 90, 108, 109, 395, 405, 1070–1079, 1087).

Fig. 1077. Osteosarcoma with undifferentiated mesenthymal elements (see Figs. 90, 108, 611, 615, 817, 819, 831, 1070–1079, 1087).

Fig. 1078. Osteosarcoma with pleomorphic fibrous histiocytoma component (×350) (see Figs. 1001, 1068, 1070–1079, 1087, 1088, 1091, 1105–1108).

Fig. 1079. Osteoblastic osteosarcoma (×350) (see Figs. 831, 843, 876, 897, 898, 958, 968, 1001, 1070–1078, 1086, 1087).

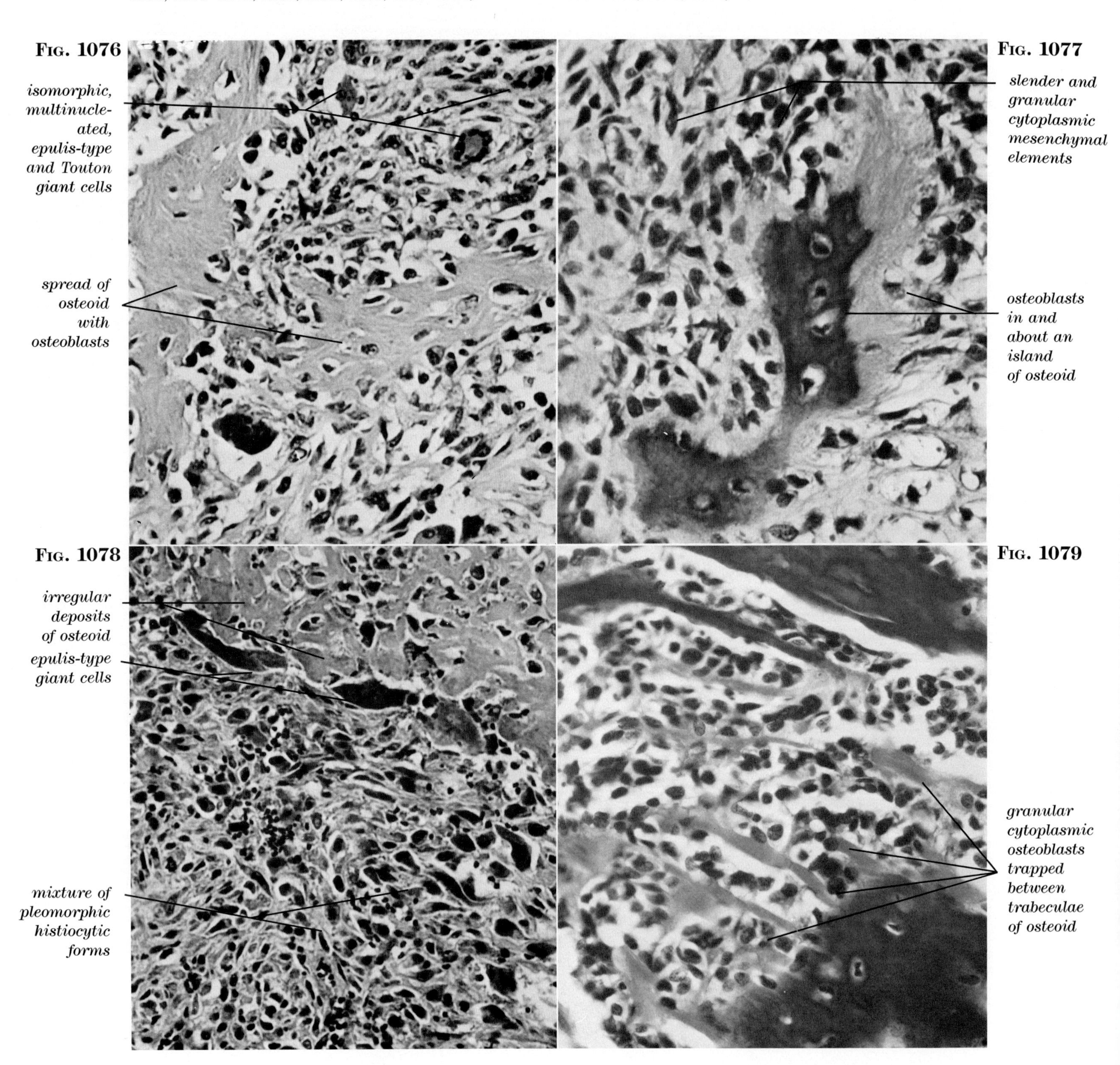

Fig. 1080. Fasciitis ossificans (×140) (see Figs. 82, 345, 605, 834, 1045).

Fig. 1081. Fibromatosis ossificans (×350) (see Figs. 1022, 1025, 1035, 1036, 1050, 1095).

Fig. 1082. Parosteal osteosarcoma (×350) (see Figs. 109, 129, 217, 224, 249, 287, 353, 354, 656, 1154).

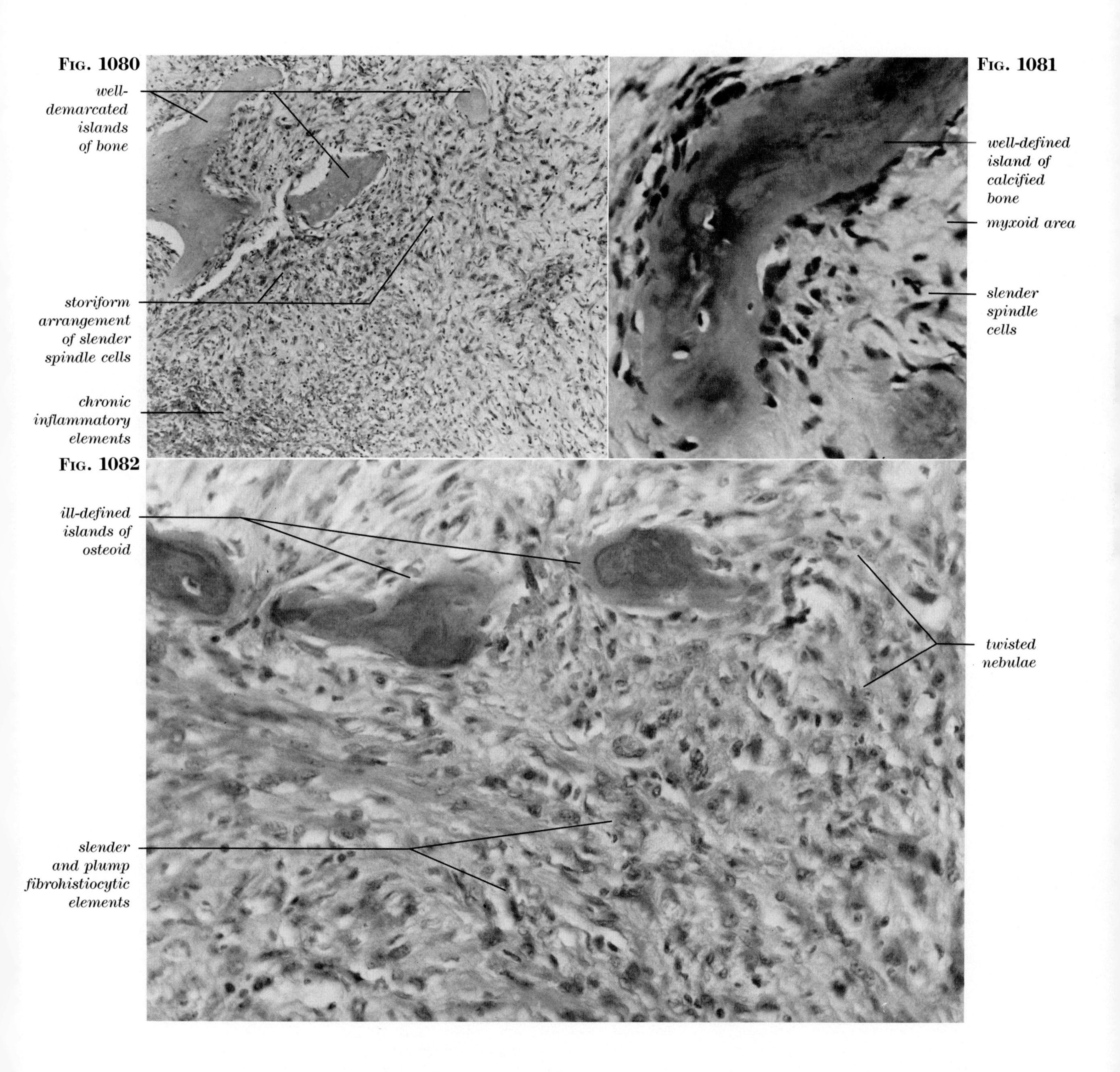

Fig. 1083. Myositis ossificans (×35) (see Figs. 145, 384, 620, 913, 1084, 1085).

Fig. 1084. Myositis ossificans (×140) (see Figs. 145, 384, 620, 913, 1083, 1085).

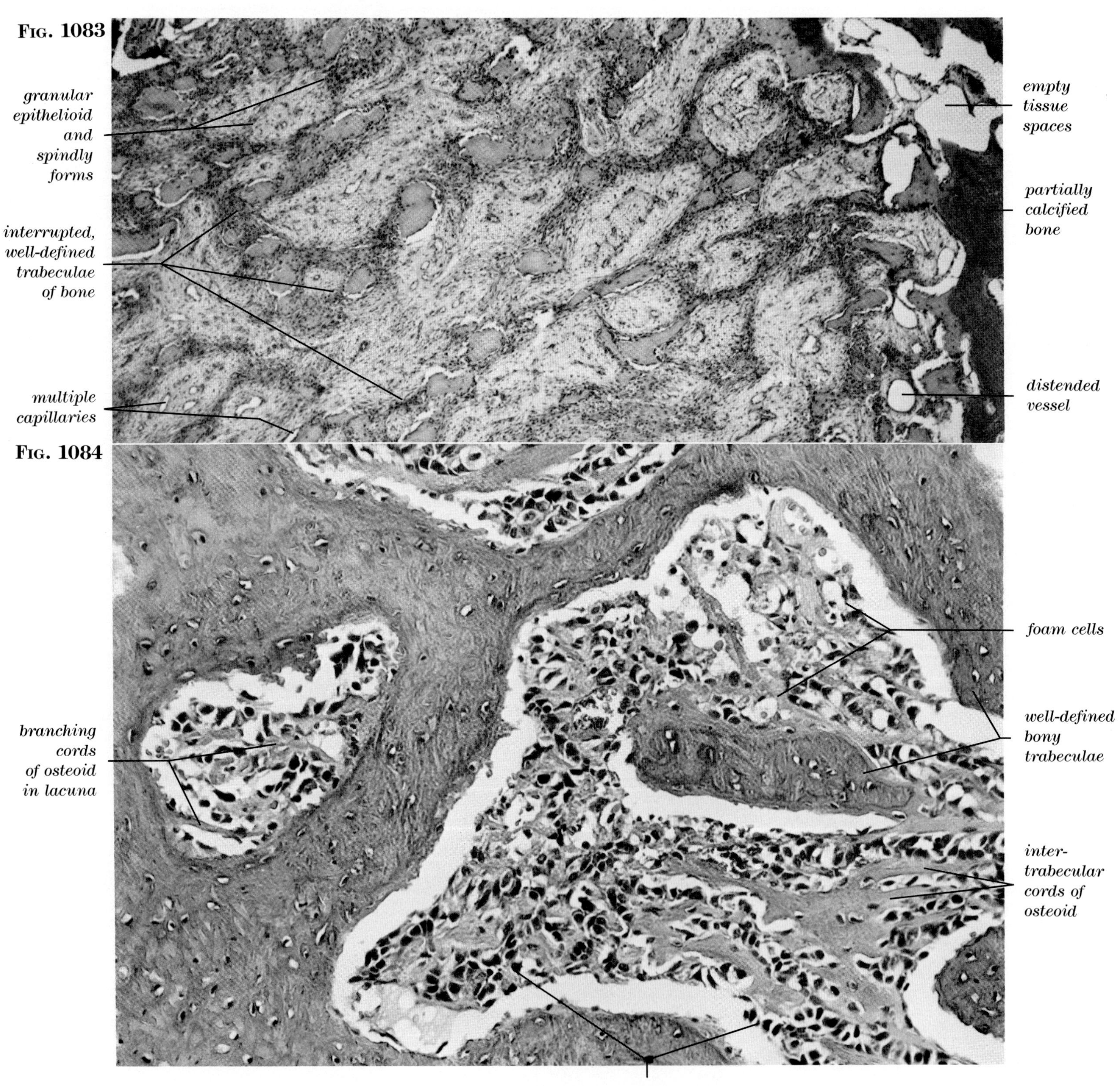

Fig. 1085. Myositis ossificans (×350) (see Figs. 145, 620, 913, 1083, 1084).

Fig. 1086. Osteosarcoma (×140) (see Figs. 90, 108, 142, 248, 395, 643, 831, 1070–1079, 1087, 1088).

Fig. 1087. Osteosarcoma (×350) (see Figs. 1070–1079, 1088, 1091).

Fig. 1088. Osteosarcoma (×350) (see Figs. 90, 108, 109, 228, 328, 615, 819, 831, 843, 876, 897, 898, 958, 968, 1001, 1068, 1070–1079, 1087, 1091).

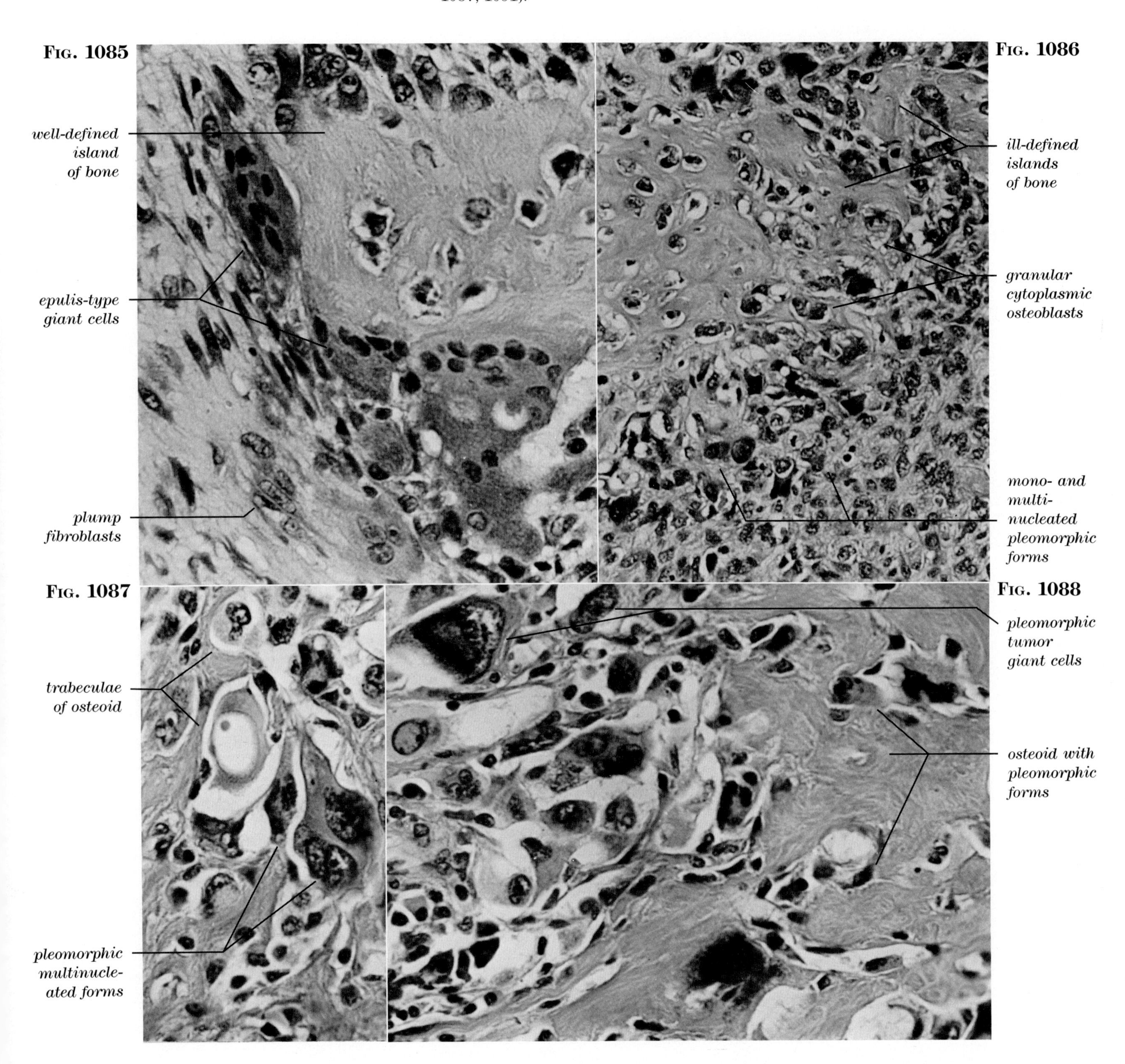

Fig. 1089. Nidus of osteoid osteoma (×140) (see Fig. 101).

Fig. 1090. Osteochondroma (×350) (see Figs. 184, 302, 303, 772, 1066, 1152).

Fig. 1091. Telangiectatic osteosarcoma (×570) (see Figs. 90, 108, 395, 405, 411, 643, 677, 730, 754, 831, 843, 876, 897, 898, 958, 968, 1001, 1068, 1070–1079, 1087, 1105–1108).

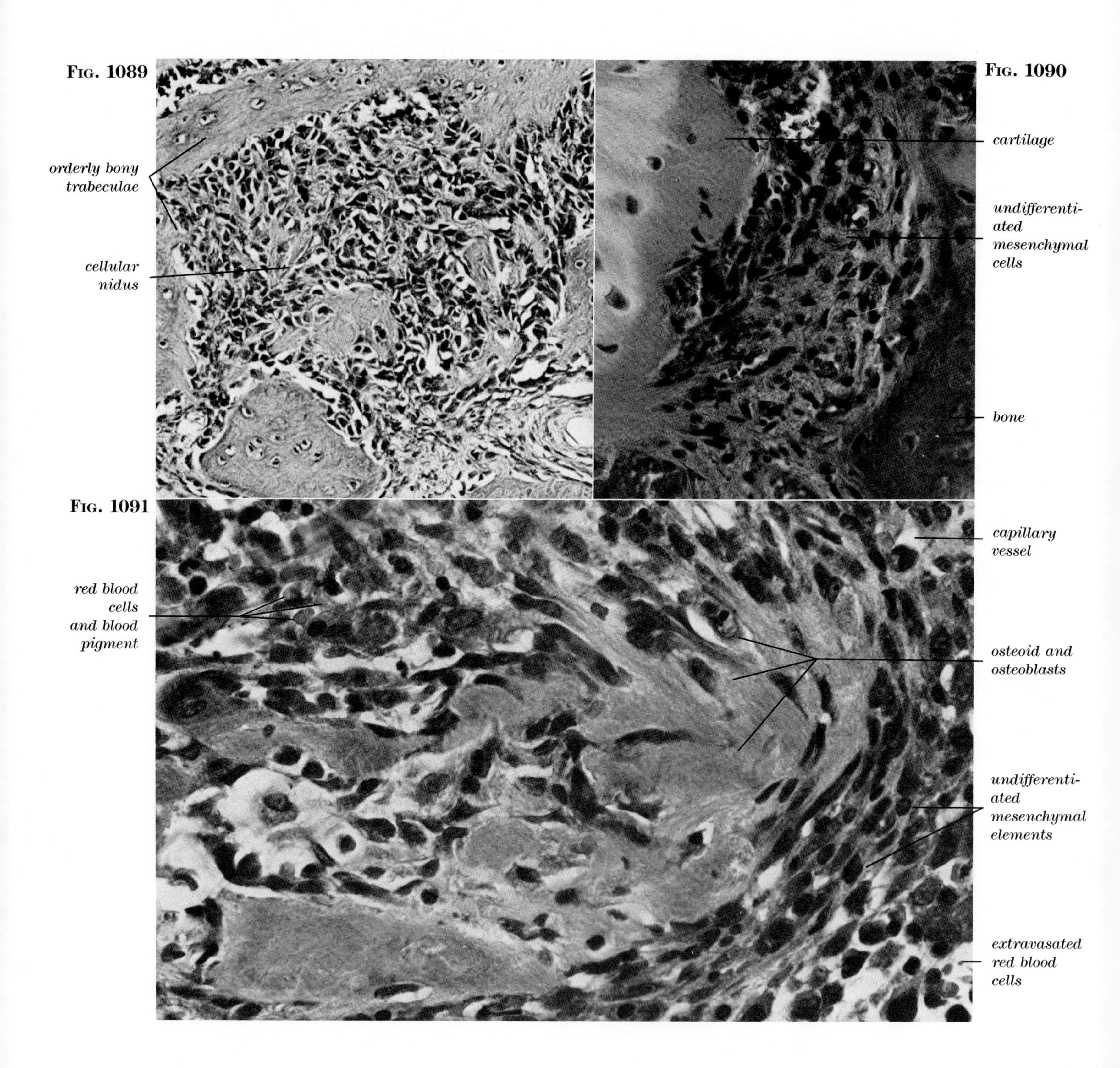

Fig. 1092. Osteoblastoma (×350) (see Figs. 97, 211, 234, 1155).

Fig. 1093. Ossified, well-differentiated chondrosarcoma (×350) (see Figs. 81, 92, 123, 285, 442, 766, 773, 1061, 1065, 1101, 1110, 1222, 1275).

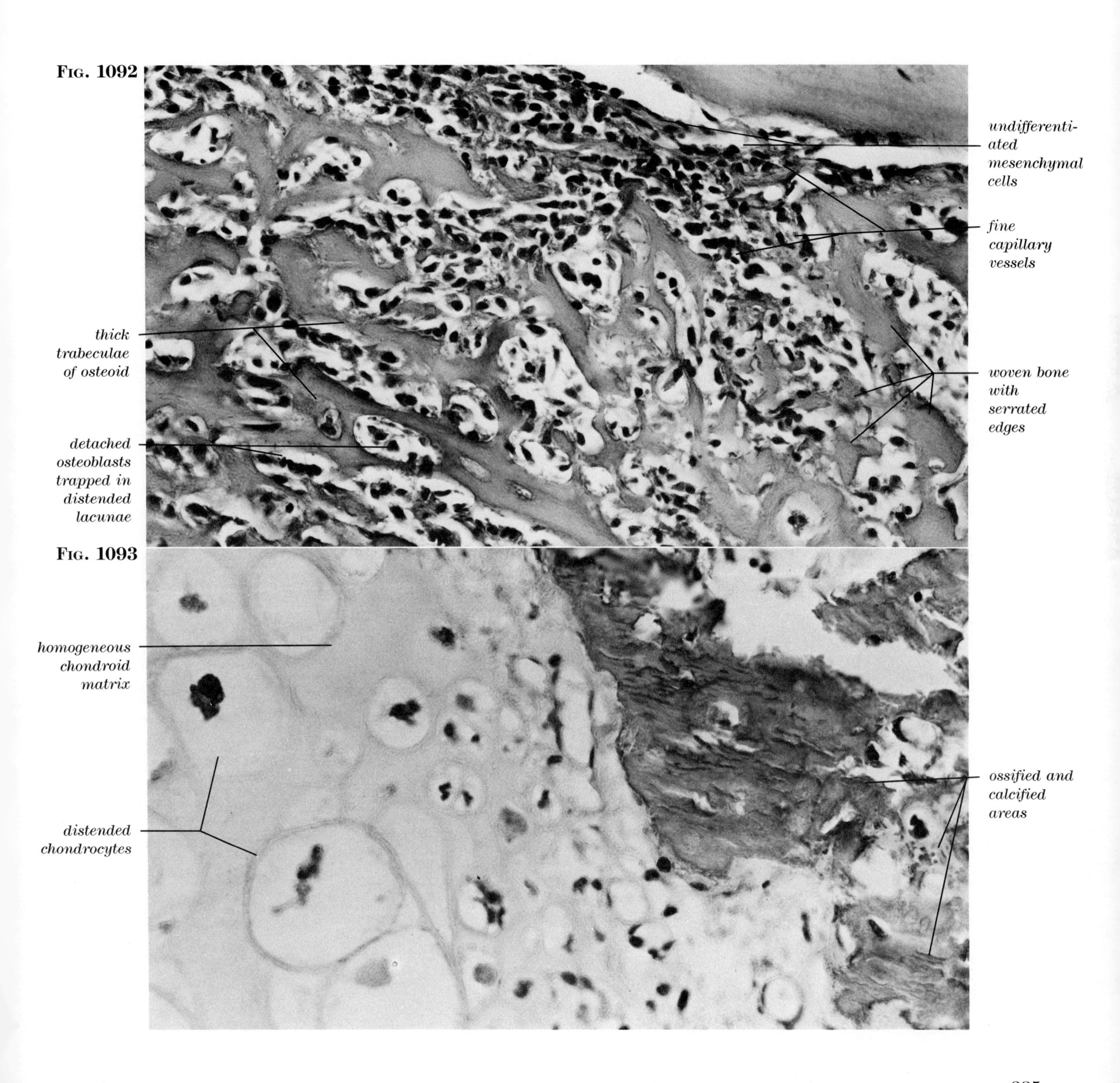

Fig. 1094. Myositis (×35) (see Figs. 1038, 1039, 1186).

Fig. 1095. Fibromatosis (×35) (see Figs. 17, 280, 582, 616, 1050, 1081).

Fig. 1096. Extracranial meningioma (×140) (see Figs. 485, 496, 537, 1225).

Fig. 1097. Poorly differentiated chondrosarcoma (×140) (see Figs. 95, 137, 317, 407, 434, 441, 445, 1064, 1102).

Fig. 1098. Lymphangiosarcoma associated with chronic lymphedema (×85) (see Figs. 84, 154, 159, 321, 424, 551, 558, 649, 954, 963, 970).

Fig. 1099. Biphasic tendosynovial sarcoma (×85) (see Figs. 85, 323, 536, 547, 727, 765, 906, 1100).

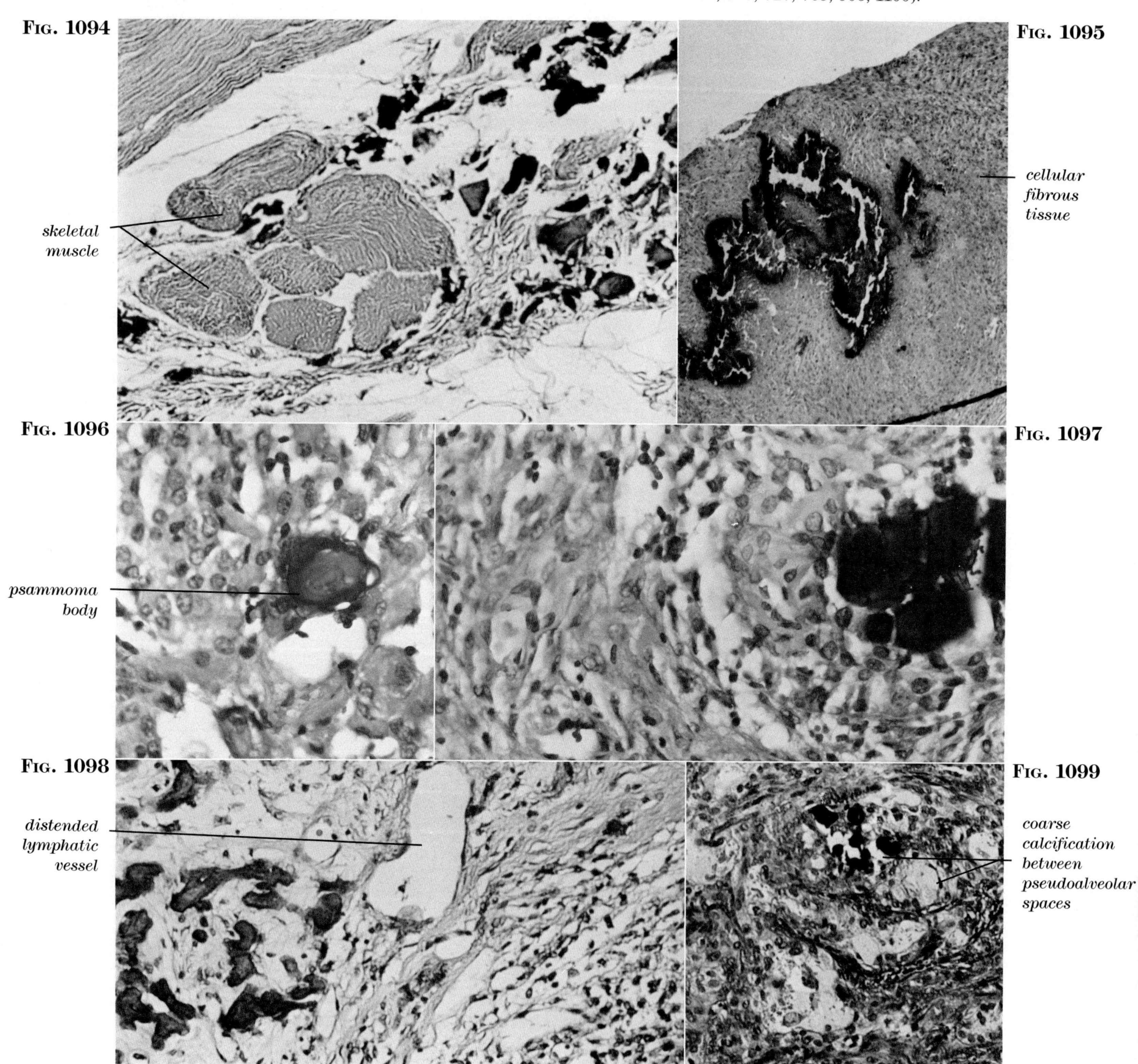

Fig. 1100. Biphasic tendosynovial sarcoma (×140) (see Figs. 85, 122, 535, 727, 1099, 1173).

Fig. 1101. (Top center) Recurrent chondrosarcoma (×140) (see Figs. 1061, 1065, 1093, 1110, 1222).

Fig. 1102. Metastatic chondrosarcoma (×140) (see Figs. 95, 124, 407, 1064, 1097).

Fig. 1103. Chondroblastic area in Paget's sarcoma (×140) (see Figs. 116, 128, 506, 1104, 1111).

Fig. 1104. Chondroblastoma (×350) (see Figs. 116, 128, 506, 918, 1103, 1111).

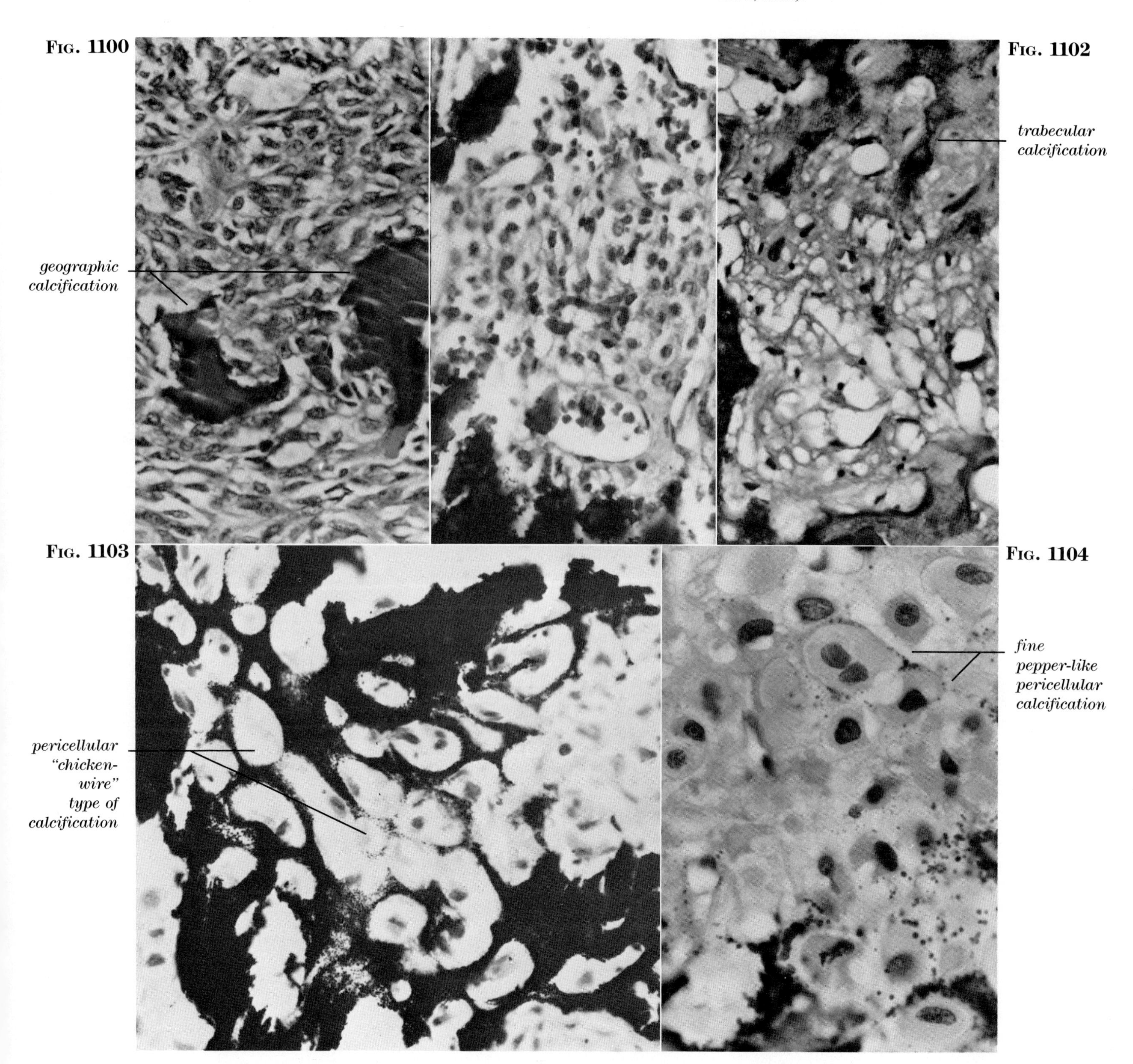

Fig. 1105. Sclerosing osteosarcoma (×85) (see Figs. 831, 843, 876, 897, 898, 958, 1070–1079, 1106–1108).

Fig. 1107. Osteoblastic osteosarcoma (×140) (see Figs. 754, 769, 777, 801, 831, 1086, 1087, 1088, 1105, 1106, 1108).

Fig. 1109. Aponeurotic fibroma (×140) (see Figs. 336, 859, 884, 886, 904, 915, 917, 1027, 1204).

Fig. 1106. Telangiectatic osteosarcoma (×140) (see Figs. 1070–1079, 1091, 1107, 1108).

Fig. 1108. Sclerosing osteosarcoma (×140) (see Figs. 90, 108, 109, 395, 643, 831, 1091, 1105–1107).

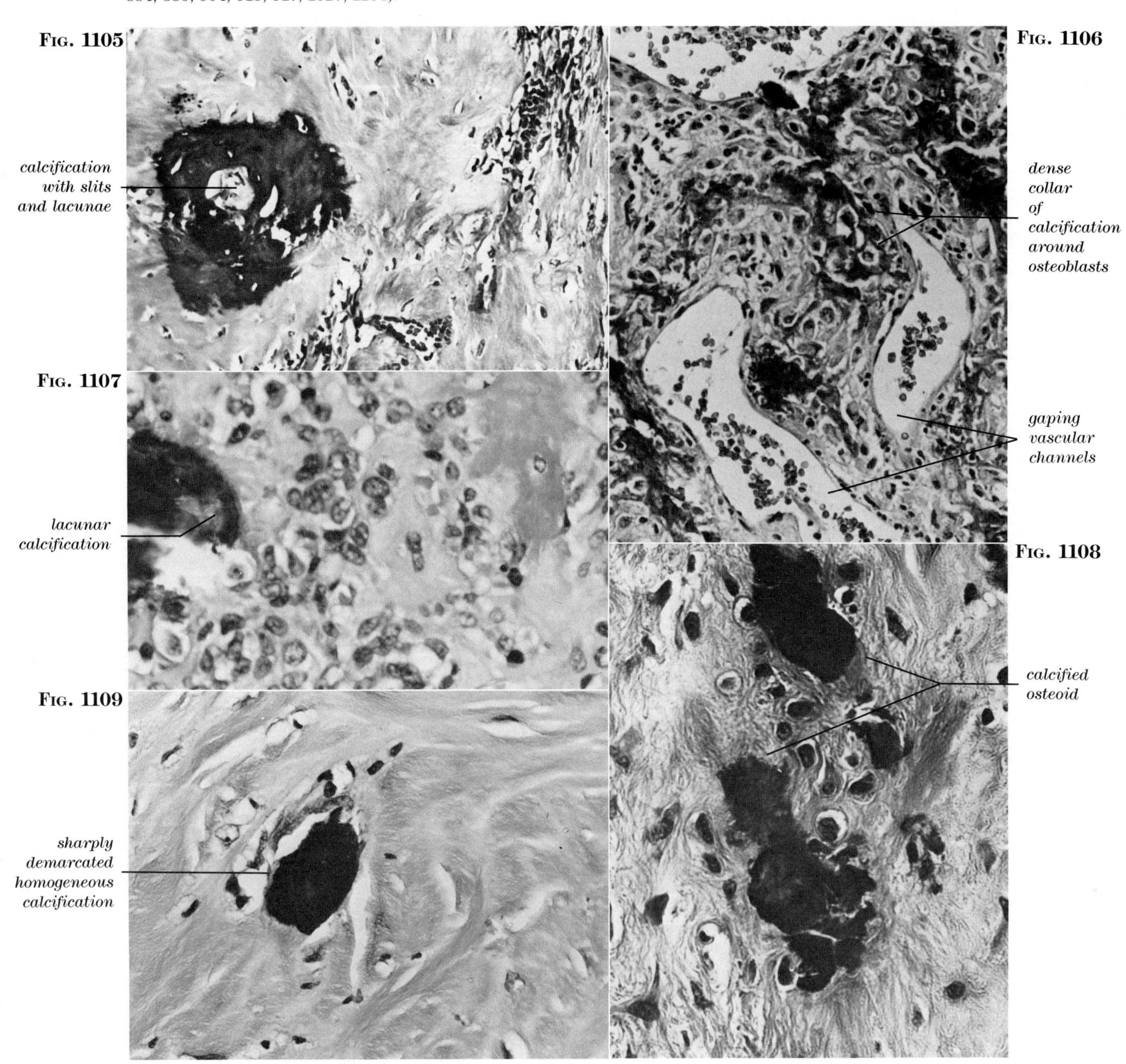

Fig. 1110. Calcified well-differentiated chondrosarcoma (×350) (see Figs. 81, 95, 223, 327, 442, 766, 773, 1061, 1065, 1093, 1101, 1222, 1275).

Fig. 1111. Chondroblastoma (×350) (see Figs. 116, 128, 506, 1103, 1104).

Fig. 1112. Post-irradiation calcification (×350).

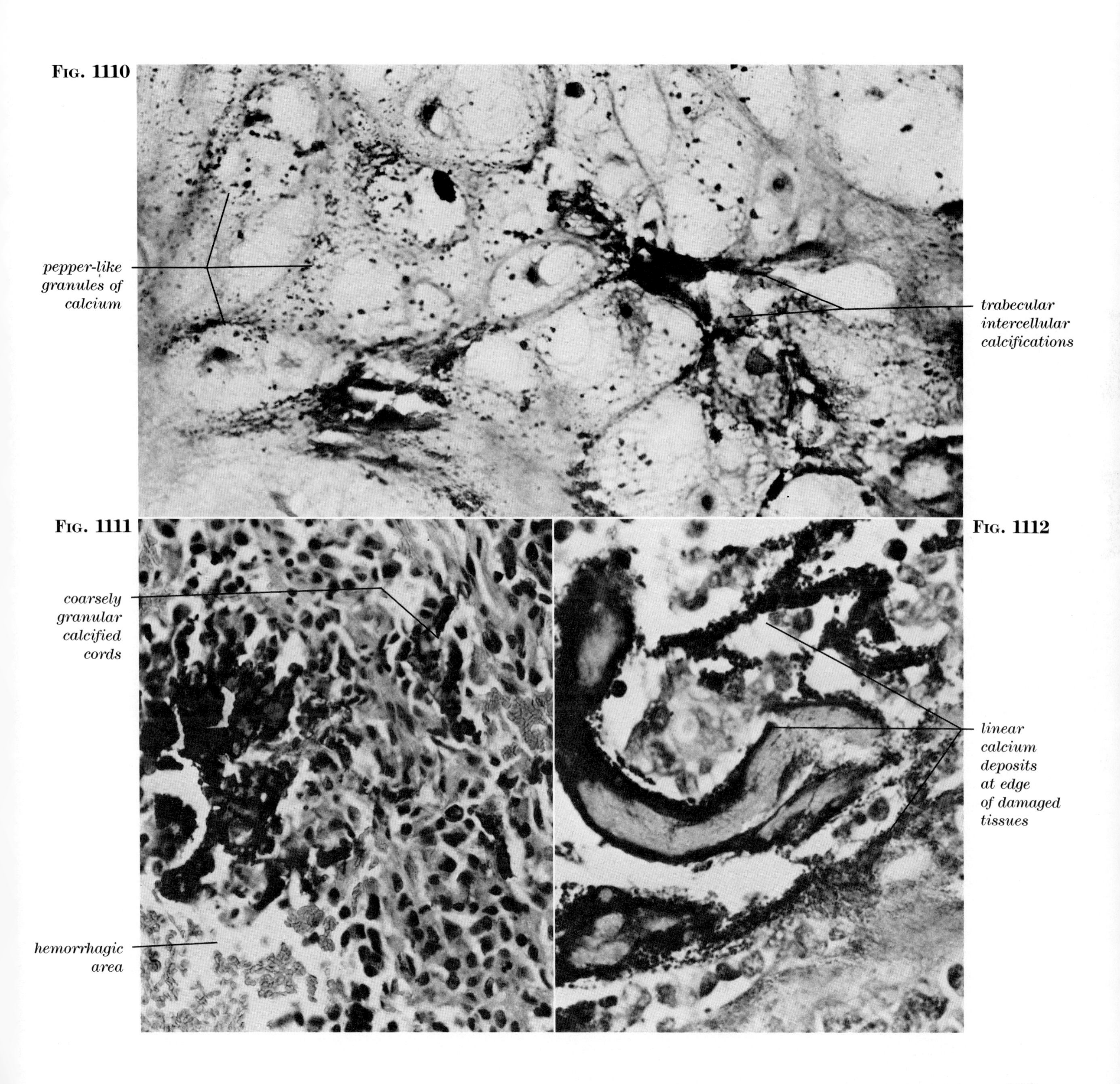

"Without biologic understanding of cancer we cannot hope for a real and lasting solution of the problem"

H. GILFORD (1861–1941)

13. Products of Cells

	NON-NEOPLASTIC LESIONS	BENIGN NEOPLASMS	MALIGNANT NEOPLASMS
1. **Collagen.** See Tables 58 to 63.			
See Figures:	1119–1123, 1125, 1127, 1136, 1140, 1153	1114, 1124, 1137, 1141–1143, 1152, 1155	1113, 1115–1118, 1126, 1128–1135, 1138, 1139, 1144–1151, 1154, 1156
2. **Glycogen.** See Tables 58 to 64.			
See Figures:		1160, 1162	1157–1159, 1161, 1163–1167
3. **Mucopolysaccharides.** See Tables 58 to 60.			
See Figures:		1168	1169–1180
4. **Fat.** See Tables 58 to 60.			
See Figures:	1185, 1186	1181–1183, 1187	1184
5. **Melanin.** See Tables 58 to 60.			
See Figures:		1191, 1193	1188, 1189 1190, 1192
6. **Secretory Granules.** See Tables 58 to 63.			
See Figures:			1194–1196
7. **Crystals.** See Tables 58 to 63.			
See Figures:	1197		1198, 1199

	NON-NEOPLASTIC LESIONS	BENIGN NEOPLASMS	MALIGNANT NEOPLASMS
8. Fine Structure. See Tables 61 to 63, 66 to 69.			
See Figures:	1201, 1206	1204, 1221, 1225, 1228, 1229, 1235	1200, 1202, 1203, 1205, 1207–1220, 1222–1224, 1226, 1227, 1230, 1231–1234, 1236–1239
9. Tissue Antigens. See Tables 58 to 60, 69, 70, 71.			
See Figures:		1252, 1254, 1258	1240–1251, 1253, 1255–1257, 1259–1275

RECENTLY, Emanuel Rubin pointed out that "The separation of structure and function is artificial and that the former determines the latter." Indeed, the advances made during the last decades in the understanding of the structure of cellular elements and the intercellular matrix of connective tissue paved the way for current functional (cytochemical and immunochemical) procedures.

Special stains may aid in demonstration of products of cells (intracellularly and after deposition in the stroma) but there are very few soft tissue or bone tumors for which special stains are diagnostic. Although many special stains exist, most of them are nonspecific. Experienced pathologists seldom resort to more than a handful of special stains, and as a rule the more experienced the pathologist is, the fewer special stains are needed to derive the diagnoses.

The three main products of connective tissue cells are collagen, carbohydrates, and fat. The collagen production is a complex process and is carried out simultaneously with the synthesis of glycogen and mucopolysaccharides. Collagen synthesized in cells is transported to the cell surface and then expelled and deposited in the intercellular matrix. It is the major supporting element of soft tissue and bone. Collagen is a polypeptide, and the quantity and quality of it varies from site to site, as well as from tumor to tumor. Six genetically distinct types of collagen have been identified. It has been shown that typing of collagen can be used for diagnostic purposes (Table 61). For example, polymerized collagen of soft tissues and bone (osteoid) is strongly refractile in *polarized light* and appears lamellar. Neoplastic collagen and osteoid, on the other hand, are much less refractile and exhibit an interrupted, criss-cross, or woven pattern.

Reticulin stain demonstrates fine collagen fibers (collagen type III) and may be of help in showing whether the fibers are around cells, for example, epithelial cells or between cells, for example, certain connective tissue cells (Table 62).

Collagen stains, such as the *trichrome,* stain collagen type I; they may also demonstrate the intracytoplasmic actin filaments of smooth muscle and other mesenchymal cells and the actin-myosin complex of rhabdomyoblasts. It is doubtful whether *phosphotungstic acid hematoxylin (PTAH) stain* has any diagnostic value in typing of connective tissue tumors.

Carbohydrate-rich macromolecules (complex carbohydrates) constitute an important part of the intercellular matrix, secretory granules, and lysosomes within cells and the surface coat of cell membranes. Complex carbohydrates are a large and widely heterogeneous group (Table 63). All the biochemically identified carbohydrates are included in the term "mucopolysaccharides (mucosubstance)" except glycogen. Simple special stains and histochemical procedures are available for demonstration of the diagnostically important components. *Periodic acid-Schiff* (PAS) stain glycoproteins red and glycosaminoglycans remain negative and allow specific identification of glycogen by obliterating PAS staining with diastase (Table 64). Granular intracytoplasmic polysaccharide material (hyalin droplets) that stains brightly with PAS after diastase digestion can be found in a number of soft tissue and bone tumors (see Table 64).

The mucosubstance found in soft tissue and bone tumors are mostly acid mucopolysaccharides (hyaluronic acid, chondroitin sulfate, and keratan sulfate). Their presence can be demonstrated with *alcian blue stain.* Alcian blue, at pH 2.5, stains nonsulfated and sulfated glycoconjugates and at pH 1.0 stains sulfated glycoconjugates only. Hyaluronic acid and chondroitin sulfates can be shown by loss of basophilia after digestion with testicular hyaluronidase (for example, in mesothelioma and in most myxoid soft tissue tumors). If, however, the mucosubstance (for example, keratan sulfate) is acid-sulfated, then at least part of the alcian blue staining will be retained after treatment with testicular hyaluronidase (for example, in epithelial tumors, certain tendosynovial tumors, and chondroid tumors). Admittedly, the value of alcian blue as a diagnostic stain is limited because the result is influenced by a number of factors, including pH. For example, highly sulfated mucopolysaccharides at low pH stain bright blue and are not digestible with hyaluronidase. At the same time, low sulfated mucopolysaccharides at low pH do not take the stain or stain pale blue and digestible with hyaluronidase (see Tables 58, 59, and 60).

Many mesenchymal tumor cells, especially the undifferentiated ones, for example, fibrous histiocytic cells, may contain intracytoplasmic globules of fat. Therefore, *fat stains* such as oil red 0 have limited diagnostic value, except perhaps in the diagnosis of adipose tissue tumors. In some of the liposarcomas, for example, one may find aggregates of fat globules eccentrically displacing the nuclei in the cytoplasm.

The value of *enzyme histochemistry* in the diagnosis of soft tissue and bone tumors is extremely limited due to lack of specificity of various enzymes as histogenetic markers. The need to use fresh, unfixed tissue sections further limits the diagnostic application of histochemical stains. The few stains that can be depended on as adjunct to routine stains are listed in Table 65. When using histochemical procedures, the pathologist should recognize that it is even more important for histochemical stains than for special stains that the preparation of the stain is understood and that adequate controls are used.

Although *transmission electron microscopic evaluation* of soft tissue and bone tumors offers little help in predicting whether a tumor is benign or malignant, it contributes to the resolution of some of the problems in differential diagnosis. It is especially helpful in determining the histogenesis of undifferentiated tumors such as small round cell neoplasms (Table 69). Careful consideration of the interrelationship between tumor cells, as well as consideration of intracytoplasmic organelles, can establish in most instances whether the tumor is epithelial, lymphoreticular, or mesenchymal in origin.

In Tables 66, 67, and 68, some of the suggestive ultrastructural features of non-neoplastic lesions, benign neoplasms, and malignant neoplasms are listed. Ultrastructural examination is a valuable adjunct to diagnosis when light microscopic studies are equivocal, but in most instances it is a confirmatory procedure. Very often, the absence of certain cell products (for example, intracytoplasmic organelles) carries more weight in determining a differential diagnosis than those that are present. Soft tissue and bone tumors are known for their heterogenous nature and wide spectrum of ultrastructural appearances. Diagnostic conclusion, particularly on the basis of an inadequate or technically suboptimal tissue sample, may not only be misleading, providing for the potential mismanagement of the patient, but may also cause unnecessary animosity between pathologists. As many tumors display a lack of specific ultrastructural features and show overlap in their fine structure, the findings of light microscopic observation should be favored if the light microscopic findings are equivocal and ultrastructural evaluation does not resolve the differential diagnostic problem.

Despite considerable limitations, correlation of the results of light microscopic examination with electron microscopic findings often extends the pathologist's diagnostic ability. Careful attention to fine detail such as composition of the basal lamina (a collection of non-cross-linked collagen that is single layered and focal in fibroblasts, single layered and uniform in smooth muscle cells, and multilayered and uniform in Schwann cells) may yield unexpected diagnostic dividends.

Recently, high resolution electron microscopy has introduced a whole new array or observations (for example, the demonstration of actomyosin in smooth muscle cells, fibroblasts, and other nonmyogenic cells) directly linked with such functions of cells as mitosis, mobility, and production of various metabolites, including immunologically active substances. Moreover, high resolution electron microscopy has resulted in a new field, *immunoelectron microscopy.*

During the last decade a variety of *immunohistochemical* and *immunocytochemical methods* have been introduced in diagnostic pathology. The application of immunologic techniques has resulted in major advances in the diagnosis and classification of lymphomas and leukemias. It seems likely that utilization of "panels" of different immunohistologic and immunocytologic markers will contribute significantly to the understanding and differential diagnosis of soft tissue and bone tumors. It must be emphasized, however, that immunohistochemical reagents are not tumor markers or sarcoma markers but tissue markers (Table 70). Therefore, one must use precaution and be absolutely certain that the cells being studied are neoplastic. It should also be remembered that fixation and embedding procedures reduce the antigenicity of cells to various degrees, and only a minor proportion of the neoplastic cells stain positive by direct and indirect immunostaining methods. A complete lack of immunoreactive substance, which regularly results in a negative stain, is often a more important diagnostic hint than the positive one.

The *immunoperoxidase methods* have several advantages over immunofluorescence procedures, including increased sensitivity, excellent preservation to allow for specific intracellular localization of the antigens being studied, the provision of a permanent record after staining, the avoidance of the need to use specialized microscopy, and the ability to study antigens after formalin fixation and paraffin embedding. It is critical, in the staining procedure of immunoperoxidase methods, to include a positive control consisting of a tissue known to contain the antigen being studied and negative controls in which the primary antibody is omitted or preabsorbed by the specific antigen being studied before staining. The major limiting factor for the diagnostic application of immunoperoxidase methods is the availability of highly specific antibodies with low cross-reactivity.

Applying immunoperoxidase techniques, it is now possible to use a number of antibodies to ascertain the histogenesis of several soft tissue and bone tumors (Table 71). *Keratin* and *carcinoembryonic antigen,* present in a vast array of epithelial (epidermoid and glandular) tumors, have not been demonstrated in connective tissue tumors. The only known exception is the focal presence of keratin in mesotheliomas, some of the tendosynovial tumors, and peripheral nerve tumors; carcinoembryonic antigen is demonstrable in granular cell tumor, chordoma, and certain osteoblastic tumors.

As more information becomes available, it becomes apparent that revision of earlier held views is necessary. For example, *Factor-VIII-related antigen,* a known endothelial cell marker, has less diagnostic potential in the diagnosis of hemangiosarcoma than was originally thought. Of the three *histiocytic markers,* alpha-1-antitrypsin, alpha-1-antichymotrypsin, and lysozyme, lysozyme is not useful as a marker for fibrous histiocytic tumors while the other two can be used as markers for lymphoreticular as well as fibrous histiocytic tumors.

Immunofluorescence and immunochemical demonstration of proteins associated with the *intermediate filament* cytoskeleton is a useful adjunct in the differential diagnosis of tumors, particularly undifferentiated tumors with equivocal histologic appearances. Although the antigenic and morphologic (ultrastructural) expression of intermediate filaments is regulated by poorly understood intracellular and extracellular factors, they remain identifiable in most benign and malignant cells. Indirect immunofluorescence microscopy, as well as immunoperoxidase methods utilizing monospecific antibody, can be used to demonstrate the five intermediate filaments: keratin, vimentin, desmin, neurofilaments, and glial filaments. These filaments can also be demonstrated by immunoperoxidase methods. It is of interest that, despite their characteristic immunologic features, intermediate filaments, especially desmin and vimentin, cannot always be distinguished ultrastructurally (see Tables 58, 59, 60, 70, and 71).

Keratin appears to be a site-specific intermediate filament and is found in association with desmosomes. However, antibodies to keratin are not specific markers of epithelial cells because the same antibodies may react with other cells, for example, mesothelial, synovial, and neuroectodermal cells. *Desmin* is a major subunit protein that is localized at the Z-line in muscle; it is demonstrable by using immunocytochemical techniques in smooth muscle and other mesenchymal cells and may play a role in regulation of cell shape. *Vimentin* is found in mesenchymal cells, as well as undifferentiated neuroectodermal cells. Vimentin is wrapped around the nucleus and seems to be limiting the nucleus to a specific space in the cytoplasm. Although vimentin is considered a marker for mesenchymal cells, its absence, e.g., in mature muscle cells, and presence in immature glial and some epithelial cells make it a less than specific marker. *Neurofilaments* are closely packed in peripheral nerve cells, possess ultrastructurally visible sidearm processes, and can be demonstrated with antibodies to S-100 protein. *Glial filaments* are randomly dispersed in the cytoplasm of glial cells and are demonstrable with antibodies to glial fibrillary acidic protein.

Today, monoclonal techniques make it possible to generate an unlimited number of monoclonal antibodies that recognize specific intracytoplasmic and stromal proteins of soft tissue and bone tumors, for example, collagen. It is foreseeable that increased judicious utilization of a panel of different immunochemical reagents and ultrastructural evaluation, coupled with special stains, will augment the ability of pathologists to come to the most appropriate conclusion in the diagnosis of soft tissue and bone tumors. However, we must be cognizant that presently there is no universal cell marker available, and there are no antibodies capable of distinguishing the more-than-200 soft tissue and bone tumors (see Fig. 6 and Table 5). Further, clonal selectivity, the level of differentiation of cytoplasmic organelles, and local tissue conditions can drastically influence the antigenicity of tumor cells. For example, metastatic nodules and neoplasms in vitro may acquire new, aberrant, or unconventional antigenicity. Therefore, blind dependence on the results of immunohistochemical stains should be avoided, and these procedures should not be used as diagnostic tests without full knowledge of their limitations.

TABLE 58. ***Product of Cells of Non-neoplastic Lesions****

	COLLAGEN[†]	POLYSACCHARIDE BEFORE AND AFTER DIGESTION[‡]	MUCOPOLYSACCHARIDE BEFORE AND AFTER DIGESTION[§]	FAT	FINE STRUCTURE	TISSUE ANTIGENS OF PRACTICAL VALUE
● Aneurysmal bone cyst	III		+−		See Tables 66–69	See Tables 70 and 71
● Giant cell granuloma		+−	+−			
● Hyperparathyroidism	III	+−				
○ Xanthogranuloma[¶]	III			+		
○ Fasciitis	III		+−			
○ Keloid	I					
○ Fibromatosis	I–III		+−			
● Fibrous dysplasia[#]	III					
○ Elastofibroma[**]	I		+−	+		
○ Collagenoma	I					
○ Tendosynovitis[#]			+−			
◑ Tendosynovial cyst	I–III		+−			
○ Tendosynovial chondromatosis	I–III		++			
○ Gouty arthritis[†]	III					
◑ Fat necrosis	III	++	+−	+		
◑ Lipogranuloma[†]			+−	+		
○ Proliferative panniculitis	I–III			+		
○ Proliferative myositis	III	+−				
○ Pyogenic granuloma	III					
○ Angiofollicular lymphoid hyperplasia	III			+		
◑ Arteriovenous malformation	I–III			+		
○ Traumatic neuroma	III					
● Callus	I–III					
◑ Eosinophilic granuloma						

*○ Primary soft tissue tumor; ● Primary bone tumor; ◑ can be primary soft tissue or bone tumor.
[†]See Tables 61 and 62.
[‡]PAS stain before and after diastase.
[§]Alcian blue stain before and after hyaluronidase.
[¶]Crystals.
[#]Calcification.
[**]Elastic fibers.

Table 59. *Product of Cells of Benign Neoplasms**

	COLLAGEN[†]	POLYSACCHARIDE BEFORE AND AFTER DIGESTION[‡]	MUCOPOLYSACCHARIDE BEFORE AND AFTER DIGESTION[§]	FAT	MELANIN	FINE STRUCTURE	TISSUE ANTIGENS OF PRACTICAL VALUE
◑ Benign fibroblastic fibrous histiocytoma[¶#]	III	+−	+−			See Tables 66–69	See Tables 70 and 71
◑ Benign histiocytic fibrous histiocytoma[¶#]	III	+−	+−				
◑ Benign pleomorphic fibrous histiocytoma[¶#**]	III	++	+−	+			
● Nonossifying fibroma	III	−−		+			
○ Fibroma[††]	I	−−	+−				
● Desmoplastic fibroma	I	−−					
◑ Well-differentiated lipoma		+−	+−	+			
◑ Myxoid lipoma		+−	+−	+			
○ Fibroblastic lipoma	III	+−	+−	+			
○ Lipoblastoma		+−	+−	+			
◑ Pleomorphic lipoma	III	+−		+			
○ Angiomyolipoma		+−		+			
○ Hibernoma		−−		+			
◑ Leiomyoma		+−					
○ Rhabdomyoma[**]		+−					
◑ Hemangioma							
◑ Benign glomus tumor							
◑ Neurofibroma[‡‡]	I–III						
○ Plexiform neurofibroma[‡‡]	III						
◑ Benign schwannoma[‡‡]	III		+−				
○ Benign nevoid schwannoma[‡‡]		++			+		
◑ Benign undifferentiated peripheral nerve tumor[‡‡§§]	III	++			+		
◑ Ganglioneuroma[‡‡]	I–III				+		
○ Benign paraganglioma[§§]	III	+−	+−				
○ Benign fibrous mesothelioma[††]			+−				
● Chondroblastoma[††]		++	++				
● Chondromyxoid fibroma	III	+−	++				
◑ Chondroma		++	++				
● Osteochondroma	I	++	++				
● Osteoblastoma	I–III						
● Osteoid osteoma	I–III						
○ Benign granular cell tumor		++					
○ Meningioma[††]	I–III	+−	+−				
○ Benign cystosarcoma phyllodes	III		+−				

*○ Primary soft tissue tumor; ● primary bone tumor; ◑ can be primary soft tissue or bone tumor.
†See Tables 61 and 62.
‡PAS stain before and after diastase.
§Alcian blue stain before and after hyaluronidase.
¶Esterases.
#Acid phosphatase.
**Crystals.
††Calcification.
‡‡May contain melanin.
§§May contain secretory granules.

Table 60. *Product of Cells of Malignant Neoplasms**

	Collagen[†]	Polysaccharide before and after digestion[‡]	Mucopolysaccharide before and after digestion[§]	Fat	Melanin	Fine structure	Tissue antigens of practical value
◑ Malignant fibroblastic fibrous histiocytoma[¶#]	III	+−	+−			See Tables 66–69	See Tables 70 and 71
◑ Malignant histiocytic fibrous histiocytoma[¶#]	III	+−	+−				
◑ Malignant pleomorphic fibrous histiocytoma[¶#]	III	++	+−	+			
◑ Desmoid tumor	I	−−					
◑ Fibroblastic fibrosarcoma	I	−−					
◑ Pleomorphic fibrosarcoma	I	−−		+			
○ Biphasic tendosynovial sarcoma[**]	III		+−				
○ Monophasic tendosynovial sarcoma, spindle cell type	III		+−				
○ Monophasic tendosynovial sarcoma pseudoglandular type[**]			+−				
○ Epithelioid sarcoma	III	+−	+−				
○ Clear cell sarcoma		+−	+−				
○ Chordoid sarcoma[**]		++	+−				
◑ Well differentiated liposarcoma		+−	+−	+			
◑ Myxoid liposarcoma	III	+−	+−	+			
◑ Lipoblastic liposarcoma		+−		+			
◑ Fibroblastic liposarcoma	III	+−		+			
◑ Pleomorphic liposarcoma	III	++		+			
◑ Leiomyoblastoma		+−					
◑ Leiomyosarcoma		++		+			
○ Embryonal rhabdomyosarcoma[††]		+−					
○ Alveolar rhabdomyosarcoma[††]		+−					
○ Myxoid rhabdomyosarcoma[††]		+−	−−				
○ Rhabdomyoblastoma[††]		+−					
○ Pleomorphic rhabdomyosarcoma[††]		+−		+			
◑ Hemangiopericytoma	III						
◑ Hemangiosarcoma[‡‡]	III	++					
○ Lymphangiosarcoma		++					
◑ Neurofibrosarcoma	I–III						
◑ Malignant schwannoma[§§]	III		+−		+		
◑ Malignant undifferentiated peripheral nerve tumor[§§¶¶]	III	++			+		
○ Malignant epithelioid mesothelioma[**]		+−	+−				
○ Malignant fibrous mesothelioma	I–III	+−	+−				
◑ Chondrosarcoma[‡‡]	II	++	++				
◑ Mesenchymal chondrosarcoma	III		++				
◑ Osteosarcoma[‡‡]	I–III						
◑ Parosteal osteosarcoma	I–III						
● Paget's sarcoma	I–III						
◑ Granulocytic sarcoma[¶¶##]		+−					
● Plasma cell myeloma[##]		+−					
○ Malignant granular cell tumor		++					
○ Alveolar soft part sarcoma[**##***]		++					
○ Kaposi's sarcoma[#]							
○ Malignant cystosarcoma phyllodes	I–III						
◑ Ewing's sarcoma	IV	+−					
● Chordoma[**]		++	++				
○ Myxopapillary ependymoma[**]		++	++				

*○ Primary soft tissue tumor; ● primary bone tumor; ◑ can be primary soft tissue or bone tumor. †See Tables 61 and 62. ‡PAS stain before and after diastase. §Alcian blue stain before and after hyaluronidase. ¶Esterases. #Acid phosphatase. **May contain calcification. ††Adenosine triphosphatase. ‡‡Alkaline phosphatase. §§May contain melanin. ¶¶May contain secretory granules. ##Crystals. ***Bowie stain positive for J-G granules. Also antirenin positive.

Table 61. *Characteristics of Collagen*

COLLAGEN	CELLS	CHARACTERISTICS
Type I	Fibroblasts	Widely distributed, well-structured large fibers. Stainable with trichrome stain.
Type II	Chondroblasts	Found in hyalin cartilages, nucleus pulposus and vitreous humor, small fibrils.
Type III	Fibroblasts, histiocytes	Widely distributed in soft tissues (for example, muscles and vessels) and parenchymal tissues. Fine reticular network. Stainable with reticulin stain.
Type IV	Epithelial and endothelial cells, cells of Ewing's sarcoma	Widely distributed, distinct basement membrane specific aggregates. Demonstrable with electron microscope.
Type V	Smooth muscle cells and myofibroblasts	Widely distributed, pericellular basement membrane. Demonstrable with electron microscope.
Type VI		Interstitial basement membrane interface.

Table 62. *Pattern of Distribution and Quantity of Reticulin Fibers*

NEOPLASM	PATTERN	QUANTITY
Fibroblastic fibrous histiocytoma	Storiform and parallel	Abundant
Histiocytic fibrous histiocytoma	Around cells and disarranged	Moderate
Fibrous tumor	Around cells and parallel	Moderate
Tendosynovial tumor	Around cells or parallel	Moderate
Striated muscle tumor	Around cells	Minimal
Smooth muscle tumor	Parallel	Minimal
Vascular tumor	Around nests of cells	Minimal
Peripheral nerve tumor	Parallel and wire-like	Abundant
Chondroid tumor	Around cells	Minimal
Histiocytic lymphoma	Around cells	Minimal
Malignant melanoma	Around nests of cells	Minimal
Epithelial tumor	Around cells	Minimal

Table 63. *Types of Complex Carbohydrates*

- Polysaccharides
  - Glycogen (homopolysaccharides)
  - Glycosaminoglycans (heteropolysaccharides)
- Glycoconjugates
  - Glycoproteins
    - Neutral
    - Acidic (sialylated, sulfated)
  - Proteoglycans
    - Nonsulfated glycosaminoglycans
    - Sulfated glycosaminoglycans
  - Glycolipids
    - Cerebroside (monosaccharide)
    - Globoside (oligosaccharides)
    - Ganglioside (oligosaccharides)

Table 64. *Role of PAS Stain as a Differential Stain*

INTRACYTOPLASMIC PAS POSITIVE DIASTASE LABILE (GLYCOGEN) MATERIAL	INTRACYTOPLASMIC PAS POSITIVE DIASTASE STABLE (GRANULAR POLYSACCHARIDE) MATERIAL
Fibrous histiocytoma	Pleomorphic fibrous histiocytoma
Liposarcoma	Liposarcoma
Clear cell sarcoma	Chordoid sarcoma
Rhabdomyosarcoma	Paraganglioma
Leiomyosarcoma	Chondroid tumors
Neuroblastoma	Alveolar soft part sarcoma
Mesothelioma	Granular cell tumor
Osteosarcoma	Mesodermal mixed tumor
Chondroid tumors	Epithelial cells
Ewing's sarcoma	
Chordoma	
Hepatoma	

Table 65. *Diagnostic Application of Histochemical Stains*

	ACID PHOSPHATASE	ALKALINE PHOSPHATASE	5-NUCLEOTIDES	ADENOSINE TRIPHOSPHATASE	ESTERASES	DIHYDROXYPHENYLALANINE OXIDASE
Epithelial tumors	Positive					
Fibrous histiocytomas	Positive					
Fibrous tumors		Positive				
Osseous tumors		Positive				
Vascular endothelium		Positive				
Lymphatic endothelium			Positive			
Striated muscle tumors				Positive		
Smooth muscle tumors				Negative		
Granulocytes					Positive	
Mast cells					Positive	
Pigmented schwannoma						Positive
Malignant melanoma						Positive

Table 66. *Suggestive Ultrastructural Features of Non-neoplastic Lesions*

Fasciitis: Fragmented basal lamina, thin filaments, myofibroblasts, dense bodies, pinocytosis
Fibromatosis: Myofibroblasts, cisternae of rough endoplasmic reticulum, thin filaments
Elastofibroma: Microfibrils, dense granular bodies
Tendosynovitis: Fine filopodia, abundant lysosomes, zonula adherens
Lipogranuloma: Lipid droplets, Langerhan's granules
Proliferative panniculitis: Fragmented basal lamina, thin filaments, dense bodies, pinocytosis
Proliferative myositis: Fragmented basal lamina, thin filaments, dense bodies, pinocytosis
Eosinophilic granuloma: Langerhan's cells.

Table 67. *Suggestive Ultrastructural Features of Benign Neoplasms*

Benign histiocytic fibrous histiocytoma: Abundant lysosomes, zonula adherens, fine filopodia, fat
Fibroma: Myofibroblasts, collagen
Hibernoma: Basement membrane, plasmalemmal dense granules, pinocytic vesicles, lipid exocytosis
Leiomyoma: Single layer basal lamina, actin filaments, attachment plaques, dense bodies, pinocytotic vesicles, scant rough endoplasmic reticulum
Rhabdomyoma: Basal lamina, abundant mitochondria, thick and thin myofilaments, membrane-bound crystalline bodies, abundant glycogen
Hemangioma: Thin basal lamina, Weibel-Palade bodies
Angiofibroma: Fragmented basal lamina, variable rough endoplasmic reticulum, attachment plaques, thin filaments, pinocytosis, intranuclear granules, collagen
Benign glomus tumor: Actin-like filaments, attachment plaques, pinocytosis, thick basal lamina, dense bodies
Lymphangiomyoma: Basal lamina, myofilaments, dense bodies, pinocytosis
Neurofibroma: Basal lamina, collagen
Benign schwannoma: Multilayer and incomplete basal lamina, microfilaments, thin cytoplasmic processes, variable collagen, sparse pinocytosis, microtubules, lysosomes, rare desmosomes, "Pi" granules
Ganglioneuroma: Dendritic cell processes
Paraganglioma: Dense core granules, sustentacular cells
Thymoma: Desmosomes, tonofilaments
Granular cell tumor: Basal lamina, abundant lysosomal granules, membrane-bound granules, angulated bodies
Meningioma: Abundant desmosomes, tonofilaments

Table 68. ***Suggestive Ultrastructural Features of Malignant Neoplasms***

Malignant fibroblastic fibrous histiocytoma: Fragmented basal lamina, variable rough endoplasmic reticulum and lysosomes, macula adherens, long cell processess, sparse collagen, intermediate filaments
Malignant histiocytic fibrous histiocytoma: Lysosomes, attachment plaques, phagocytic vacules, prominent Golgi complexes, lipid, intermediate filament, asteroid bodies
Malignant pleomorphic fibrous histiocytoma: Phagosomes, abundant rough endoplasmic reticulum and lysosomes, actin-like filaments, cytoplasmic processes, free ribosomes
Desmoid tumor: Fragmented basal lamina, attachment plaques, abundant rough endoplasmic reticulum, abundant collagen, intermediate filaments
Fibrosarcoma: Basal lamina, distended and abundant rough endoplasmic reticulum, abundant collagen hemidesmosomes, pinocytosis, no cell junction, intermediate filaments
Biphasic tendosynovial sarcoma: Basal lamina, macula adherens, microvilli-lined lumina, interdigitating filopodia, junctional complexes, prominent Golgi complexes
Epithelioid sarcoma: Actin filaments, intermediate filaments, cell junctions, filopodia-like extensions
Liposarcoma: Fragmented basal lamina, distended rough endoplasmic reticulum, abundant smooth endoplasmic reticulum, lipid droplets, pinocytosis, scant glycogen, electron-dense material, abundant mitochondria
Leiomyosarcoma: Thin basal lamina, actin filaments, fusiform dense bodies, pinocytosis, attachment plaques, vimentin, desmin
Rhabdomyosarcoma: Thick and thin myofilaments, Z-bands, glycogen, dense bodies, linear arrays of ribosomes, remnants of sarcomeres, prominent Golgi complexes, vimentin, desmin, cell junctions
Hemangiosarcoma: Fragmented basal lamina, pinocytosis, Weibel-Palade bodies, microfibrils, vimentin, tight junction
Hemangiopericytoma: Basement-membrane-like material, basal lamina, thin myofilaments, dense bodies, glycogen, tight junctions
Malignant schwannoma: Fragmented and multilayered basal lamina, desmosomes, long cell processes, microtubules, "Pi" granules, dense bodies, long-spaced collagen, enveloped axons, intermediate filaments, cell junctions
Neuroblastoma: Basal lamina, neurosecretory granules, dendritic cell processes, ribosomes, microtubules, clear vesicles, neurofilament
Neuroepithelioma: Rosettes, microtubules, sparse neurosecretory granules, junctional complexes
Malignant mesothelioma: Incomplete basal lamina, microvilli, desmosomes, tonofilaments, intermediate filaments
Granulocytic sarcoma: Nuclear blebs, Auer bodies, no filaments, no cell junctions
Malignant lymphoma: Nuclear blebs, ruffled borders, sparse rough endoplasmic reticulum, no cell processes, no cell junctions
Plasma cell myeloma: Abundant rough endoplasmic reticulum, crystals
Alveolar soft part sarcoma: Abundant smooth endoplasmic reticulum, neurosecretory-like granules, rhomboid crystalline structures, desmin and vimentin
Kaposi's sarcoma: Perithelial cells, elongated cells, lysosomes, intermediate filaments
Ewing's sarcoma: Pools of glycogen, primitive cell junctions, sparse organelles
Myxopapillary ependymoma: Basal lamina, microvilli, microfilaments, microtubules, desmosomes

Table 69. ***Some of the Differential Diagnostic Features of Small Round Cell Neoplasms***

	EMBRYONAL RHABDOMYOSARCOMA	EWING'S SARCOMA	NEUROBLASTOMA	MALIGNANT LYMPHOMA	GRANULOCYTIC SARCOMA
Nuclei*	Eccentric, pleomorphic	Uniform, round	Uniform, round	Cleaved, round	Bilobed, round
Cell Processes	Long	Absent	Dendritic	Absent	Absent
Cytoplasmic Organelles	Polysomes	Sparse	Sparse	Polysomes	Sparse
Intracytoplasmic Granules	None	None	Neurosecretory	None	Auer bodies
Intracytoplasmic filaments	Actin-myosin, sarcomeres	Intermediate	Neurofilament, microtubules	Intermediate	None
Glycogen	Uncommon	Pools	Uncommon	Rare	Rare
Cell Junctions	Primitive	Primitive	Primitive	Absent	Absent

*Many binucleated

Table 70A. *Distribution of Antigens in Mesenchymal Cells**

Cells	Antigens
Cardiac muscle	Desmin, Myoglobin, Actin, Myosin, Cathepsin B, Cathepsin D, UEA-I
Chondroblast	Vimentin, S-100, Cathepsin D
Chondrocyte	Vimentin, S-100, Cathepsin D
Endothelial	Vimentin, Factor VIII, Actin, UEA-I
Fat	S-100, UEA-I
Fibroblast	Vimentin, Desmin, Cathepsin B
Histiocyte	Vimentin, Myoglobin, S-100, Actin, Lysozyme, AACT, AAT, Myosin, UEA-I
Interdigitating	S-100, NSE
Kupffer	Actin
Leukocyte	Myosin, Vimentin
Lymphocyte	Vimentin, IgK, IgL, Lysozyme
Mast cell	AACT, AAT
Megakaryocyte	Factor VIII, Vimentin, NSE
Meningeal	Vimentin, Keratin
Mesangial	Actin
Mesothelial	Keratin, Vimentin
Myoepithelial	Keratin, Actin, S-100, NSE
Myofibroblast	Actin, Vimentin
Osteoblast	Vimentin
Pia-glial	S-100
Plasma cell	IgL, IgG, NSE
Platelets	Factor VIII
Sertoli	Vimentin
Smooth muscle, intestinal	Desmin, Actin, Myosin
Smooth muscle, uterine	Vimentin, Desmin, Actin, Myosin
Smooth muscle, vascular	Vimentin, Desmin, Actin, Myosin
Striated muscle	Vimentin, Desmin, Myoglobin, Myosin, Actin, Cathepsin B, Cathepsin D, UEA-I
Synovial	Keratin, Cathepsin D
T-cell dependent	S-100

*See Table 70C

Table 70B. *Distribution of Antigens in Neuroectodermal and Endocrine Cells**

Cells	Antigens
Adrenal cortical	Actin, Keratin
Astrocyte	Vimentin, GFAP, Actin, S-100
Astroglia	Vimentin, GFAP, Actin, Desmin
Choroidal	Keratin, GFAP
Ependymal	Vimentin, GFAP, S-100
Ganglion	S-100, Desmin
Islet	Actin, NSE, Somatostatin, Insulin, AAT, AACT, Glucagon, Keratin
Kultsitsky	Actin, NSE, Keratin, S-100, Bombesin
Langerhans	NSE, S-100, Keratin
Meissner's corpuscle	S-100
Melanocyte	S-100, NSE
Merkel cell	NSE, Keratin
Microglia	Vimentin, GFAP
Neurons	Neurofilament, NSE, S-100, Desmin
Oligodendrocyte	Vimentin, S-100, GFAP
Paccinian corpuscle	S-100
Parathyroid	Actin, NSE, Keratin
Pheochromocyte	Actin, NSE, Keratin, S-100, Chromogranin
Pineal	Keratin, NSE
Pineal interstitial	S-100
Pituitary	Keratin, NSE
Pituitary marginal	S-100
Pituitary stellate	S-100
Rosenthal fiber	Vimentin, GFAP
Schwann	Vimentin, GFAP, S-100, NSE, Myosin
Sustentacular	Actin, NSE, Keratin, S-100
Thyroid C	Actin, NSE, Calcitonin, Cathepsin B
Thyroid follicular	Actin, Thyroglobulin, Keratin
Trophoblast	HCG, Keratin
Verrocay body	S-100

*See Table 70C

Table 70C. ***Distribution of Antigens in Epithelial Cells****

Cells	Antigens
Bile duct	Keratin
Bronchial	Actin, Keratin, S-100, NSE
Duct	Actin, Keratin, S-100
Endocervical	Keratin
Epidermoid	Actin, Keratin, Vimentin, UEA-I, S-100
Epididymal	S-100
Gastric	Keratin, CEA, AACT, AAT, Gastrin
Hepatocytes	Keratin, ALPH, AFP, AACT, AAT
Intestinal	Keratin, ALPI, CEA, AACT, AAT
Mucus	Myosin
Myoepithelial	Actin, S-100, Factor VIII, Vimentin, NSE
Pancreatic acinar	Actin, S-100
Parabronchial	S-100
Prostate	Prostate specific antigen
Pulmonary alveolar	Keratin, AAT, AACT
Renal tubular cell	Keratin
Salivary gland	Keratin, GFAP, Vimentin, S-100
Seminiferous	Vimentin, ALPP
Skin adnexal	Keratin, S-100, UEA-I, Vimentin
Thymic	Keratin
Urothelial	Actin, Keratin

*UEA-I = ulex europeus agglutinin; S-100 = S-100 protein; AACT = alpha-antichymotrypsin; AAT = alpha-antitrypsin; NSE = neuron-specific enolase; IgK = immunoglobulin kappa; IgL = immunoglobulin lambda; IgG = immunoglobulin gamma; Factor VIII = factor VIII associated antigen; GFAP = glial fibrillary acidic protein; HCG = human chorionic gonadotropin; CEA = carcinoembryonic antigen; ALPH = alkaline phosphatase, hepatic; AFP = alpha-feto-protein; ALPI = alkaline phosphatase, intestinal; ALPP = alkaline phosphatase, placental.

TABLE 71. *Distribution of Certain Antigens as Tumor Markers*

Antigen	Soft Tissue and Bone Tumors*,†
Collagen	Most tumors
Vimentin†	Most tumors, including Wilms' tumor, cystosarcoma, adamantinoma, lymphomas, muscle tumors, Ewing's sarcoma, adipose tissue tumors, cartilagenous tumors, tendosynovial tumors
Desmin	Most muscle tumors, fibrous histiocytomas, many fibrous tumors
Keratin	Tendosynovial sarcoma, including clear cell sarcoma, epithelioid sarcoma; mesothelial tumors, adamantinoma, adenomatoid tumor, thymoma, Wilms' tumor
Glial fibrillary acidic protein (GFAP)	—
Neurofilament	—
Actin	Many tumors including all muscle tumors, fasciitis, angioma, angiosarcoma, fibrous histiocytomas, osteosarcoma
Myosin	Smooth muscle tumors, striated muscle tumors, cardiac muscle tumors, hemangiopericytoma
Myoglobin	Rhabdomyoma, rhabdomyoblastoma, rhabdomyosarcoma
Ulex Europaeus agglutinin (UEA-I)	Angioma, Angiosarcoma
Factor-VIII-associated antigen	Angioma, angiosarcoma, hemangioblastoma, megakaryocytic leukemia, Kaposi's sarcoma
Neuron-specific enolase (NSE)	—
S-100 Protein§	Granular cell tumor, eosinophilic granuloma, chondroma, chondrosarcoma, chordoma, chondroid tumor, smooth muscle tumors, granulosa cell tumor, histiocytosis, xanthoma, mastocytosis, synovitis
Alpha antitrypsin (AAT) and Alpha antichymotrypsin (AACT)	Fibrous histiocytomas, histiocytic lymphoma, mastocytosis, histiocytosis
Lysozyme	Lymphomas, histiocytic tumors
Immunoglobulin kappa (IgK) and immunoglobulin lambda (IgL)	Lymphomas, plasmacytomas
Alkaline phosphatase, hepatic	Osteoid- and cartilage-producing tumors, hemangiomatous tumors
Alkaline phosphatase, placental	—
Carcinoembryonic antigen (CEA)	Granular cell tumor, chordoma

*Inflammatory, phagocytic, and necrotic cells, as well as sanguinous fluid and mucus, may yield a positive reaction regardless of the antigen.
†Metastatic nodules may yield an inconsistent reaction.
‡May be present in the serum of patients with hepatitis, rheumatoid arthritis, and leprosy.
§S-100 protein level decrease in parallel with degree of malignancy.

Positive Reaction	
Neuroectodermal and Endocrine Tumors*,†	Epithelial Tumors *,†
Some tumors	Some tumors
Peripheral nerve tumors, melanoma, ganglioneuroma	Renal cell carcinoma, fibroadenoma, desmoplastic epidermoid carcinoma, salivary gland tumors, cystadenocarcinoma ovary, seminoma
Triton tumor	—
Glandular schwannoma, most endocrine tumors including carcinoid and islet cell tumor	Most epithelial tumors including renal cell carcinoma and urothelial tumors, glandular epithelial tumors, embryonal carcinoma, Brenner tumor
All gliomas, medulloblastoma, ependymoma, choroid plexus papilloma, ganglioneuroma, nevoid schwannoma	Some teratomas
Paraganglioma, schwannoma, Merkel cell tumor	—
—	Epidermoid carcinoma, adenocarcinoma, fibroadenoma
—	—
Triton tumor	—
—	—
—	—
—	Renal cell carcinoma
Paraganglioma, melanoma, Merkel cell tumor, carcinoid, insulinoma, gastrinoma, glucaginoma, somatostatinoma, medullary carcinoma of thyroid, pituitary adenoma, neuroblastoma, primitive neuroectodermal tumor	Oat cell carcinoma, adenocarcinoma, squamous carcinoma, adenosquamous carcinoma
Most gliomas, including ependymoma, retinoblastoma; peripheral nerve tumors including nevoid schwannoma; melanoma, nevus, blue nevus, ganglioneuroma	Mixed tumors of salivary glands, sclerosing adenosis, some mammary neoplasms, some salivary gland tumors, skin adnexal tumors, some renal cell carcinomas
—	—
—	—
—	—
—	Hepatoma
—	Embryonal carcinoma, seminoma
Glandular schwannoma	Many epithelial tumors

Fig. 1113. Myxoid liposarcoma (×350) (see Figs. 783, 933, 945, 948, 997, 1046, 1169, 1232).

Fig. 1114. (Top center) Plexiform neurofibroma, with wirelike cell processes (×85) (see Figs. 365, 527, 627, 842).

Fig. 1115. Malignant neuroepithelioma (×350) (see Figs. 585, 690, 1238, 1239).

Fig. 1116. Neurofibrosarcoma (×350) (see Figs. 370, 401, 584, 624, 626, 838, 1229, 1263, 1266).

Fig. 1117. (Bottom center) Fibrosarcoma (×140) (see Figs. 608, 825, 832, 901, 1205).

Fig. 1118. Desmoid tumor, low grade fibrosarcoma (×570) (see Figs. 686, 865, 899).

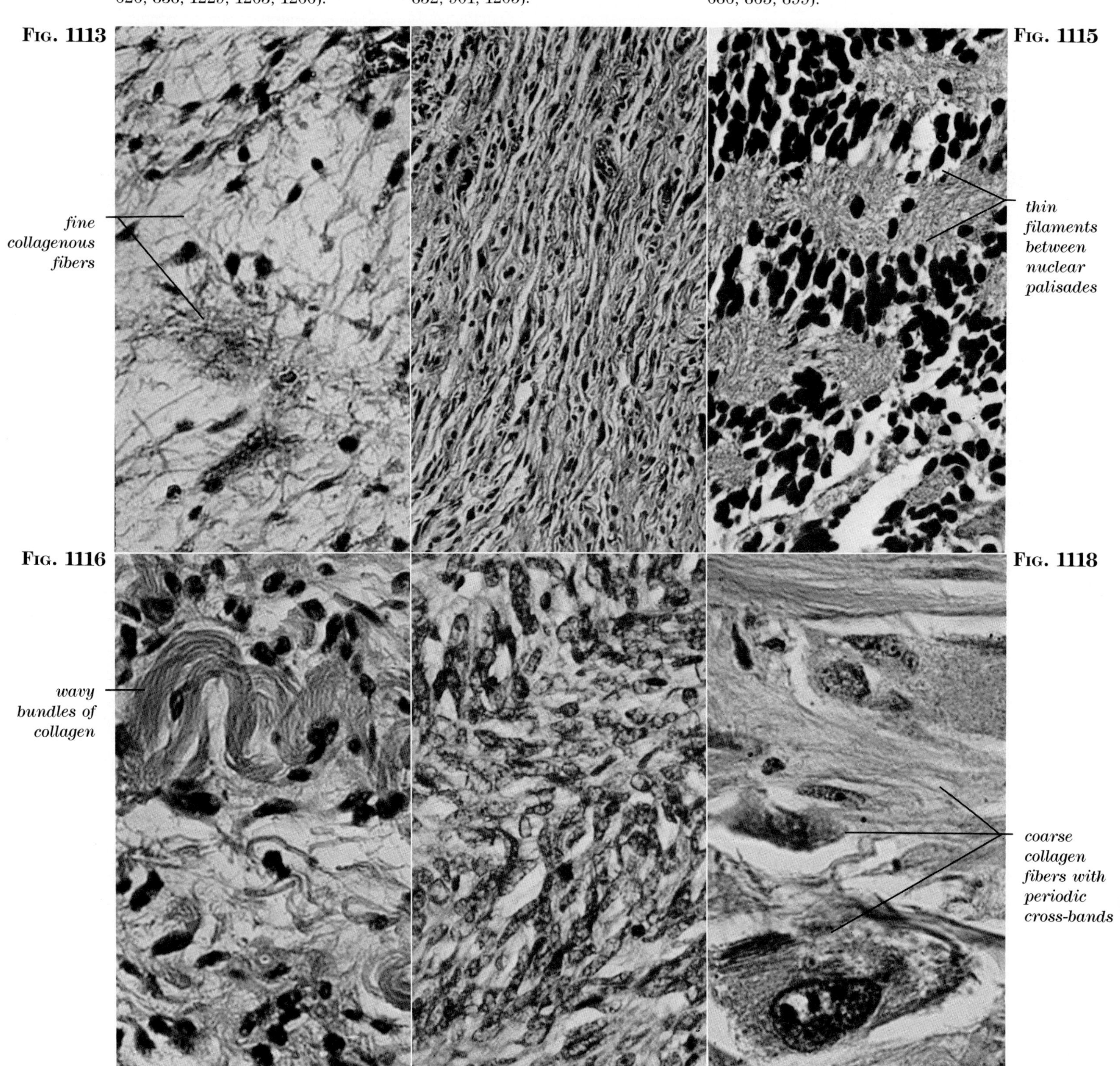

Fig. 1119. Normal dermal collagen (×350) (see Figs. 194, 1120, 1125).

Fig. 1120. (Top center) Collagenoma with scar-like collagen (see Figs. 194, 1119, 1125).

Fig. 1121. Aponeurotic fibromatosis (×350) (see Figs. 385, 581, 616, 861, 1081, 1122).

Fig. 1122. Retroperitoneal fibromatosis (×350) (see Figs. 367, 582, 616, 860, 1095, 1121).

Fig. 1123. (Bottom center) Elastofibroma with interrupted broad bundles of collagen (×350) (see Figs. 914, 1140, 1206).

Fig. 1124. Angiomyolipoma (×140) (see Fig. 959).

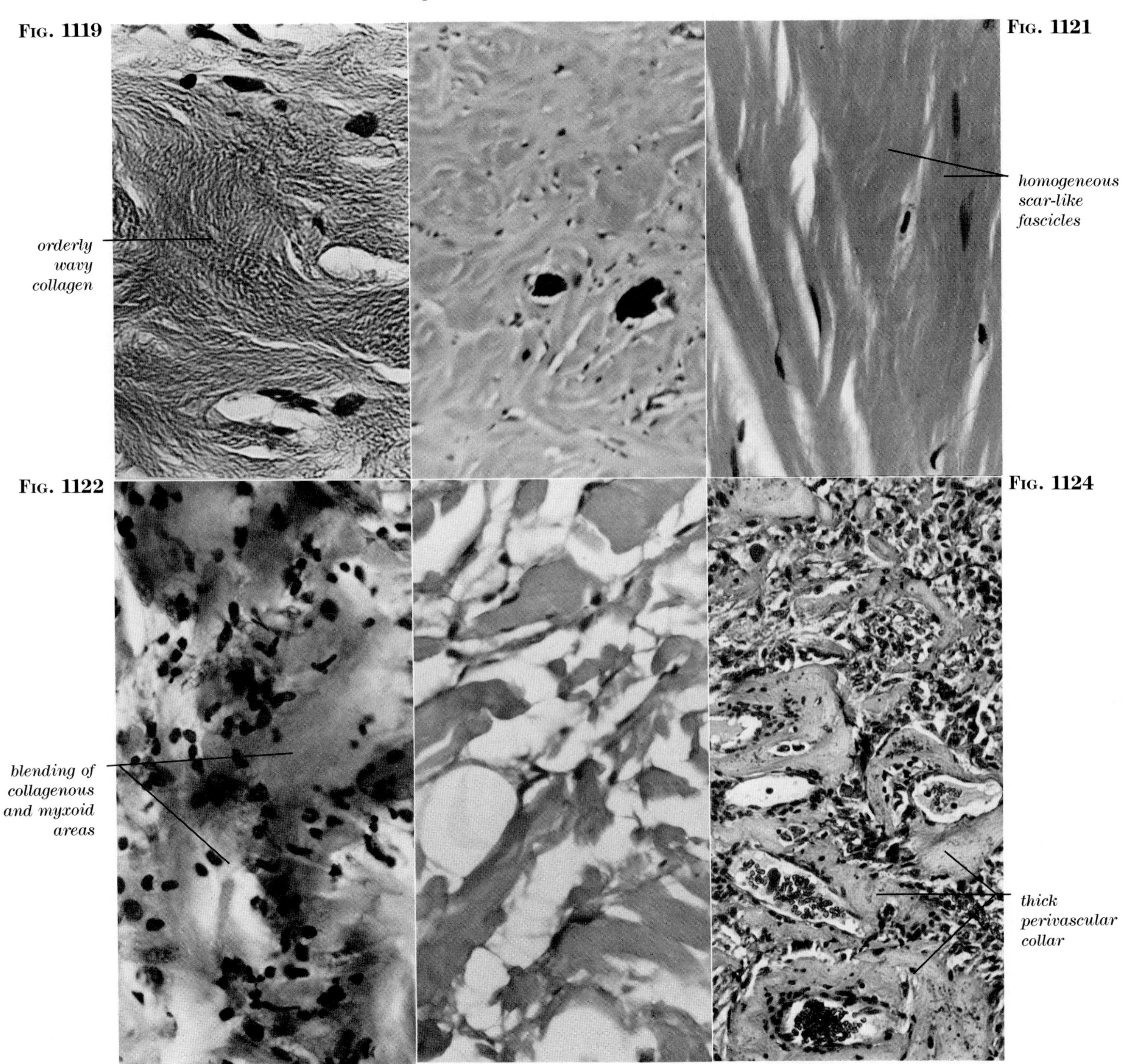

Fig. 1125. Normal dermal collagen (Reticulin, ×350) (see Figs. 194, 1119, 1120).

Fig. 1126. Kaposi's sarcoma (Reticulin, ×350) (see Figs. 63, 388, 629, 655, 658, 977, 986).

Fig. 1127. Cutaneous xanthogranuloma (Reticulin ×350) (see Figs. 607, 734, 779, 1031, 1201).

Fig. 1128. Paget's sarcoma (Reticulin, ×350) (see Figs. 107, 134, 918, 1103).

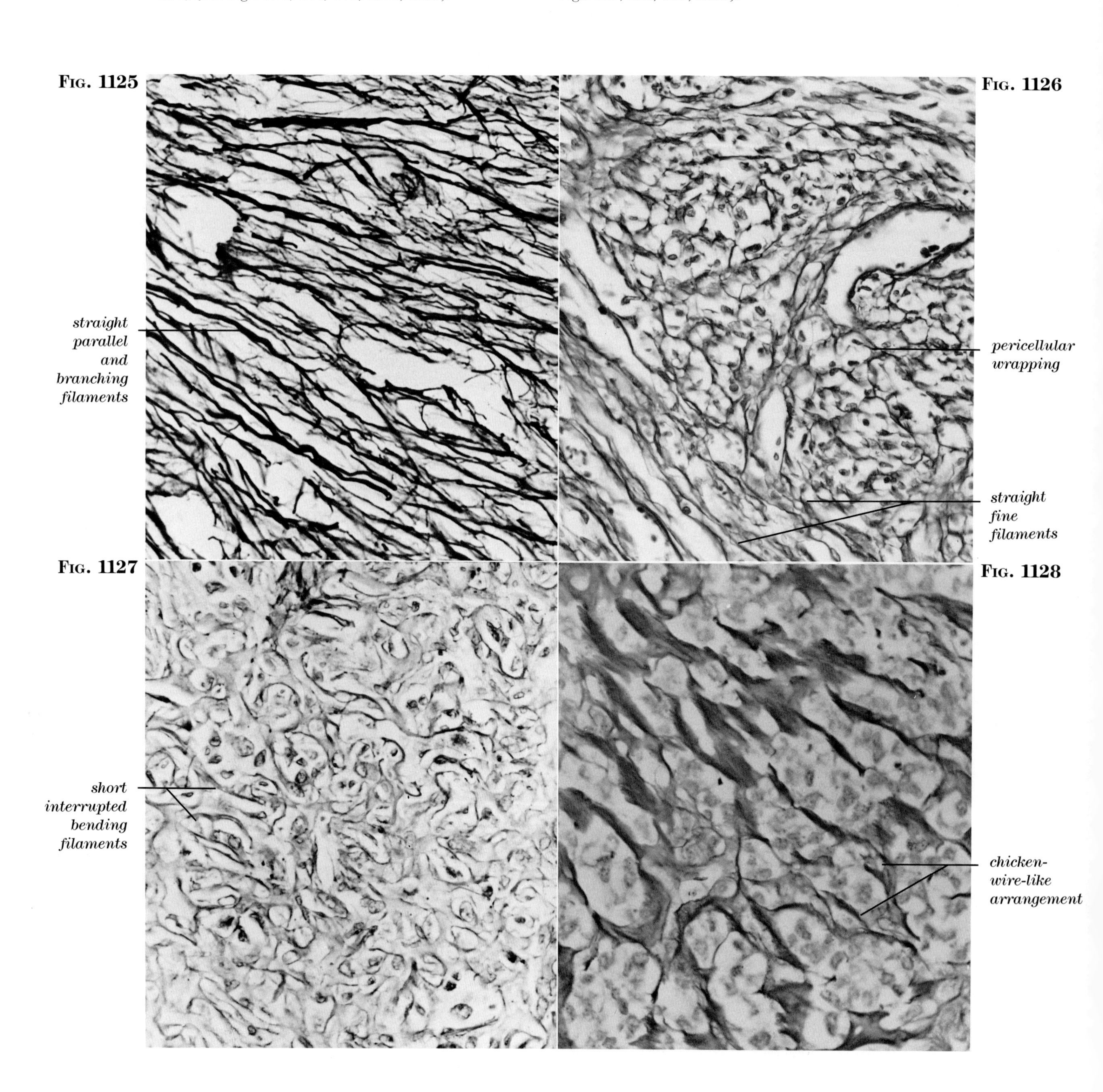

FIG. 1129. Embryonal rhabdomyosarcoma (Reticulin, ×350) (see Figs. 751, 949, 1138, 1164, 1171).

FIG. 1130. Pleomorphic rhabdomyosarcoma (Reticulin, ×140) (see Figs. 804, 1166, 1167, 1216, 1255).

FIG. 1131. Leiomyosarcoma (Trichrome, ×350) (see Figs. 803, 1051, 1132, 1163, 1211).

FIG. 1132. Leiomyosarcoma (Reticulin, ×350) (see Figs. 1051, 1131, 1163, 1211, 1212).

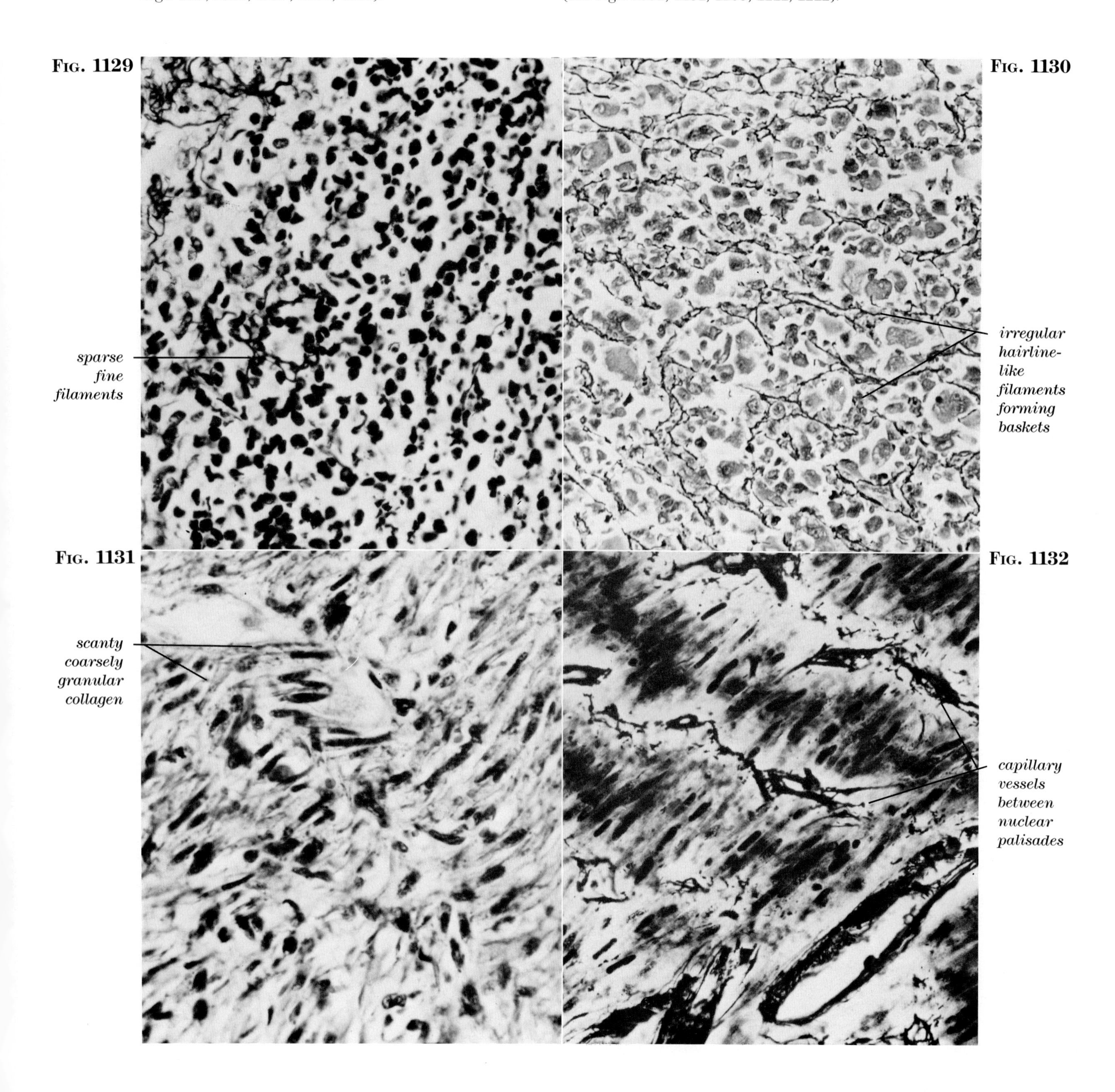

Fig. 1133. Malignant fibroblastic fibrous histiocytoma (×350) (see Figs. 76, 111, 339, 760, 833, 1134, 1135, 1147).

Fig. 1134. (Top center) Malignant fibroblastic fibrous histiocytoma showing twisted nebulae (Reticulin, ×350) (see Figs. 1133, 1135, 1147).

Fig. 1135. Malignant fibroblastic fibrous histiocytoma (PAS after digestion, ×350) (see Figs. 1133, 1134, 1147, 1178).

Fig. 1136. Tendosynovitis (Iron, ×350) (see Figs. 70, 786, 867, 1023, 1029).

Fig. 1137. Ganglioneuroma (×350) (see Figs. 486, 630, 811, 824, 857, 1191).

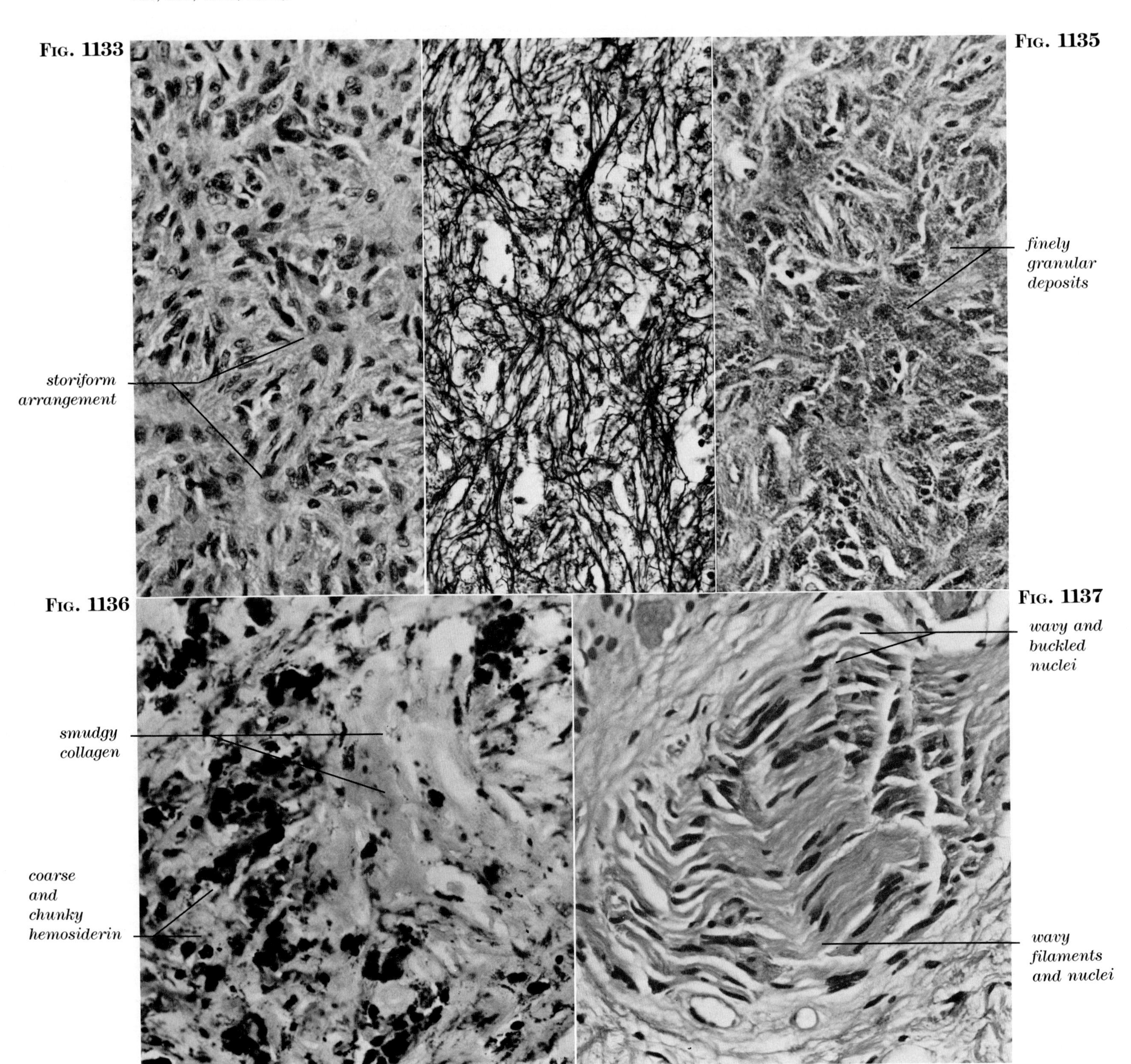

Fig. 1138. Embryonal rhabdomyosarcoma (Reticulin, ×350) (see Figs. 949, 1129, 1164, 1171, 1215).

Fig. 1139. Leiomyoblastoma (Reticulin, ×350) (see Figs. 828, 1053, 1158, 1213, 1214).

Fig. 1140. Elastofibroma (Elastic, ×350) (see Figs. 914, 1123, 1206).

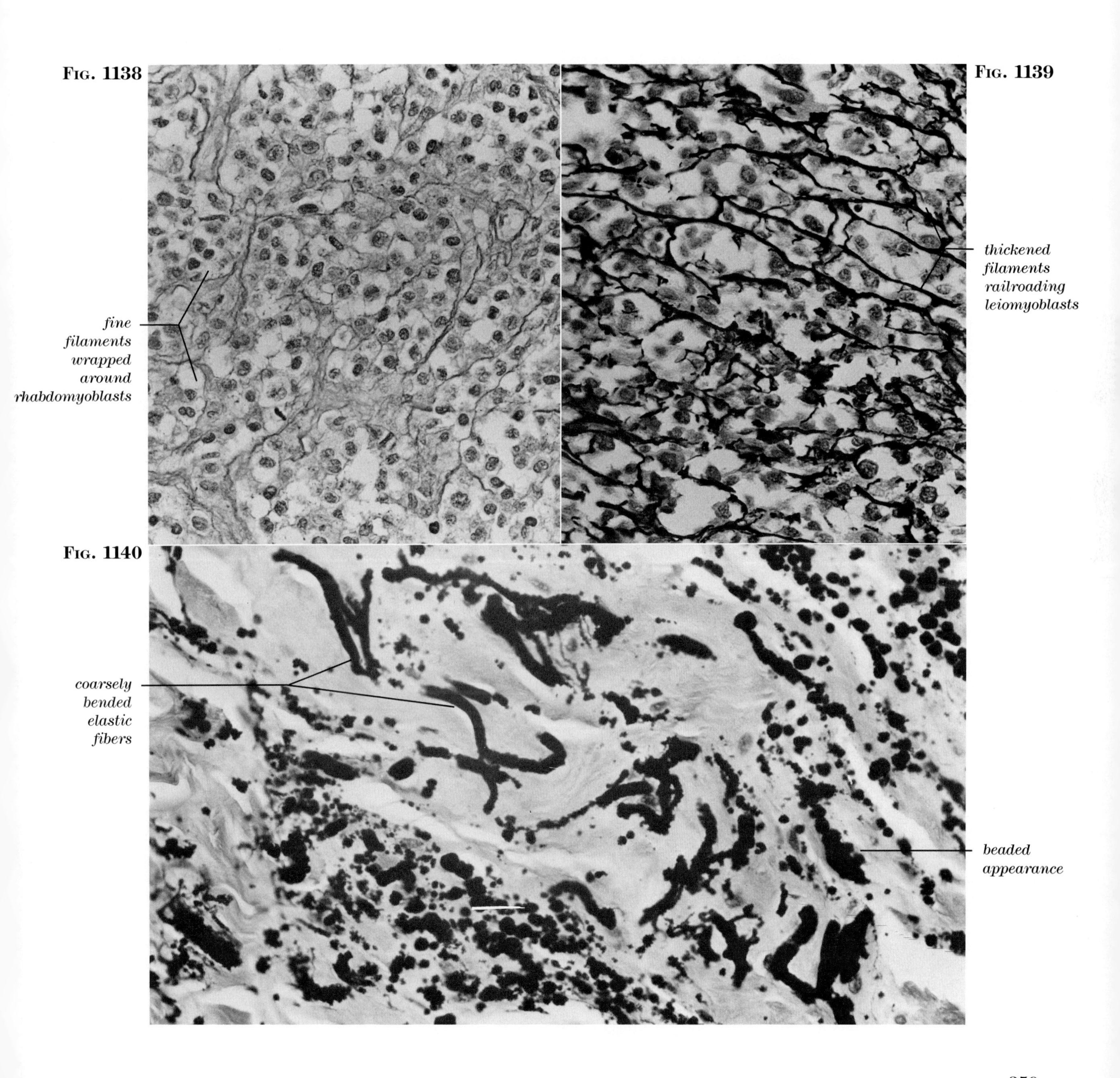

Fig. 1141. Malignant glomus tumor (Reticulin, ×570) (see Figs. 457, 482, 550, 728, 980).

Fig. 1142. Neurofibroma (Reticulin, ×350) (see Figs. 74, 253, 363, 652, 840, 871, 955, 993, 1013, 1142, 1143).

Fig. 1143. Neurofibroma (Reticulin, ×350) (see Figs. 11, 74, 245, 253, 871, 955, 993, 1013, 1142).

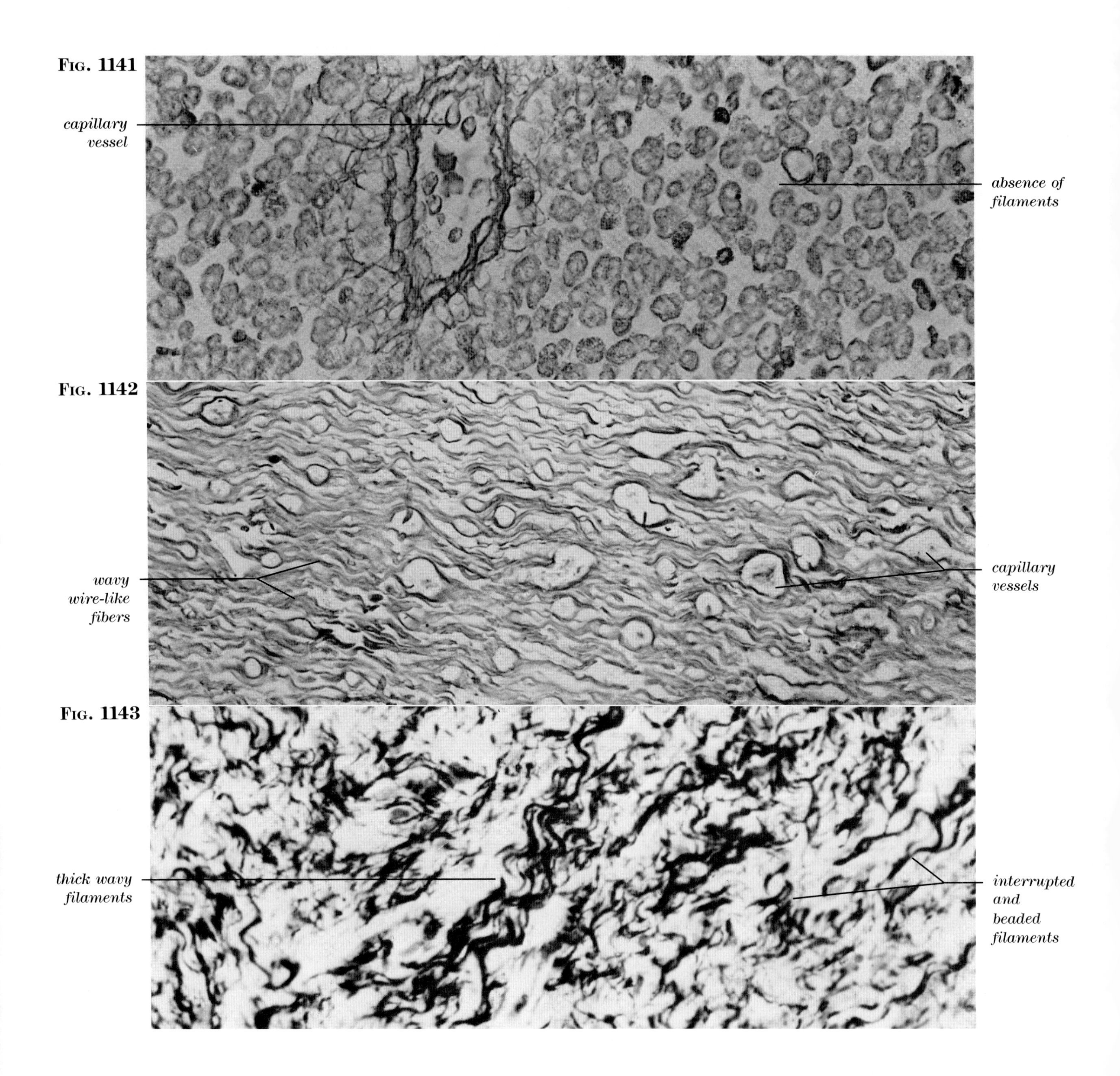

Fig. 1144. Monophasic tendosynovial sarcoma, spindle cell type (Reticulin, ×140) (see Figs. 903, 1244, 1245).

Fig. 1145. Monophasic tendosynovial sarcoma, spindle cell type with clear cell sarcoma component (Reticulin, ×350) (see Figs. 85, 684, 1144).

Fig. 1146. Epithelioid sarcoma (Reticulin, ×350) (see Figs. 676, 830, 1170, 1244).

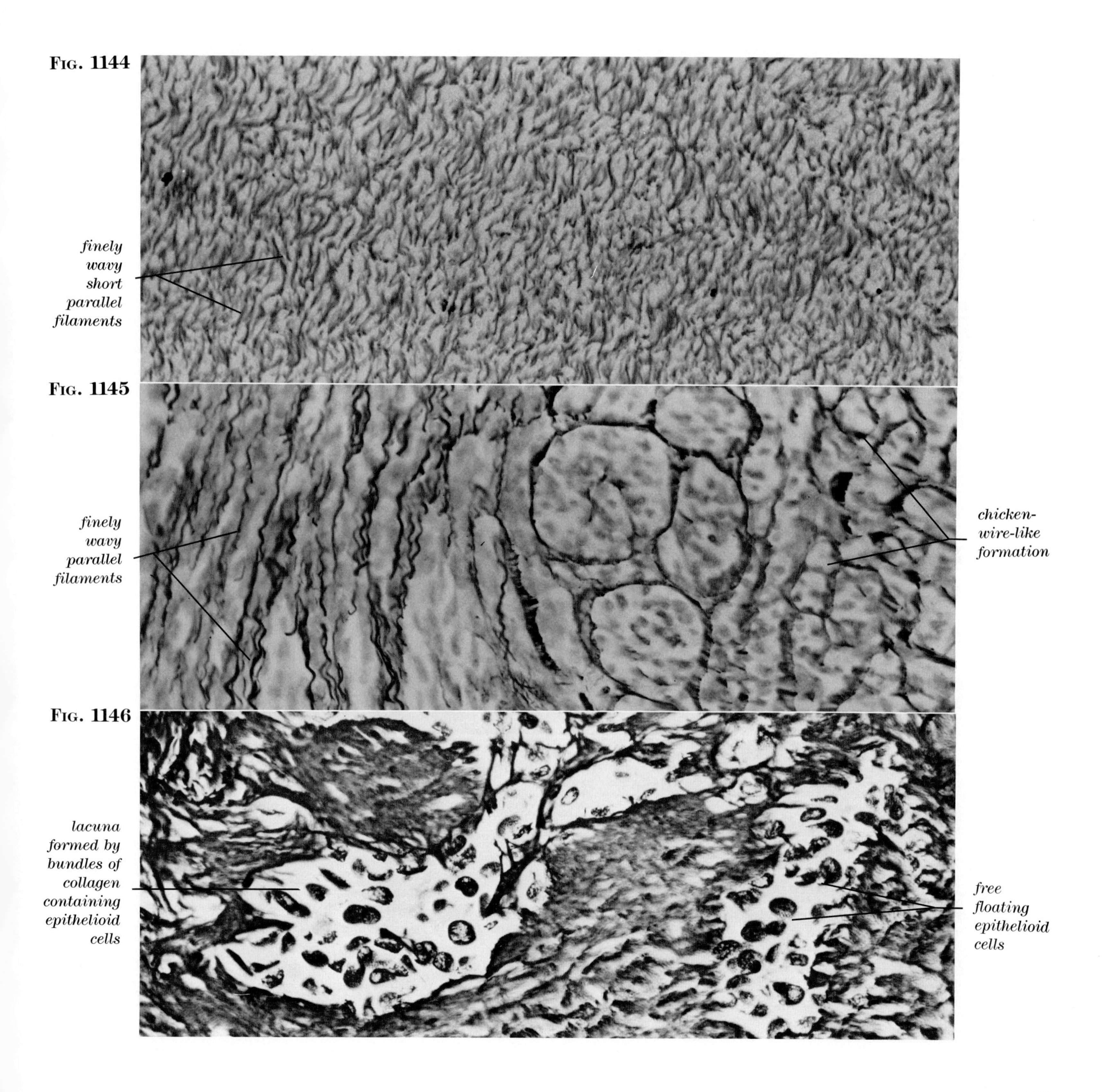

Fig. 1147. Malignant fibroblastic fibrous histiocytoma with myxoid areas (Reticulin, ×350) (see Figs. 76, 111, 339, 340, 645, 672, 833, 1026, 1133–1135, 1178).

Fig. 1148. Malignant pleomorphic fibrous histiocytoma (Reticulin, ×350) (see Figs. 88, 111, 344, 460, 679, 780, 788, 789, 792, 798, 805, 818, 1047, 1179, 1180).

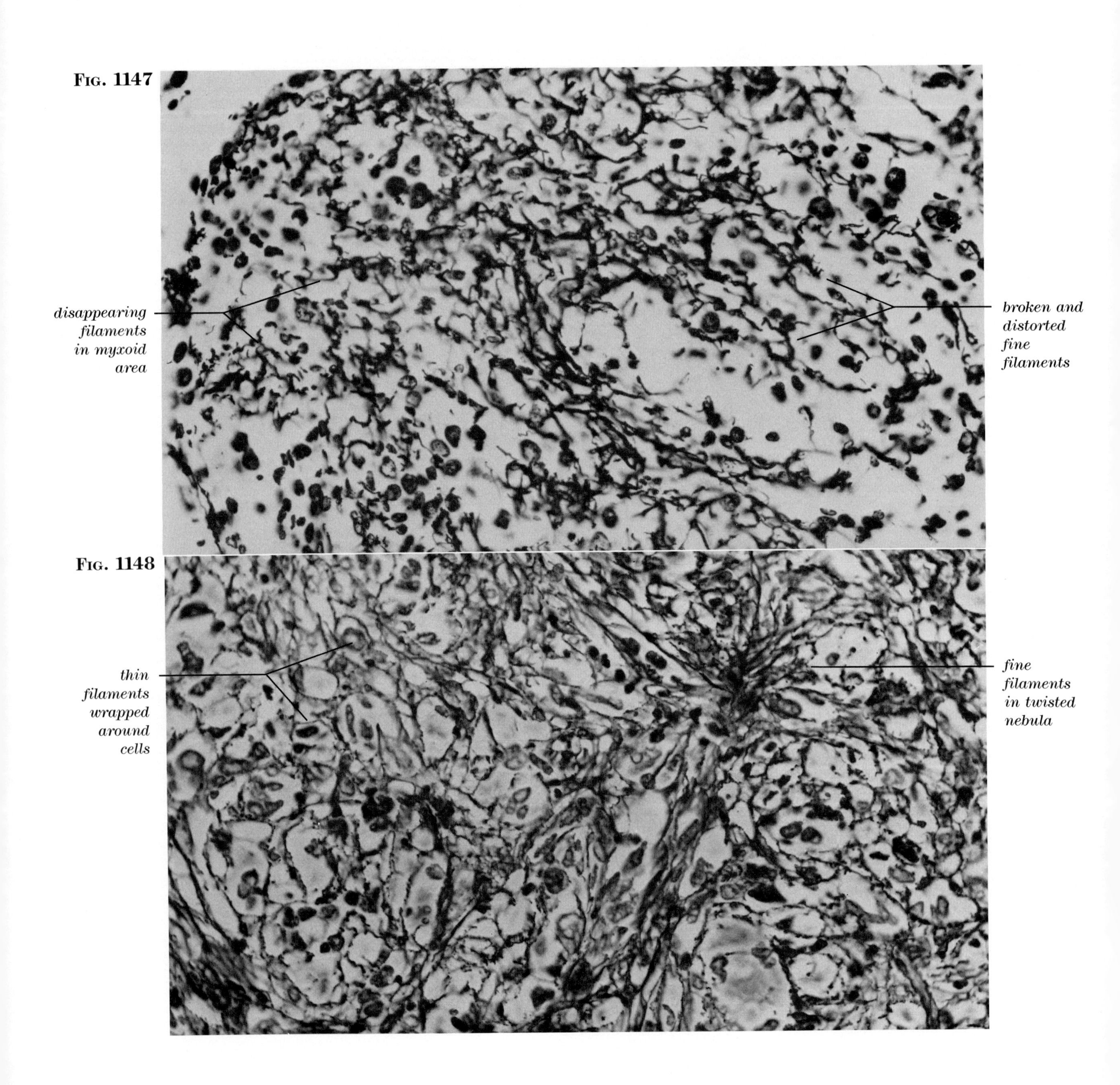

Fig. 1149. Low grade malignant histiocytic fibrous histiocytoma (Trichrome, ×570) (see Figs. 784, 791, 816, 872, 892, 893, 930, 1003, 1048, 1172, 1246, 1247).

Fig. 1150. Sclerosing osteosarcoma (×570) (see Figs. 1070–1079, 1105–1108, 1156).

Fig. 1151. Osteoblastic osteosarcoma. (×350) (see Figs. 1086, 1087, 1088, 1105, 1106, 1107, 1108).

Fig. 1152. Osteochondroma (×140) (see Figs. 772, 1063, 1066, 1090).

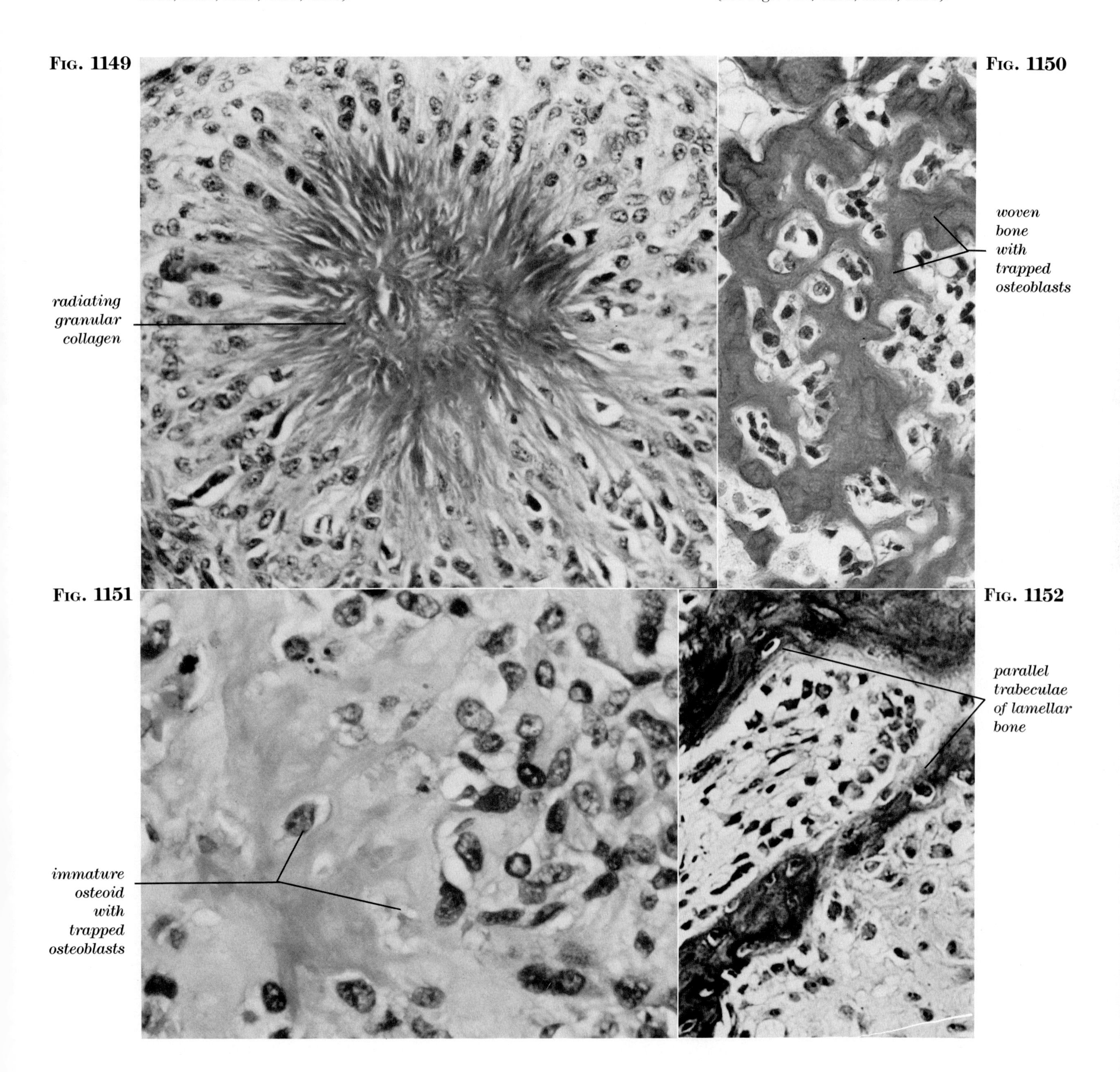

Fig. 1153. Keloid (×350) (see Fig. 877).

Fig. 1154. Parosteal osteosarcoma (×350) (see Figs. 109, 129, 249, 287, 354, 656, 1082, 1154).

Fig. 1155. Osteoblastoma (×350) (see Figs. 97, 211, 234, 1092).

Fig. 1156. Sclerosing osteosarcoma (×350) (see Figs. 108, 228, 328, 615, 819, 1070, 1105, 1150, 1156).

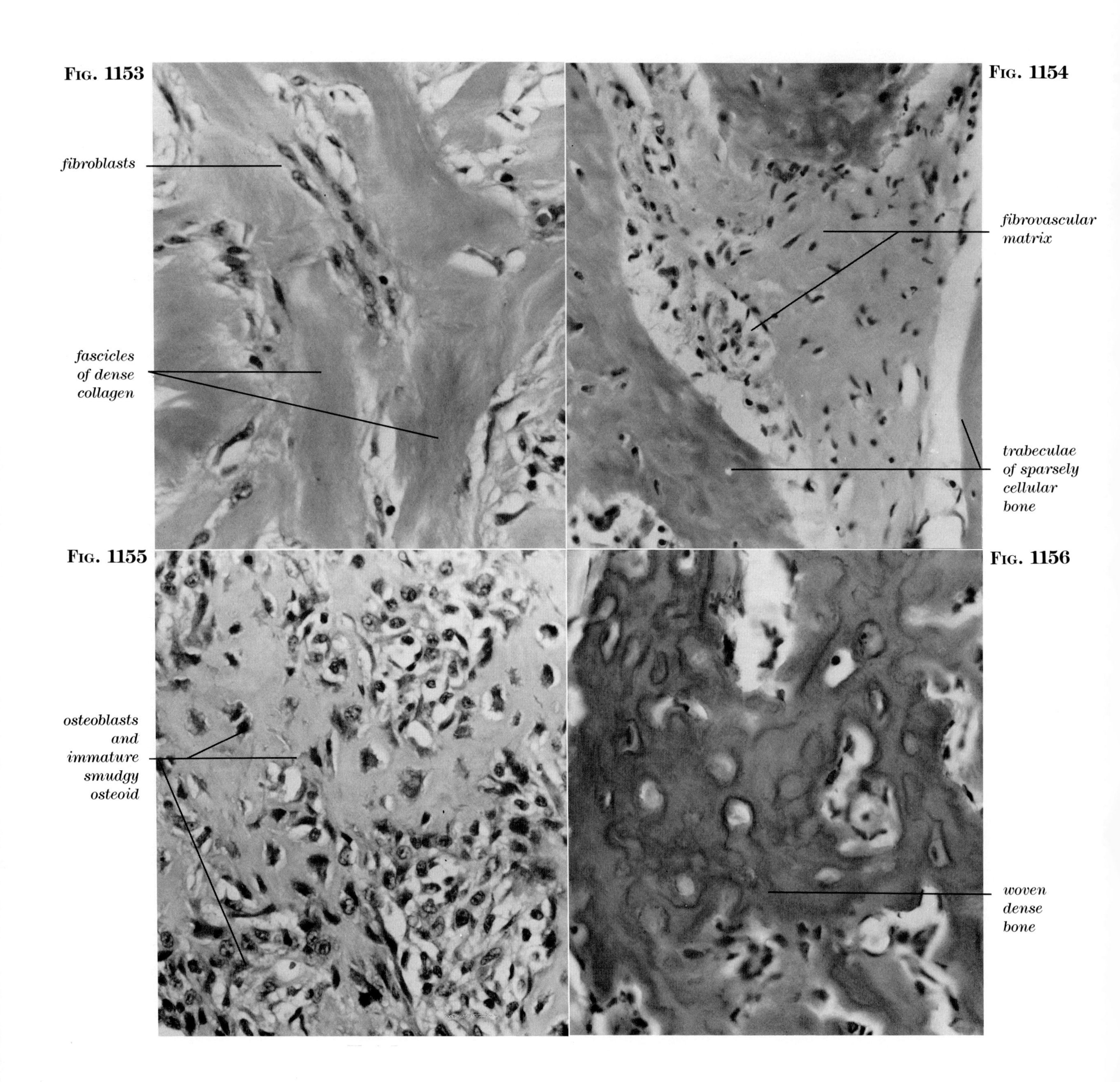

Fig. 1157. Ewing's sarcoma (PAS, ×350) (see Figs. 100, 449, 573, 698, 1230).

Fig. 1158. (Top center) Leiomyoblastoma containing coarse cytoplasmic granules (PAS, ×570) (see Figs. 1053, 1139, 1213, 1214, 1248).

Fig. 1159. Rhabdomyoblastoma (×350) (see Figs. 796, 936, 1034, 1260).

Fig. 1160. Benign granular cell tumor (PAS, ×350) (see Figs. 75, 498, 539, 736, 1235).

Fig. 1161. (Bottom center) Alveolar soft part sarcoma (×140) (see Figs. 718, 1199).

Fig. 1162. Hibernoma (PAS, ×350) (see Figs. 432, 717).

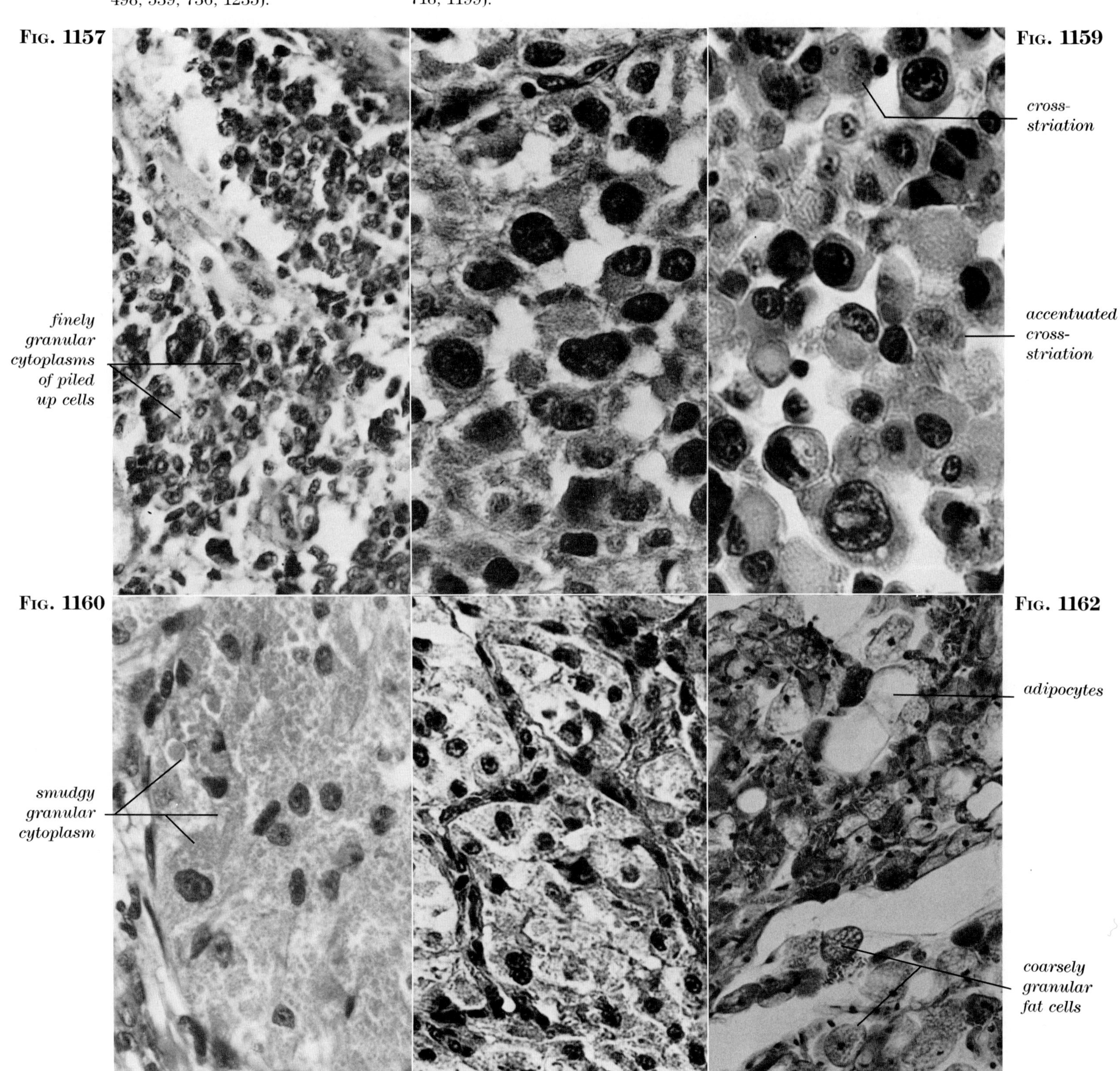

Fig. 1163. Leiomyosarcoma (×350) (see Figs. 1131, 1132, 1211, 1212, 1249).

Fig. 1164. (Top center) Embryonal rhabdomyosarcoma with flamelike filaments (×350) (see Figs. 1129, 1138, 1171, 1215, 1257).

Fig. 1165. Mesodermal mixed tumor (×570) (see Figs. 65, 589, 763, 969, 1217).

Fig. 1166. Pleomorphic rhabdomyosarcoma (×570) (see Figs. 612, 1130, 1167, 1216, 1255).

Fig. 1167. Pleomorphic rhabdomyosarcoma (×570) (see Figs. 1130, 1166, 1216, 1250, 1255).

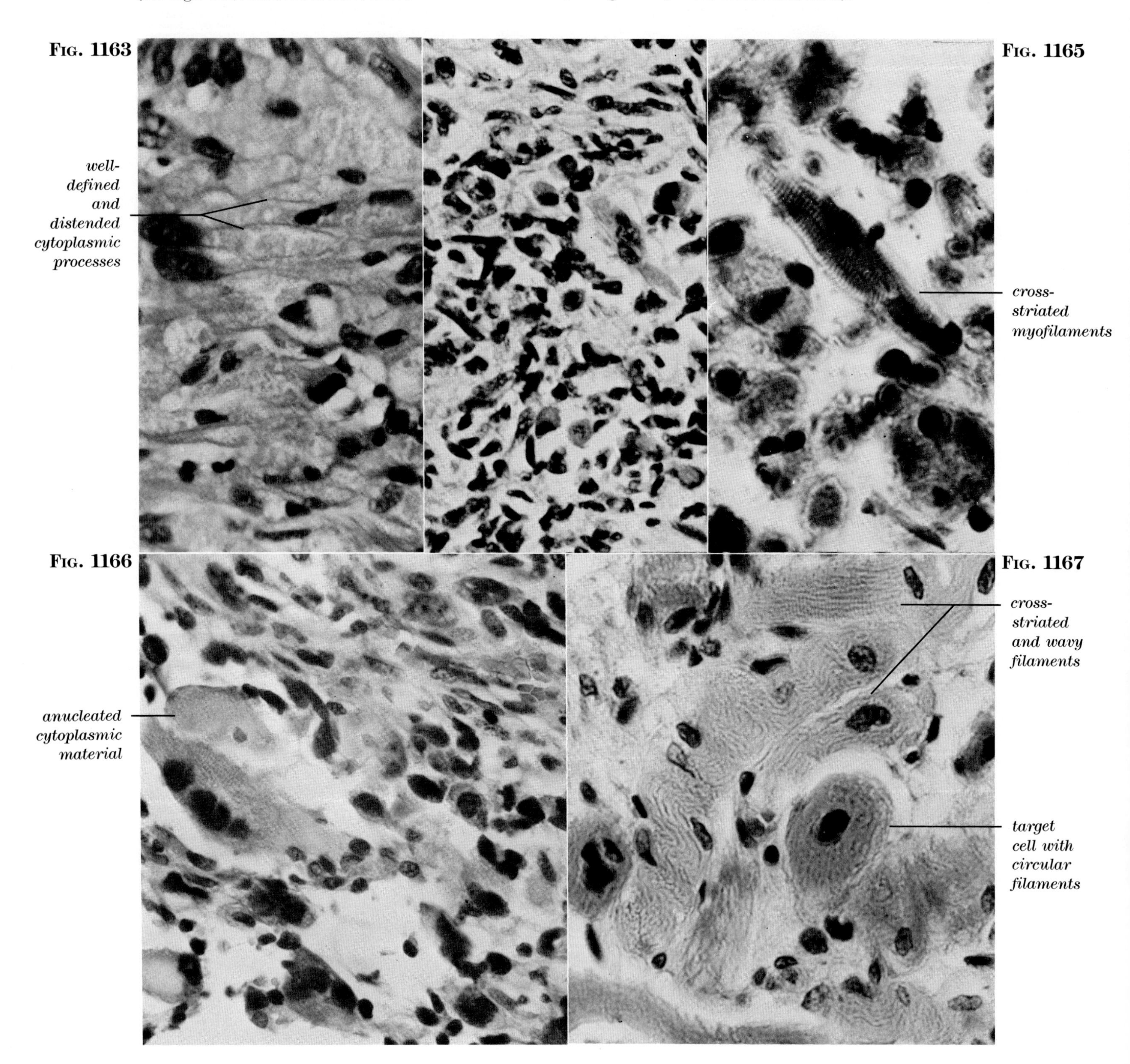

Fig. 1168. Myxoid lipoma (×350) (see Figs. 73, 635, 939, 941, 1187).

Fig. 1169. (Top center) Myxoid liposarcoma (×350) (see Figs. 1046, 1113, 1232).

Fig. 1170. Epithelioid sarcoma (×350) (see Figs. 1052, 1059, 1146, 1244, 1245).

Fig. 1171. Embryonal rhabdomyosarcoma (×350) (see Figs. 1129, 1138, 1164, 1215, 1257).

Fig. 1172. (Bottom center) Malignant histiocytic fibrous histiocytoma (×350) (see Figs. 1048, 1149, 1246–1247).

Fig. 1173. Biphasic tendosynovial sarcoma (×350) (see Figs. 765, 906, 1099, 1100, 1202).

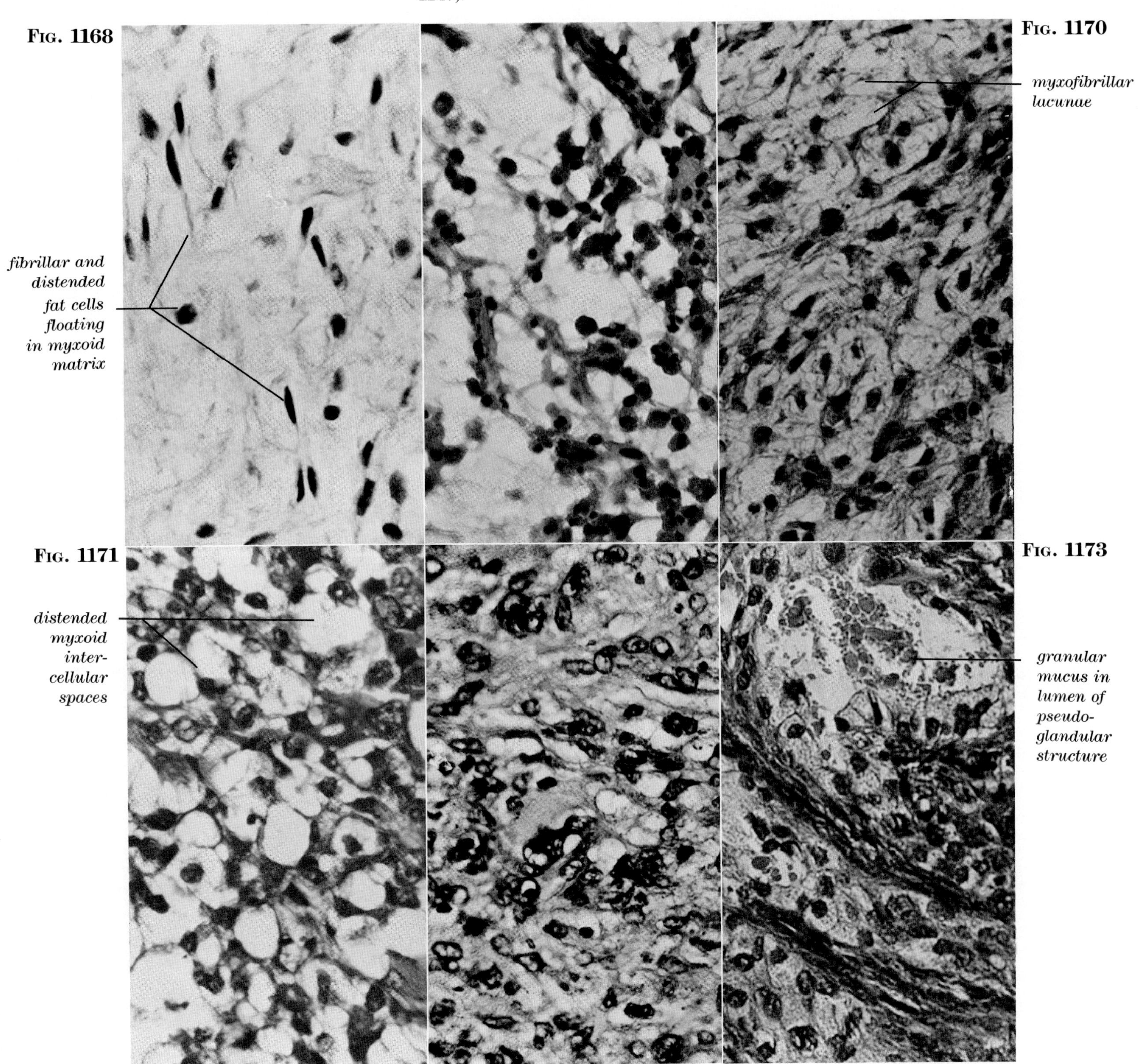

FIG. 1174. Chordoma (×350) (see Figs. 69, 215, 310, 505, 729, 767, 771, 1226).

FIG. 1175. Myxoid chondrosarcoma (×350) (see Figs. 124, 182, 213, 308, 329, 638, 932, 1175).

FIG. 1176. Chordoid sarcoma (×140) (see Figs. 456, 510, 640, 919).

FIG. 1177. Myxopapillary ependymoma (×350) (see Figs. 239, 466, 518).

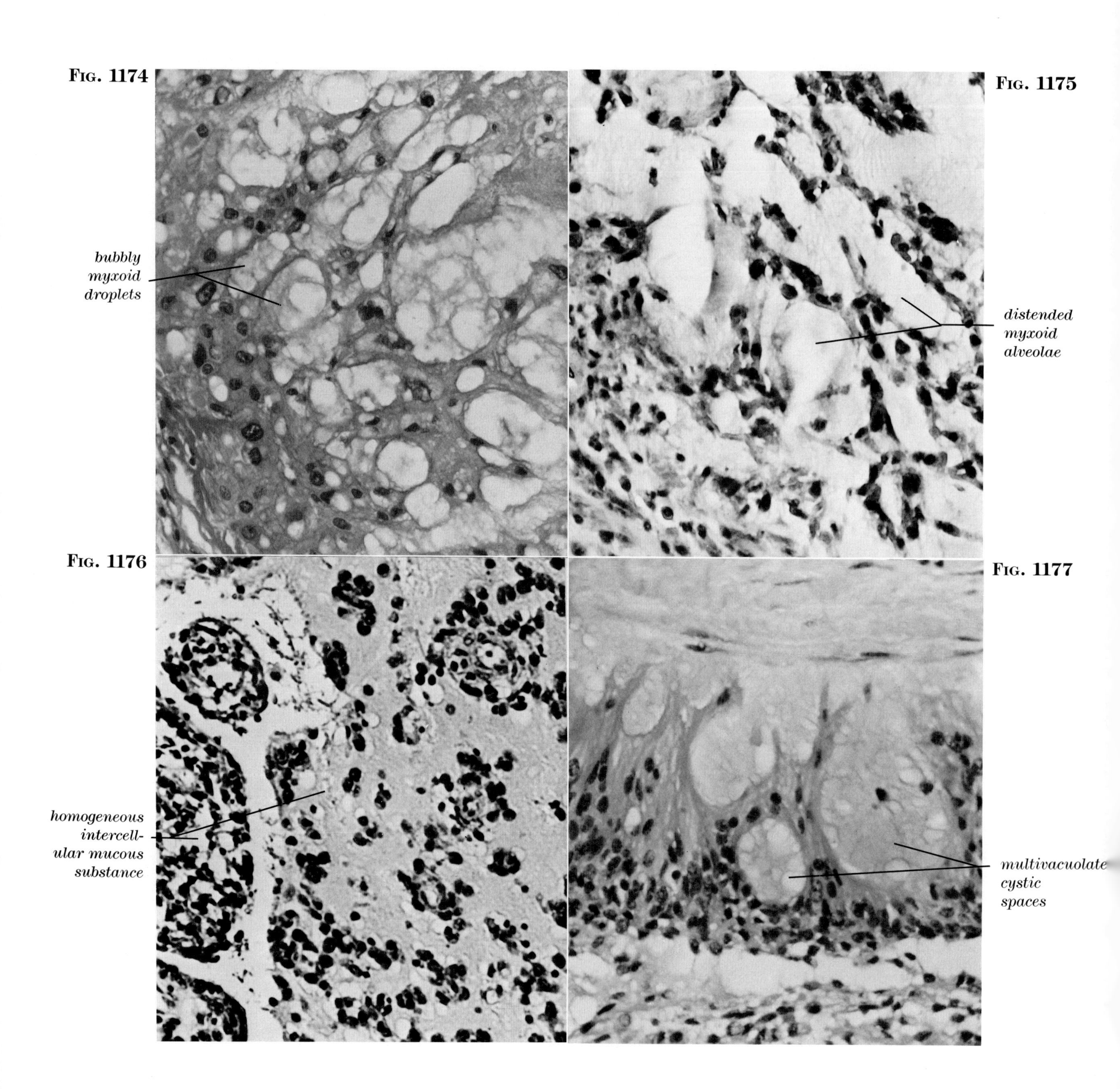

Fig. 1178. Malignant fibroblastic fibrous histiocytoma with myxoid areas (Alcian blue, ×350) (see Figs. 76, 111, 241, 339, 645, 833, 1026, 1133–1135, 1147, 1200).

Fig. 1179. Malignant pleomorphic fibrous histiocytoma (Alcian blue, ×350) (see Figs. 88, 111, 293, 598, 818, 937, 1047, 1148, 1180, 1253).

Fig. 1180. Malignant pleomorphic fibrous histiocytoma (Reticulin, ×350) (see Figs. 25, 130, 237, 344, 679, 937, 953, 1032, 1148, 1179, 1253).

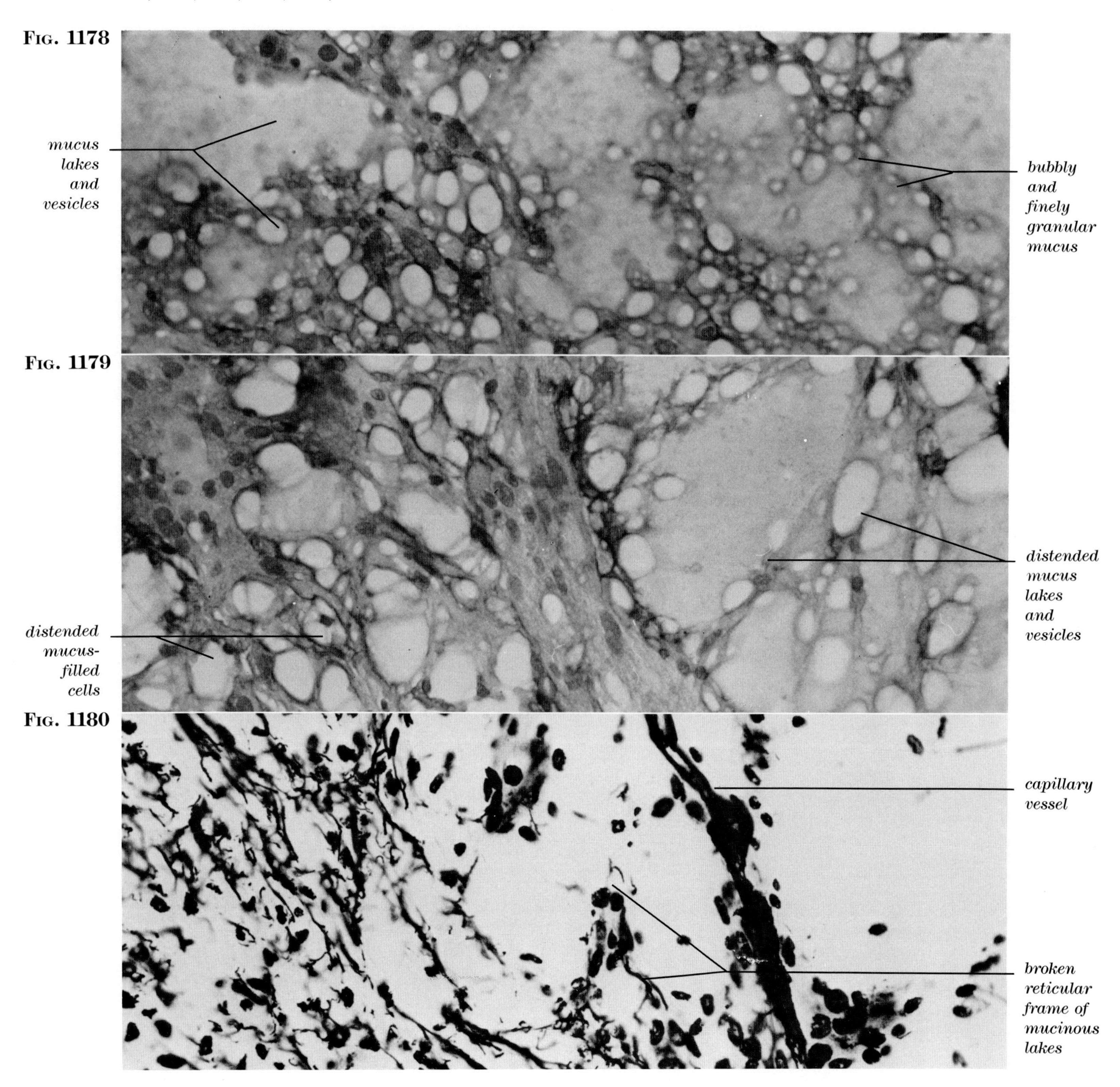

Fig. 1181. Well-differentiated lipoma (×570) (see Figs. 73, 198, 243, 455).

Fig. 1182. Well-differentiated, so-called "sclerosing liposarcoma" (Reticulin, ×350) (see Figs. 86, 219, 431, 435, 822, 999, 1184, 1233).

Fig. 1183. Pleomorphic lipoma (×350) (see Figs. 73, 613, 781, 821, 909, 957, 1182, 1183).

Fig. 1184. Well-differentiated liposarcoma (×140) (see Figs. 431, 435, 822, 999, 1233).

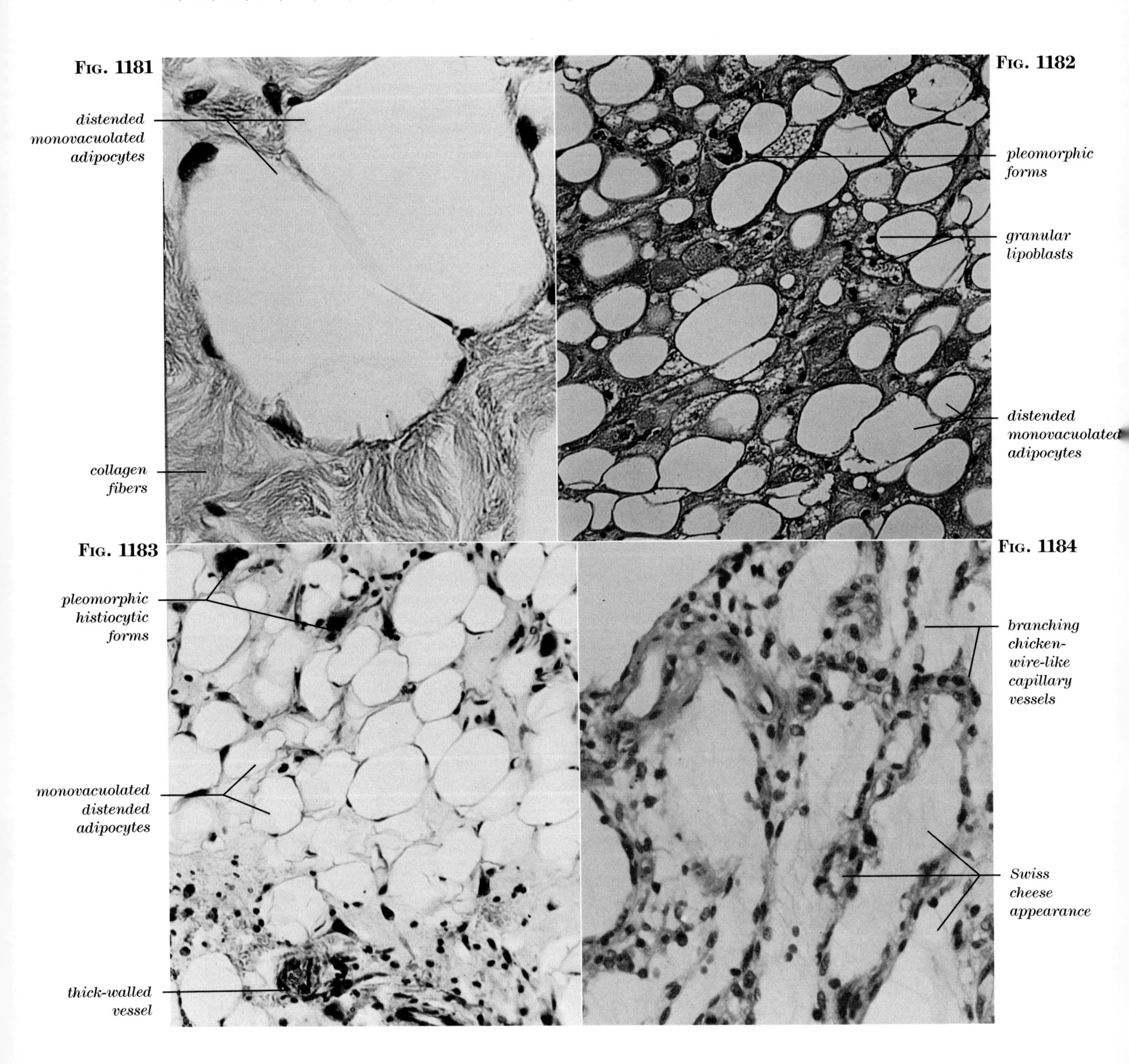

Fig. 1185. Fat necrosis (×140) (see Figs. 425, 749, 1042, 1056)

Fig. 1186. Polymyositis (×140) (see Figs. 1038, 1039, 1094).

Fig. 1187. Infiltrating lipoma (×140) (see Figs. 73, 198, 243, 455, 635, 939, 941, 950, 1168, 1181, 1187).

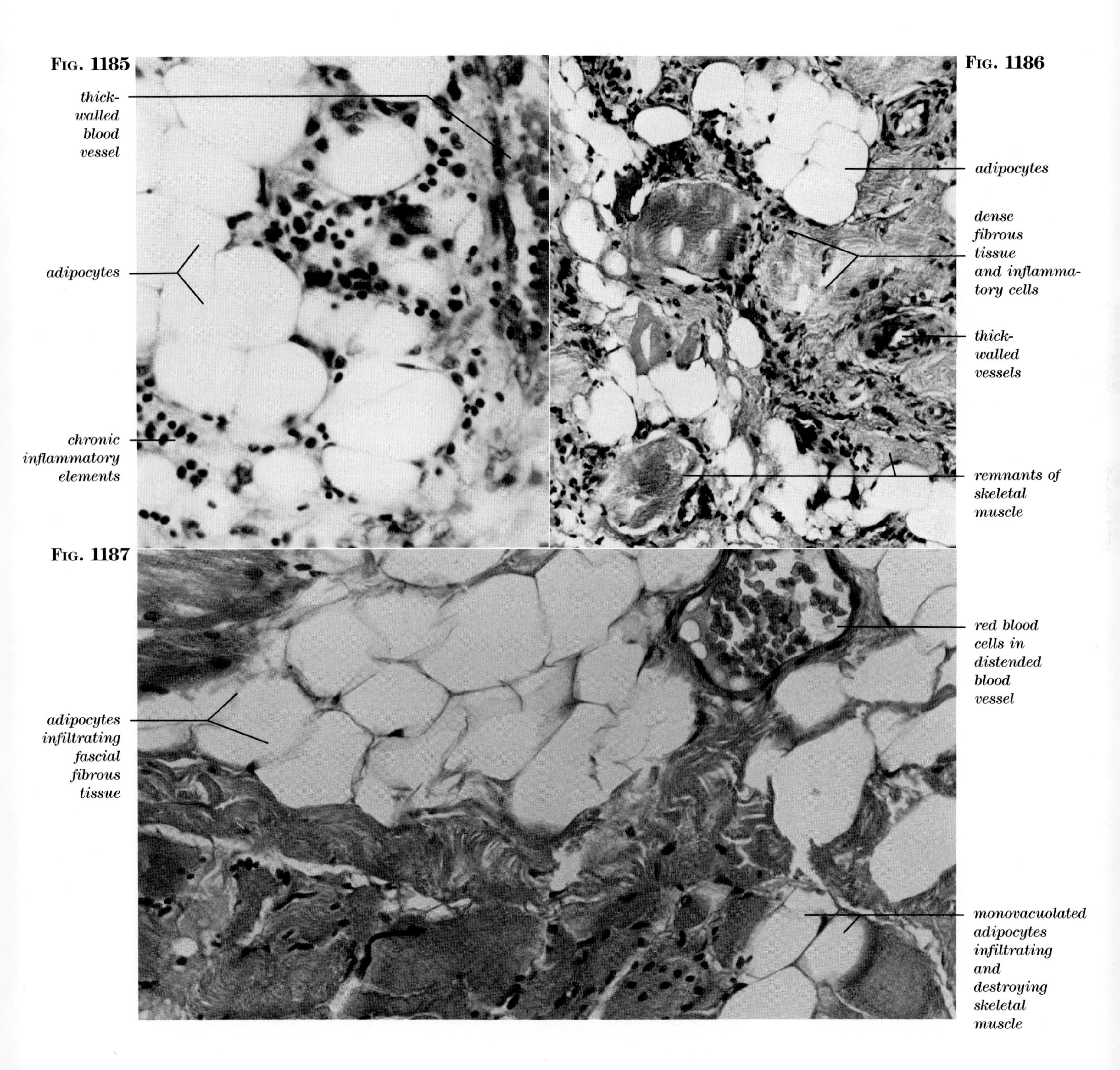

Fig. 1188. Desmoplastic melanoma (Fontana-Masson, ×350) (see Figs. 394, 409, 540, 669, 719, 748, 797, 826, 1189, 1190, 1219, 1262).

Fig. 1189. Malignant melanoma metastatic to fat (×350) (see Figs. 826, 1188, 1190, 1219).

Fig. 1190. Malignant melanoma metastatic to soft tissues (Grimelius, ×350) (see Figs. 1188, 1189, 1219).

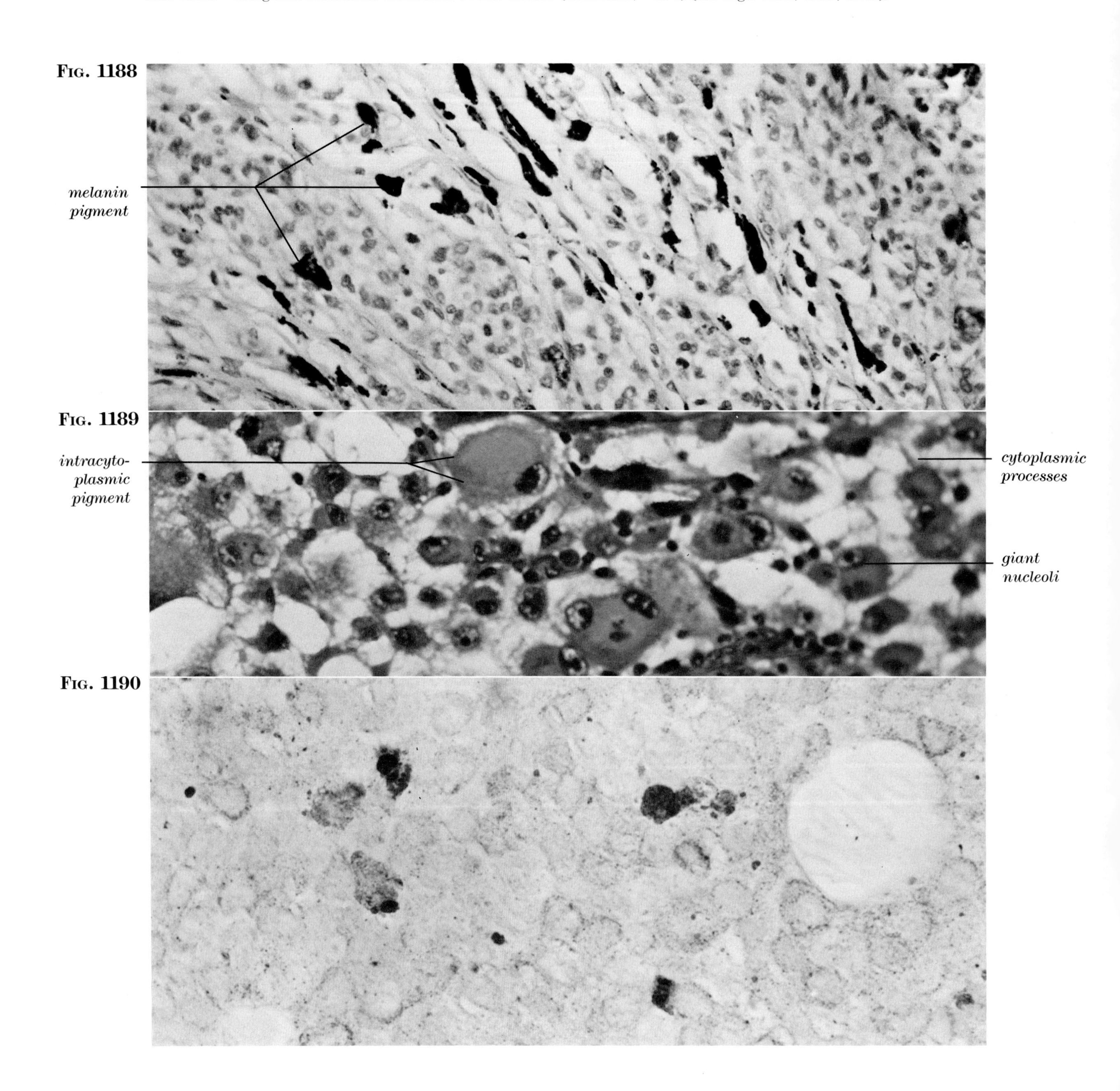

Fig. 1191. Ganglioneuroma of tibia of an infant (×140) (see Figs. 811, 824, 857, 1137, 1193).

Fig. 1192. Malignant pigmented schwannoma (×140) (see Figs. 521, 701, 735, 967).

Fig. 1193. Ganglioneuroma (×570) (see Figs. 166, 486, 630, 811, 824, 857, 1137, 1191, 1258).

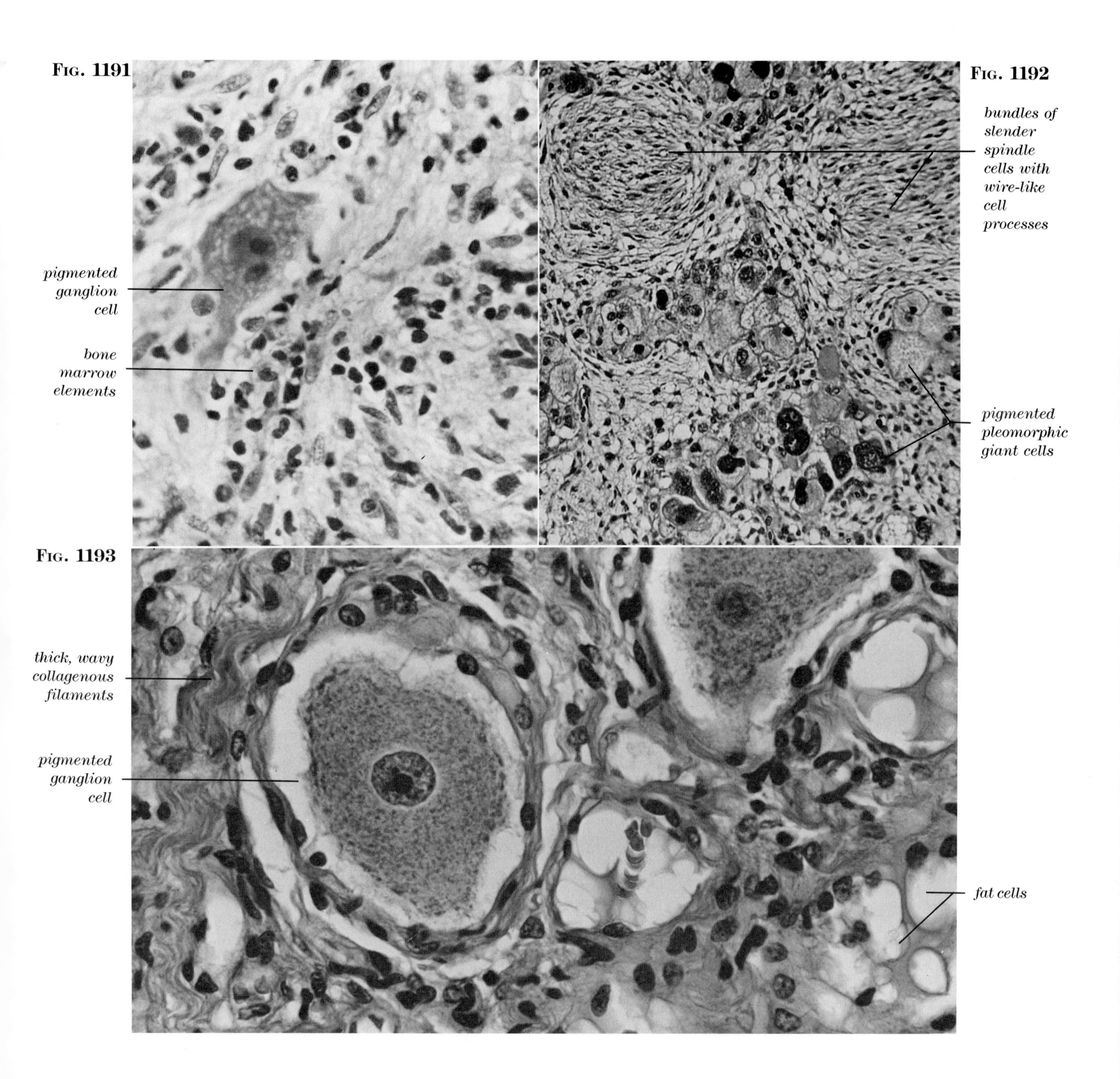

Fig. 1194. Granulocytic sarcoma (Granulocytic, ×570) (see Figs. 488, 699, 891, 1198).

Fig. 1195. Malignant islet cell tumor (Grimelius, ×570).

Fig. 1196. Malignant paraganglioma (Grimelius, ×570) (see Figs. 276, 483, 544, 714, 790, 794, 853, 1220).

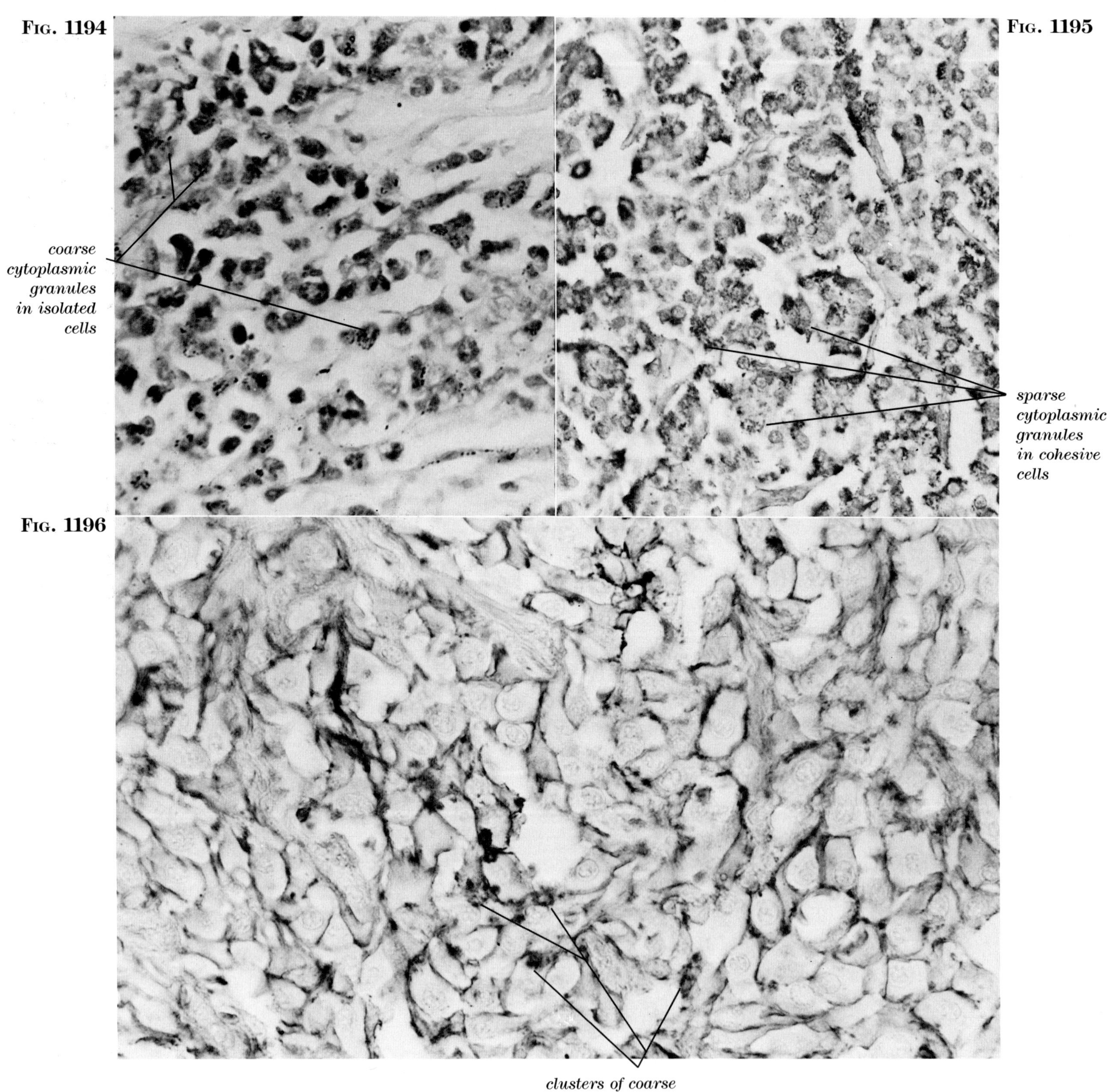

Fig. 1197. Cholesterol crystals in fat necrosis (×350) (see Figs. 1057, 1198, 1199).

Fig. 1198. Granulocytic sarcoma with Charcot-Leyden crystals (×570) (see Figs. 488, 699, 891, 1194).

Fig. 1199. Alveolar soft part sarcoma (PAS, ×350) (see Figs. 89, 538, 718, 1161, 1227).

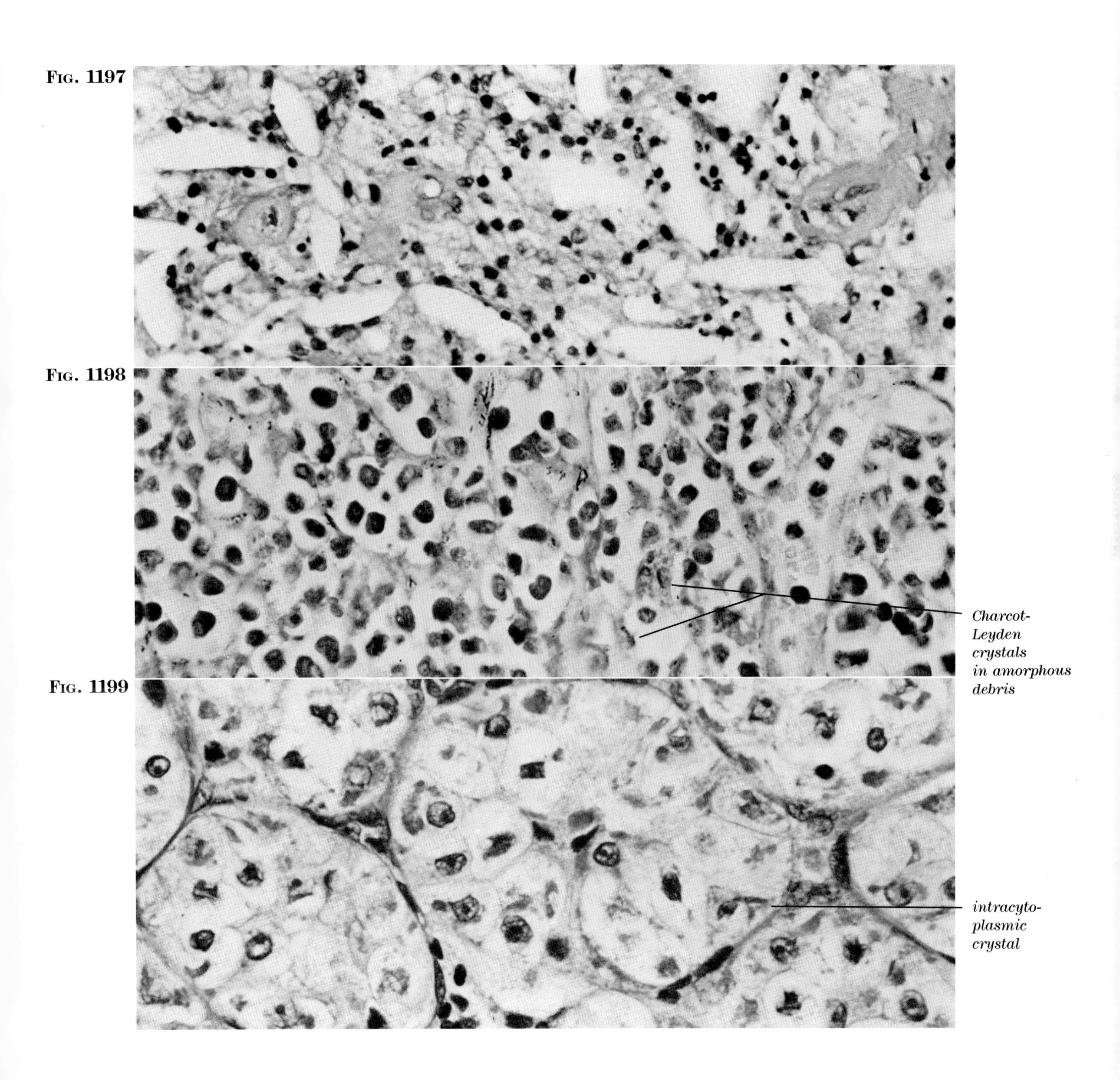

Fig. 1200. Malignant fibroblastic fibrous histiocytoma (×4,800) (see Figs. 1133–1135, 1147, 1178).

Fig. 1201. Xanthogranuloma (×16,800) (see Figs. 607, 734, 779, 1031, 1127).

Fig. 1202. Biphasic tendosynovial sarcoma (×8,000) (see Figs. 906, 1099, 1100, 1173).

Fig. 1203. Hemangiopericytoma (×8,000) (see Figs. 533, 696, 984, 995).

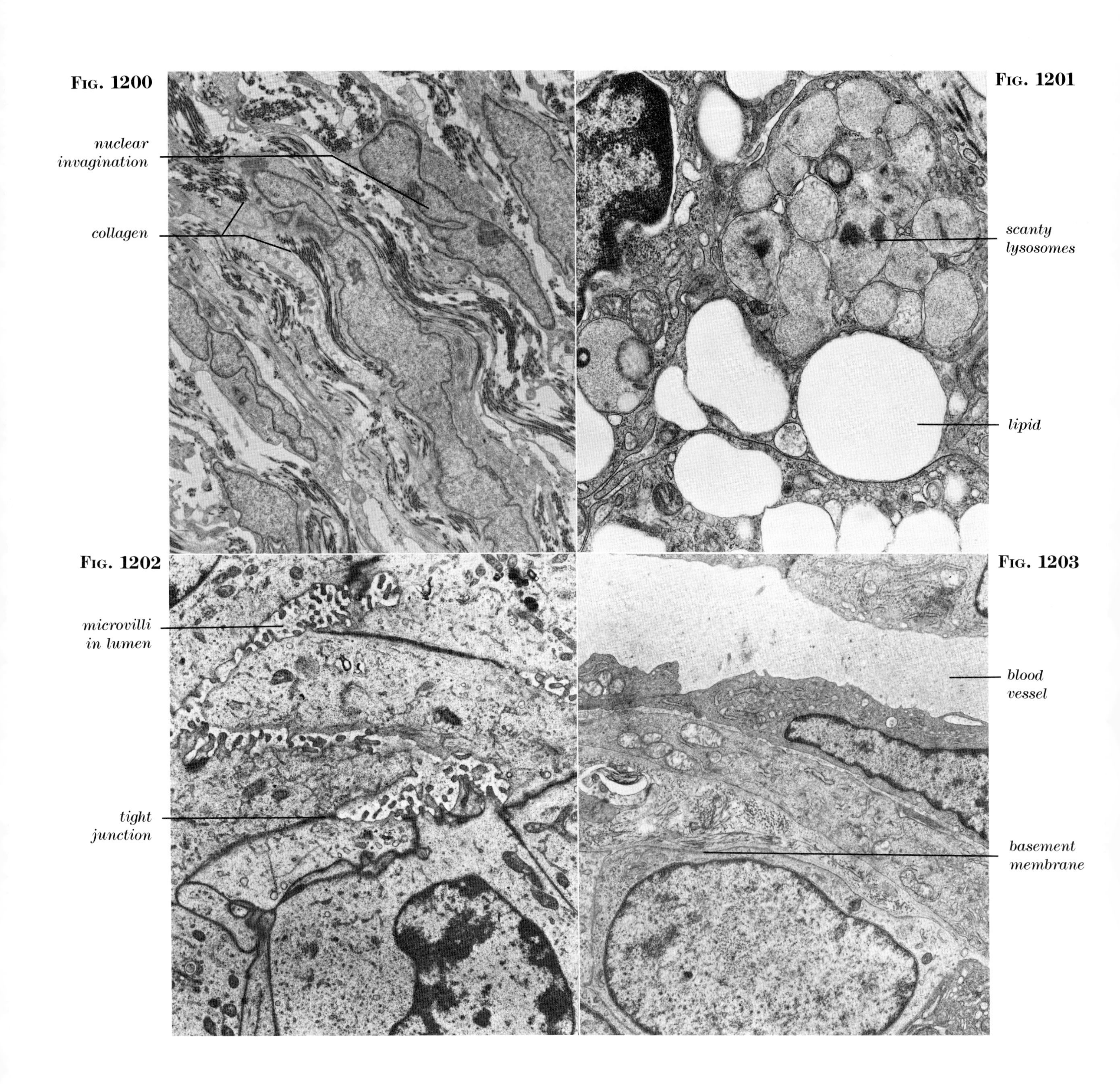

Fig. 1204. Fibroma, so-called "dermatofibroma" (×5,000) (see Figs. 336, 859, 884, 886, 1109).

Fig. 1205. Pleomorphic fibrosarcoma (×5,500) (see Figs. 91, 371, 625, 832, 901, 1117).

Fig. 1206. Elastofibroma (×8,000) (see Figs. 914, 1123, 1140).

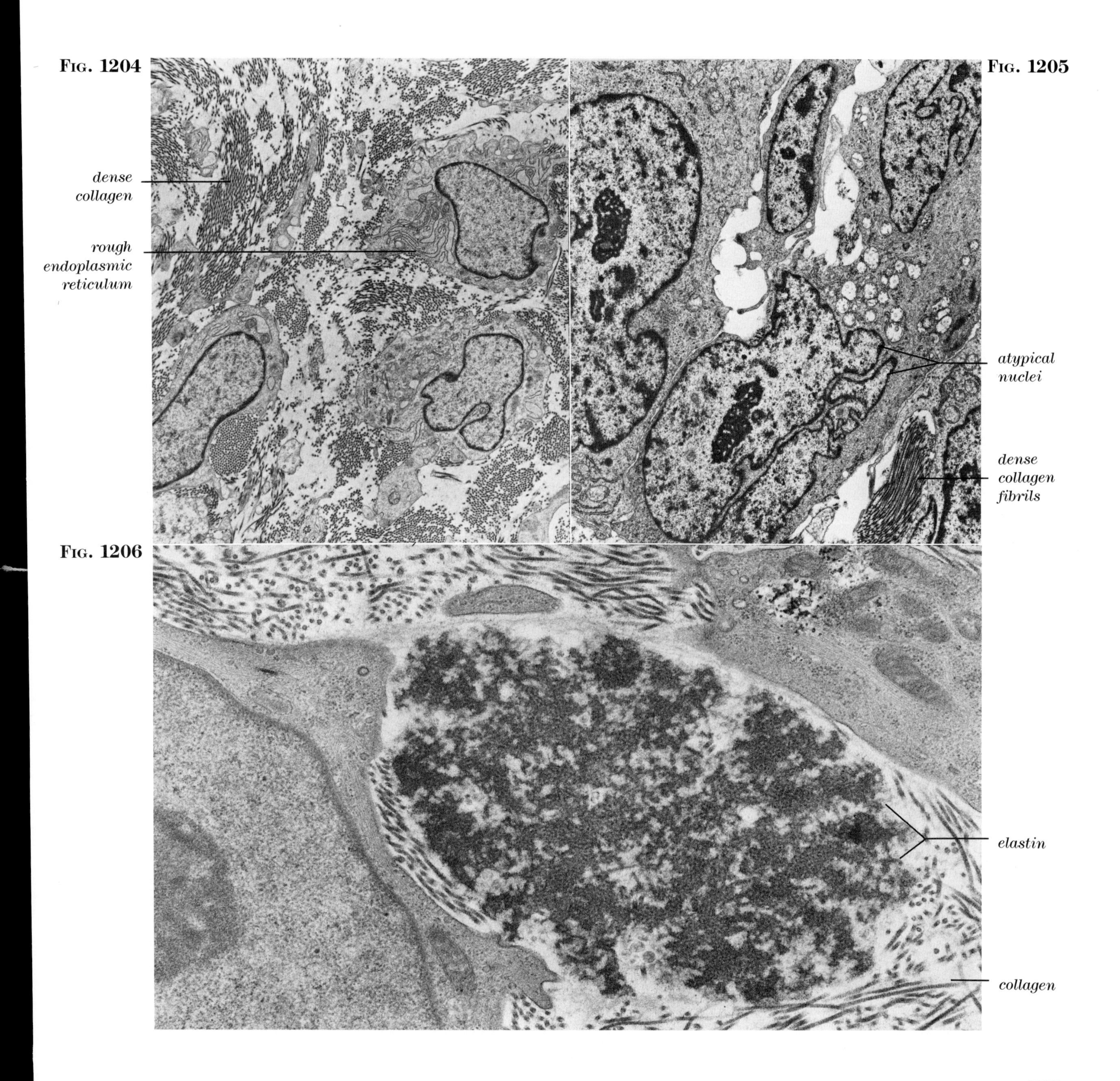

Fig. 1207. A well-differentiated area in a malignant schwannoma (×6,800) (see Figs. 815, 845, 1208).

Fig. 1208. Malignant schwannoma (×6,500) (see Figs. 603, 815, 845, 1207).

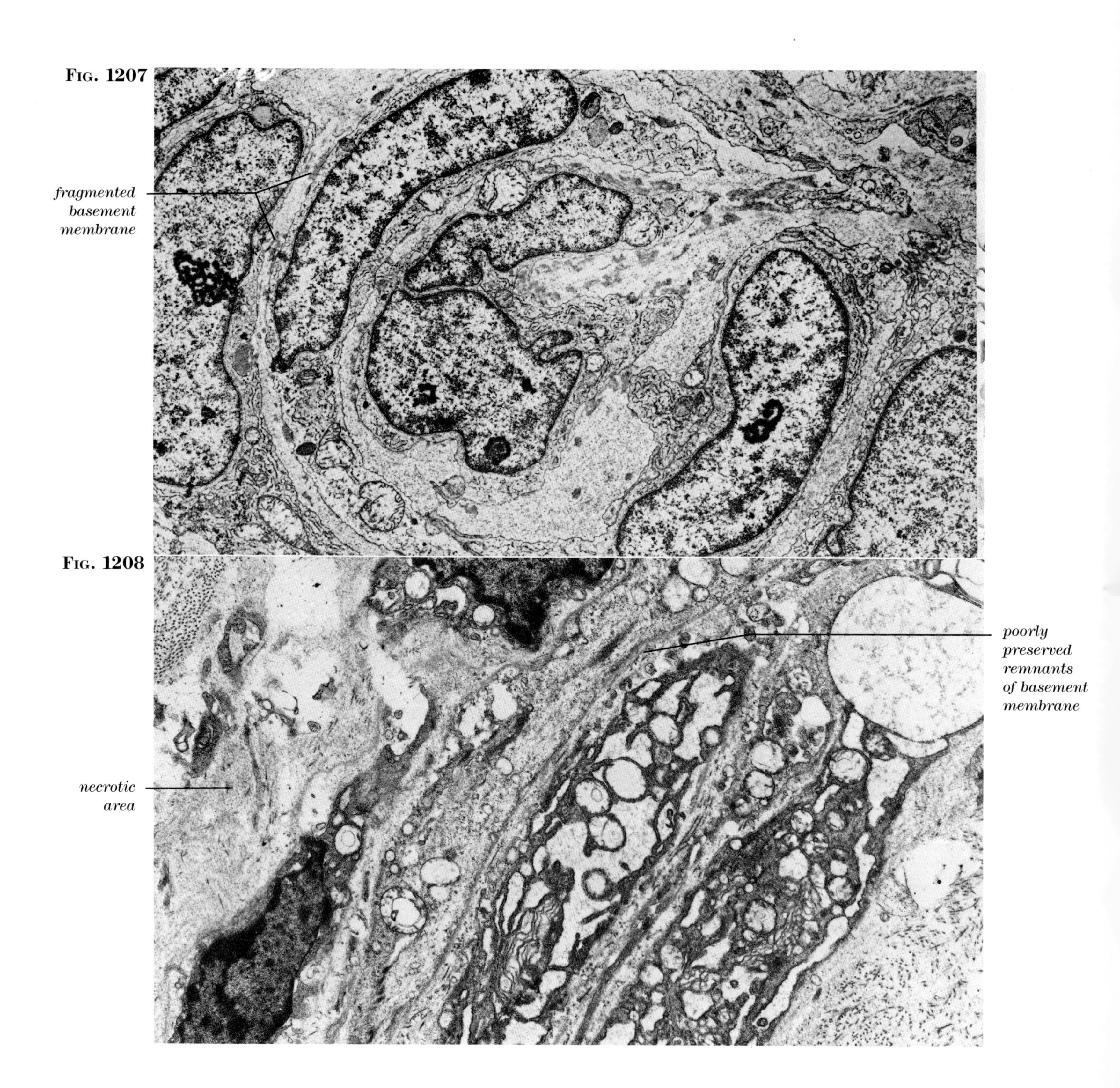

Fig. 1209. Hemangiosarcoma (×6,900) (see Figs. 576, 731, 990, 1005, 1210).

Fig. 1210. Hemangiosarcoma (×8,000) (see Figs. 720, 905, 975, 976, 1209).

Fig. 1211. Leiomyosarcoma with fusiform dense bodies *(arrow)* (×19,200) (see Figs. 1132, 1163, 1212).

Fig. 1212. Metastatic leiomyosarcoma (×9,200) (see Figs. 1131, 1132, 1163, 1211, 1249).

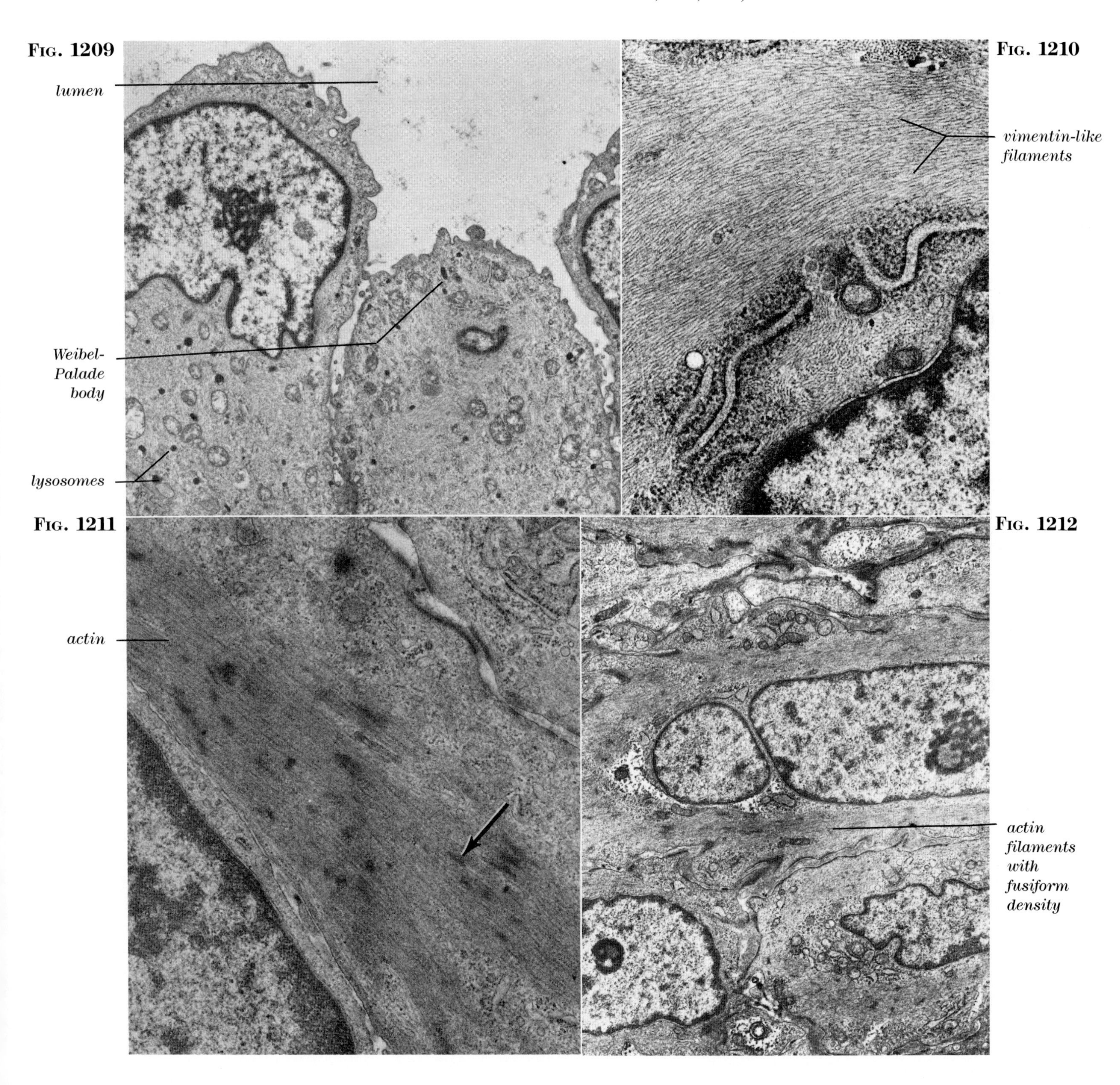

Fig. 1213. Gastric leiomyoblastoma with nuclear body (NB) and discontinuous basement membrane (*arrows*) (×5,280) (see Figs. 1139, 1158, 1214).

Fig. 1214. Gastric leiomyoblastoma (×13,200). Inset (×50,400) (see Figs. 1158, 1213, 1248).

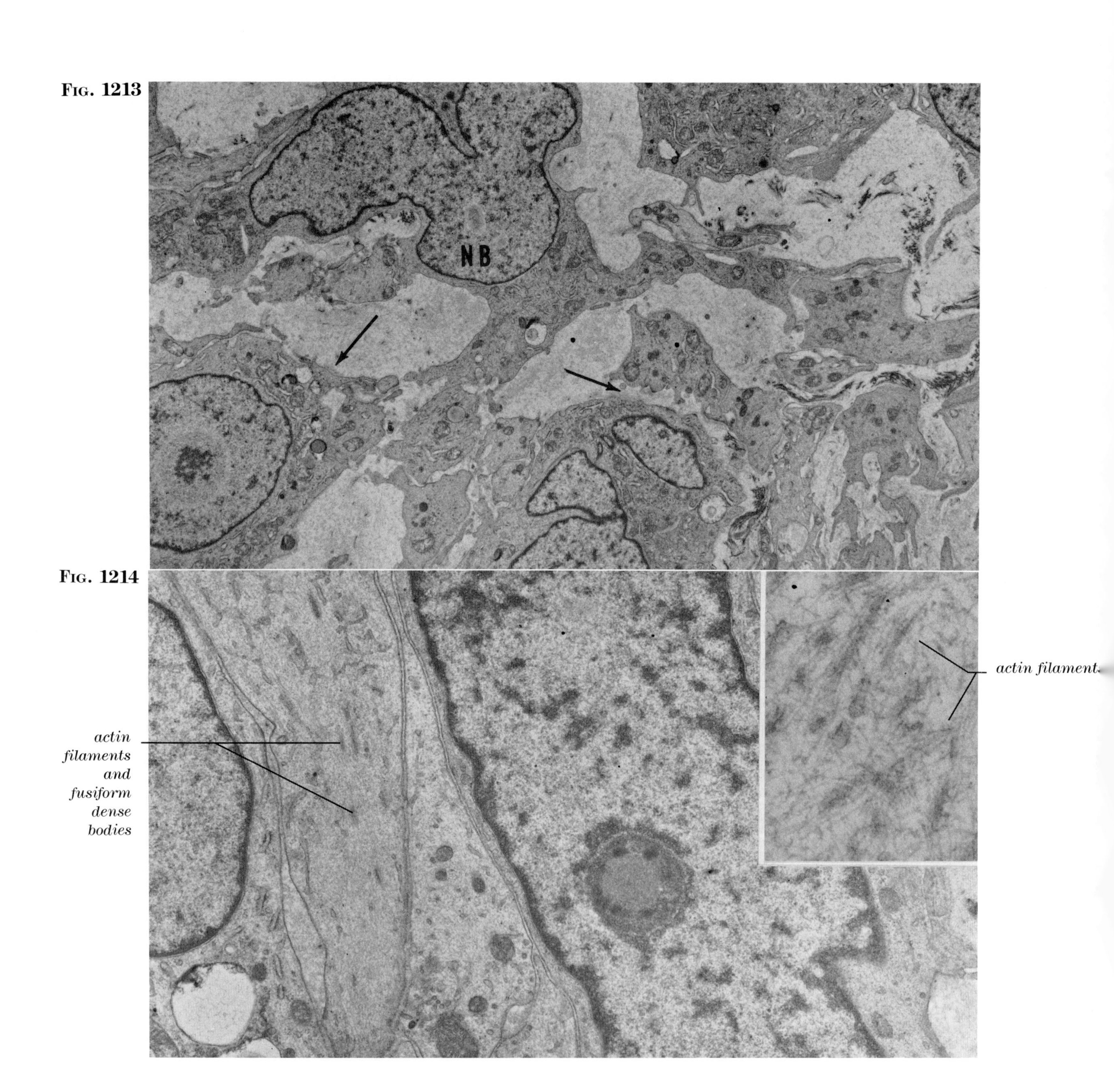

Fig. 1215. Embryonal rhabdomyosarcoma (×15,000) (see Figs. 1129, 1138, 1164, 1171, 1257).

Fig. 1216. Pleomorphic rhabdomyosarcoma (×10,000) (see Figs. 1130, 1166, 1167, 1250, 1255).

Fig. 1217. Mesodermal mixed tumor of uterus (×10,000) (see Figs. 65, 589, 763, 969, 1165).

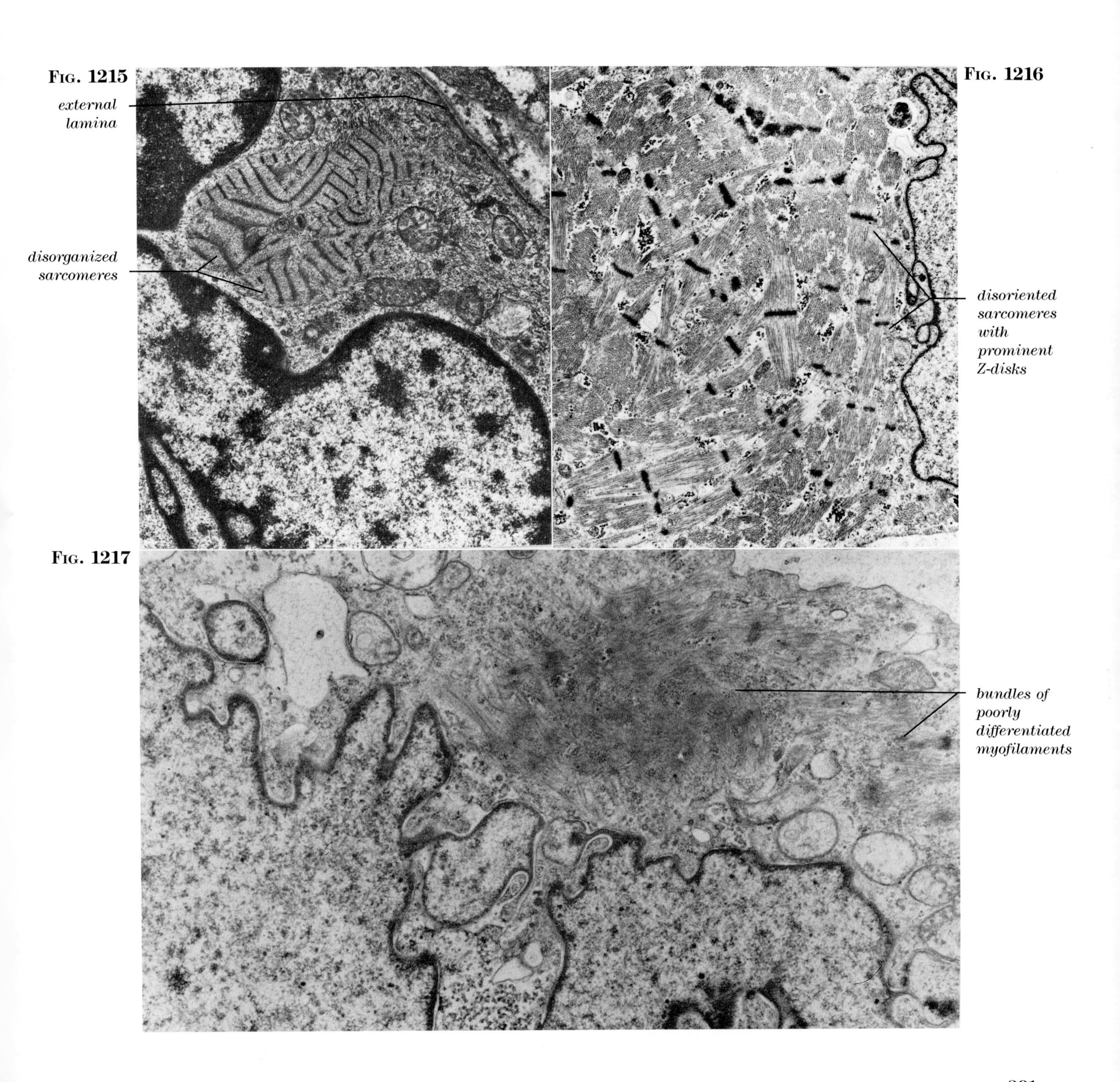

Fig. 1218. Neuroblastoma (×19,200) (see Fig. 724).

Fig. 1219. Malignant melanoma (×62,000) (see Figs. 394, 409, 540, 669, 719, 748, 797, 826, 1188, 1189, 1190, 1262).

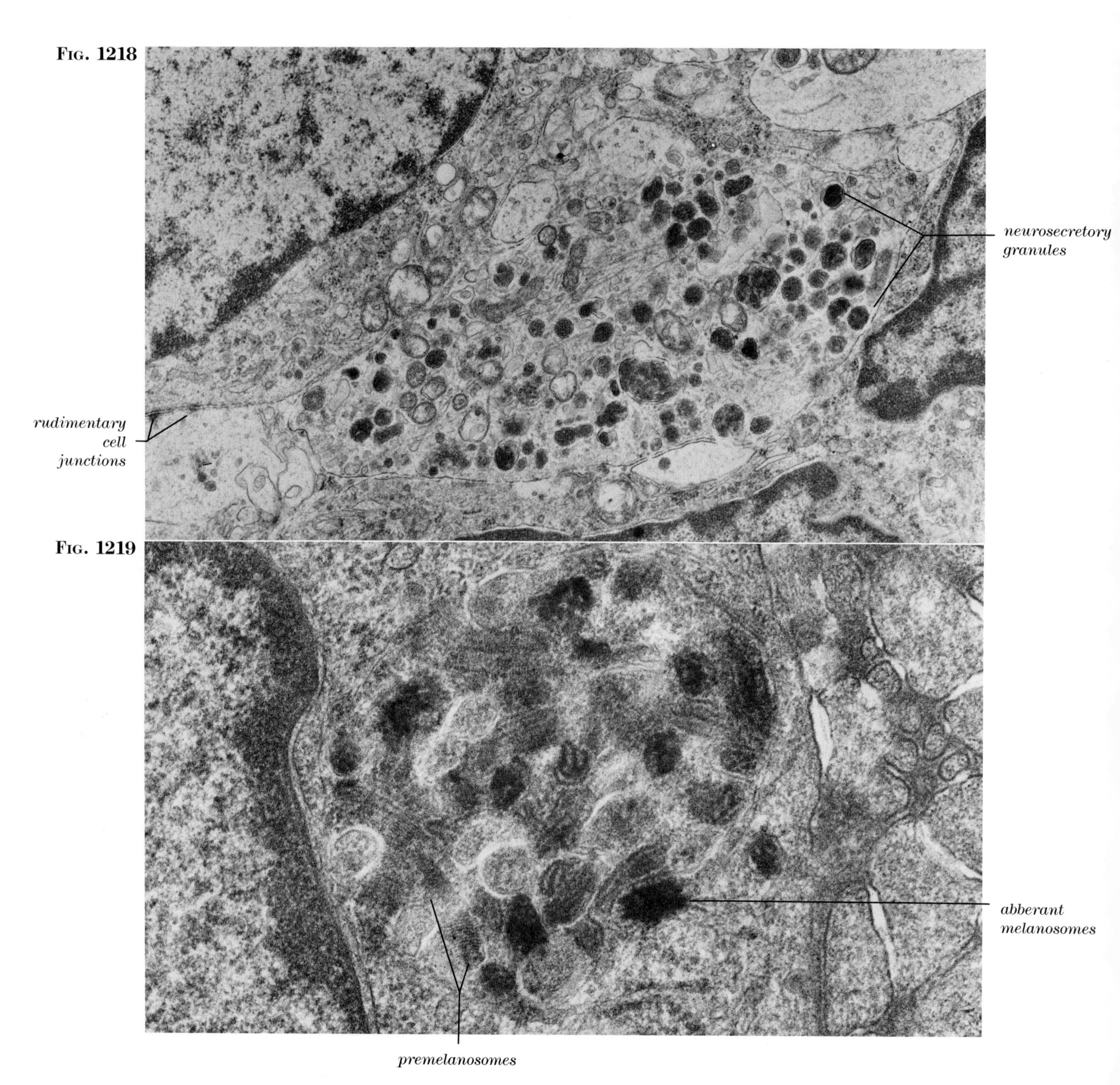

Fig. 1220. Malignant paraganglioma (×30,720) (see Figs. 790, 794, 853, 1196).

Fig. 1221. Hypertrophic hemangioma (×4,000) (see Figs. 553, 554, 974).

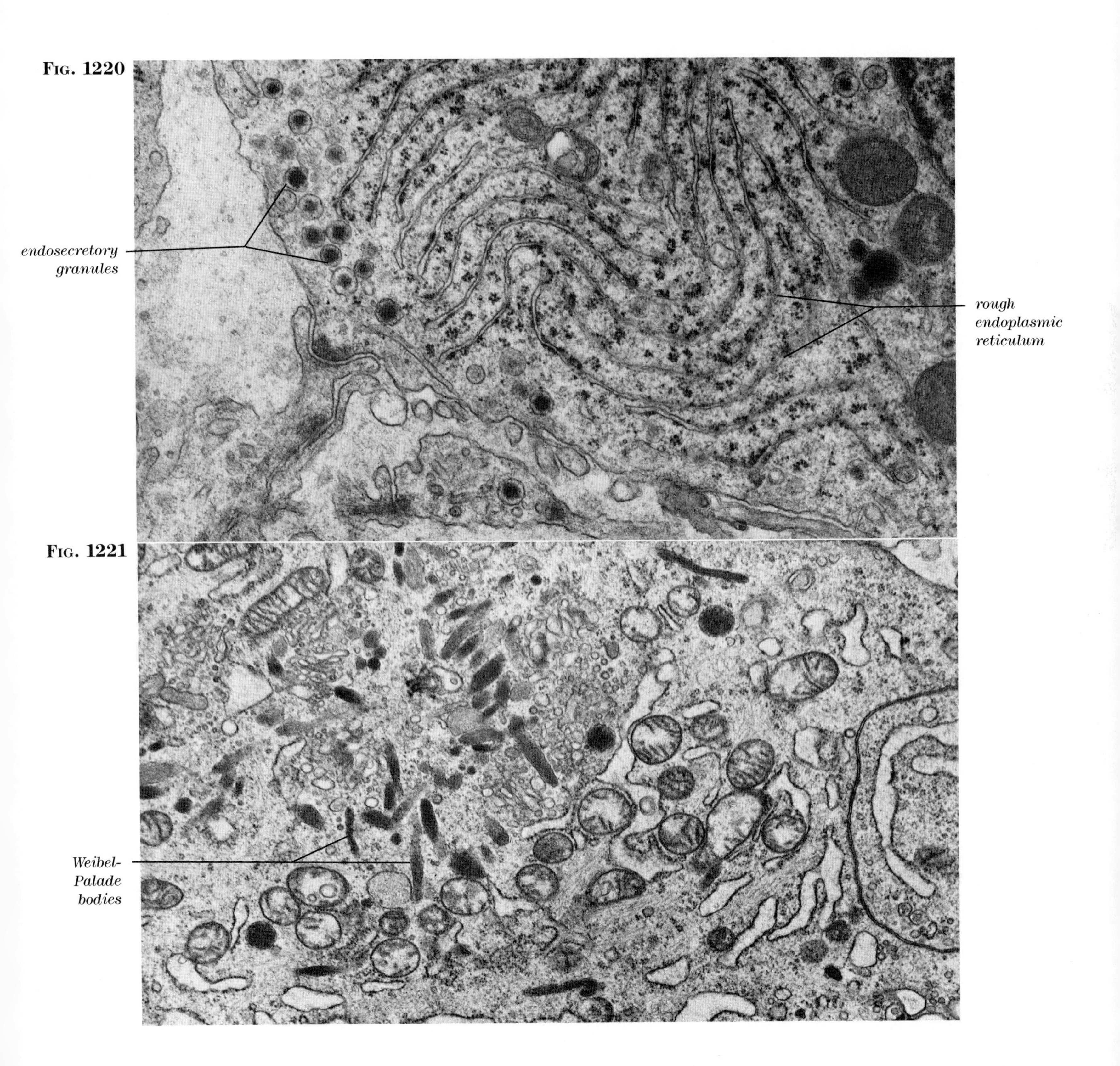

Fig. 1222. Chondrosarcoma (×4,000) (see Figs. 773, 1101, 1110, 1275).

Fig. 1223. Clear cell sarcoma of tendosynovium (×4,800) (see Figs. 927, 1145, 1244, 1245).

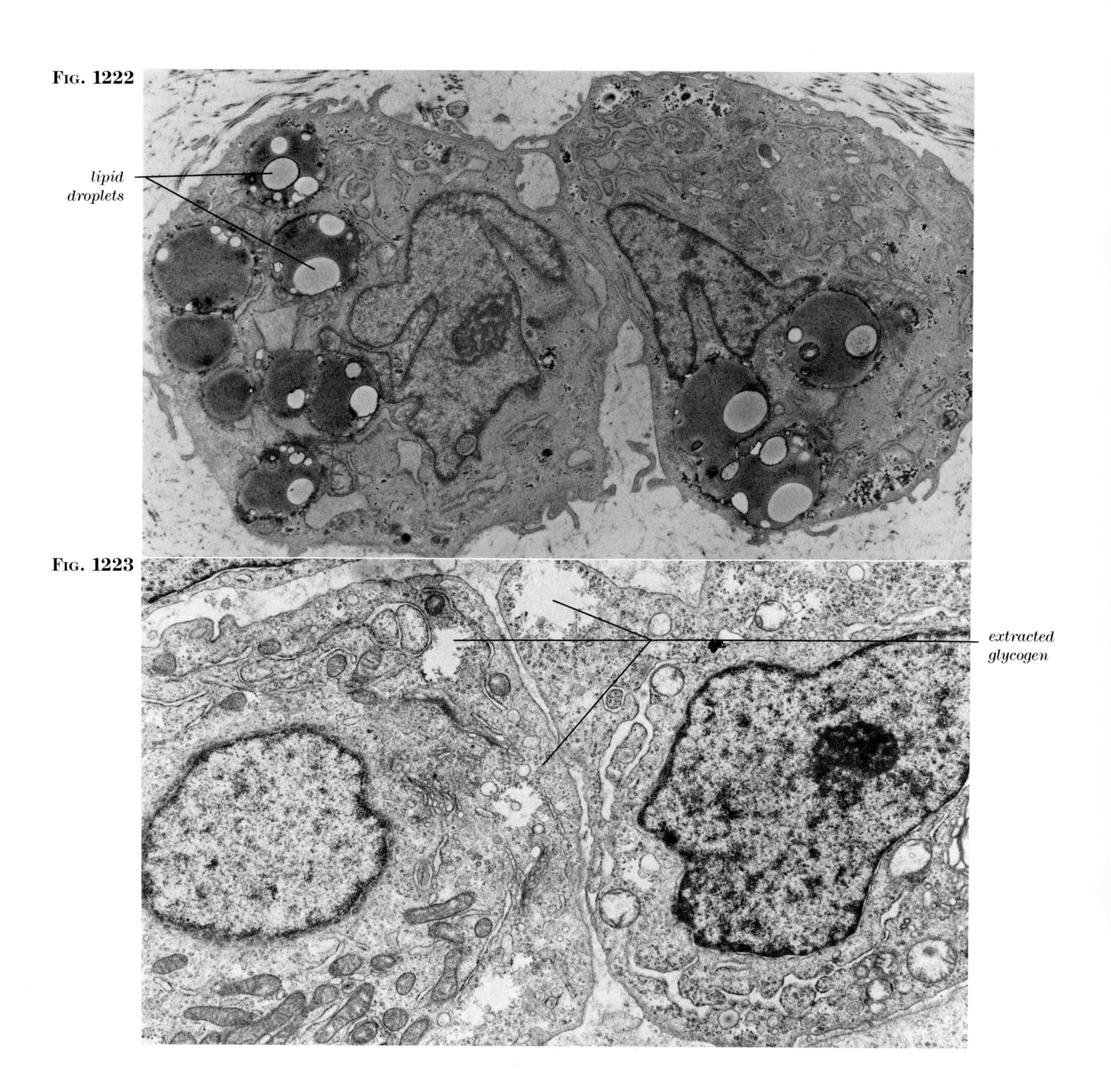

Fig. 1224. Seminoma (×7,400) (see Figs. 14, 566, 757).

Fig. 1225. Meningioma (×4,000) (see Figs. 485, 496, 537, 1096).

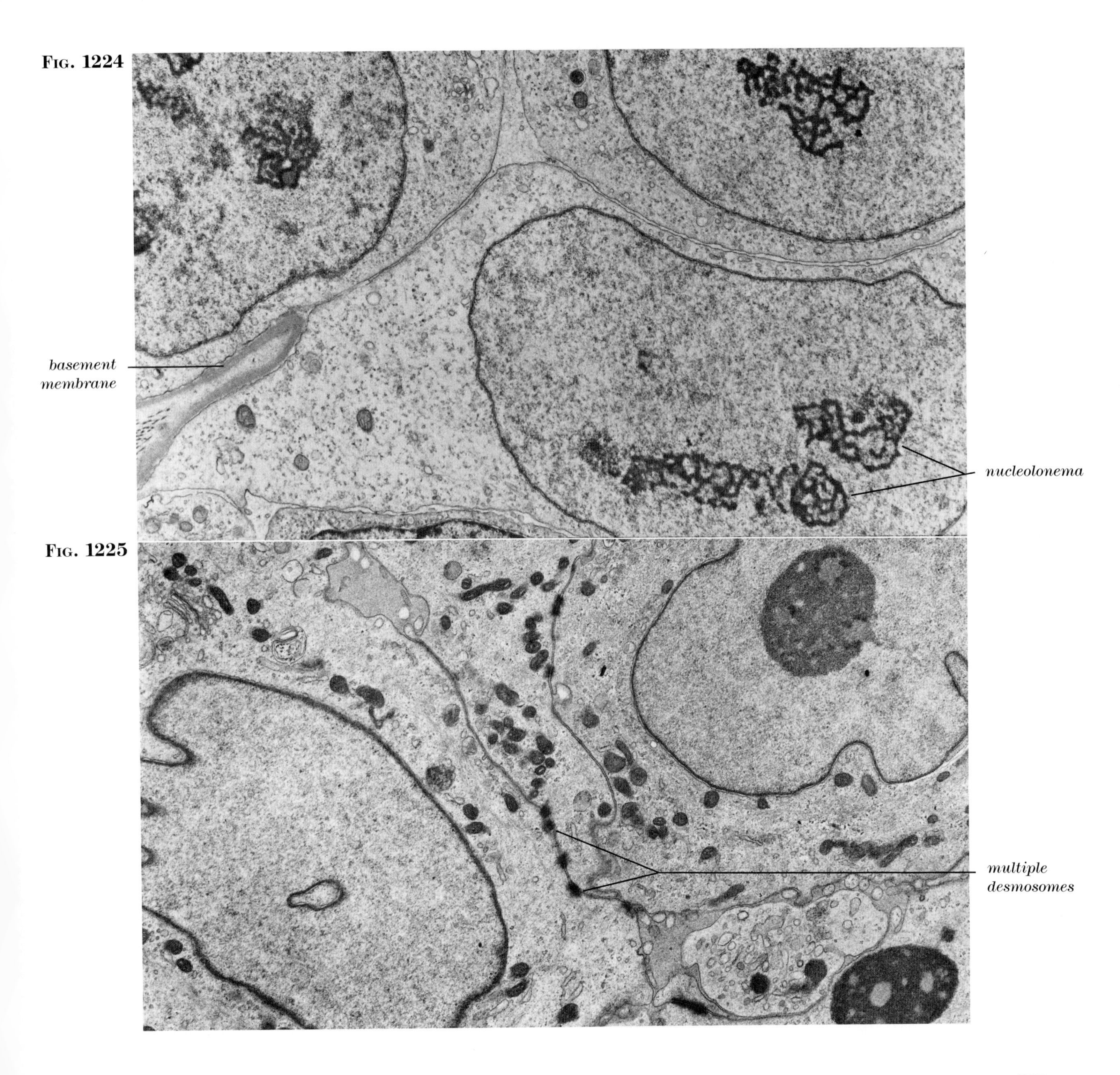

FIG. **1226.** Chordoma (×5,760) (see Figs. 729, 767, 771, 1174).

FIG. **1227.** Alveolar soft part sarcoma (×7,450) (see Figs. 697, 718, 1161, 1199).

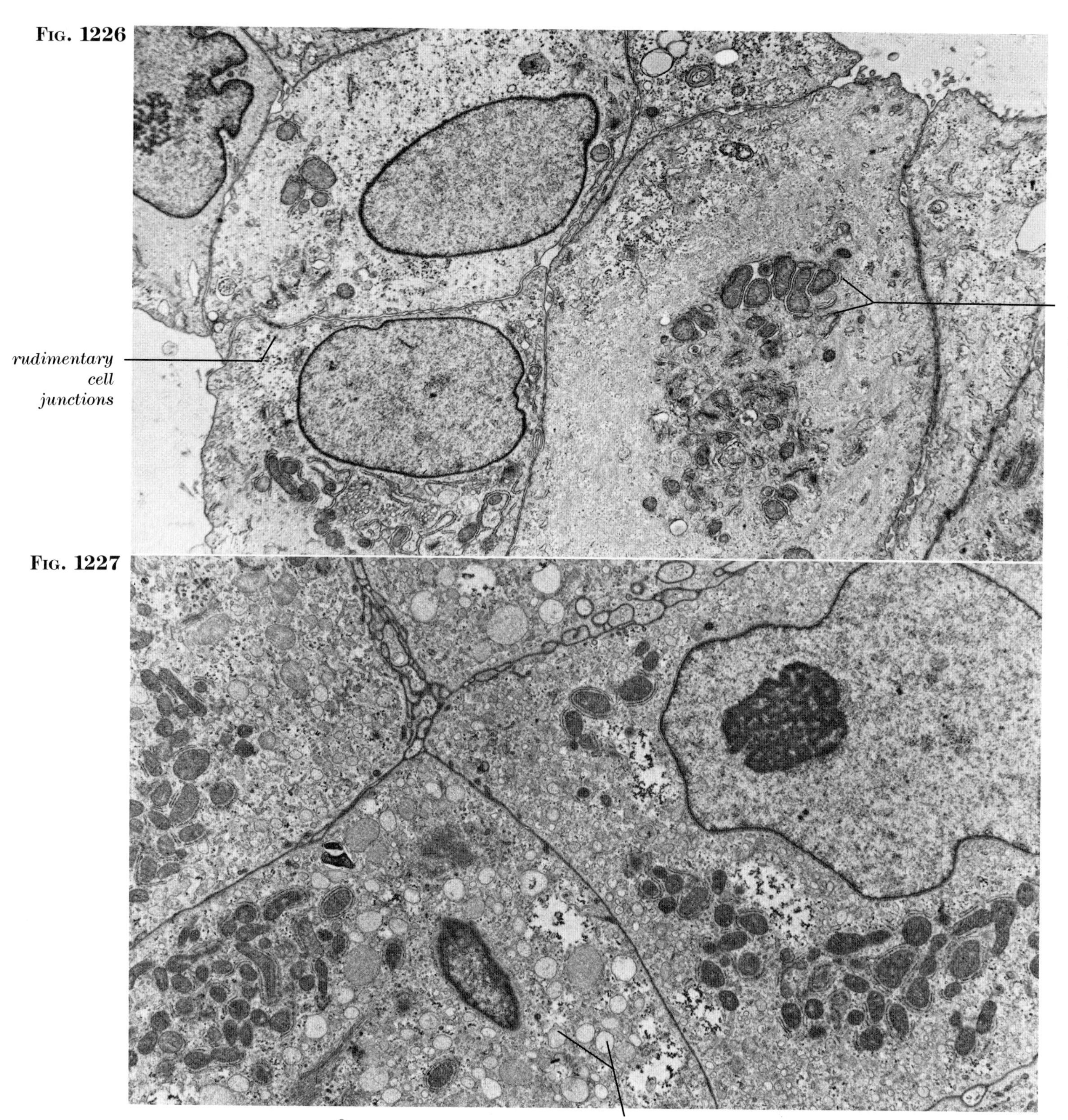

Fig. 1228. Benign schwannoma (×4,800) (see Figs. 257, 356, 850, 846).

Fig. 1229. Neurofibrosarcoma (×8,800) (see Figs. 838, 1116, 1263, 1266).

Fig. 1230. Ewing's sarcoma (×5,760) (see Figs. 449, 453, 476, 573, 698, 1157).

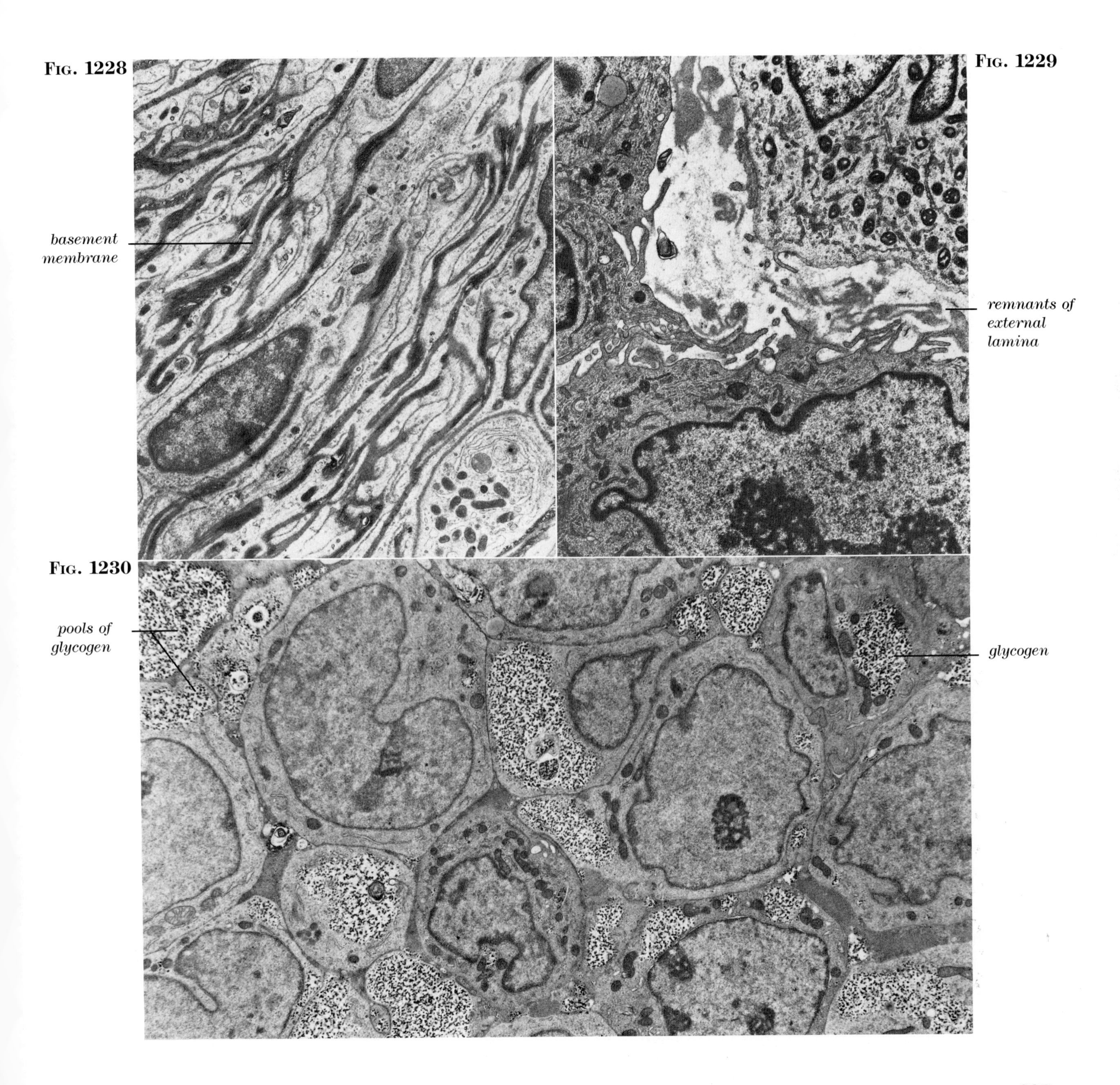

Fig. 1231. Lipoblastic liposarcoma (×9,200) (see Figs. 543, 687, 753, 934, 973).

Fig. 1232. Myxoid liposarcoma (×12,000) (see Figs. 636, 933, 1113, 1169).

Fig. 1233. Well-differentiated liposarcoma (×4,800) (see Figs. 435, 822, 999, 1184).

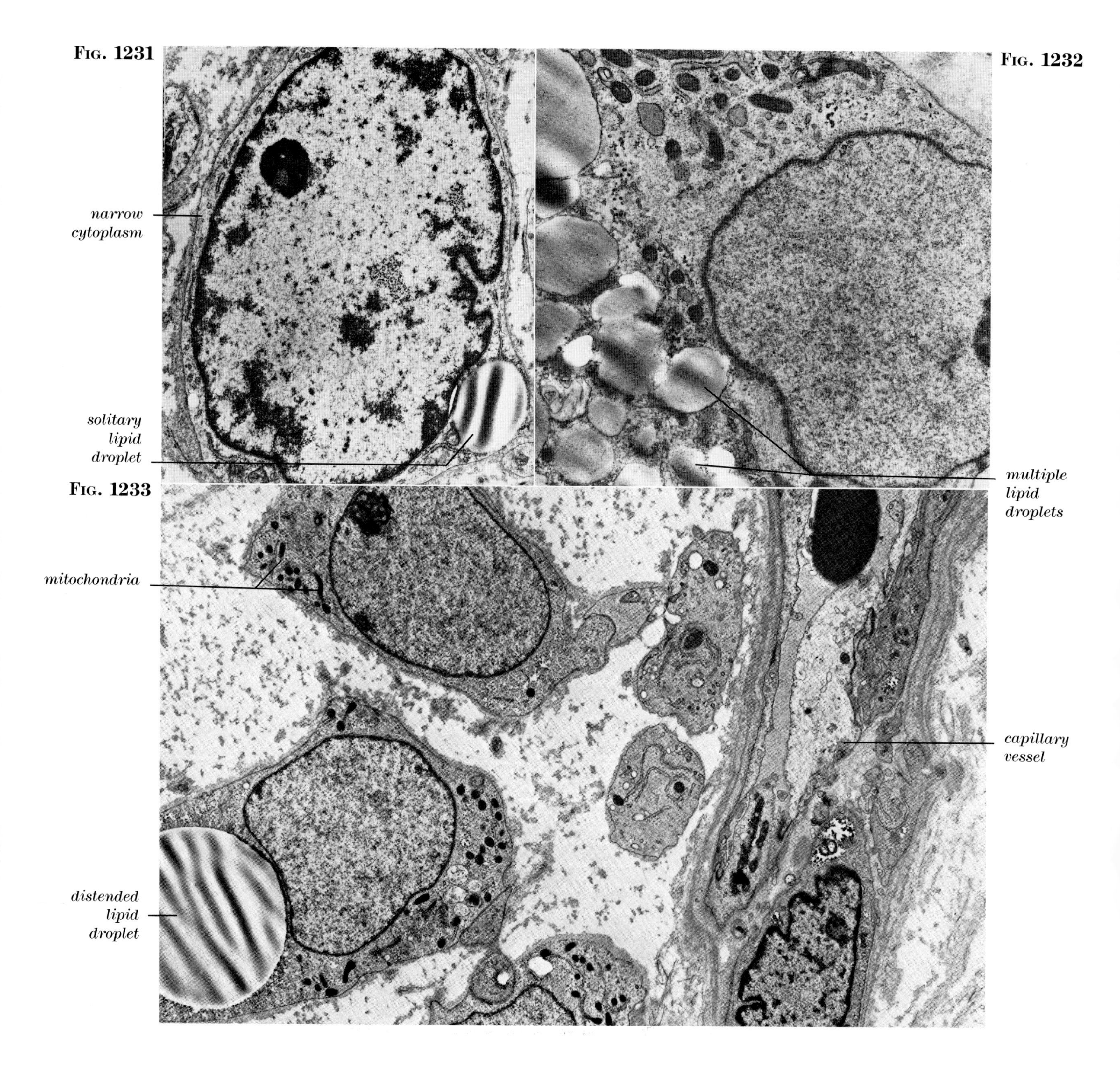

Fig. 1234. Malignant granular cell tumor (×6,000) (Courtesy of Dr. M. Marroum, Charlotte, North Carolina) (see Figs. 740, 1256).

Fig. 1235. Benign granular cell tumor (×29,000) (see Figs. 498, 539, 736, 1160).

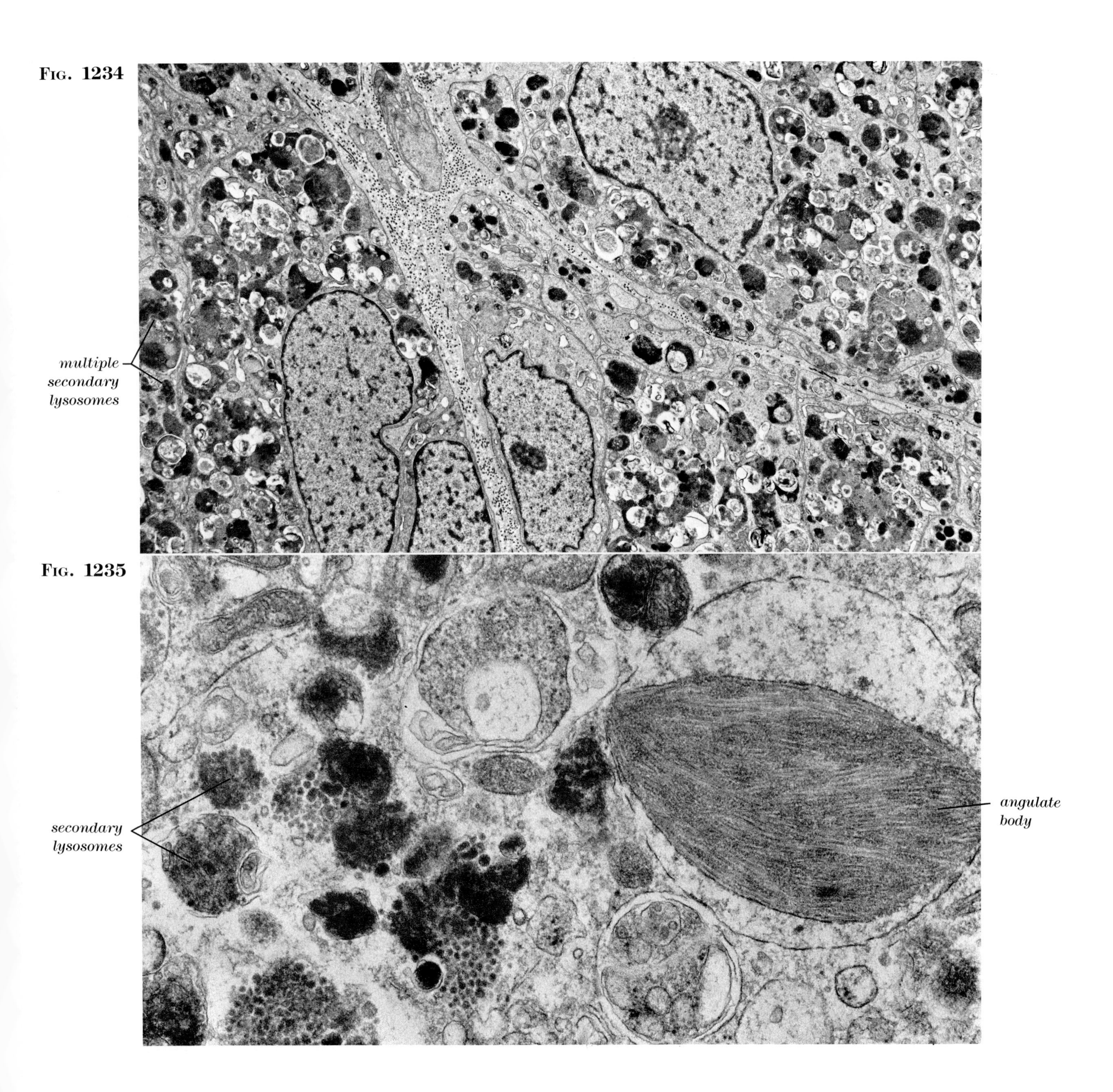

Fig. 1236. Merkel cell tumor (×14,000) (see Figs. 1112, 1237).

Fig. 1237. Merkel cell tumor (×42,000) (see Figs. 1019*B*, 1236).

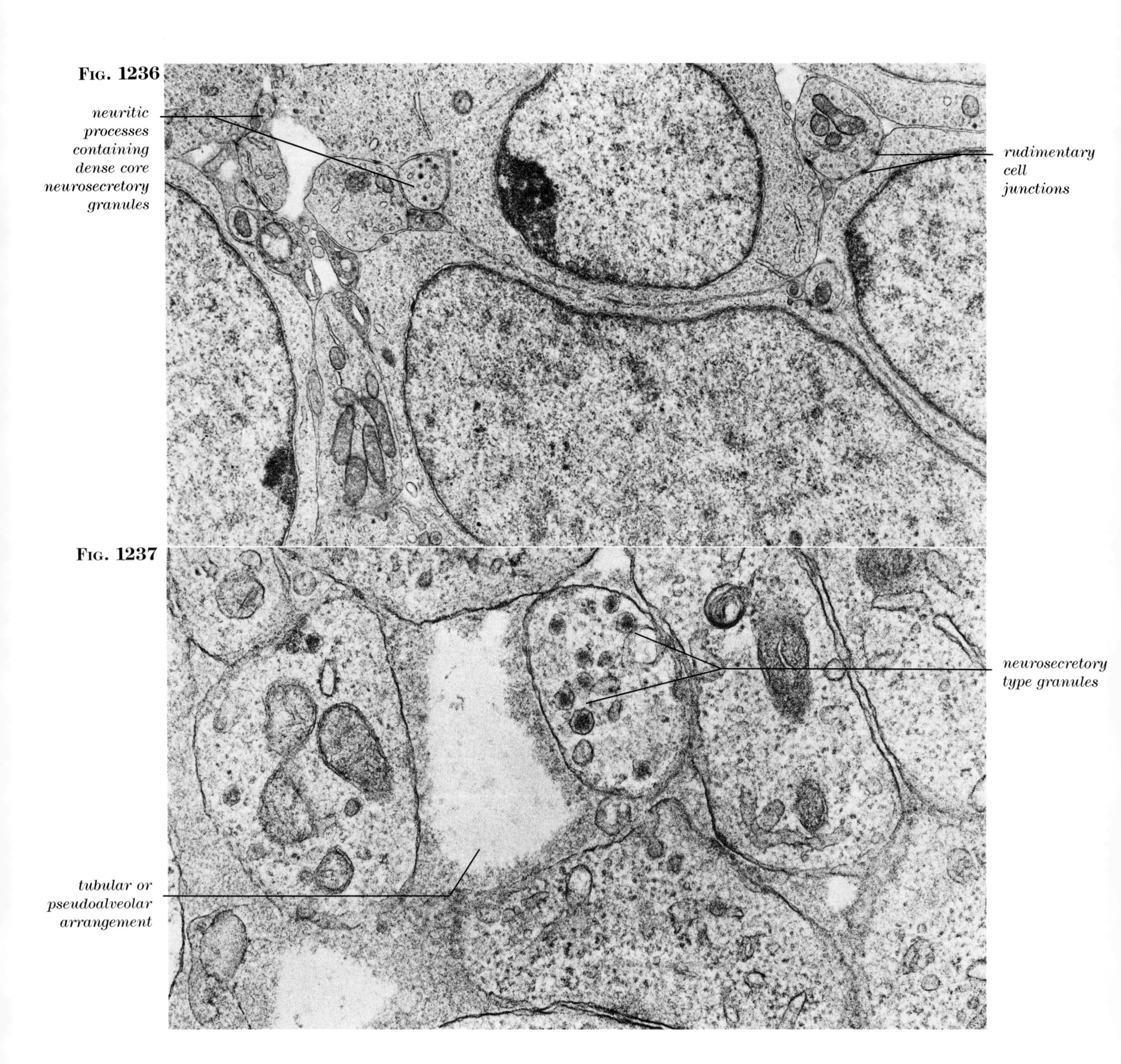

Fig. 1238. Malignant neuroepithelioma (×6,480) (see Figs. 489, 690, 1115, 1239).

Fig. 1239. Malignant neuroepithelioma (×35,000) (see Figs. 309, 585, 690, 1115, 1238).

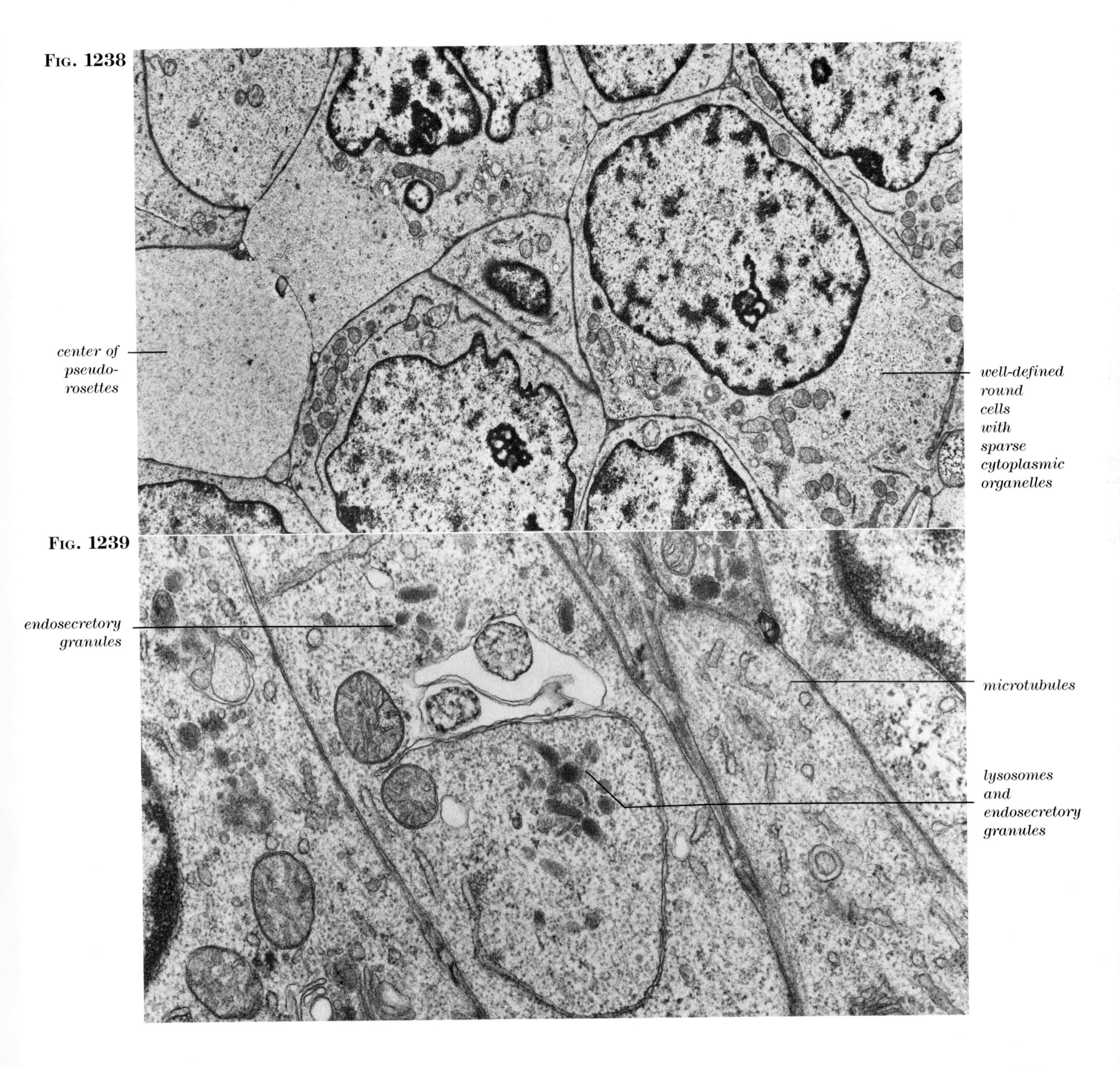

Fig. 1240. Low grade paracordal malignant fibrous mesothelioma (×350) (see Figs. 881, 1241).

Fig. 1241. Low grade paracardal malignant fibrous mesothelioma (Keratin, ×350) (see Figs. 880, 1240).

Fig. 1242. High grade malignant mesothelioma of pleura (×350) (see Figs. 511, 531, 869, 1243).

Fig. 1243. Low grade paracordal malignant mesothelioma (Keratin, ×350) (see Figs. 265, 267, 869, 1242).

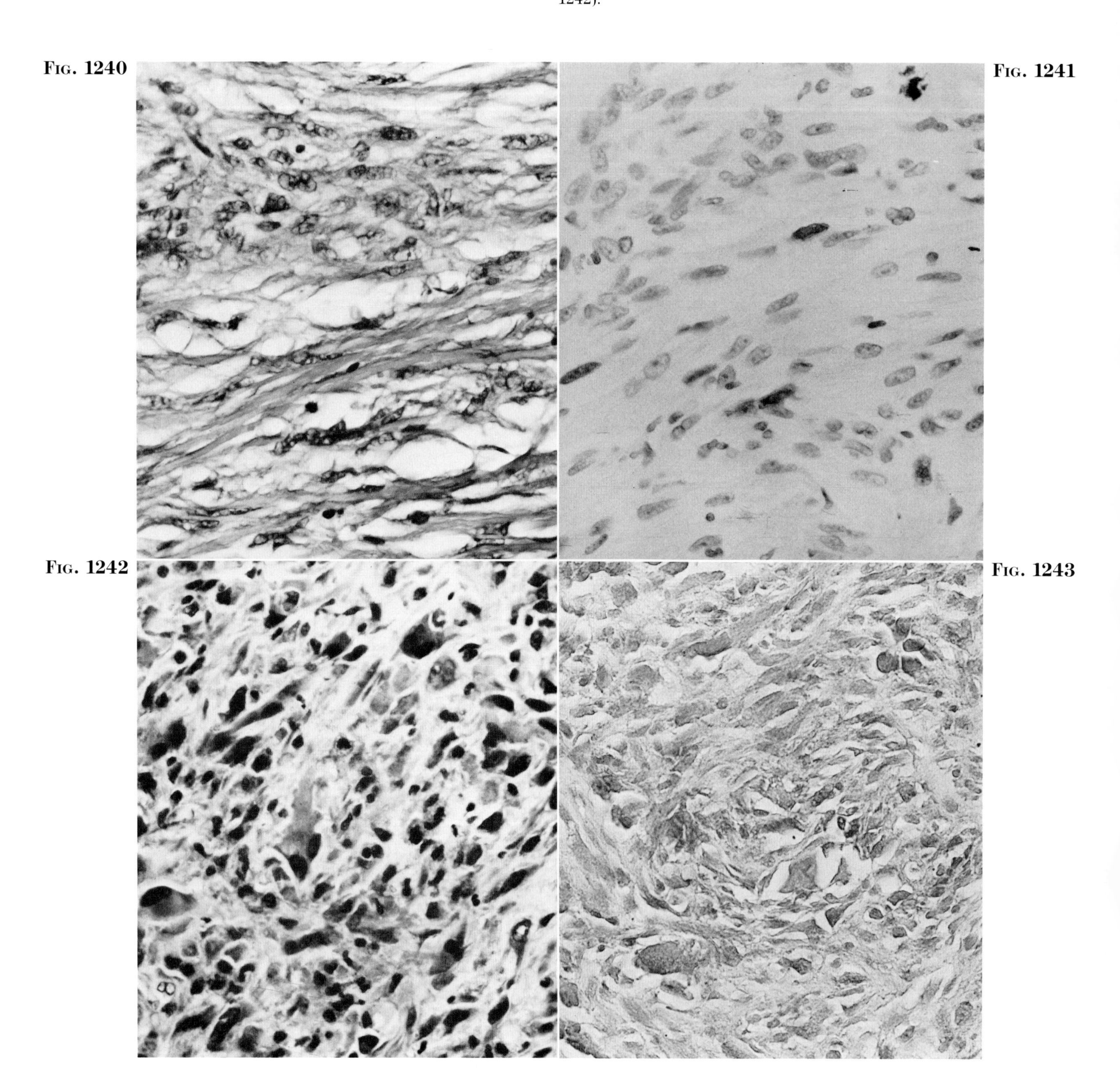
Fig. 1240

Fig. 1241

Fig. 1242

Fig. 1243

Fig. 1244. Monophasic tendosynovial sarcoma with epithelioid and clear cell features (×350) (see Figs. 1145, 1146, 1170, 1245).

Fig. 1245. Monophasic tendosynovial sarcoma with epithelioid and clear cell features (Keratin, ×350) (see Figs. 1223, 1244).

Fig. 1246. Low grade malignant histiocytic fibrous histiocytoma (×350) (see Figs. 1149, 1172, 1247).

Fig. 1247. Low grade malignant histiocytic fibrous histiocytoma (Alpha-1-antichymotrypsin, ×350) (see Figs. 1149, 1172, 1246).

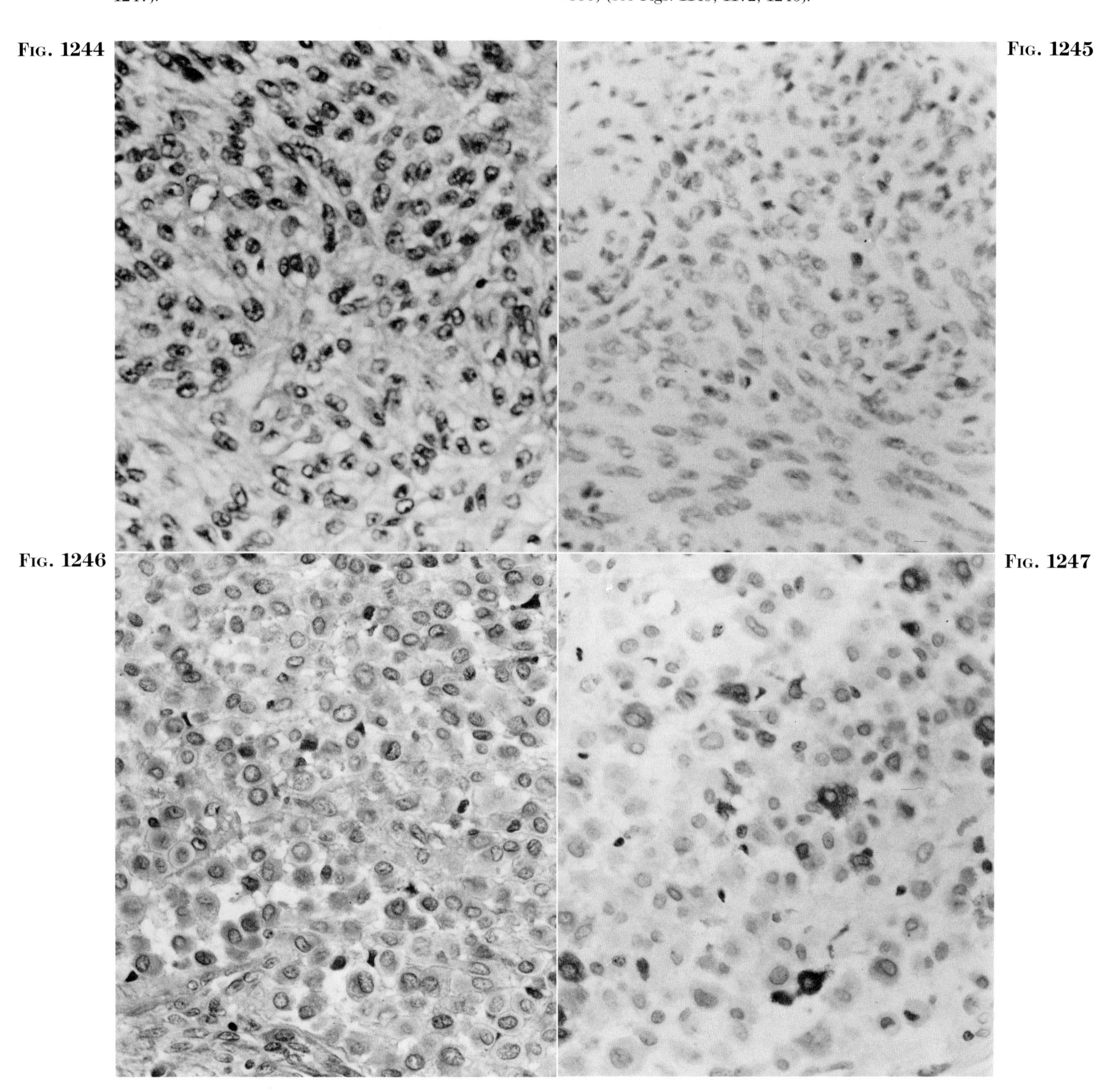

Fig. 1248. Retroperitoneal leiomyoblastoma (Actin, ×350) (see Figs. 1139, 1158, 1213, 1214).

Fig. 1249. Low grade leiomyosarcoma (Myosin, ×350) (see Figs. 1211, 1212, 1252, 1274).

Fig. 1250. Metastatic malignant teratoma partly growing as rhabdomyosarcoma (Myosin, ×350) (see Figs. 922, 1167, 1216, 1255).

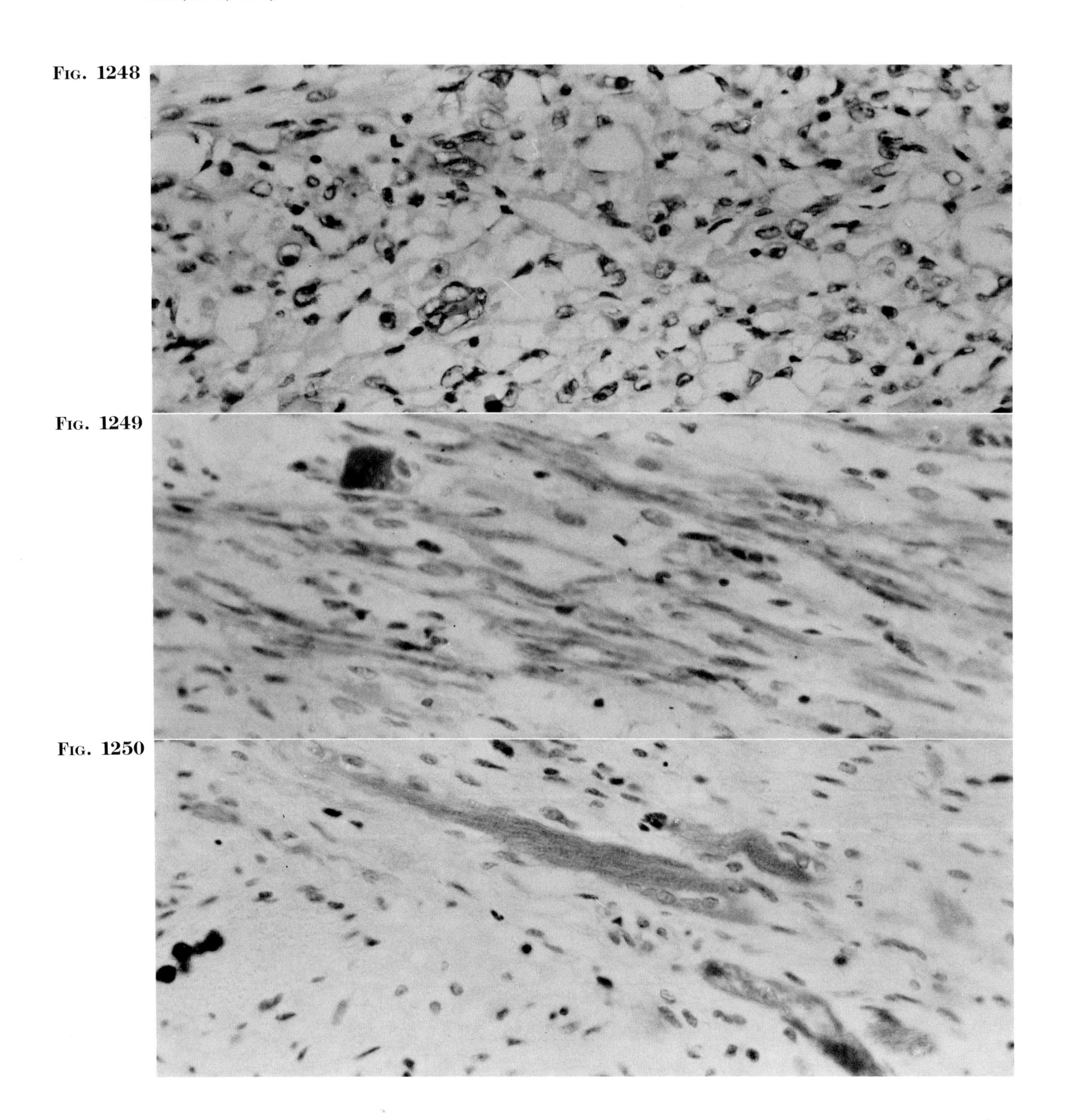

Fig. 1248

Fig. 1249

Fig. 1250

Fig. 1251. Embryonal rhabdomyosarcoma, myxoid type, so-called "botryoid sarcoma of urinary bladder" (Vimentin, ×350) (see Figs. 1033, 1129, 1257).

Fig. 1252. Borderline smooth muscle tumor of stomach (Desmin, ×350) (see Figs. 1212, 1249, 1274).

Fig. 1253. Malignant pleomorphic fibrous histiocytoma of lung (HCG, ×350) (see Figs. 1148, 1179, 1180).

Fig. 1254. Angiomyoma (Myosin, ×350) (see Figs. 653, 912, 985).

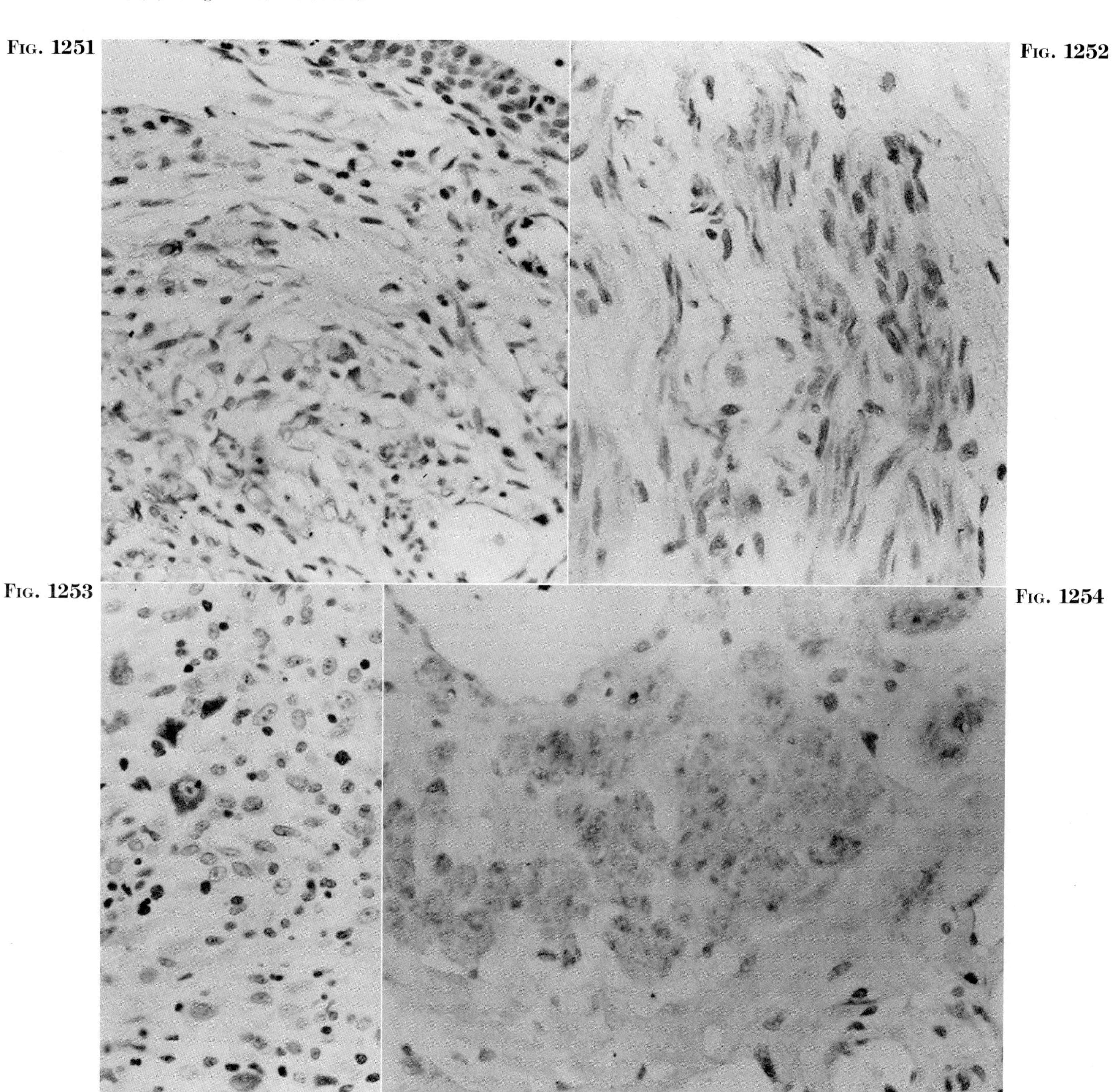

Fig. 1251

Fig. 1252

Fig. 1253

Fig. 1254

Fig. 1255. Pleomorphic rhabdomyosarcoma (Myoglobin, ×350) (see Figs. 1166, 1167, 1216, 1250, 1273).

Fig. 1256. Malignant granular cell tumor (CEA, ×570) (see Figs. 740, 1234).

Fig. 1257. Embryonal rhabdomyosarcoma (Myoglobin, ×140) (see Figs. 1138, 1164, 1171, 1115).

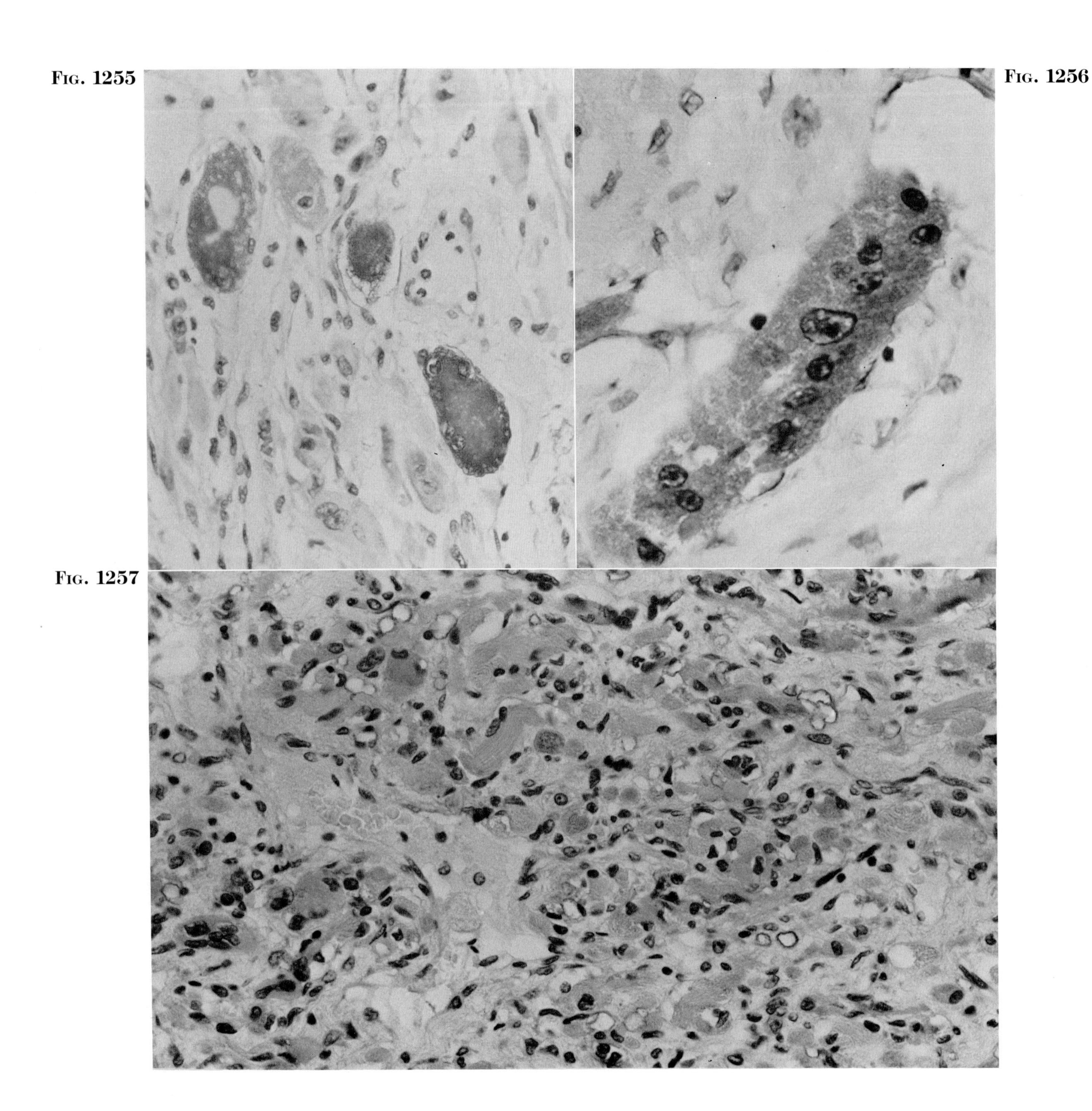

Fig. 1258. Ganglioneuroma (GFAP, ×350) (see Figs. 824, 857, 1137, 1191, 1193).

Fig. 1259. Malignant Triton tumor (Myoglobin, ×570) (see Figs. 606, 732, 812, 813).

Fig. 1260. Rhabdomyoblastoma (Myoglobin, ×350) (see Figs. 1034, 1251, 1257).

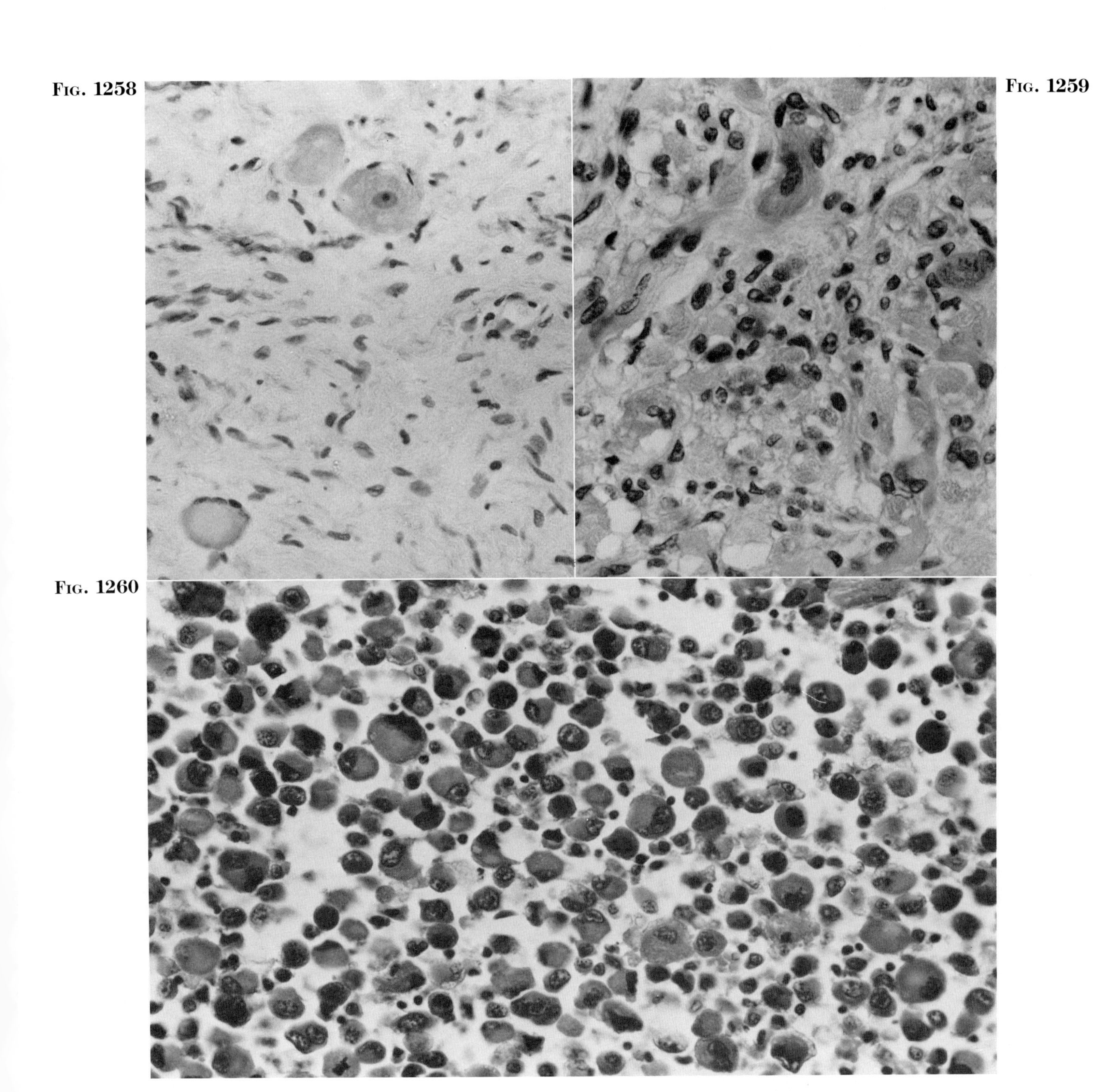

Fig. 1261. Malignant glioma (GFAP, ×350) (see Fig. 992).

Fig. 1262. Amelanotic melanoma (S-100 protein, ×350) (see Figs. 1189, 1190, 1219).

Fig. 1263. Low grade neurofibrosarcoma (S-100 protein, ×350) (see Figs. 1116, 1229, 1266).

Fig. 1264. Traumatic neuroma (S-100 protein, ×350).

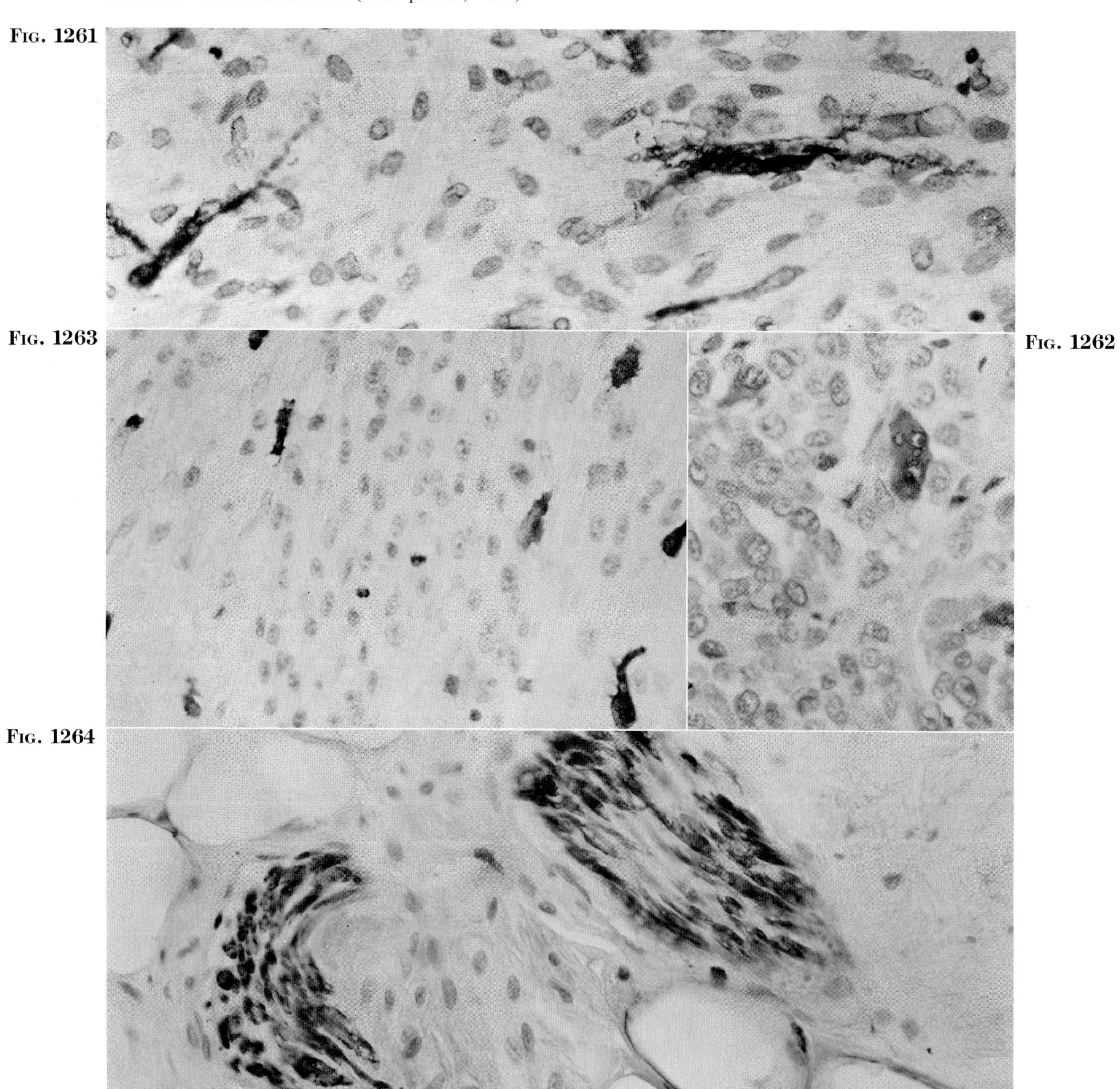

Fig. 1265. Chondrosarcoma (S-100 protein, ×350) (see Figs. 674, 707, 920, 1062).

Fig. 1266. Chordoma (S-100 protein, ×350) (see Figs. 771, 1174, 1226).

Fig. 1267. Balloon cell melanoma metastatic to lymph node (S-100 protein, ×350) (see Figs. 1189, 1219, 1262).

Fig. 1268. Benign glandular schwannoma (Keratin, ×350) (see Figs. 1229, 1259, 1263, 1266).

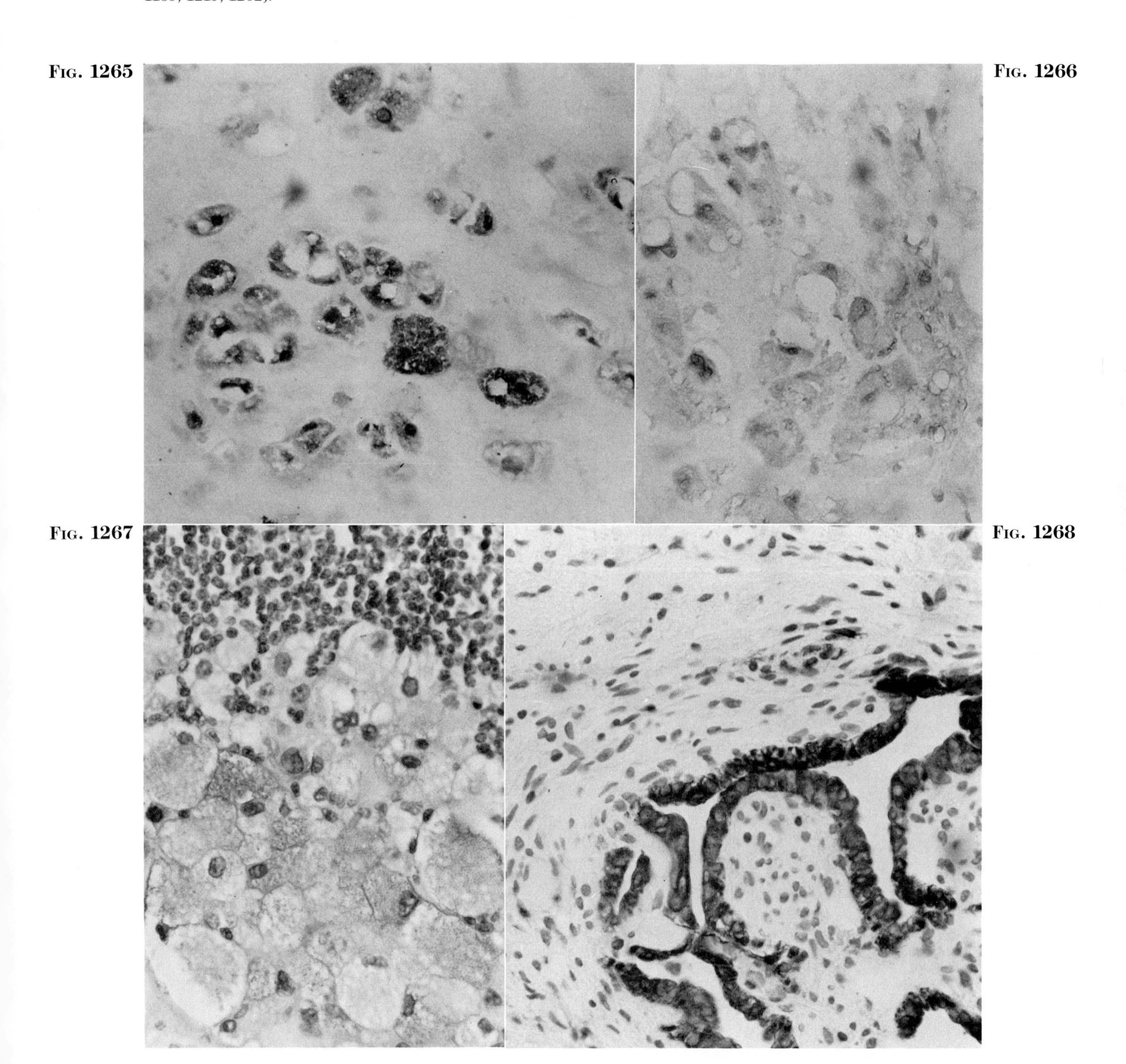

Fig. 1269. Patient had leiomyosarcoma and breast carcinoma. The metastatic tumor stains positive with CEA unequivocally, indicating an epithelial origin (CEA, ×350) (see Figs. 160, 682).

Fig. 1270. Metastatic renal cell carcinoma stains focally positive (Factor VIII) (see Figs. 764, 768, 1018A).

Fig. 1271. Large gastric neoplasm. The impression of gastric leiomyoblastoma was ruled out by positive stain (CEA, ×350).

Fig. 1272. Patient had history of gastric leiomyosarcoma. The positive stain of a pelvic mass indicates a new, prostatic primary. (Prostatic specific antigen, ×350) (see Fig. 212).

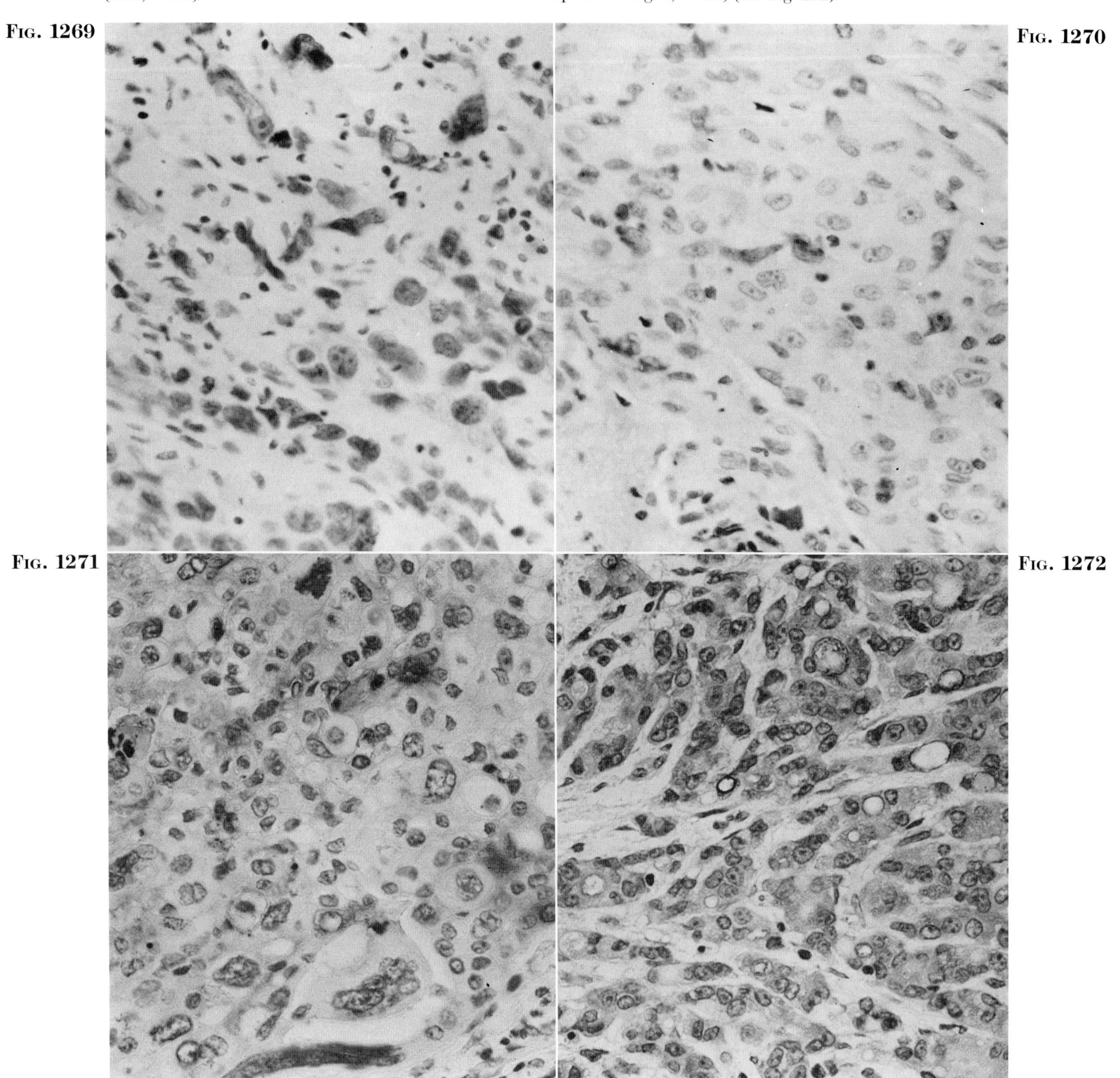

Fig. 1273. Pleomorphic rhabdomyosarcoma (Immunofluorescence stain for desmin, ×200) (see Figs. 1216, 1250, 1255).

Fig. 1274. Leiomyosarcoma (Immunofluorescence stain for actin, ×200) (see Figs. 1212, 1252, 1274).

Fig. 1275. Well-differentiated chondrosarcoma (Immunofluorescence stain for Vimentin, ×250) (see Figs. 1061, 1065, 1222).

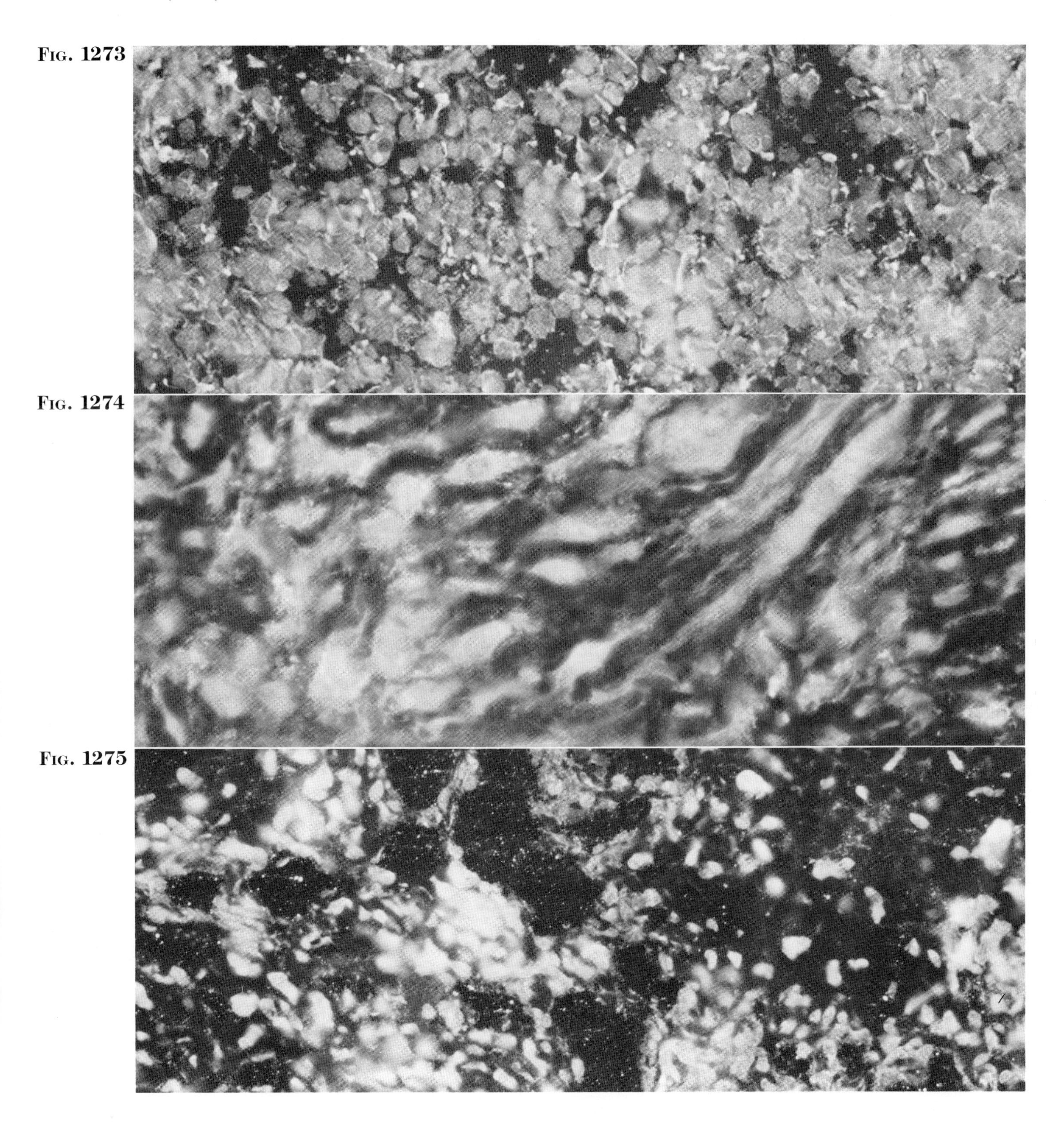

"The grade of potential malignancy is very important information regarding any tumor"
JAMES EWING (1866–1943)

14. Histologic Grade Of Sarcomas

Low Grade Sarcomas (see Table 73–75)

High Grade Sarcomas (see Table 73–75)

Either Low Grade or High Grade Sarcomas (see Tables 75 and 76)

It is generally accepted in the diagnosis of soft tissue and bone tumors that the pathologist's task is to give an accurate evaluation of sarcomas in regard to histologic type, histologic grade, and natural history. During any attempt to grade sarcomas it is mandatory that a proper diagnostic sequence is followed; the microscopic evaluation of the neoplasm requires assessment of the growth pattern, cell morphology, appearance of stroma, and products of cells. Once the histogenesis and histologic type of the sarcoma has been established, the histologic grade can be assigned. The pathologist who does not consider the histologic type of the sarcoma may end up comparing apples with oranges because sarcomas of comparable grade but different type do not have the same behavior and therapeutic response.

A major obstacle to the development of a universally acceptable grading system has been that there are no microscopic features equally applicable to all sarcomas. For example, the number of mitosis is one of the most important histologic features in determining the grade of leiomyosarcomas or malignant peripheral nerve tumors but plays a less important role in assessing malignant fibrous histiocytomas or angiosarcomas (see Table 76). It is a paradox that a number of non-neoplastic lesions and benign neoplasms may have more mitotic figures than do most high grade sarcomas (Table 72).

There is no simple formula for the grading of such a complex constellation of malignant neoplasms as the soft tissue and bone sarcomas. However, the same general principle is used in the grading of malignant epithelial neoplasms, soft tissue, and bone sarcomas. The six histologic parameters that are considered are *differentiation, cellularity, amount of stroma, vascularity, amount of necrosis,* and *number of mitosis* (Table 73).

Grading of primary sarcomas, like histologic typing, should not be attempted on suboptimal material such as a needle biopsy specimen or a frozen section. The departure from grading systems that list low grade (Grade I), intermediate grade (Grade II), and high grade (Grade III) neoplasms is done to prevent the placing of most sarcomas in the Grade II, non-decisive, category. One can only wonder how often the pathologist contributes to patient care by labelling, for example, most mammary carcinomas and colonic carcinomas as Grade II neoplasms. All available data indicate that most therapeutic decisions are made on the basis of whether the sarcoma is low grade or high grade (Table 74). Undoubtedly, a certain degree of experience and flexibility are needed to assign the most appropriate histologic grade. Fortunately, a knowledge of the histogenesis and histologic type enables the pathologist to place most soft tissue and bone sarcomas in the low grade or high grade category without particular attention to histologic parameters for grading (Table 75). Careful scrutiny of the microscopic detail of those sarcomas that can be considered either low grade or high grade almost invariably results in a definitive conclusion. In the case of unresolvable difficulty in assigning the histologic grade the size and site of the sarcoma plays an overwhelmingly important role in prognosis and staging.

Table 72. *Mitotic Figures Are Seen Very Often in the Following Lesions*

Non-neoplastic Lesions
Fasciitis
Fasciitis ossificans
Tendosynovitis
Tendosynovitis ossificans
Proliferative panniculitis
Panniculitis ossificans
Proliferative myositis
Myositis ossificans
Pyogenic granuloma
Angioblastic lymphoid hyperplasia
Vasculitis
Periostitis
Eosinophilic granuloma
Mastocytosis
Plasma cell granuloma
Benign Neoplasms
Xanthogranuloma
Benign fibrous histiocytoma
Lipoblastoma
Pleomorphic lipoma
Hypertrophic hemangioma
Papillary endothelial hyperplasia
Hemangioblastoma
Benign glomus tumor
Chondroblastoma
Osteoblastoma

Table 73. *Histologic Factors Influencing the Grade of Sarcomas*

Differentiation of malignant cells
Cellularity of sarcoma
Vascularity of sarcoma
Amount of stroma
Amount of necrosis
Number of mitosis

Table 74. *Guideline to Histologic Grading of Sarcomas*

LOW GRADE SARCOMAS	HIGH GRADE SARCOMAS
Good differentiation	**Poor differentiation**
Hypocellular	**Hypercellular**
Hypovascular	**Hypervascular**
Much stroma	**Minimal stroma**
Minimal necrosis	**Much necrosis**
Less than 5 mitoses per 10 high-power field	**More than 5 mitoses per 10 high-power field**

Table 75. *Histologic Grade of Sarcomas*

SOFT TISSUE	BONE
Low Grade Sarcomas	
Malignant fibroblastic fibrous histiocytoma	Desmoid tumor
Desmoid tumor	Well differentiated liposarcoma
Chordoid sarcoma	Myxoid liposarcoma
Well differentiated liposarcoma	Leiomyosarcoma
Myxoid liposarcoma	Well differentiated chondrosarcoma
Kaposi's sarcoma	Parosteal osteosarcoma
High Grade Sarcomas	
Monophasic tendosynovial sarcoma, spindle cell type	Malignant pleomorphic fibrous histiocytoma
Epithelioid sarcoma	Pleomorphic fibrosarcoma
Clear cell sarcoma	Poorly differentiated chondrosarcoma
Lipoblastic liposarcoma	Mesenchymal chondrosarcoma
Pleomorphic liposarcoma	Osteosarcoma
Embryonal rhabdomyosarcoma	Paget's sarcoma
Alveolar rhabdomyosarcoma	Malignant lymphoma
Myxoid rhabdomyosarcoma	Granulocytic sarcoma
Rhabdomyoblastoma	Plasma cell myeloma
Pleomorphic rhabdomyosarcoma	Ewing's sarcoma
Lymphangiosarcoma	Chordoma
Primitive neuroectodermal tumor	
Mesenchymal chondrosarcoma	
Osteosarcoma	
Granulocytic sarcoma	
Plasma cell myeloma	
Ewing's sarcoma	
Malignant granular cell tumor	
Either Low Grade or High Grade Sarcomas	
Malignant histiocytic fibrous histiocytoma (Malignant giant cell tumor)	Malignant histiocytic fibrous histiocytoma (Malignant giant cell tumor)
Malignant pleomorphic fibrous histiocytoma	Fibroblastic fibrosarcoma
Fibroblastic fibrosarcoma	Hemangiopericytoma
Fibroblastic liposarcoma	Hemangiosarcoma
Biphasic tendosynovial sarcoma	Myxoid chondrosarcoma
Monophasic tendosynovial sarcoma, pseudoglandular type	
Leiomyoblastoma	
Leiomyosarcoma	
Hemangiopericytoma	
Hemangiosarcoma	
Neurofibrosarcoma	
Malignant schwannoma	
Malignant mesothelioma	
Alveolar soft part sarcoma	
Chondrosarcoma	
Malignant cystosarcoma phyllodes	

Table 76. *Predictive Values of Histologic Factors in Grading of Sarcomas That Can Be Either Low Grade or High Grade**

	DIFFERENTIATION	CELLULARITY	VASCULARITY	STROMA	NECROSIS	MITOSIS
Malignant histiocytic fibrous histiocytoma	1	2	3	6	4	5
Malignant pleomorphic fibrous histiocytoma	1	2	5	6	3	4
Fibroblastic fibrosarcoma	1	2	5	6	4	3
Biphasic tendosynovial sarcoma	1	2	5	3	4	6
Monophasic tendosynovial sarcoma pseudoglandular type	1	2	5	3	4	6
Fibroblastic liposarcoma	1	2	5	3	4	6
Leiomyoblastoma	1	2	3	5	4	6
Leiomyosarcoma	1	3	6	5	4	2
Hemangiopericytoma	1	2	6	4	5	3
Hemangiosarcoma	1	3	6	2	4	5
Neurofibrosarcoma	2	1	6	4	5	3
Malignant schwannoma	2	1	6	4	5	3
Malignant mesothelioma	1	2	5	3	6	4
Chondrosarcoma	1	2	6	3	4	5
Malignant cystosarcoma phyllodes	1	2	6	3	5	4

*Order of importance from 1 (most important) to 6 (least important)

> *"While any attempted subdivision of a tumor type may add unwanted fuel to the conflagration which burns around the differentiation of benign and malignant forms, it yet may be of some practical value"*
>
> PRICE AND VALENTINE

15. Stage Of Sarcomas

Stage O Sarcomas (Fig. 1276, Tables 78–80)

Stage I Sarcomas (Fig. 1276, Tables 78–80)

Stage II Sarcomas (Fig. 1276, Tables 78–80)

Stage III Sarcomas (Fig. 1276, Tables 78–80)

Stage IV Sarcomas (Fig. 1276, Tables 78–80)

ANY attempt to assess accurately the stage (malignant potential) of soft tissue and bone sarcomas depends on the size, site, and histologic grade of the neoplasm. None of these three characteristics alone is an accurate determinant of prognosis. Some sarcomas with low metastatic potential can become large, for example, malignant fibroblastic fibrous histiocytoma (dermatofibrosarcoma protuberans); on the other hand, certain high grade sarcomas can be small and superficial, for example, angiosarcoma. Sarcomas that are small (less than 5 cm), superficial (not extending beyond the superficial fascia), and microscopically low grade have a favorable prognosis (Table 77), and are considered Stage O neoplasms (Table 78). Sarcomas that are large (more than 5 cm), deep (extending beyond the superficial fascia) and histologically high grade have a poor prognosis (Table 77), and are considered Stage III malignant neoplasms (Table 78). It is apparent that the pathologist who uses these criteria will list all primary bone sarcomas as deep lesions, so that there is no Stage O bone sarcoma.

Figure 1276 illustrates various combinations of prognostic signs that may occur. The upper light halves indicate favorable prognostic signs, and the lower dark halves indicate a poor prognosis. Application of this staging scheme once the histologic type, size, site, and histologic grade of the sarcoma have been established permits the listing of most sarcomas according to stage (Table 79). Furthermore, this system enables the prognostic value of the size, site, and histologic grade to be established for each sarcoma (Table 80).

All available data indicate that the stage has value and is significant in predicting the prognosis and the outcome of treatment, provided therapy has been planned and carried out according to the highest professional standards.

TABLE 77. *Factors Influencing Prognosis of Sarcomas*

SARCOMA	FAVORABLE PROGNOSTIC SIGNS	UNFAVORABLE PROGNOSTIC SIGNS
Size	Small	Big
Site	Superficial	Deep
Histologic grade	Low	High

From Hajdu, S. I.: Pathology of Soft Tissue Tumors. Philadelphia, Lea & Febiger, 1979.

TABLE 78. *Correlation between Prognostic Signs and Stage**

PROGNOSTIC SIGNS	STAGE OF SARCOMA
Three favorable signs	Stage O
Two favorable signs and one unfavorable sign	Stage I
One favorable sign and two unfavorable signs	Stage II
Three unfavorable signs	Stage III
Evidence of metastasis	Stage IV

*See Figure 1276.

Table 79. *Predictive Value of Size, Site, and Histologic Grade in Staging of Sarcomas**

	SIZE	SITE	GRADE†
Malignant fibroblastic fibrous histiocytoma	2	1	(Low)
Malignant histiocytic fibrous histiocytoma	2	3	1 (Low)
Malignant histiocytic fibrous histiocytoma	3	2	1 (High)
Malignant pleomorphic fibrous histiocytoma	2	3	1 (Low)
Malignant pleomorphic fibrous histiocytoma	3	2	1 (High)
Desmoid tumor	2	1	(Low)
Fibroblastic fibrosarcoma	2	3	1 (Low)
Fibroblastic fibrosarcoma	2	3	1 (High)
Pleomorphic fibrosarcoma	1	2	(High)
Biphasic tendosynovial sarcoma	2	3	1 (Low)
Biphasic tendosynovial sarcoma	2	3	1 (High)
Monophasic tendosynovial sarcoma, spindle cell type	1	2	(High)
Epithelioid sarcoma	1	2	(High)
Clear cell sarcoma	1	2	(High)
Chordoid sarcoma	1	2	(Low)
Well-differentiated liposarcoma	2	1	(Low)
Myxoid liposarcoma	2	1	(Low)
Lipoblastic liposarcoma	1	2	(High)
Fibroblastic liposarcoma	2	3	1 (Low)
Fibroblastic liposarcoma	3	2	1 (High)
Pleomorphic liposarcoma	1	2	(High)
Leiomyoblastoma	2	3	1 (Low)
Leiomyoblastoma	2	3	1 (High)
Leiomyosarcoma	2	3	1 (Low)
Leiomyosarcoma	3	2	1 (High)
Embryonal rhabdomyosarcoma	1	2	(High)
Alveolar rhabdomyosarcoma	2	1	(High)
Myxoid rhabdomyosarcoma	2	1	(High)
Rhabdomyoblastoma	1	2	(High)
Pleomorphic rhabdomyosarcoma	1	2	(High)
Hemangiopericytoma	2	3	1 (Low)
Hemangiopericytoma	3	2	1 (High)
Hemangiosarcoma	3	2	1 (Low)
Hemangiosarcoma	3	2	1 (High)
Lymphangiosarcoma	1	2	(High)
Neurofibrosarcoma	3	2	1 (Low)
Neurofibrosarcoma	3	2	1 (High)
Malignant schwannoma	2	3	1 (Low)
Malignant schwannoma	3	2	1 (High)
Primitive neuroectodermal tumor	2	1	(High)
Malignant mesothelioma	2	3	1 (Low)
Malignant mesothelioma	2	3	1 (High)
Well differentiated chondrosarcoma	2	1	(Low)
Myxoid chondrosarcoma	3	2	1 (Low)
Myxoid chondrosarcoma	2	3	1 (High)
Poorly differentiated chondrosarcoma	1	2	(High)
Mesenchymal chondrosarcoma	1	2	(High)
Osteosarcoma	1	2	(High)
Parosteal osteosarcoma	1	2	(Low)
Malignant granular cell tumor	1	2	(High)
Alveolar soft part sarcoma	2	3	1 (Low)
Alveolar soft part sarcoma	2	3	1 (High)
Kaposi's sarcoma	1	2	(Low)
Ewing's sarcoma	2	1	(High)
Chordoma	1	2	(High)

* Predictive value from 1 (most important) and 2 (important) to 3 (less important).

†Note that in those sarcomas that can be either low grade or high grade accurate grading is the most important. In those sarcomas that are always low grade or always high grade accurate histologic typing is the most important (see Table 75).

Table 80. *Most Common Stage of Primary Sarcomas**

SOFT TISSUE SARCOMAS	BONE SARCOMAS*
Stage 0	
Malignant fibroblastic fibrous histiocytoma	
Kaposi's sarcoma	
Stage I	
Chordoid sarcoma	Well-differentiated liposarcoma
Well-differentiated liposarcoma	Myxoid liposarcoma
	Leiomyosarcoma
Stage II	
Malignant histiocytic fibrous histiocytoma	Malignant histiocytic fibrous histiocytoma
Desmoid tumor	Desmoid tumor
Monophasic tendosynovial sarcoma, pseudoglandular type	Chondrosarcoma
	Parosteal osteosarcoma
Epithelioid sarcoma	Granulocytic sarcoma
Clear cell sarcoma	Malignant lymphoma
Myxoid liposarcoma	
Leiomyoblastoma	
Leiomyosarcoma	
Myxoid rhabdomyosarcoma	
Rhabdomyoblastoma	
Malignant schwannoma	
Chondrosarcoma	
Granulocytic sarcoma	
Malignant granular cell tumor	
Stage III	
Malignant pleomorphic fibrous histiocytoma	Malignant pleomorphic fibrous-histiocytoma
Fibrosarcoma	Fibrosarcoma
Biphasic tendosynovial sarcoma	Hemangiosarcoma
Monophasic tendosynovial sarcoma, spindle cell type	Mesenchymal chondrosarcoma
	Osteosarcoma
Lipoblastic liposarcoma	Paget's sarcoma
Fibroblastic liposarcoma	Plasma cell myeloma
Pleomorphic liposarcoma	Ewing's sarcoma
Embryonal rhabdomyosarcoma	Chordoma
Alveolar rhabdomyosarcoma	
Pleomorphic rhabdomyosarcoma	
Hemangiopericytoma	
Hemangiosarcoma	
Lymphangiosarcoma	
Neurofibrosarcoma	
Primitive neuroectodermal tumor	
Mesenchymal chondrosarcoma	
Osteosarcoma	
Granulocytic sarcoma	
Alveolar soft part sarcoma	
Ewing's sarcoma	

*See Figure 1276.

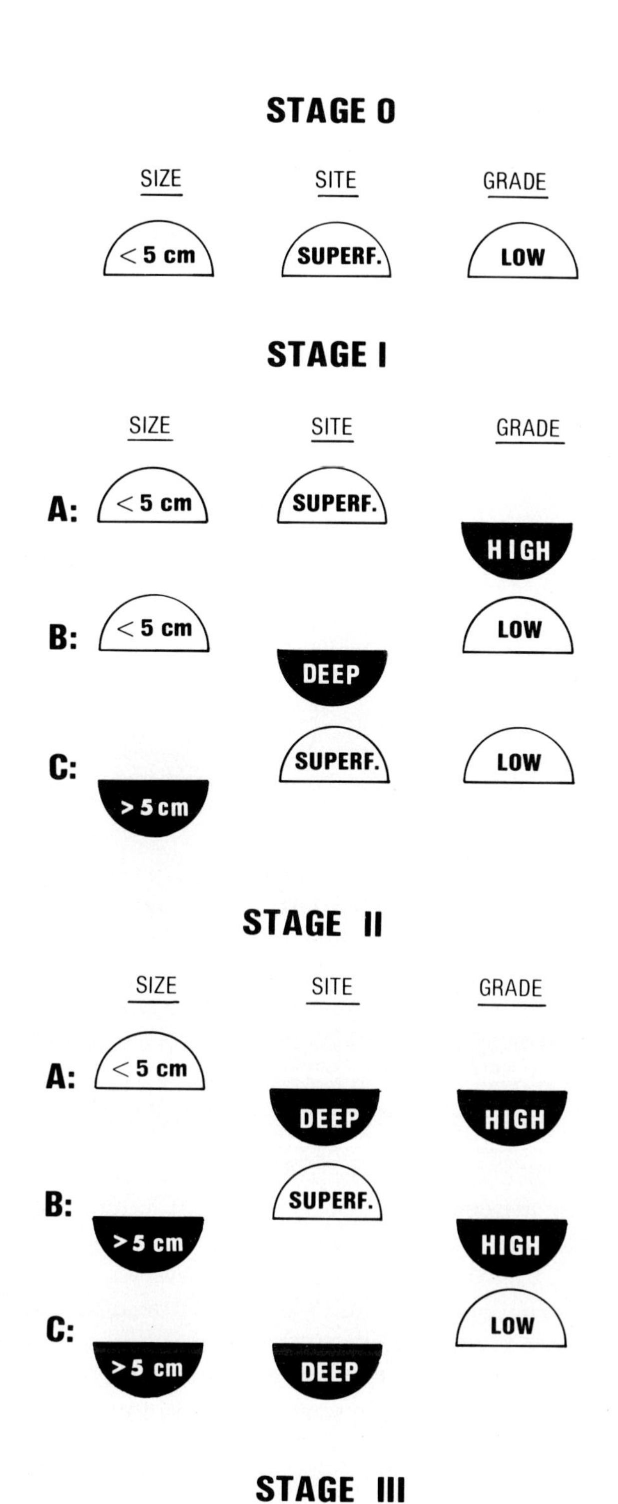

Fig. 1276. Correlation among size, site, histologic grade, and stage of the primary soft tissue sarcomas.

Three favorable prognostic signs indicate *Stage 0* sarcoma.

Two favorable prognostic signs and one unfavorable sign indicate *Stage I* sarcoma.

One favorable prognostic sign and two unfavorable ones are consistent with *Stage II* sarcoma.

Three unfavorable prognostic signs indicate *Stage III* neoplasm.

Metastatic sarcomas should be considered *Stage IV* lesions.
(From Hajdu, S. I.: Pathology of Soft Tissue Tumors. Philadelphia, Lea & Febiger, 1979).

"Classification of malignant neoplasms should be simple and should yield meaningful information as to prognosis"

American Joint Committee for Cancer Staging and End-Results Reporting (1959)

16. Prognosis

1. Non-neoplastic Lesions (see Table 81)
2. Benign Neoplasms (see Table 82)
3. Malignant Neoplasms (see Table 83)

THE prognosis of soft tissue and bone tumors is most heavily influenced by the histologic type, location, and size of the tumor. Non-neoplastic lesions and benign neoplasms, almost without exception, are self-limited. Malignant transformation of benign lesions of long duration may occur and can pose a formidable challenge to pathologists and clinicians (Tables 81 and 82).

The general prognosis of most patients with soft tissue and bone sarcomas reflects the tendency of these sarcomas to reach considerable size prior to detection, to invade surrounding tissues, to extend along anatomic planes, neurovascular bundles, and the medullary cavity, as well as their tendency for hematogenous dissemination (Table 83).

Recurrence is the sign of failure to control the disease locally (Tables 81–83). Well-planned and properly performed surgical procedures should theoretically achieve complete local control in every case. In practice, however, the rate of recurrence varies from 10% to 75%, according to the modality of therapy and histologic type of the sarcoma. About 75% of local recurrences occur in the first 2 years. Local recurrence of a primary high grade sarcoma is an unfavorable prognostic sign because many patients who develop local recurrences after treatment develop metastases (Table 83).

Less than 5% of patients treated for soft tissue or bone sarcomas develop metastases to regional lymph nodes. Sarcomas that most commonly involve the draining lymph nodes are embryonal rhabdomyosarcoma, epithelioid sarcoma, and Ewing's sarcoma. For a long time, pulmonary metastasis was viewed as a sign of an irreversible, fatal disease course. Recent progress in resection of pulmonary metastases, in the absence of extrathoracic disease, shows that from 10% to 20% of the patients so treated survive for 5 or more years (Table 83).

In addition to a variety of function-saving surgical resections, a number of adjunctive modalities, such as brachytherapy using Iridium-192 implants, external beam therapy, and both local and systemic chemotherapy, are used in an attempt to control the local growth as well as the metastatic spread of sarcomas. Despite advances made in the radiation and chemotherapy of pediatric soft tissue and bone sarcomas, clinical results indicate that adult sarcomas are less responsive to radiation and chemotherapy than those in children.

Although histologic grade is one of the most important prognostic factors, the site and size of the sarcoma should not be ignored as prognostic determinants. It has also been shown that the clinical behavior and responsiveness of the sarcoma to various therapies are significantly different and vary according to histologic type. At our present level of understanding of the natural history of soft tissue and bone sarcomas, it is apparent that these neoplasms cannot be considered only according to grade when their response to therapy is being evaluated. Other prognostic variables, including histologic types, must be considered (Table 83). The role of the pathologist in assessing sarcomas is crucial because it renders a histopathologic diagnosis that by itself is very often prognostic. Moreover, it is the pathologists' duty to supply accurate information concerning the grade and stage of the sarcoma, thereby influencing the selection of the best modality of treatment.

Table 81. ***Prognosis of Non-neoplastic Lesions***

	RECURRENCE	FRACTURE	MALIGNANT TRANSFORMATION	OTHER (SEE TABLE 19)
Aneurysmal bone cyst	20%	Rare	3%	
Giant cell granuloma	12%			
Xanthogranuloma				Spontaneous regression
Fasciitis	2%			Spontaneous regression
Keloid	30%			Familial tendency
Fibromatosis	30%		Rare	Can cause deformity, Gardner's syndrome
Fibrous dysplasia		Very uncommon	3%	Can be systemic, cutaneous pigmentation
Elastofibroma	2%			Familial tendency
Collagenoma				Can be systemic, familial
Tendosynovitis	20%	Rare		Can cause bone erosion
Tendosynovial cyst		Rare		
Tendosynovial chondromatosis	5%		1%	
Gouty arthritis		Rare		Can cause bone erosion, elevated uric acid
Proliferative panniculitis	5%			
Myositis ossificans			0.5%	Microdactylia
Angiofollicular lymphoid hyperplasia	10%			Eosinophilia
Arteriovenous malformation				Hemorrhage
Traumatic neuroma	10%			
Mesothelial hyperplasia				Effusion
Paget's disease		Very common	10%	Chronic disease, familial tendency
Osteomyelitis	5%	Rare	Very rare	Can be multifocal

Table 82. ***Prognosis of Benign Neoplasms***

	RECURRENCE	FRACTURE	MALIGNANT TRANSFORMATION	SYSTEMIC DISEASE	OTHER (SEE TABLE 20)
Benign fibroblastic fibrous histiocytoma of soft tissues	15%		2%		
Benign histiocytic fibrous histiocytoma of bone	50%	5%	10%		
Nonossifying fibroma		Common	Very rare		Familial tendency
Benign pleomorphic fibrous histiocytoma	10%		1%		
Desmoplastic fibroma		10%			
Lipoma	10%		0.1%	Can be	Familial tendency
Lipoblastoma	10%			Can be	
Angiomyolipoma	Rare				10% associated with tuberous sclerosis
Leiomyoma	10%			Can be	
Hemangioma	10%			Can be	
Glomus tumor	10%				
Neurofibroma	25%		30%*	Can be	Café-au-lait spots
Chondroblastoma	35%		2%		Joint effusion in 30%
Chondromyxoid fibroma	10%		1%		
Chondroma	20%		0.5%	Can be	
Osteochondroma			1%	Can be	Familial tendency
Osteoblastoma	25%		1%		
Osteoid osteoma	1%				
Benign granular cell tumor	10%		Very rare	Can be	

*In association with neurofibromatosis.

Table 83. *Prognosis of Malignant Neoplasms*

	Recurrence*	Fracture	Average Survival 5-Year	Average Survival 10-Year	Other (see Table 21)
Malignant fibroblastic fibrous histiocytoma of soft tissues	50%		95%	90%	
Malignant fibroblastic fibrous histiocytoma of bone		50%	60%	30%	
Malignant histiocytic fibrous histiocytoma	60%		70%	50%	
Malignant pleomorphic fibrous histiocytoma of soft tissue	40%		50%	40%	
Malignant pleomorphic fibrous histiocytoma of bone		50%	60%	30%	
Desmoid tumor	60%		98%	95%	1% associated with Gardner's syndrome
Fibrosarcoma of soft tissues	45%		60%	40%	
Fibrosarcoma of bone			40%	30%	
Biphasic tendosynovial sarcoma	40%		75%	50%	
Monophasic tendosynovial sarcoma, spindle cell type	60%		45%	25%	50% metastatic to lung at primary diagnosis
Monophasic tendosynovial sarcoma, pseudoglandular type	40%		75%	60%	
Epithelioid sarcoma	60%		60%	45%	33% metastatic to regional lymph nodes
Clear cell sarcoma	50%		55%	40%	
Chordoid sarcoma	80%		80%	75%	
Well-differentiated liposarcoma	30%		95%	92%	
Myxoid liposarcoma	50%		95%	85%	
Lipoblastic liposarcoma	65%		55%	40%	
Fibroblastic liposarcoma	60%		60%	55%	
Pleomorphic liposarcoma	55%		45%	30%	
Leiomyoblastoma			60%	50%	
Leiomyosarcoma of soft tissues	60%		45%	35%	
Leiomyosarcoma of bone		10%	80%	75%	
Embryonal rhabdomyosarcoma	30%		65%	55%	10% metastatic to regional lymph nodes
Alveolar rhabdomyosarcoma	50%		50%	30%	20% metastatic to regional lymph nodes
Myxoid rhabdomyosarcoma	30%		80%	60%	
Rhabdomyoblastoma	30%		50%	30%	
Pleomorphic rhabdomyosarcoma	30%		45%	25%	
Hemangiopericytoma	40%		60%	50%	
Hemangiosarcoma of soft tissues	50%		30%	5%	
Hemangiosarcoma of bone, solitary		10%	20%	—	
Hemangiosarcoma of bone, multifocal		10%	70%	50%	
Lymphangiosarcoma	30%		15%	5%	Often multifocal
Neurofibrosarcoma	40%		60%	50%	Often multifocal
Malignant noncollagenous peripheral nerve tumor (Malignant schwannoma)			80%	70%	
Primitive neuroectodermal tumor	40%		30%	20%	
Chondrosarcoma	30%		60%	45%	
Mesenchymal chondrosarcoma	75%		30%	20%	0.3% multifocal
Osteosarcoma of bone	10%	10%	45%	30%	
Extraskeletal osteosarcoma	70%		20%	15%	
Parosteal osteosarcoma	20%		75%	65%	
Paget's sarcoma	50%	30%	10%	—	
Granulocytic sarcoma			10%	—	
Plasma cell myeloma, systemic		50%	10%	—	
Plasma cell myeloma, solitary		50%	60%	50%	Hypercalcemia
Malignant granular cell tumor		50%	60%	40%	
Alveolar soft part sarcoma	30%		65%	45%	
Kaposi's sarcoma†	60%		90%	85%	30% associated with malignant lymphoma
Ewing's sarcoma	30%		55%	40%	
Chordoma	30%		55%	30%	

*Rate of recurrence is influenced by the modality of therapy.

†50% of patients who acquired immunodeficiency syndrome develop Kaposi's sarcoma.

"Comparison is the great organon of biological and consequently also of pathological research"

W. Roger Williams

17. Differential Diagnosis

Differential Diagnosis According to *Clinical Presentation* (see Figs. 8 to 69; Tables 17 to 23B)

Differential Diagnosis According to *Age Distribution* (see Tables 24 to 26)

Differential Diagnosis According to *Sex Distribution* (see Tables 27 to 29)

Differential Diagnosis According to *Anatomic Site* (see Figs. 70 to 117; Tables 17 to 23B; Tables 30 to 38)

Differential Diagnosis According to *Size* (see Tables 39 to 41)

Differential Diagnosis According to *Radiologic Appearance* (see Figs. 118 to 234; Tables 42 to 44)

Differential Diagnosis According to *Gross Appearance* (see Figs. 285 to 330; Tables 45 to 47)

Differential Diagnosis According to *Growth Pattern* (see Figs. 331 to 615; Tables 48 to 50)

Differential Diagnosis According to *Cell Morphology* (see Figs. 616 to 829; Tables 51 to 53)

Differential Diagnosis According to *Appearance of Stroma* (see Figs. 830 to 1112; Tables 54 to 57)

Differential Diagnosis According to *Products of Cells* (see Figs. 1113 to 1275; Tables 58 to 71)

Differential Diagnosis According to *Histologic Grade* (see Tables 72 to 76)

Differential Diagnosis According to *Stage* (see Fig. 1276; Tables 79, 80)

Differential Diagnosis According to *Prognosis* (see Tables 81 to 83)

Differential Diagnosis According to *Histologic Appearance*
- Non-neoplastic Lesions (see Tables 84, 87–92)
- Benign Neoplasms (see Tables 85–92)
- Malignant Neoplasms (see Tables 86–92)

TABLE 84. *Histologic Differential Diagnosis of Non-neoplastic Lesions**

	SHOULD BE CONSIDERED IN THE DIFFERENTIAL DIAGNOSIS		
	NON-NEOPLASTIC LESIONS	BENIGN NEOPLASMS	MALIGNANT NEOPLASMS
Aneurysmal bone cyst	Tendosynovial cyst Giant cell granuloma Hyperparathyroidism Arteriovenous malformation	Benign histiocytic fibrous histiocytoma Cavernous hemangioma Papillary endothelial hyperplasia	Malignant histiocytic fibrous histiocytoma Osteosarcoma
Giant cell granuloma	Hyperparathyroidism Aneurysmal bone cyst	Benign histiocytic fibrous histiocytoma Nonossifying fibroma	Metastatic carcinoma
Hyperparathyroidism	Aneurysmal bone cyst Giant cell granuloma Paget's disease	Benign histiocytic fibrous histiocytoma	Malignant histiocytic fibrous histiocytoma
Xanthogranuloma	Fasciitis Fat necrosis Lipogranuloma Proliferative panniculitis Foreign body granuloma Histiocytosis Gaucher's disease	Benign pleomorphic fibrous histiocytoma Hibernoma Skin adnexal adenoma	Malignant pleomorphic fibrous histiocytoma Clear cell sarcoma Malignant melanoma Clear cell carcinoma Sebaceous carcinoma
Fasciitis	Keloid Fibromatosis Proliferative myositis Pyogenic granuloma	Benign fibroblastic fibrous histiocytoma Leiomyoma Fibroma Neurofibroma Benign schwannoma	Malignant fibroblastic fibrous histiocytoma Myxoid rhabdomyosarcoma Leiomyosarcoma Fibrosarcoma Neurofibrosarcoma Malignant schwannoma Kaposi's sarcoma
Keloid	Scar Collagenoma	Neurofibroma Fibroma	Desmoid tumor
Fibromatosis	Scar Fasciitis Elastofibroma Collagenoma	Benign fibroblastic fibrous histiocytoma Leiomyoma Neurofibroma	Malignant fibroblastic fibrous histiocytoma Desmoid tumor Fibrosarcoma Leiomyosarcoma Parosteal osteosarcoma
Fibrous dysplasia	Callus Tendosynovial cyst	Desmoplastic fibroma Nonossifying fibroma	Fibrosarcoma Osteosarcoma
Elastofibroma	Fat necrosis Lipogranuloma Fibromatosis Collagenoma	Fibroblastic lipoma Pleomorphic lipoma Neurofibroma	Pleomorphic liposarcoma
Collagenoma	Keloid Fibromatosis	Benign fibrous mesothelioma Fibroma Leiomyoma	Desmoid tumor Malignant fibrous mesothelioma
Tendosynovitis	Fasciitis Fibromatosis Panniculitis	Benign histiocytic fibrous histiocytoma	Malignant histiocytic fibrous histiocytoma Monophasic tendosynovial sarcoma, spindle cell type
Tendosynovial cyst	Fasciitis Fibromatosis Tendosynovitis Aneurysmal bone cyst	Benign peripheral nerve tumor	Malignant peripheral nerve tumor Clear cell sarcoma
Tendosynovial chondromatosis	Tendosynovial cyst	Chondroma Osteochondroma	Chondrosarcoma
Fat necrosis	Lipogranuloma Proliferative panniculitis Xanthogranuloma Granulomatous synovitis Lipodystrophia	Pleomorphic lipoma Hibernoma Benign granular cell tumor	Pleomorphic liposarcoma Malignant granular cell tumor Renal cell carcinoma

*See Table 10.

TABLE 84. *Histologic Differential Diagnosis of Non-neoplastic Lesions** (*Continued*)

	SHOULD BE CONSIDERED IN THE DIFFERENTIAL DIAGNOSIS		
	NON-NEOPLASTIC LESIONS	BENIGN NEOPLASMS	MALIGNANT NEOPLASMS
Lipogranuloma	Fat necrosis Proliferative panniculitis Lipodystrophia	Pleomorphic lipoma	Pleomorphic liposarcoma Renal cell carcinoma Malignant melanoma
Proliferative panniculitis	Fat necrosis Lipogranuloma Fasciitis Tendosynovitis	Fibroblastic lipoma Hibernoma	Renal cell carcinoma Well-differentiated liposarcoma Fibroblastic liposarcoma
Proliferative myositis	Fasciitis Proliferative panniculitis Myositis ossificans Xanthogranuloma Elastofibroma	Rhabdomyoma Ganglioneuroma Benign pleomorphic fibrous histiocytoma Myxoid lipoma	Pleomorphic rhabdomyosarcoma Malignant pleomorphic fibrous histiocytoma
Myositis ossificans	Fasciitis ossificans Panniculitis ossificans Osteoid metaplasia Callus Osteogenesis imperfecta	Osteochondroma Osteoma Benign mesenchymoma	Osteosarcoma Parosteal osteosarcoma Malignant mesenchymoma
Pyogenic granuloma	Angiofollicular lymphoid hyperplasia Fasciitis Vasculitis Eosinophilic granuloma Fat necrosis Lymphocytoma cutis	Hypertrophic hemangioma Glomus tumor	Kaposi's sarcoma Hemangiosarcoma Lymphangiosarcoma Angioendotheliomatosis
Arteriovenous malformation	Vasculitis Infarct	Hemangioma Neurofibroma Angiomyoma Angiomyolipoma Lymphangioma Lymphangiomyoma	Hemangiosarcoma Lymphangiosarcoma Malignant angiomyolipoma
Traumatic neuroma	Fibromatosis	Neurofibroma Plexiform neurofibroma Ganglioneuroma Leiomyoma	
Callus	Osteoid metaplasia Myositis ossificans Panniculitis ossificans Enostosis Fasciitis ossificans Paget's disease	Osteochondroma Chondroma Chondromyxoid fibroma Osteoma	Osteosarcoma Parosteal osteosarcoma Paget's sarcoma Chondrosarcoma
Paget's disease	Hyperparathyroidism Osteomyelitis Callus	Osteoma	Osteosarcoma Paget's sarcoma
Eosinophilic granuloma	Osteomyelitis Hyperparathyroidism Histiocytosis Lymphocytoma cutis Mastocytosis Plasma cell granuloma Angiofollicular lymphoid hyperplasia		Plasmacytoma Granulocytic sarcoma Malignant lymphoma Hodgkin's disease

*See Table 10.

Table 85. *Histologic Differential Diagnosis of Benign Neoplasms**

	Should be considered in the differential diagnosis		
	Non-neoplastic lesions	Benign neoplasms	Malignant neoplasms
Benign fibroblastic fibrous histiocytoma	Fasciitis Fibromatosis Pyogenic granuloma Proliferative panniculitis	Nonossifying fibroma Benign schwannoma Fibroma Leiomyoma	Kaposi's sarcoma Malignant schwannoma Malignant fibroblastic fibrous histiocytoma Parosteal osteosarcoma Fibroblastic liposarcoma
Benign histiocytic fibrous histiocytoma	Giant cell granuloma Hyperparathyroidism Aneurysmal bone cyst Tendosynovitis Callus Myositis ossificans Foreign body granuloma	Nonossifying fibroma Osteoblastoma Osteoid osteoma	Malignant histiocytic fibrous histiocytoma Chondroblastoma Osteosarcoma Paget's sarcoma
Benign pleomorphic fibrous histiocytoma	Xanthogranuloma Storage disease Foreign body granuloma Tendosynovitis Granulomatous synovitis Tendosynovial cyst Gouty arthritis Fat necrosis Lipogranuloma	Benign histiocytic fibrous histiocytoma Papillary endothelial hyperplasia Pleomorphic lipoma Nonossifying fibroma Hibernoma	Malignant pleomorphic fibrous histiocytoma Hemangiosarcoma Pleomorphic liposarcoma Osteosarcoma Metastatic carcinoma
Nonossifying fibroma	Fibrous dysplasia Aneurysmal bone cyst Callus Infarct	Benign fibrous histiocytoma Desmoplastic fibroma Fibroblastic lipoma Neurofibroma	Parosteal osteosarcoma Fibrosarcoma Malignant fibrous histiocytoma Metastatic carcinoma Fibroblastic liposarcoma
Fibroma	Keloid Elastofibroma Fibromatosis Callus	Fibroblastic lipoma Angiofibroma Leiomyoma Neurofibroma Benign fibrous mesothelioma	Desmoid tumor Malignant fibroblastic fibrous histiocytoma Fibroblastic liposarcoma Leiomyosarcoma
Desmoplastic fibroma	Fibrous dysplasia Arteriovenous malformation Infarct Callus	Nonossifying fibroma Fibroblastic lipoma Neurofibroma	Desmoid tumor Fibrosarcoma Parosteal osteosarcoma
Well-differentiated lipoma	Normal fat Fat necrosis Lipogranuloma Proliferative panniculitis	Lipoblastoma Myxoid lipoma	Well-differentiated liposarcoma
Fibroblastic lipoma	Fasciitis Fibromatosis Proliferative panniculitis Traumatic neuroma Elastofibroma	Benign fibroblastic fibrous histiocytoma Fibroma Leiomyoma Benign schwannoma Neurofibroma	Leiomyosarcoma Malignant schwannoma Neurofibrosarcoma Well-differentiated liposarcoma
Lipoblastoma	Lipogranuloma	Myxoid lipoma	Liposarcoma Myxoid rhabdomyosarcoma

*See Table 11.

Table 85. *Histologic Differential Diagnosis of Benign Neoplasms** (*Continued*)

	Should be considered in the differential diagnosis		
	Non-neoplastic lesions	Benign neoplasms	Malignant neoplasms
Pleomorphic lipoma	Fat necrosis Lipogranuloma Proliferative panniculitis Elastofibroma Xanthogranuloma Piezogenic papule Elastofibroma	Benign pleomorphic fibrous histiocytoma Hibernoma Angiomyoma	Malignant pleomorphic fibrous histiocytoma Pleomorphic liposarcoma
Angiomyolipoma	Arteriovenous malformation Xanthogranuloma	Leiomyoma Angiomyoma	Leiomyosarcoma Liposarcoma
Hibernoma	Xanthogranuloma Fat necrosis	Benign granular cell tumor Rhabdomyoma	Alveolar soft part sarcoma Malignant melanoma Oncocytoma
Leiomyoma	Arteriovenous malformation Fibromatosis Traumatic neuroma Fasciitis	Angiomyoma Neurofibroma Fibroma Meningioma	Leiomyosarcoma Kaposi's sarcoma Fibrosarcoma Malignant schwannoma
Rhabdomyoma	Xanthogranuloma	Benign granular cell tumor Hibernoma	Rhabdomyoblastoma Alveolar soft part sarcoma
Hemangioma	Pyogenic granuloma Lymphedema Arteriovenous malformation Infarct Cystic hygroma Angiofollicular lymphoid hyperplasia Aneurysmal bone cyst	Papillary endothelial hyperplasia Benign schwannoma Angiomyoma Lymphangioma Neurofibroma Hypertrophic hemangioma	Hemangiosarcoma Lymphangiosarcoma Kaposi's sarcoma Telangiectatic osteosarcoma
Benign glomus tumor	Arteriovenous malformation	Hypertrophic hemangioma	Hemangiopericytoma Malignant glomus tumor Leiomyoblastoma
Neurofibroma	Traumatic neuroma Fasciitis Fibromatosis Keloid A-V malformation	Leiomyoma Meningioma Fibroma Nonossifying fibroma Desmoplastic fibroma	Neurofibrosarcoma Leiomyosarcoma Malignant schwannoma
Benign schwannoma	Fasciitis Fibromatosis Proliferative panniculitis Proliferative myositis Periostitis	Benign fibroblastic fibrous histiocytoma Nonossifying fibroma Desmoplastic fibroma Fibroblastic lipoma Leiomyoma Angiomyoma Myxoid neurofibroma Neuroma	Malignant fibroblastic fibrous histiocytoma Fibroblastic fibrosarcoma Monophasic tendosynovial sarcoma, spindle cell type Fibroblastic liposarcoma Leiomyosarcoma Neurofibrosarcoma Malignant fibrous mesothelioma Parosteal sarcoma Kaposi's sarcoma

*See Table 11.

TABLE 85. *Histologic Differential Diagnosis of Benign Neoplasms** (*Continued*)

	SHOULD BE CONSIDERED IN THE DIFFERENTIAL DIAGNOSIS		
	NON-NEOPLASTIC LESIONS	BENIGN NEOPLASMS	MALIGNANT NEOPLASMS
Benign paraganglioma		Blue nevus Plexiform neurofibroma Hemangioblastoma Benign glomus tumor Benign nevoid schwannoma Leiomyoma	Malignant paraganglioma Malignant nevoid schwannoma Malignant neuroepithelioma Malignant glomus tumor Neuroblastoma Clear cell sarcoma Leiomyoblastoma Malignant melanoma
Benign fibrous mesothelioma	Scar Keloid Callus Fibromatosis	Fibroma Leiomyoma Neurofibroma	Desmoid tumor Malignant fibrous mesothelioma
Chondroblastoma	Aneurysmal bone cyst Eosinophilic granuloma Chondroid metaplasia Callus	Benign histiocytic fibrous histiocytoma Chondromyxoid fibroma	Chondrosarcoma Mesenchymal chondrosarcoma Chordoid tumor Chordoma Chondrosarcomatous osteosarcoma Histiocytic lymphoma
Chondromyxoid fibroma	Fibrous dysplasia Callus	Nonossifying fibroma Chondroblastoma Desmoplastic fibroma	Mesenchymal chondrosarcoma Chordoma Chondrosarcomatous osteosarcoma
Chondroma	Chondroid metaplasia Callus Tendosynovial chondromatosis	Osteochondroma Chondromyxoid fibroma Chondroblastoma	Chondrosarcoma Chordoid sarcoma Myxoid liposarcoma Chordoma Chondrosarcomatous osteosarcoma
Osteochondroma	Callus Chondroid metaplasia Myositis ossificans Tendosynovial chondromatosis	Chondroma Ossifying fibroma	Chondrosarcoma Parosteal osteosarcoma Chordoid tumor
Osteoid osteoma	Callus Osteomyelitis Osteoid metaplasia	Osteoblastoma Osteoma Ossifying fibroma	Chondrosarcomatous osteosarcoma Osteosarcoma Parosteal osteosarcoma Paget's sarcoma
Bening granular cell tumor	Xanthogranuloma	Rhabdomyoma Benign nevoid schwannoma	Malignant granular cell tumor Alveolar soft part sarcoma

*See Table 11.

TABLE 86. *Histologic Differential Diagnosis of Malignant Neoplasms**

	SHOULD BE CONSIDERED IN THE DIFFERENTIAL DIAGNOSIS		
	NON-NEOPLASTIC LESIONS	BENIGN NEOPLASMS	MALIGNANT NEOPLASMS
Malignant fibroblastic fibrous histiocytoma	Fasciitis	Benign fibroblastic fibrous histiocytoma Benign schwannoma Nonossifying fibroma Leiomyoma Fibroma	Leiomyosarcoma Kaposi's sarcoma Malignant schwannoma Malignant tendosynovial sarcoma, spindle cell type Parosteal osteosarcoma Fibroblastic liposarcoma Desmoplastic melanoma Desmoplastic carcinoma
Malignant histiocytic fibrous histiocytoma	Giant cell granuloma Tendosynovitis Aneurysmal bone cyst Callus Myositis ossificans	Benign histiocytic fibrous histiocytoma Osteoblastoma	Leiomyoblastoma Malignant nevoid schwannoma Telangiectatic osteosarcoma Fibrous histiocytic osteosarcoma Paget's sarcoma
Malignant pleomorphic fibrous histiocytoma	Xanthogranuloma Lipogranuloma	Benign pleomorphic fibrous histiocytoma Nonossifying fibroma	Pleomorphic rhabdomyosarcoma Hemangiosarcoma Pleomorphic fibrosarcoma Pleomorphic liposarcoma Metastatic carcinoma Leiomyosarcoma Paget's sarcoma Malignant mesothelioma Neurofibrosarcoma Malignant paraganglioma Osteosarcoma
Desmoid tumor	Fibromatosis Scar Collagenoma	Desmoplastic fibroma Nonossifying fibroma Leiomyoma Neuroma Neurofibroma Fibroma Meningioma	Fibrosarcoma Leiomyosarcoma Neurofibrosarcoma
Fibroblastic fibrosarcoma	Fibromatosis Fasciitis	Nonossifying fibroma Fibroma Fibroblastic lipoma Leiomyoma Neurofibroma	Monophasic tendosynovial sarcoma, spindle cell type Desmoid tumor Fibroblastic liposarcoma Leiomyosarcoma Neurofibrosarcoma Malignant cystosarcoma phyllodes Malignant schwannoma Meningeal sarcoma Malignant fibrous mesothelioma Fibrosarcomatous osteosarcoma Parosteal sarcoma
Pleomorphic fibrosarcoma	Fibromatosis Fasciitis Proliferative myositis	Leiomyoma Fibroma Neurofibroma	Leiomyosarcoma Pleomorphic liposarcoma Neurofibrosarcoma Malignant mesothelioma Pleomorphic rhabdomyosarcoma Malignant pleomorphic fibrous histiocytoma Malignant schwannoma Malignant paraganglioma Fibrosarcomatous osteosarcoma Paget's sarcoma

*See Table 12.

Table 86. *Histologic Differential Diagnosis of Malignant Neoplasms** (*Continued*)

	SHOULD BE CONSIDERED IN THE DIFFERENTIAL DIAGNOSIS		
	NON-NEOPLASTIC LESIONS	BENIGN NEOPLASMS	MALIGNANT NEOPLASMS
Biphasic tendosynovial sarcoma	Tendosynovitis	Benign skin adnexal tumor Benign glandular schwannoma Meningioma	Adenocarcinoma Malignant epithelioid mesothelioma Hemangiosarcoma Malignant glandular schwannoma Myxopapillary ependymoma Malignant neuroepithelioma Malignant thymoma
Monophasic tendosynovial sarcoma, spindle cell type	Tendosynovitis	Leiomyoma Benign schwannoma Benign fibroblastic fibrous histiocytoma	Fibroblastic fibrosarcoma Hemangiopericytoma Malignant schwannoma Malignant fibrous mesothelioma Desmoplastic carcinoma Meningeal sarcoma Malignant fibroblastic fibrous histiocytoma
Monophasic tendosynovial sarcoma, pseudoglandular type	Tendosynovial cyst Mesothelial cyst	Skin adnexal tumor Benign glandular schwannoma	Adenocarcinoma Malignant glandular schwannoma Malignant mesothelioma
Epithelioid sarcoma	Fasciitis Fat necrosis Fibromatosis Proliferative panniculitis Foreign body granuloma Granulomatous synovitis Infarct	Benign histiocytic fibrous histiocytoma Benign nevoid schwannoma Benign neuroepithelioma Skin adnexal tumor Benign neuroepithelioma	Leiomyoblastoma Malignant histiocytic fibrous histiocytoma Malignant nevoid schwannoma Lipoblastic liposarcoma Clear cell sarcoma Adenocarcinoma Malignant glomus tumor Malignant neuroepithelioma Malignant melanoma
Clear cell sarcoma	Xanthogranuloma Tendosynovial cyst Histiocytosis	Benign nevoid schwannoma Benign pigmented schwannoma Benign neuroepithelioma Chondroblastoma	Malignant nevoid schwannoma Malignant pigmented schwannoma Malignant neuroepithelioma Leiomyoblastoma Hemangiosarcoma Chondrosarcoma Chordoma Malignant melanoma Clear cell carcinoma
Chordoid sarcoma	Tendosynovial chondromatosis	Chondroma Lipoblastoma Skin adnexal tumor	Chondrosarcoma Chordoma Lipoblastic liposarcoma Adenocarcinoma
Well-differentiated liposarcoma	Fat necrosis Proliferative panniculitis	Lipoma	Pleomorphic liposarcoma
Myxoid liposarcoma	Proliferative panniculitis	Myxoid lipoma Myxoid neurofibroma	Myxoid chondrosarcoma
Lipoblastic liposarcoma	Eosinophilic granuloma Histiocytosis Mastocytosis Plasma cell granuloma	Lipoblastoma Benign undifferentiated peripheral nerve tumor Benign paraganglioma	Embryonal rhabdomyosarcoma Malignant lymphoma Ewing's sarcoma Granulocytic sarcoma Plasmacytoma Plasma cell myeloma Malignant undifferentiated peripheral nerve tumor

*See Table 12.

Table 86. ***Histologic Differential Diagnosis of Malignant Neoplasms**** **(*Continued*)**

	Should be considered in the differential diagnosis		
	Non-neoplastic lesions	**Benign neoplasms**	**Malignant neoplasms**
Fibroblastic liposarcoma	Fasciitis Fibromatosis	Leiomyoma Benign fibroblastic fibrous histiocytoma Benign schwannoma Neurofibroma	Malignant fibroblastic fibrous histiocytoma Fibroblastic fibrosarcoma Hemangiopericytoma Leiomyosarcoma Embryonal rhabdomyosarcoma Malignant schwannoma Malignant cystosarcoma phyllodes
Pleomorphic liposarcoma	Myositis ossificans Proliferative myositis Proliferative panniculitis Xanthogranuloma Fat necrosis Lipodistrophia Elastofibroma Proliferative myositis	Pleomorphic lipoma Benign pleomorphic fibrous histiocytoma Hibernoma Rhabdomyoma Benign paraganglioma	Malignant pleomorphic fibrous histiocytoma Pleomorphic fibrosarcoma Pleomorphic rhabdomyosarcoma Well-differentiated liposarcoma Leiomyosarcoma Malignant schwannoma Fibrous histiocytic osteosarcoma Paget's sarcoma
Leiomyoblastoma	Lipogranuloma Histiocytosis Mastocytosis Plasma cell granuloma	Leiomyoma Lipoblastoma Hibernoma Hemangioblastoma Glomus tumor Benign paraganglioma Benign nevoid schwannoma Chondroblastoma	Adenocarcinoma Malignant nevoid schwannoma Epithelioid sarcoma Clear cell sarcoma Alveolar rhabdomyosarcoma Malignant glomus tumor Hemangiopericytoma Pleomorphic liposarcoma Clear cell carcinoma Malignant melanoma
Leiomyosarcoma	Traumatic neuroma Fasciitis Fibromatosis	Leiomyoma Meningioma Neurofibroma Fibroma Nonossifying fibroma Desmoplastic fibroma	Desmoid tumor Fibrosarcoma Malignant fibrous histiocytoma Malignant schwannoma Pleomorphic liposarcoma Pleomorphic rhabdomyosarcoma Meningeal sarcoma Neurofibrosarcoma Malignant fibrous mesothelioma
Embryonal rhabdomyosarcoma	Fasciitis Fibromatosis	Nonossifying fibroma Fibroblastic lipoma Benign nevoid schwannoma	Ewing's sarcoma Granulocytic sarcoma Neuroblastoma Primitive neuroectodermal tumor Small cell carcinoma Malignant teratoma
Alveolar rhabdomyosarcoma	Tendosynovitis Pyogenic granuloma	Benign histiocytic fibrous histiocytoma Benign glomus tumor Benign neuroepithelioma	Ewing's sarcoma Neuroblastoma Primitive neuroectodermal tumor Malignant glomus tumor Hemangiopericytoma Small cell carcinoma Carcinoid
Myxoid rhabdomyosarcoma	Lymphedema Fasciitis Proliferative myositis	Myxoid lipoma Myxoid neurofibroma Benign schwannoma Angiofibroma	Myxoid liposarcoma Malignant schwannoma Malignant fibroblastic fibrous histiocytoma Malignant Triton tumor

*See Table 12.

TABLE 86. *Histologic Differential Diagnosis of Malignant Neoplasms** (*Continued*)

	SHOULD BE CONSIDERED IN THE DIFFERENTIAL DIAGNOSIS		
	NON-NEOPLASTIC LESIONS	BENIGN NEOPLASMS	MALIGNANT NEOPLASMS
Rhabdomyoblastoma		Rhabdomyoma Ganglioneuroma Hibernoma Oncocytic neoplasms	Pleomorphic rhabdomyosarcoma Malignant melanoma Oncocytic neoplasms
Pleomorphic rhabdomyosarcoma	Fat necrosis Proliferative myositis Myositis ossificans	Benign pleomorphic fibrous histiocytoma Pleomorphic lipoma Rhabdomyoma Benign Triton tumor	Malignant pleomorphic fibrous histiocytoma Pleomorphic fibrosarcoma Pleomorphic liposarcoma Leiomyosarcoma Malignant teratoma Osteosarcoma Malignant melanoma
Hemangiopericytoma	Aneurysmal bone cyst Tendosynovitis Pyogenic granuloma Angiofollicular lymphoid hyperplasia Angiomatous lymphoid hamartoma	Benign glomus tumor Hypertrophic hemangioma Neurofibroma Benign schwannoma Meningioma	Monophasic tendosynovial sarcoma, spindle cell type Malignant schwannoma Embryonal rhabdomyosarcoma Malignant paraganglioma Meningeal sarcoma Malignant glomus tumor
Hemangiosarcoma	Pyogenic granuloma Angiofollicular lymphoid hyperplasia Aneurysmal bone cyst Arteriovenous malformation	Hypertrophic hemangioma Angiomyoma Papillary endothelial hyperplasia Hemangioblastoma	Malignant pleomorphic fibrous histiocytoma Lymphangiosarcoma Kaposi's sarcoma Angioendotheliomatosis Malignant glomus tumor Adenocarcinoma
Lymphangiosarcoma	Lymphedema Angiofollicular lymphoid hyperplasia	Lymphangioma Lymphangiomyoma	Hemangiosarcoma Kaposi's sarcoma Angioendotheliomatosis Adenocarcinoma
Malignant glomus tumor	Arteriovenous malformation Vasculitis Telangiectasia	Benign glomus tumor Hemangioblastoma Hypertrophic hemangioma	Leiomyoblastoma Lipoblastic liposarcoma Primitive neuroectodermal tumor Hemangiopericytoma
Neurofibrosarcoma	Fasciitis Fibromatosis Proliferative panniculitis Hypertrophy of nerve	Neurofibroma Leiomyoma Nonossifying fibroma Desmoplastic fibroma Fibroblastic lipoma Neuroma	Leiomyosarcoma Fibrosarcoma Monophasic tendosynovial sarcoma, spindle cell type Malignant fibroblastic fibrous histiocytoma Fibroblastic liposarcoma Malignant schwannoma Parosteal osteosarcoma Meningeal sarcoma Malignant fibrous mesothelioma Hemangiopericytoma

*See Table 12.

TABLE 86. ***Histologic Differential Diagnosis of Malignant Neoplasms**** **(*Continued*)**

	SHOULD BE CONSIDERED IN THE DIFFERENTIAL DIAGNOSIS		
	NON-NEOPLASTIC LESIONS	BENIGN NEOPLASMS	MALIGNANT NEOPLASMS
Malignant schwannoma	Fasciitis Proliferative panniculitis Proliferative myositis	Benign schwannoma Leiomyoma Nonossifying fibroma Fibroblastic lipoma	Monophasic tendosynovial sarcoma, spindle cell type Hemangiopericytoma Malignant fibroblastic fibrous histiocytoma Fibroblastic liposarcoma Fibrosarcoma Malignant mesothelioma Neurofibrosarcoma Meningeal sarcoma Malignant thymoma Leiomyosarcoma
Malignant undifferentiated peripheral nerve tumor		Benign schwannoma Hemangioblastoma	Neuroblastoma Embryonal rhabdomyosarcoma Ewing's sarcoma Granulocytic sarcoma Lipoblastic liposarcoma Hemangiopericytoma Leiomyoblastoma Malignant glomus tumor
Malignant paraganglioma	Xanthogranuloma Lipogranuloma	Benign paraganglioma Leiomyoma Plexiform neurofibroma Angiomyoma	Leiomyosarcoma Malignant schwannoma Hemangiopericytoma Desmoplastic malignant melanoma
Malignant epithelioid mesothelioma	Mesothelial hyperplasia Mesothelial cyst	Benign epithelioid mesothelioma Adenomatoid tumor	Adenocarcinoma Biphasic tendosynovial sarcoma Hemangiosarcoma
Malignant fibrous mesothelioma	Fibromatosis Collagenoma	Benign fibrous mesothelioma Leiomyoma Leiomyomatosis	Fibrosarcoma Neurofibrosarcoma Leiomyosarcoma
Well-differentiated chondrosarcoma	Chondroid metaplasia Tendosynovial chondromatosis Callus	Chondroma Osteochondroma Chondromyxoid fibroma Chondroblastoma	Chordoma Chordoid sarcoma Mesenchymal chondrosarcoma Osteosarcoma
Poorly differentiated chondrosarcoma	Chondroid metaplasia Tendosynovial chondromatosis	Chondroblastoma Lipoblastoma	Chordoma Chordoid sarcoma Mesenchymal chondrosarcoma Osteosarcoma Lipoblastic liposarcoma
Myxoid chondrosarcoma	Chondroid metaplasia Callus	Myxoid lipoma Chondromyxoid fibroma	Chordoid sarcoma Myxoid liposarcoma Embryonal rhabdomyosarcoma Mesenchymal chondrosarcoma
Mesenchymal chondrosarcoma	Chondroid metaplasia Tendosynovial chondromatosis Callus	Chondromyxoid fibroma Benign mesenchymoma Benign mixed tumor Osteochondroma	Ewing's sarcoma Osteosarcoma Hemangiopericytoma Malignant mixed tumor Malignant mesenchymoma

*See Table 12.

TABLE 86. *Histologic Differential Diagnosis of Malignant Neoplasms** (*Continued*)

	SHOULD BE CONSIDERED IN THE DIFFERENTIAL DIAGNOSIS		
	NON-NEOPLASTIC LESIONS	BENIGN NEOPLASMS	MALIGNANT NEOPLASMS
Osteosarcoma	Fasciitis ossificans Fibromatosis ossificans Callus Panniculitis ossificans Myositis ossificans Enostosis Osteoid metaplasia	Osteoid osteoma Osteoblastoma Ossifying fibroma Osteoma	Parosteal osteosarcoma Pleomorphic fibrosarcoma Malignant fibrous histiocytoma Chondrosarcoma Paget's sarcoma
Parosteal osteosarcoma	Callus Osteoid metaplasia Fasciitis ossificans Fibromatosis ossificans	Fibroma Leiomyoma Benign fibroblastic fibrous histiocytoma Nonossifying fibroma Ossifying fibroma	Osteosarcoma Malignant fibroblastic fibrous histiocytoma Desmoid tumor Leiomyosarcoma Fibroblastic fibrosarcoma
Paget's sarcoma	Callus Myositis ossificans Giant cell granuloma	Pleomorphic neoplasm (any type)	Osteosarcoma Chondrosarcoma Pleomorphic sarcoma (any type)
Granulocytic sarcoma	Pyogenic granuloma Eosinophilic granuloma	Benign neuroepithelioma	Malignant lymphoma Embryonal rhabdomyosarcoma Ewing's sarcoma Primitive neuroectodermal tumor
Plasma cell myeloma	Plasma cell granuloma Eosinophilic granuloma		Plasmacytoma Malignant lymphoma
Malignant granular cell tumor	Xanthogranuloma Gaucher's disease	Rhabdomyoma Hibernoma Benign granular cell tumor Oncocytoma	Alveolar soft part sarcoma Malignant paraganglioma Oncocytoma Malignant melanoma
Alveolar soft part sarcoma		Rhabdomyoma Hibernoma Oncocytoma	Malignant granular cell tumor Rhabdomyoblastoma Leiomyoblastoma Malignant melanoma Malignant paraganglioma
Kaposi's sarcoma	Pyogenic granuloma Angiofollicular lymphoid hyperplasia Fasciitis	Hemangioma Angiomyoma Blue nevus Leiomyoma	Leiomyosarcoma Hemangiosarcoma Lymphangiosarcoma
Ewing's sarcoma	Osteomyelitis Eosinophilic granuloma		Embryonal rhabdomyosarcoma Neuroblastoma Lipoblastic liposarcoma Granulocytic sarcoma Primitive neuroectodermal tumor Osteosarcoma
Chordoma	Callus Chondroid metaplasia	Chondroma Chondroblastoma Benign mixed tumor Echordosis physaliphora	Chordoid sarcoma Chondrosarcoma Malignant mixed tumor Malignant mesenchymoma Mucinous adenocarcinoma Myxopapillary ependymoma
Myxopapillary ependymoma	Mesothelial hyperplasia	Benign epithelioid mesothelioma	Chordoma Liposarcoma Biophasic tendosynovial sarcoma Adenocarcinoma Malignant mesothelioma
Adamantinoma	Osteomyelitis	Hypertrophic hemangioma	Basal cell carcinoma Metastatic carcinoma

*See Table 12.

TABLE 87. ***Differential Diagnosis of Soft Tissue and Bone Tumors with Arranged Pattern****

APPEARANCE OF STROMA	CELL MORPHOLOGY					
	SLENDER SPINDLE	PLUMP SPINDLE	GRANULAR EPITHELIOID	CLEAR EPITHELIOID	ISOMORPHIC GIANT	PLEOMORPHIC GIANT
Fibrillar	Fasciitis Xanthogranuloma Traumatic neuroma Benign fibroblastic fibrous histiocytoma Nonossifying fibroma Ossifying fibroma Panniculitis Fibroblastic lipoma Pleomorphic lipoma Neurofibroma Benign schwannoma Neuroma Ganglioneuroma Benign paraganglioma Meningioma Malignant fibroblastic fibrous histiocytoma Malignant histiocytic fibrous histiocytoma Malignant pleomorphic fibrous histiocytoma Myxoid liposarcoma Fibroblastic liposarcoma Neurofibrosarcoma Malignant schwannoma Malignant neuroepithelioma Mesenchymal chondrosarcoma Parosteal osteosarcoma Meningeal sarcoma	Fasciitis Panniculitis Fibroblastic lipoma Leiomyoma Angiomyoma Neurofibroma Neuroma Ganglioneuroma Meningioma Malignant histiocytic fibrous histiocytoma Malignant pleomorphic fibrous histiocytoma Fibroblastic fibrosarcoma Leiomyosarcoma Neurofibrosarcoma Malignant schwannoma Malignant fibrous mesothelioma Parosteal osteosarcoma Meningeal sarcoma	Benign schwannoma Benign paraganglioma Meningioma Malignant histiocytic fibrous histiocytoma Myxoid liposarcoma Malignant schwannoma Malignant neuroepithe-lioma Malignant paraganglioma Meningeal sarcoma	Xanthogranuloma Nonossifying fibroma Benign schwannoma Benign paraganglioma Malignant pleomorphic fibrous histiocytoma Leiomyosarcoma Malignant paraganglioma	Xanthogranuloma Benign histiocytic fibrous histiocytoma Nonossifying fibroma Ganglioneuroma Benign paraganglioma Malignant histiocytic fibrous histiocytoma	Pleomorphic lipoma Malignant pleomorphic fibrous histiocytoma Leiomyosarcoma Neurofibrosarcoma Malignant schwannoma Malignant paraganglioma Meningeal sarcoma

TABLE 87. *Differential Diagnosis of Soft Tissue and Bone Tumors with Arranged Pattern** (*Continued*)

APPEARANCE OF STROMA	CELL MORPHOLOGY					
	SLENDER SPINDLE	PLUMP SPINDLE	GRANULAR EPITHELIOID	CLEAR EPITHELIOID	ISOMORPHIC GIANT	PLEOMORPHIC GIANT
Sclerosed	Benign fibroblastic fibrous histiocytoma Ossifying fibroma Panniculitis Traumatic neuroma Fibroblastic lipoma Neurofibroma Benign schwannoma Neuroma Benign schwannoma Neuroma Ganglioneuroma Malignant pleomorphic fibrous histiocytoma Fibroblastic liposarcoma Neurofibrosarcoma Malignant schwannoma Parosteal osteosarcoma	Panniculitis Traumatic neuroma Ossifying fibroma Fibroblastic lipoma Leiomyoma Angiomyoma Neurofibroma Neuroma Meningioma Fibroblastic fibrosarcoma Leiomyosarcoma Neurofibrosarcoma Parosteal osteosarcoma Meningeal sarcoma	Meningioma Benign schwannoma Malignant schwannoma	Benign paraganglioma Nonossifying fibroma	Ganglioneuroma Malignant histiocytic fibrous histiocytoma	Malignant pleomorphic fibrous histiocytoma Leiomyosarcoma Neurofibrosarcoma Malignant schwannoma Meningeal sarcoma
Myxoid	Fasciitis Nonossifying fibroma Fibroblastic lipoma Pleomorphic lipoma Neurofibroma Benign schwannoma Neuroma Ganglioneuroma Malignant fibroblastic fibrous histiocytoma Fibroblastic liposarcoma Neurofibrosarcoma Malignant schwannoma Myxoid liposarcoma Malignant neuroepithelioma Mesenchymal chondrosarcoma Parosteal osteosarcoma	Fasciitis Fibroblastic lipoma Leiomyoma Neurofibroma Neuroma Malignant histiocytic fibrous histiocytoma Leiomyosarcoma Neurofibrosarcoma Malignant schwannoma	Benign schwannoma Meningioma Malignant histiocytic fibrous histiocytoma Myxoid liposarcoma Malignant schwannoma Malignant neuroepithe-lioma	Nonossifying fibroma Benign schwannoma	Nonossifying fibroma Ganglioneuroma Malignant histiocytic fibrous histiocytoma	Pleomorphic lipoma Malignant pleomorphic fibrous histiocytoma Leiomyosarcoma Neurofibrosarcoma Malignant schwannoma

Table 87. *Differential Diagnosis of Soft Tissue and Bone Tumors with Arranged Pattern** (*Continued*)

APPEARANCE OF STROMA	CELL MORPHOLOGY					
	SLENDER SPINDLE	PLUMP SPINDLE	GRANULAR EPITHELIOID	CLEAR EPITHELIOID	ISOMORPHIC GIANT	PLEOMORPHIC GIANT
Vascular	Fasciitis Benign fibroblastic fibrous hisitocytoma Benign schwannoma Fibroblastic liposarcoma Malignant schwannoma Malignant neuroepithelioma	Fasciitis Panniculitis Angiomyoma Neurofibroma Benign schwannoma Malignant histiocytic fibrous histiocytoma Leiomyosarcoma Meningioma	Benign schwannoma Benign paraganglioma Malignant histiocytic fibrous histiocytoma Myxoid liposarcoma Malignant neuroepithelioma Malignant paraganglioma	Benign schwannoma Benign paraganglioma Malignant paraganglioma	Malignant histiocytic fibrous histiocytoma	Malignant pleomorphic fibrous histiocytoma Leiomyosarcoma Malignant schwannoma Malignant paraganglioma
Inflamed	Fasciitis Benign schwannoma Malignant fibroblastic fibrous histiocytoma Parosteal osteosarcoma Meningeal sarcoma	Fasciitis Panniculitis Malignant histiocytic fibrous histiocytoma		Xanthogranuloma	Xanthogranuloma	Malignant pleomorphic fibrous histiocytoma
Necrotic	Pleomorphic lipoma Benign schwannoma Malignant histiocytic fibrous histiocytoma Fibroblastic liposarcoma Malignant schwannoma Parosteal osteosarcoma	Panniculitis Leiomyoma Malignant histiocytic fibrous histiocytoma Leiomyosarcoma	Myxoid liposarcoma Benign schwannoma Benign paraganglioma Malignant schwannoma	Benign schwannoma Benign paraganglioma	Xanthogranuloma	Pleomorphic lipoma Leiomyosarcoma Neurofibrosarcoma Malignant schwannoma Malignant paraganglioma Meningeal sarcoma
Chondrified	Benign schwannoma Malignant schwannoma Mesenchymal chondrosarcoma	Panniculitis Meningioma Meningeal sarcoma	Meningioma Myxoid liposarcoma			Pleomorphic lipoma

Table 87. *Differential Diagnosis of Soft Tissue and Bone Tumors with Arranged Pattern** (*Continued*)

APPEARANCE OF STROMA	CELL MORPHOLOGY					
	SLENDER SPINDLE	PLUMP SPINDLE	GRANULAR EPITHELIOID	CLEAR EPITHELIOID	ISOMORPHIC GIANT	PLEOMORPHIC GIANT
Ossified	Fasciitis Parosteal osteosarcoma Ossifying fibroma	Fasciitis Panniculitis Ossifying fibroma Meningioma Parosteal osteosarcoma Meningeal sarcoma	Meningioma		Ossifying fibroma	
Calcified	Benign schwannoma Neuroma Ganglioneuroma Meningioma Malignant schwannoma Parosteal osteosarcoma	Panniculitis Leiomyoma Angiomyoma Neurofibroma Neuroma Ganglioneuroma Meningioma Malignant histiocytic fibrous histiocytoma Leiomyosarcoma Parosteal osteosarcoma Meningeal sarcoma	Meningioma Malignant paraganglioma	Malignant paraganglioma	Ganglioneuroma	Leiomyosarcoma Malignant schwannoma Malignant paraganglioma

*See also Tables 48–56, 58–60, and 88–92.

TABLE 88. ***Differential Diagnosis of Soft Tissue and Bone Tumors with Spreading Pattern****

APPEARANCE OF STROMA	CELL MORPHOLOGY					
	SLENDER SPINDLE	PLUMP SPINDLE	GRANULAR EPITHELIOID	CLEAR EPITHELIOID	ISOMORPHIC GIANT	PLEOMORPHIC GIANT
Fibrillar	Fasciitis Fibromatosis Fibrous dysplasia Panniculitis Neurofibroma Benign schwannoma Neuroma Ganglioneuroma Benign cystosarcoma phyllodes Monophasic tendosynovial sarcoma, spindle cell type Embryonal rhabdomyosarcoma Hemangiopericytoma Neurofibrosarcoma Malignant schwannoma Mesenchymal chondrosarcoma Parosteal osteosarcoma Kaposi's sarcoma Meningeal sarcoma Malignant cystosarcoma phyllodes	Fasciitis Fibromatosis Fibrous dysplasia Panniculitis Desmoplastic fibroma Leiomyoma Angiomyoma Neurofibroma Benign schwannoma Neuroma Ganglioneuroma Desmoid tumor Fibroblastic fibrosarcoma Pleomorphic fibrosarcoma Pleomorphic liposarcoma Leiomyosarcoma Pleomorphic rhabdomyosarcoma Neurofibrosarcoma Malignant fibrous mesothelioma Mesenchymal chondrosarcoma Osteosarcoma Parosteal osteosarcoma Paget's sarcoma Kaposi's sarcoma Meningeal sarcoma Malignant cystosarcoma phyllodes	Malignant schwannoma Mesenchymal chondro-sarcoma	Malignant schwannoma	Ganglioneuroma Osteosarcoma	Leiomyoma Neurofibroma Benign schwannoma Pleomorphic fibrosarcoma Pleomorphic liposarcoma Leiomyosarcoma Pleomorphic rhabdomyosarcoma Neurofibrosarcoma Malignant schwannoma Malignant fibrous mesothelioma Osteosarcoma

Table 88. *Differential Diagnosis of Soft Tissue and Bone Tumors with Spreading Pattern** (*Continued*)

APPEARANCE OF STROMA	CELL MORPHOLOGY					
	SLENDER SPINDLE	PLUMP SPINDLE	GRANULAR EPITHELIOID	CLEAR EPITHELIOID	ISOMORPHIC GIANT	PLEOMORPHIC GIANT
Sclerosed	Fasciitis Panniculitis Neurofibroma Benign schwannoma Neuroma Ganglioneuroma Benign cystosarcoma phyllodes Monophasic tendosynovial sarcoma, spindle cell type Neurofibrosarcoma Parosteal osteosarcoma Meningeal sarcoma	Fasciitis Keloid Scar Fibromatosis Fibrous dysplasia Collagenoma Panniculitis Myositis ossificans Fibroma Desmoplastic fibroma Ossifying fibroma Leiomyoma Angiofibroma Angiomyoma Neurofibroma Neuroma Ganglioneuroma Benign fibrous mesothelioma Desmoid tumor Fibroblastic fibrosarcoma Pleomorphic fibrosarcoma Pleomorphic liposarcoma Leiomyosarcoma Neurofibrosarcoma Malignant fibrous mesothelioma Mesenchymal chondrosarcoma Osteosarcoma Parosteal osteosarcoma Paget's sarcoma Meningeal sarcoma	Myositis ossificans		Myositis ossificans Ganglioneuroma Osteosarcoma	Leiomyoma Neurofibroma Pleomorphic fibrosarcoma Pleomorphic liposarcoma Leiomyosarcoma Neurofibrosarcoma Malignant fibrous mesothelioma Osteosarcoma Paget's sarcoma

Table 88. *Differential Diagnosis of Soft Tissue and Bone Tumors with Spreading Pattern** (*Continued*)

APPEARANCE OF STROMA	CELL MORPHOLOGY					
	SLENDER SPINDLE	PLUMP SPINDLE	GRANULAR EPITHELIOID	CLEAR EPITHELIOID	ISOMORPHIC GIANT	PLEOMORPHIC GIANT
Myxoid	Fasciitis Fibromatosis Panniculitis Neurofibroma Benign schwannoma Neuroma Benign cystosarcoma phyllodes Monophasic tendosynovial sarcoma, spindle cell type Embryonal rhabdomyosarcoma Malignant schwannoma Mesenchymal chondrosarcoma Parosteal osteosarcoma Kaposi's sarcoma Malignant cystosarcoma phyllodes	Fasciitis Panniculitis Myositis ossificans Leiomyoma Angiomyoma Neurofibroma Benign schwannoma Pleomorphic liposarcoma Pleomorphic rhabdomyosarcoma Malignant fibrous mesothelioma Parosteal osteosarcoma Malignant cystosarcoma phyllodes	Myositis ossificans Malignant schwannoma Mesenchymal chondro-sarcoma		Myositis ossificans	Leiomyoma Neurofibroma Benign schwannoma Pleomorphic liposarcoma Pleomorphic rhabdomyo-sarcoma Malignant schwannoma Malignant fibrous mesothelioma
Vascular	Fasciitis Fibromatosis Panniculitis Neurofibroma Benign schwannoma Monophasic tendosynovial sarcoma, spindle cell type Embryonal rhabdomyosarcoma Hemangiopericytoma Malignant schwannoma Kaposi's sarcoma	Fasciitis Panniculitis Myositis ossificans Leiomyoma Angiofibroma Angiomyoma Neurofibroma Pleomorphic liposarcoma Leiomyosarcoma Kaposi's sarcoma	Myositis ossificans Malignant schwannoma		Myositis ossificans	Leiomyoma Pleomorphic liposarcoma Leiomyosarcoma Malignant schwannoma
Inflamed	Fasciitis Fibromatosis Panniculitis Benign schwannoma Monophasic tendosynovial sarcoma, spindle cell type Malignant schwannoma Kaposi's sarcoma	Fasciitis Keloid Panniculitis Myositis ossificans Angiomyoma Benign fibrous mesothelioma Pleomorphic liposarcoma Malignant fibrous mesothelioma Kaposi's sarcoma	Myositis ossificans		Myositis ossificans	Pleomorphic liposarcoma Malignant fibrous mesothelioma

TABLE 88. ***Differential Diagnosis of Soft Tissue and Bone Tumors with Spreading Pattern**** **(*Continued*)**

APPEARANCE OF STROMA	CELL MORPHOLOGY					
	SLENDER SPINDLE	PLUMP SPINDLE	GRANULAR EPITHELIOID	CLEAR EPITHELIOID	ISOMORPHIC GIANT	PLEOMORPHIC GIANT
Necrotic	Fasciitis Panniculitis Neurofibroma Benign schwannoma Monophasic tendosynovial sarcoma, spindle cell type Embryonal rhabdomyosarcoma Neurofibrosarcoma Malignant schwannoma Kaposi's sarcoma	Panniculitis Myositis ossificans Leiomyoma Neurofibroma Benign schwannoma Pleomorphic fibrosarcoma Pleomorphic liposarcoma Leiomyosarcoma Pleomorphic rhabdomyo-sarcoma Neurofibrosarcoma Malignant fibrous mesothelioma Osteosarcoma Kaposi's sarcoma	Myositis ossificans		Myositis ossificans Osteosarcoma	Leiomyoma Neurofibroma Benign schwannoma Pleomorphic fibrosarcoma Pleomorphic liposarcoma Leiomyosarcoma Pleomorphic rhabdomyo-sarcoma Neurofibrosarcoma Malignant schwannoma Malignant fibrous mesothelioma Osteosarcoma Paget's sarcoma
Chondrified	Panniculitis Mesenchymal chondrosarcoma Parosteal osteosarcoma	Scar Panniculitis Osteosarcoma Parosteal osteosarcoma	Mesenchymal chondro-sarcoma		Osteosarcoma	Osteosarcoma Paget's sarcoma
Ossified	Neuroma Benign cystosarcoma phyllodes Parosteal osteosarcoma	Fasciitis Scar Fibromatosis Panniculitis Myositic ossificans Fibroma Ossifying fibroma Leiomyoma Neuroma Ganglioneuroma Osteosarcoma Parosteal osteosarcoma	Myositis ossificans		Myositis ossificans Ganglioneuroma Osteosarcoma	Osteosarcoma Paget's sarcoma

TABLE 88. *Differential Diagnosis of Soft Tissue and Bone Tumors with Spreading Pattern* (Continued)*

APPEARANCE OF STROMA	CELL MORPHOLOGY					
	SLENDER SPINDLE	PLUMP SPINDLE	GRANULAR EPITHELIOID	CLEAR EPITHELIOID	ISOMORPHIC GIANT	PLEOMORPHIC GIANT
Calcified	Panniculitis Neurofibroma Benign schwannoma Neuroma Ganglioneuroma Benign cystosarcoma phyllodes Monophasic tendosynovial sarcoma, spindle cell type Mesenchymal chondrosarcoma Parosteal osteosarcoma	Fasciitis Keloid Scar Fibromatosis Fibrous dysplasia Collagenoma Panniculitis Myositis ossificans Fibroma Desmoplastic fibroma Ossifying fibroma Leiomyoma Angiofibroma Angiomyoma Neurofibroma Benign schwannoma Neuroma Ganglioneuroma Benign fibrous mesothelioma Pleomorphic liposarcoma Leiomyosarcoma Malignant fibrous mesothelioma Osteosarcoma Parosteal osteosarcoma Meningeal sarcoma	Myositis ossificans		Myositis ossificans Ganglioneuroma Osteosarcoma	Leiomyoma Neurofibroma Benign schwannoma Pleomorphic liposarcoma Leiomyosarcoma Osteosarcoma Paget's sarcoma

*See also Tables 48–56, 58–60, 87, and 89–92.

Table 89. *Differential Diagnosis of Soft Tissue and Bone Tumors with Lacy Pattern**

APPEARANCE OF STROMA	CELL MORPHOLOGY					
	SLENDER SPINDLE	PLUMP SPINDLE	GRANULAR EPITHELIOID	CLEAR EPITHELIOID	ISOMORPHIC GIANT	PLEOMORPHIC GIANT
Fibrillar	Elastofibroma Chondromyxoid fibroma Fibroblastic liposarcoma Myxoid rhabdomyosarcoma Mesenchymal chondrosarcoma	Elastofibroma	Xanthogranuloma Tendosynovitis Tendosynovial cyst Lipogranuloma Eosinophilic granuloma Histiocytosis Mastocytosis Plasma cell granuloma Gaucher's disease Benign histiocytic fibrous histiocytoma Hibernoma Hypertrophic hemangioma Hemangioblastoma Benign glomus tumor Benign nevoid schwannoma Benign neuroepithelioma Benign epithelioid mesothelioma Chondroblastoma Osteoblastoma Chordoid sarcoma Lipoblastic liposarcoma Leiomyoblastoma Malignant glomus tumor Malignant nevoid schwannoma Malignant neuroepithelioma	Xanthogranuloma Tendosynovial cyst Gouty arthritis Fat necrosis Chondroid metaplasia Well-differentiated lipoma Myxoid lipoma Fibroblastic lipoma Pleomorphic lipoma Benign nevoid schwannoma Chondromyxoid fibroma Chondroma Osteochondroma Clear cell sarcoma Well-differentiated liposarcoma Myxoid liposarcoma Chondrosarcoma Mesenchymal chondrosarcoma Osteosarcoma Chordoma	Tendosynovitis Benign histiocytic fibrous histiocytoma	Gouty arthritis Fat necrosis Pleomorphic lipoma Hibernoma Pleomorphic liposarcoma Leiomyoblastoma Osteosarcoma
Sclerosed	Elastofibroma Chondromyxoid fibroma	Elastofibroma	Xanthogranuloma Lipogranuloma Lipoblastoma Benign nevoid schwannoma Benign neuroepithelioma Chondroblastoma Osteoblastoma	Fibroblastic lipoma Lipoblastoma Pleomorphic lipoma Chondromyxoid fibroma Osteochondroma Clear cell sarcoma Osteosarcoma		Fat necrosis Pleomorphic lipoma Pleomorphic liposarcoma Osteosarcoma

TABLE 89. *Differential Diagnosis of Soft Tissue and Bone Tumors with Lacy Pattern** (*Continued*)

APPEARANCE OF STROMA	CELL MORPHOLOGY					
	SLENDER SPINDLE	PLUMP SPINDLE	GRANULAR EPITHELIOID	CLEAR EPITHELIOID	ISOMORPHIC GIANT	PLEOMORPHIC GIANT
Myxoid	Elastofibroma Chondromyxoid fibroma Fibroblastic liposarcoma Myxoid rhabdomyosarcoma Mesenchymal chondrosarcoma	Elastofibroma	Xanthogranuloma Tendosynovitis Tendosynovial cyst Eosinophilic granuloma Mastocytosis Gaucher's disease Benign histiocytic fibrous histiocytoma Lipoblastoma Hibernoma Hemangioblastoma Benign glomus tumor Benign nevoid schwannoma Benign neuroepithelioma Benign epithelioid mesothelioma Chondroblastoma Chordoid sarcoma Lipoblastic liposarcoma	Xanthogranuloma Tendosynovial cyst Gouty arthritis Fat necrosis Chondroid metaplasia Well-differentiated lipoma Myxoid lipoma Fibroblastic lipoma Lipoblastoma Pleomorphic lipoma Benign nevoid schwannoma Chondromyxoid fibroma Chondroma Osteochondroma Clear cell sarcoma Well-differentiated liposarcoma Myxoid liposarcoma Chondrosarcoma Mesenchymal chondrosarcoma Chordoma	Tendosynovitis Benign histiocytic fibrous histiocytoma	Gouty arthritis Fat necrosis Pleomorphic lipoma Hibernoma Pleomorphic liposarcoma
Vascular	Fibroblastic liposarcoma Myxoid rhabdomyosarcoma		Xanthogranuloma Tendosynovitis Tendosynovial cyst Lipogranuloma Eosinophilic granuloma Histiocytosis Mastocytosis Plasma cell granuloma Benign histiocytic fibrous histiocytoma Lipoblastoma Hypertrophic hemangioma Hemangioblastoma Benign glomus tumor Benign nevoid schwannoma Benign neuroepithelioma Chondroblastoma Chordoid sarcoma Lipoblastic liposarcoma Leiomyoblastoma Malignant glomus tumor	Xanthogranuloma Tendosynovial cyst Gouty arthritis Fat necrosis Lipoblastoma Clear cell sarcoma Well-differentiated liposarcoma Myxoid liposarcoma Chordoma	Tendosynovitis Benign histiocytic fibrous histiocytoma	Pleomorphic liposarcoma

TABLE 89. *Differential Diagnosis of Soft Tissue and Bone Tumors with Lacy Pattern** (*Continued*)

APPEARANCE OF STROMA	CELL MORPHOLOGY					
	SLENDER SPINDLE	PLUMP SPINDLE	GRANULAR EPITHELIOID	CLEAR EPITHELIOID	ISOMORPHIC GIANT	PLEOMORPHIC GIANT
Inflamed	Elastofibroma Myxoid rhabdomyosarcoma		Xanthogranuloma Tendosynovitis Tendosynovial cyst Lipogranuloma Eosinophilic granuloma Histiocytosis Mastocytosis Plasma cell granuloma Hibernoma	Xanthogranuloma Tendosynovial cyst Gouty arthritis Fat necrosis	Tendosynovitis	Gouty arthritis Fat necrosis
Necrotic	Elastofibroma Myxoid rhabdomyosarcoma		Xanthogranuloma Tendosynovitis Tendosynovial cyst Lipogranuloma Eosinophilic granuloma Mastocytosis Gaucher's disease Benign histiocytic fibrous histiocytoma Benign neuroepithelioma Leiomyoblastoma	Xanthogranuloma Gouty arthritis Fat necrosis Myxoid lipoma Pleomorphic lipoma Clear cell sarcoma Chondrosarcoma Osteosarcoma Chordoma	Tendosynovitis Benign histiocytic fibrous histiocytoma	Gouty arthritis Fat necrosis Pleomorphic lipoma Pleomorphic liposarcoma Leiomyoblastoma Osteosarcoma
Chondrified	Chondromyxoid fibroma Mesenchymal chondrosarcoma		Chondroblastoma	Chondroid metaplasia Myxoid lipoma Chondromyxoid fibroma Chondroma Osteochondroma Chondrosarcoma Osteosarcoma Chordoma		Osteosarcoma
Ossified			Osteoblastoma	Chondroid metaplasia Osteochondroma Osteosarcoma		Osteosarcoma
Calcified	Elastofibroma Chondromyxoid fibroma		Xanthogranuloma Tendosynovitis Tendosynovial cyst Lipogranuloma Benign histiocytic fibrous histiocytoma Benign epithelioid mesothelioma Chondroblastoma Osteoblastoma	Xanthogranuloma Tendosynovial cyst Gouty arthritis Fat necrosis Chondroid metaplasia Well-differentiated lipoma Myxoid lipoma Pleomorphic lipoma Chondromyxoid fibroma Chondroma Osteochondroma Chondrosarcoma Osteosarcoma Chordoma	Tendosynovitis Benign histiocytic fibrous histiocytoma	Gouty arthritis Fat necrosis Osteosarcoma

*See also Tables 48–56, 58–60, 87, 88, and 90–92.

Table 90. ***Differential Diagnosis of Soft Tissue and Bone Tumors with Epithelioid Pattern****

APPEARANCE OF STROMA	CELL MORPHOLOGY					
	SLENDER SPINDLE	PLUMP SPINDLE	GRANULAR EPITHELIOID	CLEAR EPITHELIOID	ISOMORPHIC GIANT	PLEOMORPHIC GIANT
Fibrillar	Panniculitis Chondromyxoid fibroma		Aneurysmal bone cyst Giant cell granuloma Hyperparathyroidism Tendosynovial cyst Pyogenic granuloma Mesothelial hyperplasia Eosinophilic granuloma Histiocytosis Benign histiocytic fibrous histiocytoma Lipoblastoma Hibernoma Rhabdomyoma Hemangioblastoma Papillary endothelial hyperplasia Benign glomus tumor Benign nevoid schwannoma Benign neuroepithelioma Benign paraganglioma Benign epithelioid mesothelioma Adenomatoid tumor Chondroblastoma Osteoblastoma Benign granular cell tumor Meningioma Malignant histiocytic fibrous histiocytoma Epithelioid sarcoma Lipoblastic liposarcoma Leiomyoblastoma Embryonal rhabdomyosarcoma Rhabdomyoblastoma Hemangiosarcoma Lymphangiosarcoma Malignant glomus tumor Angioendotheliomatosis Malignant nevoid schwannoma Malignant neuroepithelioma Malignant epithelioid mesothelioma Granulocytic sarcoma Plasmacytoma Plasma cell myeloma Ewing's sarcoma Myxopapillary ependymoma Adamantinoma	Tendosynovial cyst Panniculitis Benign nevoid schwannoma Chondromyxoid fibroma Clear cell sarcoma Leiomyoblastoma Malignant nevoid schwannoma	Aneurysmal bone cyst Giant cell granuloma Hyperparathyroidism Benign histiocytic fibrous histiocytoma Osteoblastoma Malignant histiocytic fibrous histiocytoma	Rhabdomyoma Benign paraganglioma Leiomyoblastoma Rhabdomyoblastoma Malignant epithelioid mesothelioma

TABLE 90. ***Differential Diagnosis of Soft Tissue and Bone Tumors with Epithelioid Pattern**** **(*Continued*)**

APPEARANCE OF STROMA	CELL MORPHOLOGY					
	SLENDER SPINDLE	PLUMP SPINDLE	GRANULAR EPITHELIOID	CLEAR EPITHELIOID	ISOMORPHIC GIANT	PLEOMORPHIC GIANT
Sclerosed	Panniculitis Chondromyxoid fibroma		Mesothelial hyperplasia Benign glomus tumor Benign paraganglioma Benign epithelioid mesothelioma Adenomatoid tumor Osteoblastoma Benign granular cell tumor Meningioma Epithelioid sarcoma Leiomyoblastoma Embryonal rhabdomyosarcoma Hemangiosarcoma Lymphangiosarcoma Malignant nevoid schwannoma Malignant neuroepithelioma Malignant epithelioid mesothelioma Granulocytic sarcoma Plasmacytoma Plasma cell myeloma Ewing's sarcoma Adamantinoma	Panniculitis Chondromyxoid fibroma Clear cell sarcoma Leiomyoblastoma Malignant nevoid schwannoma	Osteoblastoma	Leiomyoblastoma Malignant epithelioid mesothelioma
Myxoid	Panniculitis Chondromyxoid fibroma		Giant cell granuloma Tendosynovial cyst Mesothelial hyperplasia Mastocytosis Benign histiocytic fibrous histiocytoma Lipoblastoma Hibernoma Rhabdomyoma Benign nevoid schwannoma Benign epithelioid mesothelioma Chondroblastoma Malignant histiocytic fibrous histiocytoma Lipoblastic liposarcoma Embryonal rhabdomyosarcoma Rhabdomyoblastoma Malignant glomus tumor Malignant epithelioid mesothelioma Myxopapillary ependymoma Adamantinoma	Tendosynovial cyst Panniculitis Benign nevoid schwannoma Chondromyxoid fibroma Clear cell sarcoma Malignant nevoid schwannoma	Giant cell granuloma Benign histiocytic fibrous histiocytoma Osteoblastoma Malignant histiocytic fibrous histiocytoma	Rhabdomyoma Rhabdomyoblastoma Malignant epithelioid mesothelioma

Table 90. *Differential Diagnosis of Soft Tissue and Bone Tumors with Epithelioid Pattern** **(*Continued*)**

APPEARANCE OF STROMA	CELL MORPHOLOGY					
	SLENDER SPINDLE	PLUMP SPINDLE	GRANULAR EPITHELIOID	CLEAR EPITHELIOID	ISOMORPHIC GIANT	PLEOMORPHIC GIANT
Vascular			Aneurysmal bone cyst Giant cell granuloma Hyperparathyroidism Tendosynovial cyst Pyogenic granuloma Mesothelial hyperplasia Plasma cell granuloma Benign histiocytic fibrous histiocytoma Lipoblastoma Hemangioblastoma Papillary endothelial hyperplasia Benign glomus tumor Benign paraganglioma Benign epithelioid mesothelioma Osteoblastoma Malignant histiocytic fibrous histiocytoma Epithelioid sarcoma Lipoblastic liposarcoma Leiomyoblastoma Embryonal rhabdomyosarcoma Hemangiosarcoma Lymphangiosarcoma Malignant glomus tumor Angioendotheliomatosis Malignant epithelioid mesothelioma Ewing's sarcoma Myxopapillary ependymoma	Tendosynovial cyst Leiomyoblastoma	Aneurysmal bone cyst Giant cell granuloma Hyperparathyroidism Benign histiocytic fibrous histiocytoma Osteoblastoma Malignant histiocytic fibrous histiocytoma	Benign paraganglioma Leiomyoblastoma Malignant epithelioid mesothelioma
Inflamed	Panniculitis		Giant cell granuloma Tendosynovial cyst Pyogenic granuloma Histiocytosis Mastocytosis Plasma cell granuloma Benign epithelioid mesothelioma Epithelioid sarcoma	Tendosynovial cyst Panniculitis	Giant cell granuloma	
Necrotic			Aneurysmal bone cyst Giant cell granuloma Tendosynovial cyst Pyogenic granuloma Eosinophilic granuloma Benign histiocytic fibrous histiocytoma Epithelioid sarcoma Leiomyoblastoma Hemangiosarcoma Lymphangiosarcoma Malignant epithelioid mesothelioma Ewing's sarcoma	Tendosynovial cyst Clear cell sarcoma Leiomyoblastoma	Aneurysmal bone cyst Giant cell granuloma Benign histiocytic fibrous histiocytoma	Benign paraganglioma Leiomyoblastoma Malignant epithelioid mesothelioma

TABLE 90. *Differential Diagnosis of Soft Tissue and Bone Tumors with Epithelioid Pattern** (*Continued*)

APPEARANCE OF STROMA	CELL MORPHOLOGY					
	SLENDER SPINDLE	PLUMP SPINDLE	GRANULAR EPITHELIOID	CLEAR EPITHELIOID	ISOMORPHIC GIANT	PLEOMORPHIC GIANT
Chondrified	Chondromyxoid fibroma		Chondroblastoma	Chondromyxoid fibroma		
Ossified	Panniculitis		Osteoblastoma	Panniculitis	Osteoblastoma	
Calcified	Chondromyxoid fibroma		Tendosynovial cyst Mesothelial hyperplasia Benign epithelioid mesothelioma Chondroblastoma Osteoblastoma Meningioma Epithelioid sarcoma Malignant epithelioid mesothelioma	Tendosynovial cyst Chondromyxoid fibroma	Osteoblastoma	Benign paraganglioma Malignant epithelioid mesothelioma

*See also Tables 48–56, 58–60, 87–89, and 91–92.

TABLE 91. *Differential Diagnosis of Soft Tissue and Bone Tumors with Alveolar Pattern**

APPEARANCE OF STROMA	CELL MORPHOLOGY					
	SLENDER SPINDLE	PLUMP SPINDLE	GRANULAR EPITHELIOID	CLEAR EPITHELIOID	ISOMORPHIC GIANT	PLEOMORPHIC GIANT
Fibrillar	Benign glandular schwannoma Benign epithelioid mesothelioma Biphasic tendosynovial sarcoma Hemangiopericytoma Malignant glomus tumor Malignant glandular schwannoma	Benign epithelioid mesothelioma Malignant epithelioid mesothelioma	Aneurysmal bone cyst Pyogenic granuloma Angiofollicular lymphoid hyperplasia Arteriovenous malformation Vasculitis Mesothelial cyst Angiomatous lymphoid hamartoma Hemangioma Lymphangioma Benign glomus tumor Benign neuroepithelioma Benign paraganglioma Benign epithelioid mesothelioma Adenomatoid tumor Biphasic tendosynovial sarcoma Monophasic tendosynovial sarcoma, pseudoglandular type Epithelioid sarcoma Leiomyoblastoma Alveolar rhabdomyosarcoma Hemangiopericytoma Hemangiosarcoma Lymphangiosarcoma Malignant neuroepithelioma Neuroblastoma Malignant paraganglioma Malignant epithelioid mesothelioma Ewing's sarcoma Alveolar soft part sarcoma Myxopapillary ependymoma	Benign glandular schwannoma Benign paraganglioma Benign epithelioid mesothelioma Clear cell sarcoma Well-differentiated liposarcoma Leiomyoblastoma Malignant glandular schwannoma Chordoma	Aneurysmal bone cyst Ganglioneuroblastoma	Benign paraganglioma Hemangiosarcoma Lymphangiosarcoma Malignant paraganglioma Malignant epithelioid mesothelioma Alveolar soft part sarcoma

TABLE 91. *Differential Diagnosis of Soft Tissue and Bone Tumors with Alveolar Pattern** (*Continued*)

APPEARANCE OF STROMA	CELL MORPHOLOGY					
	SLENDER SPINDLE	PLUMP SPINDLE	GRANULAR EPITHELIOID	CLEAR EPITHELIOID	ISOMORPHIC GIANT	PLEOMORPHIC GIANT
Sclerosed	Hemangiopericytoma	Benign epithelioid mesothelioma Malignant epithelioid mesothelioma	Pyogenic granuloma Arteriovenous malformation Vasculitis Mesothelial cyst Hemangioma Lymphangioma Benign glomus tumor Benign epithelioid mesothelioma Adenomatoid tumor Epithelioid sarcoma Hemangiosarcoma Lymphangiosarcoma Malignant epithelioid mesothelioma Alveolar soft part sarcoma	Benign epithelioid mesothelioma Clear cell sarcoma Well-differentiated liposarcoma	Ganglioneuroblastoma	Hemangiosarcoma Malignant epithelioid mesothelioma
Myxoid	Benign glandular schwannoma Biphasic tendosynovial sarcoma Malignant glandular schwannoma	Benign epithelioid mesothelioma Malignant epithelioid mesothelioma	Aneurysmal bone cyst Pyogenic granuloma Angiofollicular lymphoid hyperplasia Arteriovenous malformation Vasculitis Mesothelial cyst Angiomatous lymphoid hamartoma Hemangioma Lymphangioma Benign glomus tumor Benign neuroepithelioma Benign epithelioid mesothelioma Adenomatoid tumor Biphasic tendosynovial sarcoma Monophasic tendosynovial sarcoma, pseudoglandular type Epithelioid sarcoma Alveolar rhabdomyosarcoma Hemangiosarcoma Lymphangiosarcoma Malignant neuroepithelioma Neuroblastoma Malignant epithelioid mesothelioma Alveolar soft part sarcoma Myxopapillary ependymoma	Benign glandular schwannoma Benign epithelioid mesothelioma Clear cell sarcoma Well-differentiated liposarcoma Malignant glandular schwannoma Chordoma	Aneurysmal bone cyst Ganglioneuroblastoma	Hemangiosarcoma Lymphangiosarcoma Malignant epithelioid mesothelioma

Table 91. ***Differential Diagnosis of Soft Tissue and Bone Tumors with Alveolar Pattern**** **(*Continued*)**

APPEARANCE OF STROMA	CELL MORPHOLOGY					
	SLENDER SPINDLE	PLUMP SPINDLE	GRANULAR EPITHELIOID	CLEAR EPITHELIOID	ISOMORPHIC GIANT	PLEOMORPHIC GIANT
Vascular	Hemangiopericytoma Malignant glomus tumor		Aneurysmal bone cyst Pyogenic granuloma Angiofollicular lymphoid hyperplasia Arteriovenous malformation Vasculitis Angiomatous lymphoid hyperplasia Hemangioma Lymphangioma Benign glomus tumor Benign paraganglioma Epithelioid sarcoma Leiomyoblastoma Alveolar rhabdomyosarcoma Hemangiopericytoma Hemangiosarcoma Lymphangiosarcoma Malignant glomus tumor Angioendothelioma-tosis Neuroblastoma Malignant paraganglioma Malignant epithelioid mesothelioma Ewing's sarcoma Myxopapillary ependymoma	Benign paraganglioma Clear cell sarcoma Well-differentiated liposarcoma Leiomyoblastoma Chordoma	Aneurysmal bone cyst Ganglioneuroblastoma	Benign paraganglioma Hemangiosarcoma Lymphangiosarcoma Malignant paraganglioma Malignant epithelioid mesothelioma
Inflamed		Malignant epithelioid mesothelioma	Aneurysmal bone cyst Pyogenic granuloma Angiofollicular lymphoid hyperplasia Vasculitis Angiomatous lymphoid hamartoma Epithelioid sarcoma Lymphangiosarcoma	Benign epithelioid mesothelioma	Aneurysmal bone cyst	Lymphangiosarcoma Malignant epithelioid mesothelioma

Table 91. ***Differential Diagnosis of Soft Tissue and Bone Tumors with Alveolar Pattern**** **(*Continued*)**

APPEARANCE OF STROMA	CELL MORPHOLOGY					
	SLENDER SPINDLE	PLUMP SPINDLE	GRANULAR EPITHELIOID	CLEAR EPITHELIOID	ISOMORPHIC GIANT	PLEOMORPHIC GIANT
Necrotic	Malignant glandular schwannoma	Malignant epithelioid mesothelioma	Vasculitis Epithelioid sarcoma Leiomyoblastoma Alveolar rhabdomyosarcoma Hemangiopericytoma Hemangiosarcoma Lymphangiosarcoma Neuroblastoma Malignant paraganglioma Malignant epithelioid mesothelioma Ewing's sarcoma Alveolar soft part sarcoma Myxopapillary ependymoma	Clear cell sarcoma Well-differentiated liposarcoma Leiomyoblastoma Chordoma	Aneurysmal bone cyst Ganglioneuroblastoma	Hemangiosarcoma Lymphangiosarcoma Malignant epithelioid mesothelioma
Chondrified				Well-differentiated liposarcoma Chordoma		
Ossified						
Calcified	Biphasic tendosynovial sarcoma	Benign epithelioid mesothelioma Malignant epithelioid mesothelioma	Arteriovenous malformation Mesothelial cyst Hemangioma Benign paraganglioma Benign epithelioid mesothelioma Adenomatoid tumor Biphasic tendosynovial sarcoma Monophasic tendosynovial sarcoma, pseudoglandular type Hemangiopericytoma Lymphangiosarcoma Malignant epithelioid mesothelioma Alveolar soft part sarcoma Myxopapillary ependymoma	Benign paraganglioma Benign epithelioid mesothelioma Clear cell sarcoma Well-differentiated liposarcoma Chordoma	Ganglioneuroblastoma	Benign paraganglioma Lymphangiosarcoma Malignant epithelioid mesothelioma Alveolar soft part sarcoma

*See also Tables 48–50, 58–60, 87–90, and 92.

Table 92. ***Differential Diagnosis of Soft Tissue and Bone Tumors with Disarranged Pattern****

APPEARANCE OF STROMA	CELL MORPHOLOGY					
	SLENDER SPINDLE	PLUMP SPINDLE	GRANULAR EPITHELIOID	CLEAR EPITHELIOID	ISOMORPHIC GIANT	PLEOMORPHIC GIANT
Fibrillar	Fibromatosis Ganglioneuroma Mesenchymal chondrosarcoma Osteosarcoma	Keloid Fibromatosis Elastofibroma Leiomyosarcoma Osteosarcoma Paget's sarcoma	Aneurysmal bone cyst Giant cell granuloma Hibernoma Osteoblastoma Benign granular cell tumor Embryonal rhabdomyosarcoma Rhabdomyoblastoma Hemangiosarcoma Lymphangiosarcoma Osteosarcoma Paget's sarcoma	Xanthogranuloma Lipogranuloma Chondroblastoma Leiomyoblastoma	Aneurysmal bone cyst Giant cell granuloma Xanthogranuloma Foreign body granuloma Benign pleomorphic fibrous histiocytoma Ganglioneuroma Osteoblastoma Osteoid osteoma Osteosarcoma Paget's sarcoma	Foreign body granuloma Elastofibroma Gouty arthritis Fat necrosis Lipogranuloma Proliferative myositis Myositis ossificans Callus Gaucher's disease Benign pleomorphic fibrous histiocytoma Hibernoma Rhabdomyoma Malignant pleomorphic fibrous histiocytoma Pleomorphic fibrosarcoma Leiomyoblastoma Leiomyosarcoma Rhabdomyoblastoma Pleomorphic rhabdomyosarcoma Hemangiosarcoma Lymphangiosarcoma Osteosarcoma Paget's sarcoma Malignant granular cell tumor
Sclerosed	Fibromatosis Ganglioneuroma	Keloid Fibromatosis Elastofibroma Pleomorphic fibrosarcoma Leiomyosarcoma Osteosarcoma Paget's sarcoma	Embryonal rhabdomyosarcoma Hemangiosarcoma Lymphangiosarcoma	Chondroblastoma	Foreign body granuloma Ganglioneuroma Osteoid osteoma Osteosarcoma Paget's sarcoma	Foreign body granuloma Elastofibroma Fat necrosis Lipogranuloma Myositis ossificans Callus Pleomorphic fibrosarcoma Leiomyosarcoma Hemangiosarcoma Osteosarcoma Paget's sarcoma

Table 92. *Differential Diagnosis of Soft Tissue and Bone Tumors with Disarranged Pattern** (*Continued*)

APPEARANCE OF STROMA	CELL MORPHOLOGY					
	SLENDER SPINDLE	PLUMP SPINDLE	GRANULAR EPITHELIOID	CLEAR EPITHELIOID	ISOMORPHIC GIANT	PLEOMORPHIC GIANT
Myxoid	Fibromatosis Malignant pleomorphic fibrous histiocytoma Mesenchymal chondrosarcoma Osteosarcoma	Elastofibroma Leiomyosarcoma	Aneurysmal bone cyst Giant cell granuloma Hibernoma Osteoblastoma Benign granular cell tumor Embryonal rhabdomyosarcoma Rhabdomyoblastoma Hemangiosarcoma Lymphangiosarcoma Osteosarcoma Paget's sarcoma	Xanthogranuloma Lipogranuloma Chondroblastoma Leiomyoblastoma	Aneurysmal bone cyst Giant cell granuloma Xanthogranuloma Benign pleomorphic fibrous histiocytoma Osteoblastoma Osteoid osteoma Osteosarcoma Paget's sarcoma	Elastofibroma Gouty arthritis Fat necrosis Lipogranuloma Proliferative myositis Myositis ossificans Callus Gaucher's disease Benign pleomorphic fibrous histiocytoma Hibernoma Rhabdomyoma Malignant pleomorphic fibrous histiocytoma Leiomyoblastoma Leiomyosarcoma Rhabdomyoblastoma Pleomorphic rhabdomyosarcoma Hemangiosarcoma Lymphangiosarcoma Osteosarcoma Paget's sarcoma
Vascular	Fibromatosis Malignant pleomorphic fibrous histiocytoma	Leiomyosarcoma Osteosarcoma	Aneurysmal bone cyst Giant cell granuloma Hibernoma Osteoblastoma Embryonal rhabdomyosarcoma Rhabdomyoblastoma Hemangiosarcoma Lymphangiosarcoma Lymphangiosarcoma Osteosarcoma Paget's sarcoma	Xanthogranuloma Lipogranuloma Leiomyoblastoma	Aneurysmal bone cyst Giant cell granuloma Xanthogranuloma Foreign body granuloma Benign pleomorphic fibrous histiocytoma Osteoblastoma Osteoid osteoma Osteosarcoma Paget's sarcoma	Foreign body granuloma Gouty arthritis Callus Benign pleomorphic fibrous histiocytoma Hibernoma Malignant pleomorphic fibrous histiocytoma Leiomyoblastoma Leiomyosarcoma Rhabdomyoblastoma Pleomorphic rhabdomyosarcoma Hemangiosarcoma Lymphangiosarcoma Osteosarcoma
Inflamed	Malignant pleomorphic fibrous histiocytoma	Keloid Elastofibroma	Giant cell granuloma Embryonal rhabdomyosarcoma Hemangiosarcoma Lymphangiosarcoma Paget's sarcoma	Xanthogranuloma Lipogranuloma	Giant cell granuloma Xanthogranuloma Foreign body granuloma Benign pleomorphic fibrous histiocytoma Paget's sarcoma	Foreign body granuloma Elastofibroma Gouty arthritis Fat necrosis Lipogranuloma Proliferative myositis Myositis ossificans Benign pleomorphic fibrous histiocytoma Malignant pleomorphic fibrous histiocytoma Pleomorphic rhabdomyosarcoma Hemangiosarcoma Lymphangiosarcoma Paget's sarcoma

TABLE 92. ***Differential Diagnosis of Soft Tissue and Bone Tumors with Disarranged Pattern* (Continued)***

APPEARANCE OF STROMA	CELL MORPHOLOGY					
	SLENDER SPINDLE	PLUMP SPINDLE	GRANULAR EPITHELIOID	CLEAR EPITHELIOID	ISOMORPHIC GIANT	PLEOMORPHIC GIANT
Necrotic	Malignant pleomorphic fibrous histiocytoma Mesenchymal chondrosarcoma Osteosarcoma	Elastofibroma Pleomorphic fibrosarcoma Leiomyosarcoma Osteosarcoma	Giant cell granuloma Embryonal rhabdomyosarcoma Rhabdomyoblastoma Hemangiosarcoma Lymphangiosarcoma Osteosarcoma Paget's sarcoma	Xanthogranuloma Lipogranuloma Leiomyoblastoma	Aneurysmal bone cyst Giant cell sarcoma Xanthogranuloma Foreign body granuloma Osteoblastoma Osteosarcoma Paget's sarcoma	Foreign body granuloma Elastofibroma Gouty arthritis Fat necrosis Lipogranuloma Proliferative myositis Myositis ossificans Gaucher's disease Benign pleomorphic fibrous histiocytoma Hibernoma Malignant pleomorphic fibrous histiocytoma Pleomorphic fibrosarcoma Leiomyoblastoma Leiomyosarcoma Rhabdomyoblastoma Pleomorphic rhabdomyosarcoma Hemangiosarcoma Lymphangiosarcoma Osteosarcoma Paget's sarcoma
Chondrified	Mesenchymal chondrosarcoma	Osteosarcoma	Osteosarcoma	Chondroblastoma	Foreign body granuloma Osteosarcoma Paget's sarcoma	Foreign body granuloma Myositis ossificans Callus Osteosarcoma Paget's sarcoma
Ossified		Osteosarcoma	Osteoblastoma Osteosarcoma		Foreign body granuloma Osteoblastoma Osteoid osteoma Osteosarcoma Paget's sarcoma	Foreign body granuloma Myositis ossificans Callus Osteosarcoma Paget's sarcoma
Calcified	Fibromatosis Ganglioneuroma	Keloid Fibromatosis Elastofibroma Leiomyosarcoma Osteosarcoma Paget's sarcoma	Giant cell granuloma Hibernoma Osteoblastoma Osteosarcoma Paget's sarcoma	Xanthogranuloma Lipogranuloma Chondroblastoma	Giant cell granuloma Xanthogranuloma Foreign body granuloma Benign pleomorphic fibrous histiocytoma Ganglioneuroma Osteoblastoma Osteoid osteoma Osteosarcoma Paget's sarcoma	Foreign body granuloma Elastofibroma Gouty arthritis Fat necrosis Lipogranuloma Proliferative myositis Myositis ossificans Callus Benign pleomorphic fibrous histiocytoma Hibernoma Leiomyosarcoma Osteosarcoma Paget's sarcoma

*See also Tables 48–50, 58–60, 87–91.

"If we leave out the great contribution of cytology toward early diagnosis about all the progress we have made involves the correction of old errors."

FRED W. STEWART (1894–)

Appendix: Exfoliative and Aspiration Cytology

Exfoliative Cytology (see Table 93; Figures: 1284, 1286, 1288, 1298, 1308, 1309)
Aspiration Cytology (see Table 93; Figures: 1277–1283, 1285, 1287, 1289–1297, 1299–1307, 1310–1312)

Although the history of cytology is filled with innumerable achievements and great contributions in the diagnosis of primary and metastatic cancers, cytology has its limitations. For example, cytologic techniques can be applied successfully for the diagnosis of pathologic lesions only if the lesions possess distinctive and pathognomonic cellular elements. Brushing and aspiration cytology were introduced partly to overcome the limitations of exfoliative cytology. In smears prepared from brushings and aspirates one often finds, in addition to solitary cells, tissue fragments that may show characteristic or suggestive growth patterns and stromal features (Table 93).

Brushing and aspiration cytology may become invaluable, in experienced hands, in instances when anatomic setting of the lesion or overall condition of the patient caution against tissue biopsy. It would be wrong, however, to believe that aspiration cytology eliminates the need for tissue biopsy of primary soft tissue and bone tumors. Expediency, cost containment, and availability should not preclude selection of the most appropriate and most accurate biopsy technique in the diagnosis of such complex lesions as soft tissue and bone tumors. Benign and malignant soft tissue and bone tumors are composed of a wide spectrum of bizarre, non-specific, and misleading connective tissue elements. Undoubtedly, a callus or myositis ossificans may contain some or all the cells that can be seen in osteosarcoma. Certain forms of fasciitis, proliferative myositis, and granulomatous lesions may produce cellular elements that can also be found in soft tissue sarcomas.

Readers must be aware that due to recent interest in aspiration cytology medical periodicals have published several articles on the cytology of soft tissue and bone tumors; many of these articles contain little useful information. Any cytology report on soft tissue and bone tumors that replaces factual observation with creations of a vivid imagination and does not list lesions that should be considered in the differential diagnosis does not deserve attention.

Cytologic examination of soft tissue and bone tumors is valuable but several factors including location, size, clinical presentation, and cellular composition require tissue biopsy for definitive diagnosis of primary neoplasms. Considering that soft tissue and bone tumors are uncommon, those who seldom see such tumors should be extremely cautious in interpreting smears and should not yield to pressure when a definitive diagnosis is not possible. In most cases pathologists and cytologists would do well to limit themselves to listing the pathologic entities that should be considered in the differential diagnosis.

The illustrations included in the Appendix are not all inclusive. Only lesions with cytologic features that are specific, suggestive, consistent or compatible are illustrated. Readers will notice that pattern recognition, even in smears, is of paramount importance to accurate diagnosis; therefore, many photomicrographs were selected to illustrate patterns rather than individual cells.

TABLE 93. *Comparison of Cytologic Smears and Tissue Sections*

	Exfoliative Smear	Aspiration Smear	Brush Smear	Tissue Section
Tissue Pattern	Missing	Mostly missing	Partly missing	Preserved
Cell Morphology	Preserved	Preserved	Preserved	Preserved
Intercellular Matrix	Missing	Mostly missing	Partly missing	Preserved
Intracellular Products	Mostly preserved	Mostly preserved	Mostly preserved	Mostly preserved

Fig. 1277. Benign schwannoma. Slender spindle cells with tangled wire-like cytoplasmic processes forming nuclear palisades (soft tissue aspirate, H & E, ×350).

Fig. 1278. Benign nevoid schwannoma. Admixture of granular epithelioid (nevoid) cells and slender spindle cells in fibrillar background (soft tissue aspirate, H & E, ×350).

Fig. 1279. Myxoid liposarcoma. Slender cells with pyknotic round and spindle-shaped nuclei (soft tissue aspirate, H & E, ×350).

Fig. 1277

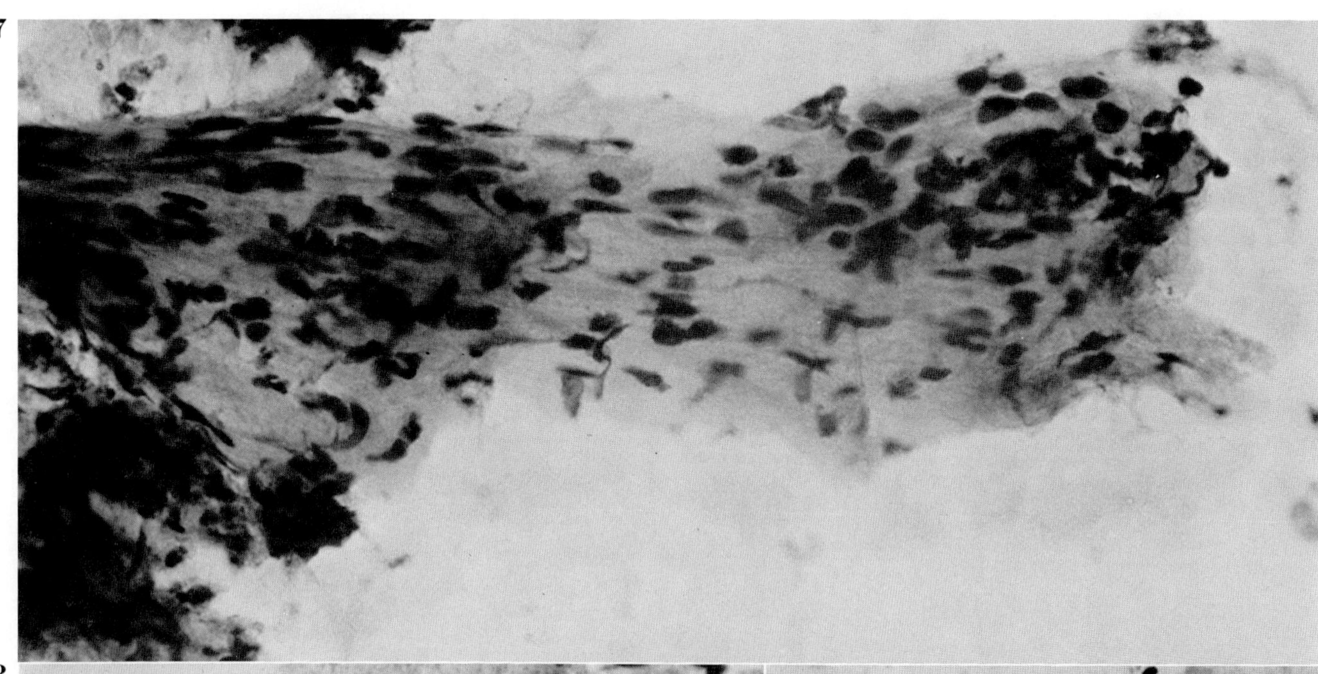

Fig. 1278

Fig. 1279

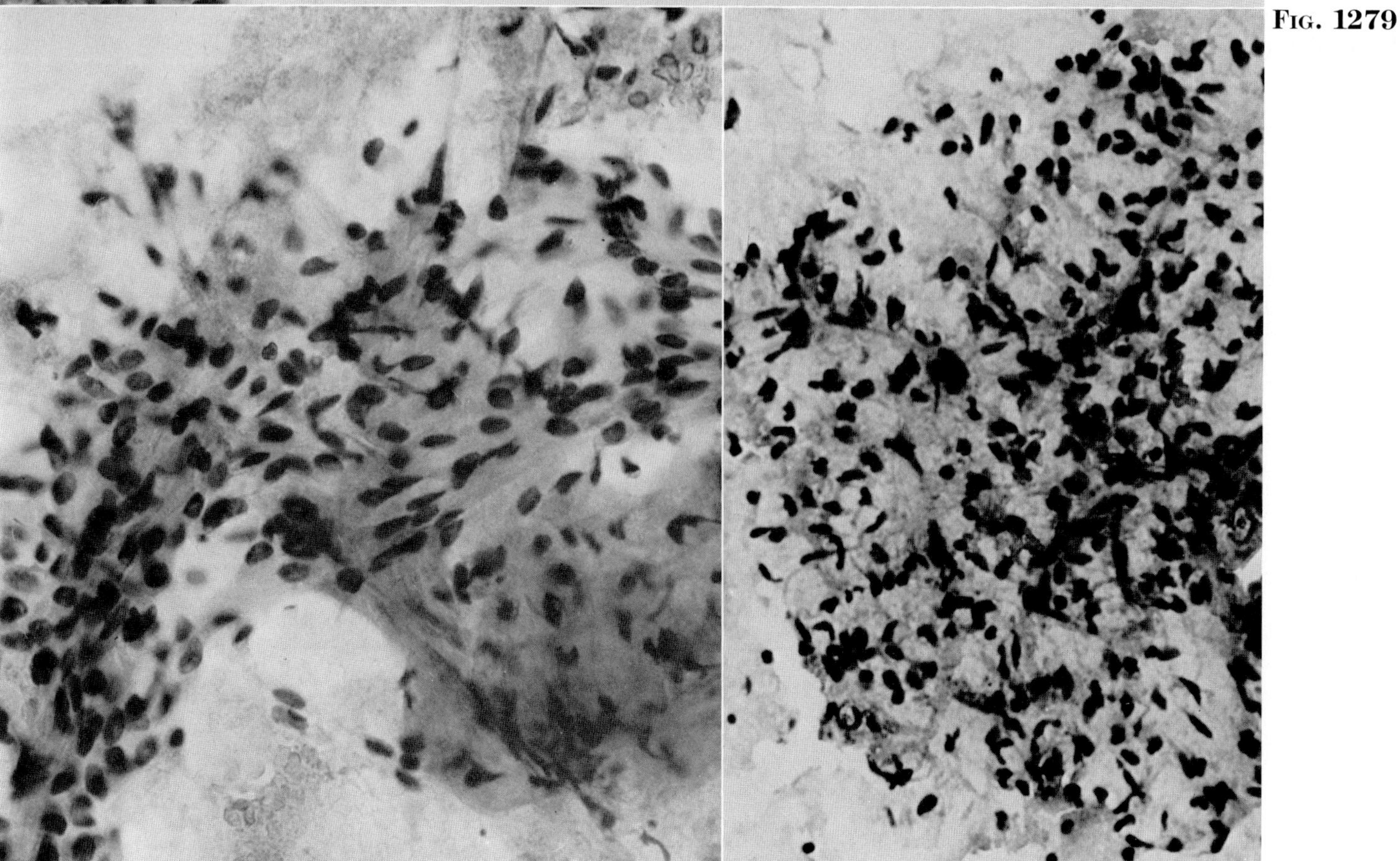

Fig. 1280. Malignant fibroblastic fibrous histiocytoma. Slender spindle cells in storiform arrangement (soft tissue aspirate, H & E, ×350).

Fig. 1281. Myxoid liposarcoma. Small cells with slender cytoplasmic processes are attached to a capillary vessel (soft tissue aspirate, H & E, ×570).

Fig. 1280

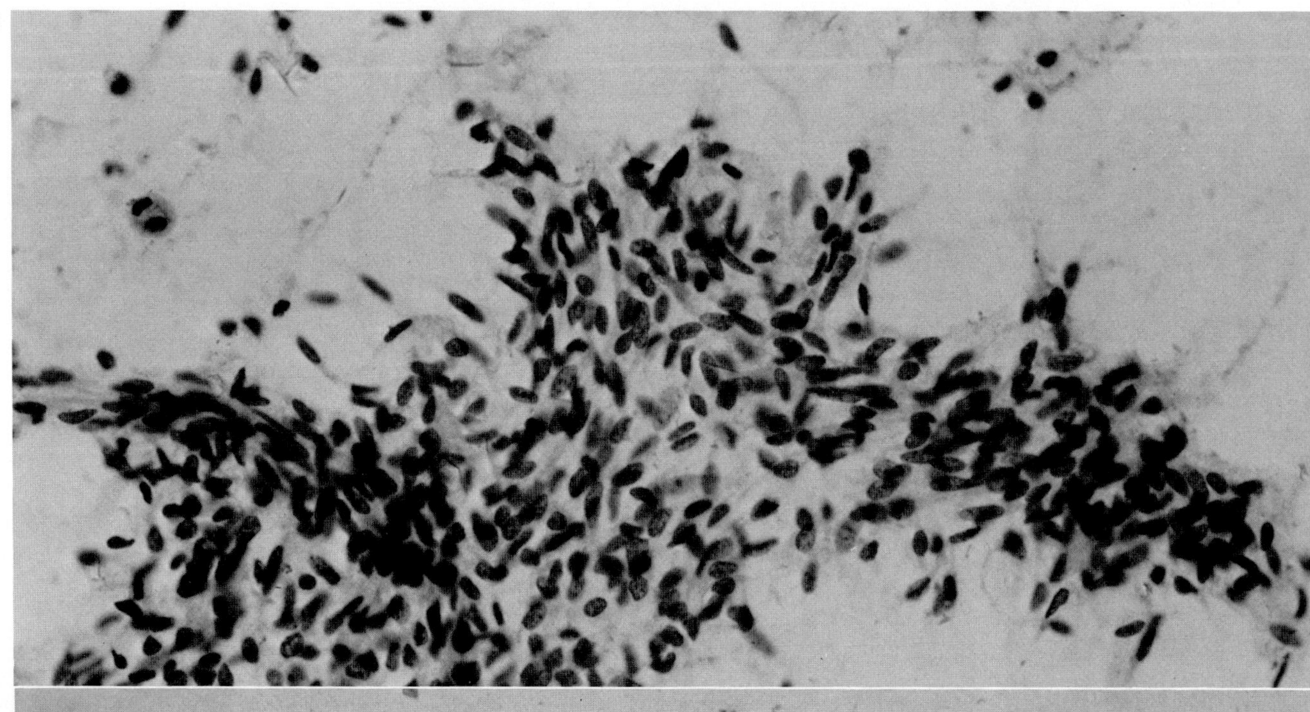

Fig. 1281

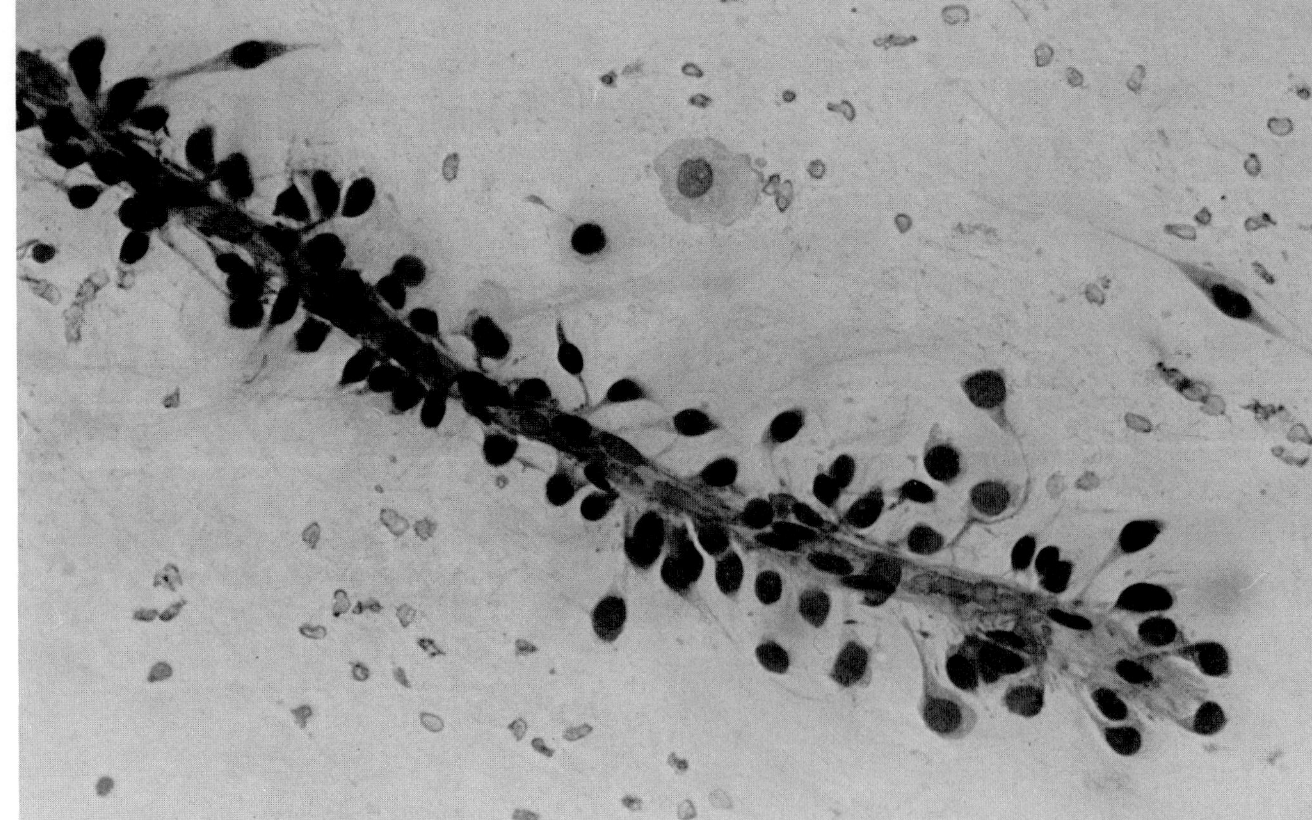

Fig. 1282. Leiomyosarcoma. Plump spindle cells with cigar-shaped nuclei showing fishbone arrangement of nuclear chromatin (soft tissue imprint, H & E, ×350).

Fig. 1283. Leiomyosarcoma. Note the cross-bared nuclear chromatin (pulmonary aspirate, Pap, ×570).

Fig. 1284. Malignant pleomorphic fibrous histiocytoma. Plump spindle cells with elongated nuclei and granular histiocytic forms with round nuclei (pleural effusion, Pap, ×450).

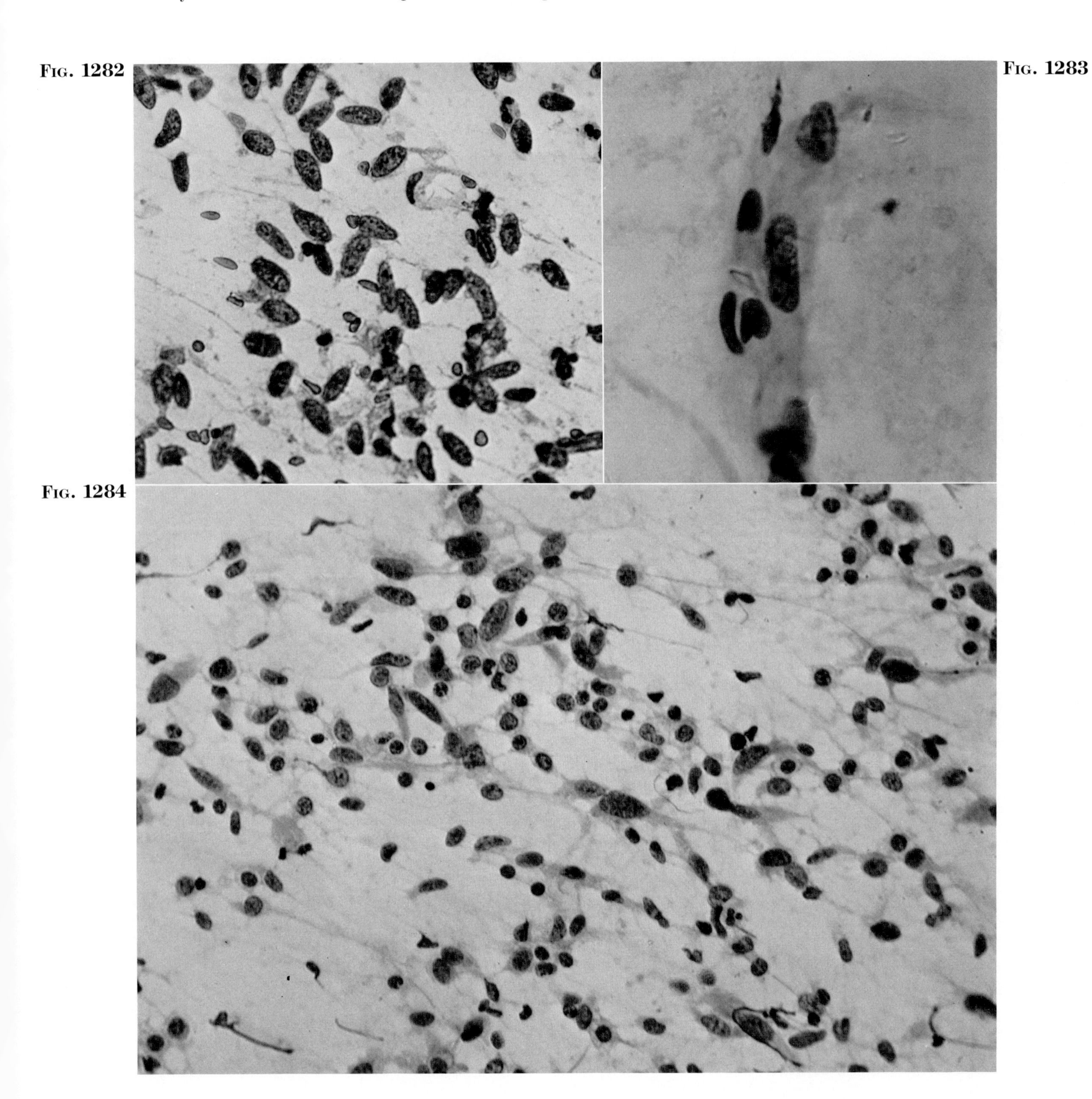

Fig. 1285. Osteomyelitis. Acute and chronic inflammatory cells and cell debris (bone aspirate, Pap, ×350).

Fig. 1286. Plasma cell myeloma (effusion, Pap, ×570).

Fig. 1287. Granuloma. Granular epithelioid cells in lymphoreticular background (lymph node aspirate, H & E, ×350).

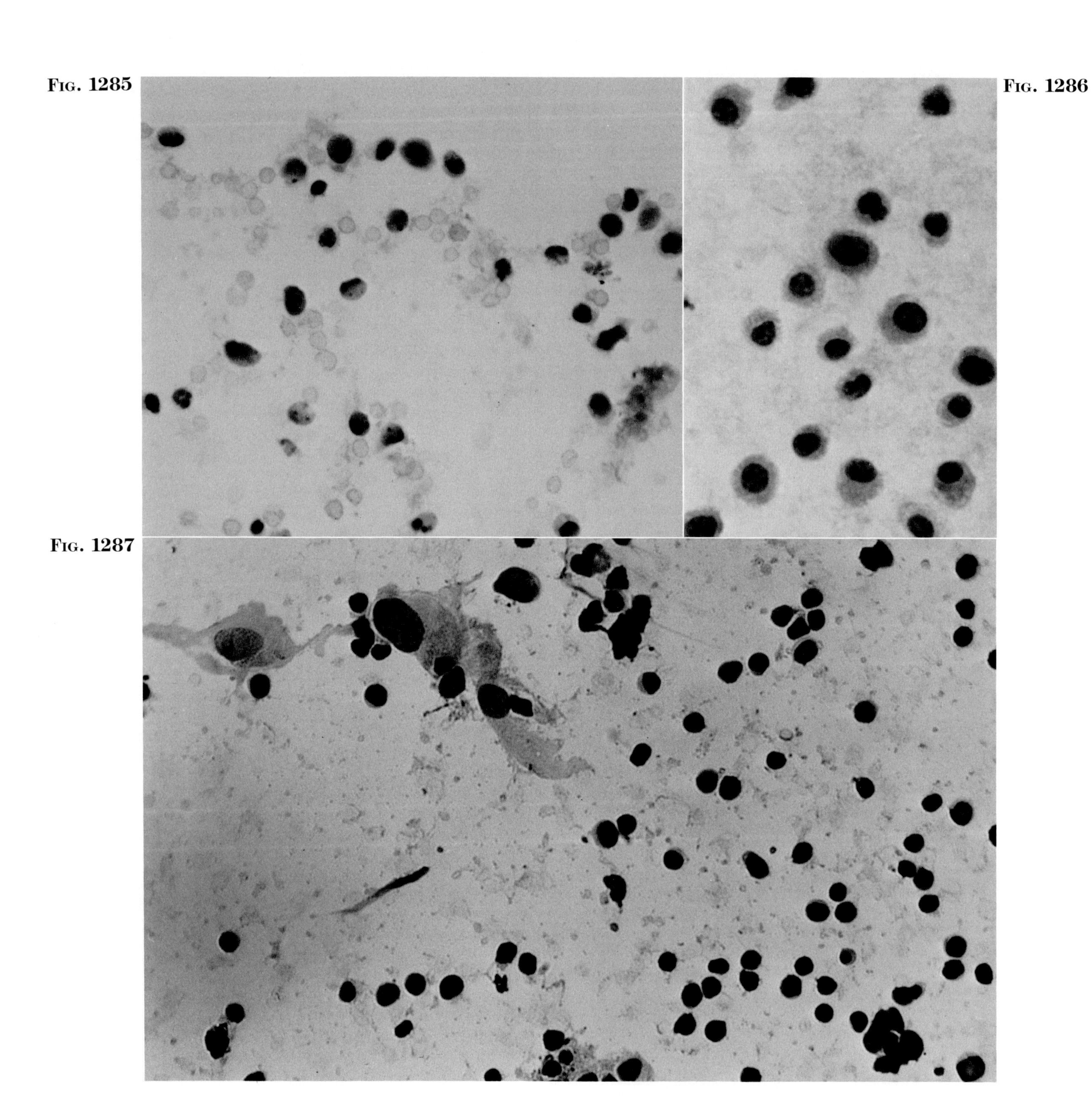

Fig. 1288. Ewing's sarcoma (sputum, Pap, ×350). **Fig. 1289.** Ewing's sarcoma (bone aspirate, H & E, ×570).

Fig. 1290. Ewing's sarcoma. Granular epithelioid cells with pyknotic nuclei shown singly and in clusters (bone aspirate, H & E, ×450).

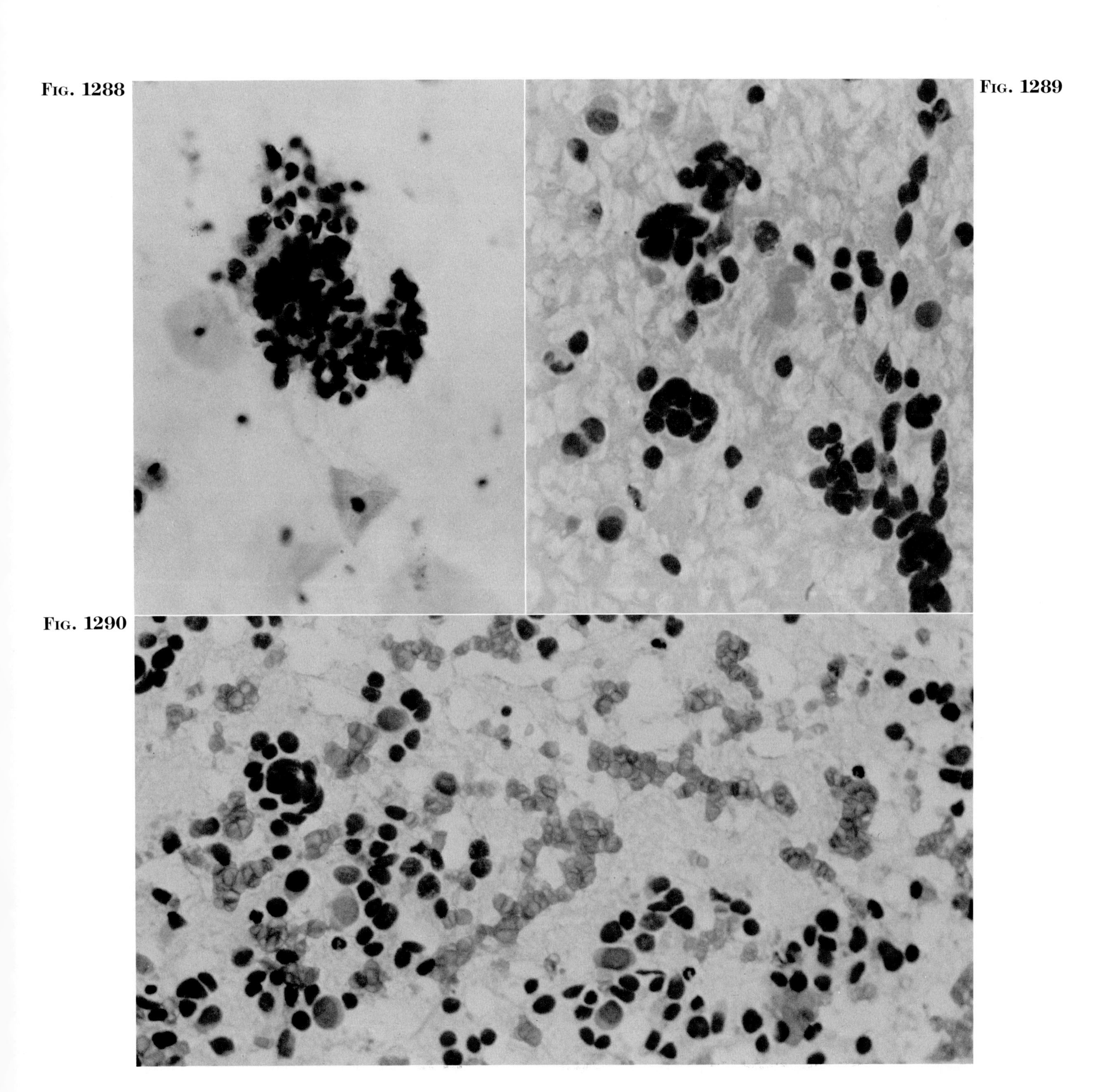

Fig. 1291. Chordoma. Granular epithelioid cells in myxomucoid matrix (bone aspirate, H & E, ×350).

Fig. 1292. Granular epithelioid, physaliferous, cells in clusters (bone aspirate, H & E, ×570).

Fig. 1293. Lipoblastic liposarcoma. Finely granular epithelioid cells in cluster surrounding a capillary vessel (soft tissue aspirate, H & E, ×570).

Fig. 1291 **Fig. 1292**

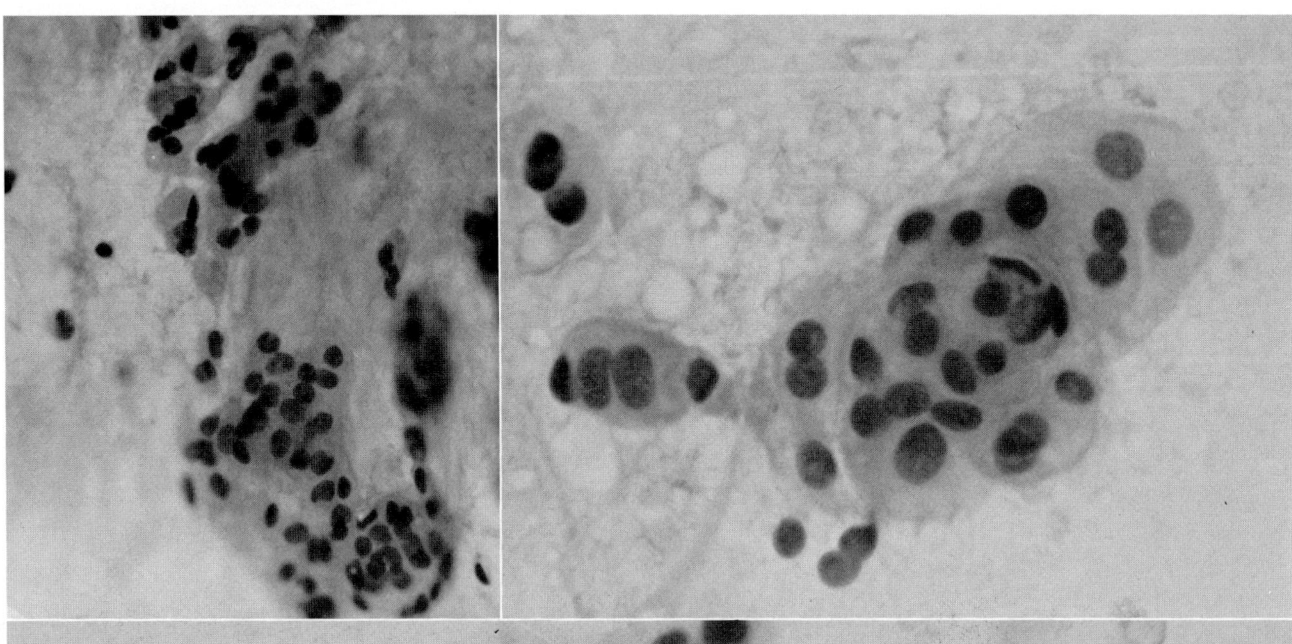

Fig. 1293

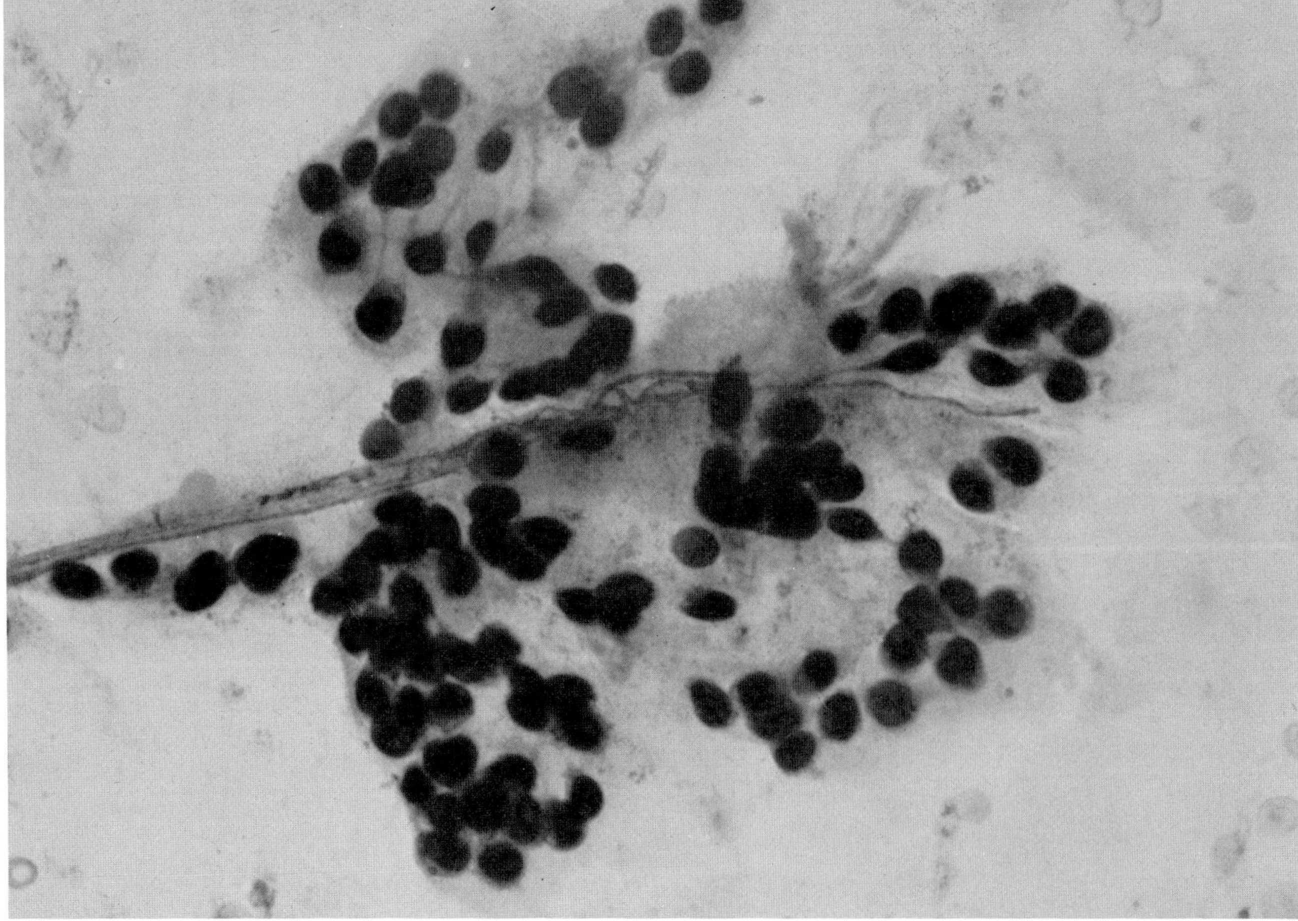

Fig. 1294. Granular cell tumor. Granular epithelioid cells with eccentrically placed uniform nuclei and ill-defined cytoplasmic borders (soft tissue aspirate, H & E, ×570).

Fig. 1295. Chondroblastoma. Mono- and binucleated granular epithelioid cells with sharply demarcated cytoplasm (bone aspirate, H & E, ×570).

Fig. 1296. Rhabdomyoblastoma. Isolated granular epithelioid cells with well-defined nuclei and cytoplasm (soft tissue aspirate, H & E, ×570).

Fig. 1294

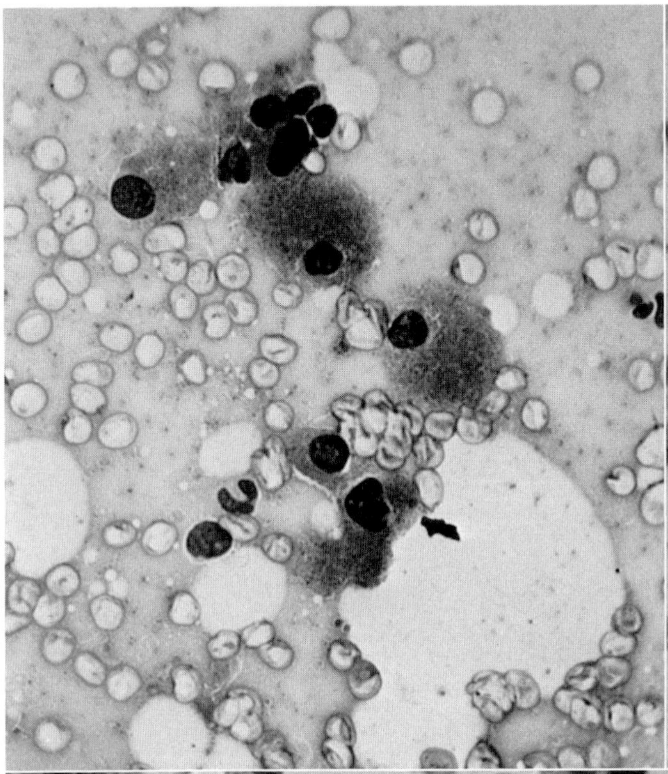

Fig. 1295

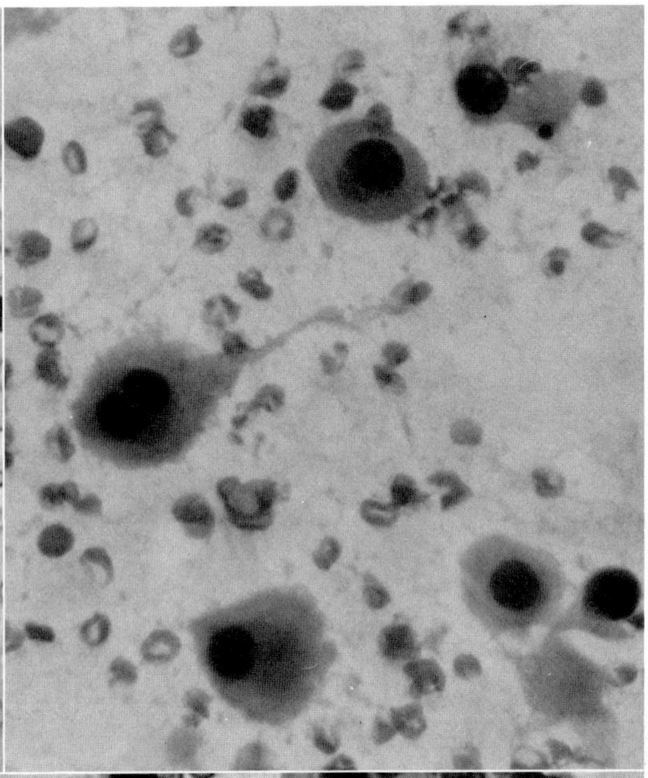

Fig. 1296

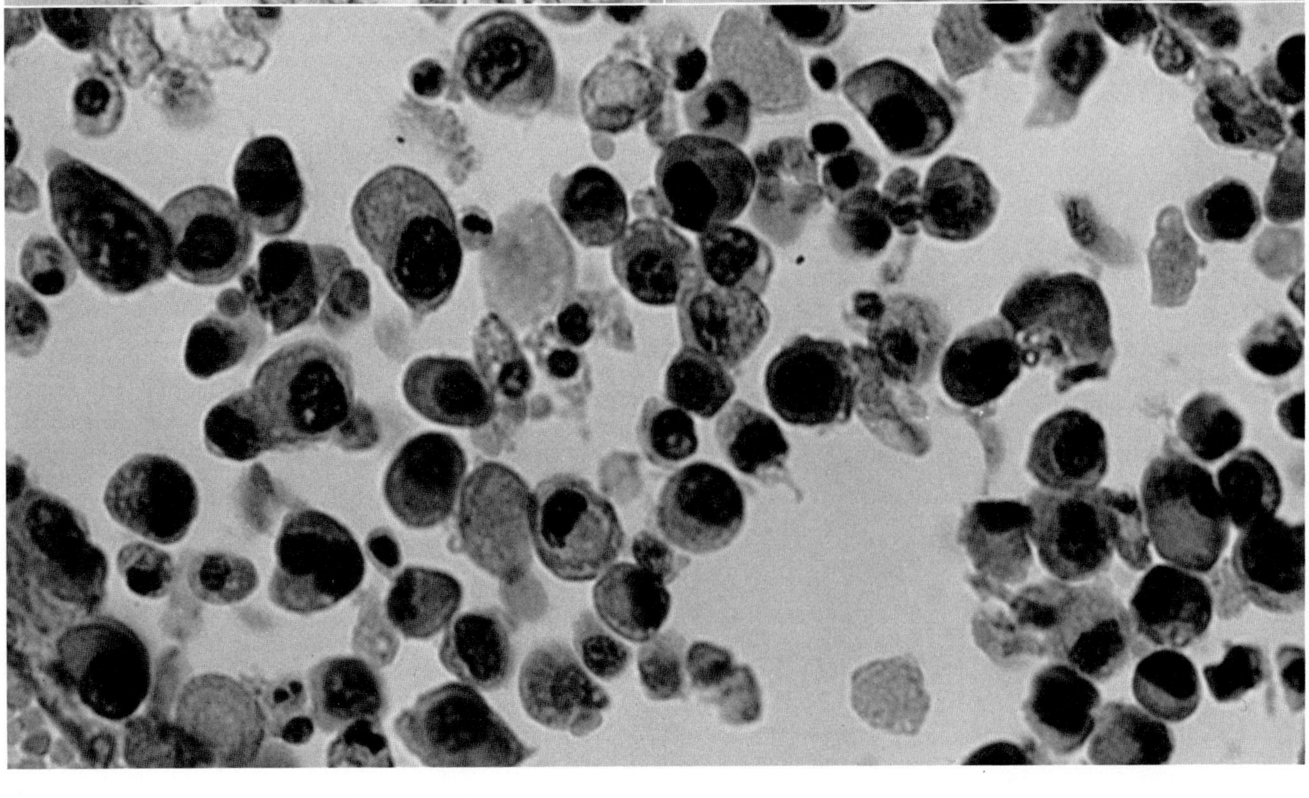

Fig. 1297. Chondrosarcoma. Clear epithelioid cells, many binucleated, in clusters and in myxochondroid background (bone aspirate, Pap, ×350).

Fig. 1298. Chondrosarcoma. Binucleated and mononucleated chondrocytes (effusion, Pap, ×450).

Fig. 1299. Clear epithelioid and foamy benign histiocytes (pulmonary infarct imprint, H & E, ×350).

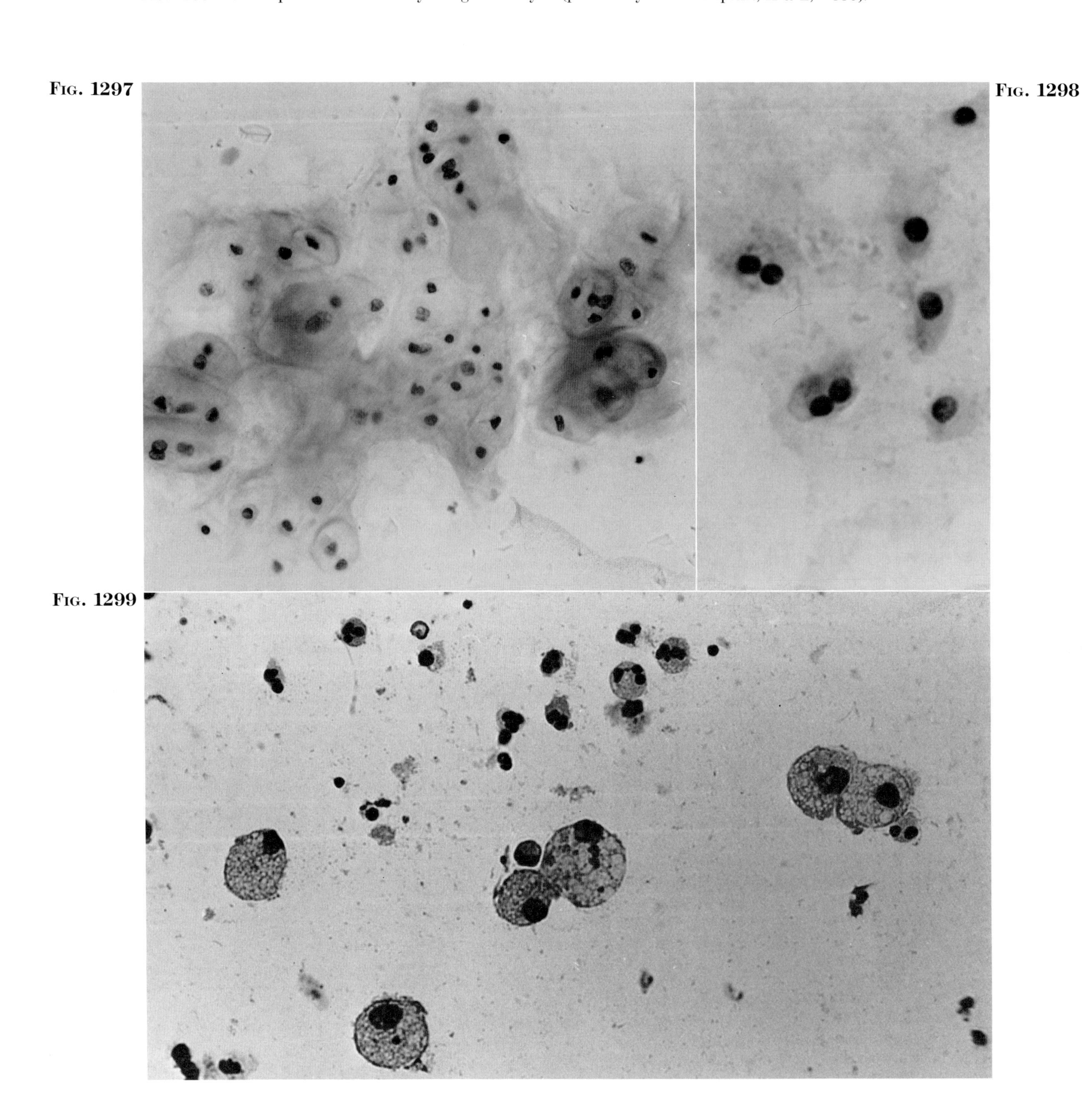

Fig. 1300. Benign histiocytic fibrous histiocytoma. Well-formed isomorphic epulis giant cells and uniform orderly stromal elements (bone aspirate, H & E, ×350).

Fig. 1301. Malignant histiocytic fibrous histiocytoma. Abortive isomorphic epulis giant cells and pleomorphic stromal elements (bone aspirate, H & E, ×350).

Fig. 1300

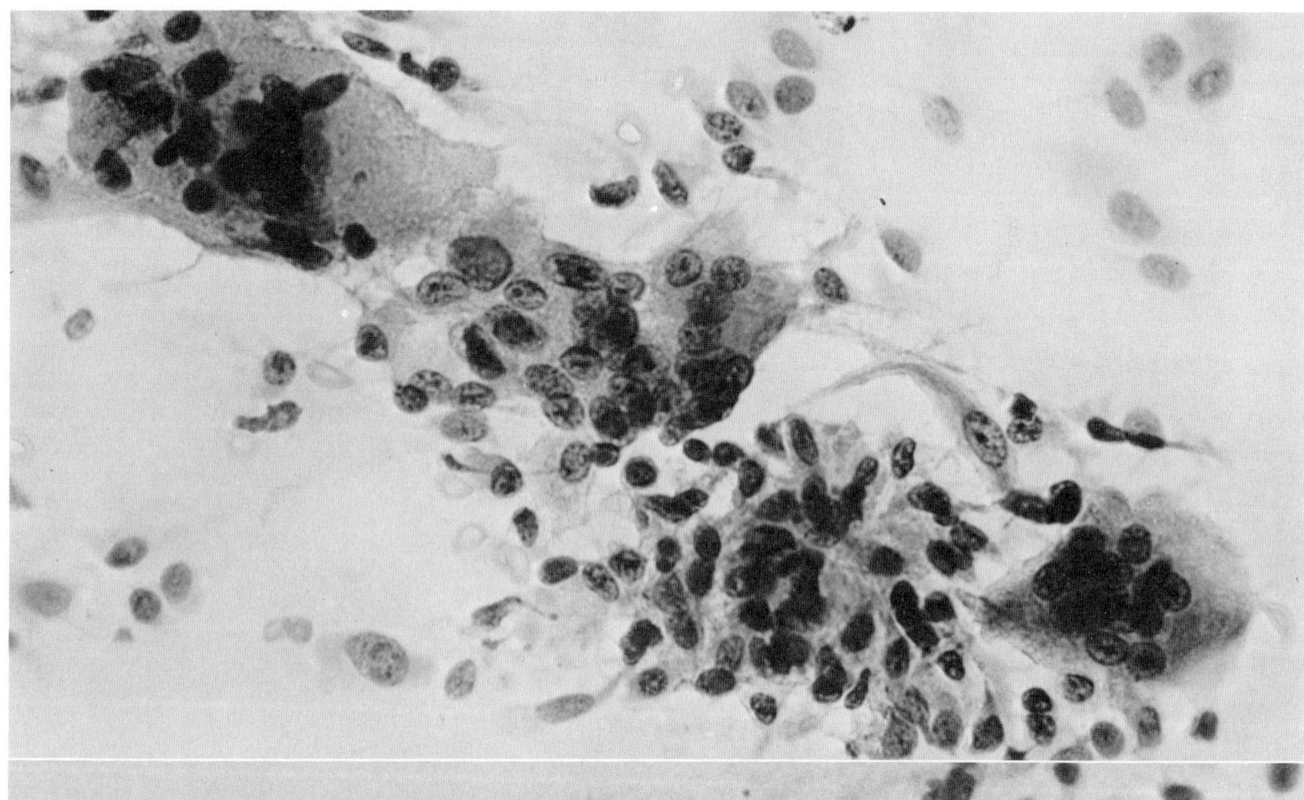

Fig. 1301

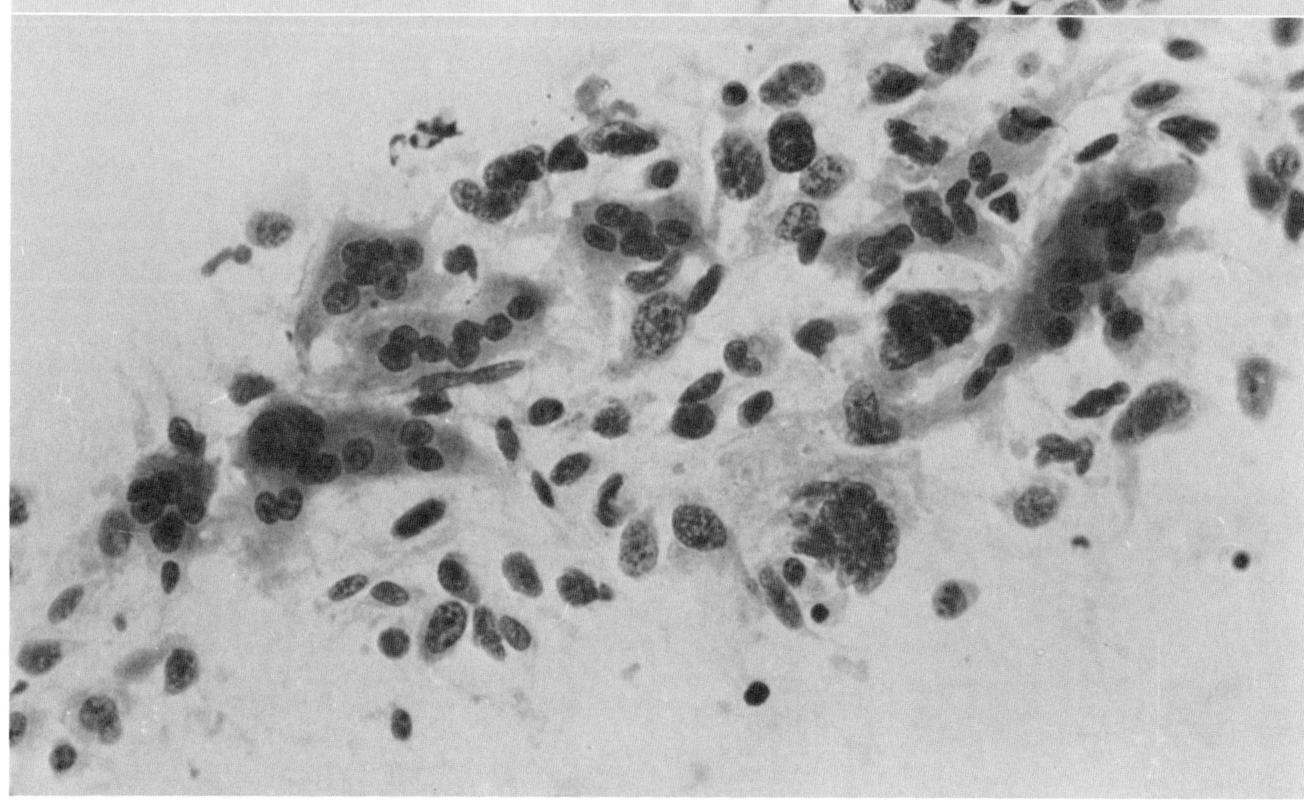

Fig. 1302. Malignant pleomorphic fibrous histiocytoma (soft tissue aspirate, H & E, ×350).

Fig. 1303. Malignant pleomorphic fibrous histiocytoma (lung aspirate, Pap, ×350).

Fig. 1304. Pleomorphic liposarcoma (soft tissue aspirate, H & E, ×350).

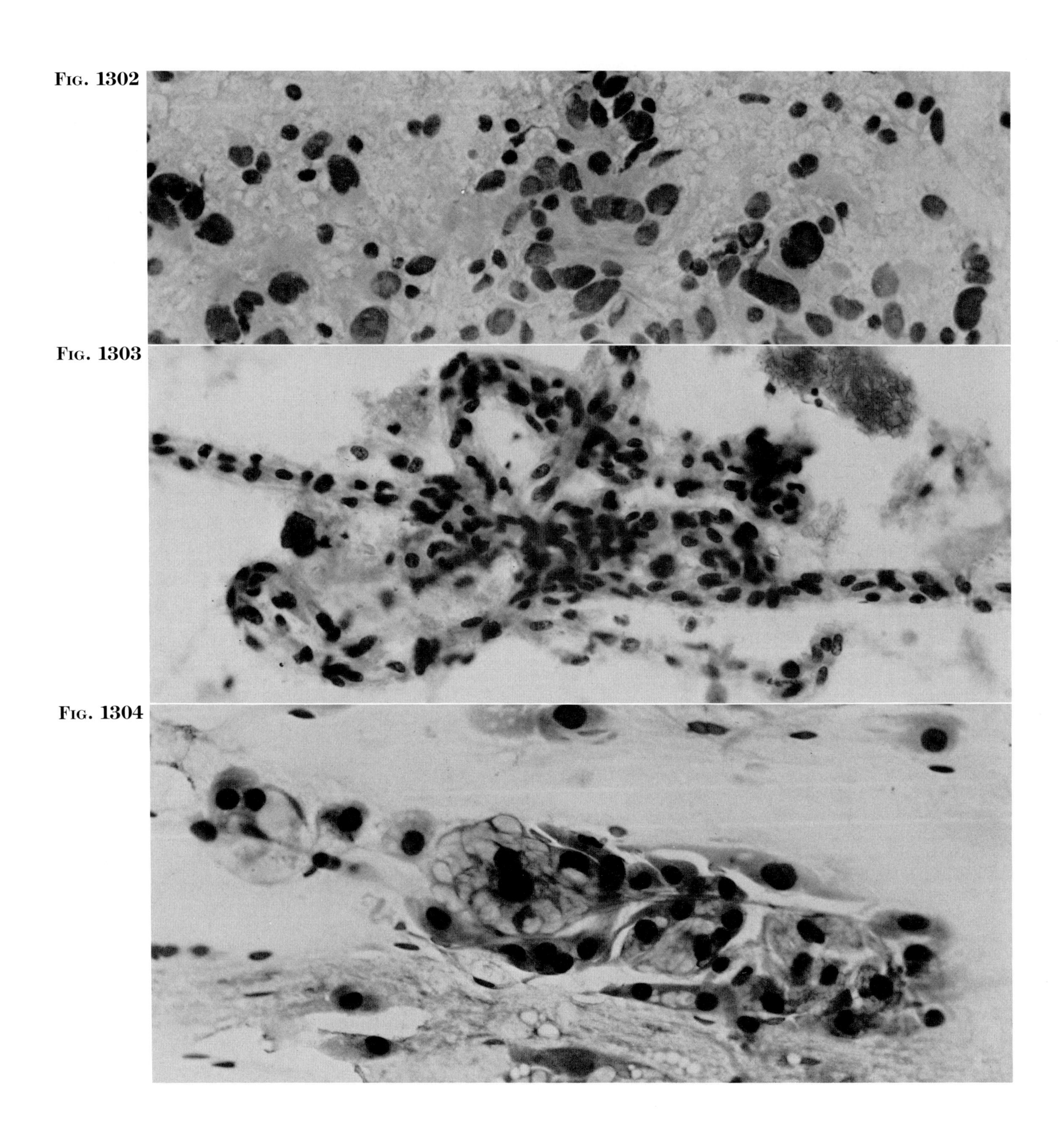

Fig. 1305. Osteosarcoma. Admixture of pleomorphic cells in granular, fragmented osteoid background (bone aspirate, H & E, ×350).

Fig. 1306. Pleomorphic rhabdomyosarcoma. Note incompletely assembled parallel filaments in several cells (soft tissue aspirate, H & E, ×570).

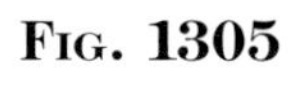

Fig. 1305

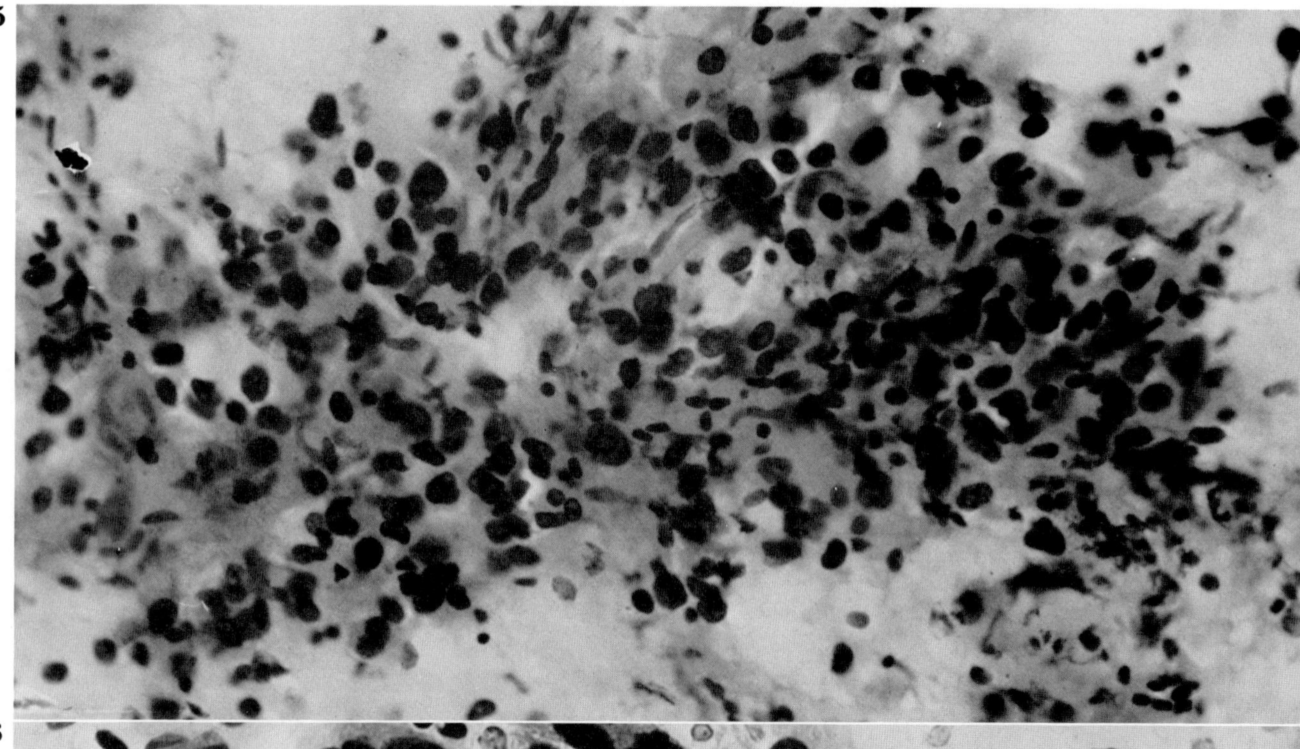

Fig. 1306

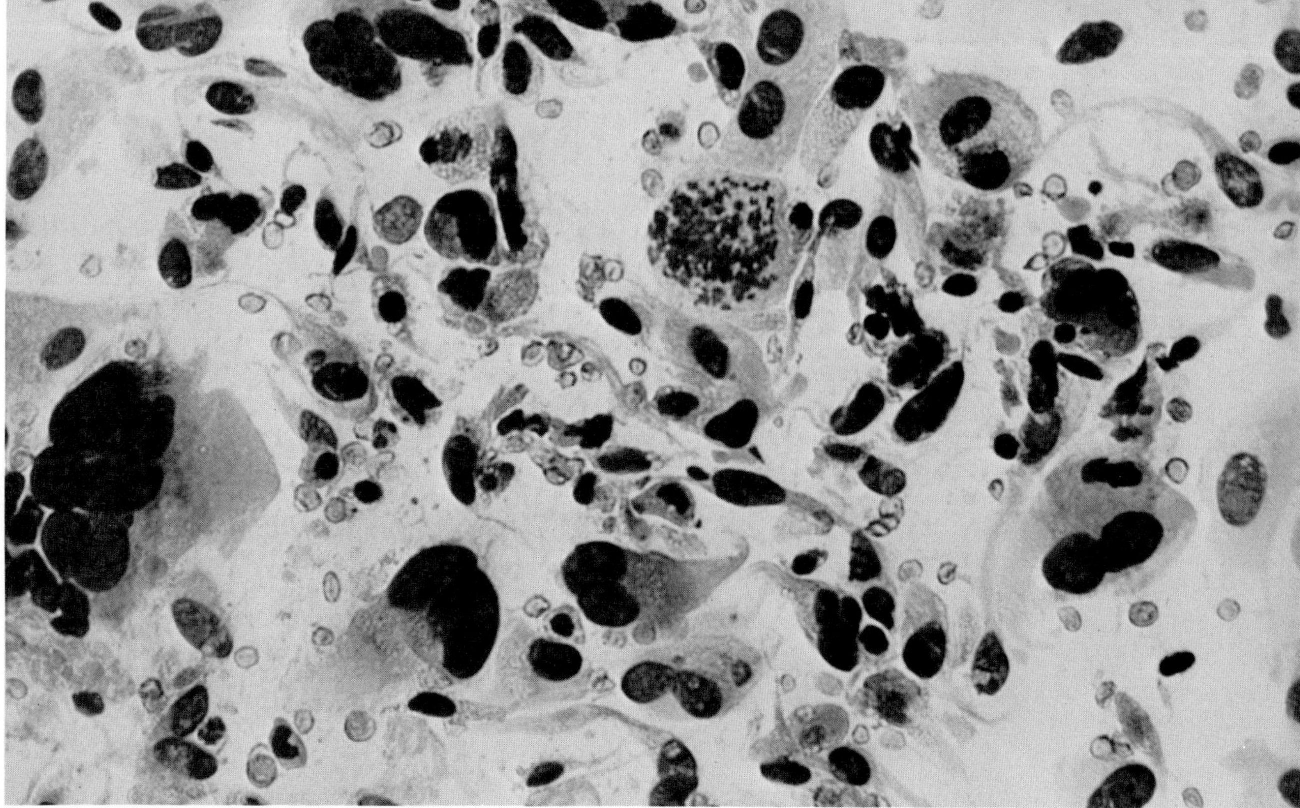

Fig. 1307. Malignant pleomorphic fibrous histiocytoma. A solitary neoplastic Touton giant cell (soft tissue aspirate, H & E, ×570).

Fig. 1308. Leiomyosarcoma. A solitary malignant cell with ill-defined tail-like process (effusion, Pap, ×570).

Fig. 1309. Osteosarcoma. A solitary neoplastic cell with two uneven nuclei and distorted cytoplasm (effusion, Pap, ×570).

Fig. 1310. Chondrosarcoma. A binucleated and mononucleated chondrocyte with vacuolated and granular cytoplasm (bone aspirate, H & E, ×570).

Fig. 1311. Malignant pleomorphic fibrous histiocytoma. A solitary multinucleated tumor giant cell with intracytoplasmic phagocytized particles (lung aspirate, H & E, ×570).

Fig. 1312. Megakaryocyte with ill-defined cytoplam and doughnut-shaped nucleus (bone aspirate, H & E, ×570).

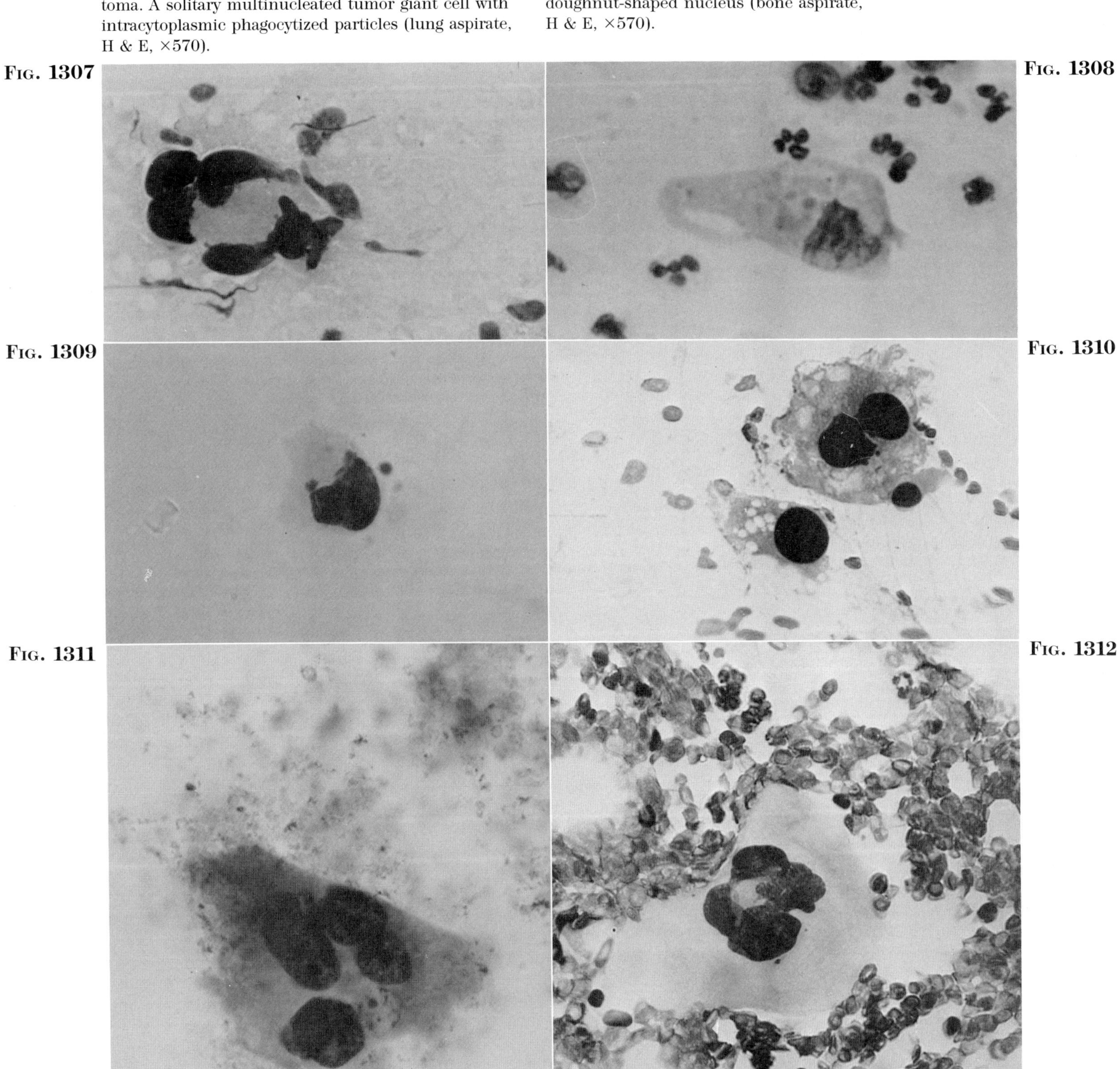

Index

Page numbers in *italics* refer to illustrations; those in **boldface** refer to definitions and differential diagnoses.

AACT, 350–353
AAT, 350–353
Aberrant antigenicity, 343
Abdominal desmoid, 26
Abdominal wall, 57–59
Abscess, 8
Acid mucopolysaccharide, 341–347
Acid phosphatase, 348
Ackerman, Lauren, 8, 82
Acquired immunodeficiency syndrome, 30, 35
Actin, 350–353
Actin–myosin complex, 341–343, 348–353
Adamantinoma, 11, 86, *99,* 140, 183, 218, **411**
  borderline, 15
  cell morphology, 218, *221*
  classification, 11
  differential diagnosis, **411, 422, 423**–445
  growth phase, 14
  radiology, 86, *99*
  site, 56, 61, 62
  size, 79, 81
Adenocarcinoma, 8
Adenofibroma, 20
Adenomatoid tumor, 11
Adenosine triphosphatase, 348
Adipocyte(s), 4
Adipose tissue, inflammation of, 18
Adipose tissue tumors, 10
Adiposis dolorosa, 10
Adult bone, 271
Adult fibrosarcoma, 26
Adult rhabdomyoma, 22
Adult Wilms' tumor, *45, 154, 272*
Age, 8, 13, 15, 50–53
  benign neoplasm, 50, 52
  malignant neoplasms, 50, 53
  non-neoplastic lesions, 50, 51
"Agent-Orange," 35
Aggressive fibromatosis, 26
AIDS, 30, 35
Alcian blue stain, 267
Alkaline phosphatase, 35, 36, 38, 348
Alpha-1-antichymotrypsin, 343, 350, 351
Alpha-1-antitrypsin, 343, 350, 351
ALPP, 351–353
Alveolar rhabdomyosarcoma, 10, 28, 140, *190, 191, 192, 201,* 218, 270, 346, 402, 405, 408, **411.** *See also* Embryonal rhabdomyosarcoma
  cell morphology, 218, *235, 259*
  complication, 408, 410
  cytology, 446
  differential diagnosis, **411, 419, 423–445**
  grade, 402–404
  gross appearance, 125
  growth pattern, 136, 137, 140
  histochemistry, 342, 348
  immunochemistry, 342, 343, 350, 353
  metastasis, 408, 410
  products, 340–343, 346–353
  prognosis, 408, 410
  stage, 405–407
  stroma, 265–267, 270
  survival, 410
  ultrastructure, 342, 348, 349
Alveolar soft part sarcoma, 11, 15, **30,** 31, 38, *69,* 86, 136, 137, 140, 218, 270, 346, 402, 405, 408, **411**
  age distribution, 50, 53
  cell morphology, 214, 215, 218, *240*
  classification, 11
  clinical appearance, 38
  cytology, 446
  definition, **30**
  differential diagnosis, **411, 422, 423–445**
  grade, 402–404
  gross appearance, 115, *193, 200*
  growth pattern, 140
  growth phase, 14
  history, 38
  illustrations, 31
  immunochemistry, 342, 343, 350, 353
  metastasis, 410
  products, 340–343, 346–353, *365, 375, 386*
  prognosis, 408, 410
  recurrence, 410
  references, 31
  sex distribution, 54, 55
  site, 38, 59, *69*
  size, 81
  special stains, 341–347
  stage, 405–407
  stroma, 270
  survival, 410
  symptoms, 38
  synonyms, 30
  ultrastructure, 342, 348, 349, *386*
Amelanotic melanoma, *398*
American Joint Committee, 408
Amniotic cavity, *3*
Amputation neuroma, 18
Amyloid tumor, 18
Amyloidoma, 11, **18,** 19, 115
Amyloidosis, 38
  localized, 18
Anatomic site, 18
Ancient neurilemoma, 22
Ancient neurofibroma, 22
Aneurysmal bone cyst, 10, **16,** 36, **71,** 84, 114, 115, 136–138, 216, 268, 344, 408, **411, 443**
  age, 50, 51
  cell morphology, 214–216, *255,* **443–445**
  clinical appearance, 36
  complication, 36, 408, 409
  cytology, 446
  definition, **16**
  differential diagnosis, **411–413, 423–445**
  gross appearance, 115
  growth pattern, 138, *187*
  histochemistry, 215
  history, 36
  illustrations, 17
  immunochemistry, 215
  products, 344
  prognosis, 408, 409
  radiology, 36, 84, *89, 97*
  recurrence, 408, 409
  references, 17
  sex distribution, 54
  site, 39, 60, 62, 71
  size, 79, 80
  stroma, 265–268, *313, 325*
  symptoms, 36
  synonym(s), 8, 16
  ultrastructure, 215
Angioblastic mesoderm, *4*
Angioendothelioma, 11, 28
Angioendotheliomatosis, 11
Angiofibroma, 10, 20, **22,** 23, *278*
Angiofollicular lymphoid hyperplasia, 10, 11, **18,** 19, 36, 39, 115, 216, 268, 344, 408, **411–413, 423–445**
  cell morphology, 214–216, **423–445**
  classification, 11
  clinical appearance, 36
  complication, 408, 409

Angiofollicular lymphoid hyperplasia (*Continued*)
cytology, 446
definition, **18**
differential diagnosis, **411–413, 423–445**
gross appearance, *41,* 115
growth pattern, 138
histochemistry, 215
history, 36
illustrations, 19
immunochemistry, 215
products, 344
radiology, 36, 84
recurrence, 408, 409
references, 19
site, 39, 40, 56, 57
size, 79, 80
stroma, 265–268, *305, 320*
survival, 408, 409
symptoms, 36
synonyms, 18
ultrastructure, 215
Angiolipoma, 10, 20, *306*
Angiomyolipoma, 9, 10, **20,** 21, 37, 40, 85, 114, 115, 139, 217, 269, 345, 408, 409, **411**
age distribution, 50, 52
cell morphology, 217
clinical appearance, 37
complication, 9, 408, 409
cytology, 446
definition, **20**
differential diagnosis, **411, 414–416, 423–445**
gross appearance, 115
growth pattern, 139
history, 37
illustrations, 21
malignant transformation, 9, 10, 409
products, 345, *355*
prognosis, 408, 409
radiology, 37
references, 21
sex distribution, 37, 54, 55
site, 37, 58
size, 79, 80
stroma, 269, *301*
symptoms, 37
synonyms, 20
Angiomatous lymphoid hamartoma, 11
Angiomyoma, 10, 22, *225, 290, 306*
Angiomyxoma, 20
Angioneurofibroma, 22
Angioneuroma, 22
Angioneuromyoma, 22
Angioreninoma, 30
Angiosarcoma, 28
Angiosarcoma multiplex, 30
Anisocytosis, 215
Anisokaryosis, 215
Antigen(s), tissue, 341–353, 354, *392–401*
Antigenicity, 342
aberrant, 343
Aponeurotic clear cell sarcoma, 26
Aponeurotic fibroma. *See* Fibroma
Aponeurotic sarcoma, 26
Appendix, 446–458
Arm, 57, 58, 59
Arranged growth pattern, 136–140, *141–152*
Arteriography, 82
Arteriovenous hemangioma, 18, 22. *See also* Arteriovenous malformation
Arteriovenous malformation, 10, **18,** 19, 36, 84, *92, 94, 105,* 114, 115, 216, 268, 344, 408, **411**
age distribution, 50, 51
cell morphology, 216, **423–445**
complication, 36, 408, 409
cytology, 446
definition, **18**
differential diagnosis, **411–413, 423–445**
gross appearance, 115
growth pattern, 138
history, 36
illustrations, 19
products, 344
prognosis, 408, 409
radiology, 36, 84, *92, 94, 105*
recurrence, 408, 409
references, 19
sex distribution, 54
site, 36, 39, 57
size, 79, 80
stroma, 265–268, *290, 311, 312*
symptoms, 36
synonyms, 18
Arteriovenous shunt, 18
Arthralgia, 36
Arthritis, 10
Articular chondrosis, 18
Asbestos, 35
Aspirate. *See* Aspiration cytology
Aspiration biopsy. *See* Aspiration cytology
Aspiration cytology, 15, 113, *446–458*
needle, 402
Aspiration smear. *See* Aspiration cytology
Asymptomatic, 12
Atrophy of muscle, 10
Atypical fibroxanthoma, 20
Atypical giant cell tumor, 16
Atypical juvenile xanthoma, 16
Atypical lipoma, 20
Atypical phase, 12, *13,* 14, 15
Atypical pyogenic granuloma, 18
Autonomic nerve tumors, 11
A-V malformation. *See* Arteriovenous malformation

Baker's cyst, 16
Balloon cell melanoma, *247, 399*
Basal cell carcinoma, *222*
Basal lamina, 342, 348, 349
Bence-Jones protein, 35, 38
Benign aortic body tumor, 22
Benign carotid body tumor, 22
Benign collagenous peripheral nerve tumor, 22
Benign collagenous schwannoma, 22
Benign counterpart(s) of malignant neoplasm(s), 12, 14
Benign cystosarcoma phyllodes, 9, 11, **24,** 25, 37, *45,* 114, 115, 136, 137, 139, 217, 269, 345, 408, 409, **411**
age distribution, 50, 52
cell morphology, 217
clinical appearance, 37, *45*
complication, 9, 408, 409
cytology, 446
definition, **24**
differential diagnosis, **411, 414–416, 423–445**
gross appearance, 115, *118*
growth pattern, 136, 137, 139, *163*
history, 37
illustrations, 25
malignant transformation, 9, 409
products, 340–343, 345, 347–353
prognosis, 408, 409
radiology, 37
references, 25
sex distribution, 54, 55
size, 79, 80
stroma, 265–267, 269, *292, 296*
symptoms, 137
synonyms, 25
Benign fibroblastic fibrous histiocytoma, 10, **20,** 21, 37, 39, 40, 85, 114, 115, 136, 137, 139, 217, 269, 345, 408, 409, **411**
age distribution, 50, 52
cell morphology, 217, *231*
clinical appearance, 37
complication, 408, 409
cytology, 446
definition, **20**
differential diagnosis, **411, 414–416, 423–445**
gross appearance, 115
growth pattern, 136, 137, 139, *141, 147*
history, 37
illustrations, 21
malignant transformation, 409
products, 340–343, 345, 347–353
prognosis, 408, 409
radiology, 37
recurrence, 408, 409
references, 21
sex distribution, 54, 55
site, 37, 56, 58
size, 79, 80
stroma, 265–267, 269, *301, 316, 317, 321*
symptoms, 37
synonyms, 20
ultrastructure, 348
Benign fibroblastic peripheral nerve tumor, 22
Benign fibrous histiocytoma, 20. *See also* Benign fibroblastic fibrous histiocytoma; Benign histiocytic fibrous histiocytoma; Benign pleomorphic fibrous histiocytoma; Nonossifying fibroma
Benign giant cell tumor, 20
Benign glomus jugulare tumor, 22
Benign hemangiopericytoma, 22
Benign histiocytic fibrous histiocytoma, 10, **20,** 21, 37, 40, *63, 78,* 85, *100, 101, 102, 109,* 114, 115, 136, 137, 139, 217, 269, 345, 408, 409, **411**
age distribution, 50, 52
atypical, *181, 185, 285, 322*
borderline, 15, *181, 185, 239, 253, 285, 310, 322*
cell morphology, *217, 231, 239, 253*
clinical appearance, 37
complication, 408, 409
cytology, 446, *455*
definition, **20**
differential diagnosis, **411, 414–416, 423–445**
gross appearance, 115, *134*
growth pattern, 136, 137, 139, *177, 181, 185*
history, 37
illustrations, 21
malignant transformation, 14, 35, 37, 409
products, 340–343, 345, 347–353
prognosis, 408, 409
radiology, 37, 85, *100, 101, 102, 109*
recurrence, 408, 409
references, 21
sex distribution, 54, 55
site, 37, 58, 60, 62, *63, 78*
size, 79, 80
stroma, 265–267, 269, *285, 310, 322*
symptoms, 37
synonyms, 20
ultrastructure, 348
Benign leiomyoblastoma, 22
Benign mediastinal paraganglioma, 22
Benign neoplasm(s), 8, 10–15, 20–25
age distribution, 50, 52
cell morphology, 214, 215, 217
classification, 10, 11, 12, 13
clinical presentation, 35, 37, 39, 40
definition(s), **20, 22, 24**

diagnosis, 12–14
differential diagnosis, **411, 414–416, 423–445**
gross appearance, 114, 115
growth pattern, 136, 137, 139
illustration(s), 21, 23, 25
incidence, 8
malignant counterparts, 12, 14
malignant transformation, 9
number, 9
products of cells, 340–343, 345, 347–354
prognosis, 408, 409
radiology, 82, 85
reference(s), 21, 23, 25
sex distribution, 54, 55
sites, 58, 60, 62
size, 79, 80
stroma, 265–267, 269, 271
synonym(s), 20, 22, 24
Benign nevoid schwannoma, *447*
Benign nonchromaffin paraganglioma, 22
Benign noncollagenous peripheral nerve tumor, 22
Benign nonfibroblastic peripheral nerve tumor, 22
Benign peripheral nerve sheath tumor, 22
Benign pleomorphic fibrous histiocytoma, 10, **20,** 21, 37, 39, 40, 85, *91,* 114, 115, 136, 137, 139, 217, 269, 345, 408, 409, **411**
age distribution, 50, 52
atypical, *164, 165, 174, 301, 312*
borderline, 15, *164, 165, 174, 231, 254, 255, 260, 301, 312*
cell morphology, 217, *231, 241, 254, 255, 260*
clinical appearance, 37
complication, 408, 409
cytology, 446
definition, **20**
differential diagnosis, 411, **414–416, 423–445**
gross appearance, 115
growth pattern, 136, 137, 139, *164, 165, 174*
history, 37
illustrations, 21
malignant transformation, 409
products, 340–343, 345, 347–353
radiology, 37, 85, *91*
recurrence, 408, 409
references, 21
sex distribution, 54, 55
site, 58, 60, 62
size, 80
stroma, 265–267, 269, *301, 312*
symptoms, 37
synonyms, 20
ultrastructure, 348
Benign retroperitoneal paraganglioma, 22
Benign schwannoma, *447*
Benign synovioma, 16
Benign undifferentiated peripheral nerve tumor, 139, 217, 269, 345, **411**
cell morphology, 217
differential diagnosis, **411**
growth pattern, 139
products, 345
stroma, 269
Benign vagal body paraganglioma, 22
Biomatrix, 266
Biopsy, 15, 82, 113, *446–458*
Biphasic synovial sarcoma, 26
Biphasic tendosynovial sarcoma, 10, **26,** 27, 38, 40, *44,* 50, 53, 54, 55, *68,* 86, 114, 116, 218, 270, 346, 402, 405, 408, **411**
cell morphology, 214, 215, 218, 224, 230, *243, 250*
clinical appearance, 38, *44*
cytology, 446
definition, **26**
differential diagnosis, **411, 418, 423–445**
grade, 402–404
gross appearance, 116
growth pattern, 140, *188–190, 192, 195, 203, 206*
growth phase, 14
history, 38
illustrations, 27
immunochemistry, 342, 343, 350, 353
products, 340–343, 346–353, *367, 376*
prognosis, 408, 410
radiology, 86, *88, 95*
recurrence, 410
references, 27
site, 38, 59, *68*
size, *81*
special stains, 341–347
stage, 405–407
stroma, 265–267, 270, *288, 336, 337*
survival, 410
symptoms, 38
synonyms, 26
ultrastructure, 342, 348, 349, *376*
Bizarre cells, 215
Bizarre leiomyosarcoma, 28
Blacks, 35, 38
Blastic lesions, 84–86
Blastoma
pulmonary, *168*
retinal, *178*
Blocking of tissue, 113
Blood vessel tumor(s), 10
Bone(s),
adult, 271
cranial, 60, 62
desmoid tumor of, 20
erosion, 87, 88, 89, 90, 92, 99, 101, 102, 107, 110
facial, 60, 61
fetal, 271
fibrosis, 16
fibrous histiocytoma with, 26
immature, 267
infarct, 8
invasion, *224*
lamellar, 267
mature, 267
pelvic, 60, 61
primitive multipotential primary sarcoma, 30
processing, 113
woven, 267
Bone cyst, aneurysmal. *See* Aneurysmal bone cyst
Bone sarcoma in Paget's disease. *See* Paget's sarcoma
Bone scan, *111*
Bone tumor(s), 3, 4, 7, 8, 10, 11. *See also* Names of neoplasms
classification, 3, 4, 10, 11
differentiation, 3, 215, 402–404
incidence, 8
malignant giant cell, 26
not considered, 18, 32
number, 8, *9*
primary, 138–140, 216–218, 268–270
size, 80, 81, 82
Bone-producing tumors, 11
Borderline lesion(s). *See* Borderline neoplasm(s)
Borderline neoplasm(s), 8, 12, *13,* 14, 15. *See also* Names of neoplasms
Borderline tumor(s). *See* Borderline neoplasm(s)
Botryoid sarcoma, 28. *See also* Myxoid rhabdomyosarcoma
Boyd, William, 8, 113
Breast carcinoma, *97, 232, 400*
Brown fat, lipoma of, 20
Brown tumor, 16
Buckled nuclei, 215
Bucy, 12
Bulging nuclei, 215
Burkitt's lymphoma, *42*
Bursal cyst, 16
Bursitis, 16
Buttock, 57–59

Café-au-lait spot(s), 37, 38
Calcification, post-irradiation, *339*
Calcified giant cell tumor, 24
Calcified lesions, 84–86
Calcified stroma, 336–339
Calcitonin, 350–353
Calcium, serum, 35, 36, 38, 267
Callus, 4, 8, 11, 115, 138, 216, 267, 268, 271, 344, 408, **411–413**
cell morphology, 408, 409
complication, 408, 409
cytology, 446
differential diagnosis, **411–413, 423–445**
gross appearance, 115
growth pattern, 138
products, 344
stroma, 268, 271
Capillary hemangioma, 22. *See also* Hemangioma
Carbohydrates, 340–347
Carcinoma,
basal cell, *222*
breast, *97, 232, 400*
clear cell, *166, 195, 224, 249, 250, 251*
desmoplastic, 142, *224, 230,* 232
embryonal, *249*
epidermoid, *142,* 230
gastric, *400*
in situ, 8
medullary, of thyroid, *186, 192, 291*
oncocytic renal cell, *240*
prostate, *108, 400*
renal cell, *195, 315,* 400
thyroid, 195
Cardiac rhabdomyoma, 22
Carpal tunnel syndrome, 10
Cartilage, 267
Cartilage tumor(s), 11
Cartilage-producing tumor(s), 11
Cathepsin B, 350–353
Cathepsin D, 350–353
Cause, 12, 35
Cavernous hemangioma, 22. *See also* Hemangioma
CEA, 343, 350, 351
Cell(s), 3, 4, 214–264. *See also* Specific names of cells
atypia, 12
changing of, 215
clear epitheliod, 214–218, 246–252
clone, 12, *13,* 215
cytology, 446–458
daughter, 4, 5
"dedifferentiation," 3
differential diagnosis, by, **423–445**
differentiation, 3, 12, 215
embryonic, 3
genetic coding, 215
mast cell, 236
maturation arrest, 3
mesenchymal, 3, 4, 5, 7, 56
morphology, 4, 12, 13, 15, 34, 137, 214–264
mutation, 12
neoplastic, 4

Cell(s) (*Continued*)
noncollagenous, 3
nuclei, 215
of embryo, 3
of origin, 4, 7, 12, 215
parent, 3, 5
phenotype, 3
pluripotential, 7, 215
poorly differentiated, 5
primitire meseuchymal, 4, 5, 7, 215
products, 3, 12, 13, 15, 34, 215, 340–401
reactive, 4
shape, 215
size, 215
smooth muscle, 347
tissue culture, 6
undifferentiated, 5, 12
well differentiated, 5
Cellular phase, 12, *13,* 14, 15
Cellularity and grade, 402–404
Cementoblastoma, 24
Cerebroside, 347
Cervical chordoma, 30
Cheloid, 16
Chemodectoma, 22
Chemotherapy, 408
Chemotherapy–induced sarcoma, 11
Chest wall, 57–59
Childhood neoplasms, 50–53
Children, 50
tumor(s), 8
Chlorophenols, 35
Cholesterol crystals, *375*
Chondrified stroma, 271, 326–327
Chondroblast(s), 4, 347
Chondroblastic osteosarcoma, 30
Chondroblastic sarcoma, 30
Chondroblastoma, 8, 11, **24,** 25, 37, 40, *78,* 85, *89,* 114, 115, 136, 137, 139, 217, 269, 345, 408, 409, **411**
age distribution, 50, 52
cell morphology, 217, *248*
classification, 11
clinical appearance, 37
complication, 408, 409
cytology, 446, *453*
definition, **24**
differential diagnosis, **411, 414–416, 423–445**
gross appearance, 115
growth pattern, 139, *186*
history, 37
illustrations, 25
malignant transformation, 9, 30, 409
products, 340–343, 345, 347–353
prognosis, 408, 409
radiology, 85, *89*
recurrence, 408, 409
references, 25
sex distribution, 54, 55
site, 40, 60, 62, *78*
size, 79, 80
stroma, 267, 269, 271, *337, 339*
symptoms, 37
synonym(s), 8, 24
Chondroid metaplasia, 8, 326
Chondroitin sulfate, 341–347
Chondroma, 11, **24,** 25, 37, 40, *71,* 85, 114, 115, 217, 269, 345, 408, 409, **411**
age distribution, 50, 52
cell morphology, 217
classification, 11
clinical appearance, 37
complication, 408, 409
cytology, 446
definition, **24**
differential diagnosis, **411, 414–416, 423–445**
gross appearance, 115
growth pattern, 139
history, 37
illustrations, 25
malignant transformation, 9, 409
products, 340–343, 345, 347–353
prognosis, 408, 409
radiology, 37, 85, *103*
references, 25
sex distribution, 54, 55
site, 60, 62, *71*
stroma, 269, 271
symptoms, 37
synonyms, 24
Chondromatosis, tendosynovial. *See* Tendosynovial chondromatosis
Chondromatous osteoelastoma, 24
Chondrometaplasia, 18
Chondromyxoid fibroma, 11, **24,** 25, 37, 40, 52, *77,* 85, 115, 139, 217, 269, 345, 408, 409, **411**
age distribution, 50, 52
cell morphology, 217
clinical appearance, 37
complication, 408, 409
cytology, 446
definition, **24**
differential diagnosis, **411, 414–416, 423–445**
gross appearance, 115
growth pattern, 139, *161*
history, 37
illustrations, 25, 409
products, 340–343, 345, 347–353
prognosis, 408, 409
radiology, 37
recurrence, 408, 409
references, 25
sex distribution, 54, 55
site, 37, 60, 62, *77*
size, 79, 80
stroma, 269, 271
symptoms, 37
synonyms, 24
Chondrosarcoma, 8, 11, **30,** 31, 38, 40, *71,* 86, 114, 116, 140, 218, 270, 271, 346, 402, 405, 408, **411**
age distribution, 50, 53
borderline, 15
cell morphology, 214, 215, 218, *222, 251, 252*
clinical appearance, 38
cytology, 446, *454, 458*
definition, **30**
differential diagnosis, **411, 421, 423–445**
differentiation, 11
grade, 402–404
gross appearance, 116, *126, 130, 133, 135*
growth pattern, 140, *168, 170*
growth phase, 14
histochemistry, 342, 348
histology, 8
history, 38
illustrations, 31
immunochemistry, 342, 343, 350, 353, *399, 401*
mesenchymal. *See* Mesenchymal chondrosarcoma
metastasis, 410
products, 340–343, 346–353, *368, 384, 399, 401*
prognosis, 410
radiology, 38, 86, *88, 89, 91, 101, 108, 109*
recurrence, 410
references, 31
sex distribution, 54, 55
site, 38, 61, 62, *71*
special stains, 341–347
stage, 405–407
stroma, 265–267, 270, 271, *294, 326, 327, 335, 336, 337, 339*
survival, 410
symptoms, 38
synonyms, 30
ultrastructure, 342, 348, 349, *384*
Chondrosarcomatous osteosarcoma. *See* Osteosarcoma
Chondrosis,
articular, 18
Chordal mesoderm, *4*
Chordoid sarcoma, 10, 15, 59, 86, 140, 218, 270, 346, 402, 405, 408, **411**
cell morphology, 218, *222*
cytology, 446
differential diagnosis, **411, 418, 423–445**
grade, 402–404
growth pattern, 140, *173, 187*
growth phase, 14
products, 346, *368*
prognosis, 408, 410
recurrence, 410
site, 59
size, 81
special stains, 341–347
stage, 405–407
stroma, 270, *292*
survival, 410
Chordoma, 11, 15, **30,** 31, 38, 40, 53, 55, 86, *108,* 116, 218, 270, 271, 346, 402, 405, 408, **411**
age distribution, 50, 53
cell morphology, 218, *243, 251, 252*
classification, 11
clinical appearance, 38, *49*
cytology, 446, *452*
definition, **30**
differential diagnosis, **411, 422, 423–445**
grade, 402–404
gross appearance, 116, *131*
growth pattern, 140 *186*
growth phase, 14
histochemistry, 342, 348
history, 38
illustrations, 31
immunochemistry, 342, 343, 350, 353, *399*
metastasis, 410
products, 340–343, 346–353, *368, 386, 399*
prognosis, 410
radiology, 38, 86, *108*
recurrence, 410
references, 31
sex distribution, 54, 55
site, 38, 61, 62
size, 79, 81
special stains, 341–347
stage, 405–407
stroma, 270, 271
survival, 410
symptoms, 38
synonyms, 30
ultrastructure, 342, 348, 349, *386*
Choriocarcinoma, *104, 264, 324*
Cicatricial fibrosarcoma, 26
Cigar-shaped nuclei, 215
Classification, 3, 9, 10, 11, 12, *13,* 14
histogenetic, 9, 10, 11
Clavicle, 60, 61
Clear cell carcinoma, *166, 195, 224, 249, 250, 251*
Clear cell chondrosarcoma, 30
Clear cell leiomyosarcoma, 28
Clear cell sarcoma, 10, 15, **26,** 27, 38, 39, *48, 68,* 86, 140, 218, 270, 346, 402, 405, 408, **411**

cell morphology, 214, 215, 218, *232, 249, 250*
clinical appearance, *48*
containing melanin, 30
cytology, 446
definition, **26**
differential diagnosis, **411, 418, 423–445**
grade, 402–404
gross appearance, *134*
growth pattern, 140, *171, 182, 191*
growth phase, 14
illustrations, 27
immunochemistry, 342, 343, 350, 353
metastasis, 410
of tendons, 26
products, 340–343, 346–353, *361, 384*
prognosis, 410
radiology, 86, *101, 102*
recurrence, 410
references, 27
site, 56, 59, *68*
size, 79, 81
special stains, 341–347
stage, 405–407
stroma, 265–267, 270, *286, 293*
survival, 410
synonyms, 26
ultrastructure, 342, 348, 349, *384*
Clear epithelioid cell(s), 214–218, 246–252
Clinical presentation, 8, 12, *13*, 15, 35–40
benign neoplasms, 35, 37, 39, 40
malignant neoplasms, 35, 38, 39, 40
non-neoplastic lesions, 35, 36, 39, 40
Codman's tumor, 24
Collagen, 340
as marker, 266
dermal, *355, 356*
production, 341
stains, 341
types, 341, 347
Collagenoma, 10, 36, 40, 57, 84, 138, 216, 268, 344
cell morphology, 216
clinical appearance, 36
cytology, 446
growth pattern, 138
history, 36
products, 344, *355*
radiology, 36, 84, *104*
site, 36, 57
size, 79
special stains, 341–347
stage, 268
symptoms, 36
Computed tomography, 82, 83
Congenital fibrosarcoma, 26
Congenital lesions, 36, 37, 38
Congenital xanthoma multiplex, 16
Consultation, 15
Cooper, S., 3
Cranial bones, 60, 62
Cranium. *See* Cranial bones; Head; Internal cranium
Cruveilhier, Jean, 54
Crystals, 340–353, *375*
CT scan, 82, 83
Cullen, William, 35
Curetting(s), 113
Cut surface, 113
Cutting of specimen, 113
Cyst(s), 16. *See also* Aneurysmal bone cyst; Tendosynovial cyst
thymic, 11
Cystic hygroma, 10, *313*
Cystic myxoma, 16
Cystic schwannoma, 22
Cysticercus cellulosae, *324*
Cystosarcoma, 24. *See* Benign cystosarcoma phyllodes; Malignant cystosarcoma phyllodes
Cystosarcoma phyllodes, malignant, 11, 140, 218, 270, 346
cell morphology, 218, *220, 228*
growth pattern, 140, *146, 153*
products, 346
stroma, 270
Cytochemical procedures, 266, 341–343, 350–354
Cytology, *446–458*
aspiration, 15, 113, *443–448*
of benign histiocytic fibrous histiocytoma, *455*, 466
of benign nevoid schwannoma, *447*
of benign schwannoma, *447*
of chondroblastoma, *453*
of chondrosarcoma, 446, *454, 458*
of chordoma, 446, *452*
of granular cell tumor, *453*
of granuloma, *450*
of infarct, *454*
of leiomyosarcoma, *449, 458*
of lipoblastic liposarcoma, *452*
of malignant fibroblastic fibrous histiocytoma, *448*
of malignant histiocytic fibrous histiocytoma, *455*
of malignant pleomorphic fibrous histiocytoma, *449, 456, 458*
of megakaryocyte, *454*
of myxoid liposarcoma, *447, 448*
of osteosarcoma, *457, 458*
of osteomyelitis, *450*
of plasma cell myelomoma, *450*
of pleomorphic liposarcoma, *456*
of rhabdomyoblastoma, *453*
Cytoplasm, 215
Cytoplasmic organelles, 3
Cytotrophoblast, 3

Darier's tumor, 26
Deafness, 36
Decalcification, 113
Decidua, *252, 293*
"Dedifferentiated" chondrosarcoma, 30
"Dedifferentiation," 3
Deep fibromatosis, 26
Deep non-neoplastic lesions, 36, 37, 38, 40
Deep sarcomas, 405–407
Definition(s), **7, 9, 12, 13, 16–31**
Definitive diagnosis, 8, 82
Dermal collagen, *355, 356*
Dermatocele, 18
Dermatofibroma, 20. *See also* Fibroma
Dermatofibrosarcoma protuberans, 26. *See also* Malignant fibroblastic fibrous histiocytoma
Dermis, 4
Desmin, 343, 350–353
Desmoplastic fibroma, 26
Desmoid fibromatosis, 26
Desmoid tumor, 8, 10, **26**, 27, 38, 40, *41, 44, 46, 53, 55, 65*, 86, 114, 116, 136, 137, 140, 218, 270, 346, 402, 405, 408, **411**
age distribution, 50, 53
cell morphology, 214, 215, 218, *227, 233*
clinical appearance, 38, *41, 44, 46*
cytology, 446
definition, **26**
differential diagnosis, **411, 417, 423–445**
grade, 402–404
gross appearance, 116, *118, 125*
growth pattern, 140, *153, 155, 158*
growth phase, 14
history, 38
illustrations, 27
immunochemistry, 342, 343, 350, 353
of bone, 20
products, 340–343, 346–353, *354*
prognosis, 410
radiology, 38, 86, *102, 105, 107*
recurrence, 410
references, 27
sex distribution, 54, 55
site, 56, 59, *65*
size, 79, 81
stage, 405–407
stroma, 265–267, 270, *279, 287*
survival, 410
symptoms, 38
synonyms, 8, 26
ultrastructure, 342, 348, 349
Desmoplastic carcinoma, 142, *224, 230, 232*
Desmoplastic fibroma, 8, 10, **20**, 21, 37, 40, *75*, 85, 114, 115, 217, 269, 345, 408, 409, **411**
age distribution, 50, 52
cell morphology, 217, *221*
clinical appearance, 37
complication, 408, 409
cytology, 466
definition, **20**
differential diagnosis, **411, 414–416, 423–445**
gross appearance, 115
growth pattern, 139
history, 37
illustrations, 21, 409
products, 345
prognosis, 408, 409
recurrence, 408, 409
references, 21
sex distribution, 54, 55
site, 40, 60, 62, *75*
size, 80
stroma, 269
symptoms, 37
synonym(s), 8, 20
Desmoplastic melanoma, *157, 162, 229*
Destructive lesions, 84–86
Diagnosis, differential. *See* Differential diagnosis
Diagnostic growth phase, 12, *13*, 14, 15
Diaphysis, 62
Diastase digestion, 341–347
Differential diagnosis, **13, 15**
benign neoplasms, **411, 417–422, 423–445**
by appearance of stroma, **423–445**
calcified, **426, 431, 434, 438, 442, 445**
chondrified, **425, 430, 442, 445**
fibrillar, **423, 427, 432, 435, 439, 443**
inflamed, **425, 429, 434, 437, 441, 444**
myxoid, **424, 429, 433, 436, 440, 444**
necrotic, **425, 430, 434, 437, 442, 445**
ossified, **426, 430, 434, 438, 445**
sclerosed, **424, 428, 432, 436, 440, 443**
vascular, **425, 429, 433, 437, 441, 444**
by growth pattern, **423–445**
aveolar, **439–442**
arranged, **423–426**
disarranged, **443–445**
epithelioid, **435–438**
lacy, **432–434**
spreading, **427–431**
by type of cells, **423–445**
malignant neoplasms, **411, 417–422, 423–445**
non-neoplastic lesions, **411–413, 423–445**
Differentiation, 3, 215
and grade, 402–404
Diffuse neurofibromatosis, 22

Dihydroxyphenylalanine oxidase, 348
Disease(s)
  Kimura's, 18
  of women, 37
  Ollier's, 24
  Paget's. *See* Paget's disease
  von Recklinghausen's, 22
    neurofibroma with, 28
  Weber-Christian, 18
Dissection of specimen(s), 113
DNA, 12
Drawings, 113
Dysplasia. *See* Fibrous dysplasia
Dystrophy of muscle, 10, *320*

Ecchordosis physaliphora, 11
Ectoderm, 3
Ectomesenchymoma, 30
Effusion, 37, 38
Elastin, 266
Elastodyplasia, 16
Elastofibroma, 10, **16,** 17, 36, 56, 84, 114, 115, 216, 268, 344, 408, 409, **411–413, 423–445**
  age distribution, 50, 51
  cell morphology, 214–216
  clinical appearance, 36
  complication, 408, 409
  cytology, 446
  definition, **16**
  differential diagnosis, **411–413, 423–445**
  gross appearance, 115
  growth pattern, 138
  histochemistry, 342, 348
  histogenesis
  history, 36
  illustrations, 17
  immunochemistry, 215
  products, 344, *355, 359, 377*
  prognosis, 408, 409
  radiology, 36, 84
  recurrence, 408, 409
  references, 17
  sex distribution, 54
  site, 40, 56, 57
  size, 79, 80
  special stains, 215, 341–347
  stroma, 268, *291*
  symptoms, 36
  synonyms, 16
  ultrastructure, 215, 342, 348, 349
Elastofibroma dorsi, 16. *See also* Elastofibroma
Elbow, 57–59
Elderly, 50
Electron microscopy, 13, 113, 215
Embryo, *3,* 4
Embryonal carcinoma, *249*
Embryonal lipoma, 20
Embryonal rhadbomyosarcoma, 4, 11, **28,** 29, 38, 40, *42, 49,* 53, 55, 65, 86, 114, 116, 218, 270, 346, 402, 405, 408, **411.** *See also* Alveolar rhabdomyosarcoma; Myxoid rhabdomyosarcoma
  age distribution, 50, 53
  cell morphology, 214, 215, 218, *222, 223, 234, 235, 236, 237, 238, 248*
  classification, 10, 11
  clinical appearance, 38, *42, 49*
  cytology, 446
  definition, **28**
  differential diagnosis, **411, 419, 423–445**
  grade, 402–404
  gross appearance, *41,* 118, *121, 125,* 127
  growth pattern, 140, 150, *152, 153, 164, 166, 170, 171, 177, 188, 190–192, 201, 206*
  growth phase, 14
  histochemistry, 342, 348
  history, 38
  illustrations, 29
  immunochemistry, 342, 343, 350, 353, *395, 396*
  metastasis, 410
  products, 340–343, 346–353, *357, 366, 367, 381, 395, 396*
  prognosis, 410
  radiology, 38, 86, *87, 106, 110*
  recurrence, 410
  references, 29
  sex distribution, 54, 55
  site, 38, 59, *65*
  size, 79, 81
  special stains, 341–347
  stage, 405–407
  stroma, 265–267, 270, *284, 293, 295, 297, 298*
  survival, 410
  symptoms, 38
  synonyms, 28
  ultrastructure, 242, 348, 349, *381*
Embryonic disc, *3,* 4
Embryonic mesoderm, 3, 4
Enchondroma, 9, 24
Endoderm, 3
Enolase, neuron specific, 350–353
Enostosis, 11
Enzyme histochemistry, 342
Eosinophilia, 36
Eosinophilic granuloma, 11, *87,* 138, 216, 268, 317, 344, 408, **411–413, 423–445**
  cell morphology, 216
  complication, 408, 409
  cytology, 446
  differential diagnosis, **411–413, 423–445**
  growth pattern, 138
  products, 344
  prognosis, 408, 409
  stroma, 268
Epidermoid carcinoma, *142,* 230
Ependymoma, *119*
  myxopapillary. *See* Myxopapillary ependymoma
Epinephrine, 35
Epiphyseal chondroblastoma, 24
Epiphyseal giant cell tumor, 24
Epiphysis, 62
Epiphysis–metaphysis, 62
Epithelial mesothelioma, 30. *See also* Mesothelioma
Epithelioid benign peripheral nerve tumor, 22
Epithelioid hemangioendothelioma, 28
Epithelioid hemangioma, 18
Epithelioid leiomyosarcoma, 28
Epithelioid liposarcoma, 28
Epithelioid rhabdomyosarcoma, 28. *See also* Embryonal rhabdomyosarcoma
Epithelioid sarcoma, 10, **26,** 27, 39, *48,* 56, 79, 86, 140, 218, 270, 346, 402, 405, 408, **411**
  cell morphology, 214, 215, 218, *230*
  clinical appearance, *48*
  cytology, 446
  definition, **26**
  differential diagnosis, **411, 418, 423–445**
  grade, 402–404
  gross appearance, *121, 134*
  growth pattern, 140, *175, 176, 182*
  growth phase, 14
  illustrations, 27
  immunochemistry, 342, 343, 350, 353
  metastasis, 410
  products, 340–343, 346–353, *361, 367*
  prognosis, 410
  radiology, 86, *101, 102*
  recurrence, 410
  references, 27
  site, 56, 59
  size, 79, 81
  stage, 405–407
  stroma, 265–267, 270, *272, 279, 286, 323, 325*
  survival, 410
  synonyms, 26
  ultrastructure, 342, 348, 349
Epitheliosarcoma, 26
Equivocal neoplasm(s), 12, *13,* 14, 15
Error(s), 8
Esterase(s), 348
Etiology, 12, 35
Ewing, James, 4, 9, 12, 402
Ewing's sarcoma, 4, 11, 15, **30,** 31, 38, 40, 53, 55, *73,* 82, 86, 114, 116, 218, 270, 346, 402, 405, 408, **411**
  age distribution, 50, 53
  cell morphology, 214, 215, 218, *235*
  classification, 11
  clinical appearance, 38
  cytology, 446, *451*
  definition, **30**
  differential diagnosis, **411, 422, 423–445**
  grade, 402–404
  gross appearance, 116, *119, 129, 130, 132, 135*
  growth pattern, 140, *171, 172, 178, 202*
  growth phase, 14
  histogenesis, 4
  history, 38
  illustrations, 31
  immunochemistry, 342, 343, 350, 353
  metastasis, 410
  products, 340–343, 346–353, *365, 387*
  prognosis, 410
  radiology, 38, 86, *99, 100, 108, 110, 111*
  recurrence, 410
  references, 31
  sex distribution, 54, 55
  site, 40, 61, 62, *73*
  size, 79, 81
  special stains, 341–347
  stage, 405–407
  stroma, 270
  symptoms, 38
  synonyms, 30
  ultrastructure, 342, 348, 349, *387*
Ewing's tumor, 30
Excisional biopsy, 82, 113
Exfoliative cytology, 15, *446–458. See also* Cytology
Exostosis, 24
Expansile lesions, 84–86
Extra-abdominal desmoid, 26
Extracellular matrix, 266
Extracranial meningioma, 139, *180,* 193. *See also* Meningioma
Extraosseous tumor(s). *See* Soft tissue tumor(s)
Extraskeletal chondrosarcoma, 30, 50, 53, 54, 55, 59, *66,* 79, 81, 116, *133. See also* Chondrosarcoma
Extraskeletal Ewing's sarcoma, 30
Extraskeletal plasmacytoma, *242*
Extraskeletal tumor(s). *See* Soft tissue tumor(s)

Face. *See* Facial bones; Head
Facial bones, 60, 61
Factor-VIII-related antigen, 343, 350, 351, 353
Familial linkage, 35, 36, 37, 38, 39, 40
Fascial fibroma, 20. *See also* Fibroma
Fasciitis, 4, 10, **16,** 17, 36, 39, *67,* 84, 114, 115, 138, 216, 268, 344, 408, 409, **411–413, 423–445**
  age distribution, 50, 51
  cell morphology, 214–216, *219*

classification, 10
clinical appearance, 36
complication, 36, 408, 409
cytology, 466
definition, **16**
differential diagnosis, **411–413, 423–445**
gross appearance, 36, 115
growth pattern, 136–138, *145, 151, 156, 160, 162, 204, 209, 211*
histochemistry, 215
illustrations, 17
immunochemistry, 215, 342, 343, 350, 353
involving fat, 18
products, 344
prognosis, 408, 409
radiology, 36
recurrence, 408, 409
references, 17
sex distribution, 54
site, 36, 39, 57, *67*
size, 79, 80
special stains, 215, 341–347
stroma, 265–268, *272, 276, 277, 278, 298, 300, 303, 308, 316, 318, 322, 331*
symptoms, 36
synonyms, 16
ultrastructure, 215, 342, 248, 249
with metaplasia, 16
Fasciitis ossificans, 10, 16, *331*
Fat, 4, 5, 340–353, 354, *370, 371*
brown, lipoma of, 20
metaplasia, 10
Fat necrosis, 8, 10, 36, 39, 40, 114, 115, 215, 268, *321, 324,* 344, 408, 409, **411–413, 423–445**
cell morphology, 214–216, *247*
clinical appearance, 36
complication, 408, 409
cytology, 446
differential diagnosis, **411–413, 423–445**
gross appearance, 115
growth pattern, 138, *166*
history, 36
immunochemistry, 215
organized, 18
products, 344, *371, 375*
radiology, 36
recurrence, 408, 409
site, 39, 57
size, 79, 80
special stains, 215
stroma, 268
symptoms, 36
ultrastructure, 215
Fat tumor(s), 10
Female predominance, 54, 55
Femur, 60, 61
Fetal bone, 271
Fetal rhabdomyoma, 22
Fever, 38
Fibrillar stroma, 272–277
Fibroadenoma
giant, 24
intracanaliculare, 24
Fibroblast(s), 3, 4, 6, 347
Fibroblastic fibrosarcoma, 26. *See also* Fibrosarcoma
Fibroblastic lipoma, 20, 139, 217, 269, 345, 408, 409, **411**
cell morphology, 217
complication, 408, 409
cytology, 466
differential diagnosis, **411, 414–416, 423–445**
growth pattern, 139
products, 340–343, 345, 347–353
prognosis, 408, 409
recurrence, 408, 409
stroma, 265–267, 269, *275, 289, 296*
Fibroblastic liposarcoma, **28,** 29, 38, 40, *42,* 53, 55, 86, 116, 218, 270, 346, 402, 405, **411**
cell morphology, 218, *222, 229*
clinical appearance, 38, *42*
cytology, 446
definition, **28**
differential diagnosis, **411, 419, 423–445**
grade, 402–404
gross appearance, 116
growth pattern, 140, *144, 169*
growth phase, 14
history, 38
illustrations, 29
metastasis, 410
products, 346
prognosis, 410
recurrence, 410
references, 29
site, 59
size, 81
stage, 405–407
stroma, 270, *273, 275, 293, 309*
survival, 410
symptoms, 38
synonyms, 28
Fibroblastoma, 20
Fibrodysplasia ossificans progressiva, 11
Fibroepithelial papilloma, 282
Fibrohistiocytic tumor, intermediate, 26
Fibroma, 8, 10, 16, **20,** 21, 37, 39, 40, 115, 136, 137, 217, 269, 345, 408, 409. *See also* Nonossifying fibroma; Ossifying fibroma
age distribution, 50, 52
cell morphology, 217
chondromyxoid. *See* Chondromyxoid fibroma
clinical appearance, 37
definition, **20**
desmoplastic, 26
gross appearance, 115
growth pattern, 139, *142*
history, 37
illustrations, 21
products, 340–343, 345, 347–353, *377*
radiology, 37
references, 21
sex distribution, 54, 55
site, 58
size, 80
stroma, 265–267, 269, *278, 283, 288, 291, 317, 338*
symptoms, 37
synonyms, 8, 20
ultrastructure, *377*
Fibroma durum, 20
Fibromatosis, 8, 10, **16,** 17, 32, 35, 36, 39, 40, *42,* 84, 114, 115, 216, 268, 344, 408, 409, **411–413, 423–445**
age distribution, 50, 51
cell morphology, 214–216
clinical appearance, 35, 36, *42*
complication, 408, 409
definition, **16**
differential diagnosis, **411–413, 423–445**
gross appearance, 115, *124*
growth pattern, 138, *151, 155, 164, 205*
histogenesis, 35
history, 36
illustrations, 17
immunochemistry, 215, 342, 343, 350, 353
malignant transformation, 9, 409
post-radiation, *281*
products, 344, *355*
prognosis, 408, 409
radiology, 36, 84
recurrence, 408, 409
references, 17
sex distribution, 54
site, 39, 40, 57, 60, 62
size, 79, 80
special stains, 215, 342, 348
stroma, 265–268, *278, 281, 282, 316, 317, 319, 323, 331, 336*
symptoms, 36
synonyms, 8, 32
types, 26
ultrastructure, 215, 242, 348, 349
Fibromatosis ossificans, 10, *331*
Fibromyxoid chondroma, 24
Fibronectin, 266
Fibrosarcoma, 7, 10, 20, **26,** 27, 38, 40, *44, 48,* 53, 55, *70, 78,* 86, 114, 116, 218, 270, 346, 402, 405, 408, **411**
age distribution, 50, 53
borderline, 15
cell morphology, 214, 215, 218, *220, 223, 227, 229, 260, 264*
classification, 10
clinical appearance, 38, *44, 48*
cytology, 446
definition, **26**
differential diagnosis, **411, 417, 423–445**
grade, 402–404
gross appearance, 116, *117, 120–122, 128*
growth pattern, 140, *152, 157, 162, 209, 212*
growth phase, 14
histochemistry, 342, 348
history, 38
illustrations, 27
immunochemistry, 342, 343, 350, 353
metastasis, 410
products, 340–343, 346–353, *354, 377*
prognosis, 410
radiology, 38, 86, *90, 98, 99*
recurrence, 410
references, 27
sex distribution, 54, 55
site, 40, 59, 61, 62, *70, 78*
size, 79, 81
special stains, 341–347
stage, 405–407
stroma, 265–267, 270, *272, 287*
survival, 410
symptoms, 38
synonyms, 26
ultrastructure, 342, 348, 349, *377*
Fibrosarcomatous mesothelioma, 30. *See also* Mesothelioma
Fibrosarcomatous osteosarcoma. *See* Osteosarcoma
Fibrosis of bone, 16
Fibrosis cortical defect, 20
Fibrous dysplasia, 8, 10, **16,** 17, 36, 40, *74,* 84, 115, 216, 268, 271, 344, 408, **411–413, 423–445**
age distribution, 50, 51
cell morphology, 216
clinical appearance, 36
complication, 408, 409
cytology, 446
definition, **16**
differential diagnosis, **411–413, 423–445**
gross appearance, 115
growth pattern, 138
histology. *See* Cell morphology; Growth pattern; Products; Stroma
history, 36
illustrations, 17
malignant transformation, 9, 409
products, 344

Fibrous dysplasia (*Continued*)
prognosis, 408, 409
radiology, 36, 84
recurrence, 408, 409
references, 17
sex distribution, 54
site, 36, 40, 60, 62, 74
size, 79, 80
stroma, 268
symptoms, 36
synonym(s), 8, 16
Fibrous histiocytic osteosarcoma, 30. *See* Osteosarcoma
Fibrous histiocytoma, 12, 20. *See also* Benign fibroblastic fibrous histiocytoma; Benign histiocytic fibrous histiocytoma; Benign pleomorphic fibrous histiocytoma; Malignant fibroblastic fibrous histiocytoma; Malignant histiocytic fibrous histiocytoma; Malignant pleomorphic fibrous histiocytoma; Nonossifying fibroma
inflammatory, 26
of bone in association with bone infarct, 26
Fibrous mesothelioma. *See* Mesothelioma
Fibrous tissue tumors, 10
Fibroxanthoma, 20
Fibroxanthosarcoma, 26
Fibula, 60, 61
Filamentous cytoplasm, 215
Fine needle biopsy. *See* Aspiration cytology
Fine structure, 341, *354, 376–391*
Finger(s). *See* Hand; Phalanges
Focal osteitis fibroma, 16
Folliculitis keloidalis, 16
Foot, 57–59
Forearm, 57–59
Foreign body granuloma, 10
Fracture, 36, 408–410
Frozen section, 15, 402
and grading, 402
Function, 341

Galen, 9, 114
Ganglion, 16
Ganglioneuroblastoma, 11
Ganglioneuroma, 11, 139, *181*, 217, 269, 408, 409
bone, *98*
cell morphology, 217, *221, 261, 264*
immunochemistry, *397*
products, 345, *358, 373, 397*
stroma, 269, *277*
Ganglioside, 347
Gardner's syndrome, 35, 36, 37, 38
Gastric carcinoma, *400*
Gastrin, 351–353
Gastrointestinal tract, 57–59
Gaucher's disease, 11
Genetic coding, 215
GFAP, 350–353
Giant cell(s)
isomorphic, 214–218, 253–255
pleomorphic, 214–218, 256–264
Giant cell granuloma, 10, **16,** 17, 36, 84, *87*, 115, 138, 216, 268, 344, 408, 409, **411–413, 423–445**
age distribution, 50, 51
cell morphology, 214–216, *221, 255*
clinical appearance, 36, 408, 409
cytology, 446
definition, **16**
differential diagnosis, **411–413, 423–445**
gross appearance, 115
growth pattern, 138
histochemistry, 215
history, 36
illustrations, 17
products, 344
radiology, 36, 84, *87*
recurrence, 408, 409
references, 17
sex distribution, 54
site, 36, 40, 60, 62
size, 79, 80
special stains, 215
stroma, 268, *310, 328*
symptoms, 36
synonyms, 16
ultrastructure, 215
Giant cell liposarcoma, 28
Giant cell predominant osteosarcoma, 30
Giant cell reparative granuloma, 16
Giant cell tumor(s), 8. *See also* Benign histiocytic fibrous histiocytoma; Malignant histiocytic fibrous histiocytoma
atypical, 16
calcified, 24
epiphyseal, 24
of bone, 26
soft tissue, 20, 26
tendon sheath, 20, 26
Giant fibroadenoma, 24
Giant osteoid osteoma, 24
Gilford, H., 340
Glandular neurofibroma, 22
Glandular schwannoma, 11, 22, 140, *188, 399*
Glial fibrillary acidic protein, 350–353
Glial filaments, 343, 350–353
Glioblastoma, *308*
Glioma, *398*
Globoid cytoplasm, 215
Globoside, 347
Glomangioma, 22
Glomus tumor
benign, 9, 10, **22,** 23, 37, 40, 139, 217, 269, 345, 408, 409, **411**
age distribution, 50, 52
atypical, *243, 305*
borderline, 15, *305*
cell morphology, 217, *243*
classification, 10
clinical appearance, 37
complication, 408, 409
cytology, 466
definition, **22**
differential diagnosis, **411, 414–416, 423–445**
gross appearance, 115
growth pattern, 139, *178, 179*
history, 37
illustrations, 23
malignant transformation, 9, 409
products, 345
prognosis, 408, 409
recurrence, 408, 409
references, 23
sex distribution, 54, 55
site, 58
size, 80
stroma, 269, *301, 305*
symptoms, 37
synonyms, 22
malignant, 10, 140, 218, 346
atypical, *243*
borderline, 15, *243*
cell morphology, 218, *243*
classification, 10
differential diagnosis, **411, 423–445**
growth pattern, 140, *173, 179, 196*
growth phase, 14
products, 346, *360*
stroma, *305*
Glucagon, 350–353
Glucoconjugates, 342, 347
Glycogen, 340, *354, 365, 366*
Glycolipids, 347
Glycoprotein(s), 266, 341–349
Glycosaminoglycans, 341–349
Gouty arthritis, 10, 36, 40, 84, 115, 138, 216, 268, *324*, 344, 408, 409
age distribution, 50, 51
cell morphology, 214–216
clinical appearance, 36
complication, 408, 409
differential diagnosis, **411–413, 423–445**
gross appearance, 115
growth pattern, 138
histochemistry, 215
history, 36
products, 344
prognosis, 408, 409
radiology, 36
recurrence, 408, 409
sex distribution, 54
site, 57
size, 80
special stains, 341–347
stroma, 268
symptoms, 36
ultrastructure, 215
Grade, 15, 113, 267, 402–404
Granular cell myoblastoma, 24
with organoid structure, 30
Granular cell neurofibroma, 22, 24
Granular cell schwannoma, 24
Granular cell tumor
benign, 9, 11, **24,** 25, 37, 39, 40, *64*, 114, 115, 217, 269, 345, 408, 409, **411**
age distribution, 50, 52
borderline, 15
cell morphology, 217, *245*
classification, 11
clinical appearance, 37
complication, 9, 408, 409
cytology, 446, *453*
definition, **24**
differential diagnosis, **411, 414–416, 423–445**
gross appearance, 121
growth pattern, 136, 137, 139, *184, 193*
history, 37
illustrations, 25
malignant transformation, 9, 409
products, 340–343, 345, 347–353, *365*
prognosis, 408, 409
recurrence, 408, 409
sex distribution, 54, 55
site, 37, 58, *64*
size, 80
stroma, 269
symptoms, 37
synonyms, 24
ultrastructure, 342, 348, 349, *389*
malignant, 11, 38, 53, 55, 140, 218, 270, 346, 402, 405, 408, **411**
age distribution, 50, 53
cell morphology, 218, *245*
clinical appearance, 38
cytology, 446
differential diagnosis, **411, 422, 423–445**
grade, 402–404
growth pattern, 140, *181*
histochemistry, 342, 348
history, 38
immunochemistry, 342, 343, 350, 353, *396*

metastasis, 410
products, 340–343, 346–353, *389, 396*
prognosis, 410
recurrence, 410
sex distribution, 54, 55
site, 59
size, 81
special stains, 341–347
stage, 405–407
stroma, 270
survival, 410
symptoms, 38
ultrastructure, 342, 348, 349, *389*
Granular epithelioid cell(s), 214–218, 234–245
Granular neuroma, 24
Granular nuclei, 215
Granulation-tissue-type hemangioma, 18
Granulocytic leukemia. *See* Granulocytic sarcoma
Granulocytic sarcoma, 8, 11, 38, 40, 53, 55, 114, 116, 140, 218, 270, 346, 402, 405, 408, **411**
age distribution, 50, 53
cell morphology, 218, *236*
clinical appearance, 38
differential diagnosis, **411, 422, 423–445**
grade, 402–404
gross appearance, 116
growth pattern, 140
growth phase, 14
histochemistry, 342, 348
history, 38
metastasis, 410
products, 340–343, 346–353, *374, 375*
prognosis, 410
recurrence, 410
sex distribution, 54, 55
site, 38
size, 81
special stains, 341–347
stage, 405–407
stroma, 270, *285*
survival, 410
symptoms, 38
synonym(s), 8
Granuloma, 4
atypical pyogenic, 18
foreign body, 10
giant cell reparative, 16
plasma cell, 11
reticulohistiocytic, 18
Granuloma gravidarum, 18
Granuloma pyogenicum, 18
Granuloma teleangiecticum, 18
Granulomatous tendosynovitis, 10
Gross appearance, 15, 114–116, *117–135*
benign neoplasms, 114, 115
malignant neoplasms, 114, 116
non-neoplastic lesions, 114, 115
Gross cutting, 113
Growth pattern, 12, 13, 15, 34, 136–213, 215
Growth phases, 12, *13,* 14, 15, 215
Gustafson, 12

Hamartoma
angiomatous lymphoid, 11
renal, 20
vascular, 18
Hand, 57–59
HCG, 350–353
Head, 57, 58, 59
Heart, rhabdomyoma of, 22
Hemangioblastoma, 58, *307*
Hemangioendothelioma, 28
Hemangioendotheliomatosis, 28
Hemangiofibroma, 22
Hemangioma, 10, 18, 20, **22,** 23, *42, 72,* 114, 115, 136, 137, 139, 217, 269, 345, 408, 409, **411**
age distribution, 50, 52
arteriovenous, 18, 22
atypical, *197, 304*
borderline, *197, 304*
capillary, 22
classification, 10, 217
clinical appearance, *42*
complication, 408, 409
cytology, 446
definition, **22**
differential diagnosis, **411, 414–416, 423–445**
gross appearance, 115
growth pattern, 139, *196–198*
hypertrophic, 10, 22, *197, 304, 383*
illustrations, 23
immunochemistry, 342, 343, 350, 353
malignant transformation, 9, 409
products, 345, *383*
prognosis, 408, 409
recurrence, 408, 409
references, 23
sex distribution, 54, 55
site, 58, 60, 62, *72*
size, 79, 80
stroma, 269, *304, 307, 311*
synonyms, 22
ultrastructure, 342, 348, 349, *383*
Hemangiomatosis, 10, 22
Hemangiopericytoma, 10, 12, **28,** 29, 38, 53, 55, *70,* 86, 116, 136, 137, 140, 218, 270, 346, 402, 405, 408, **411**
age distribution, 50, 53
benign, 22
borderline, 15
cell morphology, 214, 215, 218, *235*
classification, 10
clinical appearance, 38
cytology, 446
definition, **28**
differential diagnosis, 12, **411, 420, 423–445**
grade, 402–404
gross appearance, 116
growth pattern, 140, *177, 178, 180, 190, 192*
growth phase, 14
history, 38
illustrations, 29
immunochemistry, 342, 343, 350, 353
metastasis, 410
products, 346, *376*
prognosis, 410
radiology, 86
recurrence, 410
references, 29
sex distribution, 54, 55
site, 59, *70*
size, 81
special stains, 341–347
stage, 405–407
stroma, 265–267, 270, *303, 306, 308*
survival, 410
symptoms, 38
synonyms, 28
ultrastructure, 342, 348, 349, *376*
Hemangiosarcoma, 8, 10, **28,** 29, 35, 38, 39, 40, *44, 45,* 53, 55, *70, 73,* 86, 114, 116, 218, 270, 346, 402, 405, 408, **411**
age distribution, 50, 53
borderline, 15, *304, 307, 311*
cell morphology, 214, 215, 218, *241, 243*
classification, 10
clinical appearance, 38, *44, 45*
cytology, 446
definition, **28**
differential diagnosis, **411, 420, 423–445**
grade, 402–404
gross appearance, *42,* 116, *133*
growth pattern, 137, 140, *190, 194, 197–199, 203*
growth phase, 14
histochemistry, 342, 348
histogenesis, 35
history, 38
illustrations, 29
immunochemistry, 342, 343, 350, 353
metastasis, 410
products, 340–343, 346–353, *379*
prognosis, 410
radiology, 86
recurrence, 410
references, 29
sex distribution, 54, 55
site, 59, 61, 62, *70, 73*
size, 79, 81
special stains, 341–347
stage, 405–407
stroma, 265–267, 270, *288, 304, 307, 311*
survival, 410
symptoms, 35, 38
synonyms, 28
ultrastructure, 342, 348, 349, *379*
Hemorrhagic bone cyst, 16
Herbicides, 35
Hereditary hemorrhagic telangiectasia, 10, 57
Herring-bone pattern, 136, 137, *152*
Heuser's membrane, *3*
Hibernoma, 10, 15, **20,** 21, 37, 115, 136, 137, 139, 217, 269, 345
age distribution, 50, 52
cell morphology, 214, 215, 217, *240*
clinical appearance, 37
definition, **20**
gross appearance, 115
growth pattern, 139, *166*
history, 37
illustrations, 21
immunochemistry, 342, 343, 350, 353
products, 340–343, 345, 347–353, *365*
radiology, 37
references, 20
sex distribution, 54, 55
site, 58
size, 80
stroma, 269
symptoms, 37
synonyms, 20
ultrastructure, 342, 348, 349
High grade, 79
High grade sarcomas, 402–404
Histiocytic fibrous histiocytoma. *See* Benign histiocytic fibrous histiocytoma; Malignant histiocytic fibrous histiocytoma
Histiocytic hemangioma, 22
Histiocytic markers, 343, 350, 351
Histiocytoma. *See* Names of specific types
Histiocytosis, 11
Histochemical stain(s), 13, 342, 348
Histochemistry, *215*
Histogenesis, 3, 4, 5, 8, 9, 12, 342
Histologic factors, 403, 404
Histologic grade, 402–404
Hodgkin's disease, 11, 42, *246, 283, 317*
Human chorionic gonadotropin, 350–353
Humerus, 60, 61
Hunter, John, 13
Hyaluronic acid, 341–347
Hyaluronidase, 267
Hydroxyproline, 35, 36
Hygroma, cystic, 10, *42, 313*

Hypercalcemia, 36, 38
Hyperparathyroidism, 10, **16,** 17, 36, 115, 138, 216, 268, 344, 408, 409, **411–413, 423–445**
  cell morphology, 216
  clinical appearance, 36
  complication, 408, 409
  definition, **16**
  differential diagnosis, **411–413, 423–445**
  gross appearance, 115
  growth pattern, 138
  history, 36
  products, 344
  prognosis, 408, 409
  radiology, 36, 40
  references, 17
  site, 36, 40
  stroma, 268
  symptoms, 36
  synonyms, 16
Hyperphosphatasemia, 11
Hyperplasia
  angiofollicular. *See* Angiofollicular hyperplasia
  intravascular papillary endothelial, *198*
  lymphoid, 11
  papillary endothelial, 10, *198, 311*
  synovial, 16
Hypertrophic hemangioma, 10, 22, *197, 304, 383*
Hypophosphatasemia, 11

Idiopathic hemorrhagic sarcoma, 30
IgA, 35, 38
IgD, 35, 38
IgE, 35, 38
IgG, 35, 38, 350–353
IgK, 350–353
IgL, 350–353
IgM, 35, 38
Immature bone, 267
Immunochemical procedures, 266, 341–343, 350–354
Immunochemical stains, 340–353, *392–401*
Immunochemistry, 215
Immunocytochemical methods, 342–346, 350–353
Immunoelectron microscopy, 342
Immunofluorescence microscopy, 13
Immunofluorescence procedures, 342, 343, 401
Immunoglobulins, 35, 38, 350–353
Immunohistochemical methods, 342–346, 350–353
Immunohistochemistry, 13
Immunoperoxidase methods, 342–343, 352
Incidence, 8, *9,* 50
Incipient phase, 12, *13,* 14
Induction phase, 12, *13,* 14
Incisional biopsy, 82, 113
Infancy, malignant pigmented neuroectodermal tumor of, 30
Infantile fibrosarcoma, 26
Infarct, 10
  bone, 8, 26
Infiltrating lipoma, 20
Infiltrative fasciitis, 16
Infiltrative lesions, 84–86
Inflamed stroma, 316–322
Inflammation of adipose tissue, 18
Inflammatory fibrous histiocytoma, 26
Insulin, 350–353
Intermediate fibrohistiocytic tumor, 26
Intermediate filaments, 343, 348–353
Intermediate mesoderm, *4*
Internal cranium, 57–59
Internal thorax, 57–59
Intracanaliculare fibroadenoma, 24
Intramuscular hemangioma, 22
Intraneural lipoma, 20
Intraskeletal chondrosarcoma. *See* Chondrosarcoma
Intraskeletal tumor(s). *See* Bone tumor(s)
Intraskeletal villonodular synovitis, 16
Intravascular papillary endothelial hyperplasia, *198*
Intravenous leiomyomatosis, 22
Invasive phase, 12, *13,* 14, 15
Iridium implant, 408
Irradiation, 35
Islet cell tumor, malignant, *374*
Isomorphic giant cell(s), 214–218, 253–255

Jaffe, Henry, 9, 79, 136
Juvenile angiofibroma, 22
Juvenile aponeurotic fibroma, 20
Juvenile hemangioma, 22
Juvenile hemangiopericytoma, 28
Juvenile nasopharyngeal angiofibroma, 22
Juvenile xanthoma, 20
  atypical, 16
Juxtacortical component, 62
Juxtacortical osteogenic sarcoma, 30
Juxtacortical osteosarcoma, 30

Kaposi's sarcoma, 11, 15, **30,** 31, 35, 38, 40, *48,* 53, 55, 116, 140, 218, 270, 346, 402, 408, **411**
  age distribution, 50, 53
  cell morphology, 218, *220, 226, 227, 228*
  classification, 11
  clinical appearance, 38, *48*
  complication, 38
  definition, **30**
  differential diagnosis, **411, 422, 423–445**
  grade, 402–404
  gross appearance, 116
  growth pattern, 140, *156*
  growth phase, 14
  histogenesis, 35
  history, 35, 38
  illustrations, 31
  immunochemistry, 342, 343, 350, 353
  metastasis, 410
  products, 346, *356*
  prognosis, 410
  recurrence, 410
  references, 31
  sex distribution, 54, 55
  site, 59
  size, 81
  special stains, 341–347
  stage, 405–407
  stroma, 270, *304, 306*
  survival, 410
  symptoms, 35, 38
  synonyms, 30
  ultrastructure, 342, 348, 349
Keloid, 10, **16,** 17, 35, 36, 39, 115, 138, 216, 268, 344, 408, 409, **411,** 423
  age distribution, 50, 51
  cell morphology, 216
  clinical appearance, 35
  complication, 408, 409
  definition, **16**
  differential diagnosis, **411–413, 423–445**
  gross appearance, 39, 115
  growth pattern, 138
  histogenesis, 35
  history, 36
  illustrations, 17
  products, 340–344, 347–353, *364*
  prognosis, 408, 409
  radiology, 36
  recurrence, 36
  references, 17
  sex distribution, 54
  site, 36, 57
  size, 80
  special stains, 341–347
  stroma, 268, *281*
  symptoms, 36
  synonyms, 16
Keratan sulfate, 341–347
Keratin, 343, 350, 351
Kidney, 4
Kimura's disease, 18
Knee, 57–59
K-sarcoma, 30

Lamellar bone, 267
Laminin, 266
Large cell predominant osteosarcoma, 30
Large sarcomas, 405–407
Lateral mesoderm, *4*
Leg, 57–59
Leiomyoblast, 3
Leiomyoblastoma, 10, **28,** 29, 86, 116, 218, 270, 346, 402, 405, 408, **411**
  benign, 22
  cell morphology, 218, *234, 244, 246, 247, 248, 257, 261, 264*
  classification, 10
  cytology, 446
  definition, **28**
  differential diagnosis, **411, 419, 423–445**
  grade, 402–404
  gross appearance, 116, *123, 124*
  growth pattern, 137, 140, *164, 166, 185, 188*
  growth phase, 14
  histochemistry, 342, 348
  illustrations, 29
  immunochemistry, 342, 343, 350, 353, *394*
  metastasis, 410
  products, 340–343, 346–353, *359, 365, 380, 394*
  prognosis, 410
  radiology, 86, 104
  recurrence, 410
  references, 29
  site, 59
  size, 81
  special stains, 341–347
  stage, 405–407
  stroma, 270, *280, 294, 301, 312, 314, 324*
  survival, 410
  synonyms, 28
  ultrastructure, 342, 348, 349, *380*
Leiomyoma, 9, 10, **22,** 23, 37, 39, 40, *46,* 114, 115, 136, 137, 139, 217, 269, 345, 408, 409, **411**
  age distribution, 50, 52
  borderline, *395*
  cell morphology, 214, 215, 217, *225*
  classification, 10
  clinical appearance, 37, *46*
  complication, 408, 409
  cytology, 466
  definition, **22**
  differential diagnosis, **411, 414–416, 423–445**
  gross appearance, 115
  growth pattern, 139, *149, 150, 154, 160*
  history, 37
  illustrations, 23
  immunochemistry, 342, 343, 350, 353, *395*

malignant transformation, 9, 409
products, 340–343, 345, 347–353, *395*
prognosis, 408, 409
radiology, 37
recurrence, 408, 409
references, 23
sex distribution, 54, 55
site, 58
size, 80
stroma, 265–267, 269, *312*
symptoms, 37
synonyms, 22
ultrastructure, 342, 348, 349
Leiomyoma cutis, 22
Leiomyomatosis, 10, 22
Leiomyosarcoma
of soft tissues, 10, **28,** 29, 38, 39, 40, 53, 55, *63, 73,* 86, 114, 116, 136, 137, 140, 218, 270, 346, 402, 405, 408, **411**
age distribution, 50, 53
borderline, 15, *395*
cell morphology, 218, *219, 227, 229, 231, 233, 246, 259, 260*
classification, 10
clinical appearance, 38
cytology, 446, *449, 458*
definition, **28**
differential diagnosis, **411, 419, 423–445**
grade, 402–404
gross appearance, 116, *123*
growth pattern, 140, *143, 145, 149, 150, 151, 154, 161, 206*
growth phase, 14
histochemistry, 342, 348
history, 38
illustrations, 29
immunochemistry, 342, 343, 350, 353, *394, 395, 401*
metastasis, 410
products, 340–343, 346–353, *357, 366, 379, 394, 395, 401*
prognosis, 410
radiology, 86, *104, 109, 111*
recurrence, 410
references, 29
sex distribution, 54, 55
site, 39, 59, 61, 62, *63, 76*
size, 79, 81
special stains, 341–347
stage, 405–407
stroma, 265–267, 270, *274, 275, 277, 281, 286, 308, 311, 312, 314, 323*
survival, 410
symptoms, 38
synonyms, 28
ultrastructure, 342, 348, 349, *379*
of uterus, *109*
Leukemia, 4, 8, 11, 319
Limitation of motion, 36, 37, 38
Lipoblastic liposarcoma, 10, **28,** 29, 38, *47,* 86, 116, 218, 270, 346, 402, 405, 408, **411**
cell morphology, 218, *234, 239, 248*
clinical appearance, *47*
cytology, 446, *452*
definition, **28**
differential diagnosis, **411, 418, 423–445**
grade, 402–404
gross appearance, 116, *123, 125, 127*
growth pattern, 140, *163, 171, 172*
growth phase, 14
illustrations, 29
immunochemistry, 342, 343, 350, 353
metastasis, 410
products, 346, *388*
prognosis, 410
radiology, 86
recurrence, 410
references, 29
site, 59
size, 81
stage, 405–407
stroma, 270, *295, 303*
survival, 410
symptoms, 38
synonyms, 28
ultrastructure, 342, 348, 349, *388*
Lipoblastoma, **20,** 21, 37, 39, 115, 136, 137, 139, 217, 269, 345, 408, 409, **411**
age distribution, 50, 52
cell morphology, 217
clinical appearance, 37
complication, 408, 409
cytology, 466
definition, **20**
differential diagnosis, **411, 414–416, 423–445**
growth pattern, 139, *163, 169*
history, 37
illustrations, 21
products, 345
prognosis, 408, 409
radiology, 37
recurrence, 408, 409
references, 21
sex distribution, 54, 55
site, 58
size, 80
stroma, 265–267, 269, *309*
symptoms, 37
synonyms, 20
Lipoblastomatosis, 20
Lipodystrophia, 10
Lipogranuloma, 10, 18, 19, 36, 115, 138, 216, 268, 344, 408, 409, **411**
age distribution, 50, 51
cell morphology, 214–216, *263*
clinical appearance, 36
complication, 408, 409
cytology, 446
differential diagnosis, **411–413, 423–445**
gross appearance, 115
growth pattern, 138, *167*
histochemistry, 215
history, 36
immunochemistry, 215
products, 344
prognosis, 408, 409
radiology, 36
recurrence, 408, 409
sex distribution, 54
site, 36, 57
size, 80
stroma, 268, *289, 321*
symptoms, 36
ultrastructure, 215
Lipoid dermatoarthritis, 18
Lipoid tumor(s), 10
Lipoma, 9, 10, **20,** 21, 37, 39, 40, *42, 64,* 85, *105,* 114, 115, 136, 137, 139, 217, 269, 345, 408, 409, **411.** *See also* Fibroblastic lipoma; Pleomorphic lipoma
age distribution, 50, 52
borderline, 15
classification, 10, 214, 215, 217, *222, 254, 263*
clinical appearance, 37, *42*
complication, 408, 409
cytology, 446
definition, **20**
gross appearance, 115, *118*
growth pattern, 139, *173, 213*
history, 37
illustrations, 21
immunochemistry, 342, 343, 350, 353
malignant transformation, 9, 409
products, 340–343, 345, 347–353, *367, 370, 371*
prognosis, 408, 409
radiology, 85, *105*
recurrence, 408, 409
references, 21
sex distribution, 54, 55
site, 58, *64*
size, 79, 80
stroma, 265–267, 269, *275, 289, 296, 299, 300*
symptoms, 37
synonyms, 20
ultrastructure, 342, 348, 349
Lipoma arborescens, 20
Lipoma dolorosa, 20
Lipoma foetocellular, 20
Lipoma granulare, 20
Lipoma of brown fat, 20
Lipoma-like liposarcoma, 26
Lipomatosis, 20
Lipomyxosarcoma, 26
Liposarcoma, 28. *See also* Fibroblastic liposarcoma; Lipoblastic liposarcoma; Myxoid liposarcoma; Pleomorphic liposarcoma; Well-differentiated liposarcoma
Lobular capillary hemangioma, 18
Localized amyloid formation, 18
Localized amyloidosis, 18
Localized neurofibromatosis, 22
Loose bodies, 18
Low grade osteosarcoma, 30
Low grade sarcomas, 402–404
Lymph node(s), 113
Lymphangiography, 82
lymphangioma, 10
lymphangiomatosis, 10, 22
Lymphangiomyoma, **22,** 23, 37, 139, *199,* 269
clinical appearance, 37
definition, **22**
history, 37
illustrations, 23
references, 23
site, 58
size, 80
stroma, 269, *305*
symptoms, 37
synonyms, 22
Lymphangiopericytoma, 22
Lymphangiosarcoma, 10, 35, 38, 39, 40, *45, 48,* 53, 55, *67,* 86, 116, 218, 270, 346, 402, 405, 408, **411**
age distribution, 50, 53
cell morphology, 218, *224, 228, 229*
classification, 10
clinical appearance, 38, *45, 48*
complication, 38
cytology, 446
differential diagnosis, **411, 420, 423–445**
grade, 402–404
gross appearance, 116, *134*
growth pattern, 140, *165, 196, 198*
growth phase, 14
histogenesis, 35
history, 38
metastasis, 410
products, 346
prognosis, 410
radiology, 86, *95*
recurrence, 410
sex distribution, 54, 55
site, 59, *67*

Lymphangiosarcoma (*Continued*)
size, 81
stage, 405–407
stroma, 270, *300, 301–303, 336*
survival, 410
symptoms, 35, 38
Lymphatic vessel tumor(s), 10
Lymphedema, 9, 10, 35, 38
malignant transformation, 9
Lymphocytoma cutis, 11
Lymphocytosis, 11
Lymphoid hyperplasia, 11
Lymphoid tumor(s), 11
Lymphoma. *See* Malignant lymphoma
Lymphoreticular tumors, 11
Lysosomes, 341–343, 350, 351
Lysozyme, 343, 350, 351
Lytic lesions, 84–86

Mackenzie, D.M., 12
Magnetic resonance imaging, 83–84, *112*
Male predominance, 54, 55
Malignant angioreninoma, 30
Malignant chondroblastoma, 30
Malignant collagenous peripheral nerve tumor, 28
Malignant counterpart(s) of benign neoplasm(s), 12, 14
Malignant cutaneous neurofibroma, 28
Malignant cystosarcoma phyllodes, *45*
"Malignant degeneration," 4
Malignant fibroblastic fibrous histiocytoma, 10, **26,** 27, 38, 39, *44, 47,* 53, 55, *65, 76,* 86, 114, 116, 140, 218, 270, 346, 402, 405, 408, **411**
age distribution, 50, 53
atypical, *174, 317*
borderline, 15, *174*
cell morphology, 214, 215, 218, *223, 229, 239, 250*
clinical appearance, 38, *44, 47*
cytology, 446, *448*
definition, **26**
differential diagnosis, **411, 417, 423–445**
grade, 402–404
gross appearance, 116, *118, 129*
growth pattern, 136, 137, 140, *143, 146, 152, 174*
growth phase, 14
history, 38
illustrations, 27
immunochemistry, 342, 343, 350, 353
metastasis, 410
products, 340–343, 346–353, *358, 362, 369, 376*
prognosis, 410
recurrence, 410
references, 27
sex distribution, 54, 55
site, 59, 61, 62, *65, 76*
size, 79, 81
special stains, 341–347
stage, 405–407
stroma, 265–267, 270, *272, 317*
survival 410
symptoms, 38
synonyms, 26
ultrastructure, 342, 348, 349, *376*
Malignant fibroblastic peripheral nerve tumor, 28
Malignant fibrous histiocytoma, 26
Malignant fibrous mesothelioma, 30
Malignant giant cell tumor of bone, 26
Malignant giant cell tumor of soft tissues, 26
Malignant giant cell tumor of tendon sheath, 26
Malignant granular cell myoblastoma with organoid structure, 30
Malignant granular cell tumor. *See* Granular cell tumor, malignant
Malignant hemangioendoethelioma, 28
Malignant hemangiopericytoma, 28. *See also* Hemangiopericytoma
Malignant histiocytic fibrous histiocytoma, 10, **26,** 27, 38, *47,* 53, 55, 86, 116, 140, 218, 270, 346, 402, 405, 408, **411**
age distribution, 50, 53
atypical, *181, 185, 239, 253*
borderline, 15, *181, 185, 239, 253, 285, 310, 322*
cell morphology, 214, 215, 218, *239, 249, 253, 254, 256, 262*
clinical appearance, 38, *47*
cytology, 446, *455*
definition, **26**
differential diagnosis, **411, 417, 423–445**
grade, 402–404
gross appearance, 116, *134*
growth pattern, 136, 137, 140, *172, 181, 185, 192, 207, 209*
growth phase, 14
history, 38
illustrations, 27
immunochemistry, 342, 343, 350, 353, *393*
metastasis, 410
products, 340–343, 346–353, *363, 367, 393*
prognosis, 410
radiology, 82, 83, 86, *96, 97, 112*
recurrence, 410
references, 27
sex distribution, 54, 55
site, 38, 59, 61, 62
size, 79, 81
special stains, 341–347
stage, 405–407
stroma, 265–267, 270, *280, 285, 294, 310, 315, 322*
survival, 410
symptoms, 38
synonyms, 27
ultrastructure, 342, 348, 349
Malignant leiomyoblastoma, 28. *See also* Leiomyoblastoma
Malignant lymphoma, 4, 11, 35, *42,* 86, 140, *165, 175, 183,* 236, *237, 239, 246, 284, 349*
classification, 11
clinical appearance, 35, *42*
complication, 35
growth phase, 14
histochemistry, 342, 348
immunochemistry, 342, 343, 350, 353
radiology, 86, *90, 99*
symptoms, 35
ultrastructure, 342, 348, 349
Malignant melanoma of soft tissues, 30
Malignant mesothelioma. *See* Mesothelioma, malignant
Malignant myxoma, 26
Malignant neoplasm(s), 26–31
age distribution, 50, 53
benign counterparts, 12, 14
cell morphology, 214, 215, 218
classification, 10, 11, 12
clinical presentation, 35, 38, 39, 40
definition(s), **26, 28, 30**
diagnosis, 12
differential diagnosis, **411, 417–445**
grade, 402–404
gross appearance, 114, 116
growth pattern, 136, 137, 140
illustration(s), 27, 29, 31
incidence, 9
number, 9
products of cell, 340–343, 354–364
radiology, 82, 86
reference(s), 27, 29, 31
sex distribution, 54, 55
sites, 59, 61, 62
size, 79, 80, 81
stage, 405–407
stroma, 265–267, 270, 271
synonym(s), 26, 28, 30
Malignant neurilemoma, 30
Malignant neuroepithelioma, 30, *131,* 346, *354, 391*
Malignant neurofibromatosis, 28
Malignant nevoid schwannoma, 30, *244, 302*
Malignant nevoxanthoendothelioma, 16, 26
Malignant nonfibroblastic peripheral nerve tumor, 30
Malignant osteoclastoma, 26
Malignant peripheral nerve tumor with glandular inclusion, 30
Malignant pigmented neuroectodermal tumor of infancy, 30
Malignant pigmented neurofibromatosis, 28
Malignant pleomorphic fibrous histiocytoma, 10, **26,** 27, 38, *42,* 53, 55, *69, 76,* 86, 114, 116, 140, 218, 270, 346, 402, 405, 408, **411**
age distribution, 50, 53
atypical, *164, 165, 174, 255, 256*
borderline, 15, *164, 165, 174, 231, 255, 256, 301, 312*
cell morphology, 214, 215, 218, *231, 255, 256, 258, 260, 262*
clinical appearance, 38, *42*
cytology, 446, *449, 456, 458*
definition, **26**
differential diagnosis, **411, 417, 423–445**
grade, 402–404
gross appearance, 116, *117, 120, 121, 126, 127*
growth pattern, 136, 137, 140, *145, 164, 165, 174, 204, 209*
growth phase, 14
history, 38
illustrations, 27
immunochemistry, 342, 343, 350, 353, *395*
metastasis, 410
products, 340–343, 346–353, *362, 369, 395*
prognosis, 410
radiology, 82, 83, 86, *90, 92, 93, 94, 105, 106, 111*
recurrence, 410
references, 27
sex distribution, 54, 55
site, 56, 59, 61, 62, *69, 76*
size, 79, 81
stage, 405–407
stroma, 265–267, 270, *295, 299, 301, 310, 312, 318, 322*
survival, 410
symptoms, 38
synonyms, 26
ultrastructure, 342, 348, 349
Malignant potential. *See* Grade; Stage
Malignant retinal anlage tumor, 30
Malignant schwannoma, 28, 30–31
Malignant smooth muscle tumor, 28
Malignant synovioma, 26
Malignant thymoma, *142, 153, 219, 302*
Malignant transformation, 9
Malignant undifferentiated peripheral nerve tumor, 4, 11, **30,** 31, 140, 218, 270, 346, 402, 405, 408, **411**
cell morphology, 218, *234, 264*

classification, 11
cytology, 446
definition, **30**
differential diagnosis, **411, 421, 423–445**
grade, 402–404
gross appearance, *133*
growth pattern, 140, *144, 178,* 181, *189, 199, 200,* 206, *208*
growth phase, 14
illustrations, 31
immunochemistry, 342, 343, 350, 353
metastasis, 410
products, 346, *354, 391*
prognosis, 410
recurrence, 410
references, 31
stage, 405–407
stroma, 270, *306*
survival, 410
synonyms, 30
ultrastructure, 342, 348, 349, *391*
Malignant xanthogranuloma, 26
Malignant xanthoma, 26
Management, 8, 12, 15, 79, 408
Margin(s), excisional, 79, 113
Markers, panels of, 342
Mass, 35–40
Mast cell(s), *236*
Mastocytosis, 11
Matrix, extracellular. *See* Stroma
Maturation arrest, 3
Mature bone, 267
Medullary carcinoma of thyroid, *186, 192, 291*
Medulloblastoma, *131, 176, 242*
Medulloepithelioma, peripheral, 30
Melanim, 340–353, *372, 373*
Melanoma, *157, 162, 193, 229, 241, 247, 257, 264, 372, 382*
amelanotic, *398*
balloon cell, *247, 399*
desmoplastic, *157, 162, 229*
Melanotic neuroectodermal tumor, 140, *202*
Meningeal sarcoma, *125, 140, 183, 260, 283*
Meningioma, 9, 136, 137, 139, 217, *336, 385*
cell morphology, 217
growth pattern, 139, *180, 183*
Merkel cell tumor, *315, 390*
Mesenchymal chondrosarcoma, 11, **30,** 31, 38, 40, 53, 55, *74,* 86, 140, 218, 270, 346, 402, 405, 408, **411**
age distribution, 50, 53
cell morphology, 218, *230, 238*
classification, 11
clinical appearance, 38
cytology, 446, *454, 458*
definition, **30**
differential diagnosis, **411, 421, 423–445**
grade, 402–404
gross appearance, *133*
growth pattern, 136, 137, 140, *156, 170, 186,* 208
growth phase, 14
history, 38
illustrations, 31
immunochemistry, 342, 343, 350, 353
metastasis, 410
products, 346
prognosis, 410
radiology, 38, 82, 86, *94, 100, 101*
recurrence, 410
references, 31
sex distribution, 54, 55
site, 59, 61, 62, *74*
size, 79, 81
stage, 405–407
stroma, 265–267, 270, 271, *292, 326*
survival, 410
symptoms, 38
synonyms, 30
ultrastructure, 342, 348, 349
Mesenchymal tissue, 3
origin, 4
Mesenchymoma, 9, 11
Mesenteric fibromatosis. *See* Fibromatosis
Mesoderm, 3, *4*
Mesodermal mixed tumor, 49, *185, 207, 250, 366, 381*
Mesothelial cyst, 11
Mesothelial hyperplasia, 11
Mesothelial tumors, 11
Mesothelioma, 11, 35
benign, 9, 139, 217, 269, 345, **411**
atypical, *282*
cell morphology, 217
differential diagnosis, **411, 414–416, 423–445**
growth pattern, 139
immunochemistry, 342, 343, 350, 353
products, 340–343, 345, 347–353
special stains, 341–347
stroma, 269, *282*
ultrastructure, 342, 348, 349
malignant, 30, 38, 40, 53, 55, 86, 116, 218, 270, 346, **411**
age distribution, 50, 53
borderline, *282*
cell morphology, 218, *231*
clinical appearance, 38
complication, 38
differential diagnosis, **411, 423–445**
gross appearance, 116, *122*
growth pattern, 140, *161, 187, 192, 213*
histochemistry, 342, 348
history, 38
immunochemistry, 342, 343, 350, 353, *392*
products, 346, *392*
radiology, 86, 103
sex distribution, 54, 55
size, 81
special stains, 341–347
stroma, 270, *280, 282*
symptoms, 38
ultrastructure, 342, 348, 349
Mesothelium, 4
Metacarpal(s), 60, 61
Metanephrines, 35
Metaphyseal fibrous defect, 20
Metaphysis, 56, 62
Metaphysis-diaphysis, 62
Metaplasia, 8, 10, 11
chondroid, 8, 326
osteoid, 271, *328*
Metaplastic fasciitis, 16
Metastasis, 12, *13,* 14, 408–410
Metastatic phase, 12, *13,* 14
Metatarsal(s), 60, 61
Microdactylia, 36
Misinformation, 8
Mitosis, 4, 56, 402–404
and grade, 402–404
Mitotic count. *See* Mitosis
Mitotic figures in benign lesions, 402, 403
Molluscum fibrosum, 22
Monoclonal antibodies, 266
Monoclonal techniques, 343
Monophasic synovial sarcoma, 26
Monophasic tendosynovial sarcoma,
pseudoglandular type, 10, *140, 203*
spindle cell type, 10, **26,** 27, 38, 40, *44, 48,* 53, 55, *68,* 86, 114, 116, 140, 218, 270, 346, 402, 405, 408, **411**
cell morphology, 214, 215, 218, *220, 223, 224, 232*
clinical appearance, 38, *44,* 48
cytology, 446
definition, **26**
differential diagnosis, **411, 418, 423–445**
grade, 402–404
gross appearance, 116, 132–134
growth pattern, 136, 137, 140, *156–159, 176, 182, 188, 189*
growth phase, 14
illustrations, 27
immunochemistry, 342, 343, 350, 353, *393*
metastasis, 410
products, 340–343, 346–353, *361, 393*
prognosis, 410
radiology, 82, 83, 86, *101, 103*
recurrence, 410
references, 27
site, 59, *68*
size, 79, 81
stage, 405–407
stroma, 265–267, 270, *277, 279, 280, 285, 287, 288*
survival, 410
symptoms, 38
synonyms, 26
ultrastructure, 342, 348, 349
Morton's neuroma, 18
Motion, limitation of, 36, 37, 38
MRI, 83–84, *112*
Mucin, 267
Mucopolysaccharidoses, 11, 266, 340–353, *367–369*
Mucosubstance, 341–342
Muller, Johannes, 9
Multifocal lesions, 36, 37, 38, 40, 84–86
Multiloculated lesions, 84–86
Multiple nuclei, 215
Muscle, 4
atrophy, 10
dystrophy, 10, *320*
Muscle tumors, 10
malignant, 28
Mycosis fungoides, 11
Myelolipoma, 10, 20, *321*
Myeloma. *See* Plasma cell myeloma
Myeloplexoma, 20
Myeloxanthoma, 20
Myoblastoma, 22
granular cell, 24
Myofibroblast(s), 347
Myoglobin, 350–353
Myosarcoma. *See* Leiomyosarcoma; Rhabdomyosarcoma
Myosin, 350–353
Myositis, *320, 336. See also* Myositis ossificans; Proliferative myositis
Myositis ossificans, 4, 10, **18,** 19, 36, 40, 84, *93,* 114, 115, 216, 268, 344, 408, 409, **411**
age distribution, 50, 51
cell morphology, 214–216, *219*
clinical appearance, 36
complication, 408, 409
cytology, 446
definition, **18**
differential diagnosis, **411–413, 423–445**
gross appearance, 115
growth pattern, 138, *155*
histogenesis, 35
history, 36
illustrations, 19
malignant transformation, 9, 35, 36
products, 344

Myositis ossificans (*Continued*)
prognosis, 408, 409
radiology, 36, 84, *93*
recurrence, 408, 409
references, 19
sex distribution, 54
site, 36, 40, 56, 57
size, 79, 80
special stains, 341–347
stroma, 268, 271, *290, 332, 333*
symptoms, 35, 36
synonyms, 18
Myositis ossificans progressiva, 18
Myositis proliferans, 18
Myxofibrolipoma, 20
Myxofibroma, 20
Myxofibrosarcoma, 26
Myxoid chondroma, 24
Myxoid chondrosarcoma. *See* Chondrosarcoma
Myxoid fibrous histiocytoma, 26
Myxoid lipoma, 20, 139, 217, 269, 345
cell morphology, 217, *222*
growth pattern, 139
stroma, 269, *296, 299*
Myxoid liposarcoma, 10, **26,** 27, 38, 40, *46, 48,* 53, 55, 86, 118, 140, 218, 270, 346, 402, 405, 408, **411**
cell morphology, 214, 215, 218, *222, 254*
clinical appearance, *46, 48*
cytology, 446, *447, 448*
definition, **26**
differential diagnosis, **411, 418, 423–445**
grade, 402–404
gross appearance, 116, *127*
growth pattern, 136, 137, 140, *169*
growth phase, 14
illustrations, 27
immunochemistry, 342, 343, 350, 353
metastasis, 410
products, 340–343, 346–353, *354, 367, 388*
prognosis, 410
radiology, 82, 83, 86, *103, 107*
recurrence, 410
references, 27
site, 51
size, 81
stage, 405–407
stroma, 265–267, 270, *294, 297, 298, 309, 322*
survival, 410
symptoms, 38
synonyms, 26
ultrastructure, 342, 348, 349, *388*
Myxoid neurilemma, 22
Myxoid neurofibroma, 11, 22, *296*
Myxoid rhabdomyosarcoma, 10, 38, 53, 55, 86, 118, 140, *164, 206,* 218, 270, 346, 402, 405, 408, **411**
age distribution, 50, 53
cell morphology, 218
clinical appearance, 38
cytology, 446
differential diagnosis, **411, 419, 423–445**
grade, 402–404
gross appearance, 118
growth pattern, 136, 137, 140
histochemistry, 342, 348
history, 38
immunochemistry, 342, 343, 350, 353, *395*
metastasis, 410
products, 340–343, 346–353, *395*
prognosis, 410
radiology, 86, *87*
recurrence, 410
sex distribution, 54, 55
special stains, 8, 341–347
stage, 405–407
stroma, 270, *292, 294, 297, 299, 319*
survival, 410
symptoms, 38
ultrastructure, 342, 348, 349
Myxoid schwannoma, 22
Myxoid stroma, 292–299
Myxolipoma, 20
Myxoliposarcoma, 26
Myxoma, 20. *See also* Myxoid lipoma
cystic, 10
malignant, 26
with giant cells, 20
Myxopapillary ependymoma, 140, 175, 189, 218, 270, 346
cell morphology, 218
growth pattern, 140
immunochemistry, 342, 343, 350, 353
products, 346, *368*
special stains, 341–347
stroma, 270
ultrastructure, 342, 348, 349
Myxosarcoma, 26

Naked nuclei, 215
Nasal polyp, *250*
Neck, 57, 58, 59
Necrosis, and grade, 402–404
fat. *See* Fat necrosis.
Necrotic stroma, 323–325
Needle aspirate and grading, 402
Needle biopsy and grading, 402
Neoplasms, 4, 10, 11. *See also* Benign neoplasms; Malignant neoplasms
childhood, 50–53
equivocal, 12, *13, 14,* 15
Nerve, 10
degeneration, 11
hypertrophy, 11
Nerve tumor(s). *See* Peripheral nerve tumors
Neurilemma, myxoid, 22
Neurilemmoma
ancient, 22
malignant, 30
Neuroblastoma, 11, *242,* 349, *382*
peripheral, 30
Neuroectodermal tumor
malignant pigmented, of infancy, 30
melanotic, 140, *202*
Neuroepithelioma, 11. *See also* Malignant undifferentiated peripheral nerve tumor
malignant, 30, *131,* 346, *354, 391*
Neurofibroma, 9, 11, **22,** 23, 37, 39, 40, *45, 64,* 85, *106,* 114, 115, 136, 137, 139, 217, 269, 345, 408, 409, **411**
age distribution, 50, 52
ancient, 22
associated with von Recklinghausen's disease, 28
atypical, 159
border line, 15, *159*
cell morphology, 215, 215, 217, *220, 225*
classification, 11
clinical appearance, 37, *45*
complication, 408, 409
cytology, 446
definition, **22**
differential diagnosis, **411, 414–416, 423–445**
glandular, 22, 24
gross appearance, *41,* 115, *118, 120*
growth pattern, 139, *150, 151, 159, 181, 190*
history, 37
illustrations, 23
immunochemistry, 342, 343, 350, 353
malignant cutaneous, 28
malignant transformation, 9, 35, 37, 409
myxoid, 11, 22, *296*
plexiform. *See* Plexiform neurofibroma
products, 340–343, 345, 347–353, *354, 360*
prognosis, 408, 409
radiology, 85, *106*
recurrence, 408, 409
references, 23
sex distribution, 54, 55
site, 37, 58, *64*
size, 79, 80
special stains, 341–347
stroma, 265–267, 269, *274, 280, 296, 300, 308, 313*
symptoms, 37
synonyms, 8, 22
ultrastructure, 342, 348, 349
Neurofibromatosis, 11, 22, 28
Neurofibrosarcoma, 11, **28,** 29, 38, 39, 40, *42, 43, 44, 47,* 53, 55, *66,* 86, 114, 116, 140, 218, 270, 346, 402, 405, 408, **411**
age distribution, 50, 53
atypical, *159*
borderline, *159*
cell morphology, 214, 215, 218, *220*
classification, 11
clinical appearance, 38, *42, 43, 44, 47*
complication, 38
cytology, 446
definition, **28**
differential diagnosis, **411, 420, 423–445**
grade, 402–404
gross appearance, 116, *117, 123, 124, 127*
growth pattern, 136, 137, 140, *152, 159, 205*
growth phase, 14
history, 38
illustrations, 29
immunochemistry, 342, 343, 350, 353, *398*
metastasis, 410
products, 340–343, 346–353, *354, 387, 398*
prognosis, 410
radiology, 86
recurrence, 410
references, 29
sex distribution, 54, 55
site, 39, 59, *66*
size, 81
special stains, 341–347
stage, 405–407
stroma, 265–267, 270, *273*
survival, 410
symptoms, 38
synonyms, 28
ultrastructure, 342, 348, 349, *387*
Neurofilaments, 343, 350–353
Neurogenic sarcoma, 30
Neuroma
amputation, 18
granular, 24
Morton's, 18
plexiform, 22
traumatic. *See* Traumatic neuroma
Neuron specific enolase, 350–353
Neurosecretory granules, 340–354, 374
Neurothekoma, 22
Nevoid schwannoma, 11, 22, 139, 140, 165, *170, 197*
benign, 139, 217, 269, 345, **411**
cell morphology, 217
cytology, 446, *447*
growth pattern, 139
immunochemistry, 342, 343, 350, 353
products, 345
special stains, 341–347

stroma, 269
ultrastructure, 342, 348, 349
malignant, 30, *244, 302*
Nevoxanthoendothelioma, 16, 26
Nodular fasciitis, 16
Nodular panniculitis, 18
Nodular synovitis, 20
Noncollagenous malignant peripheral nerve tumor, 30
Nondiagnostic growth phase, 12, *13,* 14, 15
Nonfibroblastic peripheral nerve tumor, malignant, 30
Non-Hodgkin's lymphoma. *See* Malignant lymphoma
Non-neoplastic lesions, 8, *9,* 10, 11, 12
age distribution, 50, 51, 56
cell morphology, 214–216
clinical presentation, 35, 36, 39, 40
deep, 36, 37, 38, 40
definitions, **16, 18**
differential diagnosis, **411–413, 423–445**
gross appearance, 114, 115
growth pattern, 136–138
illustrations, 17, 19
not considered, 18, 32
products of cells, 340–344, 347–354
prognosis, 408, 409
radiology, 82–84
references, 17, 19
sex distribution, 54, 56
sites, 57, 60, 62
size, 79, 80, 81
stroma, 265–268, 271
synonyms, 16, 18
Nonossifying fibroma, 8, 10, **20,** 21, 37, 40, *77,* 85, *100,* 217, 269, 345, 408, 409, **411**
age distribution, 50, 52
cell morphology, 214, 215, 217, *221*
clinical appearance, 37
complication, 408, 409
cytology, 446
definition, **20**
differential diagnosis, **411, 414–416, 423–445**
gross appearance, 115
growth pattern, 139
history, 37
illustrations, 21
products, 345
prognosis, 408, 409
radiology, 37, 85, *100*
recurrence, 408, 409
references, 21
sex distribution, 54, 55
site, 37, 60, 62, *77*
size, 79, 80
stroma, 269, 271, *276*
symptoms, 37
synonym(s), 8, 20
Nonsulfated glycoconjugates, and sulfated, 342
Norepinephrine, 35
NSE, 350–353
5-Nucleotides, 348

Oil red O, 342
Olfactory neuroepithelioma, 30
Ollier's disease, 24
Oncocytic renal cell carcinoma, *240*
Organized fat necrosis, 18
Osler, Sir William, 8
Osseus metaplasia, 8, 10, 11
Ossified stroma, 271, 328–335
Ossifying fibroma, 10, 11, 16, 85
Ossifying parosteal sarcoma, 30
Osteitis deformans, 18
Osteoarthritis, 10
Osteoblast(s), 3, 4, 267
Osteoblastic sarcoma, 11, 30. *See also* Osteosarcoma
Osteoblastoma, 9, 11, **24,** 25, 37, 40, *72,* 85, *107, 111,* 114, 115, 217, 269, 345, 408, 409, **411**
age distribution, 50, 52
cell morphology, 217
clinical appearance, 37
complication, 408, 409
definition, **24**
differential diagnosis, **411, 414–416, 423–445**
gross appearance, 115
growth pattern, 139
history, 37
illustrations, 25
malignant transformation, 9, 409
products, 340–343, 345, 347–353, *364*
prognosis, 408, 409
radiology, 85, 107, 111
recurrence, 408, 409
references, 25
sex distribution, 54, 55
site, 37, 60, 62, *72*
size, 79, 80
stroma, 265–267, 269, 271, *335*
symptoms, 37
synonyms, 24
Osteocartilagenous exostosis, 24
Osteochondroma, 9, 11, **24,** 25, 37, 40, *77,* 85, 114, 115, 217, 269, 345, 408, 409, **411**
age distribution, 50, 52
cell morphology, 217, *252*
classification, 11
clinical appearance, *37*
complication, 37, 408, 409
cytology, 446
definition, **24**
differential diagnosis, **411, 414–416, 423–445**
gross appearance, 115, *129*
growth pattern, 139
history, 37
illustrations, 25
malignant transformation, 9, 409
products, 340–343, 345, 347–353, *363*
prognosis, 408, 409
radiology, 37, 85, *95, 98, 102, 108*
recurrence, 408, 409
references, 25
sex distribution, 54, 55
site, 37, 60, 62, *77*
size, 79, 80
stroma, 265–267, 269, 271, *326, 328, 334*
symptoms, 37
synonyms, 24
Osteoclast, 8
Osteoclastoma, 20
malignant, 26
Osteogenic sarcoma, 30. *See also* Osteosarcoma
Osteogenesis imperfecta, 11, 271
Osteoid, 8, 267, 341
Osteoid metaplasia, 271, *328. See also* Metaplasia
Osteoid osteoma, 11, **24,** 25, 37, 40, *73,* 85, 114, 115, 217, 269, 345, 408, 409, **411**
age distribution, 50, 52
cell morphology, 217
classification, 11
clinical appearance, 37
complication, 408, 409
cytology, 446
definition, **24**
differential diagnosis, **411, 414–416, 423–445**
gross appearance, 115
growth pattern, 139
history, 37
illustrations, 25
products, 345
prognosis, 408, 409
radiology, 37
recurrence, 408, 409
references, 25
sex distribution, 54, 55
site, 60, 62, *73*
size, 79, 80
stroma, 265–267, 269, 271, *334*
symptoms, 37
synonyms, 24
Osteoid-producing tumors, 11
Osteoma, 11, 24, 271. *See also* Osteoid osteoma
Osteomalacia, 11
Osteomyelitis, 8
malignant transformation, 9
Osteoporosis, 11
Osteosarcoma
of bone, 7, 11, **30,** 31, 38, 40, 53, 55, *75,* 82, 83, 86, 114, 116, 140, 218, 270, 346, 402, 405, 408, **411**
age distribution, 50, 53
cell morphology, 218, *223, 230, 243, 248, 251, 253, 258, 260, 262*
classification, 11
clinical appearance, 38
cytology, 446, *457, 458*
definition, **30**
differential diagnosis, **411, 422, 423–445**
grade, 402–404
gross appearance, 116, *119, 129,* 130, *132, 133, 135*
growth pattern, 136, 137, 140, *157, 161, 162, 172, 174, 209, 213, 219*
growth phase, 14
history, 38
illustrations, 31
immunochemistry, 342, 343, 350, 353
metastasis, 410
products, 340–343, 346–353, *363*
prognosis, 410
radiology, 38, 86, *89, 93, 94, 96–98, 100, 109–111*
recurrence, 410
references, 31
sex distribution, 54, 55
site, 60, 61, *75*
size, 79, 81
stage, 405–407
stroma, 265–267, 270, 271, *272, 274, 281, 286, 300, 302, 310, 328–330, 333, 334, 338*
survival, 410
symptoms, 38
synonyms, 30
ultrastructure, 342, 348, 349
of soft tissues, **30,** *69,* 82, 83, 86, *93,* 114, 116, 140, 218, 270, 346, 402, 405, 408, **411**
age distribution, 50, 52
cell morphology, 218, *223, 230, 243, 248, 251, 253, 258, 260, 262*
cytology, 446, *457, 498*
definition, **30**
differential diagnosis, **411, 422, 423–445**
grade, 402–404
gross appearance, 116, *123, 134*
growth pattern, 140, *157, 161, 162, 172, 174, 209, 212, 213*
immunochemistry, 342, 343, 350, 353
metastasis, 410
products, 340–343, 346–353, *363*
prognosis, 410
radiology, 86, *93*
recurrence, 410
site, 59, *69*

Osteosarcoma (*Continued*)
size, 79, 81
stage, 405–407
stroma, 270, 271, *272, 274, 281, 286, 300, 302, 310, 328–330, 333, 334, 338*
survival, 410
ultrastructure, 342, 348, 349
Ovarian stroma, *146, 220*

Pacinian neurofibroma, 11, 22
Paget's disease, 9, 11, **18,** 19, 35, 36, 40, *74,* 84, 211, 408, 409, **411**
age distribution, 50, 51
clinical appearance, 35
complication, 9, 36, 408, 409
definition, **18**
differential diagnosis, **411–413, 423–445**
history, 36
illustrations, 19
malignant transformation, 9, 35, 36, 409
prognosis, 408, 409
radiology, 36, 84, *91, 107, 112*
references, 19
sex distribution, 54
site, 36, 40, 60, 62, *74*
size, 80
special stains, 341–347
stroma, 271
symptoms, 35, 36
synonyms, 18
Paget's sarcoma, 11, **30,** 31, 38, 40, 53, 55, *75,* 82, 83, 86, 140, 218, 270, 346, **411**
age distribution, 50, 53
cell morphology, 218
clinical appearance, 35, 38
complication, 38
cytology, 446
definition, **30**
differential diagnosis, **411, 422, 423–445**
grade, 402–404
growth pattern, 140
growth phase, 14
history, 38
illustrations, 31
metastasis, 410
products, 346, *356, 364*
prognosis, 410
radiology, 38, 86, *91, 107, 112*
recurrence, 410
references, 31
sex distribution, 54, 55
site, 61, 62, *75*
size, 81
stage, 405–407
stroma, 265–267, 270, 271, *291, 337*
survival, 410
symptoms, 35, 38
synonyms, 30
Pain, 35, 36, 37, 38, 40
Palisading pattern, 4, 136, 137, *148–151*
Panels of markers, 342
Panniculitis. *See* Proliferative panniculitis
Panniculitis ossificans, 10
Papanicolaou, George, 6
Papillary endothelial hyperplasia, 10, *198, 311*
Papilloma, fibroepithelial, *282*
Papule, piezogenic. *See* Piezogenic papule
Paraganglioma
benign, 9, 11, **22,** 23, 37, 40, 85, 114, 115, 217, 269, 345, **411**
age distribution, 50, 52
borderline, 15
cell morphology, 214, 215, 217, *257*
clinical appearance, 37
complication, 37
definition, **22**
differential diagnosis, **411, 414–416, 423–445**
gross appearance, 115
growth pattern, 139, *179*
history, 37
illustrations, 23
immunochemistry, 342, 343, 350, 353
malignant transformation, 9
products, 345
radiology, 85
recurrence, 37
references, 23
sex distribution, 54, 55
site, 37, 58
size, 79, 80
stroma, 265–267, 269
symptoms, 37
synonyms, 22
ultrastructure, 342, 348, 349
malignant, 11, 35, 86, 116, 140, 218, 270, 346, **411**
atypical, 252
borderline, 15, *252*
cell morphology, 214, 215, 218, *239, 256, 252*
classification, 11
clinical appearance, 35
cytology, 446
differential diagnosis, **411, 423–445**
gross appearance, 116, *124*
growth pattern, 140, *195*
growth phase, 14
immunochemistry, 342, 343, 350, 353
products, 346, *374, 383*
site, 59
size, 81
special stains, 341–347
stroma, 270, *276*
symptoms, 35
ultrastructure, 342, 348, 349, *383*
Paraosteal fasciitis, 16
Parosteal osteogenic sarcoma, 30
Parosteal osteosarcoma, 11, **30,** 31, 38, 40, 53, 55, *76,* 82, 83, 86, 114, 116, 140, 218, 270, 346, 402, 405, 408, **411**
age distribution, 50, 53
cell morphology, 218, *226*
classification, 11
clinical appearance, 38
cytology, 446
definition, **30**
differential diagnosis, **411, 422, 423–445**
grade, 402–404
gross appearance, 116, *119, 126*
growth pattern, 136, 137, 140, *147*
growth phase, 14
history, 38
illustrations, 31
immunochemistry, 342, 343, 350
metastasis, 410
products, 346, *364*
prognosis, 410
radiology, 38, 82, 83, 86, *89, 108, 110*
recurrence, 410
references, 31
sex distribution, 54, 55
site, 61, 62, *76*
size, 79, 81
stage, 405–407
stroma, 265–267, 270, 271, *331*
survival, 410
symptoms, 38
synonyms, 30
ultrastructure, 342, 348, 349
Paraxial mesoderm, *4*
PAS stain, 341–349
Pathologist(s), 4, 7, 8, 12, *13,* 50, 82, 113, 137, 215, 341–343, 402, 408
Pelvic bone(s), 60, 61
Pelvis, 60, 61
Pericyte, 3, 12
Perineural hemangioma, 22
Periosteal osteoma, 24
Periosteal osteosarcoma, 30
Periostitis, 11
Peripheral leiomyosarcoma, 28
Peripheral medulloepithelioma, 30
Peripheral nerve tumor(s), 11, *13*
benign, 22
malignant, 28
fibroblastic, 28
noncollagenous, 30
nonfibroblastic, 30
undifferentiated. *See* Undifferentiated peripheral nerve tumor
with glandular inclusions, 30
Peripheral neuroblastoma, 30
Perithelioma, 28
Peritoneal leiomyomatosis. *See* Leiomyoma
Peters, Rudolph, 214
pH, 342
Phalange(s), 60, 61
Phase(s), 79
Phenotype, 3
Phenoxy acids, 35
Photography, 113
Piezogenic papule, 10, **18,** 19
definition, **18**
references, 19
site, 57
size, 80
synonyms, 18
Pigmentation, 36, 38
Pigmented neurofibroma, 22
Pigmented neurofibromatosis, 22
Pigmented retinal anlage tumor, *189*
Pigmented schwannoma, 11, 22, *236*
Pigmented villonodular synovitis, 16, 20
Pinealoma, *242*
Plasma cell granuloma, 11
Plasma cell myeloma, 11, 38, 40, 53, 55, *72,* 86, 116, 218, 270, 346, 402, 405, 508, **411**
age distribution, 50, 53
cell morphology, 218, *234, 242, 246*
classification, 11
clinical appearance, 38
complication, 38
cytology, 446, *450*
differential diagnosis, **411, 422, 423–445**
grade, 402–404
gross appearance, 116
growth pattern, 140, *172*
growth phase, 14
histochemistry, 342, 348
history, 38
immunochemistry, 38, 342, 343, 350, 353
metastasis, 410
products, 346
prognosis, 410
radiology, 38, 86, *90, 103, 107*
recurrence, 410
sex distribution, 54, 55
site, 61, 62, *72*
size, 81
special stains, 341–347
stage, 405–407
stroma, 270
survival, 410

symptoms, 38
Plasmacytoma, 11
Pleomorphic fibrosarcoma. *See* Fibrosarcoma
Pleomorphic giant cell(s), 214–218, 256–264
Pleomorphic lipoma, 20, 21, 85, 139, *213,* 217, 269, 345, **411**
cell morphology, 214, 215, 217, *254, 263*
definition, **20**
differential diagnosis, **411, 414–416, 423–445**
growth pattern, 139
illustrations, 21
immunochemistry, 342, 343, 350, 353
products, 340–343, 345, 347–353, *370*
radiology, 85
references, 21
stroma, 265–267, 269, *289, 300*
synonyms, 20
ultrastructure, 342, 348, 349
Pleomorphic liposarcoma, 10, **28,** 29, 38, 40, 53, 55, *68,* 82, 83, 86, 114, 116, 140, 218, 270, 346, 402, 405, 408, **411**
age distribution, 50, 53
cell morphology, 214, 215, 218, *258, 263*
clinical appearance, 38
cytology, 446, *456*
definition, **28**
differential diagnosis, **411, 419, 423–445**
grade, 402–404
gross appearance, 116, *127*
growth pattern, 136, 137, 140, *174, 195, 207*
growth phase, 14
illustrations, 29
immunochemistry, 342, 343, 350, 353
metastasis, 410
products, 346
prognosis, 410
radiology, 86, *92, 105*
recurrence, 410
references, 29
sex distribution, 54, 55
site, 59, *68*
size, 81
stage, 405–407
stroma, 265–267, 270, *293, 315, 323*
survival, 410
symptoms, 38
synonyms, 28
ultrastructure, 342, 348, 349
Pleomorphic myxoma, 20
Pleomorphic neurofibroma, 22
Pleomorphic rhabdomyosarcoma, 4, 10, **28,** 29, 38, 40, 42, 53, 55, 68, 82, 83, 86, 114, 116, 140, 218, 270, 346, 402, 405, 408, **411**
age distribution, 50, 53
cell morphology, 214, 215, 218, *258, 259, 261*
classification, 10
clinical appearance, 38, *42*
cytology, 446, *457*
definition, **28**
differential diagnosis, **411, 420, 423–445**
grade, 402–404
gross appearance, 116, *127, 128, 132*
growth pattern, 136, 137, 140, *209, 212, 213*
growth phase, 14
histochemistry, 342, 348
history, 38
illustrations, 29
immunochemistry, 342, 343, 350, 353, *394, 396*
metastasis, 410
products, 340–343, 346–353, *357, 366, 381, 394, 396*
prognosis, 410
radiology, 38, 86, *88*
recurrence, 410
references, 29
sex distribution, 54, 55
site, 59, *68*
size, 81
special stains, 341–347
stage, 405–407
stroma, 265–267, 270
survival, 410
symptoms, 38
synonyms, 28
ultrastructure, 342, 348, 349, *381*
Plexiform neurofibroma, 11, 139, *191,* 217, 269, 345
cell morphology, 217, *220*
growth pattern, 139
products, 345, *354*
stroma, 269, *274*
Plexiform neurofibromatosis, 22
Plexiform neuroma, 22
Plump spindle cell(s), 214–218, *222, 227, 229, 230, 231, 232*
Polarized light, 266, 341
Polyhistioma, 30
Polymerized collagen, 266
Polymyositis, 10, *371*
Polyp, 8
nasal, *250*
Polypeptide(s), 341
Polysaccharides, 347
Poorly differentiated chondrosarcoma, 30
Post-chemotherapy sarcoma, 11
Post-irradiation sarcoma, 11, 35, *106,* 140, *152, 180*
Post-radiation fibromatosis, *281*
Post-radiation sarcoma, 11, 35, *106,* 140, *152, 180*
Prechondroblast(s), 267
Pre-elastofibroma, 16
Preneoplastic phase, 12, *13,* 14
Preosteoblast(s), 267
Price and Valentine, 405
Primary bone tumor, 138–140, 216–218, 268–270
Primary soft tissue tumor, 138–140, 216–218, 268–270
Primitive multipotential primary sarcoma of bone, 30
Primitive neuroectodermal tumor. *See* Malignant undifferentiated peripheral nerve tumor
Processing
of bone, 113
of specimen, 113
Product(s) of cells, 3, 12, 13, 15, 34, 215, 340–401
Prognosis, 15, 50, 408–410, 405, 408–412
Prognostic signs, 405–407
Prolapsed disc, 11
Proliferative fasciitis, 16, 18
Proliferative myositis, **18,** 19, 36, 40, 84, 114, 115, 138, 216, 268, 344, 408, 409, **411**
age distribution, 50, 51
cell morphology, 214–216, 245
clinical appearance, 36, 408, 409
cytology, 446
definition, **18**
differential diagnosis, **411–413, 423–445**
gross appearance, 115
growth pattern, 138, *209*
history, 36
illustrations, 19
products, 344
radiology, 36
recurrence, 408, 409
references, 19
sex distribution, 54
site, 36, 40, 57
size, 80
special stains, 341–347
stroma, 268, *279, 299*
symptoms, 36
synonyms, 18
Proliferative panniculitis, 10, **18,** 19, 36, 39, 115, 138, 216, 268, 344, 408, 409, **411**
age distribution, 50, 51
cell morphology, 214–216, *241*
clinical appearance, 36
complication, 408, 409
definition, **18**
differential diagnosis, **411–413, 423–445**
gross appearance, 115
growth pattern, 138, *167, 196*
history, 36
illustrations, 19
products, 344
prognosis, 408, 409
references, 18
sex distribution, 54
site, 36, 57
size, 80
stroma, 268, *289*
symptoms, 36
synonyms, 19
Proliferative synovitis, 16
Prominent nucleoli, 215
Prostate carcinoma, *108, 400*
Prostate specific antigen, 351–353
Prostatic carcinoma, *400*
Proteoglycans, 266, 347
Pseudogout, 10
PTAH stain, 341
Pulmonary blastoma, *168*
Pyogenic granuloma, 10, **18,** 19, 36, 39, 115, 138, 216, 268, 344, 408, 409, **411**
age distribution, 50, 51
cell morphology, 214–216
classification, 10
clinical appearance, 36
complication, 408, 409
cytology, 446
definition, **18**
differential diagnosis, **411–413, 423–445**
gross appearance, 115
growth pattern, 138, *141*
history, 36
illustrations, 19
products, 344
prognosis, 408, 409
radiology, 36
recurrence, 408, 409
references, 19
sex distribution, 54
site, 36, 57
size, 80
special stains, 341–347
stroma, 268, *305*
symptoms, 36
synonyms, 18

Radiation-induced sarcoma, 11, 35, *106,* 140, *152, 180*
Radiation sarcoma, 11, 35, *106,* 140, *152, 180*
Radiation therapy, 406
Radiologic appearance, 8
Radiologic finding(s), 36, 37, 38, 40
Radiologist(s), 8, 82
Radiology, 15, *88–112*
and diagnosis, 82, 83
benign neoplasms, 82
computed tomography, 82, 83
conventional, 82

Radiology (*Continued*)
CT-scan, 82
density, 82
location, 82
magnetic resonance imaging, 83
malignant neoplasms, 82, 86
non-neoplastic lesions, 82–84
site, 82
size, 82
tomography, 82
xeroradiography, 82
ultrasound, 82
Radionuclide(s), 82, *111*
Radius, 60, 61
Reactive lesions. *See* Non-neoplastic lesions
Recurrence, 408–410
References, 7, 17–31
Renal cell carcinoma, *195, 315, 400*
Renal hamartoma, 20
Reticulin fibers, 347
Reticulin stain, 266, 341, 344–347
Reticulohistiocytic granuloma, 18
Reticulohistiocytoma, 20
Retinal anlage tumor, 30, *189*
Retinoblastoma, *178*
Retroperitoneum, 57–59
Rhabdoid tumor, 28
Rhabdomyoblastoma, 10, **28,** 29, 38, 53, 55, 86, 140, *177, 178, 184, 191, 200, 208,* 218, 270, 346, 402, 405, 408, **411**
age distribution, 50, 53
cell morphology, 218, *235, 238, 240, 245, 257*
classification, 10
clinical appearance, 38
cytology, 446, *453*
definition, **28**
differential diagnosis, **411, 420, 423–445**
grade, 402–404
growth pattern, 140
growth phase, 14
history, 38
illustrations, 29
immunochemistry, 342, 343, 350, 353
metastasis, 410
products, 340–343, 346–353, *356, 364*
prognosis, 410
recurrence, 410
references, 29
sex distribution, 54, 55
site, 59
size, 81
stage, 405–407
stroma, 270, *295, 319*
survival, 410
symptoms, 38
synonyms, 28
Rhabdomyolysis, 10
Rhabdomyoma, 10, **22,** 23, 37, 40, 85, 139, 217, 269, 345, **411**
age distribution, 50, 52
cell morphology, 217, *257*
classification, 10
clinical appearance, 37
definition, **22**
differential diagnosis, **411, 414–416, 423–445**
growth pattern, 134, *184*
history, 37
illustrations, 23
immunochemistry, 342, 343, 350, 353
of heart, 22
pleomorphic. *See* Pleomorphic rhabdomyoma
products, 345
radiology, 37
references, 23
sex distribution, 54, 55
site, 37, 58
size, 80
stroma, 269
symptoms, 37
synonyms, 22
ultrastructure, 342, 348, 349
Rhabdomyosarcoma, 28. *See* Alveolar rhabdomyosarcoma; Embryonal rhabdomyosarcoma; Myxoid rhabdomyosarcoma; Pleomorphic rhabdomyosarcoma
Rheumatoid arthritis, 10
Rib(s), 60, 61
Rindfleisch, 215
Rokitansky, Carl, 9, 50, 265
Round cell(s), 215
Round cell liposarcoma, 28
Round cell neoplasms, electron microscopy, 349
Rubin, Emanuel, 341

S-100, 350–353
Sacrococcygeal chordoma, 30
Sacrum, 60, 61
Sarcoma(s), **26, 28, 30.** *See also* Names of specific types
arising in teratoma, *233*
grades, 402–404
in situ, 8, 12, *13,* 14, 15
stages, 406
with intracytoplasmic filamentous lesions, 28
Scapula, 60, 61
Scar, 8, 9, 10
Schwannoma
benign, 9, 11, **22,** 23, 37, 39, 40, 85, 114, 115, 217, 269, 345, 408, 409, **411**
age distribution, 50, 52
borderline, 15
cell morphology, 217
classification, 11
clinical appearance, 37
complication, 37, 408, 409
cytology, 446, *447*
definition, **22**
differential diagnosis, **411, 414–416, 423–445**
glandular, 11, 22, 140, 188, *399*
gross appearance, 115, *120*
growth pattern, 139, *165, 177, 188, 197*
history, 37
illustrations, 23
immunochemistry, 342, 343, 350, 353, *399*
malignant transformation, 9, 409
products, 340–343, 345, 347–353, *387, 399*
prognosis, 408, 409
radiology, 85
recurrence, 37, 408, 409
references, 23
sex distribution, 54, 55
site, 37, 56, 58
size, 79, 80
stroma, 265–267, 269, *275, 276*
symptoms, 37
synonyms, 22
ultrastructure, 342, 348, 349, *387*
malignant, 11, **30,** 31, 38, 40, *42, 44, 46,* 53, 55, 66, 86, 114, 116, 218, 270, 346, 402, 405, 408, **411**
age distribution, 50, 53
atypical, *148*
borderline, 15, *148*
cell morphology, 214, 215, 218, *219, 220, 227, 236, 244, 261*
classification, 11
clinical appearance, 38, *42, 44, 46*
cytology, 446
definition, **30**
differential diagnosis, **411, 421, 423–445**
grade, 402–404
gross appearance, 116, *117, 118, 120, 129*
growth pattern, 136, 137, 140, *148, 154, 204, 205, 207, 209*
growth phase, 14
history, 38
illustrations, 31
immunochemistry, 342, 343, 350, 353
metastasis, 410
products, 340–343, 346–353, *373, 378*
prognosis, 410
radiology, 86, *105*
recurrence, 410
references, 31
sex distribution, 54, 55
site, 59, *66*
size, 81
special stains, 341–347
stage, 405–407
stroma, 265–267, 270, *273, 274, 302*
survival, 410
symptoms, 38
synonyms, 30
ultrastructure, 342, 348, 349, *378*
Scintigram(s), 82
Sclerosed stroma, 276–291
Sclerosing edge, 84–86
Sclerosing fibroma, 20
Sclerosing hemangioma, 20. *See* Benign fibroblastic fibrous histiocytoma
Sclerosing liposarcoma. *See* Well-differentiated liposarcoma
Sclerosing osteosarcoma, 11, 30. *See also* Osteosarcoma
Sclerotic lesions, 84–86
Second opinion, 8
Secretory granules, 340–354, 374
Sections, 113
Seminoma, *42, 200, 234, 249, 385*
Sex distribution, 15, 54, 55
Shoulder, 57–59
Site(s), anatomic, 13, 15, 56, 78
benign neoplasms, 58, 60, 62
malignant neoplasms, 59, 61, 62
non-neoplastic lesions, 57, 60, 62
Size 8, 12, *13,* 15, 79–81
benign neoplasms, 79, 80
malignant neoplasms, 79, 81
non-neoplastic lesions, 79, 80
Skeletal tissue, 3
Slender spindle cell(s), 214–218, *219, 221, 222, 224, 225, 226, 229, 230, 231, 232*
Small cell osteosarcoma, 30
Smear(s), 113
Smooth muscle, 4
Soft tissue, 3
Soft tissue tumor(s), 3, 4, 5, 6, 7
classification, 3
cells. *See* Cell(s)
incidence, 8
not considered, 18, 32
number, 8, *9*
origin, 4
primary, 138–140, 216–218, 268–270
size, 79–81
Solid rhabdomyosarcoma, 28
Solitary tumors, 40
Somatostatin, 350–353
Somites, 3
Special stain(s), 13, 113, 215, 341–347
histochemical, 13, 342, 348
immunochemical, 340–352, 392–401

reticulin, 266, 341, 344–347
Specimen, dissection of, 113
Spindle cell(s), 215. *See* Plump spindle cells; Slender spindle cells
Spindle and giant cell liposarcoma, 28
Spindle and giant cell sarcoma, 26, 28
Spindle cell lipoma, 20
Spindle cell liposarcoma, 28
Spindle cell sarcoma, 26
Stage, 15, 79, 113, 405–407
of sarcomas, 405–407
Steatopygia, 10
Sternum, 60, 61
Stewart, Fred, 448
Storiform pattern, 136, 137, *141–147*
Stout, Arthur Purdy, 9, 50, 56
Stroma, 265
amount and grade of, 402–404
calcified, 336–339
chondrified, 271, 326–327
differential diagnosis by, **423–445**
fibrillar, 272–277
inflamed, 316–332
myxoid, 292–299
necrotic, 323–325
ossified, 271
ovarian, *146, 220*
sclerosed, 276–291
vascular, 300–315
Stromal sarcoma of uterus, *142, 175, 302*
Structure, 341
Subcortical nonossifying fibroma, 20
Subcutaneous pseudosarcomatous fasciitis, 16
Superficial presentation, 36, 37, 38, 39, 405–407
Surgeon(s), 8, 82
Surgical therapy, 408
Survival, 408–410
Swelling, 36, 37, 38
Symptom(s), 35–40
Symptomatic, 12
Syncytiotrophoblast, **3**
Synonyms, 8, *13,* 16–31
Synovial chondroma, 24
Synovial chondromatosis. *See* Tendosynovial chondromatosis
Synovial chondrometaplasia, 18
Synovial cyst. *See* Tendosynovial cyst
Synovial fibroma, 20
Synovial fibrosarcoma, 26
Synovial hemangioma, 22
Synovial hyperplasia, 16
Synovial spindle cell sarcoma, 26
Synovial sarcoma, 10, 26
Synovioma, 20
benign, 16
malignant, 26
Synovitis
nodular, 20
pigmented villonodular, 16, 20
proliferative, 16

Technetium, *111*
Telangiectasia, 10, 57
Telangiectatic osteosarcoma, 11, 30. *See also* Osteosarcoma
Tenderness, 35–40
Tendinous xanthoma, 20
Tendonitis, 16
Tendosynovial chondromatosis, 8, 10, **18,** 19, 36, 40, *67,* 84, 115, 138, 216, 268, 271, 344, 408, 409, **411**
age distribution, 50, 51
cell morphology, 216
clinical appearance, 36
complication, 408, 409
definition, **18**
differential diagnosis, **411–413, 423–445**
gross appearance, 115
growth pattern, 138
history, 36
illustrations, 19
products, 344
radiology, 36, 84
recurrence, 408, 409
references, 19
sex distribution, 54
site, 36, 57, *67*
size, 80
special stains, 341–347
stroma, 268, 271
symptoms, 36
synonyms, 8, 17
Tendosynovial cyst, 10, **16,** 17, 36, 84, *93, 102,* 115, 138, 216, 268, 344, 408, 409, **411**
age distribution, 50, 51
cell morphology, 216
clinical appearance, 36
complication, 408, 409
definition, **16**
differential diagnosis, **411–413, 423–445**
gross appearance, 115
growth pattern, 138, *187*
history, 36
illustrations, 17
products, 344
prognosis, 408, 409
radiology, 36, 84, *93, 102*
references, 17
sex distribution, 54
site, 36, 57, 60, 62
size, 80
stroma, 268, *313*
symptoms, 36
synonyms, 16
Tendosynovial fibroma. *See* Fibroma
Tendosynovial sarcoma, 10, 26, 27, 38, 40, 53, 55, 82, 83, 86, 116. *See also* Biphasic tendosynovial sarcoma; Clear cell sarcoma; Epithelioid sarcoma; Monophasic tendosynovial sarcoma
age distribution, 50, 53
clinical appearance, 34, *44*
gross appearance, 116, *121, 132–134*
growth pattern, 140
growth phase, 14
history, 38
illustrations, 27
radiology, 86, *88, 101, 103*
references, 27
sex distribution, 54, 55
site, 38
synonyms, 26
Tendosynovial tumors, 10
Tendosynovitis, 8, 9, 10, **16,** 17, 36, 40, *63,* 84, 115, 138, 216, 268, 344, 408, 409, **411**
age distribution, 50, 51
cell morphology, 214–216, *255*
clinical appearance, 36
complication, 408, 409
cytology, 446
definition, **16**
differential diagnosis, **411–413, 423–445**
gross appearance, 115
growth pattern, 138
history, 36
illustrations, 17
malignant transformation, 9, 409
products, 340–344, 347–353, *358*
prognosis, 408, 409
recurrence, 36, 84
references, 17
sex distribution, 54
site, 36, 56, 57, *63*
size, 80
special stains, 341–347
stroma, 268, *316, 318*
symptoms, 36
synonyms, 16
Tendosynovitis ossificans, 10
Teratoma, *41, 117, 124, 146, 233, 292, 394*
Terminology, 7, 8
Thecagranulosa cell tumor, *188*
Therapy, 402. *See also* Treatment
Thigh, 57–59
Thinned cortex, 84–86
Thorax, internal, 57–59
Thymic cyst, 11
Thymoma, 11, *195*
benign, 9
malignant transformation, 14
malignant, *142, 153, 219, 302*
Thyroglobulin, 350–353
Thyroid carcinoma, *186, 192,* 195, *291*
Tibia, 60,61
Tissue, 3, 4
Tissue antigens, 340–353, 354, *392–401*
Tissue culture, 6
Toe(s). *See* Foot; Phalanges
Tomography, 82
computed, 82, 83
Transmission electron microscopy, 342, 348, 349
Trauma, 35
Traumatic neuroma, 10, **18,** 19, 36, 40, 115, 138, 216, 268, 344, 408, 409, **411**
cell morphology, 216
clinical appearance, 36
complication, 408, 409
definition, **18**
differential diagnosis, **411–413, 423–445**
gross appearance, 115
growth pattern, 138
history, 36
illustrations, 19
immunochemistry, *398*
products, 340–344, 347–353, *398*
radiology, 36
recurrence, 408, 409
references, 18
site, 36
size, 80
special stains, 341–347
stroma, 268
symptoms, 36
synonyms, 18
Treatment, 8, 12, 15, 79, 408
Trichrome stain, 266, 341–347
Triton tumor, malignant, 11, 28, 30, 140, *208, 211, 244, 261, 397*
Tuberous xanthoma, 20
Tubular rhabdomyosarcoma, 28
Tumor(s), 4, 11. *See also* Names of specific tumors
big, 79
borderline, 12, *13,* 14, 15
cell morphology, 214–264
classification, 9, 10, 11, 12
clinical presentation, 35–49
diagnosis, 4, 12, 215
differential **411–445**
grades, 402–404
gross appearance, 116–135
growth patterns, 4, 12, *13,* 132–213, 215
histogenesis, 10, 11, 12
incidence, 8

Tumor(s) (*Continued*)
number, 9
phases, 12, *13,* 14, 15, 215
primary, 4, 10, 11
products, 340–401
recurrent, 4
small, 79
site, 56
size, 79
stage, 405–407
Tumor matrix, 266

UEA-I, 350–353
Ulna, 60, 61
Ultrasound, 82
Ultrastructure, 340–353, *376–391. See also* Fine structure
Undifferentiated connective tissue tumors, 10
Undifferentiated sarcoma, 11
Uric acid, 36
Uterus
leiomyosarcoma of, *109*
stromal sarcoma of, *142, 175, 302*

Vacuolated cytoplasm, 215
Vascular hamartoma, 18
Vascular leiomyoma, 22
Vascular leiomyosarcoma, 28
Vascular lesions, 84–86
Vascular stroma, 300–315
Vascularity and grade, 402–404
Vasculitis, 10
Venous hemangioma. *See* Hemangioma
Vertebra(e), 60, 61
Vesicular nuclei, 215
Vessel tumors, 10
Villonodular synovitis. *See* Tendosynovitis
Vimentin, 343, 350–353
Vinylchloride, 35
Virchow, R., 9
VMA, 35
von Recklinghausen's disease, 22

Wavy nuclei, 215
Weber-Christian disease, 18
Weibel-Palade body, *379*
Well-demarcated lesions, 84–86
Well-differentiated chondrosarcoma, 30
Well-differentiated lipoma, 20, 139, 217, 269, 345, **411**
age distribution, 50, 52
cell morphology, 217
differential diagnosis, **411, 414–416, 423–445**
growth pattern, 139
immunochemistry, 342, 343, 350, 353
malignant transformation, 9, 409
products, 340–343, 345, 347–353, *370, 371*
special stains, 341–347
stroma, 269
ultrastructure, 342, 348, 349
Well-differentiated liposarcoma, 10, *26,* 27, 28, 38, 40, *47,* 53, 55, *68,* 82, 83, 86, 116, 140, 218, 270, 346, 402, 405, 408, **411**
age distribution, 50, 52
borderline, 15
cell morphology, 214, 215, 218, *263*
clinical appearance, *47*
cytology, 446
definition, **26**
differential diagnosis, **411, 418, 423–445**
grade, 402–404
gross appearance, 116
growth pattern, 136, 137, 140, *167, 168*
growth phase, 14
illustrations, 27
metastasis, 410
products, 346, *370, 388*
prognosis, 410
radiology, 86, 109
recurrence, 410
references, 27
sex distribution, 53, 54
site, 56, 59, *68*
size, 81
stage, 405–407
stroma, 265–267, 270, *309*
survival, 410
symptoms, 38
synonyms, 26
ultrastructure, 342, 348, 349, *388*
Williams, W. Roger, 13, 411
Wilms' tumor, *45, 154, 272*
Women, diseases of, 37
Woven bone, 267

Xanthogranuloma, 8, 10, **16,** 17, 26, 36, 115, 138, 216, 268, 344, 408, 409, **411**
age distribution, 50, 51
cell morphology, 214–216, *244, 254*
clinical appearance, 36
complication, 408, 409
cytology, 446
definition, **16**
differential diagnosis, **411–413, 423–445**
gross appearance, 115
growth pattern, 138, *211*
history, 36
illustrations, 17
products, 340–344, 347–353, *356, 376*
prognosis, 408, 409
radiology, 36
recurrence, 408, 409
references, 17
sex distribution, 54
site, 36, 57
size, 80
stroma, 268, *318*
symptoms, 36
synonyms, 8, 16
ultrastructure, *376*
Xanthelasma, 20
Xanthic giant cell tumor, 20
Xanthofibroma, 20
Xanthogranuloma of bone, 20
Xanthoma. *See* Benign pleomorphic fibrous histiocytoma
of bone, 20
Xanthoma multiplex
congenital, 16
juvenile, 20
atypical, 16
Xanthosarcoma, 26
Xeroradiography, 82
X-ray. *See* Radiology

Yolk sac, *3*

Z-line, 343